Linear Algebra with Applications

Fourth Edition

Otto Bretscher

Colby College

Prentice Hall
is an imprint of

Upper Saddle River, New Jersey 07458

Library of Congress Cataloging-in-Publication Data

Bretscher, Otto.
  Linear algebra with applications / Otto Bretscher.—4th ed.
    p. cm.
  Includes index.
  ISBN 978-0-13-600926-9
1. Algebras, Linear—Textbooks. I. Title.
  QA184.2.B73 2009
  512'.5–dc22   2008028930

Editor-in-Chief, Collegiate Mathematics: *Deirdre Lynch*
Senior Acquisitions Editor: *William Hoffman*
Associate Editor/Print Supplements: *Caroline Celano*
Project Manager: *Raegan Keida Heerema*
Production Coordination/Composition: *ICC Macmillan Inc.*
Associate Managing Editor: *Bayani Mendoza deLeon*
Senior Managing Editor: *Linda Mihatov Behrens*
Senior Operations Supervisor: *Alan Fischer*
Operations Specialist: *Lisa McDowell*
Executive Marketing Manager: *Jeff Weidenaar*
Marketing Assistant: *Jon Connelly*
Art Director: *Maureen Eide*
Interior Designer: *Dina Curro*
Cover Designer: *Daniel Conte*
AV Project Manager: *Thomas Benfatti*
Manager, Cover Visual Research & Permissions: *Karen Sanatar*
Art Studio: *Laserwords Private Limited*
Cover Image Credit: *Andy Burmfield/Istockphoto.com*

The cover is a close-up image of a guitar, showing four vibrating strings and the sound hole. Using Fourier analysis, we can measure the amplitudes of a plucked string's harmonics or even synthesize the guitar's sound. See pages 239–244.

Prentice Hall
is an imprint of

© 2009, 2005, 2001, 1997 Pearson Education, Inc.
Pearson Prentice Hall
Pearson Education, Inc.
Upper Saddle River, NJ 07458

Printed in the United States of America
10   9   8   7   6   5

ISBN-10:   0-13-600926-3
ISBN-13: 978-0-13-600926-9

Pearson Education Ltd., *London*
Pearson Education Singapore, Pte. Ltd
Pearson Education Canada, Inc.
Pearson Education-Japan
Pearson Education Australia PTY, Limited

Pearson Education North Asia, Ltd., *Hong Kong*
Pearson Education de Mexico, S.A. de C.V.
Pearson Education Malaysia, Pte. Ltd.
Pearson Education Upper Saddle River, New Jersey

Linear Algebra with Applications

To my parents
Otto and Margrit Bretscher-Zwicky
with love and gratitude

Contents

Text Features ix

Preface xi

1 Linear Equations   1

1.1 Introduction to Linear Systems   1
1.2 Matrices, Vectors, and Gauss–Jordan Elimination   8
1.3 On the Solutions of Linear Systems; Matrix Algebra   25

2 Linear Transformations   40

2.1 Introduction to Linear Transformations and
     Their Inverses   40
2.2 Linear Transformations in Geometry   54
2.3 Matrix Products   69
2.4 The Inverse of a Linear Transformation   79

3 Subspaces of $\mathbb{R}^n$ and Their Dimensions   101

3.1 Image and Kernel of a Linear Transformation   101
3.2 Subspaces of $\mathbb{R}^n$; Bases and Linear Independence   113
3.3 The Dimension of a Subspace of $\mathbb{R}^n$ 123
3.4 Coordinates   137

4 Linear Spaces   153

4.1 Introduction to Linear Spaces   153
4.2 Linear Transformations and Isomorphisms   165
4.3 The Matrix of a Linear Transformation   172

5 Orthogonality and Least Squares   187

5.1 Orthogonal Projections and Orthonormal Bases   187
5.2 Gram–Schmidt Process and QR Factorization   203
5.3 Orthogonal Transformations and Orthogonal Matrices   210
5.4 Least Squares and Data Fitting   220
5.5 Inner Product Spaces   233

6   Determinants   249

6.1   Introduction to Determinants   249
6.2   Properties of the Determinant   261
6.3   Geometrical Interpretations of the Determinant;
      Cramer's Rule   277

7   Eigenvalues and Eigenvectors   294

7.1   Dynamical Systems and Eigenvectors:
      An Introductory Example   294
7.2   Finding the Eigenvalues of a Matrix   308
7.3   Finding the Eigenvectors of a Matrix   319
7.4   Diagonalization   332
7.5   Complex Eigenvalues   343
7.6   Stability   357

8   Symmetric Matrices and Quadratic Forms   367

8.1   Symmetric Matrices   367
8.2   Quadratic Forms   376
8.3   Singular Values   385

9   Linear Differential Equations   397

9.1   An Introduction to Continuous Dynamical Systems   397
9.2   The Complex Case: Euler's Formula   410
9.3   Linear Differential Operators and
      Linear Differential Equations   423

Appendix A Vectors 437
Answers to Odd-Numbered Exercises 446
Subject Index 471
Name Index 478

Text Features

Continuing Text Features

- *Linear transformations* are introduced early on in the text to make the discussion of matrix operations more meaningful and easier to visualize.
- *Visualization and geometrical interpretation* are emphasized extensively throughout.
- The reader will find an abundance of *thought-provoking* (and occasionally delightful) *problems and exercises.*
- *Abstract concepts* are introduced gradually throughout the text. The major ideas are carefully developed at various levels of generality before the student is introduced to abstract vector spaces.
- *Discrete and continuous dynamical systems* are used as a motivation for eigenvectors, and as a unifying theme thereafter.

New Features in the Fourth Edition

Students and instructors generally found the third edition to be accurate and well structured. While preserving the overall organization and character of the text, some changes seemed in order.

- A large number of exercises have been added to the problem sets, from the elementary to the challenging. For example, two dozen new exercises on conic and cubic curve fitting lead up to a discussion of the Cramer-Euler Paradox on fitting a cubic through nine points.
- The section on matrix products now precedes the discussion of the inverse of a matrix, making the presentation more sensible from an algebraic point of view.
- Striking a balance between the earlier editions, the determinant is defined in terms of "patterns", a transparent way to deal with permutations. Laplace expansion and Gaussian elimination are presented as alternative approaches.
- There have been hundreds of small editorial improvements—offering a hint in a difficult problem for example—or choosing a more sensible notation in a theorem.

Preface (with David Steinsaltz)

A police officer on patrol at midnight, so runs an old joke, notices a man crawling about on his hands and knees under a streetlamp. He walks over to investigate, whereupon the man explains in a tired and somewhat slurred voice that he has lost his housekeys. The policeman offers to help, and for the next five minutes he too is searching on his hands and knees. At last he exclaims, "Are you absolutely certain that this is where you dropped the keys?"

"Here? Absolutely not. I dropped them a block down, in the middle of the street."

"Then why the devil have you got me hunting around this lamppost?"

"Because this is where the light is."

It is mathematics, and not just (as Bismarck claimed) politics, that consists in "the art of the possible." Rather than search in the darkness for solutions to problems of pressing interest, we contrive a realm of problems whose interest lies above all in the fact that solutions can conceivably be found.

Perhaps the largest patch of light surrounds the techniques of matrix arithmetic and algebra, and in particular matrix multiplication and row reduction. Here we might begin with Descartes, since it was he who discovered the conceptual meeting-point of geometry and algebra in the identification of Euclidean space with $\mathbb{R}^3$; the techniques and applications proliferated since his day. To organize and clarify those is the role of a modern linear algebra course.

Computers and Computation

An essential issue that needs to be addressed in establishing a mathematical methodology is the role of computation and of computing technology. Are the proper subjects of mathematics algorithms and calculations, or are they grand theories and abstractions that evade the need for computation? If the former, is it important that the students learn to carry out the computations with pencil and paper, or should the algorithm "press the calculator's x^{-1} button" be allowed to substitute for the traditional method of finding an inverse? If the latter, should the abstractions be taught through elaborate notational mechanisms or through computational examples and graphs?

We seek to take a consistent approach to these questions: Algorithms and computations are primary, and precisely for this reason computers are not. Again and again we examine the nitty-gritty of row reduction or matrix multiplication in order to derive new insights. Most of the proofs, whether of rank-nullity theorem, the volume-change formula for determinants, or the spectral theorem for symmetric matrices, are in this way tied to hands-on procedures.

The aim is not just to know how to compute the solution to a problem, but to *imagine* the computations. The student needs to perform enough row reductions by hand to be equipped to follow a line of argument of the form: "If we calculate the reduced row echelon form of such a matrix . . . ," and to appreciate in advance the possible outcomes of a particular computation.

In applications the solution to a problem is hardly more important than recognizing its range of validity and appreciating how sensitive it is to perturbations of the input. We emphasize the geometric and qualitative nature of the solutions, notions of approximation, stability, and "typical" matrices. The discussion of Cramer's rule, for instance, underscores the value of closed-form solutions for visualizing a system's behavior and understanding its dependence from initial conditions.

The availability of computers is, however, neither to be ignored nor regretted. Each student and instructor will have to decide how much practice is needed to be sufficiently familiar with the inner workings of the algorithm. As the explicit computations are being replaced gradually by a theoretical overview of how the algorithm works, the burden of calculation will be taken up by technology, particularly for those wishing to carry out the more numerical and applied exercises.

Examples, Exercises, Applications, and History

The exercises and examples are the heart of this book. Our objective is not just to show our readers a "patch of light" where questions may be posed and solved, but to convince them that there is indeed a great deal of useful, interesting material to be found in this area if they take the time to look around. Consequently, we have included genuine applications of the ideas and methods under discussion to a broad range of sciences: physics, chemistry, biology, economics, and, of course, mathematics itself. Often we have simplified them to sharpen the point, but they use the methods and models of contemporary scientists.

With such a large and varied set of exercises in each section, instructors should have little difficulty in designing a course that is suited to their aims and to the needs of their students. Quite a few straightforward computation problems are offered, of course. Simple (and, in a few cases, not so simple) proofs and derivations are required in some exercises. In many cases, theoretical principles that are discussed at length in more abstract linear algebra courses are here found broken up in bite-size exercises.

The examples make up a significant portion of the text; we have kept abstract exposition to a minimum. It is a matter of taste whether general theories should give rise to specific examples or be pasted together from them. In a text such as this one, attempting to keep an eye on applications, the latter is clearly preferable: The examples always precede the theorems in this book.

Scattered throughout the mathematical exposition are quite a few names and dates, some historical accounts, and anecdotes. Students of mathematics are too rarely shown that the seemingly strange and arbitrary concepts they study are the results of long and hard struggles. It will encourage the readers to know that a mere two centuries ago some of the most brilliant mathematicians were wrestling with problems such as the meaning of dimension or the interpretation of e^{it}, and to realize that the advance of time and understanding actually enables them, with some effort of their own, to see farther than those great minds.

Outline of the Text

Chapter 1   This chapter provides a careful introduction to the solution of systems of linear equations by **Gauss–Jordan elimination.** Once the concrete problem is solved, we restate it in terms of matrix formalism and discuss the geometric properties of the solutions.

Chapter 2   Here we raise the abstraction a notch and reinterpret matrices as **linear transformations.** The reader is introduced to the modern notion of a function, as an arbitrary association between an input and an output, which leads into a discussion of inverses. The traditional method for finding the inverse of a matrix is explained: It fits in naturally as a sort of automated algorithm for Gauss–Jordan elimination.

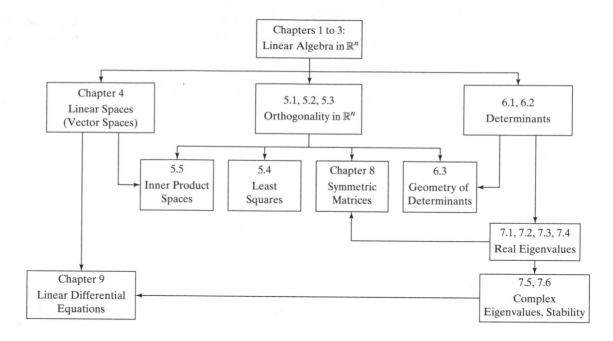

We define linear transformations primarily in terms of matrices, since that is how they are used; the abstract concept of linearity is presented as an auxiliary notion. Rotations, reflections, and orthogonal projections in $\mathbb{R}^2$ are emphasized, both as archetypal, easily visualized examples, and as preparation for future applications.

Chapter 3   We introduce the central concepts of linear algebra: subspaces, image and kernel, linear independence, bases, coordinates, and **dimension,** still firmly fixed in $\mathbb{R}^n$.

Chapter 4   Generalizing the ideas of the preceding chapter and using an abundance of examples, we introduce abstract vector spaces (which are called **linear spaces** here, to prevent the confusion some students experience with the term "vector").

Chapter 5   This chapter includes some of the most basic applications of linear algebra to **geometry** and **statistics.** We introduce orthonormal bases and the Gram–Schmidt process, along with the QR factorization. The calculation of correlation coefficients is discussed, and the important technique of least-squares approximations is explained, in a number of different contexts.

Chapter 6   Our discussion of **determinants** is algorithmic, based on the counting of patterns (a transparent way to deal with permutations). We derive the properties of the determinant from careful analysis of this procedure, tying it together with Gauss–Jordan elimination. The goal is to prepare for the main application of determinants: the computation of characteristic polynomials.

Chapter 7   This chapter introduces the central application of the latter half of the text: linear **dynamical systems.** We begin with discrete systems and are naturally led to seek **eigenvectors,** which characterize the long-term behavior of the system. Qualitative behavior is emphasized, particularly stability conditions. Complex eigenvalues are explained, without apology, and tied into earlier discussions of two-dimensional rotation matrices.

Chapter 8   The ideas and methods of Chapter 7 are applied to geometry. We discuss the spectral theorem for **symmetric matrices** and its applications to quadratic forms, conic sections, and singular values.

Chapter 9   Here we apply the methods developed for discrete dynamical systems to continuous ones, that is, to systems of first-order **linear differential equations.** Again, the cases of real and complex eigenvalues are discussed.

Solutions Manuals

- *Student's Solutions Manual*, with carefully worked solutions to all odd-numbered problems in the text (ISBN 0-13-600927-1)
- *Instructor's Solutions Manual*, with solutions to all the problems in the text (ISBN 0-13-600928-X)

Acknowledgments

I first thank my students and colleagues at Harvard University, Colby College, and Koç University (Istanbul) for the key role they have played in developing this text out of a series of rough lecture notes. The following colleagues, who have taught the course with me, have made invaluable contributions:

Attila Aşkar	Fernando Gouvêa	Barry Mazur
Persi Diaconis	Jan Holly	David Mumford
Jordan Ellenberg	Varga Kalantarov	David Steinsaltz
Matthew Emerton	David Kazhdan	Shlomo Sternberg
Edward Frenkel	Oliver Knill	Richard Taylor
Alexandru Ghitza	Leo Livshits	George Welch

I am grateful to those who have taught me algebra: Peter Gabriel, Volker Strassen, and Bartel van der Waerden at the University of Zürich; John Tate at Harvard University; Maurice Auslander at Brandeis University; and many more.

I owe special thanks to John Boller, William Calder, Devon Ducharme, and Robert Kaplan for their thoughtful review of the manuscript.

I wish to thank Sylvie Bessette, Dennis Kletzing, and Laura Lawrie for the careful preparation of the manuscript and Paul Nguyen for his well-drawn figures.

Special thanks go to Kyle Burke, who has been my assistant at Colby College for three years, having taken linear algebra with me as a first-year student. Paying close attention to the way his students understood or failed to understand the material, Kyle came up with new ways to explain many of the more troubling concepts (such as the kernel of a matrix or the notion of an isomorphism). Many of his ideas have found their way into this text. Kyle took the time to carefully review various iterations of the manuscript, and his suggestions and contributions have made this a much better text.

I am grateful to those who have contributed to the book in many ways: Marilyn Baptiste, Menoo Cung, Srdjan Divac, Devon Ducharme, Robin Gottlieb, Luke Hunsberger, George Mani, Bridget Neale, Alec van Oot, Akilesh Palanisamy, Rita Pang, Esther Silberstein, Radhika de Silva, Jonathan Tannenhauser, Ken Wada-Shiozaki, Larry Wilson, and Jingjing Zhou.

I have received valuable feedback from the book's reviewers for the various editions:

Hirotachi Abo, *University of Idaho*

Loren Argabright, *Drexel University*

Stephen Bean, *Cornell College*

Frank Beatrous, *University of Pittsburgh*

Tracy Bibelnieks, *University of Minnesota*

Jeff D. Farmer, *University of Northern* Colorado

Michael Goldberg, *Johns Hopkins University*

Herman Gollwitzer, *Drexel University*

Fernando Gouvêa, *Colby College*

David G. Handron, Jr., *Carnegie Mellon University*

Christopher Heil, *Georgia Institute of Technology*

Willy Hereman, *Colorado School of Mines*

Konrad J. Heuvers, *Michigan Technological University*

Charles Holmes, *Miami University*

Matthew Hudelson, *Washington State University*

Thomas Hunter, *Swarthmore College*

Michael Kallaher, *Washington State University*

Daniel King, *Oberlin College*

Richard Kubelka, *San Jose State University*

Michael G. Neubauer, *California State University–Northridge*

Peter C. Patton, *University of Pennsylvania*

V. Rao Potluri, *Reed College*

Jeffrey M. Rabin, *University of California, San Diego*

Daniel B. Shapiro, *Ohio State University*

David Steinsaltz, *Technische Universität Berlin*

James A. Wilson, *Iowa State University*

Darryl Yong, *Harvey Mudd College*

Jay Zimmerman, *Towson University*

I thank my editors, Caroline Celano and Bill Hoffman, for their unfailing encouragement and thoughtful advice.

The development of this text has been supported by a grant from the Instructional Innovation Fund of Harvard College.

I love to hear from the users of this text. Feel free to write to *obretsch@colby.edu* with any comments or concerns.

Otto Bretscher
www.colby.edu/~obretsch

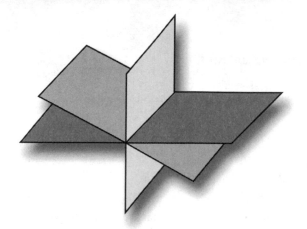

Linear Equations

1.1 Introduction to Linear Systems

Traditionally, algebra was the art of solving equations and systems of equations. The word *algebra* comes from the Arabic *al-jabr* (الجبر), which means *restoration* (of broken parts).[1] The term was first used in a mathematical sense by Mohammed al-Khowarizmi (c. 780–850), who worked at the House of Wisdom, an academy established by Caliph al-Ma'mun in Baghdad. Linear algebra, then, is the art of solving systems of linear equations.

The need to solve systems of linear equations frequently arises in mathematics, statistics, physics, astronomy, engineering, computer science, and economics.

Solving systems of linear equations is not conceptually difficult. For small systems, ad hoc methods certainly suffice. Larger systems, however, require more systematic methods. The approach generally used today was beautifully explained 2,000 years ago in a Chinese text, the *Nine Chapters on the Mathematical Art* (Jiuzhang Suanshu, 九章算術).[2] Chapter 8 of that text, called *Method of Rectangular Arrays* (Fang Cheng, 方程), contains the following problem:

> The yield of one bundle of inferior rice, two bundles of medium grade rice, and three bundles of superior rice is 39 *dou* of grain.[3] The yield of one bundle of inferior rice, three bundles of medium grade rice, and two bundles of superior rice is 34 *dou*. The yield of three bundles of inferior rice, two bundles of medium grade rice, and one bundle of superior rice is 26 *dou*. What is the yield of one bundle of each grade of rice?

In this problem the unknown quantities are the yields of one bundle of inferior, one bundle of medium grade, and one bundle of superior rice. Let us denote these quantities by x, y, and z, respectively. The problem can then be represented by the

---

[1] At one time, it was not unusual to see the sign *Algebrista y Sangrador* (bone setter and blood letter) at the entrance of a Spanish barber's shop.

[2] Shen Kangshen et al. (ed.), *The Nine Chapters on the Mathematical Art*, Companion and Commentary, Oxford University Press, 1999.

[3] The *dou* is a measure of volume, corresponding to about 2 liters at that time.

following system of linear equations:

$$\begin{vmatrix} x + 2y + 3z = 39 \\ x + 3y + 2z = 34 \\ 3x + 2y + z = 26 \end{vmatrix}.$$

To solve for x, y, and z, we need to transform this system from the form

$$\begin{vmatrix} x + 2y + 3z = 39 \\ x + 3y + 2z = 34 \\ 3x + 2y + z = 26 \end{vmatrix} \quad \text{into the form} \quad \begin{vmatrix} x = \ldots \\ y = \ldots \\ z = \ldots \end{vmatrix}.$$

In other words, we need to eliminate the terms that are off the diagonal, those circled in the following equations, and make the coefficients of the variables along the diagonal equal to 1:

$$x \ + \ \textcircled{2y} \ + \ \textcircled{3z} \ = \ 39$$
$$\textcircled{x} \ + \ 3y \ + \ \textcircled{2z} \ = \ 34$$
$$\textcircled{3x} \ + \ \textcircled{2y} \ + \ z \ = \ 26.$$

We can accomplish these goals step by step, one variable at a time. In the past, you may have simplified systems of equations by adding equations to one another or subtracting them. In this system, we can eliminate the variable x from the second equation by subtracting the first equation from the second:

$$\begin{vmatrix} x + 2y + 3z = 39 \\ x + 3y + 2z = 34 \\ 3x + 2y + z = 26 \end{vmatrix} \quad \begin{array}{c} \longrightarrow \\ -1\text{st equation} \end{array} \quad \begin{vmatrix} x + 2y + 3z = 39 \\ y - z = -5 \\ 3x + 2y + z = 26 \end{vmatrix}.$$

To eliminate the variable x from the third equation, we subtract the first equation from the third equation three times. We multiply the first equation by 3 to get

$$3x + 6y + 9z = 117 \qquad (3 \times 1\text{st equation})$$

and then subtract this result from the third equation:

$$\begin{vmatrix} x + 2y + 3z = 39 \\ y - z = -5 \\ 3x + 2y + z = 26 \end{vmatrix} \quad \begin{array}{c} \longrightarrow \\ -3 \times 1\text{st equation} \end{array} \quad \begin{vmatrix} x + 2y + 3z = 39 \\ y - z = -5 \\ {-4y} - 8z = -91 \end{vmatrix}.$$

Similarly, we eliminate the variable y above and below the diagonal:

$$\begin{vmatrix} x + 2y + 3z = 39 \\ y - z = -5 \\ {-4y} - 8z = -91 \end{vmatrix} \quad \begin{array}{c} -2 \times 2\text{nd equation} \\ \longrightarrow \\ +4 \times 2\text{nd equation} \end{array} \quad \begin{vmatrix} x + 5z = 49 \\ y - z = -5 \\ {-12z} = -111 \end{vmatrix}.$$

Before we eliminate the variable z above the diagonal, we make the coefficient of z on the diagonal equal to 1, by dividing the last equation by -12:

$$\begin{vmatrix} x + 5z = 49 \\ y - z = -5 \\ {-12z} = -111 \end{vmatrix} \quad \begin{array}{c} \longrightarrow \\ \div(-12) \end{array} \quad \begin{vmatrix} x + 5z = 49 \\ y - z = -5 \\ z = 9.25 \end{vmatrix}.$$

Finally, we eliminate the variable z above the diagonal:

$$\begin{vmatrix} x + 5z = 49 \\ y - z = -5 \\ z = 9.25 \end{vmatrix} \quad \begin{array}{c} -5 \times \text{third equation} \\ + \text{third equation} \\ \longrightarrow \end{array} \quad \begin{vmatrix} x = 2.75 \\ y = 4.25 \\ z = 9.25 \end{vmatrix}.$$

The yields of inferior, medium grade, and superior rice are 2.75, 4.25, and 9.25 *dou* per bundle, respectively.

By substituting these values, we can check that $x = 2.75$, $y = 4.25$, $z = 9.25$ is indeed the solution of the system:

$$2.75 + 2 \times 4.25 + 3 \times 9.25 = 39$$
$$2.75 + 3 \times 4.25 + 2 \times 9.25 = 34$$
$$3 \times 2.75 + 2 \times 4.25 + \qquad 9.25 = 26.$$

Happily, in linear algebra, you are almost always able to check your solutions. It will help you if you get into the habit of checking now.

Geometric Interpretation

How can we interpret this result geometrically? Each of the three equations of the system defines a plane in x–y–z-space. The solution set of the system consists of those points (x, y, z) that lie in all three planes (i.e., the intersection of the three planes). Algebraically speaking, the solution set consists of those ordered triples of numbers (x, y, z) that satisfy all three equations simultaneously. Our computations show that the system has only one solution, $(x, y, z) = (2.75, 4.25, 9.25)$. This means that the planes defined by the three equations intersect at the point $(x, y, z) = (2.75, 4.25, 9.25)$, as shown in Figure 1.

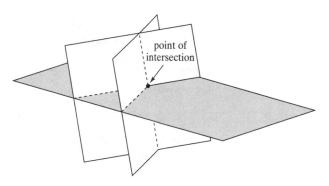

Figure 1 Three planes in space, intersecting at a point.

While three different planes in space usually intersect at a point, they may have a line in common (see Figure 2a) or may not have a common intersection at all, as shown in Figure 2b. Therefore, a system of three equations with three unknowns may have a unique solution, infinitely many solutions, or no solutions at all.

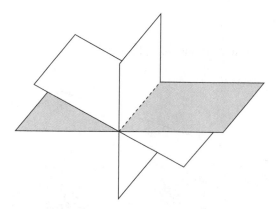

Figure 2(a) Three planes having a line in common.

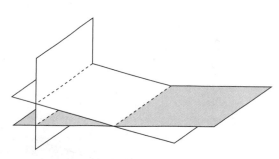

Figure 2(b) Three planes with no common intersection.

A System with Infinitely Many Solutions

Next, let's consider a system of linear equations that has infinitely many solutions:

$$\begin{vmatrix} 2x + 4y + 6z = 0 \\ 4x + 5y + 6z = 3 \\ 7x + 8y + 9z = 6 \end{vmatrix}.$$

We can solve this system using elimination as previously discussed. For simplicity, we label the equations with Roman numerals.

$$\begin{vmatrix} 2x + 4y + 6z = 0 \\ 4x + 5y + 6z = 3 \\ 7x + 8y + 9z = 6 \end{vmatrix} \begin{matrix} \div 2 \\ \longrightarrow \\ \end{matrix} \begin{vmatrix} x + 2y + 3z = 0 \\ 4x + 5y + 6z = 3 \\ 7x + 8y + 9z = 6 \end{vmatrix} \begin{matrix} \longrightarrow \\ -4\,(\mathrm{I}) \\ -7\,(\mathrm{I}) \end{matrix}$$

$$\begin{vmatrix} x + 2y + 3z = 0 \\ -3y - 6z = 3 \\ -6y - 12z = 6 \end{vmatrix} \begin{matrix} \\ \longrightarrow \\ \div(-3) \end{matrix} \begin{vmatrix} x + 2y + 3z = 0 \\ y + 2z = -1 \\ -6y - 12z = 6 \end{vmatrix} \begin{matrix} -2\,(\mathrm{II}) \\ \longrightarrow \\ +6\,(\mathrm{II}) \end{matrix}$$

$$\begin{vmatrix} x - z = 2 \\ y + 2z = -1 \\ 0 = 0 \end{vmatrix} \longrightarrow \begin{vmatrix} x - z = 2 \\ y + 2z = -1 \end{vmatrix}$$

After omitting the trivial equation $0 = 0$, we are left with only two equations with three unknowns. The solution set is the intersection of two nonparallel planes in space (i.e., a line). This system has infinitely many solutions.

The two foregoing equations can be written as follows:

$$\begin{vmatrix} x = z + 2 \\ y = -2z - 1 \end{vmatrix}.$$

We see that both x and y are determined by z. We can freely choose a value of z, an arbitrary real number; then the two preceding equations give us the values of x and y for this choice of z. For example,

- Choose $z = 1$. Then $x = z + 2 = 3$ and $y = -2z - 1 = -3$. The solution is $(x, y, z) = (3, -3, 1)$.
- Choose $z = 7$. Then $x = z + 2 = 9$ and $y = -2z - 1 = -15$. The solution is $(x, y, z) = (9, -15, 7)$.

More generally, if we choose $z = t$, an arbitrary real number, we get $x = t + 2$ and $y = -2t - 1$. Therefore, the general solution is

$$(x, y, z) = (t + 2, -2t - 1, t) = (2, -1, 0) + t(1, -2, 1).$$

This equation represents a line in space, as shown in Figure 3.

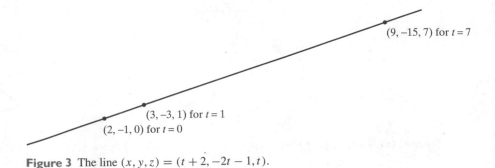

Figure 3 The line $(x, y, z) = (t + 2, -2t - 1, t)$.

A System without Solutions

In the following system, perform the eliminations yourself to obtain the result shown:

$$\begin{vmatrix} x + 2y + 3z = 0 \\ 4x + 5y + 6z = 3 \\ 7x + 8y + 9z = 0 \end{vmatrix} \longrightarrow \begin{vmatrix} x \quad - \quad z = \quad 2 \\ y + 2z = -1 \\ 0 = -6 \end{vmatrix}.$$

Whatever values we choose for x, y, and z, the equation $0 = -6$ cannot be satisfied. This system is *inconsistent*; that is, it has no solutions.

EXERCISES 1.1

GOAL *Set up and solve systems with as many as three linear equations with three unknowns, and interpret the equations and their solutions geometrically.*

In Exercises 1 through 10, find all solutions of the linear systems using elimination as discussed in this section. Then check your solutions.

1. $\begin{vmatrix} x + 2y = 1 \\ 2x + 3y = 1 \end{vmatrix}$

2. $\begin{vmatrix} 4x + 3y = 2 \\ 7x + 5y = 3 \end{vmatrix}$

3. $\begin{vmatrix} 2x + 4y = 3 \\ 3x + 6y = 2 \end{vmatrix}$

4. $\begin{vmatrix} 2x + 4y = 2 \\ 3x + 6y = 3 \end{vmatrix}$

5. $\begin{vmatrix} 2x + 3y = 0 \\ 4x + 5y = 0 \end{vmatrix}$

6. $\begin{vmatrix} x + 2y + 3z = 8 \\ x + 3y + 3z = 10 \\ x + 2y + 4z = 9 \end{vmatrix}$

7. $\begin{vmatrix} x + 2y + 3z = 1 \\ x + 3y + 4z = 3 \\ x + 4y + 5z = 4 \end{vmatrix}$

8. $\begin{vmatrix} x + 2y + 3z = 0 \\ 4x + 5y + 6z = 0 \\ 7x + 8y + 10z = 0 \end{vmatrix}$

9. $\begin{vmatrix} x + 2y + 3z = 1 \\ 3x + 2y + z = 1 \\ 7x + 2y - 3z = 1 \end{vmatrix}$

10. $\begin{vmatrix} x + 2y + 3z = 1 \\ 2x + 4y + 7z = 2 \\ 3x + 7y + 11z = 8 \end{vmatrix}$

In Exercises 11 through 13, find all solutions of the linear systems. Represent your solutions graphically, as intersections of lines in the x–y-plane.

11. $\begin{vmatrix} x - 2y = 2 \\ 3x + 5y = 17 \end{vmatrix}$

12. $\begin{vmatrix} x - 2y = 3 \\ 2x - 4y = 6 \end{vmatrix}$

13. $\begin{vmatrix} x - 2y = 3 \\ 2x - 4y = 8 \end{vmatrix}$

In Exercises 14 through 16, find all solutions of the linear systems. Describe your solutions in terms of intersecting planes. You need not sketch these planes.

14. $\begin{vmatrix} x + 4y + z = 0 \\ 4x + 13y + 7z = 0 \\ 7x + 22y + 13z = 1 \end{vmatrix}$

15. $\begin{vmatrix} x + y - z = 0 \\ 4x - y + 5z = 0 \\ 6x + y + 4z = 0 \end{vmatrix}$

16. $\begin{vmatrix} x + 4y + z = 0 \\ 4x + 13y + 7z = 0 \\ 7x + 22y + 13z = 0 \end{vmatrix}$

17. Find all solutions of the linear system

$$\begin{vmatrix} x + 2y = a \\ 3x + 5y = b \end{vmatrix},$$

where a and b are arbitrary constants.

18. Find all solutions of the linear system

$$\begin{vmatrix} x + 2y + 3z = a \\ x + 3y + 8z = b \\ x + 2y + 2z = c \end{vmatrix},$$

where a, b, and c are arbitrary constants.

19. Consider a two-commodity market. When the unit prices of the products are P_1 and P_2, the quantities demanded, D_1 and D_2, and the quantities supplied, S_1 and S_2, are given by

$$D_1 = 70 - 2P_1 + P_2, \quad S_1 = -14 + 3P_1,$$
$$D_2 = 105 + P_1 - P_2, \quad S_2 = -7 + 2P_2.$$

a. What is the relationship between the two commodities? Do they compete, as do Volvos and BMWs, or do they complement one another, as do shirts and ties?

b. Find the equilibrium prices (i.e., the prices for which supply equals demand), for both products.

20. The Russian-born U.S. economist and Nobel laureate Wassily Leontief (1906–1999) was interested in the following question: What output should each of the industries in an economy produce to satisfy the total demand for all products? Here, we consider a very simple example of input-output analysis, an economy with only two industries, A and B. Assume that the consumer demand for their products is, respectively, 1,000 and 780, in millions of dollars per year.

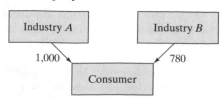

What outputs a and b (in millions of dollars per year) should the two industries generate to satisfy the demand?

You may be tempted to say 1,000 and 780, respectively, but things are not quite as simple as that. We have to take into account the interindustry demand as well. Let us say that industry A produces electricity. Of course, producing almost any product will require electric power. Suppose that industry B needs 10¢worth of electricity for each $1 of output B produces and that industry A needs 20¢worth of B's products for each $1 of output A produces. Find the outputs a and b needed to satisfy both consumer and interindustry demand.

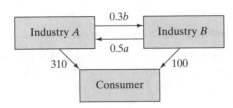

21. Find the outputs a and b needed to satisfy the consumer and interindustry demands given in the following figure (see Exercise 20):

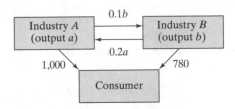

22. Consider the differential equation

$$\frac{d^2x}{dt^2} - \frac{dx}{dt} - x = \cos(t).$$

This equation could describe a forced damped oscillator, as we will see in Chapter 9. We are told that the differential equation has a solution of the form

$$x(t) = a\sin(t) + b\cos(t).$$

Find a and b, and graph the solution.

23. Find all solutions of the system

$$\begin{vmatrix} 7x - y = \lambda x \\ -6x + 8y = \lambda y \end{vmatrix}, \quad \text{for}$$

a. $\lambda = 5$ **b.** $\lambda = 10$,   and   **c.** $\lambda = 15$.

24. On your next trip to Switzerland, you should take the scenic boat ride from Rheinfall to Rheinau and back. The trip downstream from Rheinfall to Rheinau takes 20 minutes, and the return trip takes 40 minutes; the distance between Rheinfall and Rheinau along the river is 8 kilometers. How fast does the boat travel (relative to the water), and how fast does the river Rhein flow in this area? You may assume both speeds to be constant throughout the journey.

25. Consider the linear system

$$\begin{vmatrix} x + y - z = -2 \\ 3x - 5y + 13z = 18 \\ x - 2y + 5z = k \end{vmatrix},$$

where k is an arbitrary number.

a. For which value(s) of k does this system have one or infinitely many solutions?

b. For each value of k you found in part a, how many solutions does the system have?

c. Find all solutions for each value of k.

26. Consider the linear system

$$\begin{vmatrix} x + y - z = 2 \\ x + 2y + z = 3 \\ x + y + (k^2 - 5)z = k \end{vmatrix},$$

where k is an arbitrary constant. For which value(s) of k does this system have a unique solution? For which value(s) of k does the system have infinitely many solutions? For which value(s) of k is the system inconsistent?

27. Emile and Gertrude are brother and sister. Emile has twice as many sisters as brothers, and Gertrude has just as many brothers as sisters. How many children are there in this family?

28. In a grid of wires, the temperature at exterior mesh points is maintained at constant values (in °C) as shown in the accompanying figure. When the grid is in thermal equilibrium, the temperature T at each interior mesh point is the average of the temperatures at the four adjacent points. For example,

$$T_2 = \frac{T_3 + T_1 + 200 + 0}{4}.$$

Find the temperatures T_1, T_2, and T_3 when the grid is in thermal equilibrium.

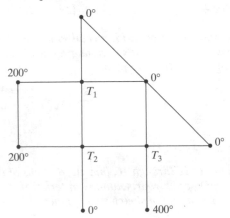

29. Find the polynomial of degree 2 [a polynomial of the form $f(t) = a + bt + ct^2$] whose graph goes through the points $(1, -1)$, $(2, 3)$, and $(3, 13)$. Sketch the graph of this polynomial.

30. Find a polynomial of degree ≤ 2 [a polynomial of the form $f(t) = a + bt + ct^2$] whose graph goes through the points $(1, p)$, $(2, q)$, $(3, r)$, where p, q, r are arbitrary constants. Does such a polynomial exist for all values of p, q, r?

31. Find all the polynomials $f(t)$ of degree ≤ 2 whose graphs run through the points $(1, 3)$ and $(2, 6)$, such that $f'(1) = 1$ [where $f'(t)$ denotes the derivative].

32. Find all the polynomials $f(t)$ of degree ≤ 2 whose graphs run through the points $(1, 1)$ and $(2, 0)$, such that $\int_1^2 f(t)\,dt = -1$.

33. Find all the polynomials $f(t)$ of degree ≤ 2 whose graphs run through the points $(1, 1)$ and $(3, 3)$, such that $f'(2) = 1$.

34. Find all the polynomials $f(t)$ of degree ≤ 2 whose graphs run through the points $(1, 1)$ and $(3, 3)$, such that $f'(2) = 3$.

35. Find the function $f(t)$ of the form $f(t) = ae^{3t} + be^{2t}$ such that $f(0) = 1$ and $f'(0) = 4$.

36. Find the function $f(t)$ of the form $f(t) = a\cos(2t) + b\sin(2t)$ such that $f''(t) + 2f'(t) + 3f(t) = 17\cos(2t)$. (This is the kind of differential equation you might have to solve when dealing with forced damped oscillators, in physics or engineering.)

37. Find the circle that runs through the points $(5, 5)$, $(4, 6)$, and $(6, 2)$. Write your equation in the form $a + bx + cy + x^2 + y^2 = 0$. Find the center and radius of this circle.

38. Find the ellipse centered at the origin that runs through the points $(1, 2)$, $(2, 2)$, and $(3, 1)$. Write your equation in the form $ax^2 + bxy + cy^2 = 1$.

39. Find all points (a, b, c) in space for which the system

$$\begin{vmatrix} x + 2y + 3z = a \\ 4x + 5y + 6z = b \\ 7x + 8y + 9z = c \end{vmatrix}$$

has at least one solution.

40. Linear systems are particularly easy to solve when they are in triangular form (i.e., all entries above or below the diagonal are zero).

    a. Solve the lower triangular system

$$\begin{vmatrix} x_1 & & & = -3 \\ -3x_1 + & x_2 & & = 14 \\ x_1 + 2x_2 + & x_3 & & = 9 \\ -x_1 + 8x_2 - & 5x_3 + x_4 & = 33 \end{vmatrix}$$

by forward substitution, finding x_1 first, then x_2, then x_3, and finally x_4.

    b. Solve the upper triangular system

$$\begin{vmatrix} x_1 + 2x_2 - & x_3 + 4x_4 = -3 \\ x_2 + 3x_3 + 7x_4 = & 5 \\ x_3 + 2x_4 = & 2 \\ x_4 = & 0 \end{vmatrix}.$$

41. Consider the linear system

$$\begin{vmatrix} x + & y = 1 \\ x + \dfrac{t}{2}y = t \end{vmatrix},$$

where t is a nonzero constant.

    a. Determine the x- and y-intercepts of the lines $x + y = 1$ and $x + (t/2)y = t$; sketch these lines. For which values of the constant t do these lines intersect? For these values of t, the point of intersection (x, y) depends on the choice of the constant t; that is, we can consider x and y as functions of t. Draw rough sketches of these functions.

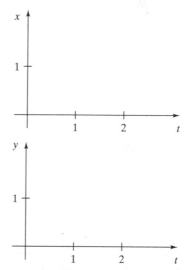

Explain briefly how you found these graphs. Argue geometrically, without solving the system algebraically.

    b. Now solve the system algebraically. Verify that the graphs you sketched in part (a) are compatible with your algebraic solution.

42. Find a system of linear equations with three unknowns whose solutions are the points on the line through $(1, 1, 1)$ and $(3, 5, 0)$.

43. Find a system of linear equations with three unknowns x, y, z whose solutions are

$$x = 6 + 5t, \quad y = 4 + 3t, \quad \text{and} \quad z = 2 + t,$$

where t is an arbitrary constant.

44. Boris and Marina are shopping for chocolate bars. Boris observes, "If I add half my money to yours, it will be enough to buy two chocolate bars." Marina naively asks, "If I add half my money to yours, how many can we buy?" Boris replies, "One chocolate bar." How much money did Boris have? (From Yuri Chernyak and Robert Rose, *The Chicken from Minsk*, Basic Books, 1995.)

45. Here is another method to solve a system of linear equations: Solve one of the equations for one of the variables, and substitute the result into the other equations. Repeat this process until you run out of variables or equations. Consider the example discussed on page 2:

$$\begin{vmatrix} x + 2y + 3z = 39 \\ x + 3y + 2z = 34 \\ 3x + 2y + z = 26 \end{vmatrix}.$$

We can solve the first equation for x:

$$x = 39 - 2y - 3z.$$

Then we substitute this equation into the other equations:

$$\begin{vmatrix} (39 - 2y - 3z) + 3y + 2z = 34 \\ 3(39 - 2y - 3z) + 2y + z = 26 \end{vmatrix}.$$

We can simplify:

$$\begin{vmatrix} y - z = -5 \\ -4y - 8z = -91 \end{vmatrix}.$$

Now, $y = z - 5$, so that $-4(z - 5) - 8z = -91$, or

$$-12z = -111.$$

We find that $z = \dfrac{111}{12} = 9.25$. Then

$$y = z - 5 = 4.25,$$

and

$$x = 39 - 2y - 3z = 2.75.$$

Explain why this method is essentially the same as the method discussed in this section; only the bookkeeping is different.

46. A hermit eats only two kinds of food: brown rice and yogurt. The rice contains 3 grams of protein and 30 grams of carbohydrates per serving, while the yogurt contains 12 grams of protein and 20 grams of carbohydrates.

  a. If the hermit wants to take in 60 grams of protein and 300 grams of carbohydrates per day, how many servings of each item should he consume?

  b. If the hermit wants to take in P grams of protein and C grams of carbohydrates per day, how many servings of each item should he consume?

47. I have 32 bills in my wallet, in the denominations of US$ 1, 5, and 10, worth $100 in total. How many do I have of each denomination?

48. Some parking meters in Milan, Italy, accept coins in the denominations of 20¢, 50¢, and € 2. As an incentive program, the city administrators offer a big reward (a brand new Ferrari Testarossa) to any meter maid who brings back exactly 1,000 coins worth exactly € 1,000 from the daily rounds. What are the odds of this reward being claimed anytime soon?

1.2 Matrices, Vectors, and Gauss–Jordan Elimination

When mathematicians in ancient China had to solve a system of simultaneous linear equations such as[4]

$$\begin{vmatrix} 3x + 21y - 3z = 0 \\ -6x - 2y - z = 62 \\ 2x - 3y + 8z = 32 \end{vmatrix},$$

they took all the numbers involved in this system and arranged them in a rectangular pattern (*Fang Cheng* in Chinese), as follows:[5]

3	21	-3	0
-6	-2	-1	62
2	-3	8	32

All the information about this system is conveniently stored in this array of numbers.

The entries were represented by bamboo rods, as shown below; red and black rods stand for positive and negative numbers, respectively. (Can you detect how this

---

[4]This example is taken from Chapter 8 of the *Nine Chapters on the Mathematical Art*; see page 1. Our source is George Gheverghese Joseph, *The Crest of the Peacock, Non-European Roots of Mathematics*, 2nd ed., Princeton University Press, 2000.

[5]Actually, the roles of rows and columns were reversed in the Chinese representation.

number system works?) The equations were then solved in a hands-on fashion, by manipulating the rods. We leave it to the reader to find the solution.

|||	=|	|||	
$\top$	||			$\perp$ ||
||	|||	|||	=||

Today, such a rectangular array of numbers,

$$\begin{bmatrix} 3 & 21 & -3 & 0 \\ -6 & -2 & -1 & 62 \\ 2 & -3 & 8 & 32 \end{bmatrix},$$

is called a *matrix*.[6] Since this particular matrix has three rows and four columns, it is called a 3×4 matrix ("three by four").

The four columns of the matrix

The three rows of the matrix $\Longleftarrow$ $\begin{bmatrix} 3 & 21 & -3 & 0 \\ -6 & -2 & -1 & 62 \\ 2 & -3 & 8 & 32 \end{bmatrix}$

Note that the first column of this matrix corresponds to the first variable of the system, while the first row corresponds to the first equation.

It is customary to label the entries of a 3×4 matrix A with double subscripts as follows:

$$A = \begin{bmatrix} a_{11} & a_{12} & a_{13} & a_{14} \\ a_{21} & a_{22} & a_{23} & a_{24} \\ a_{31} & a_{32} & a_{33} & a_{34} \end{bmatrix}.$$

The first subscript refers to the row, and the second to the column: The entry a_{ij} is located in the ith row and the jth column.

Two matrices A and B are equal if they are the same size and if corresponding entries are equal: $a_{ij} = b_{ij}$.

If the number of rows of a matrix A equals the number of columns (A is $n \times n$), then A is called a *square matrix*, and the entries $a_{11}, a_{22}, \ldots, a_{nn}$ form the (main) *diagonal* of A. A square matrix A is called *diagonal* if all its entries above and below the main diagonal are zero; that is, $a_{ij} = 0$ whenever $i \neq j$. A square matrix A is called *upper triangular* if all its entries below the main diagonal are zero; that is, $a_{ij} = 0$ whenever i exceeds j. *Lower triangular* matrices are defined analogously. A matrix whose entries are all zero is called a *zero matrix* and is denoted by 0 (regardless of its size). Consider the matrices

$$A = \begin{bmatrix} 1 & 2 & 3 \\ 4 & 5 & 6 \end{bmatrix}, \quad B = \begin{bmatrix} 1 & 2 \\ 3 & 4 \end{bmatrix}, \quad C = \begin{bmatrix} 2 & 0 & 0 \\ 0 & 3 & 0 \\ 0 & 0 & 0 \end{bmatrix},$$

$$D = \begin{bmatrix} 2 & 3 \\ 0 & 4 \end{bmatrix}, \quad E = \begin{bmatrix} 5 & 0 & 0 \\ 4 & 0 & 0 \\ 3 & 2 & 1 \end{bmatrix}.$$

---

[6]It appears that the term *matrix* was first used in this sense by the English mathematician J. J. Sylvester, in 1850.

The matrices B, C, D, and E are square, C is diagonal, C and D are upper triangular, and C and E are lower triangular.

Matrices with only one column or row are of particular interest.

Vectors and vector spaces

A matrix with only one column is called a column vector, or simply a vector. The entries of a vector are called its components. The set of all column vectors with n components is denoted by $\mathbb{R}^n$; we will refer to $\mathbb{R}^n$ as a *vector space*.

A matrix with only one row is called a row vector.

In this text, the term *vector* refers to column vectors, unless otherwise stated. The reason for our preference for column vectors will become apparent in the next section.

Examples of vectors are

$$\begin{bmatrix} 1 \\ 2 \\ 9 \\ 1 \end{bmatrix},$$

a (column) vector in $\mathbb{R}^4$, and

$$\begin{bmatrix} 1 & 5 & 5 & 3 & 7 \end{bmatrix},$$

a row vector with five components. Note that the m columns of an $n \times m$ matrix are vectors in $\mathbb{R}^n$.

In previous courses in mathematics or physics, you may have thought about vectors from a more geometric point of view. (See the Appendix for a summary of basic facts on vectors.) Let's establish some conventions regarding the geometric representation of vectors.

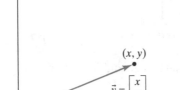

Figure 1

Standard representation of vectors

The standard representation of a vector

$$\vec{v} = \begin{bmatrix} x \\ y \end{bmatrix}$$

in the Cartesian coordinate plane is as an *arrow* (a directed line segment) from the origin to the point (x, y), as shown in Figure 1.

The standard representation of a vector in $\mathbb{R}^3$ is defined analogously.

In this text, we will consider the standard representation of vectors, unless stated otherwise.

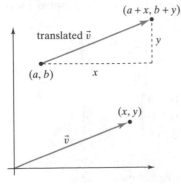

Figure 2

Occasionally, it is helpful to translate (or shift) the vector in the plane (preserving its direction and length), so that it will connect some point (a, b) to the point $(a + x, b + y)$, as shown in Figure 2.

When considering an infinite set of vectors, the arrow representation becomes impractical. In this case, it is sensible to represent the vector $\vec{v} = \begin{bmatrix} x \\ y \end{bmatrix}$ simply by the point (x, y), the head of the standard arrow representation of $\vec{v}$.

For example, the set of all vectors $\vec{v} = \begin{bmatrix} x \\ x + 1 \end{bmatrix}$ (where x is arbitrary) can be represented as the *line* $y = x + 1$. For a few special values of x we may still use the arrow representation, as illustrated in Figure 3.

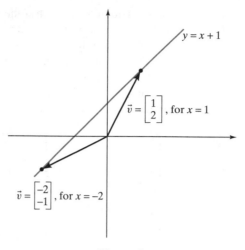

Figure 3

In this course, it will often be helpful to think about a vector numerically, as a list of numbers, which we will usually write in a column.

In our digital age, information is often transmitted and stored as a string of numbers (i.e., as a vector). A section of 10 seconds of music on a CD is stored as a vector with 440,000 components. A weather photograph taken by a satellite is transmitted to Earth as a string of numbers.

Consider the system

$$\begin{vmatrix} 2x + 8y + 4z = 2 \\ 2x + 5y + z = 5 \\ 4x + 10y - z = 1 \end{vmatrix}.$$

Sometimes we are interested in the matrix

$$\begin{bmatrix} 2 & 8 & 4 \\ 2 & 5 & 1 \\ 4 & 10 & -1 \end{bmatrix},$$

which contains the coefficients of the system, called its *coefficient matrix*.

By contrast, the matrix

$$\begin{bmatrix} 2 & 8 & 4 & 2 \\ 2 & 5 & 1 & 5 \\ 4 & 10 & -1 & 1 \end{bmatrix},$$

which displays all the numerical information contained in the system, is called its *augmented matrix*. For the sake of clarity, we will often indicate the position of the equal signs in the equations by a dotted line:

$$\begin{bmatrix} 2 & 8 & 4 & \vdots & 2 \\ 2 & 5 & 1 & \vdots & 5 \\ 4 & 10 & -1 & \vdots & 1 \end{bmatrix}.$$

To solve the system, it is more efficient to perform the elimination on the augmented matrix rather than on the equations themselves. Conceptually, the two approaches are equivalent, but working with the augmented matrix requires less writing yet is easier to read, with some practice. Instead of dividing an *equation* by a scalar,[7] you can divide a *row* by a scalar. Instead of adding a multiple of an equation to another equation, you can add a multiple of a row to another row.

As you perform elimination on the augmented matrix, you should always remember the linear system lurking behind the matrix. To illustrate this method, we perform the elimination both on the augmented matrix and on the linear system it represents:

$$
\begin{bmatrix} 2 & 8 & 4 & | & 2 \\ 2 & 5 & 1 & | & 5 \\ 4 & 10 & -1 & | & 1 \end{bmatrix} \div 2
\qquad
\begin{vmatrix} 2x + & 8y + & 4z = & 2 \\ 2x + & 5y + & z = & 5 \\ 4x + & 10y - & z = & 1 \end{vmatrix} \div 2
$$

$$\downarrow \qquad\qquad\qquad \downarrow$$

$$
\begin{bmatrix} 1 & 4 & 2 & | & 1 \\ 2 & 5 & 1 & | & 5 \\ 4 & 10 & -1 & | & 1 \end{bmatrix} \begin{matrix} \\ -2\,(\mathrm{I}) \\ -4\,(\mathrm{I}) \end{matrix}
\qquad
\begin{vmatrix} x + & 4y + & 2z = & 1 \\ 2x + & 5y + & z = & 5 \\ 4x + & 10y - & z = & 1 \end{vmatrix} \begin{matrix} \\ -2\,(\mathrm{I}) \\ -4\,(\mathrm{I}) \end{matrix}
$$

$$\downarrow \qquad\qquad\qquad \downarrow$$

$$
\begin{bmatrix} 1 & 4 & 2 & | & 1 \\ 0 & -3 & -3 & | & 3 \\ 0 & -6 & -9 & | & -3 \end{bmatrix} \begin{matrix} \\ \div(-3) \\ \ \end{matrix}
\qquad
\begin{vmatrix} x + & 4y + & 2z = & 1 \\ & -3y - & 3z = & 3 \\ & -6y - & 9z = & -3 \end{vmatrix} \div(-3)
$$

$$\downarrow \qquad\qquad\qquad \downarrow$$

$$
\begin{bmatrix} 1 & 4 & 2 & | & 1 \\ 0 & 1 & 1 & | & -1 \\ 0 & -6 & -9 & | & -3 \end{bmatrix} \begin{matrix} -4\,(\mathrm{II}) \\ \\ +6\,(\mathrm{II}) \end{matrix}
\qquad
\begin{vmatrix} x + & 4y + & 2z = & 1 \\ & y + & z = & -1 \\ & -6y - & 9z = & -3 \end{vmatrix} \begin{matrix} -4\,(\mathrm{II}) \\ \\ +6\,(\mathrm{II}) \end{matrix}
$$

$$\downarrow \qquad\qquad\qquad \downarrow$$

$$
\begin{bmatrix} 1 & 0 & -2 & | & 5 \\ 0 & 1 & 1 & | & -1 \\ 0 & 0 & -3 & | & -9 \end{bmatrix} \begin{matrix} \\ \\ \div(-3) \end{matrix}
\qquad
\begin{vmatrix} x & - & 2z = & 5 \\ & y + & z = & -1 \\ & & -3z = & -9 \end{vmatrix} \div(-3)
$$

$$\downarrow \qquad\qquad\qquad \downarrow$$

$$
\begin{bmatrix} 1 & 0 & -2 & | & 5 \\ 0 & 1 & 1 & | & -1 \\ 0 & 0 & 1 & | & 3 \end{bmatrix} \begin{matrix} +2\,(\mathrm{III}) \\ -\,(\mathrm{III}) \\ \ \end{matrix}
\qquad
\begin{vmatrix} x & - & 2z = & 5 \\ & y + & z = & -1 \\ & & z = & 3 \end{vmatrix} \begin{matrix} +2\,(\mathrm{III}) \\ -\,(\mathrm{III}) \\ \ \end{matrix}
$$

$$\downarrow \qquad\qquad\qquad \downarrow$$

$$
\begin{bmatrix} 1 & 0 & 0 & | & 11 \\ 0 & 1 & 0 & | & -4 \\ 0 & 0 & 1 & | & 3 \end{bmatrix}
\qquad
\begin{vmatrix} x & & = & 11 \\ & y & = & -4 \\ & & z = & 3 \end{vmatrix}.
$$

The solution is often represented as a vector:

$$
\begin{bmatrix} x \\ y \\ z \end{bmatrix} = \begin{bmatrix} 11 \\ -4 \\ 3 \end{bmatrix}.
$$

Thus far we have been focusing on systems of 3 linear equations with 3 unknowns. Next we will develop a technique for solving systems of linear equations of arbitrary size.

---

[7]In vector and matrix algebra, the term *scalar* is synonymous with (real) number.

Here is an example of a system of three linear equations with five unknowns:

$$\begin{vmatrix} x_1 - x_2 & & & + 4x_5 = 2 \\ & x_3 & & - x_5 = 2 \\ & & x_4 - & x_5 = 3 \end{vmatrix}$$

We can proceed as in the example on page 4. We solve each equation for the leading variable:

$$\begin{vmatrix} x_1 = 2 + x_2 - 4x_5 \\ x_3 = 2 \qquad + x_5 \\ x_4 = 3 \qquad + x_5 \end{vmatrix}.$$

Now we can freely choose values for the nonleading variables, $x_2 = t$ and $x_5 = r$, for example. The leading variables are then determined by these choices:

$$x_1 = 2 + t - 4r, \qquad x_3 = 2 + r, \qquad x_4 = 3 + r.$$

This system has infinitely many solutions; we can write the solutions in vector form as

$$\begin{bmatrix} x_1 \\ x_2 \\ x_3 \\ x_4 \\ x_5 \end{bmatrix} = \begin{bmatrix} 2 & +t & -4r \\ & t & \\ 2 & & +r \\ 3 & & +r \\ & & r \end{bmatrix}.$$

Again, you can check this answer by substituting the solutions into the original equations, for example, $x_3 - x_5 = (2 + r) - r = 2$.

What makes this system so easy to solve? The following three properties are responsible for the simplicity of the solution, with the second property playing a key role:

- P1: The leading coefficient in each equation is 1. (The leading coefficient is the coefficient of the leading variable.)
- P2: The leading variable in each equation does not appear in any of the other equations. (For example, the leading variable x_3 of the second equation appears neither in the first nor in the third equation.)
- P3: The leading variables appear in the "natural order," with increasing indices as we go down the system (x_1, x_3, x_4 as opposed to x_3, x_1, x_4, for example).

Whenever we encounter a linear system with these three properties, we can solve for the leading variables and then choose arbitrary values for the other, nonleading variables, as we did above and on page 4.

Now we are ready to tackle the case of an arbitrary system of linear equations. We will illustrate our approach by means of an example:

$$\begin{vmatrix} 2x_1 + 4x_2 - 2x_3 + 2x_4 + 4x_5 = 2 \\ x_1 + 2x_2 - x_3 + 2x_4 \qquad = 4 \\ 3x_1 + 6x_2 - 2x_3 + x_4 + 9x_5 = 1 \\ 5x_1 + 10x_2 - 4x_3 + 5x_4 + 9x_5 = 9 \end{vmatrix}.$$

We wish to reduce this system to a system satisfying the three properties (P1, P2, and P3); this reduced system will then be easy to solve.

We will proceed from equation to equation, from top to bottom. The leading variable in the first equation is x_1, with leading coefficient 2. To satisfy property P1, we will divide this equation by 2. To satisfy property P2 for the variable x_1, we will then subtract suitable multiples of the first equation from the other three equations to eliminate the variable x_1 from those equations. We will perform these operations both on the system and on the augmented matrix.

$$\begin{vmatrix} 2x_1 + & 4x_2 - & 2x_3 + & 2x_4 + & 4x_5 = & 2 \\ x_1 + & 2x_2 - & x_3 + & 2x_4 & = & 4 \\ 3x_1 + & 6x_2 - & 2x_3 + & x_4 + & 9x_5 = & 1 \\ 5x_1 + & 10x_2 - & 4x_3 + & 5x_4 + & 9x_5 = & 9 \end{vmatrix} \div 2 \qquad \begin{bmatrix} 2 & 4 & -2 & 2 & 4 & \vdots & 2 \\ 1 & 2 & -1 & 2 & 0 & \vdots & 4 \\ 3 & 6 & -2 & 1 & 9 & \vdots & 1 \\ 5 & 10 & -4 & 5 & 9 & \vdots & 9 \end{bmatrix} \div 2$$

$$\downarrow$$

$$\begin{vmatrix} x_1 + & 2x_2 - & x_3 + & x_4 + & 2x_5 = & 1 \\ x_1 + & 2x_2 - & x_3 + & 2x_4 & = & 4 \\ 3x_1 + & 6x_2 - & 2x_3 + & x_4 + & 9x_5 = & 1 \\ 5x_1 + & 10x_2 - & 4x_3 + & 5x_4 + & 9x_5 = & 9 \end{vmatrix} \begin{matrix} \\ -(I) \\ -3(I) \\ -5(I) \end{matrix} \qquad \begin{bmatrix} 1 & 2 & -1 & 1 & 2 & \vdots & 1 \\ 1 & 2 & -1 & 2 & 0 & \vdots & 4 \\ 3 & 6 & -2 & 1 & 9 & \vdots & 1 \\ 5 & 10 & -4 & 5 & 9 & \vdots & 9 \end{bmatrix} \begin{matrix} \\ -(I) \\ -3(I) \\ -5(I) \end{matrix}$$

$$\downarrow$$

$$\begin{vmatrix} x_1 + & 2x_2 - & x_3 + & x_4 + & 2x_5 = & 1 \\ & & & x_4 - & 2x_5 = & 3 \\ & & x_3 - & 2x_4 + & 3x_5 = & -2 \\ & & x_3 & & - x_5 = & 4 \end{vmatrix} \qquad \begin{bmatrix} 1 & 2 & -1 & 1 & 2 & \vdots & 1 \\ 0 & 0 & 0 & 1 & -2 & \vdots & 3 \\ 0 & 0 & 1 & -2 & 3 & \vdots & -2 \\ 0 & 0 & 1 & 0 & -1 & \vdots & 4 \end{bmatrix}$$

Now on to the second equation, with leading variable x_4 and leading coefficient 1. We could eliminate x_4 from the first and third equations and then proceed to the third equation, with leading variable x_3. However, this approach would violate our requirement P3 that the variables must be listed in the natural order, with increasing indices as we go down the system. To satisfy this requirement, we will swap the second equation with the third equation. (In the following summary, we will specify when such a swap is indicated and how it is to be performed.)

Then we can eliminate x_3 from the first and fourth equations.

$$\begin{vmatrix} x_1 + 2x_2 - & x_3 + & x_4 + & 2x_5 = & 1 \\ & x_3 - & 2x_4 + & 3x_5 = & -2 \\ & & x_4 - & 2x_5 = & 3 \\ & x_3 & & - x_5 = & 4 \end{vmatrix} \begin{matrix} +(II) \\ \\ \\ -(II) \end{matrix} \qquad \begin{bmatrix} 1 & 2 & -1 & 1 & 2 & \vdots & 1 \\ 0 & 0 & 1 & -2 & 3 & \vdots & -2 \\ 0 & 0 & 0 & 1 & -2 & \vdots & 3 \\ 0 & 0 & 1 & 0 & -1 & \vdots & 4 \end{bmatrix} \begin{matrix} +(II) \\ \\ \\ -(II) \end{matrix}$$

$$\downarrow$$

$$\begin{vmatrix} x_1 + 2x_2 & & - x_4 + & 5x_5 = & -1 \\ & x_3 - & 2x_4 + & 3x_5 = & -2 \\ & & x_4 - & 2x_5 = & 3 \\ & & 2x_4 - & 4x_5 = & 6 \end{vmatrix} \qquad \begin{bmatrix} 1 & 2 & 0 & -1 & 5 & \vdots & -1 \\ 0 & 0 & 1 & -2 & 3 & \vdots & -2 \\ 0 & 0 & 0 & 1 & -2 & \vdots & 3 \\ 0 & 0 & 0 & 2 & -4 & \vdots & 6 \end{bmatrix}$$

Now we turn our attention to the third equation, with leading variable x_4. We need to eliminate x_4 from the other three equations.

$$\begin{vmatrix} x_1 + 2x_2 & & - x_4 + & 5x_5 = & -1 \\ & x_3 - & 2x_4 & 3x_5 = & -2 \\ & & x_4 - & 2x_5 = & 3 \\ & & 2x_4 - & 4x_5 = & 6 \end{vmatrix} \begin{matrix} +(III) \\ +2(III) \\ \\ -2(III) \end{matrix} \qquad \begin{bmatrix} 1 & 2 & 0 & -1 & 5 & \vdots & -1 \\ 0 & 0 & 1 & -2 & 3 & \vdots & -2 \\ 0 & 0 & 0 & 1 & -2 & \vdots & 3 \\ 0 & 0 & 0 & 2 & -4 & \vdots & 6 \end{bmatrix} \begin{matrix} +(III) \\ +2(III) \\ \\ -2(III) \end{matrix}$$

$$\downarrow$$

$$\begin{vmatrix} x_1 + 2x_2 & & + 3x_5 = & 2 \\ & x_3 & - x_5 = & 4 \\ & & x_4 - 2x_5 = & 3 \\ & & 0 = & 0 \end{vmatrix} \qquad \begin{bmatrix} 1 & 2 & 0 & 0 & 3 & \vdots & 2 \\ 0 & 0 & 1 & 0 & -1 & \vdots & 4 \\ 0 & 0 & 0 & 1 & -2 & \vdots & 3 \\ 0 & 0 & 0 & 0 & 0 & \vdots & 0 \end{bmatrix}$$

Since there are no variables left in the fourth equation, we are done. Our system now satisfies properties P1, P2, and P3. We can solve the equations for the leading variables:

$$\begin{vmatrix} x_1 = 2 - 2x_2 - 3x_5 \\ x_3 = 4 \qquad + x_5 \\ x_4 = 3 \qquad + 2x_5 \end{vmatrix}$$

If we let $x_2 = t$ and $x_5 = r$, then the infinitely many solutions are of the form

$$\begin{bmatrix} x_1 \\ x_2 \\ x_3 \\ x_4 \\ x_5 \end{bmatrix} = \begin{bmatrix} 2 & -2t & -3r \\ & t & \\ 4 & & +r \\ 3 & & +2r \\ & & r \end{bmatrix}.$$

Let us summarize.

Solving a system of linear equations

We proceed from equation to equation, from top to bottom.

Suppose we get to the ith equation. Let x_j be the leading variable of the system consisting of the ith and all the subsequent equations. (If no variables are left in this system, then the process comes to an end.)

- If x_j does not appear in the ith equation, swap the ith equation with the first equation below that does contain x_j.

- Suppose the coefficient of x_j in the ith equation is c; thus this equation is of the form $cx_j + \cdots = \cdots$. Divide the ith equation by c.

- Eliminate x_j from all the other equations, above and below the ith, by subtracting suitable multiples of the ith equation from the others.

Now proceed to the next equation.

If an equation *zero = nonzero* emerges in this process, then the system fails to have solutions; the system is *inconsistent*.

When you are through without encountering an inconsistency, solve each equation for its leading variable. You may choose the nonleading variables freely; the leading variables are then determined by these choices.

This process can be performed on the augmented matrix. As you do so, just imagine the linear system lurking behind it.

In the preceding example, we reduced the augmented matrix

$$M = \begin{bmatrix} 2 & 4 & -2 & 2 & 4 & | & 2 \\ 1 & 2 & -1 & 2 & 0 & | & 4 \\ 3 & 6 & -2 & 1 & 9 & | & 1 \\ 5 & 10 & -4 & 5 & 9 & | & 9 \end{bmatrix} \quad \text{to} \quad E = \begin{bmatrix} 1 & 2 & 0 & 0 & 3 & | & 2 \\ 0 & 0 & 1 & 0 & -1 & | & 4 \\ 0 & 0 & 0 & 1 & -2 & | & 3 \\ 0 & 0 & 0 & 0 & 0 & | & 0 \end{bmatrix}.$$

We say that the final matrix E is in reduced row-echelon form (rref).

> **Reduced row-echelon form**
>
> A matrix is in reduced row-echelon form if it satisfies all of the following conditions:
>
> **a.** If a row has nonzero entries, then the first nonzero entry is a 1, called the *leading* 1 (or *pivot*) in this row.
>
> **b.** If a column contains a leading 1, then all the other entries in that column are 0.
>
> **c.** If a row contains a leading 1, then each row above it contains a leading 1 further to the left.
>
> Condition c implies that rows of 0's, if any, appear at the bottom of the matrix.

Conditions a, b, and c defining the reduced row-echelon form correspond to the conditions P1, P2, and P3 that we imposed on the system.

Note that the leading 1's in the matrix

$$
E = \begin{bmatrix} \textcircled{1} & 2 & 0 & 0 & 3 & \vdots & 2 \\ 0 & 0 & \textcircled{1} & 0 & -1 & \vdots & 4 \\ 0 & 0 & 0 & \textcircled{1} & -2 & \vdots & 3 \\ 0 & 0 & 0 & 0 & 0 & \vdots & 0 \end{bmatrix}
$$

correspond to the leading variables in the reduced system,

$$
\begin{array}{rrrrrr}
\textcircled{x_1} & + & 2x_2 & & + & 3x_5 & = & 2 \\
& & & \textcircled{x_3} & - & x_5 & = & 4 \\
& & & & \textcircled{x_4} & - & 2x_5 & = & 3
\end{array}.
$$

Here we draw the staircase formed by the leading variables. This is where the name *echelon form* comes from. According to Webster, an echelon is a formation "like a series of steps."

The operations we perform when bringing a matrix into reduced row-echelon form are referred to as elementary row operations. Let's review the three types of such operations.

> **Types of elementary row operations**
> - Divide a row by a nonzero scalar.
> - Subtract a multiple of a row from another row.
> - Swap two rows.

Consider the following system:

$$
\begin{array}{rrrrr}
x_1 & - & 3x_2 & & & - & 5x_4 & = & -7 \\
3x_1 & - & 12x_2 & - & 2x_3 & - & 27x_4 & = & -33 \\
-2x_1 & + & 10x_2 & + & 2x_3 & + & 24x_4 & = & 29 \\
-x_1 & + & 6x_2 & + & x_3 & + & 14x_4 & = & 17
\end{array}.
$$

The augmented matrix is

$$\begin{bmatrix} 1 & -3 & 0 & -5 & \vdots & -7 \\ 3 & -12 & -2 & -27 & \vdots & -33 \\ -2 & 10 & 2 & 24 & \vdots & 29 \\ -1 & 6 & 1 & 14 & \vdots & 17 \end{bmatrix}.$$

The reduced row-echelon form for this matrix is

$$\begin{bmatrix} 1 & 0 & 0 & 1 & \vdots & 0 \\ 0 & 1 & 0 & 2 & \vdots & 0 \\ 0 & 0 & 1 & 3 & \vdots & 0 \\ 0 & 0 & 0 & 0 & \vdots & 1 \end{bmatrix}.$$

(We leave it to you to perform the elimination.)

Since the last row of the echelon form represents the equation $0 = 1$, the system is inconsistent.

This method of solving linear systems is sometimes referred to as *Gauss–Jordan elimination*, after the German mathematician Carl Friedrich Gauss (1777–1855; see Figure 4), perhaps the greatest mathematician of modern times, and the German engineer Wilhelm Jordan (1844–1899). Gauss himself called the method *eliminatio vulgaris*. Recall that the Chinese were using this method 2,000 years ago.

AY5393908A5

Figure 4 Carl Friedrich Gauss appears on an old German 10-mark note. (In fact, this is the *mirror image* of a well-known portrait of Gauss.[8])

How Gauss developed this method is noteworthy. On January 1, 1801, the Sicilian astronomer Giuseppe Piazzi (1746–1826) discovered a planet, which he named Ceres, in honor of the patron goddess of Sicily. Today, Ceres is called a dwarf planet, because it is only about 1,000 kilometers in diameter. Piazzi was able to observe Ceres for 40 nights, but then he lost track of it. Gauss, however, at the age of 24, succeeded in calculating the orbit of Ceres, even though the task seemed hopeless on the basis of a few observations. His computations were so accurate that the German astronomer W. Olbers (1758–1840) located the asteroid on December 31, 1801. In the course of his computations, Gauss had to solve systems of 17 linear equations.[9] In dealing with this problem, Gauss also used the method of least

---

[8]Reproduced by permission of the German Bundesbank.

[9]For the mathematical details, see D. Teets and K. Whitehead, "The Discovery of Ceres: How Gauss Became Famous," *Mathematics Magazine*, 72, 2 (April 1999): 83–93.

squares, which he had developed around 1794. (See Section 5.4.) Since Gauss at first refused to reveal the methods that led to this amazing accomplishment, some even accused him of sorcery. Gauss later described his methods of orbit computation in his book *Theoria Motus Corporum Coelestium* (1809).

The method of solving a linear system by Gauss–Jordan elimination is called an *algorithm*.[10] An algorithm can be defined as "a finite procedure, written in a fixed symbolic vocabulary, governed by precise instructions, moving in discrete Steps, 1, 2, 3, …, whose execution requires no insight, cleverness, intuition, intelligence, or perspicuity, and that sooner or later comes to an end" (David Berlinski, *The Advent of the Algorithm: The Idea that Rules the World*, Harcourt Inc., 2000).

Gauss–Jordan elimination is well suited for solving linear systems on a computer, at least in principle. In practice, however, some tricky problems associated with roundoff errors can occur.

Numerical analysts tell us that we can reduce the proliferation of roundoff errors by modifying Gauss–Jordan elimination, employing more sophisticated reduction techniques.

In modifying Gauss–Jordan elimination, an interesting question arises: If we transform a matrix A into a matrix B by a sequence of elementary row operations and if B is in reduced row-echelon form, is it necessarily true that $B = \text{rref}(A)$? Fortunately (and perhaps surprisingly) this is indeed the case.

In this text, we will not utilize this fact, so there is no need to present the somewhat technical proof. If you feel ambitious, try to work out the proof yourself after studying Chapter 3. (See Exercises 3.3.84 through 3.3.87.)

---

[10] The word *algorithm* is derived from the name of the mathematician al-Khowarizmi, who introduced the term *algebra* into mathematics. (See page 1.)

EXERCISES 1.2

GOAL   *Use Gauss–Jordan elimination to solve linear systems. Do simple problems using paper and pencil, and use technology to solve more complicated problems.*

In Exercises 1 through 12, find all solutions of the equations with paper and pencil using Gauss–Jordan elimination. Show all your work. Solve the system in Exercise 8 for the variables x_1, x_2, x_3, x_4, and x_5.

1. $\begin{vmatrix} x + y - 2z = 5 \\ 2x + 3y + 4z = 2 \end{vmatrix}$

2. $\begin{vmatrix} 3x + 4y - z = 8 \\ 6x + 8y - 2z = 3 \end{vmatrix}$

3. $x + 2y + 3z = 4$

4. $\begin{vmatrix} x + y = 1 \\ 2x - y = 5 \\ 3x + 4y = 2 \end{vmatrix}$

5. $\begin{vmatrix} x_3 + x_4 = 0 \\ x_2 + x_3 = 0 \\ x_1 + x_2 = 0 \\ x_1 + x_4 = 0 \end{vmatrix}$

6. $\begin{vmatrix} x_1 - 7x_2 + x_5 = 3 \\ x_3 - 2x_5 = 2 \\ x_4 + x_5 = 1 \end{vmatrix}$

7. $\begin{vmatrix} x_1 + 2x_2 \qquad 2x_4 + 3x_5 = 0 \\ x_3 + 3x_4 + 2x_5 = 0 \\ x_3 + 4x_4 - x_5 = 0 \\ x_5 = 0 \end{vmatrix}$

8. $\begin{vmatrix} x_2 + 2x_4 + 3x_5 = 0 \\ 4x_4 + 8x_5 = 0 \end{vmatrix}$

9. $\begin{vmatrix} x_4 + 2x_5 - x_6 = 2 \\ x_1 + 2x_2 \qquad + x_5 - x_6 = 0 \\ x_1 + 2x_2 + 2x_3 \qquad - x_5 + x_6 = 2 \end{vmatrix}$

10. $\begin{vmatrix} 4x_1 + 3x_2 + 2x_3 - x_4 = 4 \\ 5x_1 + 4x_2 + 3x_3 - x_4 = 4 \\ -2x_1 - 2x_2 - x_3 + 2x_4 = -3 \\ 11x_1 + 6x_2 + 4x_3 + x_4 = 11 \end{vmatrix}$

11. $\begin{vmatrix} x_1 + 2x_3 + 4x_4 = -8 \\ x_2 - 3x_3 - x_4 = 6 \\ 3x_1 + 4x_2 - 6x_3 + 8x_4 = 0 \\ - x_2 + 3x_3 + 4x_4 = -12 \end{vmatrix}$

$0 \quad 4 \quad -12 \quad -4$

12.
$$\begin{vmatrix} 2x_1 & & -3x_3 & & +7x_5 + 7x_6 = 0 \\ -2x_1 + x_2 + 6x_3 & & -6x_5 - 12x_6 = 0 \\ & x_2 - 3x_3 & & +x_5 + 5x_6 = 0 \\ -2x_2 & & +x_4 + x_5 + x_6 = 0 \\ 2x_1 + x_2 - 3x_3 & & +8x_5 + 7x_6 = 0 \end{vmatrix}$$

Solve the linear systems in Exercises 13 through 17. You may use technology.

13.
$$\begin{vmatrix} 3x + 11y + 19z = -2 \\ 7x + 23y + 39z = 10 \\ -4x - 3y - 2z = 6 \end{vmatrix}$$

14.
$$\begin{vmatrix} 3x + 6y + 14z = 22 \\ 7x + 14y + 30z = 46 \\ 4x + 8y + 7z = 6 \end{vmatrix}$$

15.
$$\begin{vmatrix} 3x + 5y + 3z = 25 \\ 7x + 9y + 19z = 65 \\ -4x + 5y + 11z = 5 \end{vmatrix}$$

16.
$$\begin{vmatrix} 3x_1 + 6x_2 + 9x_3 + 5x_4 + 25x_5 = 53 \\ 7x_1 + 14x_2 + 21x_3 + 9x_4 + 53x_5 = 105 \\ -4x_1 - 8x_2 - 12x_3 + 5x_4 - 10x_5 = 11 \end{vmatrix}$$

17.
$$\begin{vmatrix} 2x_1 + 4x_2 + 3x_3 + 5x_4 + 6x_5 = 37 \\ 4x_1 + 8x_2 + 7x_3 + 5x_4 + 2x_5 = 74 \\ -2x_1 - 4x_2 + 3x_3 + 4x_4 - 5x_5 = 20 \\ x_1 + 2x_2 + 2x_3 - x_4 + 2x_5 = 26 \\ 5x_1 - 10x_2 + 4x_3 + 6x_4 + 4x_5 = 24 \end{vmatrix}$$

18. Determine which of the matrices below are in reduced row-echelon form:

a. $\begin{bmatrix} 1 & 2 & 0 & 2 & 0 \\ 0 & 0 & 1 & 3 & 0 \\ 0 & 0 & 1 & 4 & 0 \\ 0 & 0 & 0 & 0 & 1 \end{bmatrix}$ b. $\begin{bmatrix} 0 & 1 & 2 & 0 & 3 \\ 0 & 0 & 0 & 1 & 4 \\ 0 & 0 & 0 & 0 & 0 \end{bmatrix}$

c. $\begin{bmatrix} 1 & 2 & 0 & 3 \\ 0 & 0 & 0 & 0 \\ 0 & 0 & 1 & 2 \end{bmatrix}$ d. $\begin{bmatrix} 0 & 1 & 2 & 3 & 4 \end{bmatrix}$

19. Find all 4×1 matrices in reduced row-echelon form.

20. We say that two $n \times m$ matrices in reduced row-echelon form are of the same type if they contain the same number of leading 1's in the same positions. For example,

$$\begin{bmatrix} ① & 2 & 0 \\ 0 & 0 & ① \end{bmatrix} \text{ and } \begin{bmatrix} ① & 3 & 0 \\ 0 & 0 & ① \end{bmatrix}$$

are of the same type. How many types of 2×2 matrices in reduced row-echelon form are there?

21. How many types of 3×2 matrices in reduced row-echelon form are there? (See Exercise 20.)

22. How many types of 2×3 matrices in reduced row-echelon form are there? (See Exercise 20.)

23. Suppose you apply Gauss–Jordan elimination to a matrix. Explain how you can be sure that the resulting matrix is in reduced row-echelon form.

24. Suppose matrix A is transformed into matrix B by means of an elementary row operation. Is there an elementary row operation that transforms B into A? Explain.

25. Suppose matrix A is transformed into matrix B by a sequence of elementary row operations. Is there a sequence of elementary row operations that transforms B into A? Explain your answer. (See Exercise 24.)

26. Consider an $n \times m$ matrix A. Can you transform rref(A) into A by a sequence of elementary row operations? (See Exercise 25.)

27. Is there a sequence of elementary row operations that transforms

$$\begin{bmatrix} 1 & 2 & 3 \\ 4 & 5 & 6 \\ 7 & 8 & 9 \end{bmatrix} \text{ into } \begin{bmatrix} 1 & 0 & 0 \\ 0 & 1 & 0 \\ 0 & 0 & 0 \end{bmatrix} ?$$

Explain.

28. Suppose you subtract a multiple of an equation in a system from another equation in the system. Explain why the two systems (before and after this operation) have the same solutions.

29. *Balancing a chemical reaction.* Consider the chemical reaction

$$a\, NO_2 + b\, H_2O \rightarrow c\, HNO_2 + d\, HNO_3,$$

where a, b, c, and d are unknown positive integers. The reaction must be balanced; that is, the number of atoms of each element must be the same before and after the reaction. For example, because the number of oxygen atoms must remain the same,

$$2a + b = 2c + 3d.$$

While there are many possible values for a, b, c, and d that balance the reaction, it is customary to use the smallest possible positive integers. Balance this reaction.

30. Find the polynomial of degree 3 [a polynomial of the form $f(t) = a + bt + ct^2 + dt^3$] whose graph goes through the points $(0, 1)$, $(1, 0)$, $(-1, 0)$, and $(2, -15)$. Sketch the graph of this cubic.

31. Find the polynomial of degree 4 whose graph goes through the points $(1, 1)$, $(2, -1)$, $(3, -59)$, $(-1, 5)$, and $(-2, -29)$. Graph this polynomial.

32. *Cubic splines.* Suppose you are in charge of the design of a roller coaster ride. This simple ride will not make any left or right turns; that is, the track lies in a vertical plane. The accompanying figure shows the ride as viewed from the side. The points (a_i, b_i) are given to you, and your job is to connect the dots in a reasonably smooth way. Let $a_{i+1} > a_i$.

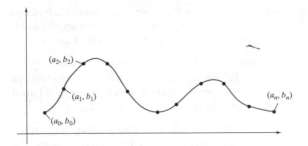

One method often employed in such design problems is the technique of cubic splines. We choose $f_i(t)$, a polynomial of degree ≤ 3, to define the shape of the ride between (a_{i-1}, b_{i-1}) and (a_i, b_i), for $i = 1, \ldots, n$.

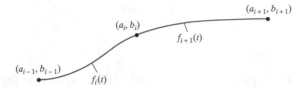

Obviously, it is required that $f_i(a_i) = b_i$ and $f_i(a_{i-1}) = b_{i-1}$, for $i = 1, \ldots, n$. To guarantee a smooth ride at the points (a_i, b_i), we want the first and the second derivatives of f_i and f_{i+1} to agree at these points:

$$f_i'(a_i) = f_{i+1}'(a_i) \quad \text{and}$$
$$f_i''(a_i) = f_{i+1}''(a_i), \quad \text{for } i = 1, \ldots, n-1.$$

Explain the practical significance of these conditions. Explain why, for the convenience of the riders, it is also required that

$$f_1'(a_0) = f_n'(a_n) = 0.$$

Show that satisfying all these conditions amounts to solving a system of linear equations. How many variables are in this system? How many equations? (*Note*: It can be shown that this system has a unique solution.)

33. Find the polynomial $f(t)$ of degree 3 such that $f(1) = 1$, $f(2) = 5$, $f'(1) = 2$, and $f'(2) = 9$, where $f'(t)$ is the derivative of $f(t)$. Graph this polynomial.

34. The *dot product* of two vectors

$$\vec{x} = \begin{bmatrix} x_1 \\ x_2 \\ \vdots \\ x_n \end{bmatrix} \quad \text{and} \quad \vec{y} = \begin{bmatrix} y_1 \\ y_2 \\ \vdots \\ y_n \end{bmatrix}$$

in $\mathbb{R}^n$ is defined by

$$\vec{x} \cdot \vec{y} = x_1 y_1 + x_2 y_2 + \cdots + x_n y_n.$$

Note that the dot product of two vectors is a scalar. We say that the vectors $\vec{x}$ and $\vec{y}$ are *perpendicular* if $\vec{x} \cdot \vec{y} = 0$.

Find all vectors in $\mathbb{R}^3$ perpendicular to

$$\begin{bmatrix} 1 \\ 3 \\ -1 \end{bmatrix}.$$

Draw a sketch.

35. Find all vectors in $\mathbb{R}^4$ that are perpendicular to the three vectors

$$\begin{bmatrix} 1 \\ 1 \\ 1 \\ 1 \end{bmatrix}, \quad \begin{bmatrix} 1 \\ 2 \\ 3 \\ 4 \end{bmatrix}, \quad \begin{bmatrix} 1 \\ 9 \\ 9 \\ 7 \end{bmatrix}.$$

(See Exercise 34.)

36. Find all solutions x_1, x_2, x_3 of the equation

$$\vec{b} = x_1 \vec{v}_1 + x_2 \vec{v}_2 + x_3 \vec{v}_3,$$

where

$$\vec{b} = \begin{bmatrix} -8 \\ -1 \\ 2 \\ 15 \end{bmatrix}, \vec{v}_1 = \begin{bmatrix} 1 \\ 4 \\ 7 \\ 5 \end{bmatrix}, \vec{v}_2 = \begin{bmatrix} 2 \\ 5 \\ 8 \\ 3 \end{bmatrix}, \vec{v}_3 = \begin{bmatrix} 4 \\ 6 \\ 9 \\ 1 \end{bmatrix}.$$

37. For some background on this exercise, see Exercise 1.1.20.

Consider an economy with three industries, I_1, I_2, I_3. What outputs x_1, x_2, x_3 should they produce to satisfy both consumer demand and interindustry demand? The demands put on the three industries are shown in the accompanying figure.

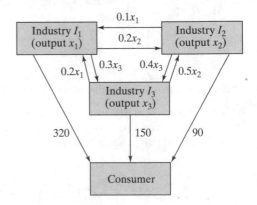

38. If we consider more than three industries in an input–output model, it is cumbersome to represent all the demands in a diagram as in Exercise 37. Suppose we have the industries $I_1, I_2, \ldots, I_n$, with outputs $x_1, x_2, \ldots, x_n$. The *output vector* is

$$\vec{x} = \begin{bmatrix} x_1 \\ x_2 \\ \vdots \\ x_n \end{bmatrix}.$$

The *consumer demand vector* is

$$\vec{b} = \begin{bmatrix} b_1 \\ b_2 \\ \vdots \\ b_n \end{bmatrix},$$

where b_i is the consumer demand on industry I_i. The demand vector for industry I_j is

$$\vec{v}_j = \begin{bmatrix} a_{1j} \\ a_{2j} \\ \vdots \\ a_{nj} \end{bmatrix},$$

where a_{ij} is the demand industry I_j puts on industry I_i, for each \$1 of output industry I_j produces. For example, $a_{32} = 0.5$ means that industry I_2 needs 50¢worth of products from industry I_3 for each \$1 worth of goods I_2 produces. The coefficient a_{ii} need not be 0: Producing a product may require goods or services from the same industry.

a. Find the four demand vectors for the economy in Exercise 37.

b. What is the meaning in economic terms of $x_j \vec{v}_j$?

c. What is the meaning in economic terms of $x_1 \vec{v}_1 + x_2 \vec{v}_2 + \cdots + x_n \vec{v}_n + \vec{b}$?

d. What is the meaning in economic terms of the equation

$$x_1 \vec{v}_1 + x_2 \vec{v}_2 + \cdots + x_n \vec{v}_n + \vec{b} = \vec{x}?$$

39. Consider the economy of Israel in 1958.[11] The three industries considered here are

$$\begin{array}{ll} I_1: & \text{agriculture,} \\ I_2: & \text{manufacturing,} \\ I_3: & \text{energy.} \end{array}$$

Outputs and demands are measured in millions of Israeli pounds, the currency of Israel at that time. We are told that

$$\vec{b} = \begin{bmatrix} 13.2 \\ 17.6 \\ 1.8 \end{bmatrix}, \qquad \vec{v}_1 = \begin{bmatrix} 0.293 \\ 0.014 \\ 0.044 \end{bmatrix},$$

$$\vec{v}_2 = \begin{bmatrix} 0 \\ 0.207 \\ 0.01 \end{bmatrix}, \qquad \vec{v}_3 = \begin{bmatrix} 0 \\ 0.017 \\ 0.216 \end{bmatrix}.$$

a. Why do the first components of $\vec{v}_2$ and $\vec{v}_3$ equal 0?

b. Find the outputs x_1, x_2, x_3 required to satisfy demand.

40. Consider some particles in the plane with position vectors $\vec{r}_1, \vec{r}_2, \ldots, \vec{r}_n$ and masses $m_1, m_2, \ldots, m_n$.

[11] W. Leontief, *Input–Output Economics*, Oxford University Press, 1966.

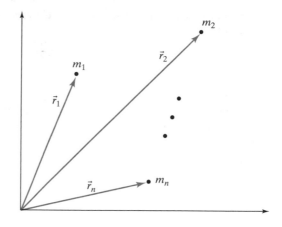

The position vector of the *center of mass* of this system is

$$\vec{r}_{cm} = \frac{1}{M}(m_1 \vec{r}_1 + m_2 \vec{r}_2 + \cdots + m_n \vec{r}_n),$$

where $M = m_1 + m_2 + \cdots + m_n$.

Consider the triangular plate shown in the accompanying sketch. How must a total mass of 1 kg be distributed among the three vertices of the plate so that the plate can be supported at the point $\begin{bmatrix} 2 \\ 2 \end{bmatrix}$; that is,

$$\vec{r}_{cm} = \begin{bmatrix} 2 \\ 2 \end{bmatrix}?$$ Assume that the mass of the plate itself is negligible.

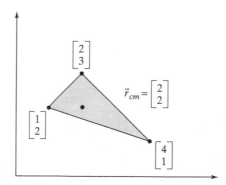

41. The *momentum* $\vec{P}$ of a system of n particles in space with masses $m_1, m_2, \ldots, m_n$ and velocities $\vec{v}_1, \vec{v}_2, \ldots, \vec{v}_n$ is defined as

$$\vec{P} = m_1 \vec{v}_1 + m_2 \vec{v}_2 + \cdots + m_n \vec{v}_n.$$

Now consider two elementary particles with velocities

$$\vec{v}_1 = \begin{bmatrix} 1 \\ 1 \\ 1 \end{bmatrix} \quad \text{and} \quad \vec{v}_2 = \begin{bmatrix} 4 \\ 7 \\ 10 \end{bmatrix}.$$

The particles collide. After the collision, their respective velocities are observed to be

$$\vec{w}_1 = \begin{bmatrix} 4 \\ 7 \\ 4 \end{bmatrix} \quad \text{and} \quad \vec{w}_2 = \begin{bmatrix} 2 \\ 3 \\ 8 \end{bmatrix}.$$

Assume that the momentum of the system is conserved throughout the collision. What does this experiment tell you about the masses of the two particles? (See the accompanying figure.)

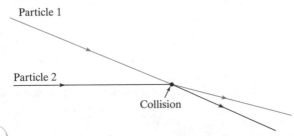

Particle 1

Particle 2

Collision

42. The accompanying sketch represents a maze of one-way streets in a city in the United States. The traffic volume through certain blocks during an hour has been measured. Suppose that the vehicles leaving the area during this hour were exactly the same as those entering it.

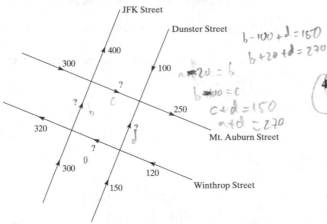

JFK Street

Dunster Street

400

300

100

?

250

?

320

?

300

120

150

Mt. Auburn Street

Winthrop Street

What can you say about the traffic volume at the four locations indicated by a question mark? Can you figure out exactly how much traffic there was on each block? If not, describe one possible scenario. For each of the four locations, find the highest and the lowest possible traffic volume.

43. Let $S(t)$ be the length of the tth day of the year 2009 in Mumbai (formerly known as Bombay), India (measured in hours, from sunrise to sunset). We are given the following values of $S(t)$:

t	$S(t)$
47	11.5
74	12
273	12

For example, $S(47) = 11.5$ means that the time from sunrise to sunset on February 16 is 11 hours and 30 minutes. For locations close to the equator, the function $S(t)$ is well approximated by a trigonometric function of the form

$$S(t) = a + b\cos\left(\frac{2\pi t}{365}\right) + c\sin\left(\frac{2\pi t}{365}\right).$$

(The period is 365 days, or 1 year.) Find this approximation for Mumbai, and graph your solution. According to this model, how long is the longest day of the year in Mumbai?

44. Kyle is getting some flowers for Olivia, his Valentine. Being of a precise analytical mind, he plans to spend exactly $24 on a bunch of exactly two dozen flowers. At the flower market they have lilies ($3 each), roses ($2 each), and daisies ($0.50 each). Kyle knows that Olivia loves lilies; what is he to do?

45. Consider the equations

$$\begin{vmatrix} x + 2y + & 3z = 4 \\ x + ky + & 4z = 6 \\ x + 2y + (k+2)z = 6 \end{vmatrix},$$

where k is an arbitrary constant.

   a. For which values of the constant k does this system have a unique solution?

   b. When is there no solution?

   c. When are there infinitely many solutions?

46. Consider the equations

$$\begin{vmatrix} y + 2kz = 0 \\ x + 2y + 6z = 2 \\ kx + 2z = 1 \end{vmatrix},$$

where k is an arbitrary constant.

   a. For which values of the constant k does this system have a unique solution?

   b. When is there no solution?

   c. When are there infinitely many solutions?

47. a. Find all solutions x_1, x_2, x_3, x_4 of the system $x_2 = \frac{1}{2}(x_1 + x_3)$, $x_3 = \frac{1}{2}(x_2 + x_4)$.

   b. In part (a), is there a solution with $x_1 = 1$ and $x_4 = 13$?

48. For an arbitrary positive integer $n \geq 3$, find all solutions $x_1, x_2, x_3, \ldots, x_n$ of the simultaneous equations $x_2 = \frac{1}{2}(x_1 + x_3)$, $x_3 = \frac{1}{2}(x_2 + x_4), \ldots, x_{n-1} = \frac{1}{2}(x_{n-2} + x_n)$. Note that we are asked to solve the simultaneous equations $x_k = \frac{1}{2}(x_{k-1} + x_{k+1})$, for $k = 2, 3, \ldots, n-1$.

49. Consider the system

$$\begin{vmatrix} 2x + y & = C \\ 3y + z = C \\ x & + 4z = C \end{vmatrix},$$

where C is a constant. Find the smallest positive integer C such that x, y, and z are all integers.

50. Find all the polynomials $f(t)$ of degree ≤ 3 such that $f(0) = 3$, $f(1) = 2$, $f(2) = 0$, and $\int_0^2 f(t)\, dt = 4$. (If you have studied Simpson's rule in calculus, explain the result.)

Exercises 51 through 60 are concerned with conics. A conic is a curve in $\mathbb{R}^2$ that can be described by an equation of the form $f(x, y) = c_1 + c_2x + c_3y + c_4x^2 + c_5xy + c_6y^2 = 0$, where at least one of the coefficients c_i is nonzero. Examples are circles, ellipses, hyperbolas, and parabolas. If k is any nonzero constant, then the equations $f(x, y) = 0$ and $kf(x, y) = 0$ describe the same conic. For example, the equation $-4 + x^2 + y^2 = 0$ and $-12 + 3x^2 + 3y^2 = 0$ both describe the circle of radius 2 centered at the origin. In Exercises 51 through 60, find all the conics through the given points, and draw a rough sketch of your solution curve(s).

51. $(0, 0)$, $(1, 0)$, $(2, 0)$, $(0, 1)$, and $(0, 2)$.

52. $(0, 0)$, $(2, 0)$, $(0, 2)$, $(2, 2)$, and $(1, 3)$.

53. $(0, 0)$, $(1, 0)$, $(2, 0)$, $(3, 0)$, and $(1, 1)$.

54. $(0, 0)$, $(1, 1)$, $(2, 2)$, $(3, 3)$, and $(1, 0)$.

55. $(0, 0)$, $(1, 0)$, $(0, 1)$, and $(1, 1)$.

56. $(0, 0)$, $(1, 0)$, $(0, 1)$, and $(1, -1)$.

57. $(5, 0)$, $(1, 2)$, $(2, 1)$, $(8, 1)$, and $(2, 9)$.

58. $(1, 0)$, $(2, 0)$, $(2, 2)$, $(5, 2)$, and $(5, 6)$.

59. $(0, 0)$, $(1, 0)$, $(2, 0)$, $(0, 1)$, $(0, 2)$, and $(1, 1)$.

60. $(0, 0)$, $(2, 0)$, $(0, 2)$, $(2, 2)$, $(1, 3)$, and $(4, 1)$.

61. Students are buying books for the new semester. Eddie buys the environmental statistics book and the set theory book for $178. Leah, who is buying books for herself and her friend, spends $319 on two environmental statistics books, one set theory book, and one educational psychology book. Mehmet buys the educational psychology book and the set theory book for $147 in total. How much does each book cost?

62. Students are buying books for the new semester. Brigitte buys the German grammar book and the German novel, *Die Leiden des jungen Werther*, for €64 in total. Claude spends €98 on the linear algebra text and the German grammar book, while Denise buys the linear algebra text and *Werther*, for €76. How much does each of the three books cost?

63. At the beginning of a political science class at a large university, the students were asked which term, *liberal* or *conservative*, best described their political views. They were asked the same question at the end of the course, to see what effect the class discussions had on their views. Of those that characterized themselves as "liberal" initially, 30% held conservative views at the end. Of those who were conservative initially, 40% moved to the liberal camp. It turned out that there were just

as many students with conservative views at the end as there had been liberal students at the beginning. Out of the 260 students in the class, how many held liberal and conservative views at the beginning of the course and at the end? (No students joined or dropped the class between the surveys, and they all participated in both surveys.)

64. At the beginning of a semester, 55 students have signed up for Linear Algebra; the course is offered in two sections that are taught at different times. Because of scheduling conflicts and personal preferences, 20% of the students in Section A switch to Section B in the first few weeks of class, while 30% of the students in Section B switch to A, resulting in a net loss of 4 students for Section B. How large were the two sections at the beginning of the semester? No students dropped Linear Algebra (why would they?) or joined the course late.

Historical Problems

65. Five cows and two sheep together cost ten *liang*[12] of silver. Two cows and five sheep together cost eight *liang* of silver. What is the cost of a cow and a sheep, respectively? (*Nine Chapters*,[13] Chapter 8, Problem 7)

66. If you sell two cows and five sheep and you buy 13 pigs, you gain 1,000 coins. If you sell three cows and three pigs and buy nine sheep, you break even. If you sell six sheep and eight pigs and you buy five cows, you lose 600 coins. What is the price of a cow, a sheep, and a pig, respectively? (*Nine Chapters*, Chapter 8, Problem 8)

67. You place five sparrows on one of the pans of a balance and six swallows on the other pan; it turns out that the sparrows are heavier. But if you exchange one sparrow and one swallow, the weights are exactly balanced. All the birds together weigh 1 *jin*. What is the weight of a sparrow and a swallow, respectively? [Give the answer in *liang*, with 1 *jin* = 16 *liang*.] (*Nine Chapters*, Chapter 8, Problem 9)

68. Consider the task of pulling a weight of 40 *dan*[14] up a hill; we have one military horse, two ordinary horses, and three weak horses at our disposal to get the job done. It turns out that the military horse and one of the ordinary horses, pulling together, are barely able to pull the

---

[12] A *liang* was about 16 grams at the time of the Han Dynasty.

[13] See page 1; we present some of the problems from the *Nine Chapters on the Mathematical Art* in a free translation, with some additional explanations, since the scenarios discussed in a few of these problems are rather unfamiliar to the modern reader.

[14] 1 *dan* = 120 *jin* = 1,920 *liang*. Thus a *dan* was about 30 kilograms at that time.

weight (but they could not pull any more). Likewise, the two ordinary horses together with one weak horse are just able to do the job, as are the three weak horses together with the military horse. How much weight can each of the horses pull alone? (*Nine Chapters*, Chapter 8, Problem 12)

69. Five households share a deep well for their water supply. Each household owns a few ropes of a certain length, which varies only from household to household. The five households, A, B, C, D, and E, own 2, 3, 4, 5, and 6 ropes, respectively. Even when tying all their ropes together, none of the households alone are able to reach the water, but A's two ropes together with one of B's ropes just reach the water. Likewise, B's three ropes with one of C's ropes, C's four ropes with one of D's ropes, D's five ropes with one of E's ropes, and E's six ropes with one of A's ropes all just reach the water. How long are the ropes of the various households, and how deep is the well?

Commentary: As stated, this problem leads to a system of 5 linear equations in 6 variables; with the given information, we are unable to determine the depth of the well. The *Nine Chapters* gives one particular solution, where the depth of the well is 7 *zhang*,[15] 2 *chi*, 1 *cun*, or 721 *cun* (since 1 *zhang* = 10 *chi* and 1 *chi* = 10 *cun*). Using this particular value for the depth of the well, find the lengths of the various ropes.

70. "A rooster is worth five coins, a hen three coins, and 3 chicks one coin. With 100 coins we buy 100 of them. How many roosters, hens, and chicks can we buy?" (From the *Mathematical Manual* by Zhang Qiujian, Chapter 3, Problem 38; 5th century A.D.)

Commentary: This famous *Hundred Fowl Problem* has reappeared in countless variations in Indian, Arabic, and European texts (see Exercises 71 through 74); it has remained popular to this day (see Exercise 44 of this section).

71. "Pigeons are sold at the rate of 5 for 3 panas, sarasabirds at the rate of 7 for 5 panas, swans at the rate of 9 for 7 panas, and peacocks at the rate of 3 for 9 panas. A man was told to bring 100 birds for 100 panas for the amusement of the King's son. What does he pay for each of the various kinds of birds that he buys?" (From the *Ganita-Sara-Sangraha* by Mahavira, India; 9th century A.D.) Find one solution to this problem.

72. "A duck costs four coins, five sparrows cost one coin, and a rooster costs one coin. Somebody buys 100 birds for 100 coins. How many birds of each kind can he buy?" (From the *Key to Arithmetic* by Al-Kashi; 15th century)

73. "A certain person buys sheep, goats, and hogs, to the number of 100, for 100 crowns; the sheep cost him $\frac{1}{2}$ a crown a-piece; the goats, $1\frac{1}{3}$ crown; and the hogs $3\frac{1}{2}$ crowns. How many had he of each?" (From the *Elements of Algebra* by Leonhard Euler, 1770)

74. "A gentleman has a household of 100 persons and orders that they be given 100 measures of grain. He directs that each man should receive three measures, each woman two measures, and each child half a measure. How many men, women, and children are there in this household?" We are told that there is at least one man, one woman, and one child. (From the *Problems for Quickening a Young Mind* by Alcuin [c. 732–804], the Abbot of St. Martins at Tours. Alcuin was a friend and tutor to Charlemagne and his family at Aachen.)

75. A father, when dying, gave to his sons 30 barrels, of which 10 were full of wine, 10 were half full, and the last 10 were empty. Divide the wine and flasks so that there will be equal division among the three sons of both wine and barrels. Find all the solutions of this problem. (From Alcuin)

76. "Make me a crown weighing 60 *minae*, mixing gold, bronze, tin, and wrought iron. Let the gold and bronze together form two-thirds, the gold and tin together three-fourths, and the gold and iron three-fifths. Tell me how much gold, tin, bronze, and iron you must put in." (From the *Greek Anthology* by Metrodorus, 6th century A.D.)

77. Three merchants find a purse lying in the road. One merchant says "If I keep the purse, I shall have twice as much money as the two of you together." "Give me the purse and I shall have three times as much as the two of you together" said the second merchant. The third merchant said "I shall be much better off than either of you if I keep the purse, I shall have five times as much as the two of you together." If there are 60 coins (of equal value) in the purse, how much money does each merchant have? (From Mahavira)

78. 3 cows graze 1 field bare in 2 days,
7 cows graze 4 fields bare in 4 days, and
3 cows graze 2 fields bare in 5 days.
It is assumed that each field initially provides the same amount, x, of grass; that the daily growth, y, of the fields remains constant; and that all the cows eat the same amount, z, each day. (Quantities x, y, and z are measured by weight.) Find all the solutions of this problem. (This is a special case of a problem discussed by Isaac Newton in his *Arithmetica Universalis*, 1707.)

---

[15] 1 *zhang* was about 2.3 meters at that time.

1.3 On the Solutions of Linear Systems; Matrix Algebra

In this final section of Chapter 1, we will discuss two rather unrelated topics:

- First, we will examine how many solutions a system of linear equations can possibly have.
- Then, we will present some definitions and rules of matrix algebra.

The Number of Solutions of a Linear System

EXAMPLE I The reduced row-echelon forms of the augmented matrices of three systems are given. How many solutions are there in each case?

$$
\textbf{a.} \begin{bmatrix} 1 & 2 & 0 & \vdots & 0 \\ 0 & 0 & 1 & \vdots & 0 \\ 0 & 0 & 0 & \vdots & 1 \\ 0 & 0 & 0 & \vdots & 0 \end{bmatrix}
\qquad
\textbf{b.} \begin{bmatrix} 1 & 2 & 0 & \vdots & 1 \\ 0 & 0 & 1 & \vdots & 2 \\ 0 & 0 & 0 & \vdots & 0 \end{bmatrix}
\qquad
\textbf{c.} \begin{bmatrix} 1 & 0 & 0 & \vdots & 1 \\ 0 & 1 & 0 & \vdots & 2 \\ 0 & 0 & 1 & \vdots & 3 \end{bmatrix}
$$

Solution

a. The third row represents the equation $0 = 1$, so that there are no solutions. We say that this system is *inconsistent*.

b. The given augmented matrix represents the system

$$
\begin{vmatrix} x_1 + 2x_2 & = 1 \\ & x_3 = 2 \end{vmatrix}, \qquad \text{or,} \qquad \begin{vmatrix} x_1 = 1 - 2x_2 \\ x_3 = 2 \end{vmatrix}.
$$

We can assign an arbitrary value, t, to the free variable x_2, so that the system has infinitely many solutions,

$$
\begin{bmatrix} x_1 \\ x_2 \\ x_3 \end{bmatrix} = \begin{bmatrix} 1 - 2t \\ t \\ 2 \end{bmatrix}, \qquad \text{where } t \text{ is an arbitrary constant.}
$$

c. Here there are no free variables, so that we have only one solution, $x_1 = 1$, $x_2 = 2$, $x_3 = 3$. ■

We can generalize our findings:[16]

Theorem 1.3.1 **Number of solutions of a linear system**

A system of equations is said to be *consistent* if there is at least one solution; it is *inconsistent* if there are no solutions.

A linear system is inconsistent if (and only if) the reduced row-echelon form of its augmented matrix contains the row $\begin{bmatrix} 0 & 0 & \cdots & 0 & \vdots & 1 \end{bmatrix}$, representing the equation $0 = 1$.

If a linear system is consistent, then it has either

- *infinitely many solutions* (if there is at least one free variable), or
- *exactly one solution* (if all the variables are leading). ■

---

[16] Starting in this section, we will number the definitions we give and the theorems we derive. The nth theorem stated in Section $p.q$ is labeled as Theorem $p.q.n$.

Example 1 illustrates what the number of leading 1's in the echelon form tells us about the number of solutions of a linear system. This observation motivates the following definition:

Definition 1.3.2 The rank of a matrix[17]

The rank of a matrix A is the number of leading 1's in rref(A).

EXAMPLE 2 rank $\begin{bmatrix} 1 & 2 & 3 \\ 4 & 5 & 6 \\ 7 & 8 & 9 \end{bmatrix} = 2$, since rref $\begin{bmatrix} 1 & 2 & 3 \\ 4 & 5 & 6 \\ 7 & 8 & 9 \end{bmatrix} = \begin{bmatrix} ① & 0 & -1 \\ 0 & ① & 2 \\ 0 & 0 & 0 \end{bmatrix}$ ∎

Note that we have defined the rank of a *matrix* rather than the rank of a linear system. When relating the concept of rank to a linear system, we must be careful to specify whether we consider the coefficient matrix or the augmented matrix of the system.

EXAMPLE 3 Consider a system of n linear equations in m variables; its coefficient matrix A has the size $n \times m$. Show that

a. The inequalities rank(A) $\leq n$ and rank(A) $\leq m$ hold.

b. If rank(A) $= n$, then the system is consistent.

c. If rank(A) $= m$, then the system has at most one solution.

d. If rank(A) $< m$, then the system has either infinitely many solutions, or none.

To get a sense for the significance of these claims, take another look at Example 1.

Solution

a. By definition of the reduced row-echelon form, there is at most one leading 1 in each of the n rows and in each of the m columns of rref(A).

b. If rank(A) $= n$, then there is a leading 1 in each row of rref(A). This implies that the echelon form of the augmented matrix cannot contain the row $\begin{bmatrix} 0 & 0 & \cdots & 0 \mid 1 \end{bmatrix}$. This system is consistent.

c. For parts c and d, it is worth noting that

$$\begin{pmatrix} \text{number of} \\ \text{free variables} \end{pmatrix} = \begin{pmatrix} \text{total number} \\ \text{of variables} \end{pmatrix} - \begin{pmatrix} \text{number of} \\ \text{leading variables} \end{pmatrix} = m - \text{rank}(A).$$

If rank(A) $= m$, then there are no free variables. Either the system is inconsistent or it has a unique solution (by Theorem 1.3.1).

d. If rank(A) $< m$, then there are free variables ($m - \text{rank } A$ of them, to be precise). Therefore, either the system is inconsistent or it has infinitely many solutions (by Theorem 1.3.1). ∎

EXAMPLE 4 Consider a linear system with fewer equations than variables. How many solutions could this system have?

Solution

Suppose there are n equations and m variables; we are told that $n < m$. Let A be the coefficient matrix of the system, of size $n \times m$. By Example 3a, we have

---

[17]This is a preliminary, rather technical definition. In Chapter 3, we will gain a better conceptual understanding of the rank.

$\text{rank}(A) \le n < m$, so that $\text{rank}(A) < m$. There are free variables ($m - \text{rank } A$ of them), so that the system will have infinitely many solutions or no solutions at all. ∎

Theorem 1.3.3   **Systems with fewer equations than variables**

A linear system with fewer equations than unknowns has either no solutions or infinitely many solutions.

To put it differently, if a linear system has a unique solution, then there must be at least as many equations as there are unknowns. ∎

To illustrate this fact, consider two linear equations in three variables, with each equation defining a plane. Two different planes in space either intersect in a line or are parallel (see Figure 1), but they will never intersect at a point! This means that a system of two linear equations with three unknowns cannot have a unique solution.

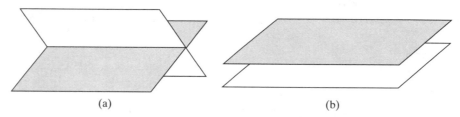

(a)   (b)

Figure 1 (a) Two planes intersect in a line. (b) Two parallel planes.

EXAMPLE 5   Consider a linear system of n equations in n variables. When does this system have a unique solution? Give your answer in terms of the rank of the coefficient matrix A.

Solution

If $\text{rank}(A) < n$, then there will be free variables ($n - \text{rank } A$ of them), so that the system has either no solutions or infinitely many solutions (see Example 3d). If $\text{rank}(A) = n$, then there are no free variables, so that there cannot be infinitely many solutions (see Example 3c). Furthermore, there must be a leading 1 in every row of $\text{rref}(A)$, so that the system is consistent (see Example 3b). We conclude that the system must have a unique solution in this case. ∎

Theorem 1.3.4   **Systems of n equations in n variables**

A linear system of n equations in n variables has a unique solution if (and only if) the rank of its coefficient matrix A is n. In this case,

$$\text{rref}(A) = \begin{bmatrix} 1 & 0 & 0 & \cdots & 0 \\ 0 & 1 & 0 & \cdots & 0 \\ 0 & 0 & 1 & \cdots & 0 \\ \vdots & \vdots & \vdots & \ddots & \vdots \\ 0 & 0 & 0 & \cdots & 1 \end{bmatrix},$$

the $n \times n$ matrix with 1's along the diagonal and 0's everywhere else. ∎

Matrix Algebra

We will now introduce some basic definitions and rules of matrix algebra. Our presentation will be somewhat lacking in motivation at first, but it will be good to have these tools available when we need them in Chapter 2.

Sums and scalar multiples of matrices are defined entry by entry, as for vectors (see Definition A.1 in the Appendix).

Definition 1.3.5 Sums of matrices

The sum of two matrices of the same size is defined entry by entry:

$$\begin{bmatrix} a_{11} & \cdots & a_{1m} \\ \vdots & & \vdots \\ a_{n1} & \cdots & a_{nm} \end{bmatrix} + \begin{bmatrix} b_{11} & \cdots & b_{1m} \\ \vdots & & \vdots \\ b_{n1} & \cdots & b_{nm} \end{bmatrix} = \begin{bmatrix} a_{11}+b_{11} & \cdots & a_{1m}+b_{1m} \\ \vdots & & \vdots \\ a_{n1}+b_{n1} & \cdots & a_{nm}+b_{nm} \end{bmatrix}.$$

Scalar Multiples of Matrices

The product of a scalar with a matrix is defined entry by entry:

$$k \begin{bmatrix} a_{11} & \cdots & a_{1m} \\ \vdots & & \vdots \\ a_{n1} & \cdots & a_{nm} \end{bmatrix} = \begin{bmatrix} ka_{11} & \cdots & ka_{1m} \\ \vdots & & \vdots \\ ka_{n1} & \cdots & ka_{nm} \end{bmatrix}.$$

EXAMPLE 6 $\begin{bmatrix} 1 & 2 & 3 \\ 4 & 5 & 6 \end{bmatrix} + \begin{bmatrix} 7 & 3 & 1 \\ 5 & 3 & -1 \end{bmatrix} = \begin{bmatrix} 8 & 5 & 4 \\ 9 & 8 & 5 \end{bmatrix}$ ∎

EXAMPLE 7 $3 \begin{bmatrix} 2 & 1 \\ -1 & 3 \end{bmatrix} = \begin{bmatrix} 6 & 3 \\ -3 & 9 \end{bmatrix}$ ∎

The definition of the product of matrices is less straightforward; we will give the general definition only in Section 2.3.

Because vectors are special matrices (with only one row or only one column), it makes sense to start with a discussion of products of vectors. The reader may be familiar with the dot product of vectors.

Definition 1.3.6 Dot product of vectors

Consider two vectors $\vec{v}$ and $\vec{w}$ with components $v_1, \ldots, v_n$ and $w_1, \ldots, w_n$, respectively. Here $\vec{v}$ and $\vec{w}$ may be column or row vectors, and they need not be of the same type. The dot product of $\vec{v}$ and $\vec{w}$ is defined to be the scalar

$$\vec{v} \cdot \vec{w} = v_1 w_1 + \cdots + v_n w_n.$$

Note that our definition of the dot product isn't row-column-sensitive. The dot product does not distinguish between row and column vectors.

EXAMPLE 8 $\begin{bmatrix} 1 & 2 & 3 \end{bmatrix} \cdot \begin{bmatrix} 3 \\ 1 \\ 2 \end{bmatrix} = 1 \cdot 3 + 2 \cdot 1 + 3 \cdot 2 = 11$ ∎

Now we are ready to define the product $A\vec{x}$, where A is a matrix and $\vec{x}$ is a vector, in terms of the dot product.

Definition 1.3.7   The product $A\vec{x}$

If A is an $n \times m$ matrix with row vectors $\vec{w}_1, \ldots, \vec{w}_n$, and $\vec{x}$ is a vector in $\mathbb{R}^m$, then

$$A\vec{x} = \begin{bmatrix} - & \vec{w}_1 & - \\ & \vdots & \\ - & \vec{w}_n & - \end{bmatrix} \vec{x} = \begin{bmatrix} \vec{w}_1 \cdot \vec{x} \\ \vdots \\ \vec{w}_n \cdot \vec{x} \end{bmatrix}.$$

In words, the ith component of $A\vec{x}$ is the dot product of the ith row of A with $\vec{x}$.
Note that $A\vec{x}$ is a column vector with n components, that is, a vector in $\mathbb{R}^n$.

EXAMPLE 9 $\begin{bmatrix} 1 & 2 & 3 \\ 1 & 0 & -1 \end{bmatrix} \begin{bmatrix} 3 \\ 1 \\ 2 \end{bmatrix} = \begin{bmatrix} 1 \cdot 3 + 2 \cdot 1 + 3 \cdot 2 \\ 1 \cdot 3 + 0 \cdot 1 + (-1) \cdot 2 \end{bmatrix} = \begin{bmatrix} 11 \\ 1 \end{bmatrix}.$ ∎

EXAMPLE 10 $\begin{bmatrix} 1 & 0 & 0 \\ 0 & 1 & 0 \\ 0 & 0 & 1 \end{bmatrix} \begin{bmatrix} x_1 \\ x_2 \\ x_3 \end{bmatrix} = \begin{bmatrix} x_1 \\ x_2 \\ x_3 \end{bmatrix}$ for all vectors $\begin{bmatrix} x_1 \\ x_2 \\ x_3 \end{bmatrix}$ in $\mathbb{R}^3$.   ∎

Note that the product $A\vec{x}$ is defined only if the number of *columns* of matrix A matches the number of components of vector $\vec{x}$:

$$\underbrace{\overbrace{A}^{n \times m} \; \overbrace{\vec{x}}^{m \times 1}}_{n \times 1}.$$

EXAMPLE 11   The product $A\vec{x} = \begin{bmatrix} 1 & 2 & 3 \\ 1 & 0 & -1 \end{bmatrix} \begin{bmatrix} 3 \\ 1 \end{bmatrix}$ is undefined, because the number of columns of matrix A fails to match the number of components of vector $\vec{x}$.   ∎

In Definition 1.3.7, we express the product $A\vec{x}$ in terms of the *rows* of the matrix A. Alternatively, the product can be expressed in terms of the *columns*.

Let's take another look at Example 9:

$$A\vec{x} = \begin{bmatrix} 1 & 2 & 3 \\ 1 & 0 & -1 \end{bmatrix} \begin{bmatrix} 3 \\ 1 \\ 2 \end{bmatrix} = \begin{bmatrix} 1 \cdot 3 + 2 \cdot 1 + 3 \cdot 2 \\ 1 \cdot 3 + 0 \cdot 1 + (-1) \cdot 2 \end{bmatrix}$$

$$= \begin{bmatrix} 1 \cdot 3 \\ 1 \cdot 3 \end{bmatrix} + \begin{bmatrix} 2 \cdot 1 \\ 0 \cdot 1 \end{bmatrix} + \begin{bmatrix} 3 \cdot 2 \\ (-1) \cdot 2 \end{bmatrix} = 3 \begin{bmatrix} 1 \\ 1 \end{bmatrix} + 1 \begin{bmatrix} 2 \\ 0 \end{bmatrix} + 2 \begin{bmatrix} 3 \\ -1 \end{bmatrix}.$$

We recognize that the expression $3 \begin{bmatrix} 1 \\ 1 \end{bmatrix} + 1 \begin{bmatrix} 2 \\ 0 \end{bmatrix} + 2 \begin{bmatrix} 3 \\ -1 \end{bmatrix}$ involves the vectors $\vec{v}_1 = \begin{bmatrix} 1 \\ 1 \end{bmatrix}, \vec{v}_2 = \begin{bmatrix} 2 \\ 0 \end{bmatrix}, \vec{v}_3 = \begin{bmatrix} 3 \\ -1 \end{bmatrix}$, the columns of A, and the scalars $x_1 = 3, x_2 = 1, x_3 = 2$, the components of $\vec{x}$. Thus we can write

$$A\vec{x} = \begin{bmatrix} | & | & | \\ \vec{v}_1 & \vec{v}_2 & \vec{v}_3 \\ | & | & | \end{bmatrix} \begin{bmatrix} x_1 \\ x_2 \\ x_3 \end{bmatrix} = x_1 \vec{v}_1 + x_2 \vec{v}_2 + x_3 \vec{v}_3.$$

We can generalize:

Theorem 1.3.8 **The product $A\vec{x}$ in terms of the columns of A**

If the column vectors of an $n \times m$ matrix A are $\vec{v}_1, \ldots, \vec{v}_m$ and $\vec{x}$ is a vector in $\mathbb{R}^m$ with components $x_1, \ldots, x_m$, then

$$A\vec{x} = \begin{bmatrix} | & & | \\ \vec{v}_1 & \cdots & \vec{v}_m \\ | & & | \end{bmatrix} \begin{bmatrix} x_1 \\ \vdots \\ x_m \end{bmatrix} = x_1\vec{v}_1 + \cdots + x_m\vec{v}_m.$$

Proof As usual, we denote the rows of A by $\vec{w}_1, \ldots, \vec{w}_n$ and the entries by a_{ij}. It suffices to show that the ith component of $A\vec{x}$ is equal to the ith component of $x_1\vec{v}_1 + \cdots + x_m\vec{v}_m$, for $i = 1, \ldots n$. Now

$$(i\text{th component of } A\vec{x}) \underbrace{=}_{\text{Step 1}} \vec{w}_i \cdot \vec{x} = a_{i1}x_1 + \cdots a_{im}x_m$$

$$= x_1(i\text{th component of } \vec{v}_1) + \cdots + x_m(i\text{th component of } \vec{v}_m)$$

$$\underbrace{=}_{\text{Step 4}} i\text{th component of } x_1\vec{v}_1 + \cdots + x_m\vec{v}_m$$

In Step 1 we are using Definition 1.3.7, and in Step 4 we are using the fact that vector addition and scalar multiplication are defined component by component. ∎

EXAMPLE 12 $A\vec{x} = \begin{bmatrix} 1 & 2 & 3 \\ 4 & 5 & 6 \\ 7 & 8 & 9 \end{bmatrix} \begin{bmatrix} 2 \\ -4 \\ 2 \end{bmatrix} = 2\begin{bmatrix} 1 \\ 4 \\ 7 \end{bmatrix} + (-4)\begin{bmatrix} 2 \\ 5 \\ 8 \end{bmatrix} + 2\begin{bmatrix} 3 \\ 6 \\ 9 \end{bmatrix}$

$= \begin{bmatrix} 2 \\ 8 \\ 14 \end{bmatrix} - \begin{bmatrix} 8 \\ 20 \\ 32 \end{bmatrix} + \begin{bmatrix} 6 \\ 12 \\ 18 \end{bmatrix} = \begin{bmatrix} 0 \\ 0 \\ 0 \end{bmatrix}.$

Note that something remarkable is happening here: Although A isn't the zero matrix and $\vec{x}$ isn't the zero vector, the product $A\vec{x}$ is the zero vector. (By contrast, the product of any two nonzero *scalars* is nonzero.) ∎

The formula for the product $A\vec{x}$ in Theorem 1.3.8 involves the expression $x_1\vec{v}_1 + \cdots + x_m\vec{v}_m$, where $\vec{v}_1, \ldots, \vec{v}_m$ are vectors in $\mathbb{R}^n$, and $x_1, \ldots, x_m$ are scalars. Such expressions come up very frequently in linear algebra; they deserve a name.

Definition 1.3.9 **Linear combinations**

A vector $\vec{b}$ in $\mathbb{R}^n$ is called a linear combination of the vectors $\vec{v}_1, \ldots, \vec{v}_m$ in $\mathbb{R}^n$ if there exist scalars $x_1, \ldots, x_m$ such that

$$\vec{b} = x_1\vec{v}_1 + \cdots + x_m\vec{v}_m.$$

Note that the product $A\vec{x}$ is the linear combination of the columns of A with the components of $\vec{x}$ as the coefficients:

$$A\vec{x} = \begin{bmatrix} | & & | \\ \vec{v}_1 & \cdots & \vec{v}_m \\ | & & | \end{bmatrix} \begin{bmatrix} x_1 \\ \vdots \\ x_m \end{bmatrix} = x_1\vec{v}_1 + \cdots + x_m\vec{v}_m.$$

Take a good look at this equation, because it is the most frequently used formula in this text. Particularly in theoretical work, it will often be useful to write the product

$A\vec{x}$ as the linear combination $x_1\vec{v}_1 + \cdots + x_m\vec{v}_m$. Conversely, when dealing with a linear combination $x_1\vec{v}_1 + \cdots + x_m\vec{v}_m$, it will often be helpful to introduce the matrix

$$A = \begin{bmatrix} | & & | \\ \vec{v}_1 & \cdots & \vec{v}_m \\ | & & | \end{bmatrix} \quad \text{and the vector} \quad \vec{x} = \begin{bmatrix} x_1 \\ \vdots \\ x_m \end{bmatrix}$$

and then write $x_1\vec{v}_1 + \cdots + x_m\vec{v}_m = A\vec{x}$.

Next we present two rules concerning the product $A\vec{x}$. In Chapter 2 we will see that these rules play a central role in linear algebra.

Theorem 1.3.10   **Algebraic rules for $A\vec{x}$**

If A is an $n \times m$ matrix; $\vec{x}$ and $\vec{y}$ are vectors in $\mathbb{R}^m$; and k is a scalar, then

    a. $A(\vec{x} + \vec{y}) = A\vec{x} + A\vec{y}$, and

    b. $A(k\vec{x}) = k(A\vec{x})$.

We will prove the first equation, leaving the second as Exercise 45. Denote the ith row of A by $\vec{w}_i$. Then

$$\left(i\text{th component of } A(\vec{x} + \vec{y})\right) = \vec{w}_i \cdot (\vec{x} + \vec{y}) \overset{\text{Step 2}}{=} \vec{w}_i \cdot \vec{x} + \vec{w}_i \cdot \vec{y}$$
$$= (i\text{th component of } A\vec{x}) + (i\text{th component of } A\vec{y})$$
$$= (i\text{th component of } A\vec{x} + A\vec{y}).$$

In Step 2 we are using a rule for dot products stated in Theorem A.5b, in the Appendix. ■

Our new tools of matrix algebra allow us to see linear systems in a new light, as illustrated in the next example. The definition of the product $A\vec{x}$ and the concept of a linear combination will be particularly helpful.

EXAMPLE 13   Consider the linear system

$$\begin{vmatrix} 3x_1 + x_2 = 7 \\ x_1 + 2x_2 = 4 \end{vmatrix}, \quad \text{with augmented matrix} \quad \begin{bmatrix} 3 & 1 & | & 7 \\ 1 & 2 & | & 4 \end{bmatrix}.$$

We can interpret the solution of this system as the intersection of two lines in the x_1x_2-plane, as illustrated in Figure 2.

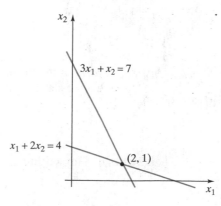

Figure 2

Alternatively, we can write the system in vector form, as

$$\begin{bmatrix} 3x_1 + x_2 \\ x_1 + 2x_2 \end{bmatrix} = \begin{bmatrix} 7 \\ 4 \end{bmatrix} \quad \text{or} \quad \begin{bmatrix} 3x_1 \\ x_1 \end{bmatrix} + \begin{bmatrix} x_2 \\ 2x_2 \end{bmatrix} = \begin{bmatrix} 7 \\ 4 \end{bmatrix} \quad \text{or} \quad x_1 \begin{bmatrix} 3 \\ 1 \end{bmatrix} + x_2 \begin{bmatrix} 1 \\ 2 \end{bmatrix} = \begin{bmatrix} 7 \\ 4 \end{bmatrix}.$$

We see that solving this system amounts to writing the vector $\begin{bmatrix} 7 \\ 4 \end{bmatrix}$ as a *linear combination* of the vectors $\begin{bmatrix} 3 \\ 1 \end{bmatrix}$ and $\begin{bmatrix} 1 \\ 2 \end{bmatrix}$ (see Definition 1.3.9). The vector equation

$$x_1 \begin{bmatrix} 3 \\ 1 \end{bmatrix} + x_2 \begin{bmatrix} 1 \\ 2 \end{bmatrix} = \begin{bmatrix} 7 \\ 4 \end{bmatrix},$$

and its solution can be represented geometrically, as shown in Figure 3. The problem amounts to resolving the vector $\begin{bmatrix} 7 \\ 4 \end{bmatrix}$ into two vectors parallel to $\begin{bmatrix} 3 \\ 1 \end{bmatrix}$ and $\begin{bmatrix} 1 \\ 2 \end{bmatrix}$, respectively, by means of a parallelogram.

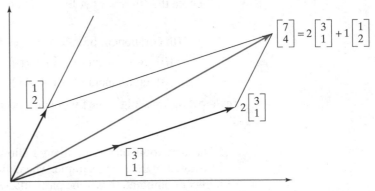

Figure 3

We can go further and write the linear combination

$$x_1 \begin{bmatrix} 3 \\ 1 \end{bmatrix} + x_2 \begin{bmatrix} 1 \\ 2 \end{bmatrix} \quad \text{as} \quad \begin{bmatrix} 3 & 1 \\ 1 & 2 \end{bmatrix} \begin{bmatrix} x_1 \\ x_2 \end{bmatrix}$$

so that the linear system

$$x_1 \begin{bmatrix} 3 \\ 1 \end{bmatrix} + x_2 \begin{bmatrix} 1 \\ 2 \end{bmatrix} = \begin{bmatrix} 7 \\ 4 \end{bmatrix} \quad \text{takes the form} \quad \begin{bmatrix} 3 & 1 \\ 1 & 2 \end{bmatrix} \begin{bmatrix} x_1 \\ x_2 \end{bmatrix} = \begin{bmatrix} 7 \\ 4 \end{bmatrix},$$

the *matrix form* of the linear system.

Note that we started out with the augmented matrix

$$\begin{bmatrix} A & \vdots & \vec{b} \end{bmatrix} = \begin{bmatrix} 3 & 1 & \vdots & 7 \\ 1 & 2 & \vdots & 4 \end{bmatrix},$$

and we ended up writing the system as

$$\underbrace{\begin{bmatrix} 3 & 1 \\ 1 & 2 \end{bmatrix}}_{A} \underbrace{\begin{bmatrix} x_1 \\ x_2 \end{bmatrix}}_{\vec{x}} = \underbrace{\begin{bmatrix} 7 \\ 4 \end{bmatrix}}_{\vec{b}}, \quad \text{or,} \quad A\vec{x} = \vec{b}. \quad \blacksquare$$

We can generalize:

Theorem 1.3.11   Matrix form of a linear system

We can write the linear system with augmented matrix $\begin{bmatrix} A & \vdots & \vec{b} \end{bmatrix}$ in matrix form as

$$A\vec{x} = \vec{b}.$$

∎

Note that the ith component of $A\vec{x}$ is $a_{i1}x_1 + \cdots + a_{im}x_m$, by Definition 1.3.7. Thus, the ith component of the equation $A\vec{x} = \vec{b}$ is

$$a_{i1}x_1 + \cdots + a_{im}x_m = b_i;$$

this is the ith equation of the system with augmented matrix $\begin{bmatrix} A & \vdots & \vec{b} \end{bmatrix}$.

EXAMPLE 14   Write the system

$$\begin{vmatrix} 2x_1 - 3x_2 + 5x_3 = 7 \\ 9x_1 + 4x_2 - 6x_3 = 8 \end{vmatrix}$$

in matrix form.

Solution

The coefficient matrix is $A = \begin{bmatrix} 2 & -3 & 5 \\ 9 & 4 & -6 \end{bmatrix}$, and $\vec{b} = \begin{bmatrix} 7 \\ 8 \end{bmatrix}$. The matrix form is

$$A\vec{x} = \vec{b}, \quad \text{or,} \quad \begin{bmatrix} 2 & -3 & 5 \\ 9 & 4 & -6 \end{bmatrix} \begin{bmatrix} x_1 \\ x_2 \\ x_3 \end{bmatrix} = \begin{bmatrix} 7 \\ 8 \end{bmatrix}.$$

∎

Now that we can write a linear system as a *single equation*, $A\vec{x} = \vec{b}$, rather than a list of simultaneous equations, we can think about it in new ways.

For example, if we have an equation $ax = b$ of *numbers*, we can divide both sides by a to find the solution x:

$$x = \frac{b}{a} = a^{-1}b \quad \text{(if } a \neq 0\text{)}.$$

It is natural to ask whether we can take an analogous approach in the case of the equation $A\vec{x} = \vec{b}$. Can we "divide by A," in some sense, and write

$$\vec{x} = \frac{\vec{b}}{A} = A^{-1}\vec{b}?$$

This issue of the invertibility of a matrix will be one of the main themes of Chapter 2.

EXERCISES 1.3

GOAL   *Use the reduced row-echelon form of the augmented matrix to find the number of solutions of a linear system. Apply the definition of the rank of a matrix. Compute the product $A\vec{x}$ in terms of the rows or the columns of A. Represent a linear system in vector or in matrix form.*

1. The reduced row-echelon forms of the augmented matrices of three systems are given below. How many solutions does each system have?

a. $\begin{bmatrix} 1 & 0 & 2 & \vdots & 0 \\ 0 & 1 & 3 & \vdots & 0 \\ 0 & 0 & 0 & \vdots & 1 \end{bmatrix}$ b. $\begin{bmatrix} 1 & 0 & \vdots & 5 \\ 0 & 1 & \vdots & 6 \end{bmatrix}$

c. $\begin{bmatrix} 0 & 1 & 0 & \vdots & 2 \\ 0 & 0 & 1 & \vdots & 3 \end{bmatrix}$

Find the rank of the matrices in Exercises 2 through 4.

2. $\begin{bmatrix} 1 & 2 & 3 \\ 0 & 1 & 2 \\ 0 & 0 & 1 \end{bmatrix}$ 3. $\begin{bmatrix} 1 & 1 & 1 \\ 1 & 1 & 1 \\ 1 & 1 & 1 \end{bmatrix}$ 4. $\begin{bmatrix} 1 & 4 & 7 \\ 2 & 5 & 8 \\ 3 & 6 & 9 \end{bmatrix}$

5. a. Write the system

$$\begin{vmatrix} x + 2y = 7 \\ 3x + y = 11 \end{vmatrix}$$

in vector form.

b. Use your answer in part (a) to represent the system geometrically. Solve the system and represent the solution geometrically.

6. Consider the vectors $\vec{v}_1$, $\vec{v}_2$, $\vec{v}_3$ in $\mathbb{R}^2$ (sketched in the accompanying figure). Vectors $\vec{v}_1$ and $\vec{v}_2$ are parallel. How many solutions x, y does the system

$$x\vec{v}_1 + y\vec{v}_2 = \vec{v}_3$$

have? Argue geometrically.

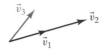

7. Consider the vectors $\vec{v}_1$, $\vec{v}_2$, $\vec{v}_3$ in $\mathbb{R}^2$ shown in the accompanying sketch. How many solutions x, y does the system

$$x\vec{v}_1 + y\vec{v}_2 = \vec{v}_3$$

have? Argue geometrically.

8. Consider the vectors $\vec{v}_1$, $\vec{v}_2$, $\vec{v}_3$, $\vec{v}_4$ in $\mathbb{R}^2$ shown in the accompanying sketch. Arguing geometrically, find two solutions x, y, z of the linear system

$$x\vec{v}_1 + y\vec{v}_2 + z\vec{v}_3 = \vec{v}_4.$$

How do you know that this system has in fact infinitely many solutions?

9. Write the system

$$\begin{vmatrix} x + 2y + 3z = 1 \\ 4x + 5y + 6z = 4 \\ 7x + 8y + 9z = 9 \end{vmatrix}$$

in matrix form.

Compute the dot products in Exercises 10 through 12 (if the products are defined).

10. $\begin{bmatrix} 1 \\ 2 \\ 3 \end{bmatrix} \cdot \begin{bmatrix} 1 \\ -2 \\ 1 \end{bmatrix}$

11. $\begin{bmatrix} 1 & 9 & 9 & 7 \end{bmatrix} \cdot \begin{bmatrix} 6 \\ 6 \\ 6 \end{bmatrix}$

12. $\begin{bmatrix} 1 & 2 & 3 & 4 \end{bmatrix} \cdot \begin{bmatrix} 5 \\ 6 \\ 7 \\ 8 \end{bmatrix}$

Compute the products $A\vec{x}$ in Exercises 13 through 15 using paper and pencil. In each case, compute the product two ways: in terms of the columns of A (Theorem 1.3.8) and in terms of the rows of A (Definition 1.3.7).

13. $\begin{bmatrix} 1 & 2 \\ 3 & 4 \end{bmatrix} \begin{bmatrix} 7 \\ 11 \end{bmatrix}$

14. $\begin{bmatrix} 1 & 2 & 3 \\ 2 & 3 & 4 \end{bmatrix} \begin{bmatrix} -1 \\ 2 \\ 1 \end{bmatrix}$

15. $\begin{bmatrix} 1 & 2 & 3 & 4 \end{bmatrix} \begin{bmatrix} 5 \\ 6 \\ 7 \\ 8 \end{bmatrix}$

Compute the products $A\vec{x}$ in Exercises 16 through 19 using paper and pencil (if the products are defined).

16. $\begin{bmatrix} 0 & 1 \\ 3 & 2 \end{bmatrix} \begin{bmatrix} 2 \\ -3 \end{bmatrix}$

17. $\begin{bmatrix} 1 & 2 & 3 \\ 4 & 5 & 6 \end{bmatrix} \begin{bmatrix} 7 \\ 8 \end{bmatrix}$

18. $\begin{bmatrix} 1 & 2 \\ 3 & 4 \\ 5 & 6 \end{bmatrix} \begin{bmatrix} 1 \\ 2 \end{bmatrix}$

19. $\begin{bmatrix} 1 & 1 & -1 \\ -5 & 1 & 1 \\ 1 & -5 & 3 \end{bmatrix} \begin{bmatrix} 1 \\ 2 \\ 3 \end{bmatrix}$

20. a. Find $\begin{bmatrix} 2 & 3 \\ 4 & 5 \\ 6 & 7 \end{bmatrix} + \begin{bmatrix} 7 & 5 \\ 3 & 1 \\ 0 & -1 \end{bmatrix}$.

b. Find $9 \begin{bmatrix} 1 & -1 & 2 \\ 3 & 4 & 5 \end{bmatrix}$.

21. Use technology to compute the product

$$\begin{bmatrix} 1 & 7 & 8 & 9 \\ 1 & 2 & 9 & 1 \\ 1 & 5 & 1 & 5 \\ 1 & 6 & 4 & 8 \end{bmatrix} \begin{bmatrix} 1 \\ 9 \\ 5 \\ 6 \end{bmatrix}.$$

22. Consider a linear system of three equations with three unknowns. We are told that the system has a unique solution. What does the reduced row-echelon form of the coefficient matrix of this system look like? Explain your answer.

23. Consider a linear system of four equations with three unknowns. We are told that the system has a unique solution. What does the reduced row-echelon form of the coefficient matrix of this system look like? Explain your answer.

24. Let A be a 4×4 matrix, and let $\vec{b}$ and $\vec{c}$ be two vectors in $\mathbb{R}^4$. We are told that the system $A\vec{x} = \vec{b}$ has a unique solution. What can you say about the number of solutions of the system $A\vec{x} = \vec{c}$?

25. Let A be a 4×4 matrix, and let $\vec{b}$ and $\vec{c}$ be two vectors in $\mathbb{R}^4$. We are told that the system $A\vec{x} = \vec{b}$ is inconsistent. What can you say about the number of solutions of the system $A\vec{x} = \vec{c}$?

26. Let A be a 4×3 matrix, and let $\vec{b}$ and $\vec{c}$ be two vectors in $\mathbb{R}^4$. We are told that the system $A\vec{x} = \vec{b}$ has a unique solution. What can you say about the number of solutions of the system $A\vec{x} = \vec{c}$?

27. If the rank of a 4×4 matrix A is 4, what is rref(A)?

28. If the rank of a 5×3 matrix A is 3, what is rref(A)?

In Problems 29 through 32, let $\vec{x} = \begin{bmatrix} 5 \\ 3 \\ -9 \end{bmatrix}$ *and* $\vec{y} = \begin{bmatrix} 2 \\ 0 \\ 1 \end{bmatrix}$.

29. Find a *diagonal* matrix A such that $A\vec{x} = \vec{y}$.

30. Find a matrix A of rank 1 such that $A\vec{x} = \vec{y}$.

31. Find an *upper triangular* matrix A such that $A\vec{x} = \vec{y}$. Also, it is required that all the entries of A on and above the diagonal be nonzero.

32. Find a matrix A with all nonzero entries such that $A\vec{x} = \vec{y}$.

33. Let A be the $n \times n$ matrix with all 1's on the diagonal and all 0's above and below the diagonal. What is $A\vec{x}$, where $\vec{x}$ is a vector in $\mathbb{R}^n$?

34. We define the vectors
$$\vec{e}_1 = \begin{bmatrix} 1 \\ 0 \\ 0 \end{bmatrix}, \quad \vec{e}_2 = \begin{bmatrix} 0 \\ 1 \\ 0 \end{bmatrix}, \quad \vec{e}_3 = \begin{bmatrix} 0 \\ 0 \\ 1 \end{bmatrix}$$
in $\mathbb{R}^3$.

a. For
$$A = \begin{bmatrix} a & b & c \\ d & e & f \\ g & h & k \end{bmatrix},$$
compute $A\vec{e}_1$, $A\vec{e}_2$, and $A\vec{e}_3$.

b. If B is an $n \times 3$ matrix with columns $\vec{v}_1$, $\vec{v}_2$, and $\vec{v}_3$, what is $B\vec{e}_1$, $B\vec{e}_2$, $B\vec{e}_3$?

35. In $\mathbb{R}^m$, we define
$$\vec{e}_i = \begin{bmatrix} 0 \\ 0 \\ \vdots \\ 1 \\ \vdots \\ 0 \end{bmatrix} \leftarrow i\text{th component.}$$

If A is an $n \times m$ matrix, what is $A\vec{e}_i$?

36. Find a 3×3 matrix A such that
$$A \begin{bmatrix} 1 \\ 0 \\ 0 \end{bmatrix} = \begin{bmatrix} 1 \\ 2 \\ 3 \end{bmatrix}, \quad A \begin{bmatrix} 0 \\ 1 \\ 0 \end{bmatrix} = \begin{bmatrix} 4 \\ 5 \\ 6 \end{bmatrix},$$
$$\text{and} \quad A \begin{bmatrix} 0 \\ 0 \\ 1 \end{bmatrix} = \begin{bmatrix} 7 \\ 8 \\ 9 \end{bmatrix}.$$

37. Find all vectors $\vec{x}$ such that $A\vec{x} = \vec{b}$, where
$$A = \begin{bmatrix} 1 & 2 & 0 \\ 0 & 0 & 1 \\ 0 & 0 & 0 \end{bmatrix} \quad \text{and} \quad \vec{b} = \begin{bmatrix} 2 \\ 1 \\ 0 \end{bmatrix}.$$

38. a. Using technology, generate a random 3×3 matrix A. (The entries may be either single-digit integers or numbers between 0 and 1, depending on the technology you are using.) Find rref(A). Repeat this experiment a few times.

b. What does the reduced row-echelon form of most 3×3 matrices look like? Explain.

39. Repeat Exercise 38 for 3×4 matrices.

40. Repeat Exercise 38 for 4×3 matrices.

41. How many solutions do most systems of three linear equations with three unknowns have? Explain in terms of your work in Exercise 38.

42. How many solutions do most systems of three linear equations with four unknowns have? Explain in terms of your work in Exercise 39.

43. How many solutions do most systems of four linear equations with three unknowns have? Explain in terms of your work in Exercise 40.

44. Consider an $n \times m$ matrix A with more rows than columns ($n > m$). Show that there is a vector $\vec{b}$ in $\mathbb{R}^n$ such that the system $A\vec{x} = \vec{b}$ is inconsistent.

45. Consider an $n \times m$ matrix A, a vector $\vec{x}$ in $\mathbb{R}^m$, and a scalar k. Show that
$$A(k\vec{x}) = k(A\vec{x}).$$

46. Find the rank of the matrix
$$\begin{bmatrix} a & b & c \\ 0 & d & e \\ 0 & 0 & f \end{bmatrix},$$
where a, d, and f are nonzero, and b, c, and e are arbitrary numbers.

47. A linear system of the form
$$A\vec{x} = \vec{0}$$
is called *homogeneous*. Justify the following facts:
a. All homogeneous systems are consistent.

b. A homogeneous system with fewer equations than unknowns has infinitely many solutions.

c. If $\vec{x}_1$ and $\vec{x}_2$ are solutions of the homogeneous system $A\vec{x} = \vec{0}$, then $\vec{x}_1 + \vec{x}_2$ is a solution as well.

d. If $\vec{x}$ is a solution of the homogeneous system $A\vec{x} = \vec{0}$ and k is an arbitrary constant, then $k\vec{x}$ is a solution as well.

48. Consider a solution $\vec{x}_1$ of the linear system $A\vec{x} = \vec{b}$. Justify the facts stated in parts (a) and (b):

a. If $\vec{x}_h$ is a solution of the system $A\vec{x} = \vec{0}$, then $\vec{x}_1 + \vec{x}_h$ is a solution of the system $A\vec{x} = \vec{b}$.

b. If $\vec{x}_2$ is another solution of the system $A\vec{x} = \vec{b}$, then $\vec{x}_2 - \vec{x}_1$ is a solution of the system $A\vec{x} = \vec{0}$.

c. Now suppose A is a 2×2 matrix. A solution vector $\vec{x}_1$ of the system $A\vec{x} = \vec{b}$ is shown in the accompanying figure. We are told that the solutions of the system $A\vec{x} = \vec{0}$ form the line shown in the sketch. Draw the line consisting of all solutions of the system $A\vec{x} = \vec{b}$.

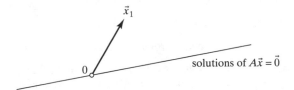

If you are puzzled by the generality of this problem, think about an example first:

$$A = \begin{bmatrix} 1 & 2 \\ 3 & 6 \end{bmatrix}, \quad \vec{b} = \begin{bmatrix} 3 \\ 9 \end{bmatrix}, \quad \text{and} \quad \vec{x}_1 = \begin{bmatrix} 1 \\ 1 \end{bmatrix}.$$

49. Consider the accompanying table. For some linear systems $A\vec{x} = \vec{b}$, you are given either the rank of the coefficient matrix A, or the rank of the augmented matrix $[A \mid \vec{b}]$. In each case, state whether the system could have no solution, one solution, or infinitely many solutions. There may be more than one possibility for some systems. Justify your answers.

	Number of Equations	Number of Unknowns	Rank of A	Rank of $[A \mid \vec{b}]$
a.	3	4	—	2
b.	4	3	3	—
c.	4	3	—	4
d.	3	4	3	—

50. Consider a linear system $A\vec{x} = \vec{b}$, where A is a 4×3 matrix. We are told that rank $[A \mid \vec{b}] = 4$. How many solutions does this system have?

51. Consider an $n \times m$ matrix A, an $r \times s$ matrix B, and a vector $\vec{x}$ in $\mathbb{R}^p$. For which values of $n, m, r, s,$ and p is the product

$$A(B\vec{x})$$

defined?

52. Consider the matrices

$$A = \begin{bmatrix} 1 & 0 \\ 1 & 2 \end{bmatrix} \quad \text{and} \quad B = \begin{bmatrix} 0 & -1 \\ 1 & 0 \end{bmatrix}.$$

Can you find a 2×2 matrix C such that

$$A(B\vec{x}) = C\vec{x},$$

for all vectors $\vec{x}$ in $\mathbb{R}^2$?

53. If A and B are two $n \times m$ matrices, is

$$(A + B)\vec{x} = A\vec{x} + B\vec{x}$$

for all $\vec{x}$ in $\mathbb{R}^m$?

54. Consider two vectors $\vec{v}_1$ and $\vec{v}_2$ in $\mathbb{R}^3$ that are not parallel. Which vectors in $\mathbb{R}^3$ are linear combinations of $\vec{v}_1$ and $\vec{v}_2$? Describe the set of these vectors geometrically. Include a sketch in your answer.

55. Is the vector $\begin{bmatrix} 7 \\ 8 \\ 9 \end{bmatrix}$ a linear combination of

$$\begin{bmatrix} 1 \\ 2 \\ 3 \end{bmatrix} \quad \text{and} \quad \begin{bmatrix} 4 \\ 5 \\ 6 \end{bmatrix}?$$

56. Is the vector

$$\begin{bmatrix} 30 \\ -1 \\ 38 \\ 56 \\ 62 \end{bmatrix}$$

a linear combination of

$$\begin{bmatrix} 1 \\ 7 \\ 1 \\ 9 \\ 4 \end{bmatrix}, \quad \begin{bmatrix} 5 \\ 6 \\ 3 \\ 2 \\ 8 \end{bmatrix}, \quad \begin{bmatrix} 9 \\ 2 \\ 3 \\ 5 \\ 2 \end{bmatrix}, \quad \begin{bmatrix} -2 \\ -5 \\ 4 \\ 7 \\ 9 \end{bmatrix}?$$

57. Express the vector $\begin{bmatrix} 7 \\ 11 \end{bmatrix}$ as the sum of a vector on the line $y = 3x$ and a vector on the line $y = x/2$.

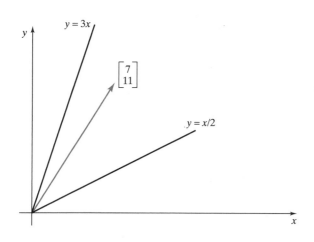

58. For which values of the constants b and c is the vector

$\begin{bmatrix} 3 \\ b \\ c \end{bmatrix}$ a linear combination of $\begin{bmatrix} 1 \\ 3 \\ 2 \end{bmatrix}$, $\begin{bmatrix} 2 \\ 6 \\ 4 \end{bmatrix}$, and $\begin{bmatrix} -1 \\ -3 \\ -2 \end{bmatrix}$?

59. For which values of the constants c and d is $\begin{bmatrix} 5 \\ 7 \\ c \\ d \end{bmatrix}$ a linear

combination of $\begin{bmatrix} 1 \\ 1 \\ 1 \\ 1 \end{bmatrix}$ and $\begin{bmatrix} 1 \\ 2 \\ 3 \\ 4 \end{bmatrix}$?

60. For which values of the constants a, b, c, and d is $\begin{bmatrix} a \\ b \\ c \\ d \end{bmatrix}$

a linear combination of $\begin{bmatrix} 0 \\ 0 \\ 3 \\ 0 \end{bmatrix}$, $\begin{bmatrix} 1 \\ 0 \\ 4 \\ 0 \end{bmatrix}$, and $\begin{bmatrix} 2 \\ 0 \\ 5 \\ 6 \end{bmatrix}$?

61. For which values of the constant c is $\begin{bmatrix} 1 \\ c \\ c^2 \end{bmatrix}$ a linear

combination of $\begin{bmatrix} 1 \\ 2 \\ 4 \end{bmatrix}$ and $\begin{bmatrix} 1 \\ 3 \\ 9 \end{bmatrix}$?

62. For which values of the constant c is $\begin{bmatrix} 1 \\ c \\ c^2 \end{bmatrix}$ a linear

combination of $\begin{bmatrix} 1 \\ a \\ a^2 \end{bmatrix}$ and $\begin{bmatrix} 1 \\ b \\ b^2 \end{bmatrix}$, where a and b are

arbitrary constants?

In Exercises 63 through 67, consider the vectors $\vec{v}$ and $\vec{w}$ in the accompanying figure.

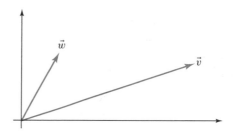

63. Give a geometrical description of the set of all vectors of the form $\vec{v} + c\vec{w}$, where c is an arbitrary real number.

64. Give a geometrical description of the set of all vectors of the form $\vec{v} + c\vec{w}$, where $0 \leq c \leq 1$.

65. Give a geometrical description of the set of all vectors of the form $a\vec{v} + b\vec{w}$, where $0 \leq a \leq 1$ and $0 \leq b \leq 1$.

66. Give a geometrical description of the set of all vectors of the form $a\vec{v} + b\vec{w}$, where $a + b = 1$.

67. Give a geometrical description of the set of all vectors of the form $a\vec{v} + b\vec{w}$, where $0 \leq a$, $0 \leq b$, and $a + b \leq 1$.

68. Give a geometrical description of the set of all vectors $\vec{u}$ in $\mathbb{R}^2$ such that $\vec{u} \cdot \vec{v} = \vec{u} \cdot \vec{w}$.

69. Solve the linear system

$$\begin{vmatrix} y + z = a \\ x \quad\ + z = b \\ x + y \quad\ = c \end{vmatrix},$$

where a, b, and c are arbitrary constants.

70. Let A be the $n \times n$ matrix with 0's on the main diagonal, and 1's everywhere else. For an arbitrary vector $\vec{b}$ in $\mathbb{R}^n$, solve the linear system $A\vec{x} = \vec{b}$, expressing the components $x_1, \ldots, x_n$ of $\vec{x}$ in terms of the components of $\vec{b}$. (See Exercise 69 for the case $n = 3$.)

Chapter One Exercises

TRUE OR FALSE?[18]

Determine whether the statements that follow are true or false, and justify your answer.

1. There exists a 3×4 matrix with rank 4.

2. If A is a 3×4 matrix and vector $\vec{v}$ is in $\mathbb{R}^4$, then vector $A\vec{v}$ is in $\mathbb{R}^3$.

3. If the 4×4 matrix A has rank 4, then any linear system with coefficient matrix A will have a unique solution.

4. There exists a system of three linear equations with three unknowns that has exactly three solutions.

5. There exists a 5×5 matrix A of rank 4 such that the system $A\vec{x} = \vec{0}$ has only the solution $\vec{x} = \vec{0}$.

6. If matrix A is in rref, then at least one of the entries in each column must be 1.

7. If A is an $n \times n$ matrix and $\vec{x}$ is a vector in $\mathbb{R}^n$, then the product $A\vec{x}$ is a linear combination of the columns of matrix A.

8. If vector $\vec{u}$ is a linear combination of vectors $\vec{v}$ and $\vec{w}$, then we can write $\vec{u} = a\vec{v} + b\vec{w}$ for some scalars a and b.

9. Matrix $\begin{bmatrix} 1 & 2 & 0 \\ 0 & 0 & 1 \\ 0 & 0 & 0 \end{bmatrix}$ is in rref.

10. A system of four linear equations in three unknowns is always inconsistent.

11. If A is a nonzero matrix of the form $\begin{bmatrix} a & -b \\ b & a \end{bmatrix}$, then the rank of A must be 2.

12. $\operatorname{rank} \begin{bmatrix} 1 & 1 & 1 \\ 1 & 2 & 3 \\ 1 & 3 & 6 \end{bmatrix} = 3$

13. The system $A\vec{x} = \begin{bmatrix} 0 \\ 0 \\ 0 \\ 1 \end{bmatrix}$ is inconsistent for all 4×3 matrices A.

14. There exists a 2×2 matrix A such that
$$A \begin{bmatrix} 1 \\ 1 \end{bmatrix} = \begin{bmatrix} 1 \\ 2 \end{bmatrix} \quad \text{and} \quad A \begin{bmatrix} 2 \\ 2 \end{bmatrix} = \begin{bmatrix} 2 \\ 1 \end{bmatrix}.$$

15. $\operatorname{rank} \begin{bmatrix} 2 & 2 & 2 \\ 2 & 2 & 2 \\ 2 & 2 & 2 \end{bmatrix} = 2$

16. $\begin{bmatrix} 11 & 13 & 15 \\ 17 & 19 & 21 \end{bmatrix} \begin{bmatrix} -1 \\ 3 \\ -1 \end{bmatrix} = \begin{bmatrix} 13 \\ 19 \\ 21 \end{bmatrix}$

17. There exists a matrix A such that $A \begin{bmatrix} -1 \\ 2 \end{bmatrix} = \begin{bmatrix} 3 \\ 5 \\ 7 \end{bmatrix}$.

18. Vector $\begin{bmatrix} 1 \\ 2 \\ 3 \end{bmatrix}$ is a linear combination of vectors
$$\begin{bmatrix} 4 \\ 5 \\ 6 \end{bmatrix} \quad \text{and} \quad \begin{bmatrix} 7 \\ 8 \\ 9 \end{bmatrix}.$$

19. The system $\begin{bmatrix} 1 & 2 & 3 \\ 4 & 5 & 6 \\ 0 & 0 & 0 \end{bmatrix} \vec{x} = \begin{bmatrix} 1 \\ 2 \\ 3 \end{bmatrix}$ is inconsistent.

20. There exists a 2×2 matrix A such that $A \begin{bmatrix} 1 \\ 2 \end{bmatrix} = \begin{bmatrix} 3 \\ 4 \end{bmatrix}$.

21. If A and B are any two 3×3 matrices of rank 2, then A can be transformed into B by means of elementary row operations.

22. If vector $\vec{u}$ is a linear combination of vectors $\vec{v}$ and $\vec{w}$, and $\vec{v}$ is a linear combination of vectors $\vec{p}, \vec{q}$, and $\vec{r}$, then $\vec{u}$ must be a linear combination of $\vec{p}, \vec{q}, \vec{r}$, and $\vec{w}$.

23. A linear system with fewer unknowns than equations must have infinitely many solutions or none.

24. The rank of any upper triangular matrix is the number of nonzero entries on its diagonal.

25. If the system $A\vec{x} = \vec{b}$ has a unique solution, then A must be a square matrix.

26. If A is any 4×3 matrix, then there exists a vector $\vec{b}$ in $\mathbb{R}^4$ such that the system $A\vec{x} = \vec{b}$ is inconsistent.

27. There exist scalars a and b such that matrix
$$\begin{bmatrix} 0 & 1 & a \\ -1 & 0 & b \\ -a & -b & 0 \end{bmatrix}$$
has rank 3.

---

[18] We will conclude each chapter (except for Chapter 9) with some true–false questions, over 400 in all. We will start with a group of about 10 straightforward statements that refer directly to definitions and theorems given in the chapter. Then there may be some computational exercises, and the remaining ones are more conceptual, calling for independent reasoning. In some chapters, a few of the problems toward the end can be quite challenging. Don't expect a balanced coverage of all the topics; some concepts are better suited for this kind of questioning than others.

28. If $\vec{v}$ and $\vec{w}$ are vectors in $\mathbb{R}^4$, then $\vec{v}$ must be a linear combination of $\vec{v}$ and $\vec{w}$.

29. If $\vec{u}$, $\vec{v}$, and $\vec{w}$ are nonzero vectors in $\mathbb{R}^2$, then $\vec{w}$ must be a linear combination of $\vec{u}$ and $\vec{v}$.

30. If $\vec{v}$ and $\vec{w}$ are vectors in $\mathbb{R}^4$, then the zero vector in $\mathbb{R}^4$ must be a linear combination of $\vec{v}$ and $\vec{w}$.

31. There exists a 4×3 matrix A of rank 3 such that $A \begin{bmatrix} 1 \\ 2 \\ 3 \end{bmatrix} = \vec{0}$.

32. The system $A\vec{x} = \vec{b}$ is inconsistent if (and only if) rref(A) contains a row of zeros.

33. If A is a 4×3 matrix of rank 3 and $A\vec{v} = A\vec{w}$ for two vectors $\vec{v}$ and $\vec{w}$ in $\mathbb{R}^3$, then vectors $\vec{v}$ and $\vec{w}$ must be equal.

34. If A is a 4×4 matrix and the system $A\vec{x} = \begin{bmatrix} 2 \\ 3 \\ 4 \\ 5 \end{bmatrix}$ has a unique solution, then the system $A\vec{x} = \vec{0}$ has only the solution $\vec{x} = \vec{0}$.

35. If vector $\vec{u}$ is a linear combination of vectors $\vec{v}$ and $\vec{w}$, then $\vec{w}$ must be a linear combination of $\vec{u}$ and $\vec{v}$.

36. If $A = \begin{bmatrix} \vec{u} & \vec{v} & \vec{w} \end{bmatrix}$ and rref$(A) = \begin{bmatrix} 1 & 0 & 2 \\ 0 & 1 & 3 \\ 0 & 0 & 0 \end{bmatrix}$, then the equation $\vec{w} = 2\vec{u} + 3\vec{v}$ must hold.

37. If A and B are matrices of the same size, then the formula rank$(A + B) = $ rank$(A) + $ rank(B) must hold.

38. If A and B are any two $n \times n$ matrices of rank n, then A can be transformed into B by means of elementary row operations.

39. If a vector $\vec{v}$ in $\mathbb{R}^4$ is a linear combination of $\vec{u}$ and $\vec{w}$, and if A is a 5×4 matrix, then $A\vec{v}$ must be a linear combination of $A\vec{u}$ and $A\vec{w}$.

40. If matrix E is in reduced row-echelon form, and if we omit a row of E, then the remaining matrix must be in reduced row-echelon form as well.

41. The linear system $A\vec{x} = \vec{b}$ is consistent if (and only if) rank$(A) = $ rank $\begin{bmatrix} A & \vdots & \vec{b} \end{bmatrix}$.

42. If A is a 3×4 matrix of rank 3, then the system $A\vec{x} = \begin{bmatrix} 1 \\ 2 \\ 3 \end{bmatrix}$ must have infinitely many solutions.

43. If two matrices A and B have the same reduced row-echelon form, then the equations $A\vec{x} = \vec{0}$ and $B\vec{x} = \vec{0}$ must have the same solutions.

44. If matrix E is in reduced row-echelon form, and if we omit a column of E, then the remaining matrix must be in reduced row-echelon form as well.

45. If A and B are two 2×2 matrices such that the equations $A\vec{x} = \vec{0}$ and $B\vec{x} = \vec{0}$ have the same solutions, then rref(A) must be equal to rref(B).

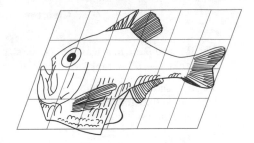

Linear Transformations

2.1 Introduction to Linear Transformations and Their Inverses

Imagine yourself cruising in the Mediterranean as a crew member on a French coast guard boat, looking for evildoers. Periodically, your boat radios its position to headquarters in Marseille. You expect that communications will be intercepted. So, before you broadcast anything, you have to transform the actual position of the boat,

$$\begin{bmatrix} x_1 \\ x_2 \end{bmatrix}$$

(x_1 for Eastern longitude, x_2 for Northern latitude), into an encoded position

$$\begin{bmatrix} y_1 \\ y_2 \end{bmatrix}.$$

You use the following code:

$$\begin{aligned} y_1 &= \ x_1 + 3x_2 \\ y_2 &= 2x_1 + 5x_2. \end{aligned}$$

For example, when the actual position of your boat is 5° E, 42° N, or

$$\vec{x} = \begin{bmatrix} x_1 \\ x_2 \end{bmatrix} = \begin{bmatrix} 5 \\ 42 \end{bmatrix},$$

your encoded position will be

$$\vec{y} = \begin{bmatrix} y_1 \\ y_2 \end{bmatrix} = \begin{bmatrix} x_1 + 3x_2 \\ 2x_1 + 5x_2 \end{bmatrix} = \begin{bmatrix} 5 + 3 \cdot 42 \\ 2 \cdot 5 + 5 \cdot 42 \end{bmatrix} = \begin{bmatrix} 131 \\ 220 \end{bmatrix}.$$

(See Figure 1.)

The coding transformation can be represented as

$$\underbrace{\begin{bmatrix} y_1 \\ y_2 \end{bmatrix}}_{\vec{y}} = \begin{bmatrix} x_1 + 3x_2 \\ 2x_1 + 5x_2 \end{bmatrix} = \underbrace{\begin{bmatrix} 1 & 3 \\ 2 & 5 \end{bmatrix}}_{A} \underbrace{\begin{bmatrix} x_1 \\ x_2 \end{bmatrix}}_{\vec{x}},$$

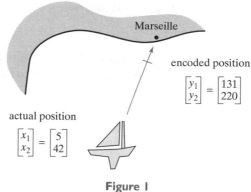

actual position

$$\begin{bmatrix} x_1 \\ x_2 \end{bmatrix} = \begin{bmatrix} 5 \\ 42 \end{bmatrix}$$

encoded position

$$\begin{bmatrix} y_1 \\ y_2 \end{bmatrix} = \begin{bmatrix} 131 \\ 220 \end{bmatrix}$$

Figure I

or, more succinctly, as

$$\vec{y} = A\vec{x}.$$

The matrix A is called the (*coefficient*) *matrix* of the transformation.

A transformation of the form

$$\vec{y} = A\vec{x}$$

is called a *linear transformation*. We will discuss this important concept in greater detail later in this section and throughout this chapter.

As the ship reaches a new position, the sailor on duty at headquarters in Marseille receives the encoded message

$$\vec{b} = \begin{bmatrix} 133 \\ 223 \end{bmatrix}.$$

He must determine the actual position of the boat. He will have to solve the linear system

$$A\vec{x} = \vec{b},$$

or, more explicitly,

$$\begin{vmatrix} x_1 + 3x_2 = 133 \\ 2x_1 + 5x_2 = 223 \end{vmatrix}.$$

Here is his solution. Is it correct?

$$\vec{x} = \begin{bmatrix} x_1 \\ x_2 \end{bmatrix} = \begin{bmatrix} 4 \\ 43 \end{bmatrix}$$

As the boat travels on and dozens of positions are radioed in, the sailor gets a little tired of solving all those linear systems, and he thinks there must be a general formula to simplify the task. He wants to solve the system

$$\begin{vmatrix} x_1 + 3x_2 = y_1 \\ 2x_1 + 5x_2 = y_2 \end{vmatrix}$$

when y_1 and y_2 are arbitrary constants, rather than particular numerical values. He is looking for the *decoding transformation*

$$\vec{y} \to \vec{x},$$

which is the *inverse*[1] of the coding transformation

$$\vec{x} \rightarrow \vec{y}.$$

The method of finding this solution is nothing new. We apply elimination as we have for a linear system with known values y_1 and y_2:

$$
\begin{vmatrix} x_1 + 3x_2 = y_1 \\ 2x_1 + 5x_2 = y_2 \end{vmatrix}
\begin{array}{c} \longrightarrow \\ -2 \text{ (I)} \end{array}
\begin{vmatrix} x_1 + 3x_2 = y_1 \\ -x_2 = -2y_1 + y_2 \end{vmatrix}
\begin{array}{c} \longrightarrow \\ \div(-1) \end{array}
$$

$$
\begin{vmatrix} x_1 + 3x_2 = y_1 \\ x_2 = 2y_1 - y_2 \end{vmatrix}
\begin{array}{c} -3 \text{ (II)} \\ \longrightarrow \end{array}
\begin{vmatrix} x_1 = -5y_1 + 3y_2 \\ x_2 = 2y_1 - y_2 \end{vmatrix}.
$$

The formula for the decoding transformation is

$$x_1 = -5y_1 + 3y_2,$$
$$x_2 = 2y_1 - y_2,$$

or

$$\vec{x} = B\vec{y}, \quad \text{where } B = \begin{bmatrix} -5 & 3 \\ 2 & -1 \end{bmatrix}.$$

Note that the decoding transformation is linear and that its coefficient matrix is

$$B = \begin{bmatrix} -5 & 3 \\ 2 & -1 \end{bmatrix}.$$

The relationship between the two matrices A and B is shown in Figure 2.

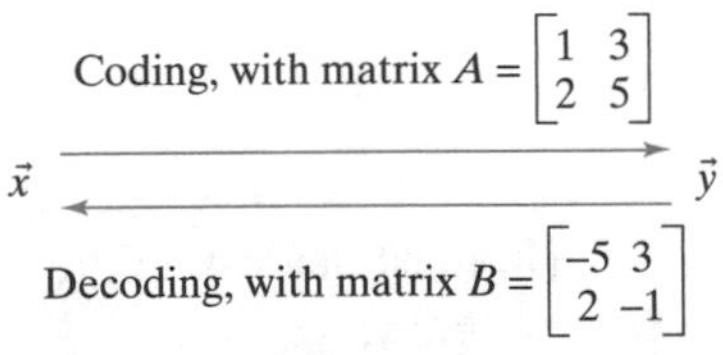

Coding, with matrix $A = \begin{bmatrix} 1 & 3 \\ 2 & 5 \end{bmatrix}$

$\vec{x}$ $\vec{y}$

Decoding, with matrix $B = \begin{bmatrix} -5 & 3 \\ 2 & -1 \end{bmatrix}$

Figure 2

Since the decoding transformation $\vec{x} = B\vec{y}$ is the inverse of the coding transformation $\vec{y} = A\vec{x}$, we say that the matrix B is the *inverse* of the matrix A. We can write this as $B = A^{-1}$.

Not all linear transformations

$$\begin{bmatrix} x_1 \\ x_2 \end{bmatrix} \rightarrow \begin{bmatrix} y_1 \\ y_2 \end{bmatrix}$$

are invertible. Suppose some ignorant officer chooses the code

$$\begin{array}{l} y_1 = x_1 + 2x_2 \\ y_2 = 2x_1 + 4x_2 \end{array} \quad \text{with matrix} \quad A = \begin{bmatrix} 1 & 2 \\ 2 & 4 \end{bmatrix}$$

for the French coast guard boats. When the sailor in Marseille has to decode a position, for example,

$$\vec{b} = \begin{bmatrix} 89 \\ 178 \end{bmatrix},$$

---

[1] We will discuss the concept of the inverse of a transformation more systematically in Section 2.4.

he will be chagrined to discover that the system

$$\begin{vmatrix} x_1 + 2x_2 = & 89 \\ 2x_1 + 4x_2 = & 178 \end{vmatrix}$$

has infinitely many solutions, namely,

$$\begin{bmatrix} x_1 \\ x_2 \end{bmatrix} = \begin{bmatrix} 89 - 2t \\ t \end{bmatrix},$$

where t is an arbitrary number.

Because this system does not have a unique solution, it is impossible to recover the actual position from the encoded position: The coding transformation and the coding matrix A are *noninvertible*. This code is useless!

Now let us discuss the important concept of *linear transformations* in greater detail. Since linear transformations are a special class of functions, it may be helpful to review the concept of a *function* first.

Consider two sets X and Y. A function T from X to Y is a rule that associates with each element x of X a unique element y of Y. The set X is called the *domain* of the function, and Y is its *target space*. We will sometimes refer to x as the *input* of the function and to y as its *output*. Figure 3 shows an example where domain X and target space Y are finite.

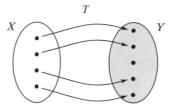

Figure 3 Domain X and target space Y of a function T.

In precalculus and calculus, you studied functions whose input and output are scalars (i.e., whose domain and target space are the real numbers $\mathbb{R}$ or subsets of $\mathbb{R}$); for example,

$$y = x^2, \quad f(x) = e^x, \quad g(t) = \frac{t^2 - 2}{t - 1}.$$

In multivariable calculus, you may have encountered functions whose input or output were vectors.

EXAMPLE I
$$y = x_1^2 + x_2^2 + x_3^2$$

This formula defines a function from the vector space $\mathbb{R}^3$ to $\mathbb{R}$. The input is the vector $\vec{x} = \begin{bmatrix} x_1 \\ x_2 \\ x_3 \end{bmatrix}$, and the output is the scalar y. ∎

EXAMPLE 2
$$\vec{r} = \begin{bmatrix} \cos(t) \\ \sin(t) \\ t \end{bmatrix}$$

This formula defines a function from $\mathbb{R}$ to the vector space $\mathbb{R}^3$, with input t and output $\vec{r}$. ∎

We now return to the topic of linear transformations.

Definition 2.1.1 Linear transformations[2]

A function T from $\mathbb{R}^m$ to $\mathbb{R}^n$ is called a *linear transformation* if there exists an $n \times m$ matrix A such that

$$T(\vec{x}) = A\vec{x},$$

for all $\vec{x}$ in the vector space $\mathbb{R}^m$.

It is important to note that a linear transformation is a special kind of *function*. The input and the output are both vectors. If we denote the output vector $T(\vec{x})$ by $\vec{y}$, we can write

$$\vec{y} = A\vec{x}.$$

Let us write this equation in terms of its components:

$$\begin{bmatrix} y_1 \\ y_2 \\ \vdots \\ y_n \end{bmatrix} = \begin{bmatrix} a_{11} & a_{12} & \cdots & a_{1m} \\ a_{21} & a_{22} & \cdots & a_{2m} \\ \vdots & \vdots & & \vdots \\ a_{n1} & a_{n2} & \cdots & a_{nm} \end{bmatrix} \begin{bmatrix} x_1 \\ x_2 \\ \vdots \\ x_m \end{bmatrix} = \begin{bmatrix} a_{11}x_1 + a_{12}x_2 + \cdots + a_{1m}x_m \\ a_{21}x_1 + a_{22}x_2 + \cdots + a_{2m}x_m \\ \vdots & \vdots & & \vdots \\ a_{n1}x_1 + a_{n2}x_2 + \cdots + a_{nm}x_m \end{bmatrix},$$

or

$$\begin{aligned}
y_1 &= a_{11}x_1 + a_{12}x_2 + \cdots + a_{1m}x_m \\
y_2 &= a_{21}x_1 + a_{22}x_2 + \cdots + a_{2m}x_m \\
\vdots &= \vdots \quad\quad \vdots \quad\quad\quad\quad \vdots \\
y_n &= a_{n1}x_1 + a_{n2}x_2 + \cdots + a_{nm}x_m.
\end{aligned}$$

The output variables y_i are linear functions of the input variables x_j. In some branches of mathematics, a first-order function with a constant term, such as $y = 3x_1 - 7x_2 + 5x_3 + 8$, is called linear. Not so in linear algebra: The linear functions of m variables are those of the form $y = c_1 x_1 + c_2 x_2 + \cdots + c_m x_m$, for some coefficients $c_1, c_2, \ldots, c_m$. By contrast, a function such as $y = 3x_1 - 7x_2 + 5x_3 + 8$ is called *affine*.

EXAMPLE 3 The linear transformation

$$\begin{aligned}
y_1 &= 7x_1 + 3x_2 - 9x_3 + 8x_4 \\
y_2 &= 6x_1 + 2x_2 - 8x_3 + 7x_4 \\
y_3 &= 8x_1 + 4x_2 \quad\quad\quad + 7x_4
\end{aligned}$$

(a function from $\mathbb{R}^4$ to $\mathbb{R}^3$) is represented by the 3×4 matrix

$$A = \begin{bmatrix} 7 & 3 & -9 & 8 \\ 6 & 2 & -8 & 7 \\ 8 & 4 & 0 & 7 \end{bmatrix}. \quad\blacksquare$$

---

[2]This is one of several possible definitions of a linear transformation; we could just as well have chosen the statement of Theorem 2.1.3 as the definition (as many texts do). This will be a recurring theme in this text: Most of the central concepts of linear algebra can be characterized in two or more ways. Each of these characterizations can serve as a possible definition; the other characterizations will then be stated as theorems, since we need to prove that they are equivalent to the chosen definition. Among these multiple characterizations, there is no "correct" definition (although mathematicians may have their favorite). Each characterization will be best suited for certain purposes and problems, while it is inadequate for others.

EXAMPLE 4   The coefficient matrix of the *identity transformation*

$$\begin{aligned} y_1 &= x_1 \\ y_2 &= \quad x_2 \\ &\vdots \qquad \qquad \ddots \\ y_n &= \qquad \qquad x_n \end{aligned}$$

(a linear transformation from $\mathbb{R}^n$ to $\mathbb{R}^n$ whose output equals its input) is the $n \times n$ matrix

$$\begin{bmatrix} 1 & 0 & \cdots & 0 \\ 0 & 1 & \cdots & 0 \\ \vdots & \vdots & \ddots & \vdots \\ 0 & 0 & \cdots & 1 \end{bmatrix}.$$

All entries on the main diagonal are 1, and all other entries are 0. This matrix is called the *identity matrix* and is denoted by I_n:

$$I_2 = \begin{bmatrix} 1 & 0 \\ 0 & 1 \end{bmatrix}, \quad I_3 = \begin{bmatrix} 1 & 0 & 0 \\ 0 & 1 & 0 \\ 0 & 0 & 1 \end{bmatrix}, \quad \text{and so on.} \qquad \blacksquare$$

We have already seen the identity matrix in other contexts. For example, we have shown that a linear system $A\vec{x} = \vec{b}$ of n equations with n unknowns has a unique solution if and only if $\text{rref}(A) = I_n$. (See Theorem 1.3.4.)

EXAMPLE 5   Consider the letter **L** (for Linear?) in Figure 4, made up of the vectors $\begin{bmatrix} 1 \\ 0 \end{bmatrix}$ and $\begin{bmatrix} 0 \\ 2 \end{bmatrix}$. Show the effect of the linear transformation

$$T(\vec{x}) = \begin{bmatrix} 0 & -1 \\ 1 & 0 \end{bmatrix} \vec{x}$$

on this letter, and describe the transformation in words.

Solution
We have

$$T\begin{bmatrix} 1 \\ 0 \end{bmatrix} = \begin{bmatrix} 0 & -1 \\ 1 & 0 \end{bmatrix} \begin{bmatrix} 1 \\ 0 \end{bmatrix} = \begin{bmatrix} 0 \\ 1 \end{bmatrix} \quad \text{and} \quad T\begin{bmatrix} 0 \\ 2 \end{bmatrix} = \begin{bmatrix} 0 & -1 \\ 1 & 0 \end{bmatrix} \begin{bmatrix} 0 \\ 2 \end{bmatrix} = \begin{bmatrix} -2 \\ 0 \end{bmatrix},$$

as shown in Figure 5.

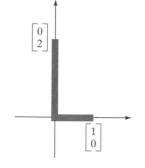

Figure 4

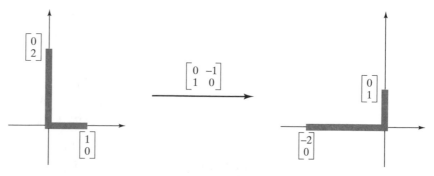

Figure 5

The **L** is rotated through an angle of 90° in the counterclockwise direction. Let's examine the effect of transformation T on an arbitrary vector $\vec{x} = \begin{bmatrix} x_1 \\ x_2 \end{bmatrix}$:

$$T(\vec{x}) = \begin{bmatrix} 0 & -1 \\ 1 & 0 \end{bmatrix} \vec{x} = \begin{bmatrix} 0 & -1 \\ 1 & 0 \end{bmatrix} \begin{bmatrix} x_1 \\ x_2 \end{bmatrix} = \begin{bmatrix} -x_2 \\ x_1 \end{bmatrix}.$$

We observe that the vectors $\vec{x}$ and $T(\vec{x})$ have the same length,

$$\sqrt{x_1^2 + x_2^2} = \sqrt{(-x_2)^2 + x_1^2},$$

and that they are perpendicular to one another, since the dot product equals zero (see Definition A.8 in the appendix):

$$\vec{x} \cdot T(\vec{x}) = \begin{bmatrix} x_1 \\ x_2 \end{bmatrix} \cdot \begin{bmatrix} -x_2 \\ x_1 \end{bmatrix} = -x_1 x_2 + x_2 x_1 = 0.$$

Paying attention to the signs of the components, we see that if $\vec{x}$ is in the first quadrant (meaning that x_1 and x_2 are both positive), then $T(\vec{x}) = \begin{bmatrix} -x_2 \\ x_1 \end{bmatrix}$ is in the second quadrant. See Figure 6.

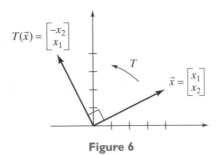

Figure 6

We can conclude that $T(\vec{x})$ is obtained by rotating vector $\vec{x}$ through an angle of 90° in the counterclockwise direction, as in the special cases $\vec{x} = \begin{bmatrix} 1 \\ 0 \end{bmatrix}$ and $\vec{x} = \begin{bmatrix} 0 \\ 2 \end{bmatrix}$ considered earlier. (Check that the rotation is indeed counterclockwise when $\vec{x}$ is in the second, third, or fourth quadrant.) ∎

EXAMPLE 6   Consider the linear transformation $T(\vec{x}) = A\vec{x}$, with

$$A = \begin{bmatrix} 1 & 2 & 3 \\ 4 & 5 & 6 \\ 7 & 8 & 9 \end{bmatrix}.$$

Find

$$T\begin{bmatrix} 1 \\ 0 \\ 0 \end{bmatrix} \quad \text{and} \quad T\begin{bmatrix} 0 \\ 0 \\ 1 \end{bmatrix},$$

where for simplicity we write $T\begin{bmatrix} 1 \\ 0 \\ 0 \end{bmatrix}$ instead of $T\left(\begin{bmatrix} 1 \\ 0 \\ 0 \end{bmatrix}\right)$.

Solution

A straightforward computation shows that

$$T \begin{bmatrix} 1 \\ 0 \\ 0 \end{bmatrix} = \begin{bmatrix} 1 & 2 & 3 \\ 4 & 5 & 6 \\ 7 & 8 & 9 \end{bmatrix} \begin{bmatrix} 1 \\ 0 \\ 0 \end{bmatrix} = \begin{bmatrix} 1 \\ 4 \\ 7 \end{bmatrix}$$

and

$$T \begin{bmatrix} 0 \\ 0 \\ 1 \end{bmatrix} = \begin{bmatrix} 1 & 2 & 3 \\ 4 & 5 & 6 \\ 7 & 8 & 9 \end{bmatrix} \begin{bmatrix} 0 \\ 0 \\ 1 \end{bmatrix} = \begin{bmatrix} 3 \\ 6 \\ 9 \end{bmatrix}.$$

Note that $T \begin{bmatrix} 1 \\ 0 \\ 0 \end{bmatrix}$ is the first column of the matrix A and that $T \begin{bmatrix} 0 \\ 0 \\ 1 \end{bmatrix}$ is its third column. ∎

We can generalize this observation:

Theorem 2.1.2   The columns of the matrix of a linear transformation

Consider a linear transformation T from $\mathbb{R}^m$ to $\mathbb{R}^n$. Then, the matrix of T is

$$A = \begin{bmatrix} | & | & & | \\ T(\vec{e}_1) & T(\vec{e}_2) & \cdots & T(\vec{e}_m) \\ | & | & & | \end{bmatrix}, \quad \text{where} \quad \vec{e}_i = \begin{bmatrix} 0 \\ 0 \\ \vdots \\ 1 \\ \vdots \\ 0 \end{bmatrix} \leftarrow i\text{th} \quad . \qquad ∎$$

To justify this result, write

$$A = \begin{bmatrix} | & | & & | \\ \vec{v}_1 & \vec{v}_2 & \cdots & \vec{v}_m \\ | & | & & | \end{bmatrix}.$$

Then

$$T(\vec{e}_i) = A\vec{e}_i = \begin{bmatrix} | & | & & | & & | \\ \vec{v}_1 & \vec{v}_2 & \cdots & \vec{v}_i & \cdots & \vec{v}_m \\ | & | & & | & & | \end{bmatrix} \begin{bmatrix} 0 \\ 0 \\ \vdots \\ 1 \\ \vdots \\ 0 \end{bmatrix} = \vec{v}_i,$$

by Theorem 1.3.8.

The vectors $\vec{e}_1, \vec{e}_2, \ldots, \vec{e}_m$ in the vector space $\mathbb{R}^m$ are sometimes referred to as the *standard vectors* in $\mathbb{R}^m$. The standard vectors $\vec{e}_1, \vec{e}_2, \vec{e}_3$ in $\mathbb{R}^3$ are often denoted by $\vec{i}, \vec{j}, \vec{k}$.

EXAMPLE 7   Consider a linear transformation $T(\vec{x}) = A\vec{x}$ from $\mathbb{R}^m$ to $\mathbb{R}^n$.

a. What is the relationship between $T(\vec{v})$, $T(\vec{w})$, and $T(\vec{v} + \vec{w})$, where $\vec{v}$ and $\vec{w}$ are vectors in $\mathbb{R}^m$?

b. What is the relationship between $T(\vec{v})$ and $T(k\vec{v})$, where $\vec{v}$ is a vector in $\mathbb{R}^m$ and k is a scalar?

Solution

a. Applying Theorem 1.3.10, we find that

$$T(\vec{v} + \vec{w}) = A(\vec{v} + \vec{w}) = A\vec{v} + A\vec{w} = T(\vec{v}) + T(\vec{w}).$$

In words, the transform of the sum of two vectors equals the sum of the transforms.

b. Again, apply Theorem 1.3.10:

$$T(k\vec{v}) = A(k\vec{v}) = kA\vec{v} = kT(\vec{v}).$$

In words, the transform of a scalar multiple of a vector is the scalar multiple of the transform.   ∎

Figure 7 illustrates these two properties in the case of the linear transformation T from $\mathbb{R}^2$ to $\mathbb{R}^2$ that rotates a vector through an angle of $90°$ in the counterclockwise direction. (Compare this with Example 5.)

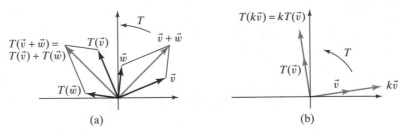

Figure 7   (a) Illustrating the property $T(\vec{v} + \vec{w}) = T(\vec{v}) + T(\vec{w})$.
(b) Illustrating the property $T(k\vec{v}) = kT(\vec{v})$.

In Example 7, we saw that a linear transformation satisfies the two equations $T(\vec{v} + \vec{w}) = T(\vec{v}) + T(\vec{w})$ and $T(k\vec{v}) = kT(\vec{v})$. Now we will show that the converse is true as well: Any transformation from $\mathbb{R}^m$ to $\mathbb{R}^n$ that satisfies these two equations is a linear transformation.

Theorem 2.1.3   Linear transformations

A transformation T from $\mathbb{R}^m$ to $\mathbb{R}^n$ is linear if (and only if)

a. $T(\vec{v} + \vec{w}) = T(\vec{v}) + T(\vec{w})$, for all vectors $\vec{v}$ and $\vec{w}$ in $\mathbb{R}^m$, and

b. $T(k\vec{v}) = kT(\vec{v})$, for all vectors $\vec{v}$ in $\mathbb{R}^m$ and all scalars k.

Proof   In Example 7, we saw that a linear transformation satisfies the equations in (a) and (b). To prove the converse, consider a transformation T from $\mathbb{R}^m$ to $\mathbb{R}^n$ that satisfies equations (a) and (b). We must show that there exists a matrix A such that

$T(\vec{x}) = A\vec{x}$, for all $\vec{x}$ in the vector space $\mathbb{R}^m$. Let $\vec{e}_1, \ldots, \vec{e}_m$ be the standard vectors introduced in Theorem 2.1.2.

$$T(\vec{x}) = T\begin{bmatrix} x_1 \\ x_2 \\ \vdots \\ x_m \end{bmatrix} = T(x_1\vec{e}_1 + x_2\vec{e}_2 + \cdots + x_m\vec{e}_m)$$

$$= T(x_1\vec{e}_1) + T(x_2\vec{e}_2) + \cdots + T(x_m\vec{e}_m) \quad \text{(by property a)}$$

$$= x_1T(\vec{e}_1) + x_2T(\vec{e}_2) + \cdots + x_mT(\vec{e}_m) \quad \text{(by property b)}$$

$$= \begin{bmatrix} | & | & & | \\ T(\vec{e}_1) & T(\vec{e}_2) & \cdots & T(\vec{e}_m) \\ | & | & & | \end{bmatrix} \begin{bmatrix} x_1 \\ x_2 \\ \vdots \\ x_m \end{bmatrix} = A\vec{x} \qquad \blacksquare$$

Here is an example illustrating Theorem 2.1.3.

EXAMPLE 8 Consider a linear transformation T from $\mathbb{R}^2$ to $\mathbb{R}^2$ such that $T(\vec{v}_1) = \frac{1}{2}\vec{v}_1$ and $T(\vec{v}_2) = 2\vec{v}_2$, for the vectors $\vec{v}_1$ and $\vec{v}_2$ sketched in Figure 8. On the same axes, sketch $T(\vec{x})$, for the given vector $\vec{x}$. Explain your solution.

Solution

Using a parallelogram, we can represent $\vec{x}$ as a linear combination of $\vec{v}_1$ and $\vec{v}_2$, as shown in Figure 9:

$$\vec{x} = c_1\vec{v}_1 + c_2\vec{v}_2.$$

By Theorem 2.1.3,

$$T(\vec{x}) = T(c_1\vec{v}_1 + c_2\vec{v}_2) = c_1T(\vec{v}_1) + c_2T(\vec{v}_2) = \frac{1}{2}c_1\vec{v}_1 + 2c_2\vec{v}_2.$$

The vector $c_1\vec{v}_1$ is cut in half, and the vector $c_2\vec{v}_2$ is doubled, as shown in Figure 10.

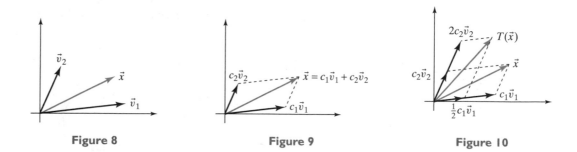

Figure 8   Figure 9   Figure 10

Imagine that vector $\vec{x}$ is drawn on a rubber sheet. Transformation T expands this sheet by a factor of 2 in the $\vec{v}_2$-direction and contracts it by a factor of 2 in the $\vec{v}_1$-direction. (We prefer "contracts by a factor of 2" to the awkward "expands by a factor of $\frac{1}{2}$.") $\qquad \blacksquare$

EXERCISES 2.1

GOAL *Use the concept of a linear transformation in terms of the formula $\vec{y} = A\vec{x}$, and interpret simple linear transformations geometrically. Find the inverse of a linear transformation from $\mathbb{R}^2$ to $\mathbb{R}^2$ (if it exists). Find the matrix of a linear transformation column by column.*

Consider the transformations from $\mathbb{R}^3$ to $\mathbb{R}^3$ defined in Exercises 1 through 3. Which of these transformations are linear?

1. $y_1 = 2x_2$
$y_2 = x_2 + 2$
$y_3 = 2x_2$

2. $y_1 = 2x_2$
$y_2 = 3x_3$
$y_3 = x_1$

3. $y_1 = x_2 - x_3$
$y_2 = x_1 x_3$
$y_3 = x_1 - x_2$

4. Find the matrix of the linear transformation
$$y_1 = 9x_1 + 3x_2 - 3x_3$$
$$y_2 = 2x_1 - .9x_2 + x_3$$
$$y_3 = 4x_1 - 9x_2 - 2x_3$$
$$y_4 = 5x_1 + x_2 + 5x_3.$$

5. Consider the linear transformation T from $\mathbb{R}^3$ to $\mathbb{R}^2$ with
$$T\begin{bmatrix} 1 \\ 0 \\ 0 \end{bmatrix} = \begin{bmatrix} 7 \\ 11 \end{bmatrix}, \quad T\begin{bmatrix} 0 \\ 1 \\ 0 \end{bmatrix} = \begin{bmatrix} 6 \\ 9 \end{bmatrix},$$
and $T\begin{bmatrix} 0 \\ 0 \\ 1 \end{bmatrix} = \begin{bmatrix} -13 \\ 17 \end{bmatrix}$.

Find the matrix A of T.

6. Consider the transformation T from $\mathbb{R}^2$ to $\mathbb{R}^3$ given by
$$T\begin{bmatrix} x_1 \\ x_2 \end{bmatrix} = x_1 \begin{bmatrix} 1 \\ 2 \\ 3 \end{bmatrix} + x_2 \begin{bmatrix} 4 \\ 5 \\ 6 \end{bmatrix}.$$

Is this transformation linear? If so, find its matrix.

7. Suppose $\vec{v}_1, \vec{v}_2, \ldots, \vec{v}_m$ are arbitrary vectors in $\mathbb{R}^n$. Consider the transformation from $\mathbb{R}^m$ to $\mathbb{R}^n$ given by
$$T\begin{bmatrix} x_1 \\ x_2 \\ \vdots \\ x_m \end{bmatrix} = x_1 \vec{v}_1 + x_2 \vec{v}_2 + \cdots + x_m \vec{v}_m.$$

Is this transformation linear? If so, find its matrix A in terms of the vectors $\vec{v}_1, \vec{v}_2, \ldots, \vec{v}_m$.

8. Find the inverse of the linear transformation
$$y_1 = x_1 + 7x_2$$
$$y_2 = 3x_1 + 20x_2.$$

In Exercises 9 through 12, decide whether the given matrix is invertible. Find the inverse if it exists. In Exercise 12, the constant k is arbitrary.

9. $\begin{bmatrix} 2 & 3 \\ 6 & 9 \end{bmatrix}$

10. $\begin{bmatrix} 1 & 2 \\ 4 & 9 \end{bmatrix}$

11. $\begin{bmatrix} 1 & 2 \\ 3 & 9 \end{bmatrix}$

12. $\begin{bmatrix} 1 & k \\ 0 & 1 \end{bmatrix}$

13. Prove the following facts:

a. The 2×2 matrix
$$A = \begin{bmatrix} a & b \\ c & d \end{bmatrix}$$
is invertible if and only if $ad - bc \neq 0$. (*Hint*: Consider the cases $a \neq 0$ and $a = 0$ separately.)

b. If
$$\begin{bmatrix} a & b \\ c & d \end{bmatrix}$$
is invertible, then
$$\begin{bmatrix} a & b \\ c & d \end{bmatrix}^{-1} = \frac{1}{ad - bc} \begin{bmatrix} d & -b \\ -c & a \end{bmatrix}.$$

[The formula in part (b) is worth memorizing.]

14. a. For which values of the constant k is the matrix $\begin{bmatrix} 2 & 3 \\ 5 & k \end{bmatrix}$ invertible?

b. For which values of the constant k are all entries of $\begin{bmatrix} 2 & 3 \\ 5 & k \end{bmatrix}^{-1}$ integers?

(See Exercise 13.)

15. For which values of the constants a and b is the matrix
$$A = \begin{bmatrix} a & -b \\ b & a \end{bmatrix}$$
invertible? What is the inverse in this case? (See Exercise 13.)

Give a geometric interpretation of the linear transformations defined by the matrices in Exercises 16 through 23. Show the effect of these transformations on the letter L *considered in Example 5. In each case, decide whether the transformation is invertible. Find the inverse if it exists, and interpret it geometrically. See Exercise 13.*

16. $\begin{bmatrix} 3 & 0 \\ 0 & 3 \end{bmatrix}$

17. $\begin{bmatrix} -1 & 0 \\ 0 & -1 \end{bmatrix}$

18. $\begin{bmatrix} 0.5 & 0 \\ 0 & 0.5 \end{bmatrix}$

19. $\begin{bmatrix} 1 & 0 \\ 0 & 0 \end{bmatrix}$

20. $\begin{bmatrix} 0 & 1 \\ 1 & 0 \end{bmatrix}$

21. $\begin{bmatrix} 0 & 1 \\ -1 & 0 \end{bmatrix}$

22. $\begin{bmatrix} 1 & 0 \\ 0 & -1 \end{bmatrix}$

23. $\begin{bmatrix} 0 & 2 \\ -2 & 0 \end{bmatrix}$

Consider the circular face in the accompanying figure. For each of the matrices A in Exercises 24 through 30, draw a sketch showing the effect of the linear transformation $T(\vec{x}) = A\vec{x}$ on this face.

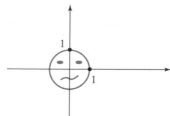

24. $\begin{bmatrix} 0 & -1 \\ 1 & 0 \end{bmatrix}$ **25.** $\begin{bmatrix} 2 & 0 \\ 0 & 2 \end{bmatrix}$ **26.** $\begin{bmatrix} 0 & 1 \\ 1 & 0 \end{bmatrix}$

27. $\begin{bmatrix} 1 & 0 \\ 0 & -1 \end{bmatrix}$ **28.** $\begin{bmatrix} 1 & 0 \\ 0 & 2 \end{bmatrix}$ **29.** $\begin{bmatrix} -1 & 0 \\ 0 & -1 \end{bmatrix}$

30. $\begin{bmatrix} 0 & 0 \\ 0 & 1 \end{bmatrix}$

31. In Chapter 1, we mentioned that an old German bill shows the mirror image of Gauss's likeness. What linear transformation T can you apply to get the actual picture back?

32. Find an $n \times n$ matrix A such that $A\vec{x} = 3\vec{x}$, for all $\vec{x}$ in $\mathbb{R}^n$.

33. Consider the transformation T from $\mathbb{R}^2$ to $\mathbb{R}^2$ that rotates any vector $\vec{x}$ through an angle of $45°$ in the counterclockwise direction, as shown in the following figure:

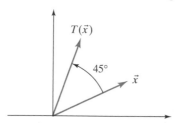

You are told that T is a linear transformation. (This will be shown in the next section.) Find the matrix of T.

34. Consider the transformation T from $\mathbb{R}^2$ to $\mathbb{R}^2$ that rotates any vector $\vec{x}$ through a given angle θ in the counterclockwise direction. (Compare this with Exercise 33.) You are told that T is linear. Find the matrix of T in terms of θ.

35. In the example about the French coast guard in this section, suppose you are a spy watching the boat and listening in on the radio messages from the boat. You collect the following data:

- When the actual position is $\begin{bmatrix} 5 \\ 42 \end{bmatrix}$, they radio $\begin{bmatrix} 89 \\ 52 \end{bmatrix}$.

- When the actual position is $\begin{bmatrix} 6 \\ 41 \end{bmatrix}$, they radio $\begin{bmatrix} 88 \\ 53 \end{bmatrix}$.

Can you crack their code (i.e., find the coding matrix), assuming that the code is linear?

36. Let T be a linear transformation from $\mathbb{R}^2$ to $\mathbb{R}^2$. Let $\vec{v}_1$, $\vec{v}_2$, and $\vec{w}$ be three vectors in $\mathbb{R}^2$, as shown below. We are told that $T(\vec{v}_1) = \vec{v}_1$ and $T(\vec{v}_2) = 3\vec{v}_2$. On the same axes, sketch $T(\vec{w})$.

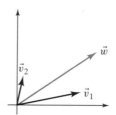

37. Consider a linear transformation T from $\mathbb{R}^2$ to $\mathbb{R}^2$. Suppose that $\vec{v}$ and $\vec{w}$ are two arbitrary vectors in $\mathbb{R}^2$ and that $\vec{x}$ is a third vector whose endpoint is on the line segment connecting the endpoints of $\vec{v}$ and $\vec{w}$. Is the endpoint of the vector $T(\vec{x})$ necessarily on the line segment connecting the endpoints of $T(\vec{v})$ and $T(\vec{w})$? Justify your answer.

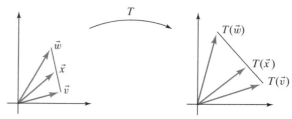

[*Hint*: We can write $\vec{x} = \vec{v} + k(\vec{w} - \vec{v})$, for some scalar k between 0 and 1.]

We can summarize this exercise by saying that a linear transformation maps a line onto a line.

38. The two column vectors $\vec{v}_1$ and $\vec{v}_2$ of a 2×2 matrix A are shown in the accompanying sketch. Consider the linear transformation $T(\vec{x}) = A\vec{x}$, from $\mathbb{R}^2$ to $\mathbb{R}^2$. Sketch the vector

$$T \begin{bmatrix} 2 \\ -1 \end{bmatrix}.$$

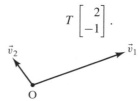

39. Show that if T is a linear transformation from $\mathbb{R}^m$ to $\mathbb{R}^n$, then

$$T \begin{bmatrix} x_1 \\ x_2 \\ \vdots \\ x_m \end{bmatrix} = x_1 T(\vec{e}_1) + x_2 T(\vec{e}_2) + \cdots + x_m T(\vec{e}_m),$$

where $\vec{e}_1, \vec{e}_2, \ldots, \vec{e}_m$ are the standard vectors in $\mathbb{R}^m$.

40. Describe all linear transformations from $\mathbb{R}\,(=\mathbb{R}^1)$ to $\mathbb{R}$. What do their graphs look like?

41. Describe all linear transformations from $\mathbb{R}^2$ to $\mathbb{R}\,(=\mathbb{R}^1)$. What do their graphs look like?

42. When you represent a three-dimensional object graphically in the plane (on paper, the blackboard, or a computer screen), you have to transform spatial coordinates,
$$\begin{bmatrix} x_1 \\ x_2 \\ x_3 \end{bmatrix},$$
into plane coordinates, $\begin{bmatrix} y_1 \\ y_2 \end{bmatrix}$. The simplest choice is a linear transformation, for example, the one given by the matrix
$$\begin{bmatrix} -\frac{1}{2} & 1 & 0 \\ -\frac{1}{2} & 0 & 1 \end{bmatrix}.$$

a. Use this transformation to represent the unit cube with corner points
$$\begin{bmatrix} 0 \\ 0 \\ 0 \end{bmatrix}, \quad \begin{bmatrix} 1 \\ 0 \\ 0 \end{bmatrix}, \quad \begin{bmatrix} 0 \\ 1 \\ 0 \end{bmatrix}, \quad \begin{bmatrix} 0 \\ 0 \\ 1 \end{bmatrix},$$
$$\begin{bmatrix} 1 \\ 1 \\ 0 \end{bmatrix}, \quad \begin{bmatrix} 0 \\ 1 \\ 1 \end{bmatrix}, \quad \begin{bmatrix} 1 \\ 0 \\ 1 \end{bmatrix}, \quad \begin{bmatrix} 1 \\ 1 \\ 1 \end{bmatrix}.$$
Include the images of the x_1, x_2, and x_3 axes in your sketch:

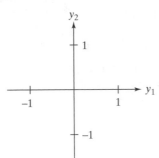

b. Represent the image of the point $\begin{bmatrix} 1 \\ \frac{1}{2} \\ \frac{1}{2} \end{bmatrix}$ in your figure in part (a). Explain.

c. Find all the points
$$\begin{bmatrix} x_1 \\ x_2 \\ x_3 \end{bmatrix} \text{ in } \mathbb{R}^3$$
that are transformed to $\begin{bmatrix} 0 \\ 0 \end{bmatrix}$. Explain.

43. a. Consider the vector $\vec{v} = \begin{bmatrix} 2 \\ 3 \\ 4 \end{bmatrix}$. Is the transformation $T(\vec{x}) = \vec{v} \cdot \vec{x}$ (the dot product) from $\mathbb{R}^3$ to $\mathbb{R}$ linear? If so, find the matrix of T.

b. Consider an arbitrary vector $\vec{v}$ in $\mathbb{R}^3$. Is the transformation $T(\vec{x}) = \vec{v} \cdot \vec{x}$ linear? If so, find the matrix of T (in terms of the components of $\vec{v}$).

c. Conversely, consider a linear transformation T from $\mathbb{R}^3$ to $\mathbb{R}$. Show that there exists a vector $\vec{v}$ in $\mathbb{R}^3$ such that $T(\vec{x}) = \vec{v} \cdot \vec{x}$, for all $\vec{x}$ in $\mathbb{R}^3$.

44. The cross product of two vectors in $\mathbb{R}^3$ is given by
$$\begin{bmatrix} a_1 \\ a_2 \\ a_3 \end{bmatrix} \times \begin{bmatrix} b_1 \\ b_2 \\ b_3 \end{bmatrix} = \begin{bmatrix} a_2 b_3 - a_3 b_2 \\ a_3 b_1 - a_1 b_3 \\ a_1 b_2 - a_2 b_1 \end{bmatrix}.$$
(See Definition A.9 and Theorem A.11 in the Appendix.) Consider an arbitrary vector $\vec{v}$ in $\mathbb{R}^3$. Is the transformation $T(\vec{x}) = \vec{v} \times \vec{x}$ from $\mathbb{R}^3$ to $\mathbb{R}^3$ linear? If so, find its matrix in terms of the components of the vector $\vec{v}$.

45. Consider two linear transformations $\vec{y} = T(\vec{x})$ and $\vec{z} = L(\vec{y})$, where T goes from $\mathbb{R}^m$ to $\mathbb{R}^p$ and L goes from $\mathbb{R}^p$ to $\mathbb{R}^n$. Is the transformation $\vec{z} = L(T(\vec{x}))$ linear as well? [The transformation $\vec{z} = L(T(\vec{x}))$ is called the *composite* of T and L.]

46. Let
$$A = \begin{bmatrix} a & b \\ c & d \end{bmatrix} \quad \text{and} \quad B = \begin{bmatrix} p & q \\ r & s \end{bmatrix}.$$
Find the matrix of the linear transformation $T(\vec{x}) = B(A\vec{x})$. (See Exercise 45.) [*Hint*: Find $T(\vec{e}_1)$ and $T(\vec{e}_2)$.]

47. Let T be a linear transformation from $\mathbb{R}^2$ to $\mathbb{R}^2$. Three vectors $\vec{v}_1$, $\vec{v}_2$, $\vec{w}$ in $\mathbb{R}^2$ and the vectors $T(\vec{v}_1)$, $T(\vec{v}_2)$ are shown in the accompanying figure. Sketch $T(\vec{w})$. Explain your answer.

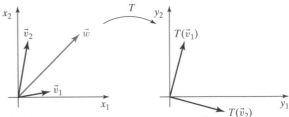

48. Consider two linear transformations T and L from $\mathbb{R}^2$ to $\mathbb{R}^2$. We are told that $T(\vec{v}_1) = L(\vec{v}_1)$ and $T(\vec{v}_2) = L(\vec{v}_2)$ for the vectors $\vec{v}_1$ and $\vec{v}_2$ sketched below. Show that $T(\vec{x}) = L(\vec{x})$, for all vectors $\vec{x}$ in $\mathbb{R}^2$.

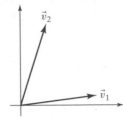

49. Some parking meters in downtown Geneva, Switzerland, accept 2 Franc and 5 Franc coins.

   a. A parking officer collects 51 coins worth 144 Francs. How many coins are there of each kind?

   b. Find the matrix A that transforms the vector

$$\begin{bmatrix} \text{number of 2 Franc coins} \\ \text{number of 5 Franc coins} \end{bmatrix}$$

    into the vector

$$\begin{bmatrix} \text{total value of coins} \\ \text{total number of coins} \end{bmatrix}.$$

   c. Is the matrix A in part (b) invertible? If so, find the inverse (use Exercise 13). Use the result to check your answer in part (a).

50. A goldsmith uses a platinum alloy and a silver alloy to make jewelry; the densities of these alloys are exactly 20 and 10 grams per cubic centimeter, respectively.

   a. King Hiero of Syracuse orders a crown from this goldsmith, with a total mass of 5 kilograms (or 5,000 grams), with the stipulation that the platinum alloy must make up at least 90% of the mass. The goldsmith delivers a beautiful piece, but the king's friend Archimedes has doubts about its purity. While taking a bath, he comes up with a method to check the composition of the crown (famously shouting "Eureka!" in the process, and running to the king's palace naked). Submerging the crown in water, he finds its volume to be 370 cubic centimeters. How much of each alloy went into this piece (by mass)? Is this goldsmith a crook?

   b. Find the matrix A that transforms the vector

$$\begin{bmatrix} \text{mass of platinum alloy} \\ \cdot \text{ mass of silver alloy} \end{bmatrix}$$

    into the vector

$$\begin{bmatrix} \text{total mass} \\ \text{total volume} \end{bmatrix},$$

    for any piece of jewelry this goldsmith makes.

   c. Is the matrix A in part (b) invertible? If so, find the inverse (use Exercise 13). Use the result to check your answer in part (a).

51. The conversion formula $C = \frac{5}{9}(F-32)$ from Fahrenheit to Celsius (as measures of temperature) is nonlinear, in the sense of linear algebra (why?). Still, there is a technique that allows us to use a matrix to represent this conversion.

   a. Find the 2×2 matrix A that transforms the vector $\begin{bmatrix} F \\ 1 \end{bmatrix}$ into the vector $\begin{bmatrix} C \\ 1 \end{bmatrix}$. (The second row of A will be $\begin{bmatrix} 0 & 1 \end{bmatrix}$.)

   b. Is the matrix A in part (a) invertible? If so, find the inverse (use Exercise 13). Use the result to write a formula expressing F in terms of C.

52. In the financial pages of a newspaper, one can sometimes find a table (or matrix) listing the exchange rates between currencies. In this exercise we will consider a miniature version of such a table, involving only the Canadian dollar (C\$) and the South African Rand (ZAR). Consider the matrix

$$A = \begin{bmatrix} 1 & 1/8 \\ 8 & 1 \end{bmatrix} \begin{matrix} \text{C\$} \\ \text{ZAR} \end{matrix}$$

with columns labeled C\$ and ZAR,

representing the fact that C\$1 is worth ZAR8 (as of June 2008).

   a. After a trip you have C\$100 and ZAR1,600 in your pocket. We represent these two values in the vector $\vec{x} = \begin{bmatrix} 100 \\ 1,600 \end{bmatrix}$. Compute $A\vec{x}$. What is the practical significance of the two components of the vector $A\vec{x}$?

   b. Verify that matrix A fails to be invertible. For which vectors $\vec{b}$ is the system $A\vec{x} = \vec{b}$ consistent? What is the practical significance of your answer? If the system $A\vec{x} = \vec{b}$ is consistent, how many solutions $\vec{x}$ are there? Again, what is the practical significance of the answer?

53. Consider a larger currency exchange matrix (see Exercise 52), involving four of the world's leading currencies: Euro (€), U.S. dollar (\$), Japanese yen (¥), and British pound (£).

$$A = \begin{bmatrix} * & \frac{5}{8} & * & * \\ * & * & * & 2 \\ 170 & * & * & * \\ * & * & * & * \end{bmatrix} \begin{matrix} \text{€} \\ \text{\$} \\ \text{¥} \\ \text{£} \end{matrix}$$

with columns labeled €, \$, ¥, £.

The entry a_{ij} gives the value of one unit of the jth currency, expressed in terms of the ith currency. For example, $a_{31} = 170$ means that €1 = ¥170 (as of June 2008). Find the exact values of the 13 missing entries of A (expressed as fractions).

54. Consider an arbitrary currency exchange matrix A (see Exercises 52 and 53).

   a. What are the diagonal entries a_{ii} of A?

   b. What is the relationship between a_{ij} and a_{ji}?

   c. What is the relationship between a_{ik}, a_{kj}, and a_{ij}?

   d. What is the rank of A? What is the relationship between A and rref(A)?

2.2 Linear Transformations in Geometry

In Example 2.1.5 we saw that the matrix $\begin{bmatrix} 0 & -1 \\ 1 & 0 \end{bmatrix}$ represents a counterclockwise rotation through 90° in the coordinate plane. Many other 2×2 matrices define simple geometrical transformations as well; this section is dedicated to a discussion of some of those transformations.

EXAMPLE 1　Consider the matrices

$$A = \begin{bmatrix} 2 & 0 \\ 0 & 2 \end{bmatrix}, \quad B = \begin{bmatrix} 1 & 0 \\ 0 & 0 \end{bmatrix}, \quad C = \begin{bmatrix} -1 & 0 \\ 0 & 1 \end{bmatrix},$$

$$D = \begin{bmatrix} 0 & 1 \\ -1 & 0 \end{bmatrix}, \quad E = \begin{bmatrix} 1 & 0.2 \\ 0 & 1 \end{bmatrix}, \quad \text{and} \quad F = \begin{bmatrix} 1 & -1 \\ 1 & 1 \end{bmatrix}.$$

Show the effect of each of these matrices on our standard letter L,[3] and describe each transformation in words.

a.

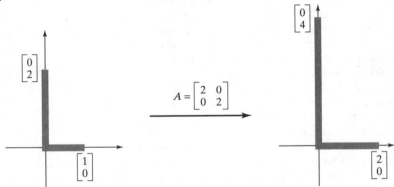

The L gets enlarged by a factor of 2; we will call this transformation a *scaling by 2*.

b.

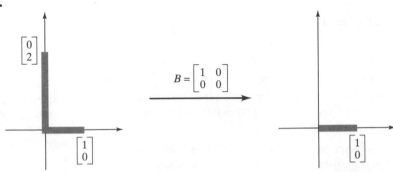

The L gets smashed into the horizontal axis. We will call this transformation the *orthogonal projection onto the horizontal axis*.

---

[3]See Example 2.1.5. Recall that vector $\begin{bmatrix} 1 \\ 0 \end{bmatrix}$ is the foot of our standard L, and $\begin{bmatrix} 0 \\ 2 \end{bmatrix}$ is its back.

c.

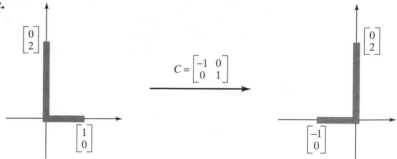

$$C = \begin{bmatrix} -1 & 0 \\ 0 & 1 \end{bmatrix}$$

The L gets flipped over the vertical axis. We will call this the *reflection about the vertical axis*.

d.

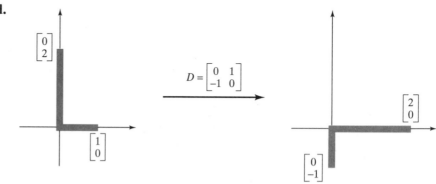

$$D = \begin{bmatrix} 0 & 1 \\ -1 & 0 \end{bmatrix}$$

The L is *rotated* through 90°, in the clockwise direction (this amounts to a rotation through −90°). The result is the opposite of what we got in Example 2.1.5.

e.

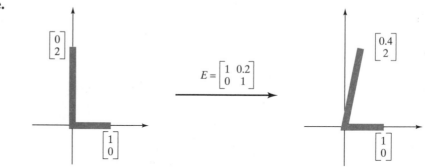

$$E = \begin{bmatrix} 1 & 0.2 \\ 0 & 1 \end{bmatrix}$$

The foot of the L remains unchanged, while the back is shifted horizontally to the right; the L is italicized, becoming *L*. We will call this transformation a *horizontal shear*.

f.

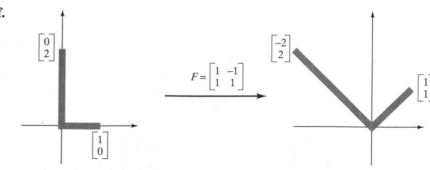

$$F = \begin{bmatrix} 1 & -1 \\ 1 & 1 \end{bmatrix}$$

There are two things going on here: The L is rotated through 45° and also enlarged (scaled) by a factor of $\sqrt{2}$. This is a *rotation combined with a scaling* (you may perform the two transformations in either order). Among all the possible composites of the transformations considered in parts (a) through (e), this one is particularly important in applications as well as in pure mathematics (see Theorem 7.5.3, for example). ■

We will now take a closer look at the six types of transformations we encountered in Example 1.

Scalings

For any positive constant k, the matrix $\begin{bmatrix} k & 0 \\ 0 & k \end{bmatrix}$ defines a scaling by k, since

$$\begin{bmatrix} k & 0 \\ 0 & k \end{bmatrix} \vec{x} = \begin{bmatrix} k & 0 \\ 0 & k \end{bmatrix} \begin{bmatrix} x_1 \\ x_2 \end{bmatrix} = \begin{bmatrix} kx_1 \\ kx_2 \end{bmatrix} = k \begin{bmatrix} x_1 \\ x_2 \end{bmatrix} = k\vec{x}.$$

This is a *dilation* (or enlargement) if k exceeds 1, and it is a *contraction* (or shrinking) for values of k between 0 and 1. (What happens when k is negative or zero?)

Orthogonal Projections[4]

Consider a line L in the plane, running through the origin. Any vector $\vec{x}$ in $\mathbb{R}^2$ can be written uniquely as

$$\vec{x} = \vec{x}^{\parallel} + \vec{x}^{\perp},$$

where $\vec{x}^{\parallel}$ is parallel to line L, and $\vec{x}^{\perp}$ is perpendicular to L. See Figure 1.

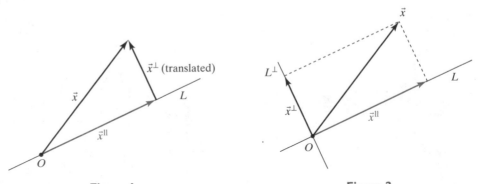

Figure 1   **Figure 2**

The transformation $T(\vec{x}) = \vec{x}^{\parallel}$ from $\mathbb{R}^2$ to $\mathbb{R}^2$ is called the *orthogonal projection of $\vec{x}$ onto L*, often denoted by $\text{proj}_L(\vec{x})$:

$$\text{proj}_L(\vec{x}) = \vec{x}^{\parallel}.$$

You can think of $\text{proj}_L(\vec{x})$ as the shadow vector $\vec{x}$ casts on L if you shine a light straight down on L.

Let $L^{\perp}$ be the line through the origin perpendicular to L. Note that $\vec{x}^{\perp}$ is parallel to $L^{\perp}$, and we can interpret $\vec{x}^{\perp}$ as the orthogonal projection of $\vec{x}$ onto $L^{\perp}$, as illustrated in Figure 2.

---

[4]The term *orthogonal* is synonymous with perpendicular. For a more general discussion of projections, see Exercise 33.

We can use the dot product to write a formula for an orthogonal projection. Before proceeding, you may want to review the section "Dot Product, Length, Orthogonality" in the appendix.

To find a formula for $\vec{x}^{\parallel}$, let $\vec{w}$ be a nonzero vector parallel to L. Since $\vec{x}^{\parallel}$ is parallel to $\vec{w}$, we can write

$$\vec{x}^{\parallel} = k\vec{w},$$

for some scalar k about to be determined. Now $\vec{x}^{\perp} = \vec{x} - \vec{x}^{\parallel} = \vec{x} - k\vec{w}$ is perpendicular to line L, that is, perpendicular to $\vec{w}$, meaning that

$$(\vec{x} - k\vec{w}) \cdot \vec{w} = 0.$$

It follows that

$$\vec{x} \cdot \vec{w} - k(\vec{w} \cdot \vec{w}) = 0, \quad \text{or,} \quad k = \frac{\vec{x} \cdot \vec{w}}{\vec{w} \cdot \vec{w}}.$$

We can conclude that

$$\text{proj}_L(\vec{x}) = \vec{x}^{\parallel} = k\vec{w} = \left(\frac{\vec{x} \cdot \vec{w}}{\vec{w} \cdot \vec{w}} \right) \vec{w}.$$

See Figure 3. Consider the special case of a *unit* vector $\vec{u}$ parallel to L. Then the formula for projection simplifies to

$$\text{proj}_L(\vec{x}) = \left(\frac{\vec{x} \cdot \vec{u}}{\vec{u} \cdot \vec{u}} \right) \vec{u} = (\vec{x} \cdot \vec{u})\vec{u}$$

since $\vec{u} \cdot \vec{u} = \|\vec{u}\|^2 = 1$ for a unit vector $\vec{u}$.

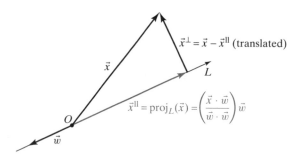

Figure 3

Is the transformation $T(\vec{x}) = \text{proj}_L(\vec{x})$ linear? If so, what is its matrix? If we write

$$\vec{x} = \begin{bmatrix} x_1 \\ x_2 \end{bmatrix} \quad \text{and} \quad \vec{u} = \begin{bmatrix} u_1 \\ u_2 \end{bmatrix},$$

then

$$\text{proj}_L(\vec{x}) = (\vec{x} \cdot \vec{u})\vec{u} = \left(\begin{bmatrix} x_1 \\ x_2 \end{bmatrix} \cdot \begin{bmatrix} u_1 \\ u_2 \end{bmatrix} \right) \begin{bmatrix} u_1 \\ u_2 \end{bmatrix}$$

$$= (x_1 u_1 + x_2 u_2) \begin{bmatrix} u_1 \\ u_2 \end{bmatrix}$$

$$= \begin{bmatrix} u_1^2 x_1 + u_1 u_2 x_2 \\ u_1 u_2 x_1 + u_2^2 x_2 \end{bmatrix}$$

$$= \begin{bmatrix} u_1^2 & u_1 u_2 \\ u_1 u_2 & u_2^2 \end{bmatrix} \begin{bmatrix} x_1 \\ x_2 \end{bmatrix}$$

$$= \begin{bmatrix} u_1^2 & u_1 u_2 \\ u_1 u_2 & u_2^2 \end{bmatrix} \vec{x}.$$

It turns out that $T(\vec{x}) = \text{proj}_L(\vec{x})$ is indeed a linear transformation, with matrix $\begin{bmatrix} u_1^2 & u_1 u_2 \\ u_1 u_2 & u_2^2 \end{bmatrix}$. More generally, if $\vec{w}$ is a nonzero vector parallel to L, then the matrix is $\dfrac{1}{w_1^2 + w_2^2} \begin{bmatrix} w_1^2 & w_1 w_2 \\ w_1 w_2 & w_2^2 \end{bmatrix}$. (See Exercise 12.)

EXAMPLE 2   Find the matrix A of the orthogonal projection onto the line L spanned by $\vec{w} = \begin{bmatrix} 4 \\ 3 \end{bmatrix}$.

Solution

$$A = \frac{1}{w_1^2 + w_2^2} \begin{bmatrix} w_1^2 & w_1 w_2 \\ w_1 w_2 & w_2^2 \end{bmatrix} = \frac{1}{25} \begin{bmatrix} 16 & 12 \\ 12 & 9 \end{bmatrix} = \begin{bmatrix} 0.64 & 0.48 \\ 0.48 & 0.36 \end{bmatrix} \qquad \blacksquare$$

Let us summarize our findings.

Definition 2.2.1   Orthogonal Projections

Consider a line L in the coordinate plane, running through the origin. Any vector $\vec{x}$ in $\mathbb{R}^2$ can be written uniquely as

$$\vec{x} = \vec{x}^{\parallel} + \vec{x}^{\perp},$$

where $\vec{x}^{\parallel}$ is parallel to line L, and $\vec{x}^{\perp}$ is perpendicular to L.

The transformation $T(\vec{x}) = \vec{x}^{\parallel}$ from $\mathbb{R}^2$ to $\mathbb{R}^2$ is called the *orthogonal projection of $\vec{x}$ onto L*, often denoted by $\text{proj}_L(\vec{x})$. If $\vec{w}$ is a nonzero vector parallel to L, then

$$\text{proj}_L(\vec{x}) = \left(\frac{\vec{x} \cdot \vec{w}}{\vec{w} \cdot \vec{w}} \right) \vec{w}.$$

In particular, if $\vec{u} = \begin{bmatrix} u_1 \\ u_2 \end{bmatrix}$ is a *unit* vector parallel to L, then

$$\text{proj}_L(\vec{x}) = (\vec{x} \cdot \vec{u}) \vec{u}.$$

The transformation $T(\vec{x}) = \text{proj}_L(\vec{x})$ is linear, with matrix

$$\frac{1}{w_1^2 + w_2^2} \begin{bmatrix} w_1^2 & w_1 w_2 \\ w_1 w_2 & w_2^2 \end{bmatrix} = \begin{bmatrix} u_1^2 & u_1 u_2 \\ u_1 u_2 & u_2^2 \end{bmatrix}.$$

Reflections

Again, consider a line L in the coordinate plane, running through the origin, and let $\vec{x}$ be a vector in $\mathbb{R}^2$. The reflection $\text{ref}_L(\vec{x})$ of $\vec{x}$ about L is shown in Figure 4: We are flipping vector $\vec{x}$ over the line L. The line segment joining the tips of vectors $\vec{x}$ and $\text{ref}_L \vec{x}$ is perpendicular to line L and bisected by L. In previous math courses you have surely seen examples of reflections about the horizontal and vertical axes [when comparing the graphs of $y = f(x)$, $y = -f(x)$, and $y = f(-x)$, for example].

We can use the representation $\vec{x} = \vec{x}^{\parallel} + \vec{x}^{\perp}$ to write a formula for $\text{ref}_L(\vec{x})$. See Figure 5.

We can see that

$$\text{ref}_L(\vec{x}) = \vec{x}^{\parallel} - \vec{x}^{\perp}.$$

Alternatively, we can express $\text{ref}_L(\vec{x})$ in terms of $\vec{x}^{\perp}$ alone or in terms of $\vec{x}^{\parallel}$ alone:

$$\text{ref}_L(\vec{x}) = \vec{x}^{\parallel} - \vec{x}^{\perp} = (\vec{x} - \vec{x}^{\perp}) - \vec{x}^{\perp} = \vec{x} - 2\vec{x}^{\perp}$$

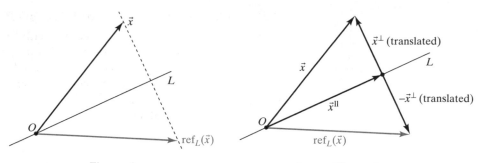

| Figure 4 | Figure 5 |

(use Figure 5 to explain this formula geometrically) and

$$\text{ref}_L(\vec{x}) = \vec{x}^{\parallel} - \vec{x}^{\perp} = \vec{x}^{\parallel} - (\vec{x} - \vec{x}^{\parallel}) = 2\vec{x}^{\parallel} - \vec{x}$$
$$= 2\text{proj}_L(\vec{x}) - \vec{x} = 2(\vec{x} \cdot \vec{u})\vec{u} - \vec{x},$$

where $\vec{u}$ is a unit vector parallel to L.

The formula $\text{ref}(\vec{x}) = 2(\vec{x} \cdot \vec{u})\vec{u} - \vec{x}$ allows us to find the matrix of a reflection. It turns out that this matrix is of the form $\begin{bmatrix} a & b \\ b & -a \end{bmatrix}$, where $a^2 + b^2 = 1$ (see Exercise 13), and that, conversely, any 2×2 matrix of this form represents a reflection about a line (see Exercise 17).

Definition 2.2.2 Reflections

Consider a line L in the coordinate plane, running through the origin, and let $\vec{x} = \vec{x}^{\parallel} + \vec{x}^{\perp}$ be a vector in $\mathbb{R}^2$. The linear transformation $T(\vec{x}) = \vec{x}^{\parallel} - \vec{x}^{\perp}$ is called the *reflection of $\vec{x}$ about L*, often denoted by $\text{ref}_L(\vec{x})$:

$$\text{ref}_L(\vec{x}) = \vec{x}^{\parallel} - \vec{x}^{\perp}.$$

We have a formula relating $\text{ref}_L(\vec{x})$ to $\text{proj}_L(\vec{x})$:

$$\text{ref}_L(\vec{x}) = 2\text{proj}_L(\vec{x}) - \vec{x} = 2(\vec{x} \cdot \vec{u})\vec{u} - \vec{x}.$$

The matrix of T is of the form $\begin{bmatrix} a & b \\ b & -a \end{bmatrix}$, where $a^2 + b^2 = 1$. Conversely, any matrix of this form represents a reflection about a line.

Use Figure 6 to explain the formula $\text{ref}_L(\vec{x}) = 2\text{proj}_L(\vec{x}) - \vec{x}$ geometrically.

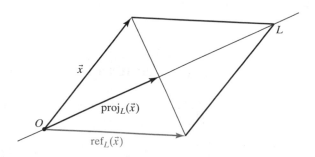

Figure 6

Orthogonal Projections and Reflections in Space

Although this section is mostly concerned with linear transformations from $\mathbb{R}^2$ to $\mathbb{R}^2$, we will take a quick look at orthogonal projections and reflections in space, since this theory is analogous to the case of two dimensions.

Let L be a line in coordinate space, running through the origin. Any vector $\vec{x}$ in $\mathbb{R}^3$ can be written uniquely as $\vec{x} = \vec{x}^{\parallel} + \vec{x}^{\perp}$, where $\vec{x}^{\parallel}$ is parallel to L, and $\vec{x}^{\perp}$ is perpendicular to L. We define

$$\text{proj}_L(\vec{x}) = \vec{x}^{\parallel},$$

and we have the formula

$$\text{proj}_L(\vec{x}) = \vec{x}^{\parallel} = (\vec{x} \cdot \vec{u})\vec{u},$$

where $\vec{u}$ is a unit vector parallel to L. See Definition 2.2.1.

Let $L^{\perp} = V$ be the plane through the origin perpendicular to L; note that the vector $\vec{x}^{\perp}$ will be parallel to $L^{\perp} = V$. We can give formulas for the orthogonal projection onto V, as well as for the reflections about V and L, in terms of the orthogonal projection onto L:

$$\text{proj}_V(\vec{x}) = \vec{x} - \text{proj}_L(\vec{x}) = \vec{x} - (\vec{x} \cdot \vec{u})\vec{u},$$

$$\text{ref}_L(\vec{x}) = \text{proj}_L(\vec{x}) - \text{proj}_V(\vec{x}) = 2\text{proj}_L(\vec{x}) - \vec{x} = 2(\vec{x} \cdot \vec{u})\vec{u} - \vec{x}, \quad \text{and}$$

$$\text{ref}_V(\vec{x}) = \text{proj}_V(\vec{x}) - \text{proj}_L(\vec{x}) = -\text{ref}_L(\vec{x}) = \vec{x} - 2(\vec{x} \cdot \vec{u})\vec{u}.$$

See Figure 7, and compare with Definition 2.2.2.

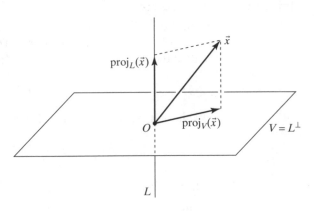

Figure 7

EXAMPLE 3   Let V be the plane defined by $2x_1 + x_2 - 2x_3 = 0$, and let $\vec{x} = \begin{bmatrix} 5 \\ 4 \\ -2 \end{bmatrix}$. Find $\text{ref}_V(\vec{x})$.

Solution
Note that the vector $\vec{v} = \begin{bmatrix} 2 \\ 1 \\ -2 \end{bmatrix}$ is perpendicular to plane V (the components of $\vec{v}$ are the coefficients of the variables in the given equation of the plane: 2, 1, and -2). Thus

$$\vec{u} = \frac{1}{\|\vec{v}\|}\vec{v} = \frac{1}{3}\begin{bmatrix} 2 \\ 1 \\ -2 \end{bmatrix}$$

is a unit vector perpendicular to V, and we can use the formula we derived earlier:

$$\text{ref}_V(\vec{x}) = \vec{x} - 2(\vec{x} \cdot \vec{u})\vec{u} = \begin{bmatrix} 5 \\ 4 \\ -2 \end{bmatrix} - \frac{2}{9}\left(\begin{bmatrix} 5 \\ 4 \\ -2 \end{bmatrix} \cdot \begin{bmatrix} 2 \\ 1 \\ -2 \end{bmatrix}\right)\begin{bmatrix} 2 \\ 1 \\ -2 \end{bmatrix}$$

$$= \begin{bmatrix} 5 \\ 4 \\ -2 \end{bmatrix} - 4\begin{bmatrix} 2 \\ 1 \\ -2 \end{bmatrix}$$

$$= \begin{bmatrix} 5 \\ 4 \\ -2 \end{bmatrix} - \begin{bmatrix} 8 \\ 4 \\ -8 \end{bmatrix}$$

$$= \begin{bmatrix} -3 \\ 0 \\ 6 \end{bmatrix}. \qquad \blacksquare$$

Rotations

Consider the linear transformation T from $\mathbb{R}^2$ to $\mathbb{R}^2$ that rotates any vector $\vec{x}$ through a fixed angle θ in the counterclockwise direction,[5] as shown in Figure 8. Recall Example 2.1.5, where we studied a rotation through $\theta = \pi/2$.

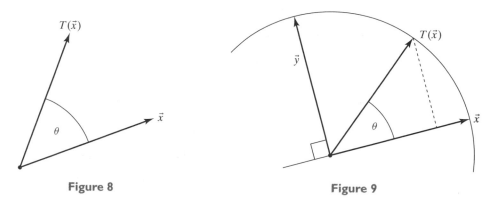

Figure 8   **Figure 9**

Now consider Figure 9, where we introduce the auxiliary vector $\vec{y}$, obtained by rotating $\vec{x}$ through $\pi/2$. From Example 2.1.5 we know that if $\vec{x} = \begin{bmatrix} x_1 \\ x_2 \end{bmatrix}$, then $\vec{y} = \begin{bmatrix} -x_2 \\ x_1 \end{bmatrix}$. Using basic trigonometry, we find that

$$T(\vec{x}) = (\cos\theta)\vec{x} + (\sin\theta)\vec{y} = (\cos\theta)\begin{bmatrix} x_1 \\ x_2 \end{bmatrix} + (\sin\theta)\begin{bmatrix} -x_2 \\ x_1 \end{bmatrix}$$

$$= \begin{bmatrix} (\cos\theta)x_1 - (\sin\theta)x_2 \\ (\sin\theta)x_1 + (\cos\theta)x_2 \end{bmatrix}$$

$$= \begin{bmatrix} \cos\theta & -\sin\theta \\ \sin\theta & \cos\theta \end{bmatrix}\begin{bmatrix} x_1 \\ x_2 \end{bmatrix}$$

$$= \begin{bmatrix} \cos\theta & -\sin\theta \\ \sin\theta & \cos\theta \end{bmatrix}\vec{x}.$$

---

[5]We can define a rotation more formally in terms of the polar coordinates of $\vec{x}$. The length of $T(\vec{x})$ equals the length of $\vec{x}$, and the polar angle (or argument) of $T(\vec{x})$ exceeds the polar angle of $\vec{x}$ by θ.

This computation shows that a rotation through θ is indeed a linear transformation, with the matrix

$$\begin{bmatrix} \cos\theta & -\sin\theta \\ \sin\theta & \cos\theta \end{bmatrix}.$$

Theorem 2.2.3   Rotations

The matrix of a counterclockwise rotation in $\mathbb{R}^2$ through an angle θ is

$$\begin{bmatrix} \cos\theta & -\sin\theta \\ \sin\theta & \cos\theta \end{bmatrix}.$$

Note that this matrix is of the form $\begin{bmatrix} a & -b \\ b & a \end{bmatrix}$, where $a^2 + b^2 = 1$. Conversely, any matrix of this form represents a rotation.   ■

EXAMPLE 4   The matrix of a counterclockwise rotation through $\pi/6$ (or $30°$) is

$$\begin{bmatrix} \cos(\pi/6) & -\sin(\pi/6) \\ \sin(\pi/6) & \cos(\pi/6) \end{bmatrix} = \frac{1}{2}\begin{bmatrix} \sqrt{3} & -1 \\ 1 & \sqrt{3} \end{bmatrix}.$$ ■

Rotations Combined with a Scaling

EXAMPLE 5   Examine how the linear transformation

$$T(\vec{x}) = \begin{bmatrix} a & -b \\ b & a \end{bmatrix}\vec{x}$$

affects our standard letter L. Here a and b are arbitrary constants.

Solution

Figure 10 suggests that T represents a *rotation combined with a scaling*. Think polar coordinates: This is a rotation through the phase angle θ of vector $\begin{bmatrix} a \\ b \end{bmatrix}$, combined with a scaling by the magnitude $r = \sqrt{a^2 + b^2}$ of vector $\begin{bmatrix} a \\ b \end{bmatrix}$. To verify this claim algebraically, we can write the vector $\begin{bmatrix} a \\ b \end{bmatrix}$ in polar coordinates, as

$$\begin{bmatrix} a \\ b \end{bmatrix} = \begin{bmatrix} r\cos\theta \\ r\sin\theta \end{bmatrix},$$

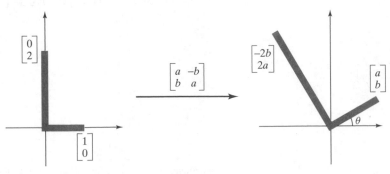

Figure 10

Figure 11

as illustrated in Figure 11. Then

$$\begin{bmatrix} a & -b \\ b & a \end{bmatrix} = \begin{bmatrix} r\cos\theta & -r\sin\theta \\ r\sin\theta & r\cos\theta \end{bmatrix} = r\begin{bmatrix} \cos\theta & -\sin\theta \\ \sin\theta & \cos\theta \end{bmatrix}.$$

It turns out that matrix $\begin{bmatrix} a & -b \\ b & a \end{bmatrix}$ is a scalar multiple of a rotation matrix, as claimed. ∎

Theorem 2.2.4   Rotations combined with a scaling

A matrix of the form $\begin{bmatrix} a & -b \\ b & a \end{bmatrix}$ represents a rotation combined with a scaling.

More precisely, if r and θ are the polar coordinates of vector $\begin{bmatrix} a \\ b \end{bmatrix}$, then $\begin{bmatrix} a & -b \\ b & a \end{bmatrix}$ represents a rotation through θ combined with a scaling by r. ∎

Shears

We will introduce shears by means of some simple experiments involving a ruler and a deck of cards.[6]

In the first experiment, we place the deck of cards on the ruler, as shown in Figure 12. Note that the 2 of diamonds is placed on one of the short edges of the ruler. That edge will stay in place throughout the experiment. Now we lift the other short edge of the ruler up, keeping the cards in vertical position at all times. The cards will slide up, being "fanned out," without any horizontal displacement.

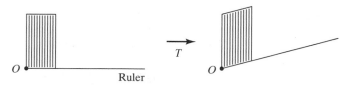

Figure 12   ·

Figure 13 shows a side view of this transformation. The origin represents the ruler's short edge that is staying in place.

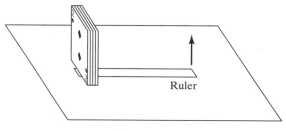

Figure 13

Such a transformation T is called a *vertical shear*. If we focus on the side view only, we have a vertical shear in $\mathbb{R}^2$ (although in reality the experiment takes place in space).

---

[6]Two hints for instructors:
- Use several decks of cards for dramatic effect.
- Hold the decks together with a rubber band to avoid embarrassing accidents.

Now let's draw a vector $\vec{x} = \begin{bmatrix} x_1 \\ x_2 \end{bmatrix}$ on the side of our deck of cards, and let's find a formula for the sheared vector $T(\vec{x})$, using Figure 14 as a guide. Here, k denotes the slope of the ruler after the transformation:

$$T(\vec{x}) = T\left(\begin{bmatrix} x_1 \\ x_2 \end{bmatrix}\right) = \begin{bmatrix} x_1 \\ kx_1 + x_2 \end{bmatrix} = \begin{bmatrix} 1 & 0 \\ k & 1 \end{bmatrix}\begin{bmatrix} x_1 \\ x_2 \end{bmatrix} = \begin{bmatrix} 1 & 0 \\ k & 1 \end{bmatrix}\vec{x}.$$

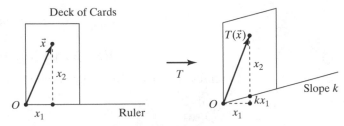

Figure 14

We find that the matrix of a vertical shear is of the form $\begin{bmatrix} 1 & 0 \\ k & 1 \end{bmatrix}$, where k is an arbitrary constant.

Horizontal shears are defined analogously; consider Figure 15.

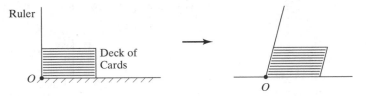

Figure 15

We leave it as an exercise to the reader to verify that the matrix of a horizontal shear is of the form $\begin{bmatrix} 1 & k \\ 0 & 1 \end{bmatrix}$. Take another look at part (e) of Example 1.

Oblique shears are far less important in applications, and we will not consider them in this introductory text.

Theorem 2.2.5   Horizontal and vertical shears

The matrix of a *horizontal shear* is of the form $\begin{bmatrix} 1 & k \\ 0 & 1 \end{bmatrix}$, and the matrix of a *vertical shear* is of the form $\begin{bmatrix} 1 & 0 \\ k & 1 \end{bmatrix}$, where k is an arbitrary constant.   ■

The Scottish scholar d'Arcy Thompson showed how the shapes of related species of plants and animals can often be transformed into one another, using linear as well as nonlinear transformations.[7] In Figure 16 he uses a horizontal shear to transform the shape of one species of fish into another.

---

[7]Thompson, d'Arcy W., *On Growth and Form*, Cambridge University Press, 1917. P. B. Medawar calls this "the finest work of literature in all the annals of science that have been recorded in the English tongue."

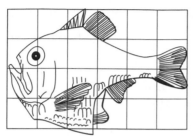

Argyropelecus olfersi.   *Sternoptyx diaphana.*

Figure 16

EXERCISES 2.2

GOAL *Use the matrices of orthogonal projections, re-flections, and rotations. Apply the definitions of shears, orthogonal projections, and reflections.*

1. Sketch the image of the standard L under the linear transformation

$$T(\vec{x}) = \begin{bmatrix} 3 & 1 \\ 1 & 2 \end{bmatrix} \vec{x}.$$

(See Example 1.)

2. Find the matrix of a rotation through an angle of 60° in the counterclockwise direction.

3. Consider a linear transformation T from $\mathbb{R}^2$ to $\mathbb{R}^3$. Use $T(\vec{e}_1)$ and $T(\vec{e}_2)$ to describe the image of the unit square geometrically.

4. Interpret the following linear transformation geometrically:

$$T(\vec{x}) = \begin{bmatrix} 1 & 1 \\ -1 & 1 \end{bmatrix} \vec{x}.$$

5. The matrix

$$\begin{bmatrix} -0.8 & -0.6 \\ 0.6 & -0.8 \end{bmatrix}$$

represents a rotation. Find the angle of rotation (in radians).

6. Let L be the line in $\mathbb{R}^3$ that consists of all scalar multiples of the vector $\begin{bmatrix} 2 \\ 1 \\ 2 \end{bmatrix}$. Find the orthogonal projection of the vector $\begin{bmatrix} 1 \\ 1 \\ 1 \end{bmatrix}$ onto L.

7. Let L be the line in $\mathbb{R}^3$ that consists of all scalar multiples of $\begin{bmatrix} 2 \\ 1 \\ 2 \end{bmatrix}$. Find the reflection of the vector $\begin{bmatrix} 1 \\ 1 \\ 1 \end{bmatrix}$ about the line L.

8. Interpret the following linear transformation geometrically:

$$T(\vec{x}) = \begin{bmatrix} 0 & -1 \\ -1 & 0 \end{bmatrix} \vec{x}.$$

9. Interpret the following linear transformation geometrically:

$$T(\vec{x}) = \begin{bmatrix} 1 & 0 \\ 1 & 1 \end{bmatrix} \vec{x}.$$

10. Find the matrix of the orthogonal projection onto the line L in $\mathbb{R}^2$ shown in the accompanying figure:

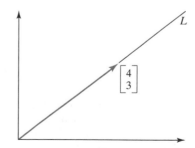

11. Refer to Exercise 10. Find the matrix of the reflection about the line L.

12. Consider a line L in the plane, running through the origin. If $\vec{w} = \begin{bmatrix} w_1 \\ w_2 \end{bmatrix}$ is a nonzero vector parallel to L, show that the matrix of $\text{proj}_L(\vec{x})$ is

$$\frac{1}{w_1^2 + w_2^2} \begin{bmatrix} w_1^2 & w_1 w_2 \\ w_1 w_2 & w_2^2 \end{bmatrix}.$$

13. Suppose a line L in $\mathbb{R}^2$ contains the unit vector

$$\vec{u} = \begin{bmatrix} u_1 \\ u_2 \end{bmatrix}.$$

Find the matrix A of the linear transformation $T(\vec{x}) = \text{ref}_L(\vec{x})$. Give the entries of A in terms of u_1 and u_2. Show that A is of the form $\begin{bmatrix} a & b \\ b & -a \end{bmatrix}$, where $a^2 + b^2 = 1$.

14. Suppose a line L in $\mathbb{R}^3$ contains the unit vector

$$\vec{u} = \begin{bmatrix} u_1 \\ u_2 \\ u_3 \end{bmatrix}.$$

a. Find the matrix A of the linear transformation $T(\vec{x}) = \text{proj}_L(\vec{x})$. Give the entries of A in terms of the components u_1, u_2, u_3 of $\vec{u}$.

b. What is the sum of the diagonal entries of the matrix A you found in part (a)?

15. Suppose a line L in $\mathbb{R}^3$ contains the unit vector

$$\vec{u} = \begin{bmatrix} u_1 \\ u_2 \\ u_3 \end{bmatrix}.$$

Find the matrix A of the linear transformation $T(\vec{x}) = \text{ref}_L(\vec{x})$. Give the entries of A in terms of the components u_1, u_2, u_3 of $\vec{u}$.

16. Let $T(\vec{x}) = \text{ref}_L(\vec{x})$ be the reflection about the line L in $\mathbb{R}^2$ shown in the accompanying figure.

a. Draw sketches to illustrate that T is linear.

b. Find the matrix of T in terms of θ.

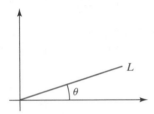

17. Consider a matrix A of the form $A = \begin{bmatrix} a & b \\ b & -a \end{bmatrix}$, where $a^2 + b^2 = 1$. Find two nonzero perpendicular vectors $\vec{v}$ and $\vec{w}$ such that $A\vec{v} = \vec{v}$ and $A\vec{w} = -\vec{w}$ (write the entries of $\vec{v}$ and $\vec{w}$ in terms of a and b). Conclude that $T(\vec{x}) = A\vec{x}$ represents the reflection about the line L spanned by $\vec{v}$.

18. The linear transformation $T(\vec{x}) = \begin{bmatrix} 0.6 & 0.8 \\ 0.8 & -0.6 \end{bmatrix} \vec{x}$ is a reflection about a line L (see Exercise 17). Find the equation of line L (in the form $y = mx$).

Find the matrices of the linear transformations from $\mathbb{R}^3$ to $\mathbb{R}^3$ given in Exercises 19 through 23. Some of these transformations have not been formally defined in the text. Use common sense. You may assume that all these transformations are linear.

19. The orthogonal projection onto the x–y-plane.

20. The reflection about the x–z-plane.

21. The rotation about the z-axis through an angle of $\pi/2$, counterclockwise as viewed from the positive z-axis.

22. The rotation about the y-axis through an angle θ, counterclockwise as viewed from the positive y-axis.

23. The reflection about the plane $y = z$.

24. Rotations and reflections have two remarkable properties: They preserve the length of vectors and the angle between vectors. (Draw figures illustrating these properties.) We will show that, conversely, any linear transformation T from $\mathbb{R}^2$ to $\mathbb{R}^2$ that preserves length and angles is either a rotation or a reflection (about a line).

a. Show that if $T(\vec{x}) = A\vec{x}$ preserves length and angles, then the two column vectors $\vec{v}$ and $\vec{w}$ of A must be perpendicular unit vectors.

b. Write the first column vector of A as $\vec{v} = \begin{bmatrix} a \\ b \end{bmatrix}$; note that $a^2 + b^2 = 1$, since $\vec{v}$ is a unit vector. Show that for a given $\vec{v}$ there are two possibilities for $\vec{w}$, the second column vector of A. Draw a sketch showing $\vec{v}$ and the two possible vectors $\vec{w}$. Write the components of $\vec{w}$ in terms of a and b.

c. Show that if a linear transformation T from $\mathbb{R}^2$ to $\mathbb{R}^2$ preserves length and angles, then T is either a rotation or a reflection (about a line). See Exercise 17.

25. Find the inverse of the matrix $\begin{bmatrix} 1 & k \\ 0 & 1 \end{bmatrix}$, where k is an arbitrary constant. Interpret your result geometrically.

26. a. Find the scaling matrix A that transforms $\begin{bmatrix} 2 \\ -1 \end{bmatrix}$ into $\begin{bmatrix} 8 \\ -4 \end{bmatrix}$.

b. Find the orthogonal projection matrix B that transforms $\begin{bmatrix} 2 \\ 3 \end{bmatrix}$ into $\begin{bmatrix} 2 \\ 0 \end{bmatrix}$.

c. Find the rotation matrix C that transforms $\begin{bmatrix} 0 \\ 5 \end{bmatrix}$ into $\begin{bmatrix} 3 \\ 4 \end{bmatrix}$.

d. Find the shear matrix D that transforms $\begin{bmatrix} 1 \\ 3 \end{bmatrix}$ into $\begin{bmatrix} 7 \\ 3 \end{bmatrix}$.

e. Find the reflection matrix E that transforms $\begin{bmatrix} 7 \\ 1 \end{bmatrix}$ into $\begin{bmatrix} -5 \\ 5 \end{bmatrix}$.

27. Consider the matrices A through E below.

$$A = \begin{bmatrix} 0.6 & 0.8 \\ 0.8 & -0.6 \end{bmatrix}, \quad B = \begin{bmatrix} 3 & 0 \\ 0 & 3 \end{bmatrix},$$

$$C = \begin{bmatrix} 0.36 & -0.48 \\ -0.48 & 0.64 \end{bmatrix} \quad D = \begin{bmatrix} -0.8 & 0.6 \\ -0.6 & -0.8 \end{bmatrix},$$

$$E = \begin{bmatrix} 1 & 0 \\ -1 & 1 \end{bmatrix}$$

Fill in the blanks in the sentences below.
We are told that there is a solution in each case.
Matrix _____ represents a scaling.
Matrix _____ represents an orthogonal projection.
Matrix _____ represents a shear.
Matrix _____ represents a reflection.
Matrix _____ represents a rotation.

28. Each of the linear transformations in parts (a) through (e) corresponds to one (and only one) of the matrices A through J. Match them up.

a. Scaling   **b.** Shear   **c.** Rotation

d. Orthogonal Projection   **e.** Reflection

$$A = \begin{bmatrix} 0 & 0 \\ 0 & 1 \end{bmatrix} \quad B = \begin{bmatrix} 2 & 1 \\ 1 & 0 \end{bmatrix} \quad C = \begin{bmatrix} -0.6 & 0.8 \\ -0.8 & -0.6 \end{bmatrix}$$

$$D = \begin{bmatrix} 7 & 0 \\ 0 & 7 \end{bmatrix} \quad E = \begin{bmatrix} 1 & 0 \\ -3 & 1 \end{bmatrix} \quad F = \begin{bmatrix} 0.6 & 0.8 \\ 0.8 & -0.6 \end{bmatrix}$$

$$G = \begin{bmatrix} 0.6 & 0.6 \\ 0.8 & 0.8 \end{bmatrix} \quad H = \begin{bmatrix} 2 & -1 \\ 1 & 2 \end{bmatrix} \quad I = \begin{bmatrix} 0 & 0 \\ 1 & 0 \end{bmatrix}$$

$$J = \begin{bmatrix} 0.8 & -0.6 \\ 0.6 & -0.8 \end{bmatrix}$$

29. Let T be a function from $\mathbb{R}^m$ to $\mathbb{R}^n$, and let L be a function from $\mathbb{R}^n$ to $\mathbb{R}^m$. Suppose that $L\left(T(\vec{x})\right) = \vec{x}$ for all $\vec{x}$ in $\mathbb{R}^m$ and $T\left(L(\vec{y})\right) = \vec{y}$ for all $\vec{y}$ in $\mathbb{R}^n$. If T is a linear transformation, show that L is linear as well. [*Hint:* $\vec{v} + \vec{w} = T\left(L(\vec{v})\right) + T\left(L(\vec{w})\right) = T\left(L(\vec{v}) + L(\vec{w})\right)$ since T is linear. Now apply L on both sides.]

30. Find a nonzero 2×2 matrix A such that $A\vec{x}$ is parallel to the vector $\begin{bmatrix} 1 \\ 2 \end{bmatrix}$, for all $\vec{x}$ in $\mathbb{R}^2$.

31. Find a nonzero 3×3 matrix A such that $A\vec{x}$ is perpendicular to $\begin{bmatrix} 1 \\ 2 \\ 3 \end{bmatrix}$, for all $\vec{x}$ in $\mathbb{R}^3$.

32. Consider the rotation matrix $D = \begin{bmatrix} \cos\alpha & -\sin\alpha \\ \sin\alpha & \cos\alpha \end{bmatrix}$ and the vector $\vec{v} = \begin{bmatrix} \cos\beta \\ \sin\beta \end{bmatrix}$, where α and β are arbitrary angles.

a. Draw a sketch to explain why $D\vec{v} = \begin{bmatrix} \cos(\alpha + \beta) \\ \sin(\alpha + \beta) \end{bmatrix}$.

b. Compute $D\vec{v}$. Use the result to derive the addition theorems for sine and cosine:

$$\cos(\alpha + \beta) = \dots, \quad \sin(\alpha + \beta) = \dots.$$

33. Consider two nonparallel lines L_1 and L_2 in $\mathbb{R}^2$. Explain why a vector $\vec{v}$ in $\mathbb{R}^2$ can be expressed uniquely as

$$\vec{v} = \vec{v}_1 + \vec{v}_2,$$

where $\vec{v}_1$ is on L_1 and $\vec{v}_2$ on L_2. Draw a sketch. The transformation $T(\vec{v}) = \vec{v}_1$ is called the *projection onto L_1 along L_2.* Show algebraically that T is linear.

34. One of the five given matrices represents an orthogonal projection onto a line and another represents a reflection about a line. Identify both and briefly justify your choice.

$$A = \frac{1}{3}\begin{bmatrix} 1 & 2 & 2 \\ 2 & 1 & 2 \\ 2 & 2 & 1 \end{bmatrix}, \quad B = \frac{1}{3}\begin{bmatrix} 1 & 1 & 1 \\ 1 & 1 & 1 \\ 1 & 1 & 1 \end{bmatrix},$$

$$C = \frac{1}{3}\begin{bmatrix} 2 & 1 & 1 \\ 1 & 2 & 1 \\ 1 & 1 & 2 \end{bmatrix}, \quad D = -\frac{1}{3}\begin{bmatrix} 1 & 2 & 2 \\ 2 & 1 & 2 \\ 2 & 2 & 1 \end{bmatrix},$$

$$E = \frac{1}{3}\begin{bmatrix} -1 & 2 & 2 \\ 2 & -1 & 2 \\ 2 & 2 & -1 \end{bmatrix}$$

35. Let T be an invertible linear transformation from $\mathbb{R}^2$ to $\mathbb{R}^2$. Let P be a parallelogram in $\mathbb{R}^2$ with one vertex at the origin. Is the image of P a parallelogram as well? Explain. Draw a sketch of the image.

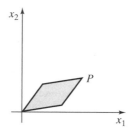

36. Let T be an invertible linear transformation from $\mathbb{R}^2$ to $\mathbb{R}^2$. Let P be a parallelogram in $\mathbb{R}^2$. Is the image of P a parallelogram as well? Explain.

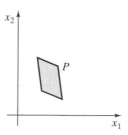

37. The *trace* of a matrix $\begin{bmatrix} a & b \\ c & d \end{bmatrix}$ is the sum $a + d$ of its diagonal entries. What can you say about the trace of a 2×2 matrix that represents a(n)

a. orthogonal projection   **b.** reflection about a line

c. rotation   **d.** (horizontal or vertical) shear.

In three cases, give the exact value of the trace, and in one case, give an interval of possible values.

38. The *determinant* of a matrix $\begin{bmatrix} a & b \\ c & d \end{bmatrix}$ is $ad - bc$ (we have seen this quantity in Exercise 2.1.13 already). Find the determinant of a matrix that represents a(n)

a. orthogonal projection   **b.** reflection about a line

c. rotation   **d.** (horizontal or vertical) shear.

What do your answers tell you about the invertibility of these matrices?

39. Describe each of the linear transformations defined by the matrices in parts (a) through (c) geometrically, as a well-known transformation combined with a scaling. Give the scaling factor in each case.

a. $\begin{bmatrix} 1 & 1 \\ 1 & 1 \end{bmatrix}$ **b.** $\begin{bmatrix} 3 & 0 \\ -1 & 3 \end{bmatrix}$

c. $\begin{bmatrix} 3 & 4 \\ 4 & -3 \end{bmatrix}$

40. Let P and Q be two perpendicular lines in $\mathbb{R}^2$. For a vector $\vec{x}$ in $\mathbb{R}^2$, what is $\text{proj}_P(\vec{x}) + \text{proj}_Q(\vec{x})$? Give your answer in terms of $\vec{x}$. Draw a sketch to justify your answer.

41. Let P and Q be two perpendicular lines in $\mathbb{R}^2$. For a vector $\vec{x}$ in $\mathbb{R}^2$, what is the relationship between $\text{ref}_P(\vec{x})$ and $\text{ref}_Q(\vec{x})$? Draw a sketch to justify your answer.

42. Let $T(\vec{x}) = \text{proj}_L(\vec{x})$ be the orthogonal projection onto a line in $\mathbb{R}^2$. What is the relationship between $T(\vec{x})$ and $T(T(\vec{x}))$? Justify your answer carefully.

43. Use the formula derived in Exercise 2.1.13 to find the inverse of the rotation matrix

$$A = \begin{bmatrix} \cos\theta & -\sin\theta \\ \sin\theta & \cos\theta \end{bmatrix}.$$

Interpret the linear transformation defined by A^{-1} geometrically. Explain.

44. A nonzero matrix of the form $A = \begin{bmatrix} a & -b \\ b & a \end{bmatrix}$ represents a rotation combined with a scaling. Use the formula derived in Exercise 2.1.13 to find the inverse of A. Interpret the linear transformation defined by A^{-1} geometrically. Explain.

45. A matrix of the form $A = \begin{bmatrix} a & b \\ b & -a \end{bmatrix}$, where $a^2 + b^2 = 1$, represents a reflection about a line (see Exercise 17). Use the formula derived in Exercise 2.1.13 to find the inverse of A. Explain.

46. A nonzero matrix of the form $A = \begin{bmatrix} a & b \\ b & -a \end{bmatrix}$ represents a reflection about a line L combined with a scaling. (Why? What is the scaling factor?) Use the formula derived in Exercise 2.1.13 to find the inverse of A. Interpret

the linear transformation defined by A^{-1} geometrically. Explain.

47. Let T be a linear transformation from $\mathbb{R}^2$ to $\mathbb{R}^2$. Consider the function

$$f(t) = \left(T \begin{bmatrix} \cos(t) \\ \sin(t) \end{bmatrix} \right) \cdot \left(T \begin{bmatrix} -\sin(t) \\ \cos(t) \end{bmatrix} \right),$$

from $\mathbb{R}$ to $\mathbb{R}$. Show the following:

a. The function $f(t)$ is continuous. You may take for granted that the functions $\sin(t)$ and $\cos(t)$ are continuous, and also that sums and products of continuous functions are continuous.

b. $f(\pi/2) = -f(0)$.

c. There exists a number c between 0 and $\pi/2$ such that $f(c) = 0$. Use the intermediate value theorem of calculus, which tells us the following: If a function $g(t)$ is continuous for $a \leq t \leq b$, and L is a number between $g(a)$ and $g(b)$, then there exists at least one number c between a and b such that $g(c) = L$.

d. There exist two perpendicular unit vectors $\vec{v}_1$ and $\vec{v}_2$ in $\mathbb{R}^2$ such that the vectors $T(\vec{v}_1)$ and $T(\vec{v}_2)$ are perpendicular as well. See the accompanying figure. (Compare with Theorem 8.3.3 for a generalization.)

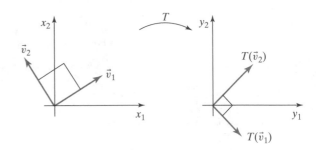

48. Refer to Exercise 47. Consider the linear transformation

$$T(\vec{x}) = \begin{bmatrix} 0 & 4 \\ 5 & -3 \end{bmatrix} \vec{x}.$$

Find the function $f(t)$ defined in Exercise 47, graph it (using technology), and find a number c between 0 and $\pi/2$ such that $f(c) = 0$. Use your answer to find two perpendicular unit vectors $\vec{v}_1$ and $\vec{v}_2$ such that $T(\vec{v}_1)$ and $T(\vec{v}_2)$ are perpendicular. Draw a sketch.

49. Sketch the image of the unit circle under the linear transformation

$$T(\vec{x}) = \begin{bmatrix} 5 & 0 \\ 0 & 2 \end{bmatrix} \vec{x}.$$

50. Let T be an invertible linear transformation from $\mathbb{R}^2$ to $\mathbb{R}^2$. Show that the image of the unit circle is an ellipse centered at the origin.[8] [*Hint*: Consider two perpendicular unit vectors $\vec{v}_1$ and $\vec{v}_2$ such that $T(\vec{v}_1)$ and $T(\vec{v}_2)$ are perpendicular.] (See Exercise 47d.) The unit circle consists of all vectors of the form

$$\vec{v} = \cos(t)\vec{v}_1 + \sin(t)\vec{v}_2,$$

where t is a parameter.

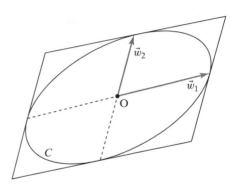

51. Let $\vec{w}_1$ and $\vec{w}_2$ be two nonparallel vectors in $\mathbb{R}^2$. Consider the curve C in $\mathbb{R}^2$ that consists of all vectors of the form $\cos(t)\vec{w}_1 + \sin(t)\vec{w}_2$, where t is a parameter. Show that C is an ellipse. (*Hint*: You can interpret C as the image of the unit circle under a suitable linear transformation; then use Exercise 50.)

52. Consider an invertible linear transformation T from $\mathbb{R}^2$ to $\mathbb{R}^2$. Let C be an ellipse in $\mathbb{R}^2$. Show that the image of C under T is an ellipse as well. (*Hint*: Use the result of Exercise 51.)

2.3 Matrix Products

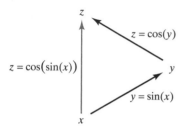

Figure 1

Recall the *composition* of two functions: The composite of the functions $y = \sin(x)$ and $z = \cos(y)$ is $z = \cos(\sin(x))$, as illustrated in Figure 1.

Similarly, we can compose two linear transformations.

To understand this concept, let's return to the coding example discussed in Section 2.1. Recall that the position $\vec{x} = \begin{bmatrix} x_1 \\ x_2 \end{bmatrix}$ of your boat is encoded and that you radio the encoded position $\vec{y} = \begin{bmatrix} y_1 \\ y_2 \end{bmatrix}$ to Marseille. The coding transformation is

$$\vec{y} = A\vec{x}, \quad \text{with} \quad A = \begin{bmatrix} 1 & 2 \\ 3 & 5 \end{bmatrix}.$$

In Section 2.1, we left out one detail: Your position is radioed on to Paris, as you would expect in a centrally governed country such as France. Before broadcasting to Paris, the position $\vec{y}$ is again encoded, using the linear transformation

$$\vec{z} = B\vec{y}, \quad \text{with} \quad B = \begin{bmatrix} 6 & 7 \\ 8 & 9 \end{bmatrix}$$

---

[8]An ellipse in $\mathbb{R}^2$ centered at the origin may be defined as a curve that can be parametrized as

$$\cos(t)\vec{w}_1 + \sin(t)\vec{w}_2,$$

for two perpendicular vectors $\vec{w}_1$ and $\vec{w}_2$. Suppose the length of $\vec{w}_1$ exceeds the length of $\vec{w}_2$. Then we call the vectors $\pm\vec{w}_1$ the semimajor axes of the ellipse and $\pm\vec{w}_2$ the semiminor axes.

Convention: All ellipses considered in this text are centered at the origin unless stated otherwise.

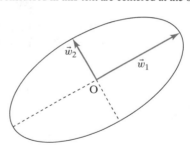

this time, and the sailor in Marseille radios the encoded position $\vec{z}$ to Paris. (See Figure 2.)

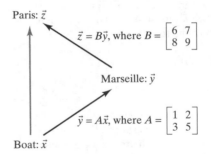

Figure 2

We can think of the message $\vec{z}$ received in Paris as a function of the actual position $\vec{x}$ of the boat,

$$\vec{z} = B(A\vec{x}),$$

the composite of the two transformations $\vec{y} = A\vec{x}$ and $\vec{z} = B\vec{y}$. Is this transformation $\vec{z} = T(\vec{x})$ linear, and, if so, what is its matrix? We will show two approaches to these important questions: (a) using brute force, and (b) using some theory.

a. We write the components of the two transformations and substitute.

$$\begin{matrix} z_1 = 6y_1 + 7y_2 \\ z_2 = 8y_1 + 9y_2 \end{matrix} \quad \text{and} \quad \begin{matrix} y_1 = x_1 + 2x_2 \\ y_2 = 3x_1 + 5x_2 \end{matrix}$$

so that

$$z_1 = 6(x_1 + 2x_2) + 7(3x_1 + 5x_2) = (6 \cdot 1 + 7 \cdot 3)x_1 + (6 \cdot 2 + 7 \cdot 5)x_2$$
$$= 27x_1 + 47x_2,$$
$$z_2 = 8(x_1 + 2x_2) + 9(3x_1 + 5x_2) = (8 \cdot 1 + 9 \cdot 3)x_1 + (8 \cdot 2 + 9 \cdot 5)x_2$$
$$= 35x_1 + 61x_2.$$

This shows that the composite is indeed linear, with matrix

$$\begin{bmatrix} 6 \cdot 1 + 7 \cdot 3 & 6 \cdot 2 + 7 \cdot 5 \\ 8 \cdot 1 + 9 \cdot 3 & 8 \cdot 2 + 9 \cdot 5 \end{bmatrix} = \begin{bmatrix} 27 & 47 \\ 35 & 61 \end{bmatrix}.$$

b. We can use Theorem 1.3.10 to show that the transformation $T(\vec{x}) = B(A\vec{x})$ is linear:

$$T(\vec{v} + \vec{w}) = B\big(A(\vec{v} + \vec{w})\big) = B(A\vec{v} + A\vec{w})$$
$$= B(A\vec{v}) + B(A\vec{w}) = T(\vec{v}) + T(\vec{w}),$$
$$T(k\vec{v}) = B\big(A(k\vec{v})\big) = B\big(k(A\vec{v})\big) = k\big(B(A\vec{v})\big) = kT(\vec{v}).$$

Once we know that T is linear, we can find its matrix by computing the vectors $T(\vec{e}_1) = B(A\vec{e}_1)$ and $T(\vec{e}_2) = B(A\vec{e}_2)$; the matrix of T is then $[T(\vec{e}_1) \quad T(\vec{e}_2)]$, by Theorem 2.1.2:

$$T(\vec{e}_1) = B(A\vec{e}_1) = B(\text{first column of } A) = \begin{bmatrix} 6 & 7 \\ 8 & 9 \end{bmatrix} \begin{bmatrix} 1 \\ 3 \end{bmatrix} = \begin{bmatrix} 27 \\ 35 \end{bmatrix},$$

$$T(\vec{e}_2) = B(A\vec{e}_2) = B(\text{second column of } A) = \begin{bmatrix} 6 & 7 \\ 8 & 9 \end{bmatrix} \begin{bmatrix} 2 \\ 5 \end{bmatrix} = \begin{bmatrix} 47 \\ 61 \end{bmatrix}.$$

We find that the matrix of the linear transformation $T(\vec{x}) = B(A\vec{x})$ is

$$\begin{bmatrix} | & | \\ T(\vec{e}_1) & T(\vec{e}_2) \\ | & | \end{bmatrix} = \begin{bmatrix} 27 & 47 \\ 35 & 61 \end{bmatrix}.$$

This result agrees with the result in (a), of course.

The matrix of the linear transformation $T(\vec{x}) = B(A\vec{x})$ is called the *product* of the matrices B and A, written as BA. This means that

$$T(\vec{x}) = B(A\vec{x}) = (BA)\vec{x},$$

for all vectors $\vec{x}$ in $\mathbb{R}^2$. (See Figure 3.)

Now let's look at the product of larger matrices. Let B be an $n \times p$ matrix and A a $p \times m$ matrix. These matrices represent linear transformations as shown in Figure 4.

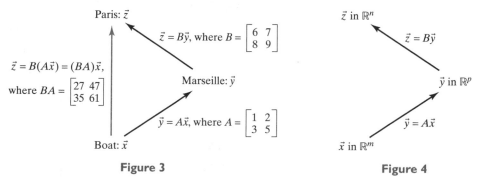

Figure 3   **Figure 4**

Again, the composite transformation $\vec{z} = B(A\vec{x})$ is linear. (The foregoing justification applies in this more general case as well.) The matrix of the linear transformation $\vec{z} = B(A\vec{x})$ is called the *product* of the matrices B and A, written as BA. Note that BA is an $n \times m$ matrix (as it represents a linear transformation from $\mathbb{R}^m$ to $\mathbb{R}^n$). As in the case of $\mathbb{R}^2$, the equation

$$\vec{z} = B(A\vec{x}) = (BA)\vec{x}$$

holds for all vectors $\vec{x}$ in $\mathbb{R}^m$, by definition of the product BA. (See Figure 5.)

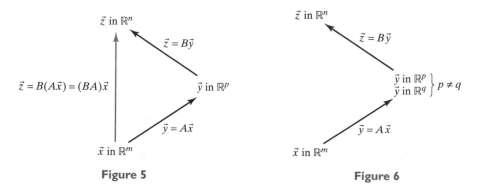

Figure 5   **Figure 6**

In the definition of the matrix product BA, the number of columns of B matches the number of rows of A. What happens if these two numbers are different? Suppose B is an $n \times p$ matrix and A is a $q \times m$ matrix, with $p \neq q$.

In this case, the transformations $\vec{z} = B\vec{y}$ and $\vec{y} = A\vec{x}$ cannot be composed, since the target space of $\vec{y} = A\vec{x}$ is different from the domain of $\vec{z} = B\vec{y}$. (See Figure 6.) To put it more plainly: The output of $\vec{y} = A\vec{x}$ is not an acceptable input for the transformation $\vec{z} = B\vec{y}$. In this case, the matrix product BA is undefined.

Definition 2.3.1 Matrix multiplication

    a. Let B be an $n \times p$ matrix and A a $q \times m$ matrix. The product BA is defined if (and only if) $p = q$.

    b. If B is an $n \times p$ matrix and A a $p \times m$ matrix, then the product BA is defined as the matrix of the linear transformation $T(\vec{x}) = B(A\vec{x})$. This means that $T(\vec{x}) = B(A\vec{x}) = (BA)\vec{x}$, for all $\vec{x}$ in the vector space $\mathbb{R}^m$. The product BA is an $n \times m$ matrix.

Although this definition of matrix multiplication does not give us concrete instructions for computing the product of two numerically given matrices, such instructions can be derived easily from the definition.

As in Definition 2.3.1, let B be an $n \times p$ matrix and A a $p \times m$ matrix. Let's think about the columns of the matrix BA:

$$(i\text{th column of } BA) = (BA)\vec{e}_i$$
$$= B(A\vec{e}_i)$$
$$= B(i\text{th column of } A).$$

If we denote the columns of A by $\vec{v}_1, \vec{v}_2, \ldots, \vec{v}_m$, we can write

$$BA = B \begin{bmatrix} | & | & & | \\ \vec{v}_1 & \vec{v}_2 & \cdots & \vec{v}_m \\ | & | & & | \end{bmatrix} = \begin{bmatrix} | & | & & | \\ B\vec{v}_1 & B\vec{v}_2 & \cdots & B\vec{v}_m \\ | & | & & | \end{bmatrix}.$$

Theorem 2.3.2 The columns of the matrix product

Let B be an $n \times p$ matrix and A a $p \times m$ matrix with columns $\vec{v}_1, \vec{v}_2, \ldots, \vec{v}_m$. Then, the product BA is

$$BA = B \begin{bmatrix} | & | & & | \\ \vec{v}_1 & \vec{v}_2 & \cdots & \vec{v}_m \\ | & | & & | \end{bmatrix} = \begin{bmatrix} | & | & & | \\ B\vec{v}_1 & B\vec{v}_2 & \cdots & B\vec{v}_m \\ | & | & & | \end{bmatrix}.$$

To find BA, we can multiply B with the columns of A and combine the resulting vectors. ∎

This is exactly how we computed the product

$$BA = \begin{bmatrix} 6 & 7 \\ 8 & 9 \end{bmatrix} \begin{bmatrix} 1 & 2 \\ 3 & 5 \end{bmatrix} = \begin{bmatrix} 27 & 47 \\ 35 & 61 \end{bmatrix}$$

on page 70, using approach (b).

For practice, let us multiply the same matrices in the reverse order. The first column of AB is $\begin{bmatrix} 1 & 2 \\ 3 & 5 \end{bmatrix} \begin{bmatrix} 6 \\ 8 \end{bmatrix} = \begin{bmatrix} 22 \\ 58 \end{bmatrix}$; the second is $\begin{bmatrix} 1 & 2 \\ 3 & 5 \end{bmatrix} \begin{bmatrix} 7 \\ 9 \end{bmatrix} = \begin{bmatrix} 25 \\ 66 \end{bmatrix}$. Thus,

$$AB = \begin{bmatrix} 1 & 2 \\ 3 & 5 \end{bmatrix} \begin{bmatrix} 6 & 7 \\ 8 & 9 \end{bmatrix} = \begin{bmatrix} 22 & 25 \\ 58 & 66 \end{bmatrix}.$$

Compare the two previous displays to see that $AB \neq BA$: Matrix multiplication is *noncommutative*. This should come as no surprise, in view of the fact that the matrix product represents a composite of transformations. Even for functions of one variable, the order in which we compose matters. Refer to the first example in this section and note that the functions $\cos\left(\sin(x)\right)$ and $\sin\left(\cos(x)\right)$ are different.

Theorem 2.3.3   Matrix multiplication is noncommutative

$AB \neq BA$, in general. However, at times it does happen that $AB = BA$; then we say that the matrices A and B *commute*.   ∎

It is useful to have a formula for the ijth entry of the product BA of an $n \times p$ matrix B and a $p \times m$ matrix A.

Let $\vec{v}_1, \vec{v}_2, \ldots, \vec{v}_m$ be the columns of A. Then, by Theorem 2.3.2,

$$BA = B \begin{bmatrix} | & | & & | & & | \\ \vec{v}_1 & \vec{v}_2 & \cdots & \vec{v}_j & \cdots & \vec{v}_m \\ | & | & & | & & | \end{bmatrix} = \begin{bmatrix} | & | & & | & & | \\ B\vec{v}_1 & B\vec{v}_2 & \cdots & B\vec{v}_j & \cdots & B\vec{v}_m \\ | & | & & | & & | \end{bmatrix}.$$

The ijth entry of the product BA is the ith component of the vector $B\vec{v}_j$, which is the dot product of the ith row of B and $\vec{v}_j$, by Definition 1.3.7.

Theorem 2.3.4   The entries of the matrix product

Let B be an $n \times p$ matrix and A a $p \times m$ matrix. The ijth entry of BA is the dot product of the ith row of B with the jth column of A.

$$BA = \begin{bmatrix} b_{11} & b_{12} & \cdots & b_{1p} \\ b_{21} & b_{22} & \cdots & b_{2p} \\ \vdots & \vdots & \ddots & \vdots \\ b_{i1} & b_{i2} & \cdots & b_{ip} \\ \vdots & \vdots & \ddots & \vdots \\ b_{n1} & b_{n2} & \cdots & b_{np} \end{bmatrix} \begin{bmatrix} a_{11} & a_{12} & \cdots & a_{1j} & \cdots & a_{1m} \\ a_{21} & a_{22} & \cdots & a_{2j} & \cdots & a_{2m} \\ \vdots & \vdots & \ddots & \vdots & \ddots & \vdots \\ a_{p1} & a_{p2} & \cdots & a_{pj} & \cdots & a_{pm} \end{bmatrix}$$

is the $n \times m$ matrix whose ijth entry is

$$b_{i1}a_{1j} + b_{i2}a_{2j} + \cdots + b_{ip}a_{pj} = \sum_{k=1}^{p} b_{ik}a_{kj}.$$ ∎

EXAMPLE 1

$$\begin{bmatrix} 6 & 7 \\ 8 & 9 \end{bmatrix} \begin{bmatrix} 1 & 2 \\ 3 & 5 \end{bmatrix} = \begin{bmatrix} 6 \cdot 1 + 7 \cdot 3 & 6 \cdot 2 + 7 \cdot 5 \\ 8 \cdot 1 + 9 \cdot 3 & 8 \cdot 2 + 9 \cdot 5 \end{bmatrix} = \begin{bmatrix} 27 & 47 \\ 35 & 61 \end{bmatrix}$$

We have done these computations before. (Where?)   ∎

EXAMPLE 2   Compute the products BA and AB for $A = \begin{bmatrix} 0 & 1 \\ 1 & 0 \end{bmatrix}$ and $B = \begin{bmatrix} -1 & 0 \\ 0 & 1 \end{bmatrix}$. Interpret your answers geometrically, as composites of linear transformation. Draw composition diagrams.

Solution

$$BA = \begin{bmatrix} -1 & 0 \\ 0 & 1 \end{bmatrix} \begin{bmatrix} 0 & 1 \\ 1 & 0 \end{bmatrix} = \begin{bmatrix} 0 & -1 \\ 1 & 0 \end{bmatrix} \quad \text{and} \quad AB = \begin{bmatrix} 0 & 1 \\ 1 & 0 \end{bmatrix} \begin{bmatrix} -1 & 0 \\ 0 & 1 \end{bmatrix} = \begin{bmatrix} 0 & 1 \\ -1 & 0 \end{bmatrix}.$$

Note that in this special example it turns out that $BA = -AB$.

From Section 2.2 we recall the following geometrical interpretations:

$$A = \begin{bmatrix} 0 & 1 \\ 1 & 0 \end{bmatrix} \text{ represents the reflection about the vector } \begin{bmatrix} 1 \\ 1 \end{bmatrix};$$

$$B = \begin{bmatrix} -1 & 0 \\ 0 & 1 \end{bmatrix} \text{ represents the reflection about } \begin{bmatrix} 0 \\ 1 \end{bmatrix};$$

$$BA = \begin{bmatrix} 0 & -1 \\ 1 & 0 \end{bmatrix} \text{ represents the rotation through } \frac{\pi}{2}; \text{ and}$$

$$AB = \begin{bmatrix} 0 & 1 \\ -1 & 0 \end{bmatrix} \text{ represents the rotation through } -\frac{\pi}{2}.$$

Let's use our standard L to show the effect of these transformations. See Figures 7 and 8. ■

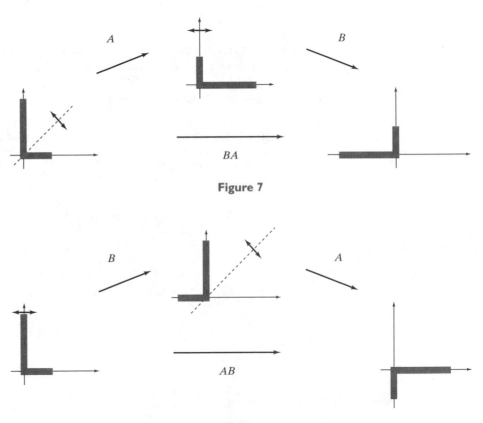

Figure 7

Figure 8

Matrix Algebra

Next let's discuss some algebraic rules for matrix multiplication.

- Composing a linear transformation with the identity transformation, on either side, leaves the transformation unchanged. (See Example 2.1.4.)

Theorem 2.3.5 **Multiplying with the identity matrix**

For an $n \times m$ matrix A,

$$AI_m = I_n A = A.$$

■

- If A is an $n \times p$ matrix, B a $p \times q$ matrix, and C a $q \times m$ matrix, what is the relationship between $(AB)C$ and $A(BC)$?

  One way to think about this problem (although perhaps not the most elegant one) is to write C in terms of its columns: $C = \begin{bmatrix} \vec{v}_1 & \vec{v}_2 & \cdots & \vec{v}_m \end{bmatrix}$. Then

  $$(AB)C = (AB) \begin{bmatrix} \vec{v}_1 & \vec{v}_2 & \cdots & \vec{v}_m \end{bmatrix} = \begin{bmatrix} (AB)\vec{v}_1 & (AB)\vec{v}_2 & \cdots & (AB)\vec{v}_m \end{bmatrix},$$

  and

  $$A(BC) = A \begin{bmatrix} B\vec{v}_1 & B\vec{v}_2 & \cdots & B\vec{v}_m \end{bmatrix} = \begin{bmatrix} A(B\vec{v}_1) & A(B\vec{v}_2) & \cdots & A(B\vec{v}_m) \end{bmatrix}.$$

  Since $(AB)\vec{v}_i = A(B\vec{v}_i)$, by definition of the matrix product, we find that $(AB)C = A(BC)$.

Theorem 2.3.6   Matrix multiplication is associative

$$(AB)C = A(BC)$$

We can simply write ABC for the product $(AB)C = A(BC)$.   ∎

A more conceptual proof is based on the fact that the composition of functions is associative. The two linear transformations

$$T(\vec{x}) = \big((AB)C\big)\vec{x} \quad \text{and} \quad L(\vec{x}) = \big(A(BC)\big)\vec{x}$$

are identical because, by definition of matrix multiplication,

$$T(\vec{x}) = \big((AB)C\big)\vec{x} = (AB)(C\vec{x}) = A\big(B(C\vec{x})\big)$$

and

$$L(\vec{x}) = \big(A(BC)\big)\vec{x} = A\big((BC)\vec{x}\big) = A\big(B(C\vec{x})\big).$$

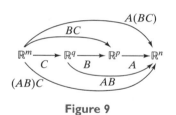

Figure 9

The domains and target spaces of the linear transformations defined by the matrices $A, B, C, BC, AB, A(BC)$, and $(AB)C$ are shown in Figure 9.

Theorem 2.3.7   Distributive property for matrices

If A and B are $n \times p$ matrices, and C and D are $p \times m$ matrices, then

$$A(C + D) = AC + AD, \quad \text{and}$$
$$(A + B)C = AC + BC.$$ ∎

You will be asked to verify this property in Exercise 27.

Theorem 2.3.8   If A is an $n \times p$ matrix, B is a $p \times m$ matrix, and k is a scalar, then

$$(kA)B = A(kB) = k(AB).$$ ∎

You will be asked to verify this property in Exercise 28.

Block Matrices (Optional)

In the popular puzzle Sudoku, one considers a 9×9 matrix A that is subdivided into nine 3×3 matrices called *blocks*. The puzzle setter provides some of the 81 entries of matrix A, and the objective is to fill in the remaining entries so that each row of A, each column of A, and each block contains each of the digits 1 through 9 exactly once.

5	3			7				
6			1	9	5			
	9	8					6	
8				6				3
4			8		3			1
7				2				6
	6					2	8	
			4	1	9			5
				8			7	9

This is an example of a *block matrix* (or *partitioned matrix*), that is, a matrix that is partitioned into rectangular submatrices, called blocks, by means of horizontal and vertical lines that go all the way through the matrix.

The blocks need not be of equal size.

For example, we can partition the matrix

$$B = \begin{bmatrix} 1 & 2 & 3 \\ 4 & 5 & 6 \\ 6 & 7 & 9 \end{bmatrix} \quad \text{as} \quad B = \left[\begin{array}{cc|c} 1 & 2 & 3 \\ 4 & 5 & 6 \\ \hline 6 & 7 & 9 \end{array}\right] = \begin{bmatrix} B_{11} & B_{12} \\ B_{21} & B_{22} \end{bmatrix},$$

where $B_{11} = \begin{bmatrix} 1 & 2 \\ 4 & 5 \end{bmatrix}$, $B_{12} = \begin{bmatrix} 3 \\ 6 \end{bmatrix}$, $B_{21} = \begin{bmatrix} 6 & 7 \end{bmatrix}$, and $B_{22} = [9]$.

A useful property of block matrices is the following:

Theorem 2.3.9   Multiplying block matrices

Block matrices can be multiplied as though the blocks were scalars (i.e., using the formula in Theorem 2.3.4):

$$AB = \begin{bmatrix} A_{11} & A_{12} & \cdots & A_{1p} \\ A_{21} & A_{22} & \cdots & A_{2p} \\ \vdots & \vdots & \ddots & \vdots \\ A_{i1} & A_{i2} & \cdots & A_{ip} \\ \vdots & \vdots & \ddots & \vdots \\ A_{n1} & A_{n2} & \cdots & A_{np} \end{bmatrix} \begin{bmatrix} B_{11} & B_{12} & \cdots & B_{1j} & \cdots & B_{1m} \\ B_{21} & B_{22} & \cdots & B_{2j} & \cdots & B_{2m} \\ \vdots & \vdots & \ddots & \vdots & \ddots & \vdots \\ B_{p1} & B_{p2} & \cdots & B_{pj} & \cdots & B_{pm} \end{bmatrix}$$

is the block matrix whose ijth block is the matrix

$$A_{i1}B_{1j} + A_{i2}B_{2j} + \cdots + A_{ip}B_{pj} = \sum_{k=1}^{p} A_{ik}B_{kj},$$

provided that all the products $A_{ik}B_{kj}$ are defined. ∎

Verifying this fact is left as an exercise. A numerical example follows.

EXAMPLE 3 $\left[\begin{array}{cc|c} 0 & 1 & -1 \\ 1 & 0 & 1 \end{array}\right] \left[\begin{array}{cc|c} 1 & 2 & 3 \\ 4 & 5 & 6 \\ \hline 7 & 8 & 9 \end{array}\right]$

$$= \left[\begin{array}{c|c} \begin{bmatrix} 0 & 1 \\ 1 & 0 \end{bmatrix}\begin{bmatrix} 1 & 2 \\ 4 & 5 \end{bmatrix} + \begin{bmatrix} -1 \\ 1 \end{bmatrix}\begin{bmatrix} 7 & 8 \end{bmatrix} & \begin{bmatrix} 0 & 1 \\ 1 & 0 \end{bmatrix}\begin{bmatrix} 3 \\ 6 \end{bmatrix} + \begin{bmatrix} -1 \\ 1 \end{bmatrix}[9] \end{array}\right]$$

$$= \left[\begin{array}{cc|c} -3 & -3 & -3 \\ 8 & 10 & 12 \end{array}\right].$$

Compute this product without using a partition, and see whether or not you find the same result. ∎

In this simple example, using blocks is somewhat pointless. Example 3 merely illustrates Theorem 2.3.9. In Example 2.4.7, we will see a more sensible application of the concept of block matrices.

EXERCISES 2.3

GOAL *Compute matrix products column by column and entry by entry. Interpret matrix multiplication in terms of the underlying linear transformations. Use the rules of matrix algebra. Multiply block matrices.*

If possible, compute the matrix products in Exercises 1 through 13, using paper and pencil.

1. $\begin{bmatrix} 1 & 1 \\ 0 & 1 \end{bmatrix} \begin{bmatrix} 1 & 2 \\ 3 & 4 \end{bmatrix}$

2. $\begin{bmatrix} 1 & -1 \\ -2 & 2 \end{bmatrix} \begin{bmatrix} 7 & 5 \\ 3 & 1 \end{bmatrix}$

3. $\begin{bmatrix} 1 & 2 & 3 \\ 4 & 5 & 6 \end{bmatrix} \begin{bmatrix} 1 & 2 \\ 3 & 4 \end{bmatrix}$

4. $\begin{bmatrix} 1 & -1 \\ 0 & 2 \\ 2 & 1 \end{bmatrix} \begin{bmatrix} 3 & 2 \\ 1 & 0 \end{bmatrix}$

5. $\begin{bmatrix} 1 & 0 \\ 0 & 1 \\ 0 & 0 \end{bmatrix} \begin{bmatrix} a & b \\ c & d \end{bmatrix}$

6. $\begin{bmatrix} a & b \\ c & d \end{bmatrix} \begin{bmatrix} d & -b \\ -c & a \end{bmatrix}$

7. $\begin{bmatrix} 1 & 0 & -1 \\ 0 & 1 & 1 \\ 1 & -1 & -2 \end{bmatrix} \begin{bmatrix} 1 & 2 & 3 \\ 3 & 2 & 1 \\ 2 & 1 & 3 \end{bmatrix}$

8. $\begin{bmatrix} 0 & 1 \\ 0 & 0 \end{bmatrix} \begin{bmatrix} 0 & 1 \\ 0 & 0 \end{bmatrix}$

9. $\begin{bmatrix} 1 & 2 \\ 2 & 4 \end{bmatrix} \begin{bmatrix} -6 & 8 \\ 3 & -4 \end{bmatrix}$

10. $\begin{bmatrix} 1 & 0 & -1 \end{bmatrix} \begin{bmatrix} 1 & 2 \\ 2 & 1 \\ 1 & 1 \end{bmatrix}$

11. $\begin{bmatrix} 1 & 2 & 3 \end{bmatrix} \begin{bmatrix} 3 \\ 2 \\ 1 \end{bmatrix}$

12. $\begin{bmatrix} 1 \\ 2 \\ 3 \end{bmatrix} \begin{bmatrix} 1 & 2 & 3 \end{bmatrix}$

13. $\begin{bmatrix} 0 & 0 & 1 \end{bmatrix} \begin{bmatrix} a & b & c \\ d & e & f \\ g & h & k \end{bmatrix} \begin{bmatrix} 0 \\ 1 \\ 0 \end{bmatrix}$

14. For the matrices

$$A = \begin{bmatrix} 1 & 1 \\ 1 & 1 \end{bmatrix}, \quad B = \begin{bmatrix} 1 & 2 & 3 \end{bmatrix},$$

$$C = \begin{bmatrix} 1 & 0 & -1 \\ 2 & 1 & 0 \\ 3 & 2 & 1 \end{bmatrix}, \quad D = \begin{bmatrix} 1 \\ 1 \\ 1 \end{bmatrix}, \quad E = \begin{bmatrix} 5 \end{bmatrix},$$

determine which of the 25 matrix products AA, AB, $AC, \ldots, ED$, EE are defined, and compute those that are defined.

Use the given partitions to compute the products in Exercises 15 and 16. Check your work by computing the same products without using a partition. Show all your work.

15. $\left[\begin{array}{cc|c} 1 & 0 & 0 \\ 0 & 1 & 0 \\ \hline 1 & 3 & 4 \end{array}\right] \left[\begin{array}{c|c} 1 & 0 \\ 2 & 0 \\ \hline 3 & 4 \end{array}\right]$

16. $\left[\begin{array}{cc|cc} 1 & 0 & 1 & 0 \\ 0 & 1 & 0 & 1 \\ \hline 0 & 0 & 1 & 0 \\ 0 & 0 & 0 & 1 \end{array}\right] \left[\begin{array}{cc|cc} 1 & 2 & 2 & 3 \\ 3 & 4 & 4 & 5 \\ \hline 0 & 0 & 1 & 2 \\ 0 & 0 & 3 & 4 \end{array}\right]$

In the Exercises 17 through 26, find all matrices that commute with the given matrix A.

17. $A = \begin{bmatrix} 1 & 0 \\ 0 & 2 \end{bmatrix}$

18. $A = \begin{bmatrix} 1 & 2 \\ 0 & 1 \end{bmatrix}$

19. $A = \begin{bmatrix} 0 & -2 \\ 2 & 0 \end{bmatrix}$

20. $A = \begin{bmatrix} 2 & 3 \\ -3 & 2 \end{bmatrix}$

21. $A = \begin{bmatrix} 1 & 2 \\ 2 & -1 \end{bmatrix}$

22. $A = \begin{bmatrix} 1 & 1 \\ 1 & 1 \end{bmatrix}$

23. $A = \begin{bmatrix} 1 & 3 \\ 2 & 6 \end{bmatrix}$

24. $A = \begin{bmatrix} 2 & 0 & 0 \\ 0 & 3 & 0 \\ 0 & 0 & 4 \end{bmatrix}$

25. $A = \begin{bmatrix} 2 & 0 & 0 \\ 0 & 3 & 0 \\ 0 & 0 & 2 \end{bmatrix}$

26. $A = \begin{bmatrix} 2 & 0 & 0 \\ 0 & 2 & 0 \\ 0 & 0 & 3 \end{bmatrix}$

27. Prove the *distributive laws* for matrices:

$$A(C + D) = AC + AD$$

and

$$(A + B)C = AC + BC.$$

28. Consider an $n \times p$ matrix A, a $p \times m$ matrix B, and a scalar k. Show that

$$(kA)B = A(kB) = k(AB).$$

29. Consider the matrix

$$D_\alpha = \begin{bmatrix} \cos\alpha & -\sin\alpha \\ \sin\alpha & \cos\alpha \end{bmatrix}.$$

We know that the linear transformation $T(\vec{x}) = D_\alpha \vec{x}$ is a counterclockwise rotation through an angle α.

a. For two angles, α and β, consider the products $D_\alpha D_\beta$ and $D_\beta D_\alpha$. Arguing geometrically, describe the linear transformations $\vec{y} = D_\alpha D_\beta \vec{x}$ and $\vec{y} = D_\beta D_\alpha \vec{x}$. Are the two transformations the same?

b. Now compute the products $D_\alpha D_\beta$ and $D_\beta D_\alpha$. Do the results make sense in terms of your answer in part (a)? Recall the trigonometric identities

$$\sin(\alpha \pm \beta) = \sin \alpha \cos \beta \pm \cos \alpha \sin \beta$$
$$\cos(\alpha \pm \beta) = \cos \alpha \cos \beta \mp \sin \alpha \sin \beta.$$

30. Consider the lines P and Q in $\mathbb{R}^2$ sketched below. Consider the linear transformation $T(\vec{x}) = \text{ref}_Q(\text{ref}_P(\vec{x}))$, that is, we first reflect $\vec{x}$ about P and then we reflect the result about Q.

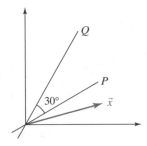

a. For the vector $\vec{x}$ given in the figure, sketch $T(\vec{x})$. What angle do the vectors $\vec{x}$ and $T(\vec{x})$ enclose? What is the relationship between the lengths of $\vec{x}$ and $T(\vec{x})$?

b. Use your answer in part (a) to describe the transformation T geometrically, as a reflection, rotation, shear, or projection.

c. Find the matrix of T.

d. Give a geometrical interpretation of the linear transformation $L(\vec{x}) = \text{ref}_P(\text{ref}_Q(\vec{x}))$, and find the matrix of L.

31. Consider two matrices A and B whose product AB is defined. Describe the ith row of the product AB in terms of the rows of A and the matrix B.

32. Find all 2×2 matrices X such that $AX = XA$ for all 2×2 matrices A.

For the matrices A in Exercises 33 through 42, compute $A^2 = AA$, $A^3 = AAA$, and A^4. Describe the pattern that emerges, and use this pattern to find $A^{1,001}$. Interpret your answers geometrically, in terms of rotations, reflections, shears, and orthogonal projections.

33. $\begin{bmatrix} -1 & 0 \\ 0 & -1 \end{bmatrix}$ 34. $\begin{bmatrix} 1 & 0 \\ 0 & -1 \end{bmatrix}$ 35. $\begin{bmatrix} 0 & 1 \\ 1 & 0 \end{bmatrix}$

36. $\begin{bmatrix} 1 & 1 \\ 0 & 1 \end{bmatrix}$ 37. $\begin{bmatrix} 1 & 0 \\ -1 & 1 \end{bmatrix}$ 38. $\begin{bmatrix} 0 & -1 \\ 1 & 0 \end{bmatrix}$

39. $\frac{1}{\sqrt{2}} \begin{bmatrix} 1 & 1 \\ -1 & 1 \end{bmatrix}$ 40. $\frac{1}{2} \begin{bmatrix} -1 & -\sqrt{3} \\ \sqrt{3} & -1 \end{bmatrix}$

41. $\frac{1}{\sqrt{2}} \begin{bmatrix} 1 & 1 \\ 1 & -1 \end{bmatrix}$ 42. $\frac{1}{2} \begin{bmatrix} 1 & 1 \\ 1 & 1 \end{bmatrix}$

In Exercises 43 through 48, find a 2×2 matrix A with the given properties. (Hint: It helps to think of geometrical examples.)

43. $A \neq I_2$, $A^2 = I_2$ 44. $A^2 \neq I_2$, $A^4 = I_2$

45. $A^2 \neq I_2$, $A^3 = I_2$

46. $A^2 = A$, all entries of A are nonzero.

47. $A^3 = A$, all entries of A are nonzero.

48. $A^{10} = \begin{bmatrix} 1 & 1 \\ 0 & 1 \end{bmatrix}$

In Exercises 49 through 54, consider the matrices

$$A = \begin{bmatrix} 0 & 1 \\ 1 & 0 \end{bmatrix}, \quad B = \begin{bmatrix} -1 & 0 \\ 0 & 1 \end{bmatrix}, \quad C = \begin{bmatrix} 1 & 0 \\ 0 & -1 \end{bmatrix},$$

$$D = \begin{bmatrix} 0 & -1 \\ -1 & 0 \end{bmatrix}, \quad E = \begin{bmatrix} 0.6 & 0.8 \\ 0.8 & -0.6 \end{bmatrix}, \quad F = \begin{bmatrix} 0 & -1 \\ 1 & 0 \end{bmatrix},$$

$$G = \begin{bmatrix} 0 & 1 \\ -1 & 0 \end{bmatrix}, \quad H = \begin{bmatrix} 0.8 & -0.6 \\ 0.6 & 0.8 \end{bmatrix}, \quad J = \begin{bmatrix} 1 & -1 \\ 1 & 1 \end{bmatrix}.$$

Compute the indicated products. Interpret these products geometrically, and draw composition diagrams, as in Example 2.

49. AF and FA 50. CG and CG

51. FJ and JF 52. JH and HJ

53. CD and DC 54. BE and EB.

In Exercises 55 through 64, find all matrices X that satisfy the given matrix equation.

55. $\begin{bmatrix} 1 & 2 \\ 2 & 4 \end{bmatrix} X = \begin{bmatrix} 0 & 0 \\ 0 & 0 \end{bmatrix}$

56. $X \begin{bmatrix} 2 & 1 \\ 4 & 2 \end{bmatrix} = \begin{bmatrix} 0 & 0 \\ 0 & 0 \end{bmatrix}$ 57. $\begin{bmatrix} 1 & 2 \\ 3 & 5 \end{bmatrix} X = I_2$

58. $X \begin{bmatrix} 1 & 2 \\ 3 & 5 \end{bmatrix} = I_2$ 59. $X \begin{bmatrix} 2 & 1 \\ 4 & 2 \end{bmatrix} = I_2$

60. $\begin{bmatrix} 1 & 2 \\ 2 & 4 \end{bmatrix} X = I_2$ 61. $\begin{bmatrix} 1 & 2 & 3 \\ 0 & 1 & 2 \end{bmatrix} X = I_2$

62. $\begin{bmatrix} 1 & 2 & 3 \\ 4 & 5 & 6 \end{bmatrix} X = I_2$ 63. $\begin{bmatrix} 1 & 4 \\ 2 & 5 \\ 3 & 6 \end{bmatrix} X = I_3$

64. $\begin{bmatrix} 1 & 0 \\ 2 & 1 \\ 3 & 2 \end{bmatrix} X = I_3$

65. Find all upper triangular 2×2 matrices X such that X^2 is the zero matrix.

66. Find all lower triangular 3×3 matrices X such that X^3 is the zero matrix.

67. Consider any 2×2 matrix A that represents a horizontal or vertical shear. Compute $(A - I_2)^2$. Explain your answer geometrically: If $\vec{x}$ is any vector in $\mathbb{R}^2$, what is $(A - I_2)^2 \vec{x} = A^2 \vec{x} - 2A\vec{x} + \vec{x}$?

68. Consider an $n \times m$ matrix A of rank n. Show that there exists an $m \times n$ matrix X such that $AX = I_n$. If $n < m$, how many such matrices X are there?

69. Consider an $n \times n$ matrix A of rank n. How many $n \times n$ matrices X are there such that $AX = I_n$?

2.4 The Inverse of a Linear Transformation

Let's first review the concept of an invertible function. As you read these abstract definitions, consider the examples in Figures 1 and 2, where X and Y are finite sets.

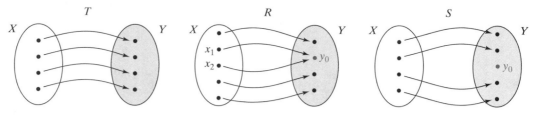

Figure 1 T is invertible. R is not invertible: The equation $R(x) = y_0$ has two solutions, x_1 and x_2.
S is not invertible: There is no x such that $S(x) = y_0$.

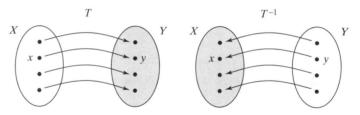

Figure 2 A function T and its inverse T^{-1}.

Definition 2.4.1   Invertible Functions

A function T from X to Y is called invertible if the equation $T(x) = y$ has a unique solution x in X for each y in Y.

In this case, the inverse T^{-1} from Y to X is defined by

$$T^{-1}(y) = \big(\text{the unique } x \text{ in } X \text{ such that } T(x) = y\big).$$

To put it differently, the equation

$$x = T^{-1}(y) \qquad \text{means that} \qquad y = T(x).$$

Note that

$$T^{-1}\big(T(x)\big) = x \qquad \text{and} \qquad T\big(T^{-1}(y)\big) = y$$

for all x in X and for all y in Y.

Conversely, if L is a function from Y to X such that

$$L\big(T(x)\big) = x \qquad \text{and} \qquad T\big(L(y)\big) = y$$

for all x in X and for all y in Y, then T is invertible and $T^{-1} = L$.

If a function T is invertible, then so is T^{-1} and $(T^{-1})^{-1} = T$.

If a function is given by a formula, we may be able to find the inverse by solving the formula for the input variable(s). For example, the inverse of the function

$$y = \frac{x^3 - 1}{5} \quad \text{(from } \mathbb{R} \text{ to } \mathbb{R})$$

is

$$x = \sqrt[3]{5y + 1}.$$

Now consider the case of a linear transformation T from $\mathbb{R}^n$ to $\mathbb{R}^n$ given by

$$\vec{y} = T(\vec{x}) = A\vec{x},$$

where A is an $n \times n$ matrix. (The case of an $n \times m$ matrix will be discussed in Exercise 48.)

According to Definition 2.4.1, the linear transformation $\vec{y} = T(\vec{x}) = A\vec{x}$ is invertible if the *linear system*

$$A\vec{x} = \vec{y}$$

has a unique solution $\vec{x}$ in $\mathbb{R}^n$ for all $\vec{y}$ in the vector space $\mathbb{R}^n$. By Theorem 1.3.4, this is the case if (and only if) rank$(A) = n$, or equivalently, if

$$\text{rref}(A) = \begin{bmatrix} 1 & 0 & 0 & \cdots & 0 \\ 0 & 1 & 0 & \cdots & 0 \\ 0 & 0 & 1 & \cdots & 0 \\ \vdots & \vdots & \vdots & \ddots & \vdots \\ 0 & 0 & 0 & \cdots & 1 \end{bmatrix} = I_n.$$

Definition 2.4.2 Invertible matrices

A square matrix A is said to be *invertible* if the linear transformation $\vec{y} = T(\vec{x}) = A\vec{x}$ is invertible. In this case, the matrix[9] of T^{-1} is denoted by A^{-1}. If the linear transformation $\vec{y} = T(\vec{x}) = A\vec{x}$ is invertible, then its inverse is $\vec{x} = T^{-1}(\vec{y}) = A^{-1}\vec{y}$.

Theorem 2.4.3 Invertibility

An $n \times n$ matrix A is invertible if (and only if)

$$\text{rref}(A) = I_n$$

or, equivalently, if

$$\text{rank}(A) = n.$$ ∎

The following proposition follows directly from Theorem 1.3.4 and Example 1.3.3d.

Theorem 2.4.4 Invertibility and linear systems

Let A be an $n \times n$ matrix.

a. Consider a vector $\vec{b}$ in $\mathbb{R}^n$. If A is invertible, then the system $A\vec{x} = \vec{b}$ has the unique solution $\vec{x} = A^{-1}\vec{b}$. If A is noninvertible, then the system $A\vec{x} = \vec{b}$ has infinitely many solutions or none.

---

[9]The inverse transformation is linear. (See Exercise 2.2.29.)

b. Consider the special case when $\vec{b} = \vec{0}$. The system $A\vec{x} = \vec{0}$ has $\vec{x} = \vec{0}$ as a solution. If A is invertible, then this is the only solution. If A is noninvertible, then the system $A\vec{x} = \vec{0}$ has infinitely many solutions. ■

EXAMPLE 1 Is the matrix

$$A = \begin{bmatrix} 1 & 1 & 1 \\ 2 & 3 & 2 \\ 3 & 8 & 2 \end{bmatrix}$$

invertible?

Solution

$$\begin{bmatrix} 1 & 1 & 1 \\ 2 & 3 & 2 \\ 3 & 8 & 2 \end{bmatrix} \begin{matrix} \\ -2(I) \\ -3(I) \end{matrix} \rightarrow \begin{bmatrix} 1 & 1 & 1 \\ 0 & 1 & 0 \\ 0 & 5 & -1 \end{bmatrix} \begin{matrix} -(II) \\ \\ -5(II) \end{matrix} \rightarrow$$

$$\begin{bmatrix} 1 & 0 & 1 \\ 0 & 1 & 0 \\ 0 & 0 & -1 \end{bmatrix} \begin{matrix} \\ \\ \div(-1) \end{matrix} \rightarrow \begin{bmatrix} 1 & 0 & 1 \\ 0 & 1 & 0 \\ 0 & 0 & 1 \end{bmatrix} \begin{matrix} -(III) \\ \\ \end{matrix} \rightarrow \begin{bmatrix} 1 & 0 & 0 \\ 0 & 1 & 0 \\ 0 & 0 & 1 \end{bmatrix} = I_3 = \text{rref}(A)$$

Matrix A is invertible since $\text{rref}(A) = I_3$. ■

Let's find the inverse of the matrix

$$A = \begin{bmatrix} 1 & 1 & 1 \\ 2 & 3 & 2 \\ 3 & 8 & 2 \end{bmatrix}$$

in Example 1, or, equivalently, the inverse of the linear transformation

$$\vec{y} = A\vec{x} \qquad \text{or} \qquad \begin{bmatrix} y_1 \\ y_2 \\ y_3 \end{bmatrix} = \begin{bmatrix} x_1 + x_2 + x_3 \\ 2x_1 + 3x_2 + 2x_3 \\ 3x_1 + 8x_2 + 2x_3 \end{bmatrix}.$$

To find the inverse transformation, we solve this system for the input variables x_1, x_2, and x_3:

$$\begin{vmatrix} x_1 + x_2 + x_3 = & y_1 & \\ 2x_1 + 3x_2 + 2x_3 = & y_2 & \\ 3x_1 + 8x_2 + 2x_3 = & & y_3 \end{vmatrix} \begin{matrix} \longrightarrow \\ -2 \,(I) \\ -3 \,(I) \end{matrix}$$

$$\begin{vmatrix} x_1 + x_2 + x_3 = & y_1 & \\ x_2 & = -2y_1 + y_2 & \\ 5x_2 - x_3 = -3y_1 & + y_3 \end{vmatrix} \begin{matrix} -\,(II) \\ \longrightarrow \\ -5\,(II) \end{matrix}$$

$$\begin{vmatrix} x_1 & + x_3 = & 3y_1 - y_2 & \\ x_2 & = -2y_1 + y_2 & \\ & - x_3 = & 7y_1 - 5y_2 + y_3 \end{vmatrix} \begin{matrix} \longrightarrow \\ \\ \div(-1) \end{matrix}$$

$$\begin{vmatrix} x_1 & + x_3 = & 3y_1 - y_2 & \\ x_2 & = -2y_1 + y_2 & \\ & x_3 = -7y_1 + 5y_2 - y_3 \end{vmatrix} \begin{matrix} -\,(III) \\ \longrightarrow \\ \end{matrix}$$

$$\begin{vmatrix} x_1 & = & 10y_1 - 6y_2 + y_3 \\ x_2 & = -2y_1 + y_2 \\ & x_3 = -7y_1 + 5y_2 - y_3 \end{vmatrix}.$$

We have found the inverse transformation; its matrix is

$$B = A^{-1} = \begin{bmatrix} 10 & -6 & 1 \\ -2 & 1 & 0 \\ -7 & 5 & -1 \end{bmatrix}.$$

We can write the preceding computations in matrix form:

$$\begin{bmatrix} 1 & 1 & 1 & \vdots & 1 & 0 & 0 \\ 2 & 3 & 2 & \vdots & 0 & 1 & 0 \\ 3 & 8 & 2 & \vdots & 0 & 0 & 1 \end{bmatrix} \begin{matrix} \\ -2\,(I) \\ -3\,(I) \end{matrix} \longrightarrow \begin{bmatrix} 1 & 1 & 1 & \vdots & 1 & 0 & 0 \\ 0 & 1 & 0 & \vdots & -2 & 1 & 0 \\ 0 & 5 & -1 & \vdots & -3 & 0 & 1 \end{bmatrix} \begin{matrix} -(II) \\ \longrightarrow \\ -5\,(II) \end{matrix}$$

$$\begin{bmatrix} 1 & 0 & 1 & \vdots & 3 & -1 & 0 \\ 0 & 1 & 0 & \vdots & -2 & 1 & 0 \\ 0 & 0 & -1 & \vdots & 7 & -5 & 1 \end{bmatrix} \begin{matrix} \\ \\ \div(-1) \end{matrix} \longrightarrow \begin{bmatrix} 1 & 0 & 1 & \vdots & 3 & -1 & 0 \\ 0 & 1 & 0 & \vdots & -2 & 1 & 0 \\ 0 & 0 & 1 & \vdots & -7 & 5 & -1 \end{bmatrix} \begin{matrix} -(III) \\ \longrightarrow \\ \end{matrix}$$

$$\begin{bmatrix} 1 & 0 & 0 & \vdots & 10 & -6 & 1 \\ 0 & 1 & 0 & \vdots & -2 & 1 & 0 \\ 0 & 0 & 1 & \vdots & -7 & 5 & -1 \end{bmatrix}.$$

This process can be described succinctly as follows.

Theorem 2.4.5   **Finding the inverse of a matrix**

To find the *inverse* of an $n \times n$ matrix A, form the $n \times (2n)$ matrix $\begin{bmatrix} A & \vdots & I_n \end{bmatrix}$ and compute rref $\begin{bmatrix} A & \vdots & I_n \end{bmatrix}$.

- If rref $\begin{bmatrix} A & \vdots & I_n \end{bmatrix}$ is of the form $\begin{bmatrix} I_n & \vdots & B \end{bmatrix}$, then A is invertible, and $A^{-1} = B$.
- If rref $\begin{bmatrix} A & \vdots & I_n \end{bmatrix}$ is of another form (i.e., its left half fails to be I_n), then A is not invertible. Note that the left half of rref $\begin{bmatrix} A & \vdots & I_n \end{bmatrix}$ is rref(A).   ∎

Next let's discuss some algebraic rules for matrix inversion.

- Consider an invertible linear transformation $T(\vec{x}) = A\vec{x}$ from $\mathbb{R}^n$ to $\mathbb{R}^n$. By Definition 2.4.1, the equation $T^{-1}(T(\vec{x})) = \vec{x}$ holds for all $\vec{x}$ in $\mathbb{R}^n$. Written in matrix form, this equation reads $A^{-1}A\vec{x} = \vec{x} = I_n\vec{x}$. It follows that $A^{-1}A = I_n$. Likewise we can show that $AA^{-1} = I_n$.

Theorem 2.4.6   **Multiplying with the inverse**

For an invertible $n \times n$ matrix A,

$$A^{-1}A = I_n \qquad \text{and} \qquad AA^{-1} = I_n.$$ ∎

- If A and B are invertible $n \times n$ matrices, is BA invertible as well? If so, what is its inverse?
  To find the inverse of the linear transformation

$$\vec{y} = BA\vec{x},$$

we solve the equation for $\vec{x}$ in two steps. First, we multiply both sides of the equation by B^{-1} from the left:

$$B^{-1}\vec{y} = B^{-1}BA\vec{x} = I_n A\vec{x} = A\vec{x}.$$

Now, we multiply by A^{-1} from the left:

$$A^{-1}B^{-1}\vec{y} = A^{-1}A\vec{x} = \vec{x}.$$

This computation shows that the linear transformation

$$\vec{y} = BA\vec{x}$$

is invertible and that its inverse is

$$\vec{x} = A^{-1}B^{-1}\vec{y}.$$

Theorem 2.4.7   The inverse of a product of matrices

If A and B are invertible $n \times n$ matrices, then BA is invertible as well, and

$$(BA)^{-1} = A^{-1}B^{-1}.$$

Pay attention to the order of the matrices. (Order matters!)

To verify this result, we can multiply $A^{-1}B^{-1}$ by BA (in either order), and check that the result is I_n:

$$BAA^{-1}B^{-1} = BI_nB^{-1} = BB^{-1} = I_n, \text{ and}$$
$$A^{-1}B^{-1}BA = A^{-1}A = I_n.$$

Everything works out!

To understand the order of the factors in the formula $(BA)^{-1} = A^{-1}B^{-1}$, think about our French coast guard story again.

To recover the actual position $\vec{x}$ from the doubly encoded position $\vec{z}$, you *first* apply the decoding transformation $\vec{y} = B^{-1}\vec{z}$ and *then* the decoding transformation $\vec{x} = A^{-1}\vec{y}$. The inverse of $\vec{z} = BA\vec{x}$ is therefore $\vec{x} = A^{-1}B^{-1}\vec{z}$, as illustrated in Figure 3.

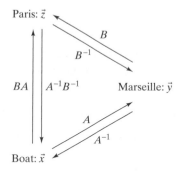

Figure 3

The following result is often useful in finding inverses:

Theorem 2.4.8   A criterion for invertibility

Let A and B be two $n \times n$ matrices such that

$$BA = I_n.$$

Then

    a. A and B are both invertible,

    b. $A^{-1} = B$ and $B^{-1} = A$, and

    c. $AB = I_n$.

It follows from the definition of an invertible function that if $AB = I_n$ *and* $BA = I_n$, then A and B are inverses, that is, $A = B^{-1}$ and $B = A^{-1}$. Theorem 2.4.8 makes the point that the equation $BA = I_n$ *alone* guarantees that A and B are inverses. Exercise 107 illustrates the significance of this claim.

Proof   To demonstrate that A is invertible, it suffices to show that the linear system $A\vec{x} = \vec{0}$ has only the solution $\vec{x} = \vec{0}$ (by Theorem 2.4.4b). If we multiply the equation $A\vec{x} = \vec{0}$ by B from the left, we find that $BA\vec{x} = B\vec{0} = \vec{0}$. It follows that $\vec{x} = I_n\vec{x} = BA\vec{x} = \vec{0}$, as claimed. Therefore, A is invertible. If we multiply the equation $BA = I_n$ by A^{-1} from the right, we find that $B = A^{-1}$. Matrix B, being the inverse of A, is itself invertible, and $B^{-1} = (A^{-1})^{-1} = A$. (See Definition 2.4.1.) Finally, $AB = AA^{-1} = I_n$.

You can use Theorem 2.4.8 to check your work when computing the inverse of a matrix. Earlier in this section we claimed that

$$B = \begin{bmatrix} 10 & -6 & 1 \\ -2 & 1 & 0 \\ -7 & 5 & -1 \end{bmatrix} \quad \text{is the inverse of} \quad A = \begin{bmatrix} 1 & 1 & 1 \\ 2 & 3 & 2 \\ 3 & 8 & 2 \end{bmatrix}.$$

Let's use Theorem 2.4.8b to check our work:

$$BA = \begin{bmatrix} 10 & -6 & 1 \\ -2 & 1 & 0 \\ -7 & 5 & -1 \end{bmatrix} \begin{bmatrix} 1 & 1 & 1 \\ 2 & 3 & 2 \\ 3 & 8 & 2 \end{bmatrix} = \begin{bmatrix} 1 & 0 & 0 \\ 0 & 1 & 0 \\ 0 & 0 & 1 \end{bmatrix} = I_3. \qquad \blacksquare$$

EXAMPLE 2   Suppose A, B, and C are three $n \times n$ matrices such that $ABC = I_n$. Show that B is invertible, and express B^{-1} in terms of A and C.

Solution

Write $ABC = (AB)C = I_n$. We have $C(AB) = I_n$, by Theorem 2.4.8c. Since matrix multiplication is associative, we can write $(CA)B = I_n$. Applying Theorem 2.4.8 again, we conclude that B is invertible, and $B^{-1} = CA$. $\qquad \blacksquare$

EXAMPLE 3   For an arbitrary 2×2 matrix $A = \begin{bmatrix} a & b \\ c & d \end{bmatrix}$, compute the product $\begin{bmatrix} d & -b \\ -c & a \end{bmatrix} \begin{bmatrix} a & b \\ c & d \end{bmatrix}$. When is A invertible? If so, what is A^{-1}?

Solution

$$\begin{bmatrix} d & -b \\ -c & a \end{bmatrix} \begin{bmatrix} a & b \\ c & d \end{bmatrix} = \begin{bmatrix} ad - bc & 0 \\ 0 & ad - bc \end{bmatrix} = (ad - bc)I_2.$$

If $ad - bc \neq 0$, we can write $\underbrace{\left(\dfrac{1}{ad - bc} \begin{bmatrix} d & -b \\ -c & a \end{bmatrix} \right)}_{B} \underbrace{\begin{bmatrix} a & b \\ c & d \end{bmatrix}}_{A} = I_2.$

It now follows from Theorem 2.4.8 that A is invertible, with $A^{-1} = \dfrac{1}{ad - bc} \begin{bmatrix} d & -b \\ -c & a \end{bmatrix}$. Conversely, if A is invertible, then we can multiply the equation $\begin{bmatrix} d & -b \\ -c & a \end{bmatrix} \begin{bmatrix} a & b \\ c & d \end{bmatrix} = (ad - bc)I_2$ with A^{-1} from the right, finding $\begin{bmatrix} d & -b \\ -c & a \end{bmatrix} = (ad - bc)A^{-1}$. Since some of the scalars a, b, c, d are nonzero (being the entries of the invertible matrix A), it follows that $ad - bc \neq 0$. $\qquad \blacksquare$

Theorem 2.4.9 Inverse and determinant of a 2 × 2 matrix

a. The 2 × 2 matrix

$$A = \begin{bmatrix} a & b \\ c & d \end{bmatrix}$$

is invertible if (and only if) $ad - bc \neq 0$.

Quantity $ad - bc$ is called the *determinant* of A, written $\det(A)$:

$$\det(A) = \det \begin{bmatrix} a & b \\ c & d \end{bmatrix} = ad - bc.$$

b. If

$$A = \begin{bmatrix} a & b \\ c & d \end{bmatrix}$$

is invertible, then

$$\begin{bmatrix} a & b \\ c & d \end{bmatrix}^{-1} = \frac{1}{ad - bc} \begin{bmatrix} d & -b \\ -c & a \end{bmatrix} = \frac{1}{\det(A)} \begin{bmatrix} d & -b \\ -c & a \end{bmatrix}. \qquad \blacksquare$$

In Chapter 6 we will introduce the determinant of a square matrix of arbitrary size, and we will generalize the results of Theorem 2.4.9 to $n \times n$ matrices. See Theorems 6.2.4 and 6.3.9.

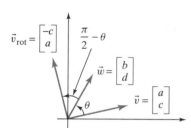

Figure 4

What is the geometrical interpretation of the determinant of a 2 × 2 matrix A? Write $A = \begin{bmatrix} a & b \\ c & d \end{bmatrix}$, and consider the column vectors $\vec{v} = \begin{bmatrix} a \\ c \end{bmatrix}$ and $\vec{w} = \begin{bmatrix} b \\ d \end{bmatrix}$. It turns out to be helpful to introduce the auxiliary vector $\vec{v}_{rot} = \begin{bmatrix} -c \\ a \end{bmatrix}$, obtained by rotating vector $\vec{v} = \begin{bmatrix} a \\ c \end{bmatrix}$ through an angle of $\frac{\pi}{2}$. Let θ be the (oriented) angle from $\vec{v}$ to $\vec{w}$, with $-\pi < \theta \leq \pi$. See Figure 4. Then

$$\det A = ad - bc \underset{\text{Step 2}}{=} \vec{v}_{rot} \cdot \vec{w} \underset{\text{Step 3}}{=} \|\vec{v}_{rot}\| \cos\left(\frac{\pi}{2} - \theta\right) \|\vec{w}\| = \|\vec{v}\| \sin\theta \|\vec{w}\|.$$

In Steps 2 and 3 we use the definition of the dot product and its geometrical interpretation. See Definition A.4 in the Appendix.

Theorem 2.4.10 Geometrical interpretation of the determinant of a 2 × 2 matrix

If $A = \begin{bmatrix} \vec{v} & \vec{w} \end{bmatrix}$ is a 2 × 2 matrix with nonzero columns $\vec{v}$ and $\vec{w}$, then

$$\det A = \det \begin{bmatrix} \vec{v} & \vec{w} \end{bmatrix} = \|\vec{v}\| \sin\theta \|\vec{w}\|,$$

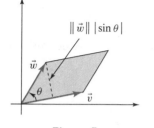

Figure 5

where θ is the oriented angle from $\vec{v}$ to $\vec{w}$, with $-\pi < \theta \leq \pi$. It follows that

- $|\det A| = \|\vec{v}\| \, |\sin\theta| \, \|\vec{w}\|$ is the *area of the parallelogram* spanned by $\vec{v}$ and $\vec{w}$ (see Figure 5),
- $\det A = 0$ if $\vec{v}$ and $\vec{w}$ are *parallel*, meaning that $\theta = 0$ or $\theta = \pi$,
- $\det A > 0$ if $0 < \theta < \pi$, and
- $\det A < 0$ if $-\pi < \theta < 0$. $\qquad \blacksquare$

In Chapter 6 we will go a step further and interpret $\det A$ in terms of the linear transformation $T(\vec{x}) = A\vec{x}$.

EXAMPLE 4 Is the matrix $A = \begin{bmatrix} 1 & 3 \\ 2 & 1 \end{bmatrix}$ invertible? If so, find the inverse. Interpret $\det A$ geometrically.

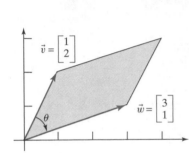

Figure 6

Solution

We find the determinant $\det(A) = 1 \cdot 1 - 3 \cdot 2 = -5 \neq 0$, so that A is indeed invertible, by Theorem 2.4.9a. Then

$$A^{-1} = \frac{1}{\det A} \begin{bmatrix} d & -b \\ -c & a \end{bmatrix} = \frac{1}{(-5)} \begin{bmatrix} 1 & -3 \\ -2 & 1 \end{bmatrix} = \begin{bmatrix} -\frac{1}{5} & \frac{3}{5} \\ \frac{2}{5} & -\frac{1}{5} \end{bmatrix},$$

by Theorem 2.4.9b.

Furthermore, $|\det A| = 5$ is the area of the shaded parallelogram in Figure 6, and $\det A$ is negative since the angle θ from $\vec{v}$ to $\vec{w}$ is negative. ■

EXAMPLE 5 For which values of the constant k is the matrix $A = \begin{bmatrix} 1-k & 2 \\ 4 & 3-k \end{bmatrix}$ invertible?

Solution

By Theorem 2.4.9a, the matrix A fails to be invertible if $\det A = 0$. Now

$$\det A = \det \begin{bmatrix} 1-k & 2 \\ 4 & 3-k \end{bmatrix} = (1-k)(3-k) - 2 \cdot 4$$
$$= k^2 - 4k - 5 = (k-5)(k+1) = 0$$

when $k = 5$ or $k = -1$. Thus A is invertible for all values of k *except* $k = 5$ and $k = -1$. ■

EXAMPLE 6 Consider a matrix A that represents the reflection about a line L in the plane. Use the determinant to verify that A is invertible. Find A^{-1}. Explain your answer conceptually, and interpret the determinant geometrically.

Solution

By Definition 2.2.2, a reflection matrix is of the form $A = \begin{bmatrix} a & b \\ b & -a \end{bmatrix}$, where $a^2 + b^2 = 1$. Now $\det A = \det \begin{bmatrix} a & b \\ b & -a \end{bmatrix} = -a^2 - b^2 = -1$. It turns out that A is invertible, and $A^{-1} = \frac{1}{(-1)} \begin{bmatrix} -a & -b \\ -b & a \end{bmatrix} = \begin{bmatrix} a & b \\ b & -a \end{bmatrix} = A$. It makes good sense that A is its own inverse, since $A(A\vec{x}) = \vec{x}$ for all $\vec{x}$ in $\mathbb{R}^2$, by definition of a reflection. See Figure 7.

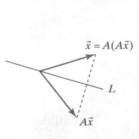

Figure 7

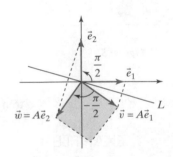

Figure 8

To interpret the determinant geometrically, recall that $\vec{v} = \begin{bmatrix} a \\ b \end{bmatrix} = A\vec{e}_1$ and $\vec{w} = \begin{bmatrix} b \\ -a \end{bmatrix} = A\vec{e}_2$. The parallelogram spanned by $\vec{v}$ and $\vec{w}$ is actually a unit square, with area $1 = |\det A|$, and θ is $-\dfrac{\pi}{2}$ since the reflection about L reverses the orientation of an angle. See Figure 8. ∎

The Inverse of a Block Matrix (Optional)

We will conclude this chapter with two examples involving block matrices. To refresh your memory, take another look at Theorem 2.3.9.

EXAMPLE 7 Let A be a block matrix

$$A = \begin{bmatrix} A_{11} & A_{12} \\ 0 & A_{22} \end{bmatrix},$$

where A_{11} is an $n \times n$ matrix, A_{22} is an $m \times m$ matrix, and A_{12} is an $n \times m$ matrix.

a. For which choices of A_{11}, A_{12}, and A_{22} is A invertible?

b. If A is invertible, what is A^{-1} (in terms of A_{11}, A_{12}, A_{22})?

Solution

We are looking for an $(n + m) \times (n + m)$ matrix B such that

$$BA = I_{n+m} = \begin{bmatrix} I_n & 0 \\ 0 & I_m \end{bmatrix}.$$

Let us partition B in the same way as A:

$$B = \begin{bmatrix} B_{11} & B_{12} \\ B_{21} & B_{22} \end{bmatrix},$$

where B_{11} is $n \times n$, B_{22} is $m \times m$, and so on. The fact that B is the inverse of A means that

$$\begin{bmatrix} B_{11} & B_{12} \\ B_{21} & B_{22} \end{bmatrix}\begin{bmatrix} A_{11} & A_{12} \\ 0 & A_{22} \end{bmatrix} = \begin{bmatrix} I_n & 0 \\ 0 & I_m \end{bmatrix},$$

or, using Theorem 2.3.9,

$$\begin{vmatrix} B_{11}A_{11} = I_n \\ B_{11}A_{12} + B_{12}A_{22} = 0 \\ B_{21}A_{11} = 0 \\ B_{21}A_{12} + B_{22}A_{22} = I_m \end{vmatrix}.$$

We have to solve for the blocks B_{ij}. Applying Theorem 2.4.8 to the equation $B_{11}A_{11} = I_n$, we find that A_{11} is invertible, and $B_{11} = A_{11}^{-1}$. Equation 3 now implies that $B_{21} = 0A_{11}^{-1} = 0$. Next, Equation 4 simplifies to $B_{22}A_{22} = I_m$. By Theorem 2.4.8, A_{22} is invertible, and $B_{22} = A_{22}^{-1}$. Lastly, Equation 2 becomes $A_{11}^{-1}A_{12} + B_{12}A_{22} = 0$, or $B_{12}A_{22} = -A_{11}^{-1}A_{12}$, or $B_{12} = -A_{11}^{-1}A_{12}A_{22}^{-1}$. We conclude that

a. A is invertible if (and only if) both A_{11} and A_{22} are invertible (no condition is imposed on A_{12}), and

b. If A is invertible, then its inverse is

$$A^{-1} = \begin{bmatrix} A_{11}^{-1} & -A_{11}^{-1}A_{12}A_{22}^{-1} \\ 0 & A_{22}^{-1} \end{bmatrix}. \qquad ∎$$

Verify this result for the following example:

EXAMPLE 8

$$\begin{bmatrix} 1 & 1 & | & 1 & 2 & 3 \\ 1 & 2 & | & 4 & 5 & 6 \\ \hline 0 & 0 & | & 1 & 0 & 0 \\ 0 & 0 & | & 0 & 1 & 0 \\ 0 & 0 & | & 0 & 0 & 1 \end{bmatrix}^{-1} = \begin{bmatrix} 2 & -1 & | & 2 & 1 & 0 \\ -1 & 1 & | & -3 & -3 & -3 \\ \hline 0 & 0 & | & 1 & 0 & 0 \\ 0 & 0 & | & 0 & 1 & 0 \\ 0 & 0 & | & 0 & 0 & 1 \end{bmatrix}$$ ∎

EXERCISES 2.4

GOAL   *Apply the concept of an invertible function. Determine whether a matrix (or a linear transformation) is invertible, and find the inverse if it exists.*

Decide whether the matrices in Exercises 1 through 15 are invertible. If they are, find the inverse. Do the computations with paper and pencil. Show all your work.

1. $\begin{bmatrix} 2 & 3 \\ 5 & 8 \end{bmatrix}$

2. $\begin{bmatrix} 1 & 1 \\ 1 & 1 \end{bmatrix}$

3. $\begin{bmatrix} 0 & 2 \\ 1 & 1 \end{bmatrix}$

4. $\begin{bmatrix} 1 & 2 & 1 \\ 1 & 3 & 2 \\ 1 & 0 & 1 \end{bmatrix}$

5. $\begin{bmatrix} 1 & 2 & 2 \\ 1 & 3 & 1 \\ 1 & 1 & 3 \end{bmatrix}$

6. $\begin{bmatrix} 1 & 2 & 3 \\ 0 & 1 & 2 \\ 0 & 0 & 1 \end{bmatrix}$

7. $\begin{bmatrix} 1 & 2 & 3 \\ 0 & 0 & 2 \\ 0 & 0 & 3 \end{bmatrix}$

8. $\begin{bmatrix} 0 & 0 & 1 \\ 0 & 1 & 0 \\ 1 & 0 & 0 \end{bmatrix}$

9. $\begin{bmatrix} 1 & 1 & 1 \\ 1 & 1 & 1 \\ 1 & 1 & 1 \end{bmatrix}$

10. $\begin{bmatrix} 1 & 1 & 1 \\ 1 & 2 & 3 \\ 1 & 3 & 6 \end{bmatrix}$

11. $\begin{bmatrix} 1 & 0 & 1 \\ 0 & 1 & 0 \\ 0 & 0 & 1 \end{bmatrix}$

12. $\begin{bmatrix} 1 & 1 & 2 & 3 \\ 0 & -1 & 0 & 0 \\ 2 & 2 & 5 & 4 \\ 0 & 3 & 0 & 1 \end{bmatrix}$

13. $\begin{bmatrix} 1 & 0 & 0 & 0 \\ 2 & 1 & 0 & 0 \\ 3 & 2 & 1 & 0 \\ 4 & 3 & 2 & 1 \end{bmatrix}$

14. $\begin{bmatrix} 2 & 5 & 0 & 0 \\ 1 & 3 & 0 & 0 \\ 0 & 0 & 1 & 2 \\ 0 & 0 & 2 & 5 \end{bmatrix}$

15. $\begin{bmatrix} 1 & 2 & 3 & 4 \\ 2 & 4 & 7 & 11 \\ 3 & 7 & 14 & 25 \\ 4 & 11 & 25 & 50 \end{bmatrix}$

Decide whether the linear transformations in Exercises 16 through 20 are invertible. Find the inverse transformation if it exists. Do the computations with paper and pencil. Show all your work.

16. $\begin{aligned} y_1 &= 3x_1 + 5x_2, \\ y_2 &= 5x_1 + 8x_2 \end{aligned}$

17. $\begin{aligned} y_1 &= x_1 + 2x_2, \\ y_2 &= 4x_1 + 8x_2 \end{aligned}$

18. $\begin{aligned} y_1 &= x_2, \\ y_2 &= x_3, \\ y_3 &= x_1 \end{aligned}$

19. $\begin{aligned} y_1 &= x_1 + x_2 + x_3, \\ y_2 &= x_1 + 2x_2 + 3x_3, \\ y_3 &= x_1 + 4x_2 + 9x_3 \end{aligned}$

20. $\begin{aligned} y_1 &= x_1 + 3x_2 + 3x_3, \\ y_2 &= x_1 + 4x_2 + 8x_3, \\ y_3 &= 2x_1 + 7x_2 + 12x_3 \end{aligned}$

Which of the functions f from $\mathbb{R}$ to $\mathbb{R}$ in Exercises 21 through 24 are invertible?

21. $f(x) = x^2$

22. $f(x) = 2^x$

23. $f(x) = x^3 + x$

24. $f(x) = x^3 - x$

Which of the (nonlinear) transformations from $\mathbb{R}^2$ to $\mathbb{R}^2$ in Exercises 25 through 27 are invertible? Find the inverse if it exists.

25. $\begin{bmatrix} y_1 \\ y_2 \end{bmatrix} = \begin{bmatrix} x_1^3 \\ x_2 \end{bmatrix}$

26. $\begin{bmatrix} y_1 \\ y_2 \end{bmatrix} = \begin{bmatrix} x_2 \\ x_1^3 + x_2 \end{bmatrix}$

27. $\begin{bmatrix} y_1 \\ y_2 \end{bmatrix} = \begin{bmatrix} x_1 + x_2 \\ x_1 \cdot x_2 \end{bmatrix}$

28. Find the inverse of the linear transformation

$$T\begin{bmatrix} x_1 \\ x_2 \\ x_3 \\ x_4 \end{bmatrix} = x_1 \begin{bmatrix} 22 \\ -16 \\ 8 \\ 5 \end{bmatrix} + x_2 \begin{bmatrix} 13 \\ -3 \\ 9 \\ 4 \end{bmatrix}$$
$$+ x_3 \begin{bmatrix} 8 \\ -2 \\ 7 \\ 3 \end{bmatrix} + x_4 \begin{bmatrix} 3 \\ -2 \\ 2 \\ 1 \end{bmatrix}$$

from $\mathbb{R}^4$ to $\mathbb{R}^4$.

29. For which values of the constant k is the following matrix invertible?

$$\begin{bmatrix} 1 & 1 & 1 \\ 1 & 2 & k \\ 1 & 4 & k^2 \end{bmatrix}$$

30. For which values of the constants b and c is the following matrix invertible?

$$\begin{bmatrix} 0 & 1 & b \\ -1 & 0 & c \\ -b & -c & 0 \end{bmatrix}$$

31. For which values of the constants a, b, and c is the following matrix invertible?

$$\begin{bmatrix} 0 & a & b \\ -a & 0 & c \\ -b & -c & 0 \end{bmatrix}$$

32. Find all matrices $\begin{bmatrix} a & b \\ c & d \end{bmatrix}$ such that $ad - bc = 1$ and $A^{-1} = A$.

33. Consider the matrices of the form $A = \begin{bmatrix} a & b \\ b & -a \end{bmatrix}$, where a and b are arbitrary constants. For which values of a and b is $A^{-1} = A$?

34. Consider the diagonal matrix

$$A = \begin{bmatrix} a & 0 & 0 \\ 0 & b & 0 \\ 0 & 0 & c \end{bmatrix}.$$

a. For which values of a, b, and c is A invertible? If it is invertible, what is A^{-1}?

b. For which values of the diagonal elements is a diagonal matrix (of arbitrary size) invertible?

35. Consider the upper triangular 3×3 matrix

$$A = \begin{bmatrix} a & b & c \\ 0 & d & e \\ 0 & 0 & f \end{bmatrix}.$$

a. For which values of a, b, c, d, e, and f is A invertible?

b. More generally, when is an upper triangular matrix (of arbitrary size) invertible?

c. If an upper triangular matrix is invertible, is its inverse an upper triangular matrix as well?

d. When is a lower triangular matrix invertible?

36. To determine whether a square matrix A is invertible, it is not always necessary to bring it into reduced row-echelon form. Justify the following rule: To determine whether a square matrix A is invertible, reduce it to triangular form (upper or lower), using elementary row operations. A is invertible if (and only if) all entries on the diagonal of this triangular form are nonzero.

37. If A is an invertible matrix and c is a nonzero scalar, is the matrix cA invertible? If so, what is the relationship between A^{-1} and $(cA)^{-1}$?

38. Find A^{-1} for $A = \begin{bmatrix} 1 & k \\ 0 & -1 \end{bmatrix}$.

39. Consider a square matrix that differs from the identity matrix at just one entry, off the diagonal, for example,

$$\begin{bmatrix} 1 & 0 & 0 \\ 0 & 1 & 0 \\ -\frac{1}{2} & 0 & 1 \end{bmatrix}.$$

In general, is a matrix M of this form invertible? If so, what is the M^{-1}?

40. Show that if a square matrix A has two equal columns, then A is not invertible.

41. Which of the following linear transformations T from $\mathbb{R}^3$ to $\mathbb{R}^3$ are invertible? Find the inverse if it exists.

a. Reflection about a plane

b. Orthogonal projection onto a plane

c. Scaling by a factor of 5 [i.e., $T(\vec{v}) = 5\vec{v}$, for all vectors $\vec{v}$]

d. Rotation about an axis

42. A square matrix is called a *permutation matrix* if it contains a 1 exactly once in each row and in each column, with all other entries being 0. Examples are I_n and

$$\begin{bmatrix} 0 & 0 & 1 \\ 1 & 0 & 0 \\ 0 & 1 & 0 \end{bmatrix}.$$

Are permutation matrices invertible? If so, is the inverse a permutation matrix as well?

43. Consider two invertible $n \times n$ matrices A and B. Is the linear transformation $\vec{y} = A(B\vec{x})$ invertible? If so, what is the inverse? [*Hint*: Solve the equation $\vec{y} = A(B\vec{x})$ first for $B\vec{x}$ and then for $\vec{x}$.]

44. Consider the $n \times n$ matrix M_n that contains all integers $1, 2, 3, \ldots, n^2$ as its entries, written in sequence, column by column; for example,

$$M_4 = \begin{bmatrix} 1 & 5 & 9 & 13 \\ 2 & 6 & 10 & 14 \\ 3 & 7 & 11 & 15 \\ 4 & 8 & 12 & 16 \end{bmatrix}.$$

a. Determine the rank of M_4.

b. Determine the rank of M_n, for an arbitrary $n \geq 2$.

c. For which integers n is M_n invertible?

45. To gauge the complexity of a computational task, mathematicians and computer scientists count the number of elementary operations (additions, subtractions, multiplications, and divisions) required. For a rough count, we will sometimes consider multiplications and divisions only, referring to those jointly as *multiplicative operations*. As an example, we examine the process of inverting a 2×2 matrix by elimination.

$$\begin{bmatrix} a & b & \vdots & 1 & 0 \\ c & d & \vdots & 0 & 1 \end{bmatrix} \quad \div a, \text{ requires 2 multiplicative}$$
operations: b/a and $1/a$

↓

$$\begin{bmatrix} 1 & b' & \vdots & e & 0 \\ c & d & \vdots & 0 & 1 \end{bmatrix} \quad \begin{array}{l}(\text{where } b' = b/a, \text{ and } e = 1/a) \\ -c \text{ (I), requires 2 multiplicative} \\ \text{operations: } cb' \text{ and } ce\end{array}$$

↓

$$\begin{bmatrix} 1 & b' & \vdots & e & 0 \\ 0 & d' & \vdots & g & 1 \end{bmatrix} \quad \begin{array}{l} \div d', \text{ requires 2 multiplicative} \\ \text{operations}\end{array}$$

↓

$$\begin{bmatrix} 1 & b' & \vdots & e & 0 \\ 0 & 1 & \vdots & g' & h \end{bmatrix} \quad \begin{array}{l} -b' \text{ (II), requires 2 multiplicative} \\ \text{operations}\end{array}$$

↓

$$\begin{bmatrix} 1 & 0 & \vdots & e' & f \\ 0 & 1 & \vdots & g' & h \end{bmatrix}$$

The whole process requires 8 multiplicative operations. Note that we do not count operations with predictable results, such as $1a$, $0a$, a/a, $0/a$.

a. How many multiplicative operations are required to invert a 3×3 matrix by elimination?

b. How many multiplicative operations are required to invert an $n \times n$ matrix by elimination?

c. If it takes a slow hand-held calculator 1 second to invert a 3×3 matrix, how long will it take the same calculator to invert a 12×12 matrix? Assume that the matrices are inverted by Gauss–Jordan elimination and that the duration of the computation is proportional to the number of multiplications and divisions involved.

46. Consider the linear system

$$A\vec{x} = \vec{b},$$

where A is an invertible matrix. We can solve this system in two different ways:

- By finding the reduced row-echelon form of the augmented matrix $\begin{bmatrix} A & \vdots & \vec{b} \end{bmatrix}$
- By computing A^{-1} and using the formula $\vec{x} = A^{-1}\vec{b}$

In general, which approach requires fewer multiplicative operations? See Exercise 45.

47. Give an example of a noninvertible function f from $\mathbb{R}$ to $\mathbb{R}$ and a number b such that the equation

$$f(x) = b$$

has a unique solution.

48. Consider an invertible linear transformation $T(\vec{x}) = A\vec{x}$ from $\mathbb{R}^m$ to $\mathbb{R}^n$, with inverse $L = T^{-1}$ from $\mathbb{R}^n$ to $\mathbb{R}^m$. In Exercise 2.2.29 we show that L is a linear transformation, so that $L(\vec{y}) = B\vec{y}$ for some $m \times n$ matrix B. Use the equations $BA = I_n$ and $AB = I_m$ to show that

$n = m$. (*Hint:* Think about the number of solutions of the linear systems $A\vec{x} = \vec{0}$ and $B\vec{y} = \vec{0}$.)

49. *Input–Output Analysis.* (This exercise builds on Exercises 1.1.20, 1.2.37, 1.2.38, and 1.2.39.) Consider the industries $J_1, J_2, \ldots, J_n$ in an economy. Suppose the consumer demand vector is $\vec{b}$, the output vector is $\vec{x}$ and the demand vector of the jth industry is $\vec{v}_j$. (The ith component a_{ij} of $\vec{v}_j$ is the demand industry J_j puts on industry J_i, per unit of output of J_j.) As we have seen in Exercise 1.2.38, the output $\vec{x}$ just meets the aggregate demand if

$$\underbrace{x_1\vec{v}_1 + x_2\vec{v}_2 + \cdots + x_n\vec{v}_n + \vec{b}}_{\text{aggregate demand}} = \underbrace{\vec{x}}_{\text{output}}.$$

This equation can be written more succinctly as

$$\begin{bmatrix} \vert & \vert & & \vert \\ \vec{v}_1 & \vec{v}_2 & \cdots & \vec{v}_n \\ \vert & \vert & & \vert \end{bmatrix} \begin{bmatrix} x_1 \\ x_2 \\ \vdots \\ x_n \end{bmatrix} + \vec{b} = \vec{x},$$

or $A\vec{x} + \vec{b} = \vec{x}$. The matrix A is called the *technology matrix* of this economy; its coefficients a_{ij} describe the interindustry demand, which depends on the technology used in the production process. The equation

$$A\vec{x} + \vec{b} = \vec{x}$$

describes a linear system, which we can write in the customary form:

$$\vec{x} - A\vec{x} = \vec{b},$$
$$I_n\vec{x} - A\vec{x} = \vec{b},$$
$$(I_n - A)\vec{x} = \vec{b}.$$

If we want to know the output $\vec{x}$ required to satisfy a given consumer demand $\vec{b}$ (this was our objective in the previous exercises), we can solve this linear system, preferably via the augmented matrix.

In economics, however, we often ask other questions: If $\vec{b}$ *changes*, how will $\vec{x}$ change in response? If the consumer demand on one industry increases by 1 unit and the consumer demand on the other industries remains unchanged, how will $\vec{x}$ change?[10] If we

---

[10] The relevance of questions like these became particularly clear during World War II, when the demand on certain industries suddenly changed dramatically. When U.S. President F. D. Roosevelt asked for 50,000 airplanes to be built, it was easy enough to predict that the country would have to produce more aluminum. Unexpectedly, the demand for copper dramatically increased (why?). A copper shortage then occurred, which was solved by borrowing silver from Fort Knox. People realized that input–output analysis can be effective in modeling and predicting chains of increased demand like this. After World War II, this technique rapidly gained acceptance and was soon used to model the economies of more than 50 countries.

ask questions like these, we think of the output $\vec{x}$ as a *function* of the consumer demand $\vec{b}$.

If the matrix $(I_n - A)$ is invertible,[11] we can express $\vec{x}$ as a function of $\vec{b}$ (in fact, as a linear transformation):

$$\vec{x} = (I_n - A)^{-1}\vec{b}.$$

a. Consider the example of the economy of Israel in 1958 (discussed in Exercise 1.2.39). Find the technology matrix A, the matrix $(I_n - A)$, and its inverse $(I_n - A)^{-1}$.

b. In the example discussed in part (a), suppose the consumer demand on agriculture (Industry 1) is 1 unit (1 million pounds), and the demands on the other two industries are zero. What output $\vec{x}$ is required in this case? How does your answer relate to the matrix $(I_n - A)^{-1}$?

c. Explain, in terms of economics, why the diagonal elements of the matrix $(I_n - A)^{-1}$ you found in part a must be at least 1.

d. If the consumer demand on manufacturing increases by 1 (from whatever it was), and the consumer demand on the two other industries remains the same, how will the output have to change? How does your answer relate to the matrix $(I_n - A)^{-1}$?

e. Using your answers in parts (a) through (d) as a guide, explain in general (not just for this example) what the columns and the entries of the matrix $(I_n - A)^{-1}$ tell you, in terms of economics. Those who have studied multivariable calculus may wish to consider the partial derivatives

$$\frac{\partial x_i}{\partial b_j}.$$

50. This exercise refers to Exercise 49a. Consider the entry $k = a_{11} = 0.293$ of the technology matrix A. Verify that the entry in the first row and the first column of $(I_n - A)^{-1}$ is the value of the geometrical series $1 + k + k^2 + \cdots$. Interpret this observation in terms of economics.

51. a. Consider an $n \times m$ matrix A with $\text{rank}(A) < n$. Show that there exists a vector $\vec{b}$ in $\mathbb{R}^n$ such that the system $A\vec{x} = \vec{b}$ is inconsistent. [*Hint:* For $E = \text{rref}(A)$, show that there exists a vector $\vec{c}$ in $\mathbb{R}^n$ such that the system $E\vec{x} = \vec{c}$ is inconsistent; then, "work backward."]

b. Consider an $n \times m$ matrix A with $n > m$. Show that there exists a vector $\vec{b}$ in $\mathbb{R}^n$ such that the system $A\vec{x} = \vec{b}$ is inconsistent.

52. For

$$A = \begin{bmatrix} 0 & 1 & 2 \\ 0 & 2 & 4 \\ 0 & 3 & 6 \\ 1 & 4 & 8 \end{bmatrix},$$

find a vector $\vec{b}$ in $\mathbb{R}^4$ such that the system $A\vec{x} = \vec{b}$ is inconsistent. See Exercise 51.

53. Let $A = \begin{bmatrix} 3 & 1 \\ 3 & 5 \end{bmatrix}$ in all parts of this problem.

a. Find a scalar λ (lambda) such that the matrix $A - \lambda I_2$ fails to be invertible. There are two solutions; choose one and use it in parts (b) and (c).

b. For the λ you chose in part (a), find the matrix $A - \lambda I_2$; then find a nonzero vector $\vec{x}$ such that $(A - \lambda I_2)\vec{x} = \vec{0}$. (This can be done, since $A - \lambda I_2$ fails to be invertible.)

c. Note that the equation $(A - \lambda I_2)\vec{x} = \vec{0}$ can be written as $A\vec{x} - \lambda \vec{x} = \vec{0}$, or, $A\vec{x} = \lambda \vec{x}$. Check that the equation $A\vec{x} = \lambda \vec{x}$ holds for your λ from part (a) and your $\vec{x}$ from part (b).

54. Let $A = \begin{bmatrix} 1 & 10 \\ -3 & 12 \end{bmatrix}$. Using Exercise 53 as a guide, find a scalar λ and a nonzero vector $\vec{x}$ such that $A\vec{x} = \lambda \vec{x}$.

In Exercises 55 through 65, show that the given matrix A is invertible, and find the inverse. Interpret the linear transformation $T(\vec{x}) = A\vec{x}$ and the inverse transformation $T^{-1}(\vec{y}) = A^{-1}\vec{y}$ geometrically. Interpret det A geometrically. In your figure, show the angle θ and the vectors $\vec{v}$ and $\vec{w}$ introduced in Theorem 2.4.10.

55. $\begin{bmatrix} 2 & 0 \\ 0 & 2 \end{bmatrix}$

56. $\begin{bmatrix} -3 & 0 \\ 0 & -3 \end{bmatrix}$

57. $\begin{bmatrix} \cos\alpha & \sin\alpha \\ \sin\alpha & -\cos\alpha \end{bmatrix}$

58. $\begin{bmatrix} \cos\alpha & -\sin\alpha \\ \sin\alpha & \cos\alpha \end{bmatrix}$

59. $\begin{bmatrix} 0.6 & -0.8 \\ 0.8 & 0.6 \end{bmatrix}$

60. $\begin{bmatrix} -0.8 & 0.6 \\ 0.6 & 0.8 \end{bmatrix}$

61. $\begin{bmatrix} 1 & 1 \\ -1 & 1 \end{bmatrix}$

62. $\begin{bmatrix} 3 & 4 \\ -4 & 3 \end{bmatrix}$

63. $\begin{bmatrix} -3 & 4 \\ 4 & 3 \end{bmatrix}$

64. $\begin{bmatrix} 1 & -1 \\ 0 & 1 \end{bmatrix}$

65. $\begin{bmatrix} 1 & 0 \\ 1 & 1 \end{bmatrix}$

66. Consider two $n \times n$ matrices A and B such that the product AB is invertible. Show that the matrices A and B are both invertible. [*Hint:* $AB(AB)^{-1} = I_n$ and $(AB)^{-1}AB = I_n$. Use Theorem 2.4.8.]

For two invertible $n \times n$ matrices A and B, determine which of the formulas stated in Exercises 67 through 75 are necessarily true.

67. $(A + B)^2 = A^2 + 2AB + B^2$

68. A^2 is invertible, and $(A^2)^{-1} = (A^{-1})^2$

69. $A + B$ is invertible, and $(A + B)^{-1} = A^{-1} + B^{-1}$

70. $(A - B)(A + B) = A^2 - B^2$

71. $ABB^{-1}A^{-1} = I_n$

[11] This will always be the case for a "productive" economy. See Exercise 103.

72. $ABA^{-1} = B$

73. $(ABA^{-1})^3 = AB^3A^{-1}$

74. $(I_n + A)(I_n + A^{-1}) = 2I_n + A + A^{-1}$

75. $A^{-1}B$ is invertible, and $(A^{-1}B)^{-1} = B^{-1}A$

76. Find all linear transformations T from $\mathbb{R}^2$ to $\mathbb{R}^2$ such that

$$T\begin{bmatrix} 1 \\ 2 \end{bmatrix} = \begin{bmatrix} 2 \\ 1 \end{bmatrix} \quad \text{and} \quad T\begin{bmatrix} 2 \\ 5 \end{bmatrix} = \begin{bmatrix} 1 \\ 3 \end{bmatrix}.$$

[*Hint*: We are looking for the 2×2 matrices A such that

$$A\begin{bmatrix} 1 \\ 2 \end{bmatrix} = \begin{bmatrix} 2 \\ 1 \end{bmatrix} \quad \text{and} \quad A\begin{bmatrix} 2 \\ 5 \end{bmatrix} = \begin{bmatrix} 1 \\ 3 \end{bmatrix}.$$

These two equations can be combined to form the matrix equation

$$A\begin{bmatrix} 1 & 2 \\ 2 & 5 \end{bmatrix} = \begin{bmatrix} 2 & 1 \\ 1 & 3 \end{bmatrix}.]$$

77. Using the last exercise as a guide, justify the following statement:
Let $\vec{v}_1, \vec{v}_2, \ldots, \vec{v}_m$ be vectors in $\mathbb{R}^m$ such that the matrix

$$S = \begin{bmatrix} | & | & & | \\ \vec{v}_1 & \vec{v}_2 & \cdots & \vec{v}_m \\ | & | & & | \end{bmatrix}$$

is invertible. Let $\vec{w}_1, \vec{w}_2, \ldots, \vec{w}_m$ be arbitrary vectors in $\mathbb{R}^n$. Then there exists a unique linear transformation T from $\mathbb{R}^m$ to $\mathbb{R}^n$ such that $T(\vec{v}_i) = \vec{w}_i$, for all $i = 1, \ldots, m$. Find the matrix A of this transformation in terms of S and

$$B = \begin{bmatrix} | & | & & | \\ \vec{w}_1 & \vec{w}_2 & \cdots & \vec{w}_m \\ | & | & & | \end{bmatrix}.$$

78. Find the matrix A of the linear transformation T from $\mathbb{R}^2$ to $\mathbb{R}^3$ with

$$T\begin{bmatrix} 1 \\ 2 \end{bmatrix} = \begin{bmatrix} 7 \\ 5 \\ 3 \end{bmatrix} \quad \text{and} \quad T\begin{bmatrix} 2 \\ 5 \end{bmatrix} = \begin{bmatrix} 1 \\ 2 \\ 3 \end{bmatrix}.$$

(compare with Exercise 77).

79. Find the matrix A of the linear transformation T from $\mathbb{R}^2$ to $\mathbb{R}^2$ with

$$T\begin{bmatrix} 3 \\ 1 \end{bmatrix} = 2\begin{bmatrix} 3 \\ 1 \end{bmatrix} \quad \text{and} \quad T\begin{bmatrix} 1 \\ 2 \end{bmatrix} = 3\begin{bmatrix} 1 \\ 2 \end{bmatrix}.$$

(compare with Exercise 77).

80. Consider the regular tetrahedron sketched below, whose center is at the origin.

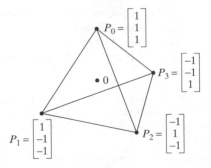

Let T from $\mathbb{R}^3$ to $\mathbb{R}^3$ be the rotation about the axis through the points 0 and P_2 that transforms P_1 into P_3. Find the images of the four corners of the tetrahedron under this transformation.

$$P_0 \overset{T}{\to}$$
$$P_1 \to P_3$$
$$P_2 \to$$
$$P_3 \to$$

Let L from $\mathbb{R}^3$ to $\mathbb{R}^3$ be the reflection about the plane through the points 0, P_0, and P_3. Find the images of the four corners of the tetrahedron under this transformation.

$$P_0 \overset{L}{\to}$$
$$P_1 \to$$
$$P_2 \to$$
$$P_3 \to$$

Describe the transformations in parts (a) through (c) geometrically.

a. T^{-1} **b.** L^{-1}

c. $T^2 = T \circ T$ (the composite of T with itself)

d. Find the images of the four corners under the transformations $T \circ L$ and $L \circ T$. Are the two transformations the same?

$$P_0 \overset{T \circ L}{\to} \qquad\qquad P_0 \overset{L \circ T}{\to}$$
$$P_1 \to \qquad\qquad\qquad P_1 \to$$
$$P_2 \to \qquad\qquad\qquad P_2 \to$$
$$P_3 \to \qquad\qquad\qquad P_3 \to$$

e. Find the images of the four corners under the transformation $L \circ T \circ L$. Describe this transformation geometrically.

81. Find the matrices of the transformations T and L defined in Exercise 80.

82. Consider the matrix

$$E = \begin{bmatrix} 1 & 0 & 0 \\ -3 & 1 & 0 \\ 0 & 0 & 1 \end{bmatrix}$$

and an arbitrary 3×3 matrix

$$A = \begin{bmatrix} a & b & c \\ d & e & f \\ g & h & k \end{bmatrix}.$$

a. Compute EA. Comment on the relationship between A and EA, in terms of the technique of elimination we learned in Section 1.2.

b. Consider the matrix

$$E = \begin{bmatrix} 1 & 0 & 0 \\ 0 & \frac{1}{4} & 0 \\ 0 & 0 & 1 \end{bmatrix}$$

and an arbitrary 3×3 matrix A. Compute EA. Comment on the relationship between A and EA.

c. Can you think of a 3×3 matrix E such that EA is obtained from A by swapping the last two rows (for any 3×3 matrix A)?

d. The matrices of the forms introduced in parts (a), (b), and (c) are called *elementary*: An $n \times n$ matrix E is elementary if it can be obtained from I_n by performing one of the three elementary row operations on I_n. Describe the format of the three types of elementary matrices.

83. Are elementary matrices invertible? If so, is the inverse of an elementary matrix elementary as well? Explain the significance of your answers in terms of elementary row operations.

84. a. Justify the following: If A is an $n \times m$ matrix, then there exist elementary $n \times n$ matrices E_1, E_2, ..., E_p such that

$$\text{rref}(A) = E_1 E_2 \cdots E_p A.$$

b. Find such elementary matrices E_1, E_2, ..., E_p for

$$A = \begin{bmatrix} 0 & 2 \\ 1 & 3 \end{bmatrix}.$$

85. a. Justify the following: If A is an $n \times m$ matrix, then there exists an invertible $n \times n$ matrix S such that

$$\text{rref}(A) = SA.$$

b. Find such an invertible matrix S for

$$A = \begin{bmatrix} 2 & 4 \\ 4 & 8 \end{bmatrix}.$$

86. a. Justify the following: Any invertible matrix is a product of elementary matrices.

b. Write $A = \begin{bmatrix} 0 & 2 \\ 1 & 3 \end{bmatrix}$ as a product of elementary matrices.

87. Write all possible forms of elementary 2×2 matrices E. In each case, describe the transformation $\vec{y} = E\vec{x}$ geometrically.

88. Consider an invertible $n \times n$ matrix A and an $n \times n$ matrix B. A certain sequence of elementary row operations transforms A into I_n.

a. What do you get when you apply the same row operations in the same order to the matrix AB?

b. What do you get when you apply the same row operations to I_n?

89. Is the product of two lower triangular matrices a lower triangular matrix as well? Explain your answer.

90. Consider the matrix

$$A = \begin{bmatrix} 1 & 2 & 3 \\ 2 & 6 & 7 \\ 2 & 2 & 4 \end{bmatrix}.$$

a. Find lower triangular elementary matrices E_1, E_2, ..., E_m such that the product

$$E_m \cdots E_2 E_1 A$$

is an upper triangular matrix U. *Hint*: Use elementary row operations to eliminate the entries below the diagonal of A.

b. Find lower triangular elementary matrices M_1, M_2, ..., M_m and an upper triangular matrix U such that

$$A = M_1 M_2 \cdots M_m U.$$

c. Find a lower triangular matrix L and an upper triangular matrix U such that

$$A = LU.$$

Such a representation of an invertible matrix is called an *LU-factorization*. The method outlined in this exercise to find an *LU*-factorization can be streamlined somewhat, but we have seen the major ideas. An *LU*-factorization (as introduced here) does not always exist (see Exercise 92).

d. Find a lower triangular matrix L with 1's on the diagonal, an upper triangular matrix U with 1's on the diagonal, and a diagonal matrix D such that $A = LDU$. Such a representation of an invertible matrix is called an *LDU-factorization*.

91. Knowing an *LU*-factorization of a matrix A makes it much easier to solve a linear system

$$A\vec{x} = \vec{b}.$$

Consider the *LU*-factorization

$$A = \begin{bmatrix} 1 & 2 & -1 & 4 \\ -3 & -5 & 6 & -5 \\ 1 & 4 & 6 & 20 \\ -1 & 6 & 20 & 43 \end{bmatrix}$$

$$= \begin{bmatrix} 1 & 0 & 0 & 0 \\ -3 & 1 & 0 & 0 \\ 1 & 2 & 1 & 0 \\ -1 & 8 & -5 & 1 \end{bmatrix} \begin{bmatrix} 1 & 2 & -1 & 4 \\ 0 & 1 & 3 & 7 \\ 0 & 0 & 1 & 2 \\ 0 & 0 & 0 & 1 \end{bmatrix}$$

$$= LU.$$

Suppose we have to solve the system $A\vec{x} = LU\vec{x} = \vec{b}$, where

$$\vec{b} = \begin{bmatrix} -3 \\ 14 \\ 9 \\ 33 \end{bmatrix}.$$

a. Set $\vec{y} = U\vec{x}$, and solve the system $L\vec{y} = \vec{b}$, by forward substitution (finding first y_1, then y_2, etc.). Do this using paper and pencil. Show all your work.

b. Solve the system $U\vec{x} = \vec{y}$, using back substitution, to find the solution $\vec{x}$ of the system $A\vec{x} = \vec{b}$. Do this using paper and pencil. Show all your work.

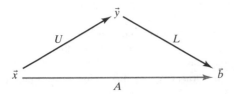

92. Show that the matrix $A = \begin{bmatrix} 0 & 1 \\ 1 & 0 \end{bmatrix}$ cannot be written in the form $A = LU$, where L is lower triangular and U is upper triangular.

93. In this exercise we will examine which invertible $n \times n$ matrices A admit an LU-factorization $A = LU$, as discussed in Exercise 90. The following definition will be useful: For $m = 1, \ldots, n$ the *principal submatrix* $A^{(m)}$ of A is obtained by omitting all rows and columns of A past the mth. For example, the matrix

$$A = \begin{bmatrix} 1 & 2 & 3 \\ 4 & 5 & 6 \\ 7 & 8 & 7 \end{bmatrix}$$

has the principal submatrices

$$A^{(1)} = [1], \quad A^{(2)} = \begin{bmatrix} 1 & 2 \\ 4 & 5 \end{bmatrix}, \quad A^{(3)} = A = \begin{bmatrix} 1 & 2 & 3 \\ 4 & 5 & 6 \\ 7 & 8 & 7 \end{bmatrix}.$$

We will show that an invertible $n \times n$ matrix A admits an LU-factorization $A = LU$ if (and only if) all its principal submatrices are invertible.

a. Let $A = LU$ be an LU-factorization of an $n \times n$ matrix A. Use block matrices to show that $A^{(m)} = L^{(m)}U^{(m)}$ for $m = 1, \ldots, n$.

b. Use part (a) to show that if an invertible $n \times n$ matrix A has an LU-factorization, then all its principal submatrices $A^{(m)}$ are invertible.

c. Consider an $n \times n$ matrix A whose principal submatrices are all invertible. Show that A admits an LU-factorization. (*Hint*: By induction, you can assume that $A^{(n-1)}$ has an LU-factorization $A^{(n-1)} = L'U'$.) Use block matrices to find an LU-factorization for A. Alternatively, you can explain this result in terms of Gauss–Jordan elimination (if

all principal submatrices are invertible, then no row swaps are required).

94. a. Show that if an invertible $n \times n$ matrix A admits an LU-factorization, then it admits an LDU-factorization (see Exercise 90 d).

b. Show that if an invertible $n \times n$ matrix A admits an LDU-factorization, then this factorization is unique. (*Hint*: Suppose that $A = L_1 D_1 U_1 = L_2 D_2 U_2$.) Then $U_2 U_1^{-1} = D_2^{-1} L_2^{-1} L_1 D_1$ is diagonal (why?). Conclude that $U_2 = U_1$.

95. Consider a block matrix

$$A = \begin{bmatrix} A_{11} & 0 \\ 0 & A_{22} \end{bmatrix},$$

where A_{11} and A_{22} are square matrices. For which choices of A_{11} and A_{22} is A invertible? In these cases, what is A^{-1}?

96. Consider a block matrix

$$A = \begin{bmatrix} A_{11} & 0 \\ A_{21} & A_{22} \end{bmatrix},$$

where A_{11} and A_{22} are square matrices. For which choices of A_{11}, A_{21}, and A_{22} is A invertible? In these cases, what is A^{-1}?

97. Consider the block matrix

$$A = \begin{bmatrix} A_{11} & A_{12} & A_{13} \\ 0 & 0 & A_{23} \end{bmatrix},$$

where A_{11} is an invertible matrix. Determine the rank of A in terms of the ranks of the blocks A_{11}, A_{12}, A_{13}, and A_{23}.

98. Consider the block matrix

$$A = \begin{bmatrix} I_n & \vec{v} \\ \vec{w} & 1 \end{bmatrix},$$

where $\vec{v}$ is a vector in $\mathbb{R}^n$, and $\vec{w}$ is a row vector with n components. For which choices of $\vec{v}$ and $\vec{w}$ is A invertible? In these cases, what is A^{-1}?

99. Find all invertible $n \times n$ matrices A such that $A^2 = A$.

100. Find a nonzero $n \times n$ matrix A with identical entries such that $A^2 = A$.

101. Consider two $n \times n$ matrices A and B whose entries are positive or zero. Suppose that all entries of A are less than or equal to s, and all column sums of B are less than or equal to r (the jth column sum of a matrix is the sum of all the entries in its jth column). Show that all entries of the matrix AB are less than or equal to sr.

102. (This exercise builds on Exercise 101.) Consider an $n \times n$ matrix A whose entries are positive or zero. Suppose that all column sums of A are less than 1. Let r be the largest column sum of A.

a. Show that the entries of A^m are less than or equal to r^m, for all positive integers m.

b. Show that

$$\lim_{m\to\infty} A^m = 0$$

(meaning that all entries of A^m approach zero).

c. Show that the infinite series

$$I_n + A + A^2 + \cdots + A^m + \cdots$$

converges (entry by entry).

d. Compute the product

$$(I_n - A)(I_n + A + A^2 + \cdots + A^m).$$

Simplify the result. Then let m go to infinity, and thus show that

$$(I_n - A)^{-1} = I_n + A + A^2 + \cdots + A^m + \cdots.$$

103. (This exercise builds on Exercises 49, 101, and 102.)

a. Consider the industries $J_1, \ldots, J_n$ in an economy. We say that industry J_j is *productive* if the jth column sum of the technology matrix A is less than 1. What does this mean in terms of economics?

b. We say that an economy is productive if all of its industries are productive. Exercise 102 shows that if A is the technology matrix of a productive economy, then the matrix $I_n - A$ is invertible. What does this result tell you about the ability of a productive economy to satisfy consumer demand?

c. Interpret the formula

$$(I_n - A)^{-1} = I_n + A + A^2 + \cdots + A^m + \cdots$$

derived in Exercise 102d in terms of economics.

104. The color of light can be represented in a vector

$$\begin{bmatrix} R \\ G \\ B \end{bmatrix},$$

where R = amount of red, G = amount of green, and B = amount of blue. The human eye and the brain transform the incoming signal into the signal

$$\begin{bmatrix} I \\ L \\ S \end{bmatrix},$$

where

$$\text{intensity} \quad I = \frac{R + G + B}{3}$$

$$\text{long-wave signal} \quad L = R - G$$

$$\text{short-wave signal} \quad S = B - \frac{R + G}{2}.$$

a. Find the matrix P representing the transformation from

$$\begin{bmatrix} R \\ G \\ B \end{bmatrix} \quad \text{to} \quad \begin{bmatrix} I \\ L \\ S \end{bmatrix}.$$

b. Consider a pair of yellow sunglasses for water sports that cuts out all blue light and passes all red and green light. Find the 3×3 matrix A that represents the transformation incoming light undergoes as it passes through the sunglasses.

c. Find the matrix for the composite transformation that light undergoes as it first passes through the sunglasses and then the eye.

d. As you put on the sunglasses, the signal you receive (intensity, long- and short-wave signals) undergoes a transformation. Find the matrix M of this transformation.

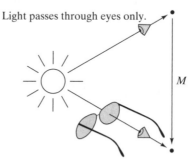

Light passes through eyes only.

Light passes through glasses and then through eyes.

105. A village is divided into three mutually exclusive groups called *clans*. Each person in the village belongs to a clan, and this identification is permanent. There are rigid rules concerning marriage: A person from one clan can only marry a person from one other clan. These rules are encoded in the matrix A below. The fact that the 2–3 entry is 1 indicates that marriage between a man from clan III and a woman from clan II is allowed. The clan of a child is determined by the mother's clan, as indicated by the matrix B. According to this scheme siblings belong to the same clan.

Husband's clan

$$A = \begin{bmatrix} 0 & 1 & 0 \\ 0 & 0 & 1 \\ 1 & 0 & 0 \end{bmatrix} \begin{matrix} \text{I} \\ \text{II} \\ \text{III} \end{matrix} \quad \begin{matrix} \text{Wife's} \\ \text{clan} \end{matrix}$$

with columns labeled I, II, III.

Mother's clan

$$B = \begin{bmatrix} 1 & 0 & 0 \\ 0 & 0 & 1 \\ 0 & 1 & 0 \end{bmatrix} \begin{matrix} \text{I} \\ \text{II} \\ \text{III} \end{matrix} \quad \begin{matrix} \text{Child's} \\ \text{clan} \end{matrix}$$

with columns labeled I, II, III.

The identification of a person with clan I can be represented by the vector

$$\vec{e}_1 = \begin{bmatrix} 1 \\ 0 \\ 0 \end{bmatrix},$$

and likewise for the two other clans. Matrix A transforms the husband's clan into the wife's clan (if $\vec{x}$ represents the husband's clan, then $A\vec{x}$ represents the wife's clan).

a. Are the matrices A and B invertible? Find the inverses if they exist. What do your answers mean, in practical terms?

b. What is the meaning of B^2, in terms of the rules of the community?

c. What is the meaning of AB and BA, in terms of the rules of the community? Are AB and BA the same?

d. Bueya is a young woman who has many male first cousins, both on her mother's and on her father's sides. The kinship between Bueya and each of her male cousins can be represented by one of the four diagrams below:

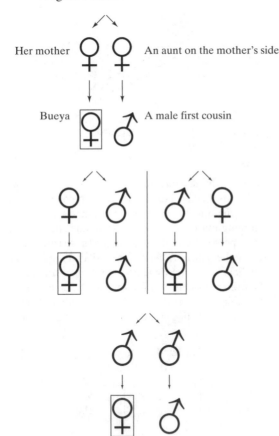

In each of the four cases, find the matrix which gives you the cousin's clan in terms of Bueya's clan.

e. According to the rules of the village, could Bueya marry a first cousin? (We do not know Bueya's clan.)

106. As background to this exercise, see Exercise 45.

a. If you use Theorem 2.3.4, how many multiplications of scalars are necessary to multiply two 2×2 matrices?

b. If you use Theorem 2.3.4, how many multiplications are needed to multiply an $n \times p$ and a $p \times m$ matrix?

In 1969, the German mathematician Volker Strassen surprised the mathematical community by showing that two 2×2 matrices can be multiplied with only seven multiplications of numbers. Here is his trick: Suppose you have to find AB for $A = \begin{bmatrix} a & b \\ c & d \end{bmatrix}$ and $B = \begin{bmatrix} p & q \\ r & s \end{bmatrix}$. First compute

$$
\begin{aligned}
h_1 &= (a+d)(p+s) \\
h_2 &= (c+d)p \\
h_3 &= a(q-s) \\
h_4 &= d(r-p). \\
h_5 &= (a+b)s \\
h_6 &= (c-a)(p+q) \\
h_7 &= (b-d)(r+s)
\end{aligned}
$$

Then

$$
AB = \begin{bmatrix} h_1 + h_4 - h_5 + h_7 & h_3 + h_5 \\ h_2 + h_4 & h_1 + h_3 - h_2 + h_6 \end{bmatrix}.
$$

107. Let $\mathbb{N}$ be the set of all positive integers, $1, 2, 3, \ldots$. We define two functions f and g from $\mathbb{N}$ to $\mathbb{N}$:

$$f(x) = 2x, \quad \text{for all } x \text{ in } \mathbb{N}$$

$$g(x) = \begin{cases} x/2 & \text{if } x \text{ is even} \\ (x+1)/2 & \text{if } x \text{ is odd} \end{cases}$$

Find formulas for the composite functions $g\big(f(x)\big)$ and $f\big(g(x)\big)$. Is one of them the identity transformation from $\mathbb{N}$ to $\mathbb{N}$? Are the functions f and g invertible?

108. *Geometrical optics.* Consider a thin biconvex lens with two spherical faces.

This is a good model for the lens of the human eye and for the lenses used in many optical instruments, such as reading glasses, cameras, microscopes, and telescopes. The line through the centers of the spheres defining the two faces is called the *optical axis* of the lens.

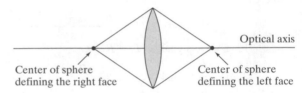

In this exercise, we learn how we can track the path of a ray of light as it passes through the lens, provided that the following conditions are satisfied:

• The ray lies in a plane with the optical axis.

• The angle the ray makes with the optical axis is small.

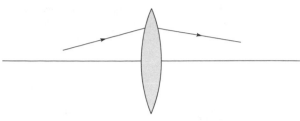

To keep track of the ray, we introduce two *reference planes* perpendicular to the optical axis, to the left and to the right of the lens.

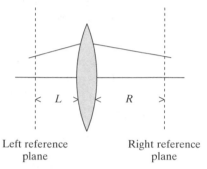

Left reference plane

Right reference plane

We can characterize the incoming ray by its slope m and its intercept x with the left reference plane. Likewise, we characterize the outgoing ray by slope n and intercept y.

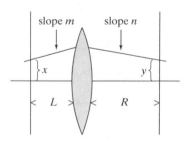

We want to know how the outgoing ray depends on the incoming ray, that is, we are interested in the transformation

$$T: \mathbb{R}^2 \to \mathbb{R}^2; \quad \begin{bmatrix} x \\ m \end{bmatrix} \to \begin{bmatrix} y \\ n \end{bmatrix}.$$

We will see that T can be approximated by a linear transformation provided that m is small, as we assumed. To study this transformation, we divide the path of the ray into three segments, as shown in the following figure:

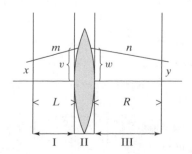

We have introduced two auxiliary reference planes, directly to the left and to the right of the lens. Our transformation

$$\begin{bmatrix} x \\ m \end{bmatrix} \to \begin{bmatrix} y \\ n \end{bmatrix}$$

can now be represented as the composite of three simpler transformations:

$$\begin{bmatrix} x \\ m \end{bmatrix} \to \begin{bmatrix} v \\ m \end{bmatrix} \to \begin{bmatrix} w \\ n \end{bmatrix} \to \begin{bmatrix} y \\ n \end{bmatrix}.$$

From the definition of the slope of a line we get the relations $v = x + Lm$ and $y = w + Rn$.

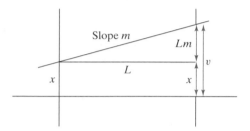

$$\begin{bmatrix} v \\ m \end{bmatrix} = \begin{bmatrix} x + Lm \\ m \end{bmatrix} = \begin{bmatrix} 1 & L \\ 0 & 1 \end{bmatrix} \begin{bmatrix} x \\ m \end{bmatrix};$$

$$\begin{bmatrix} y \\ n \end{bmatrix} = \begin{bmatrix} 1 & R \\ 0 & 1 \end{bmatrix} \begin{bmatrix} w \\ n \end{bmatrix}$$

$$\begin{bmatrix} x \\ m \end{bmatrix} \xrightarrow[\begin{bmatrix} 1 & L \\ 0 & 1 \end{bmatrix}]{} \begin{bmatrix} v \\ m \end{bmatrix} \longrightarrow \begin{bmatrix} w \\ n \end{bmatrix} \xrightarrow[\begin{bmatrix} 1 & R \\ 0 & 1 \end{bmatrix}]{} \begin{bmatrix} y \\ n \end{bmatrix}$$

It would lead us too far into physics to derive a formula for the transformation

$$\begin{bmatrix} v \\ m \end{bmatrix} \to \begin{bmatrix} w \\ n \end{bmatrix}$$

here.[12] Under the assumptions we have made, the transformation is well approximated by

$$\begin{bmatrix} w \\ n \end{bmatrix} = \begin{bmatrix} 1 & 0 \\ -k & 1 \end{bmatrix} \begin{bmatrix} v \\ m \end{bmatrix},$$

for some positive constant k (this formula implies that $w = v$).

$$\begin{bmatrix} x \\ m \end{bmatrix} \xrightarrow[\begin{bmatrix} 1 & L \\ 0 & 1 \end{bmatrix}]{} \begin{bmatrix} v \\ m \end{bmatrix} \xrightarrow[\begin{bmatrix} 1 & 0 \\ -k & 1 \end{bmatrix}]{} \begin{bmatrix} w \\ n \end{bmatrix} \xrightarrow[\begin{bmatrix} 1 & R \\ 0 & 1 \end{bmatrix}]{} \begin{bmatrix} y \\ n \end{bmatrix}$$

---

[12] See, for example, Paul Bamberg and Shlomo Sternberg, *A Course in Mathematics for Students of Physics 1*, Cambridge University Press, 1991.

The transformation $\begin{bmatrix} x \\ m \end{bmatrix} \rightarrow \begin{bmatrix} y \\ n \end{bmatrix}$ is represented by the matrix product

$$\begin{bmatrix} 1 & R \\ 0 & 1 \end{bmatrix} \begin{bmatrix} 1 & 0 \\ -k & 1 \end{bmatrix} \begin{bmatrix} 1 & L \\ 0 & 1 \end{bmatrix}$$
$$= \begin{bmatrix} 1 - Rk & L + R - kLR \\ -k & 1 - kL \end{bmatrix}.$$

a. *Focusing parallel rays.* Consider the lens in the human eye, with the retina as the right reference plane. In an adult, the distance R is about 0.025 meters (about 1 inch). The ciliary muscles allow you to vary the shape of the lens and thus the lens constant k, within a certain range. What value of k enables you to focus parallel incoming rays, as shown in the figure? This value of k will allow you to see a distant object clearly. (The customary unit of measurement for k is 1 diopter $= \frac{1}{1 \text{ meter}}$.)

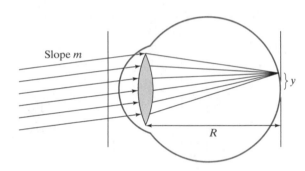

(*Hint*: In terms of the transformation

$$\begin{bmatrix} x \\ m \end{bmatrix} \rightarrow \begin{bmatrix} y \\ n \end{bmatrix},$$

you want y to be independent of x (y must depend on the slope m alone). Explain why $1/k$ is called the *focal length* of the lens.)

b. What value of k enables you to read this text from a distance of $L = 0.3$ meters? Consider the following figure (which is not to scale):

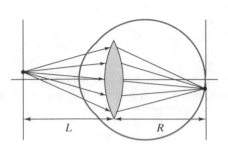

c. *The telescope.* An astronomical telescope consists of two lenses with the same optical axis.

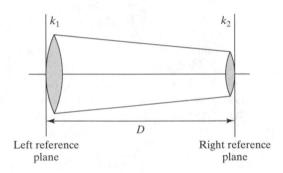

Left reference plane

Right reference plane

Find the matrix of the transformation

$$\begin{bmatrix} x \\ m \end{bmatrix} \rightarrow \begin{bmatrix} y \\ n \end{bmatrix},$$

in terms of k_1, k_2, and D. For given values of k_1 and k_2, how do you choose D so that parallel incoming rays are converted into parallel outgoing rays? What is the relationship between D and the focal lengths of the two lenses, $1/k_1$ and $1/k_2$?

Chapter Two Exercises

TRUE OR FALSE?

1. If A is any invertible $n \times n$ matrix, then $\text{rref}(A) = I_n$.

2. The formula $(A^2)^{-1} = (A^{-1})^2$ holds for all invertible matrices A.

3. The formula $AB = BA$ holds for all $n \times n$ matrices A and B.

4. If $AB = I_n$ for two $n \times n$ matrices A and B, then A must be the inverse of B.

5. If A is a 3×4 matrix and B is a 4×5 matrix, then AB will be a 5×3 matrix.

6. The function $T \begin{bmatrix} x \\ y \end{bmatrix} = \begin{bmatrix} y \\ 1 \end{bmatrix}$ is a linear transformation.

7. The matrix $\begin{bmatrix} 5 & 6 \\ -6 & 5 \end{bmatrix}$ represents a rotation combined with a scaling.

8. If A is any invertible $n \times n$ matrix, then A commutes with A^{-1}.

9. The function $T \begin{bmatrix} x \\ y \end{bmatrix} = \begin{bmatrix} x - y \\ y - x \end{bmatrix}$ is a linear transformation.

10. Matrix $\begin{bmatrix} 1/2 & -1/2 \\ 1/2 & 1/2 \end{bmatrix}$ represents a rotation.

11. There exists a real number k such that the matrix $\begin{bmatrix} k-2 & 3 \\ -3 & k-2 \end{bmatrix}$ fails to be invertible.

12. Matrix $\begin{bmatrix} -0.6 & 0.8 \\ -0.8 & -0.6 \end{bmatrix}$ represents a rotation.

13. The formula $\det(2A) = 2\det(A)$ holds for all 2×2 matrices A.

14. There exists a matrix A such that
$\begin{bmatrix} 1 & 2 \\ 3 & 4 \end{bmatrix} A \begin{bmatrix} 5 & 6 \\ 7 & 8 \end{bmatrix} = \begin{bmatrix} 1 & 1 \\ 1 & 1 \end{bmatrix}$.

15. Matrix $\begin{bmatrix} 1 & 2 \\ 3 & 6 \end{bmatrix}$ is invertible.

16. Matrix $\begin{bmatrix} 1 & 1 & 1 \\ 1 & 0 & 1 \\ 1 & 1 & 0 \end{bmatrix}$ is invertible.

17. There exists an upper triangular 2×2 matrix A such that
$A^2 = \begin{bmatrix} 1 & 1 \\ 0 & 1 \end{bmatrix}$.

18. The function $T \begin{bmatrix} x \\ y \end{bmatrix} = \begin{bmatrix} (y+1)^2 - (y-1)^2 \\ (x-3)^2 - (x+3)^2 \end{bmatrix}$ is a linear transformation.

19. Matrix $\begin{bmatrix} k & -2 \\ 5 & k-6 \end{bmatrix}$ is invertible for all real numbers k.

20. There exists a real number k such that the matrix $\begin{bmatrix} k-1 & -2 \\ -4 & k-3 \end{bmatrix}$ fails to be invertible.

21. The matrix product $\begin{bmatrix} a & b \\ c & d \end{bmatrix} \begin{bmatrix} d & -b \\ -c & a \end{bmatrix}$ is always a scalar multiple of I_2.

22. There exists a nonzero upper triangular 2×2 matrix A such that $A^2 = \begin{bmatrix} 0 & 0 \\ 0 & 0 \end{bmatrix}$.

23. There exists a positive integer n such that
$\begin{bmatrix} 0 & -1 \\ 1 & 0 \end{bmatrix}^n = I_2$.

24. There exists an invertible 2×2 matrix A such that
$A^{-1} = \begin{bmatrix} 1 & 1 \\ 1 & 1 \end{bmatrix}$.

25. There exists an invertible $n \times n$ matrix with two identical rows.

26. If $A^2 = I_n$, then matrix A must be invertible.

27. There exists a matrix A such that $A \begin{bmatrix} 1 & 1 \\ 1 & 1 \end{bmatrix} = \begin{bmatrix} 1 & 2 \\ 1 & 2 \end{bmatrix}$.

28. There exists a matrix A such that $\begin{bmatrix} 1 & 2 \\ 1 & 2 \end{bmatrix} A = \begin{bmatrix} 1 & 1 \\ 1 & 1 \end{bmatrix}$.

29. The matrix $\begin{bmatrix} 1 & 1 \\ 1 & -1 \end{bmatrix}$ represents a reflection about a line.

30. $\begin{bmatrix} 1 & k \\ 0 & 1 \end{bmatrix}^3 = \begin{bmatrix} 1 & 3k \\ 0 & 1 \end{bmatrix}$ for all real numbers k.

31. If matrix $\begin{bmatrix} a & b & c \\ d & e & f \\ g & h & i \end{bmatrix}$ is invertible, then matrix $\begin{bmatrix} a & b \\ d & e \end{bmatrix}$ must be invertible as well.

32. If A^2 is invertible, then matrix A itself must be invertible.

33. If $A^{17} = I_2$, then matrix A must be I_2.

34. If $A^2 = I_2$, then matrix A must be either I_2 or $-I_2$.

35. If matrix A is invertible, then matrix $5A$ must be invertible as well.

36. If A and B are two 4×3 matrices such that $A\vec{v} = B\vec{v}$ for all vectors $\vec{v}$ in $\mathbb{R}^3$, then matrices A and B must be equal.

37. If matrices A and B commute, then the formula $A^2B = BA^2$ must hold.

38. If $A^2 = A$ for an invertible $n \times n$ matrix A, then A must be I_n.

39. If matrices A and B are both invertible, then matrix $A + B$ must be invertible as well.

40. The equation $A^2 = A$ holds for all 2×2 matrices A representing a projection.

41. The equation $A^{-1} = A$ holds for all 2×2 matrices A representing a reflection.

42. The formula $(A\vec{v}) \cdot (A\vec{w}) = \vec{v} \cdot \vec{w}$ holds for all invertible 2×2 matrices A and for all vectors $\vec{v}$ and $\vec{w}$ in $\mathbb{R}^2$.

43. There exist a 2×3 matrix A and a 3×2 matrix B such that $AB = I_2$.

44. There exist a 3×2 matrix A and a 2×3 matrix B such that $AB = I_3$.

45. If $A^2 + 3A + 4I_3 = 0$ for a 3×3 matrix A, then A must be invertible.

46. If A is an $n \times n$ matrix such that $A^2 = 0$, then matrix $I_n + A$ must be invertible.

47. If matrix A commutes with B, and B commutes with C, then matrix A must commute with C.

48. If T is any linear transformation from $\mathbb{R}^3$ to $\mathbb{R}^3$, then $T(\vec{v} \times \vec{w}) = T(\vec{v}) \times T(\vec{w})$ for all vectors $\vec{v}$ and $\vec{w}$ in $\mathbb{R}^3$.

49. There exists an invertible 10×10 matrix that has 92 ones among its entries.

50. The formula $\text{rref}(AB) = \text{rref}(A)\,\text{rref}(B)$ holds for all $n \times p$ matrices A and for all $p \times m$ matrices B.

51. There exists an invertible matrix S such that $S^{-1} \begin{bmatrix} 0 & 1 \\ 0 & 0 \end{bmatrix} S$ is a diagonal matrix.

52. If the linear system $A^2 \vec{x} = \vec{b}$ is consistent, then the system $A\vec{x} = \vec{b}$ must be consistent as well.

53. There exists an invertible 2×2 matrix A such that $A^{-1} = -A$.

54. There exists an invertible 2×2 matrix A such that $A^2 = \begin{bmatrix} 1 & 0 \\ 0 & -1 \end{bmatrix}$.

55. If a matrix $A = \begin{bmatrix} a & b \\ c & d \end{bmatrix}$ represents the orthogonal projection onto a line L, then the equation $a^2 + b^2 + c^2 + d^2 = 1$ must hold.

56. If A is an invertible 2×2 matrix and B is any 2×2 matrix, then the formula $\text{rref}(AB) = \text{rref}(B)$ must hold.

3

Subspaces of $\mathbb{R}^n$ and Their Dimensions

3.1 Image and Kernel of a Linear Transformation

You may be familiar with the notion of the *image*[1] of a function.

Definition 3.1.1 Image of a function

The *image* of a function consists of all the values the function takes in its target space. If f is a function from X to Y, then

$$\text{image}(f) = \{f(x): x \text{ in } X\}$$
$$= \{b \text{ in } Y: b = f(x), \text{ for some } x \text{ in } X\}.$$

EXAMPLE 1 A group X of students and a group Y of professors stand in the yard. Each student throws a tomato at one of the professors (and each tomato hits its intended target). Consider the function $y = f(x)$ from X to Y that associates with each student x the target y of his or her tomato. The image of f consists of those professors that are hit. See Figure 1. ∎

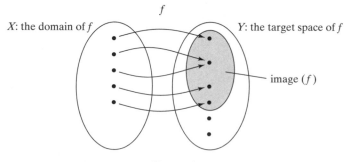

Figure 1

---

[1]Some authors use the term *range* for what we call the image, while others use the term *range* for what we call the target space. Because of this ambiguity, we will not use the term *range* at all. Make sure to check which definition is used when you encounter the term *range* in a text.

EXAMPLE 2 The image of the exponential function $f(x) = e^x$ from $\mathbb{R}$ to $\mathbb{R}$ consists of all positive numbers. Indeed, $f(x) = e^x$ is positive for all x, and every positive number b can be written as $b = e^{\ln b} = f(\ln b)$. See Figure 2. ∎

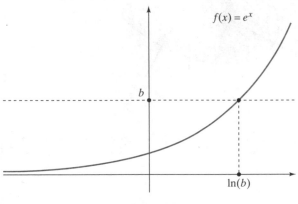

Figure 2

More generally, the image of a function $f(x)$ from $\mathbb{R}$ to $\mathbb{R}$ consists of all numbers b such that the line $y = b$ intersects the graph of f (take another look at Figure 2). The image of f is the orthogonal projection of the graph of f onto the vertical axis.

EXAMPLE 3 The image of the function

$$f(t) = \begin{bmatrix} \cos t \\ \sin t \end{bmatrix} \quad \text{from } \mathbb{R} \text{ to } \mathbb{R}^2$$

is the unit circle centered at the origin (see Figure 3). Indeed, $\begin{bmatrix} \cos t \\ \sin t \end{bmatrix}$ is a unit vector for all t, since $\cos^2 t + \sin^2 t = 1$, and, conversely, any unit vector $\vec{u}$ in $\mathbb{R}^2$ can be written in polar coordinates as $\vec{u} = \begin{bmatrix} \cos t \\ \sin t \end{bmatrix}$, where t is its polar angle. ∎

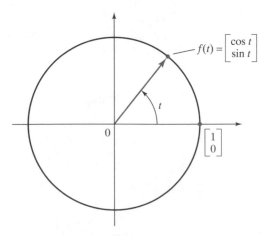

Figure 3

The function f in Example 3 is called a *parametrization* of the unit circle. More generally, a parametrization of a curve C in $\mathbb{R}^2$ is a function g from $\mathbb{R}$ to $\mathbb{R}^2$ whose image is C.

EXAMPLE 4 If the function f from X to Y is invertible, then the image of f is Y. Indeed, for every b in Y there exists one (and only one) x in X such that $b = f(x)$, namely, $x = f^{-1}(b)$:

$$b = f\left(f^{-1}(b)\right).$$

See Figure 4. ■

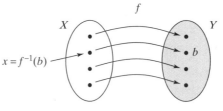

Figure 4

EXAMPLE 5 Consider the linear transformation T from $\mathbb{R}^3$ to $\mathbb{R}^3$ that projects a vector $\vec{x}$ orthogonally into the x_1–x_2-plane, meaning that $T\begin{bmatrix} x_1 \\ x_2 \\ x_3 \end{bmatrix} = \begin{bmatrix} x_1 \\ x_2 \\ 0 \end{bmatrix}$. See Figure 5.

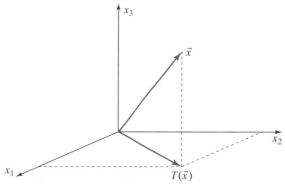

Figure 5

The image of T is the x_1–x_2-plane in $\mathbb{R}^3$, consisting of all vectors of the form $\begin{bmatrix} x_1 \\ x_2 \\ 0 \end{bmatrix}$. ■

EXAMPLE 6 Describe the image of the linear transformation

$$T(\vec{x}) = A\vec{x} \quad \text{from } \mathbb{R}^2 \text{ to } \mathbb{R}^2, \text{ where} \quad A = \begin{bmatrix} 1 & 3 \\ 2 & 6 \end{bmatrix}.$$

Solution

The image of T consists of all the values of T, that is, all vectors of the form

$$T\begin{bmatrix} x_1 \\ x_2 \end{bmatrix} = A\begin{bmatrix} x_1 \\ x_2 \end{bmatrix} = \begin{bmatrix} 1 & 3 \\ 2 & 6 \end{bmatrix}\begin{bmatrix} x_1 \\ x_2 \end{bmatrix} = x_1\begin{bmatrix} 1 \\ 2 \end{bmatrix} + x_2\begin{bmatrix} 3 \\ 6 \end{bmatrix}$$

$$= x_1\begin{bmatrix} 1 \\ 2 \end{bmatrix} + 3x_2\begin{bmatrix} 1 \\ 2 \end{bmatrix} = (x_1 + 3x_2)\begin{bmatrix} 1 \\ 2 \end{bmatrix}.$$

Since the vectors $\begin{bmatrix} 1 \\ 2 \end{bmatrix}$ and $\begin{bmatrix} 3 \\ 6 \end{bmatrix}$ are parallel, the image of T is the line of all scalar multiples of the vector $\begin{bmatrix} 1 \\ 2 \end{bmatrix}$, as illustrated in Figure 6. ∎

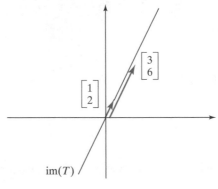

Figure 6

EXAMPLE 7   Describe the image of the linear transformation

$$T(\vec{x}) = A\vec{x} \quad \text{from } \mathbb{R}^2 \text{ to } \mathbb{R}^3, \text{ where} \quad A = \begin{bmatrix} 1 & 1 \\ 1 & 2 \\ 1 & 3 \end{bmatrix}.$$

Solution

The image of T consists of all vectors of the form

$$T\begin{bmatrix} x_1 \\ x_2 \end{bmatrix} = \begin{bmatrix} 1 & 1 \\ 1 & 2 \\ 1 & 3 \end{bmatrix} \begin{bmatrix} x_1 \\ x_2 \end{bmatrix} = x_1 \begin{bmatrix} 1 \\ 1 \\ 1 \end{bmatrix} + x_2 \begin{bmatrix} 1 \\ 2 \\ 3 \end{bmatrix},$$

that is, all linear combinations of the column vectors of A,

$$\vec{v}_1 = \begin{bmatrix} 1 \\ 1 \\ 1 \end{bmatrix} \quad \text{and} \quad \vec{v}_2 = \begin{bmatrix} 1 \\ 2 \\ 3 \end{bmatrix}.$$

The image of T is the plane V "spanned" by the two vectors $\vec{v}_1$ and $\vec{v}_2$, that is, the plane through the origin and the end points $(1, 1, 1)$ and $(1, 2, 3)$ of $\vec{v}_1$ and $\vec{v}_2$, respectively. See Figure 7. ∎

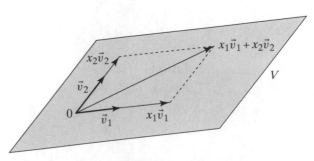

Figure 7

The observations we made in Examples 6 and 7 motivate the following definition.

Definition 3.1.2   Span

Consider the vectors $\vec{v}_1, \ldots, \vec{v}_m$ in $\mathbb{R}^n$. The set of all linear combinations $c_1\vec{v}_1 + \cdots + c_m\vec{v}_m$ of the vectors $\vec{v}_1, \ldots, \vec{v}_m$ is called their *span*:

$$\text{span}(\vec{v}_1, \ldots, \vec{v}_m) = \{c_1\vec{v}_1 + \cdots + c_m\vec{v}_m : c_1, \ldots, c_m \text{ in } \mathbb{R}\}.$$

Theorem 3.1.3   Image of a linear transformation

The image of a linear transformation $T(\vec{x}) = A\vec{x}$ is the span of the column vectors of A.[2] We denote the image of T by $\text{im}(T)$ or $\text{im}(A)$.   ■

To justify this fact, we write the transformation T in vector form as in Examples 6 and 7:

$$T(\vec{x}) = A\vec{x} = \begin{bmatrix} \vec{v}_1 & \cdots & \vec{v}_m \end{bmatrix} \begin{bmatrix} x_1 \\ \vdots \\ x_m \end{bmatrix} = x_1\vec{v}_1 + \cdots + x_m\vec{v}_m.$$

This shows that the image of T consists of all linear combinations of the column vectors $\vec{v}_1, \ldots, \vec{v}_m$ of matrix A. Thus $\text{im}(T)$ is the span of the vectors $\vec{v}_1, \ldots, \vec{v}_m$.

The image of a linear transformation has some noteworthy properties.

Theorem 3.1.4   Some properties of the image

The image of a linear transformation T (from $\mathbb{R}^m$ to $\mathbb{R}^n$) has the following properties:

  a. The zero vector $\vec{0}$ in $\mathbb{R}^n$ is in the image of T.
  b. The image of T is *closed under addition*: If $\vec{v}_1$ and $\vec{v}_2$ are in the image of T, then so is $\vec{v}_1 + \vec{v}_2$.
  c. The image of T is *closed under scalar multiplication*: If $\vec{v}$ is in the image of T and k is an arbitrary scalar, then $k\vec{v}$ is in the image of T as well.

Proof
  a. $\vec{0} = A\vec{0} = T(\vec{0})$.
  b. There exist vectors $\vec{w}_1$ and $\vec{w}_2$ in $\mathbb{R}^m$ such that $\vec{v}_1 = T(\vec{w}_1)$ and $\vec{v}_2 = T(\vec{w}_2)$. Then $\vec{v}_1 + \vec{v}_2 = T(\vec{w}_1) + T(\vec{w}_2) = T(\vec{w}_1 + \vec{w}_2)$, so that $\vec{v}_1 + \vec{v}_2$ is in the image of T as well.
  c. If $\vec{v} = T(\vec{w})$, then $k\vec{v} = kT(\vec{w}) = T(k\vec{w})$.   ■

It follows from properties (b) and (c) that the image of T is *closed under linear combinations*: If some vectors $\vec{v}_1, \ldots, \vec{v}_p$ are in the image and $c_1, \ldots, c_p$ are arbitrary scalars, then $c_1\vec{v}_1 + \cdots + c_p\vec{v}_p$ is in the image as well. In Figure 8 we illustrate this property in the case when $p = 2$ and $n = 3$ (that is, the target space of T is $\mathbb{R}^3$).

EXAMPLE 8   Consider an $n \times n$ matrix A. Show that $\text{im}(A^2)$ is a subset of $\text{im}(A)$, that is, each vector in $\text{im}(A^2)$ is also in $\text{im}(A)$.

Solution

Consider a vector $\vec{b} = A^2\vec{v} = AA\vec{v}$ in the image of A^2. We can write $\vec{b} = A(A\vec{v}) = A\vec{w}$, where $\vec{w} = A\vec{v}$. The equation $\vec{b} = A\vec{w}$ shows that $\vec{b}$ is in the image of A. See Figure 9.   ■

---

[2]The image of T is also called the *column space* of A.

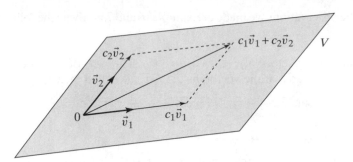

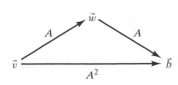

Figure 8 If $\vec{v}_1$ and $\vec{v}_2$ are in the image, then so are all vectors in the plane V in $\mathbb{R}^3$ spanned by $\vec{v}_1$ and $\vec{v}_2$.

Figure 9

The Kernel of a Linear Transformation

When you study functions $y = f(x)$ of one variable, you are often interested in the *zeros* of $f(x)$, that is, the solutions of the equation $f(x) = 0$. For example, the function $y = \sin(x)$ has infinitely many zeros, namely, all integer multiples of π.

The zeros of a linear transformation are of interest as well.

Definition 3.1.5   Kernel

The *kernel*[3] of a linear transformation $T(\vec{x}) = A\vec{x}$ from $\mathbb{R}^m$ to $\mathbb{R}^n$ consists of all zeros of the transformation, that is, the solutions of the equation $T(\vec{x}) = A\vec{x} = \vec{0}$. See Figure 10, where we show the kernel along with the image.

In other words, the kernel of T is the solution set of the linear system

$$A\vec{x} = \vec{0}.$$

We denote the kernel of T by $\ker(T)$ or $\ker(A)$.

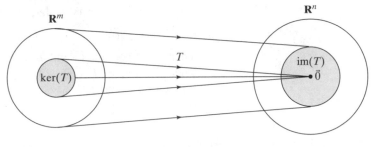

Figure 10

For a linear transformation T from $\mathbb{R}^m$ to $\mathbb{R}^n$,

- $\operatorname{im}(T) = \{T(\vec{x}) : \vec{x} \text{ in } \mathbb{R}^m\}$ is a subset of the *target space* $\mathbb{R}^n$ of T, and
- $\ker(T) = \{\vec{x} \text{ in } \mathbb{R}^m : T(\vec{x}) = \vec{0}\}$ is a subset of the *domain* $\mathbb{R}^m$ of T.

EXAMPLE 9   Consider the linear transformation T from $\mathbb{R}^3$ to $\mathbb{R}^3$ that projects a vector orthogonally into the x_1–x_2-plane (see Example 5 and Figure 5).

The kernel of T consists of the solutions of the equation $T(\vec{x}) = \vec{0}$, that is, the vectors whose orthogonal projection onto the x_1–x_2-plane is zero. Those are the vectors on the x_3-axis, that is, the scalar multiples of $\vec{e}_3$. See Figure 11. ∎

---

[3]The kernel of T is also called the *null space* of A.

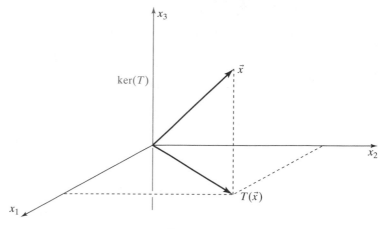

Figure 11

EXAMPLE 10   Find the kernel of the linear transformation

$$T(\vec{x}) = \begin{bmatrix} 1 & 1 & 1 \\ 1 & 2 & 3 \end{bmatrix} \vec{x} \quad \text{from } \mathbb{R}^3 \text{ to } \mathbb{R}^2.$$

Solution

We have to solve the linear system

$$T(\vec{x}) = \begin{bmatrix} 1 & 1 & 1 \\ 1 & 2 & 3 \end{bmatrix} \vec{x} = \vec{0}.$$

Since we have studied this kind of problem carefully in Section 1.2, we can be brief.

$$\text{rref} \begin{bmatrix} 1 & 1 & 1 & | & 0 \\ 1 & 2 & 3 & | & 0 \end{bmatrix} = \begin{bmatrix} 1 & 0 & -1 & | & 0 \\ 0 & 1 & 2 & | & 0 \end{bmatrix}$$

$$\begin{vmatrix} x_1 & - & x_3 = 0 \\ & x_2 + 2x_3 = 0 \end{vmatrix}, \quad \text{or} \quad \begin{vmatrix} x_1 = & x_3 \\ x_2 = & -2x_3 \end{vmatrix}$$

$$\begin{bmatrix} x_1 \\ x_2 \\ x_3 \end{bmatrix} = \begin{bmatrix} t \\ -2t \\ t \end{bmatrix} = t \begin{bmatrix} 1 \\ -2 \\ 1 \end{bmatrix}, \qquad \text{where } t \text{ is an arbitrary constant.}$$

The kernel of T is the line spanned by $\begin{bmatrix} 1 \\ -2 \\ 1 \end{bmatrix}$ in $\mathbb{R}^3$. ■

Consider a linear transformation $T(\vec{x}) = A\vec{x}$ from $\mathbb{R}^m$ to $\mathbb{R}^n$, where m exceeds n (as in Example 10, where $m = 3$ and $n = 2$). There will be free variables for the equation $T(\vec{x}) = A\vec{x} = \vec{0}$; that is, this system has infinitely many solutions. Therefore the kernel of T consists of infinitely many vectors. This agrees with our intuition: We expect some collapsing to take place as we transform the "large" vector space $\mathbb{R}^m$ into the "smaller" $\mathbb{R}^n$. (Recall that the kernel consists of everything that "collapses to zero.")

EXAMPLE 11   Find the kernel of the linear transformation $T(\vec{x}) = A\vec{x}$ from $\mathbb{R}^5$ to $\mathbb{R}^4$, where

$$A = \begin{bmatrix} 1 & 2 & 2 & -5 & 6 \\ -1 & -2 & -1 & 1 & -1 \\ 4 & 8 & 5 & -8 & 9 \\ 3 & 6 & 1 & 5 & -7 \end{bmatrix}.$$

Solution

We have to solve the linear system

$$T(\vec{x}) = A\vec{x} = \vec{0}.$$

We leave it to the reader to verify that

$$\text{rref}(A) = \begin{bmatrix} 1 & 2 & 0 & 3 & -4 \\ 0 & 0 & 1 & -4 & 5 \\ 0 & 0 & 0 & 0 & 0 \\ 0 & 0 & 0 & 0 & 0 \end{bmatrix}.$$

There is no need to write the zeros in the last column of the augmented matrix. The kernel of T consists of the solutions of the system

$$\begin{vmatrix} x_1 + 2x_2 & + 3x_4 - 4x_5 = 0 \\ x_3 - 4x_4 + 5x_5 = 0 \end{vmatrix}, \quad \text{or,} \quad \begin{vmatrix} x_1 = -2x_2 - 3x_4 + 4x_5 \\ x_3 = \phantom{-2x_2 - {}} 4x_4 - 5x_5 \end{vmatrix}$$

The solutions are the vectors of the form

$$\vec{x} = \begin{bmatrix} x_1 \\ x_2 \\ x_3 \\ x_4 \\ x_5 \end{bmatrix} = \begin{bmatrix} -2s & -3t & +4r \\ s & & \\ & 4t & -5r \\ & t & \\ & & r \end{bmatrix} = s\begin{bmatrix} -2 \\ 1 \\ 0 \\ 0 \\ 0 \end{bmatrix} + t\begin{bmatrix} -3 \\ 0 \\ 4 \\ 1 \\ 0 \end{bmatrix} + r\begin{bmatrix} 4 \\ 0 \\ -5 \\ 0 \\ 1 \end{bmatrix}$$

where s, t, and r are arbitrary constants. Using the concept of the *span* introduced in Definition 3.1.2, we can write

$$\ker(T) = \text{span}\left(\begin{bmatrix} -2 \\ 1 \\ 0 \\ 0 \\ 0 \end{bmatrix}, \begin{bmatrix} -3 \\ 0 \\ 4 \\ 1 \\ 0 \end{bmatrix}, \begin{bmatrix} 4 \\ 0 \\ -5 \\ 0 \\ 1 \end{bmatrix} \right). \qquad \blacksquare$$

The kernel has some remarkable properties, analogous to the properties of the image listed in Theorem 3.1.4.

Theorem 3.1.6 **Some properties of the kernel**

Consider a linear transfomration T from $\mathbb{R}^m$ to $\mathbb{R}^n$.

a. The zero vector $\vec{0}$ in $\mathbb{R}^m$ is in the kernel of T.

b. The kernel is closed under addition.

c. The kernel is closed under scalar multiplication. $\qquad \blacksquare$

The verification of Theorem 3.1.6 is left as Exercise 49.

EXAMPLE 12 For an invertible $n \times n$ matrix A, find $\ker(A)$.

Solution

By Theorem 2.4.4b, the system

$$A\vec{x} = \vec{0}$$

has the sole solution $\vec{x} = \vec{0}$, so that $\ker(A) = \{\vec{0}\}$. $\qquad \blacksquare$

Conversely, if A is a noninvertible $n \times n$ matrix, then $\ker(A) \neq \{\vec{0}\}$, meaning that the kernel consists of more than just the zero vector (again by Theorem 2.4.4b).

EXAMPLE 13   For which $n \times m$ matrices A is $\ker(A) = \{\vec{0}\}$? Give your answer in terms of the rank of A.

Solution

It is required that there be no free variables for the system $A\vec{x} = \vec{0}$, meaning that all m variables are leading variables. Thus we want $\text{rank}(A) = m$, since the rank is the number of leading variables.   ∎

Let us summarize the results of Examples 12 and 13.

Theorem 3.1.7   When is $\ker(A) = \{\vec{0}\}$?

    a. Consider an $n \times m$ matrix A. Then $\ker(A) = \{\vec{0}\}$ if (and only if) $\text{rank}(A) = m$.

    b. Consider an $n \times m$ matrix A. If $\ker(A) = \{\vec{0}\}$, then $m \leq n$. Equivalently, if $m > n$, then there are nonzero vectors in the kernel of A.

    c. For a *square* matrix A, we have $\ker(A) = \{\vec{0}\}$ if (and only if) A is invertible.   ∎

We conclude this section with a summary that relates many concepts we have introduced thus far.

SUMMARY 3.1.8 | **Various characterizations of invertible matrices**

For an $n \times n$ matrix A, the following statements are equivalent; that is, for a given A they are either all true or all false.

    i. A is invertible.

    ii. The linear system $A\vec{x} = \vec{b}$ has a unique solution $\vec{x}$, for all $\vec{b}$ in $\mathbb{R}^n$.

    iii. $\text{rref}(A) = I_n$.

    iv. $\text{rank}(A) = n$.

    v. $\text{im}(A) = \mathbb{R}^n$.

    vi. $\ker(A) = \{\vec{0}\}$.

In Figure 12 we briefly recall the justification for these equivalences.

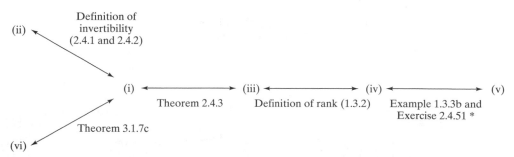

*Note that $\vec{b}$ is in the image of A if and only if the system $A\vec{x} = \vec{b}$ is consistent (by definition of the image).

Figure 12

EXERCISES 3.1

GOAL   *Use the concepts of the image and the kernel of a linear transformation (or a matrix). Express the image and the kernel of any matrix as the span of some vectors. Use kernel and image to determine whether a matrix is invertible.*

For each matrix A in Exercises 1 through 13, find vectors that span the kernel of A. Use paper and pencil.

1. $A = \begin{bmatrix} 1 & 2 \\ 3 & 4 \end{bmatrix}$ **2.** $A = \begin{bmatrix} 2 & 3 \\ 6 & 9 \end{bmatrix}$

3. $A = \begin{bmatrix} 0 & 0 \\ 0 & 0 \end{bmatrix}$ **4.** $A = \begin{bmatrix} 1 & 2 & 3 \end{bmatrix}$

5. $A = \begin{bmatrix} 1 & 1 & 1 \\ 1 & 2 & 3 \\ 1 & 3 & 5 \end{bmatrix}$ **6.** $A = \begin{bmatrix} 1 & 1 & 1 \\ 1 & 1 & 1 \\ 1 & 1 & 1 \end{bmatrix}$

7. $A = \begin{bmatrix} 1 & 2 & 3 \\ 1 & 3 & 2 \\ 3 & 2 & 1 \end{bmatrix}$ **8.** $A = \begin{bmatrix} 1 & 1 & 1 \\ 1 & 2 & 3 \end{bmatrix}$

9. $A = \begin{bmatrix} 1 & 1 \\ 1 & 2 \\ 1 & 3 \end{bmatrix}$ **10.** $A = \begin{bmatrix} 1 & 2 & 3 & 4 \\ 0 & 1 & 2 & 3 \\ 0 & 0 & 0 & 1 \end{bmatrix}$

11. $A = \begin{bmatrix} 1 & 0 & 2 & 4 \\ 0 & 1 & -3 & -1 \\ 3 & 4 & -6 & 8 \\ 0 & -1 & 3 & 4 \end{bmatrix}$

12. $A = \begin{bmatrix} 1 & -1 & -1 & 1 & 1 \\ -1 & 1 & 0 & -2 & 2 \\ 1 & -1 & -2 & 0 & 3 \\ 2 & -2 & -1 & 3 & 4 \end{bmatrix}$

13. $A = \begin{bmatrix} 1 & 2 & 0 & 0 & 3 & 0 \\ 0 & 0 & 1 & 0 & 2 & 0 \\ 0 & 0 & 0 & 1 & 1 & 0 \\ 0 & 0 & 0 & 0 & 0 & 0 \\ 0 & 0 & 0 & 0 & 0 & 0 \end{bmatrix}$

For each matrix A in Exercises 14 through 16, find vectors that span the image of A. Give as few vectors as possible. Use paper and pencil.

14. $A = \begin{bmatrix} 1 & 1 \\ 1 & 2 \\ 1 & 3 \\ 1 & 4 \end{bmatrix}$ **15.** $A = \begin{bmatrix} 1 & 1 & 1 & 1 \\ 1 & 2 & 3 & 4 \end{bmatrix}$

16. $A = \begin{bmatrix} 1 & 2 & 3 \\ 1 & 2 & 3 \\ 1 & 2 & 3 \end{bmatrix}$

For each matrix A in Exercises 17 through 22, describe the image of the transformation $T(\vec{x}) = A\vec{x}$ geometrically (as a line, plane, etc. in $\mathbb{R}^2$ or $\mathbb{R}^3$).

17. $A = \begin{bmatrix} 1 & 2 \\ 3 & 4 \end{bmatrix}$ **18.** $A = \begin{bmatrix} 1 & 4 \\ 3 & 12 \end{bmatrix}$

19. $A = \begin{bmatrix} 1 & 2 & 3 & 4 \\ -2 & -4 & -6 & -8 \end{bmatrix}$

20. $A = \begin{bmatrix} 1 & 1 & 1 \\ 1 & 1 & 1 \\ 1 & 1 & 1 \end{bmatrix}$ **21.** $A = \begin{bmatrix} 4 & 7 & 3 \\ 1 & 9 & 2 \\ 5 & 6 & 8 \end{bmatrix}$

22. $A = \begin{bmatrix} 2 & 1 & 3 \\ 3 & 4 & 2 \\ 6 & 5 & 7 \end{bmatrix}$

Describe the images and kernels of the transformations in Exercises 23 through 25 geometrically.

23. Reflection about the line $y = x/3$ in $\mathbb{R}^2$

24. Orthogonal projection onto the plane $x + 2y + 3z = 0$ in $\mathbb{R}^3$

25. Rotation through an angle of $\pi/4$ in the counterclockwise direction (in $\mathbb{R}^2$)

26. What is the image of a function f from $\mathbb{R}$ to $\mathbb{R}$ given by

$$f(t) = t^3 + at^2 + bt + c,$$

where a, b, c are arbitrary scalars?

27. Give an example of a noninvertible function f from $\mathbb{R}$ to $\mathbb{R}$ with $\operatorname{im}(f) = \mathbb{R}$.

28. Give an example of a parametrization of the ellipse

$$x^2 + \frac{y^2}{4} = 1$$

in $\mathbb{R}^2$. (See Example 3.)

29. Give an example of a function whose image is the unit sphere

$$x^2 + y^2 + z^2 = 1$$

in $\mathbb{R}^3$.

30. Give an example of a matrix A such that $\operatorname{im}(A)$ is spanned by the vector $\begin{bmatrix} 1 \\ 5 \end{bmatrix}$.

31. Give an example of a matrix A such that $\operatorname{im}(A)$ is the plane with normal vector $\begin{bmatrix} 1 \\ 3 \\ 2 \end{bmatrix}$ in $\mathbb{R}^3$.

32. Give an example of a linear transformation whose image is the line spanned by

$$\begin{bmatrix} 7 \\ 6 \\ 5 \end{bmatrix}$$

in $\mathbb{R}^3$.

33. Give an example of a linear transformation whose kernel is the plane $x + 2y + 3z = 0$ in $\mathbb{R}^3$.

34. Give an example of a linear transformation whose kernel is the line spanned by

$$\begin{bmatrix} -1 \\ 1 \\ 2 \end{bmatrix}$$

in $\mathbb{R}^3$.

35. Consider a nonzero vector $\vec{v}$ in $\mathbb{R}^3$. Arguing geometrically, describe the image and the kernel of the linear transformation T from $\mathbb{R}^3$ to $\mathbb{R}$ given by

$$T(\vec{x}) = \vec{v} \cdot \vec{x}.$$

36. Consider a nonzero vector $\vec{v}$ in $\mathbb{R}^3$. Arguing geometrically, describe the image and the kernel of the linear transformation T from $\mathbb{R}^3$ to $\mathbb{R}^3$ given by

$$T(\vec{x}) = \vec{v} \times \vec{x}.$$

See Definition A.9 in the Appendix.

37. For the matrix

$$A = \begin{bmatrix} 0 & 1 & 0 \\ 0 & 0 & 1 \\ 0 & 0 & 0 \end{bmatrix},$$

describe the images and kernels of the matrices A, A^2, and A^3 geometrically.

38. Consider a square matrix A.
  a. What is the relationship between $\ker(A)$ and $\ker(A^2)$? Are they necessarily equal? Is one of them necessarily contained in the other? More generally, what can you say about $\ker(A)$, $\ker(A^2)$, $\ker(A^3)$, $\ker(A^4)$, ...?
  b. What can you say about $\text{im}(A)$, $\text{im}(A^2)$, $\text{im}(A^3)$, ...?
  (*Hint*: Exercise 37 is helpful.)

39. Consider an $n \times p$ matrix A and a $p \times m$ matrix B.
  a. What is the relationship between $\ker(AB)$ and $\ker(B)$? Are they always equal? Is one of them always contained in the other?
  b. What is the relationship between $\text{im}(A)$ and $\text{im}(AB)$?

40. Consider an $n \times p$ matrix A and a $p \times m$ matrix B. If $\ker(A) = \text{im}(B)$, what can you say about the product AB?

41. Consider the matrix $A = \begin{bmatrix} 0.36 & 0.48 \\ 0.48 & 0.64 \end{bmatrix}$.
  a. Describe $\ker(A)$ and $\text{im}(A)$ geometrically.
  b. Find A^2. If $\vec{v}$ is in the image of A, what can you say about $A\vec{v}$?
  c. Describe the transformation $T(\vec{x}) = A\vec{x}$ geometrically.

42. Express the image of the matrix

$$A = \begin{bmatrix} 1 & 1 & 1 & 6 \\ 1 & 2 & 3 & 4 \\ 1 & 3 & 5 & 2 \\ 1 & 4 & 7 & 0 \end{bmatrix}$$

as the kernel of a matrix B. *Hint*: The image of A consists of all vectors $\vec{y}$ in $\mathbb{R}^4$ such that the system $A\vec{x} = \vec{y}$ is consistent. Write this system more explicitly:

$$\begin{vmatrix} x_1 + x_2 + x_3 + 6x_4 = y_1 \\ x_1 + 2x_2 + 3x_3 + 4x_4 = y_2 \\ x_1 + 3x_2 + 5x_3 + 2x_4 = y_3 \\ x_1 + 4x_2 + 7x_3 \qquad = y_4 \end{vmatrix}.$$

Now, reduce rows:

$$\begin{vmatrix} x_1 \quad - \quad x_3 + 8x_4 = \quad 4y_3 - 3y_4 \\ x_2 + 2x_3 - 2x_4 = \quad - y_3 + y_4 \\ 0 \qquad\qquad = y_1 \quad - 3y_3 + 2y_4 \\ 0 \qquad\qquad = \quad y_2 - 2y_3 + y_4 \end{vmatrix}.$$

For which vectors $\vec{y}$ is this system consistent? The answer allows you to express $\text{im}(A)$ as the kernel of a 2×4 matrix B.

43. Using your work in Exercise 42 as a guide, explain how you can write the image of any matrix A as the kernel of some matrix B.

44. Consider a matrix A, and let $B = \text{rref}(A)$.
  a. Is $\ker(A)$ necessarily equal to $\ker(B)$? Explain.
  b. Is $\text{im}(A)$ necessarily equal to $\text{im}(B)$? Explain.

45. Consider an $n \times m$ matrix A with $\text{rank}(A) = r < m$. Explain how you can write $\ker(A)$ as the span of $m - r$ vectors.

46. Consider a 3×4 matrix A in reduced row-echelon form. What can you say about the image of A? Describe all cases in terms of $\text{rank}(A)$, and draw a sketch for each.

47. Let T be the projection along a line L_1 onto a line L_2. (See Exercise 2.2.33.) Describe the image and the kernel of T geometrically.

48. Consider a 2×2 matrix A with $A^2 = A$.

   a. If $\vec{w}$ is in the image of A, what is the relationship between $\vec{w}$ and $A\vec{w}$?

   b. What can you say about A if rank$(A) = 2$? What if rank$(A) = 0$?

   c. If rank$(A) = 1$, show that the linear transformation $T(\vec{x}) = A\vec{x}$ is the projection onto im(A) along ker(A). (See Exercise 2.2.33.)

49. Verify that the kernel of a linear transformation is closed under addition and scalar multiplication. (See Theorem 3.1.6.)

50. Consider a square matrix A with ker$(A^2) = $ ker(A^3). Is ker$(A^3) = $ ker(A^4)? Justify your answer.

51. Consider an $n \times p$ matrix A and a $p \times m$ matrix B such that ker$(A) = \{\vec{0}\}$ and ker$(B) = \{\vec{0}\}$. Find ker(AB).

52. Consider a $p \times m$ matrix A and a $q \times m$ matrix B, and form the block matrix

$$C = \begin{bmatrix} A \\ B \end{bmatrix}.$$

What is the relationship between ker(A), ker(B), and ker(C)?

53. In Exercises 53 and 54, we will work with the binary digits (or bits) 0 and 1, instead of the real numbers $\mathbb{R}$. Addition and multiplication in this system are defined as usual, except for the rule $1 + 1 = 0$. We denote this number system with $\mathbb{F}_2$, or simply $\mathbb{F}$. The set of all vectors with n components in $\mathbb{F}$ is denoted by $\mathbb{F}^n$; note that $\mathbb{F}^n$ consists of 2^n vectors. (Why?) In information technology, a vector in $\mathbb{F}^8$ is called a *byte*. (A byte is a string of 8 binary digits.)

The basic ideas of linear algebra introduced so far (for the real numbers) apply to $\mathbb{F}$ without modifications.

A *Hamming matrix* with n rows is a matrix that contains all nonzero vectors in $\mathbb{F}^n$ as its columns (in any order). Note that there are $2^n - 1$ columns. Here is an example:

$$H = \begin{bmatrix} 1 & 0 & 0 & 1 & 0 & 1 & 1 \\ 0 & 1 & 0 & 1 & 1 & 0 & 1 \\ 0 & 0 & 1 & 1 & 1 & 1 & 0 \end{bmatrix}, \quad \begin{matrix} 3 \text{ rows} \\ 2^3 - 1 = 7 \text{ columns.} \end{matrix}$$

   a. Express the kernel of H as the span of four vectors in $\mathbb{F}^7$ of the form

$$\vec{v}_1 = \begin{bmatrix} * \\ * \\ * \\ 1 \\ 0 \\ 0 \\ 0 \end{bmatrix}, \ \vec{v}_2 = \begin{bmatrix} * \\ * \\ * \\ 0 \\ 1 \\ 0 \\ 0 \end{bmatrix}, \ \vec{v}_3 = \begin{bmatrix} * \\ * \\ * \\ 0 \\ 0 \\ 1 \\ 0 \end{bmatrix}, \ \vec{v}_4 = \begin{bmatrix} * \\ * \\ * \\ 0 \\ 0 \\ 0 \\ 1 \end{bmatrix}.$$

   b. Form the 7×4 matrix

$$M = \begin{bmatrix} | & | & | & | \\ \vec{v}_1 & \vec{v}_2 & \vec{v}_3 & \vec{v}_4 \\ | & | & | & | \end{bmatrix}.$$

Explain why im$(M) = $ ker(H). If $\vec{x}$ is an arbitrary vector in $\mathbb{F}^4$, what is $H(M\vec{x})$?

54. (See Exercise 53 for some background.) When information is transmitted, there may be some errors in the communication. We present a method of adding extra information to messages so that most errors that occur during transmission can be detected and corrected. Such methods are referred to as *error-correcting codes*. (Compare these with codes whose purpose is to conceal information.) The pictures of man's first landing on the moon (in 1969) were televised just as they had been received and were not very clear, since they contained many errors induced during transmission. On later missions, much clearer error-corrected pictures were obtained.

In computers, information is stored and processed in the form of strings of binary digits, 0 and 1. This stream of binary digits is often broken up into "blocks" of eight binary digits (bytes). For the sake of simplicity, we will work with blocks of only four binary digits (i.e., with vectors in $\mathbb{F}^4$), for example,

$$\cdots \mid \ 1 \ 0 \ 1 \ 1 \mid 1 \ 0 \ 0 \ 1 \mid 1 \ 0 \ 1 \ 0 \mid 1 \ 0 \ 1 \ 1 \mid$$
$$1 \ 0 \ 0 \ 0 \mid \cdots.$$

Suppose these vectors in $\mathbb{F}^4$ have to be transmitted from one computer to another, say, from a satellite to ground control in Kourou, French Guiana (the station of the European Space Agency). A vector $\vec{u}$ in $\mathbb{F}^4$ is first transformed into a vector $\vec{v} = M\vec{u}$ in $\mathbb{F}^7$, where M is the matrix you found in Exercise 53. The last four entries of $\vec{v}$ are just the entries of $\vec{u}$; the first three entries of $\vec{v}$ are added to detect errors. The vector $\vec{v}$ is now transmitted to Kourou. We assume that at most one error will occur during transmission; that is, the vector $\vec{w}$ received in Kourou will be either $\vec{v}$ (if no error has occurred) or $\vec{w} = \vec{v} + \vec{e}_i$ (if there is an error in the ith component of the vector).

   a. Let H be the Hamming matrix introduced in Exercise 53. How can the computer in Kourou use $H\vec{w}$ to determine whether there was an error in the transmission? If there was no error, what is $H\vec{w}$? If there was an error, how can the computer determine in which component the error was made?

b. Suppose the vector

$$\vec{w} = \begin{bmatrix} 1 \\ 0 \\ 1 \\ 0 \\ 1 \\ 0 \\ 0 \end{bmatrix}$$

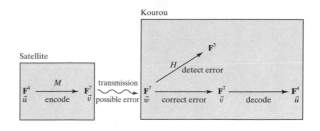

is received in Kourou. Determine whether an error was made in the transmission and, if so, correct it. (That is, find $\vec{v}$ and $\vec{u}$.)

3.2 Subspaces of $\mathbb{R}^n$; Bases and Linear Independence

In the last section, we saw that both the image and the kernel of a linear transformation contain the zero vector (of the target space and the domain, respectively), are closed under addition, and are closed under scalar multiplication. Subsets of the vector space $\mathbb{R}^n$ with these three properties are called (linear) *subspaces* of $\mathbb{R}^n$.

Definition 3.2.1   Subspaces of $\mathbb{R}^n$

A subset W of the vector space $\mathbb{R}^n$ is called a (linear) *subspace of* $\mathbb{R}^n$ if it has the following three properties:

a. W contains the zero vector in $\mathbb{R}^n$.

b. W is closed under addition: If $\vec{w}_1$ and $\vec{w}_2$ are both in W, then so is $\vec{w}_1 + \vec{w}_2$.

c. W is closed under scalar multiplication: If $\vec{w}$ is in W and k is an arbitrary scalar, then $k\vec{w}$ is in W.

Properties (b) and (c) together mean that W is *closed under linear combinations*: If vectors $\vec{w}_1, \ldots, \vec{w}_m$ are in W and $k_1, \ldots, k_m$ are scalars, then the linear combination $k_1\vec{w}_1 + \cdots + k_m\vec{w}_m$ is in W as well.

Theorems 3.1.4 and 3.1.6 tell us the following:

Theorem 3.2.2   Image and kernel are subspaces

If $T(\vec{x}) = A\vec{x}$ is a linear transformation from $\mathbb{R}^m$ to $\mathbb{R}^n$, then

- $\ker(T) = \ker(A)$ is a subspace of $\mathbb{R}^m$, and
- $\text{image}(T) = \text{im}(A)$ is a subspace of $\mathbb{R}^n$. ∎

EXAMPLE 1   Is $W = \left\{ \begin{bmatrix} x \\ y \end{bmatrix} \text{ in } \mathbb{R}^2 : x \geq 0 \text{ and } y \geq 0 \right\}$ a subspace of $\mathbb{R}^2$?

Solution

Note that W consists of all vectors in the first quadrant of the x–y-plane, including the positive axes and the origin, as illustrated in Figure 1.

W contains the zero vector and is closed under addition, but it is not closed under multiplication with a *negative* scalar. (See Figure 2.) Thus W fails to be a subspace of $\mathbb{R}^2$. ∎

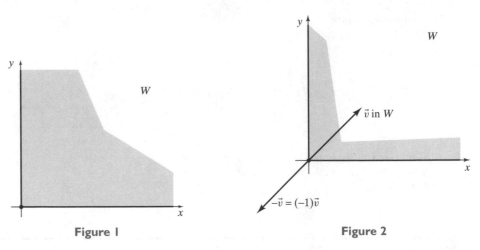

Figure I   Figure 2

EXAMPLE 2   Show that the only subspaces of $\mathbb{R}^2$ are $\mathbb{R}^2$ itself, the set $\{\vec{0}\}$, and any of the lines through the origin.

Solution

Suppose W is a subspace of $\mathbb{R}^2$ that is neither $\{\vec{0}\}$ nor a line through the origin. We have to show that W must equal $\mathbb{R}^2$. Consider a nonzero vector $\vec{v}_1$ in W. (We can find such a vector since $W \neq \{\vec{0}\}$.) The line L spanned by $\vec{v}_1$ is a subset of W, since W is closed under scalar multiplication; but W does not equal L, since W isn't a line. Consider a vector $\vec{v}_2$ in W that isn't on L (see Figure 3). Using a parallelogram, we can express any vector $\vec{v}$ in $\mathbb{R}^2$ as a linear combination of $\vec{v}_1$ and $\vec{v}_2$. Therefore, $\vec{v}$ belongs to W, since W is closed under linear combinations. This shows that $W = \mathbb{R}^2$, as claimed.   ∎

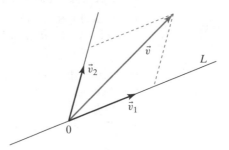

Figure 3

Similarly, the only subspaces of $\mathbb{R}^3$ are $\mathbb{R}^3$ itself, the planes through the origin, the lines through the origin, and the set $\{\vec{0}\}$. (See Exercise 5.) Note the hierarchy of subspaces, arranged according to their dimensions. (The concept of dimension will be made precise in the next section.)

	Subspaces of $\mathbb{R}^2$	Subspaces of $\mathbb{R}^3$
Dimension 3		$\mathbb{R}^3$
Dimension 2	$\mathbb{R}^2$	Planes through $\vec{0}$
Dimension 1	Lines through $\vec{0}$	Lines through $\vec{0}$
Dimension 0	$\{\vec{0}\}$	$\{\vec{0}\}$

We have seen that both the kernel and the image of a linear transformation are subspaces (of domain and target space respectively). Conversely, can we express any subspace V of $\mathbb{R}^n$ as the kernel or the image of a linear transformation (or, equivalently, of a matrix)?

Let us consider an example.

EXAMPLE 3 Consider the plane V in $\mathbb{R}^3$ given by the equation $x_1 + 2x_2 + 3x_3 = 0$.

a. Find a matrix A such that $V = \ker(A)$.

b. Find a matrix B such that $V = \operatorname{im}(B)$.

Solution

a. We can write the equation $x_1 + 2x_2 + 3x_3 = 0$ as $\begin{bmatrix} 1 & 2 & 3 \end{bmatrix} \begin{bmatrix} x_1 \\ x_2 \\ x_3 \end{bmatrix} = 0$, so that $V = \ker \begin{bmatrix} 1 & 2 & 3 \end{bmatrix}$.

b. Since the image of a matrix is the span of its columns, we need to describe V as the span of some vectors. For the plane V, any two nonparallel vectors will do, for example, $\begin{bmatrix} -2 \\ 1 \\ 0 \end{bmatrix}$ and $\begin{bmatrix} -3 \\ 0 \\ 1 \end{bmatrix}$. Thus $V = \operatorname{im} \begin{bmatrix} -2 & -3 \\ 1 & 0 \\ 0 & 1 \end{bmatrix}$. ∎

A subspace of $\mathbb{R}^n$ is usually given either as the solution set of a homogeneous linear system (that is, as a kernel), as in Example 3, or as the span of some vectors (that is, as an image). Sometimes, a subspace that has been defined as a kernel must be given as an image (as in part b of Example 3), or vice versa. The transition from kernel to image is straightforward: Using Gaussian elimination, we can represent the solution set as the span of some vectors. (See Examples 10 and 11 of Section 3.1.) A method of writing the image of a matrix as a kernel is discussed in Exercises 3.1.42 and 3.1.43.

Bases and Linear Independence

EXAMPLE 4 Consider the matrix

$$A = \begin{bmatrix} 1 & 2 & 1 & 2 \\ 1 & 2 & 2 & 3 \\ 1 & 2 & 3 & 4 \end{bmatrix}.$$

Find vectors in $\mathbb{R}^3$ that span the image of A. What is the smallest number of vectors needed to span the image of A?

Solution

We know from Theorem 3.1.3 that the image of A is spanned by the four column vectors of A,

$$\vec{v}_1 = \begin{bmatrix} 1 \\ 1 \\ 1 \end{bmatrix}, \quad \vec{v}_2 = \begin{bmatrix} 2 \\ 2 \\ 2 \end{bmatrix}, \quad \vec{v}_3 = \begin{bmatrix} 1 \\ 2 \\ 3 \end{bmatrix}, \quad \vec{v}_4 = \begin{bmatrix} 2 \\ 3 \\ 4 \end{bmatrix}.$$

Figure 4 illustrates that the image of A is a plane; we don't need all four vectors to span $\operatorname{im}(A)$. We observe that $\vec{v}_2 = 2\vec{v}_1$ and $\vec{v}_4 = \vec{v}_1 + \vec{v}_3$, so that the vectors $\vec{v}_2$ and $\vec{v}_4$ are "redundant" as far as the span is concerned:

$$\operatorname{im} \begin{bmatrix} 1 & 2 & 1 & 2 \\ 1 & 2 & 2 & 3 \\ 1 & 2 & 3 & 4 \end{bmatrix} = \operatorname{span}(\vec{v}_1, \vec{v}_2, \vec{v}_3, \vec{v}_4) = \operatorname{span}(\vec{v}_1, \vec{v}_3).$$

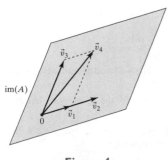
Figure 4

The image of A can be spanned by two vectors, but not by one vector alone.

im(A)

Let us verify the equation span$(\vec{v}_1, \vec{v}_2, \vec{v}_3, \vec{v}_4) = $ span$(\vec{v}_1, \vec{v}_3)$ algebraically. If a vector $\vec{v}$ is in span$(\vec{v}_1, \vec{v}_2, \vec{v}_3, \vec{v}_4)$, then

$$\begin{aligned}
\vec{v} &= c_1\vec{v}_1 + c_2\vec{v}_2 + c_3\vec{v}_3 + c_4\vec{v}_4 \\
&= c_1\vec{v}_1 + c_2(2\vec{v}_1) + c_3\vec{v}_3 + c_4(\vec{v}_1 + \vec{v}_3) \\
&= (c_1 + 2c_2 + c_4)\vec{v}_1 + (c_3 + c_4)\vec{v}_3,
\end{aligned}$$

showing that $\vec{v}$ is in span$(\vec{v}_1, \vec{v}_3)$, as claimed.   ∎

The preceding example motivates the following important definitions.

Definition 3.2.3   Redundant vectors[4]; linear independence; basis

Consider vectors $\vec{v}_1, \ldots, \vec{v}_m$ in $\mathbb{R}^n$.

a. We say that a vector $\vec{v}_i$ in the list $\vec{v}_1, \ldots, \vec{v}_m$ is *redundant* if $\vec{v}_i$ is a linear combination of the preceding vectors $\vec{v}_1, \ldots, \vec{v}_{i-1}$.[5]

b. The vectors $\vec{v}_1, \ldots, \vec{v}_m$ are called *linearly independent* if none of them is redundant. Otherwise, the vectors are called *linearly dependent* (meaning that at least one of them is redundant).[6]

c. We say that the vectors $\vec{v}_1, \ldots, \vec{v}_m$ form a *basis* of a subspace V of $\mathbb{R}^n$ if they span V and are linearly independent. (Also, it is required that vectors $\vec{v}_1, \ldots, \vec{v}_m$ be in V.)[7]

Let's take another look at Example 4: In the list

$$\vec{v}_1 = \begin{bmatrix} 1 \\ 1 \\ 1 \end{bmatrix}, \quad \vec{v}_2 = \begin{bmatrix} 2 \\ 2 \\ 2 \end{bmatrix}, \quad \vec{v}_3 = \begin{bmatrix} 1 \\ 2 \\ 3 \end{bmatrix}, \quad \vec{v}_4 = \begin{bmatrix} 2 \\ 3 \\ 4 \end{bmatrix}$$

of column vectors of A, the vectors $\vec{v}_2$ and $\vec{v}_4$ are redundant, since $\vec{v}_2 = 2\vec{v}_1$ and $\vec{v}_4 = \vec{v}_1 + \vec{v}_3$. If we omit the redundant vectors from the list, then the remaining vectors

$$\vec{v}_1 = \begin{bmatrix} 1 \\ 1 \\ 1 \end{bmatrix}, \quad \vec{v}_3 = \begin{bmatrix} 1 \\ 2 \\ 3 \end{bmatrix}$$

are linearly independent; they form a basis of $V = $ image (A).

We can generalize the result of Example 4.

Theorem 3.2.4   Basis of the image

To construct a basis of the image of a matrix A, list all the column vectors of A, and omit the redundant vectors from this list.   ∎

But how can we identify the redundant column vectors? In simple cases, this can often be done by inspection (as in Example 4); in the next section we will develop a general algorithm, based on Gaussian elimination.

---

[4]The notion of a redundant vector is not part of the established vocabulary of linear algebra. However, we will find this concept quite useful in discussing linear independence.

[5]We call the first vector, $\vec{v}_1$, redundant if it is the zero vector. This agrees with the convention that the *empty linear combination* of vectors is the zero vector.

[6]A list of alternative characterizations of linear independence will be presented in Summary 3.2.9. In many texts, characterization (iv) of that list is used to define linear independence.

[7]An alternative characterization of a basis will be presented in Theorem 3.2.10.

EXAMPLE 5   Are the following vectors in $\mathbb{R}^7$ linearly independent?

$$\vec{v}_1 = \begin{bmatrix} 7 \\ 0 \\ 4 \\ 0 \\ 1 \\ 9 \\ 0 \end{bmatrix}, \quad \vec{v}_2 = \begin{bmatrix} 6 \\ 0 \\ 7 \\ 1 \\ 4 \\ 8 \\ 0 \end{bmatrix}, \quad \vec{v}_3 = \begin{bmatrix} 5 \\ 0 \\ 6 \\ 2 \\ 3 \\ 1 \\ 7 \end{bmatrix}, \quad \vec{v}_4 = \begin{bmatrix} 4 \\ 5 \\ 3 \\ 3 \\ 2 \\ 2 \\ 4 \end{bmatrix}$$

Solution

Let's look for redundant vectors in this list. Vectors $\vec{v}_1$ and $\vec{v}_2$ are clearly nonredundant, since $\vec{v}_1$ is nonzero and $\vec{v}_2$ fails to be a scalar multiple of $\vec{v}_1$ (look at the fourth components). Looking at the last components, we realize that $\vec{v}_3$ cannot be a linear combination of $\vec{v}_1$ and $\vec{v}_2$, since any linear combination of $\vec{v}_1$ and $\vec{v}_2$ will have a 0 in the last component, while the last component of $\vec{v}_3$ is 7. Looking at the second components, we can see that $\vec{v}_4$ isn't a linear combination of $\vec{v}_1, \vec{v}_2, \vec{v}_3$. Thus the vectors $\vec{v}_1, \vec{v}_2, \vec{v}_3, \vec{v}_4$ are linearly independent.   ∎

We will frequently use the approach of Example 5 to show linear independence.

Theorem 3.2.5   Linear independence and zero components

Consider vectors $\vec{v}_1, \ldots, \vec{v}_m$ in $\mathbb{R}^n$. If $\vec{v}_1$ is nonzero, and if each of the vectors $\vec{v}_i$ (for $i \geq 2$) has a nonzero entry in a component where all the preceding vectors $\vec{v}_1, \ldots, \vec{v}_{i-1}$ have a 0, then the vectors $\vec{v}_1, \ldots, \vec{v}_m$ are linearly independent.   ∎

To understand what we are trying to say in Theorem 3.2.5, take another look at the vectors in Example 5.

$$\begin{bmatrix} 7 \\ 0 \\ 4 \\ 0 \\ 1 \\ 9 \\ 0 \end{bmatrix}, \quad \begin{bmatrix} 6 \\ 0 \\ 7 \\ \textcircled{1} \\ 4 \\ 8 \\ 0 \end{bmatrix}, \quad \begin{bmatrix} 5 \\ 0 \\ 6 \\ 2 \\ 3 \\ 1 \\ \textcircled{7} \end{bmatrix}, \quad \begin{bmatrix} 4 \\ \textcircled{5} \\ 3 \\ 3 \\ 2 \\ 2 \\ 4 \end{bmatrix}$$

EXAMPLE 6   Are the vectors

$$\vec{v}_1 = \begin{bmatrix} 1 \\ 2 \\ 3 \end{bmatrix}, \quad \vec{v}_2 = \begin{bmatrix} 4 \\ 5 \\ 6 \end{bmatrix}, \quad \vec{v}_3 = \begin{bmatrix} 7 \\ 8 \\ 9 \end{bmatrix}$$

linearly independent?

Solution

Theorem 3.2.5 doesn't help here, since these vectors don't have any zero components. The vectors $\vec{v}_1$ and $\vec{v}_2$ are clearly nonredundant, as $\vec{v}_1$ is nonzero and $\vec{v}_2$ fails to be a scalar multiple of $\vec{v}_1$. To see whether vector $\vec{v}_3$ is redundant, we need to examine whether $\vec{v}_3$ can be written as $\vec{v}_3 = c_1\vec{v}_1 + c_2\vec{v}_2$. Considering the augmented matrix

$$M = \begin{bmatrix} 1 & 4 & 7 \\ 2 & 5 & 8 \\ 3 & 6 & 9 \end{bmatrix}, \quad \text{with} \quad \text{rref}(M) = \begin{bmatrix} 1 & 0 & -1 \\ 0 & 1 & 2 \\ 0 & 0 & 0 \end{bmatrix},$$

we find the unique solution $c_1 = -1$, $c_2 = 2$, so that

$$\vec{v}_3 = -\vec{v}_1 + 2\vec{v}_2.$$

It turns out that vector $\vec{v}_3$ is redundant, making vectors $\vec{v}_1, \vec{v}_2, \vec{v}_3$ linearly dependent. ∎

For good reasons, mathematicians like to write their equations in the form

$$(\text{Something}) = 0.$$

Applying this principle,[8] we can write the equation $\vec{v}_3 = -\vec{v}_1 + 2\vec{v}_2$ from Example 6 as

$$\vec{v}_1 - 2\vec{v}_2 + \vec{v}_3 = \vec{0}.$$

This equation is called a *linear relation* among the vectors $\vec{v}_1$, $\vec{v}_2$, and $\vec{v}_3$.

Definition 3.2.6   Linear Relations

Consider the vectors $\vec{v}_1, \ldots, \vec{v}_m$ in $\mathbb{R}^n$. An equation of the form

$$c_1\vec{v}_1 + \cdots + c_m\vec{v}_m = \vec{0}$$

is called a (linear) *relation* among the vectors $\vec{v}_1, \ldots, \vec{v}_m$. There is always the *trivial relation*, with $c_1 = \cdots = c_m = 0$. *Nontrivial relations* (where at least one coefficient c_i is nonzero) may or may not exist among the vectors $\vec{v}_1, \ldots, \vec{v}_m$.

Example 6 suggests the following result.

Theorem 3.2.7   Relations and linear dependence

The vectors $\vec{v}_1, \ldots, \vec{v}_m$ in $\mathbb{R}^n$ are linearly dependent if (and only if) there are non-trivial relations among them.

Proof
- Suppose vectors $\vec{v}_1, \ldots, \vec{v}_m$ are linearly dependent, and $\vec{v}_i = c_1\vec{v}_1 + \cdots + c_{i-1}\vec{v}_{i-1}$ is a redundant vector in this list. Then we can generate a nontrivial relation by subtracting $\vec{v}_i$ from both sides: $c_1\vec{v}_1 + \cdots + c_{i-1}\vec{v}_{i-1} + (-1)\vec{v}_i = \vec{0}$.
- Conversely, if there is a nontrivial relation $c_1\vec{v}_1 + \cdots + c_i\vec{v}_i + \cdots + c_m\vec{v}_m = \vec{0}$, where i is the highest index such that $c_i \neq 0$, then we can solve for $\vec{v}_i$ and thus express $\vec{v}_i$ as a linear combination of the preceding vectors:

$$\vec{v}_i = -\frac{c_1}{c_i}\vec{v}_1 - \cdots - \frac{c_{i-1}}{c_i}\vec{v}_{i-1}.$$

This shows that vector $\vec{v}_i$ is redundant, so that vectors $\vec{v}_1, \ldots, \vec{v}_m$ are linearly dependent, as claimed. ∎

---

[8]This method was popularized by Descartes, and is often credited to him, but it was used earlier by the English geographer Thomas Harriot (1560–1621). For more on "Harriot's Principle," see W. P. Berlinghoff and F.Q. Gouvêa, *Math Through the Ages*, Oxton House Publishers and MAA, 2004.

EXAMPLE 7   Suppose the column vectors of an $n \times m$ matrix A are linearly independent. Find the kernel of matrix A.

Solution

We need to solve the equation

$$A\vec{x} = \vec{0} \quad \text{or} \quad \begin{bmatrix} | & & | \\ \vec{v}_1 & \cdots & \vec{v}_m \\ | & & | \end{bmatrix} \begin{bmatrix} x_1 \\ \vdots \\ x_m \end{bmatrix} = \vec{0} \quad \text{or} \quad x_1\vec{v}_1 + \cdots + x_m\vec{v}_m = \vec{0}.$$

We see that finding the kernel of A amounts to finding the relations among the column vectors of A. By Theorem 3.2.7, there is only the trivial relation, with $x_1 = \cdots = x_m = 0$, so that $\ker(A) = \{\vec{0}\}$.   ■

Let us summarize the findings of Example 7.

Theorem 3.2.8   Kernel and relations

The vectors in the kernel of an $n \times m$ matrix A correspond to the linear relations among the column vectors $\vec{v}_1, \ldots, \vec{v}_m$ of A: The equation

$$A\vec{x} = \vec{0} \quad \text{means that} \quad x_1\vec{v}_1 + \cdots + x_m\vec{v}_m = \vec{0}.$$

In particular, the column vectors of A are linearly independent if (and only if) $\ker(A) = \{\vec{0}\}$, or, equivalently, if $\operatorname{rank}(A) = m$. This condition implies that $m \leq n$.

Thus we can find at most n linearly independent vectors in $\mathbb{R}^n$.   ■

EXAMPLE 8   Consider the matrix

$$A = \begin{bmatrix} 1 & 4 & 7 \\ 2 & 5 & 8 \\ 3 & 6 & 9 \end{bmatrix}$$

to illustrate the connection between redundant column vectors, relations among the column vectors, and the kernel. See Example 6.

$$\text{Redundant column vector:} \quad \begin{bmatrix} 7 \\ 8 \\ 9 \end{bmatrix} = -\begin{bmatrix} 1 \\ 2 \\ 3 \end{bmatrix} + 2\begin{bmatrix} 4 \\ 5 \\ 6 \end{bmatrix}$$

$$\updownarrow$$

$$\text{Relation among column vectors:} \quad 1\begin{bmatrix} 1 \\ 2 \\ 3 \end{bmatrix} - 2\begin{bmatrix} 4 \\ 5 \\ 6 \end{bmatrix} + 1\begin{bmatrix} 7 \\ 8 \\ 9 \end{bmatrix} = \vec{0}$$

$$\updownarrow$$

$$\text{Vector} \begin{bmatrix} 1 \\ -2 \\ 1 \end{bmatrix} \text{ is in } \ker(A), \text{ since} \quad \begin{bmatrix} 1 & 4 & 7 \\ 2 & 5 & 8 \\ 3 & 6 & 9 \end{bmatrix} \begin{bmatrix} 1 \\ -2 \\ 1 \end{bmatrix} = \begin{bmatrix} 0 \\ 0 \\ 0 \end{bmatrix}$$

    ■

In the following summary we list the various characterizations of linear independence discussed thus far (in Definition 3.2.3b, Theorem 3.2.7, and Theorem 3.2.8). We include one new characterization, (iii). The proof of the equivalence of statements (iii) and (iv) is left to the reader as Exercise 35; it is analogous to the proof of Theorem 3.2.7.

SUMMARY 3.2.9 | **Various characterizations of linear independence**

For a list $\vec{v}_1, \ldots, \vec{v}_m$ of vectors in $\mathbb{R}^n$, the following statements are equivalent:

i. Vectors $\vec{v}_1, \ldots, \vec{v}_m$ are linearly independent.

ii. None of the vectors $\vec{v}_1, \ldots, \vec{v}_m$ is redundant, meaning that none of them is a linear combination of preceding vectors.

iii. None of the vectors $\vec{v}_i$ is a linear combination of the other vectors $\vec{v}_1, \ldots, \vec{v}_{i-1}, \vec{v}_{i+1}, \ldots, \vec{v}_m$ in the list.

iv. There is only the trivial relation among the vectors $\vec{v}_1, \ldots, \vec{v}_m$, meaning that the equation $c_1 \vec{v}_1 + \cdots + c_m \vec{v}_m = \vec{0}$ has only the solution $c_1 = \cdots = c_m = 0$.

v. $\ker \begin{bmatrix} \vec{v}_1 & \cdots & \vec{v}_m \end{bmatrix} = \{\vec{0}\}$.

vi. $\operatorname{rank} \begin{bmatrix} \vec{v}_1 & \cdots & \vec{v}_m \end{bmatrix} = m$.

We conclude this section with an important alternative characterization of a basis. (See Definition 3.2.3c.)

EXAMPLE 9　If $\vec{v}_1, \ldots, \vec{v}_m$ is a basis of a subspace V of $\mathbb{R}^n$, and if $\vec{v}$ is a vector in V, how many solutions $c_1, \ldots, c_m$ does the equation

$$\vec{v} = c_1 \vec{v}_1 + \cdots + c_m \vec{v}_m$$

have?

Solution

There is at least one solution, since the vectors $\vec{v}_1, \ldots, \vec{v}_m$ span V (that's part of the definition of a basis). Suppose we have two representations

$$\vec{v} = c_1 \vec{v}_1 + \cdots + c_m \vec{v}_m$$
$$= d_1 \vec{v}_1 + \cdots + d_m \vec{v}_m.$$

By subtraction, we find

$$(c_1 - d_1)\vec{v}_1 + \cdots + (c_m - d_m)\vec{v}_m = \vec{0},$$

a relation among the vectors $\vec{v}_1, \ldots, \vec{v}_m$. Since the vectors $\vec{v}_1, \ldots, \vec{v}_m$ are linearly independent, this must be the trivial relation, and we have $c_1 - d_1 = 0, \ldots,$ $c_m - d_m = 0$, or $c_1 = d_1, \ldots, c_m = d_m$. It turns out that the two representations $\vec{v} = c_1 \vec{v}_1 + \cdots + c_m \vec{v}_m$ and $\vec{v} = d_1 \vec{v}_1 + \cdots + d_m \vec{v}_m$ are identical. We have shown that there is one and only one way to write $\vec{v}$ as a linear combination of the basis vectors $\vec{v}_1, \ldots, \vec{v}_m$. ■

Let us summarize.

Theorem 3.2.10　**Basis and unique representation**

Consider the vectors $\vec{v}_1, \ldots, \vec{v}_m$ in a subspace V of $\mathbb{R}^n$.

The vectors $\vec{v}_1, \ldots, \vec{v}_m$ form a basis of V if (and only if) every vector $\vec{v}$ in V can be expressed *uniquely* as a linear combination

$$\vec{v} = c_1 \vec{v}_1 + \cdots + c_m \vec{v}_m.$$

(In Section 3.4, we will call the coefficients $c_1, \ldots, c_m$ the *coordinates* of $\vec{v}$ with respect to the basis $\vec{v}_1, \ldots, \vec{v}_m$.)

Proof   In Example 9 we have shown only one part of Theorem 3.2.10; we still need to verify that the uniqueness of the representation $\vec{v} = c_1\vec{v}_1 + \cdots + c_m\vec{v}_m$ (for every $\vec{v}$ in V) implies that $\vec{v}_1, \ldots, \vec{v}_m$ is a basis of V. Clearly, the vectors $\vec{v}_1, \ldots, \vec{v}_m$ span V, since every $\vec{v}$ in V can be written as a linear combination of $\vec{v}_1, \ldots, \vec{v}_m$.

To show the linear independence of vectors $\vec{v}_1, \ldots, \vec{v}_m$, consider a relation $c_1\vec{v}_1 + \cdots + c_m\vec{v}_m = \vec{0}$. This relation is a representation of the zero vector as a linear combination of $\vec{v}_1, \ldots, \vec{v}_m$. But this representation is unique, with $c_1 = \cdots = c_m = 0$, so that $c_1\vec{v}_1 + \cdots + c_m\vec{v}_m = \vec{0}$ must be the trivial relation. We have shown that vectors $\vec{v}_1, \ldots, \vec{v}_m$ are linearly independent. ∎

Consider the plane $V = \text{image}(A) = \text{span}(\vec{v}_1, \vec{v}_2, \vec{v}_3, \vec{v}_4)$ introduced in Example 4. (Take another look at Figure 4.)

We can write

$$\vec{v}_4 = 1\vec{v}_1 + 0\vec{v}_2 + 1\vec{v}_3 + 0\vec{v}_4$$
$$= 0\vec{v}_1 + 0\vec{v}_2 + 0\vec{v}_3 + 1\vec{v}_4,$$

illustrating the fact that the vectors $\vec{v}_1, \vec{v}_2, \vec{v}_3, \vec{v}_4$ do not form a basis of V. However, every vector $\vec{v}$ in V can be expressed *uniquely* as a linear combination of $\vec{v}_1$ and $\vec{v}_3$ alone, meaning that the vectors $\vec{v}_1, \vec{v}_3$ do form a basis of V.

EXERCISES 3.2

GOAL   *Check whether or not a subset of $\mathbb{R}^n$ is a subspace. Apply the concept of linear independence (in terms of Definition 3.2.3, Theorem 3.2.7, and Theorem 3.2.8). Apply the concept of a basis, both in terms of Definition 3.2.3 and in terms of Theorem 3.2.10.*

Which of the sets W in Exercises 1 through 3 are subspaces of $\mathbb{R}^3$?

1. $W = \left\{ \begin{bmatrix} x \\ y \\ z \end{bmatrix} : x + y + z = 1 \right\}$

2. $W = \left\{ \begin{bmatrix} x \\ y \\ z \end{bmatrix} : x \leq y \leq z \right\}$

3. $W = \left\{ \begin{bmatrix} x + 2y + 3z \\ 4x + 5y + 6z \\ 7x + 8y + 9z \end{bmatrix} : x, y, z \text{ arbitrary constants} \right\}$

4. Consider the vectors $\vec{v}_1, \vec{v}_2, \ldots, \vec{v}_m$ in $\mathbb{R}^n$. Is $\text{span}(\vec{v}_1, \ldots, \vec{v}_m)$ necessarily a subspace of $\mathbb{R}^n$? Justify your answer.

5. Give a geometrical description of all subspaces of $\mathbb{R}^3$. Justify your answer.

6. Consider two subspaces V and W of $\mathbb{R}^n$.
   a. Is the intersection $V \cap W$ necessarily a subspace of $\mathbb{R}^n$?
   b. Is the union $V \cup W$ necessarily a subspace of $\mathbb{R}^n$?

7. Consider a nonempty subset W of $\mathbb{R}^n$ that is closed under addition and under scalar multiplication. Is W necessarily a subspace of $\mathbb{R}^n$? Explain.

8. Find a nontrivial relation among the following vectors:

$$\begin{bmatrix} 1 \\ 2 \end{bmatrix}, \quad \begin{bmatrix} 2 \\ 3 \end{bmatrix}, \quad \begin{bmatrix} 3 \\ 4 \end{bmatrix}.$$

9. Consider the vectors $\vec{v}_1, \vec{v}_2, \ldots, \vec{v}_m$ in $\mathbb{R}^n$, with $\vec{v}_m = \vec{0}$. Are these vectors linearly independent?

In Exercises 10 through 20, use paper and pencil to identify the redundant vectors. Thus determine whether the given vectors are linearly independent.

10. $\begin{bmatrix} 2 \\ 1 \end{bmatrix}, \begin{bmatrix} 6 \\ 3 \end{bmatrix}$ **11.** $\begin{bmatrix} 7 \\ 11 \end{bmatrix}, \begin{bmatrix} 11 \\ 7 \end{bmatrix}$

12. $\begin{bmatrix} 7 \\ 11 \end{bmatrix}, \begin{bmatrix} 0 \\ 0 \end{bmatrix}$ **13.** $\begin{bmatrix} 1 \\ 2 \end{bmatrix}, \begin{bmatrix} 1 \\ 2 \end{bmatrix}$

14. $\begin{bmatrix} 1 \\ 0 \\ 0 \end{bmatrix}, \begin{bmatrix} 1 \\ 2 \\ 0 \end{bmatrix}, \begin{bmatrix} 1 \\ 2 \\ 3 \end{bmatrix}$ **15.** $\begin{bmatrix} 1 \\ 2 \end{bmatrix}, \begin{bmatrix} 2 \\ 3 \end{bmatrix}, \begin{bmatrix} 3 \\ 4 \end{bmatrix}$

16. $\begin{bmatrix} 1 \\ 1 \\ 1 \end{bmatrix}, \begin{bmatrix} 3 \\ 2 \\ 1 \end{bmatrix}, \begin{bmatrix} 6 \\ 5 \\ 4 \end{bmatrix}$ **17.** $\begin{bmatrix} 1 \\ 1 \\ 1 \end{bmatrix}, \begin{bmatrix} 1 \\ 2 \\ 3 \end{bmatrix}, \begin{bmatrix} 1 \\ 3 \\ 6 \end{bmatrix}$

18. $\begin{bmatrix} 1 \\ 1 \\ 1 \\ 1 \end{bmatrix}, \begin{bmatrix} 1 \\ 2 \\ 3 \\ 4 \end{bmatrix}, \begin{bmatrix} 1 \\ 4 \\ 7 \\ 10 \end{bmatrix}$

19. $\begin{bmatrix} 1 \\ 0 \\ 0 \\ 0 \end{bmatrix}, \begin{bmatrix} 2 \\ 0 \\ 0 \\ 0 \end{bmatrix}, \begin{bmatrix} 0 \\ 1 \\ 0 \\ 0 \end{bmatrix}, \begin{bmatrix} 0 \\ 0 \\ 1 \\ 0 \end{bmatrix}, \begin{bmatrix} 3 \\ 4 \\ 5 \\ 0 \end{bmatrix}$

20. $\begin{bmatrix} 0 \\ 0 \\ 0 \end{bmatrix}, \begin{bmatrix} 1 \\ 0 \\ 0 \end{bmatrix}, \begin{bmatrix} 3 \\ 0 \\ 0 \end{bmatrix}, \begin{bmatrix} 0 \\ 1 \\ 0 \end{bmatrix}, \begin{bmatrix} 4 \\ 5 \\ 0 \end{bmatrix}, \begin{bmatrix} 6 \\ 7 \\ 0 \end{bmatrix}, \begin{bmatrix} 0 \\ 0 \\ 1 \end{bmatrix}$

In Exercises 21 through 26, find a redundant column vector of the given matrix A, and write it as a linear combination of preceding columns. Use this representation to write a nontrivial relation among the columns, and thus find a nonzero vector in the kernel of A. (This procedure is illustrated in Example 8.)

21. $\begin{bmatrix} 1 & 1 \\ 1 & 1 \end{bmatrix}$ **22.** $\begin{bmatrix} 1 & 3 \\ 2 & 6 \end{bmatrix}$.   **23.** $\begin{bmatrix} 0 & 1 \\ 0 & 2 \end{bmatrix}$

24. $\begin{bmatrix} 1 & 3 & 6 \\ 1 & 2 & 5 \\ 1 & 1 & 4 \end{bmatrix}$ **25.** $\begin{bmatrix} 1 & 0 & 1 \\ 1 & 1 & 1 \\ 1 & 0 & 1 \end{bmatrix}$

26. $\begin{bmatrix} 1 & 0 & 2 & 0 \\ 0 & 1 & 3 & 0 \\ 0 & 0 & 0 & 1 \end{bmatrix}$

Find a basis of the image of the matrices in Exercises 27 through 33.

27. $\begin{bmatrix} 1 & 1 \\ 1 & 2 \\ 1 & 3 \end{bmatrix}$ **28.** $\begin{bmatrix} 1 & 1 & 1 \\ 1 & 2 & 5 \\ 1 & 3 & 7 \end{bmatrix}$ **29.** $\begin{bmatrix} 1 & 2 & 3 \\ 4 & 5 & 6 \end{bmatrix}$

30. $\begin{bmatrix} 0 & 1 & 0 \\ 0 & 0 & 1 \\ 0 & 0 & 0 \end{bmatrix}$ **31.** $\begin{bmatrix} 1 & 5 \\ 2 & 6 \\ 3 & 7 \\ 5 & 8 \end{bmatrix}$

32. $\begin{bmatrix} 0 & 1 & 2 & 0 & 0 & 3 \\ 0 & 0 & 0 & 1 & 0 & 4 \\ 0 & 0 & 0 & 0 & 1 & 5 \\ 0 & 0 & 0 & 0 & 0 & 0 \end{bmatrix}$

33. $\begin{bmatrix} 0 & 1 & 2 & 0 & 3 & 0 \\ 0 & 0 & 0 & 1 & 4 & 0 \\ 0 & 0 & 0 & 0 & 0 & 1 \\ 0 & 0 & 0 & 0 & 0 & 0 \end{bmatrix}$

34. Consider the 5×4 matrix

$$A = \begin{bmatrix} | & | & | & | \\ \vec{v}_1 & \vec{v}_2 & \vec{v}_3 & \vec{v}_4 \\ | & | & | & | \end{bmatrix}.$$

We are told that the vector $\begin{bmatrix} 1 \\ 2 \\ 3 \\ 4 \end{bmatrix}$ is in the kernel of A.

Write $\vec{v}_4$ as a linear combination of $\vec{v}_1, \vec{v}_2, \vec{v}_3$.

35. Show that there is a nontrivial relation among the vectors $\vec{v}_1, \ldots, \vec{v}_m$ if (and only if) at least one of the vectors $\vec{v}_i$ is a linear combination of the other vectors $\vec{v}_1, \ldots, \vec{v}_{i-1}, \vec{v}_{i+1}, \ldots, \vec{v}_m$.

36. Consider a linear transformation T from $\mathbb{R}^n$ to $\mathbb{R}^p$ and some linearly dependent vectors $\vec{v}_1, \vec{v}_2, \ldots, \vec{v}_m$ in $\mathbb{R}^n$. Are the vectors $T(\vec{v}_1), T(\vec{v}_2), \ldots, T(\vec{v}_m)$ necessarily linearly dependent? How can you tell?

37. Consider a linear transformation T from $\mathbb{R}^n$ to $\mathbb{R}^p$ and some linearly independent vectors $\vec{v}_1, \vec{v}_2, \ldots, \vec{v}_m$ in $\mathbb{R}^n$. Are the vectors $T(\vec{v}_1), T(\vec{v}_2), \ldots, T(\vec{v}_m)$ necessarily linearly independent? How can you tell?

38. a. Let V be a subspace of $\mathbb{R}^n$. Let m be the largest number of linearly independent vectors we can find in V. (Note that $m \leq n$, by Theorem 3.2.8.) Choose linearly independent vectors $\vec{v}_1, \vec{v}_2, \ldots, \vec{v}_m$ in V. Show that the vectors $\vec{v}_1, \vec{v}_2, \ldots, \vec{v}_m$ span V and are therefore a basis of V. This exercise shows that any subspace of $\mathbb{R}^n$ has a basis.

   If you are puzzled, think first about the special case when V is a plane in $\mathbb{R}^3$. What is m in this case?

 b. Show that any subspace V of $\mathbb{R}^n$ can be represented as the image of a matrix.

39. Consider some linearly independent vectors $\vec{v}_1, \vec{v}_2, \ldots, \vec{v}_m$ in $\mathbb{R}^n$ and a vector $\vec{v}$ in $\mathbb{R}^n$ that is not contained in the span of $\vec{v}_1, \vec{v}_2, \ldots, \vec{v}_m$. Are the vectors $\vec{v}_1, \vec{v}_2, \ldots, \vec{v}_m, \vec{v}$ necessarily linearly independent? Justify your answer.

40. Consider an $n \times p$ matrix A and a $p \times m$ matrix B. We are told that the columns of A and the columns of B are linearly independent. Are the columns of the product AB linearly independent as well? *Hint*: Exercise 3.1.51 is useful.

41. Consider an $m \times n$ matrix A and an $n \times m$ matrix B (with $n \neq m$) such that $AB = I_m$. (We say that A is a *left inverse* of B.) Are the columns of B linearly independent? What about the columns of A?

42. Consider some perpendicular unit vectors $\vec{v}_1, \vec{v}_2, \ldots, \vec{v}_m$ in $\mathbb{R}^n$. Show that these vectors are necessarily linearly independent. (*Hint*: Form the dot product of $\vec{v}_i$ and both sides of the equation

$$c_1 \vec{v}_1 + c_2 \vec{v}_2 + \cdots + c_i \vec{v}_i + \cdots + c_m \vec{v}_m = \vec{0}.)$$

43. Consider three linearly independent vectors $\vec{v}_1, \vec{v}_2, \vec{v}_3$ in $\mathbb{R}^n$. Are the vectors $\vec{v}_1, \vec{v}_1 + \vec{v}_2, \vec{v}_1 + \vec{v}_2 + \vec{v}_3$ linearly independent as well? How can you tell?

44. Consider linearly independent vectors $\vec{v}_1, \vec{v}_2, \ldots, \vec{v}_m$ in $\mathbb{R}^n$, and let A be an invertible $m \times m$ matrix. Are the columns of the following matrix linearly independent?

$$\begin{bmatrix} | & | & & | \\ \vec{v}_1 & \vec{v}_2 & \cdots & \vec{v}_m \\ | & | & & | \end{bmatrix} A$$

45. Are the columns of an invertible matrix linearly independent?

46. Find a basis of the kernel of the matrix

$$\begin{bmatrix} 1 & 2 & 0 & 3 & 5 \\ 0 & 0 & 1 & 4 & 6 \end{bmatrix}.$$

Justify your answer carefully; that is, explain how you know that the vectors you found are linearly independent and span the kernel.

47. Consider three linearly independent vectors $\vec{v}_1, \vec{v}_2, \vec{v}_3$ in $\mathbb{R}^4$. Find

$$\text{rref} \begin{bmatrix} | & | & | \\ \vec{v}_1 & \vec{v}_2 & \vec{v}_3 \\ | & | & | \end{bmatrix}.$$

48. Express the plane V in $\mathbb{R}^3$ with equation $3x_1 + 4x_2 + 5x_3 = 0$ as the kernel of a matrix A and as the image of a matrix B.

49. Express the line L in $\mathbb{R}^3$ spanned by the vector $\begin{bmatrix} 1 \\ 1 \\ 1 \end{bmatrix}$ as the image of a matrix A and as the kernel of a matrix B.

50. Consider two subspaces V and W of $\mathbb{R}^n$. Let $V + W$ be the set of all vectors in $\mathbb{R}^n$ of the form $\vec{v} + \vec{w}$, where $\vec{v}$ is in V and $\vec{w}$ in W. Is $V + W$ necessarily a subspace of $\mathbb{R}^n$?

If V and W are two distinct lines in $\mathbb{R}^3$, what is $V + W$? Draw a sketch.

51. Consider two subspaces V and W of $\mathbb{R}^n$ whose intersection consists only of the vector $\vec{0}$.
   a. Consider linearly independent vectors $\vec{v}_1, \vec{v}_2, \ldots,$ $\vec{v}_p$ in V and $\vec{w}_1, \vec{w}_2, \ldots, \vec{w}_q$ in W. Explain why the vectors $\vec{v}_1, \vec{v}_2, \ldots, \vec{v}_p, \vec{w}_1, \vec{w}_2, \ldots, \vec{w}_q$ are linearly independent.
   b. Consider a basis $\vec{v}_1, \vec{v}_2, \ldots, \vec{v}_p$ of V and a basis $\vec{w}_1, \vec{w}_2, \ldots, \vec{w}_q$ of W. Explain why $\vec{v}_1, \vec{v}_2, \ldots,$ $\vec{v}_p, \vec{w}_1, \vec{w}_2, \ldots, \vec{w}_q$ is a basis of $V + W$. (See Exercise 50.)

52. For which values of the constants $a, b, c, d, e,$ and f are the following vectors linearly independent? Justify your answer.

$$\begin{bmatrix} a \\ 0 \\ 0 \\ 0 \end{bmatrix}, \begin{bmatrix} b \\ c \\ 0 \\ 0 \end{bmatrix}, \begin{bmatrix} d \\ e \\ f \\ 0 \end{bmatrix}$$

53. Consider a subspace V of $\mathbb{R}^n$. We define the *orthogonal complement* $V^\perp$ of V as the set of those vectors $\vec{w}$ in $\mathbb{R}^n$ that are perpendicular to all vectors in V; that is, $\vec{w} \cdot \vec{v} = 0$, for all $\vec{v}$ in V. Show that $V^\perp$ is a subspace of $\mathbb{R}^n$.

54. Consider the line L spanned by $\begin{bmatrix} 1 \\ 2 \\ 3 \end{bmatrix}$ in $\mathbb{R}^3$. Find a basis of $L^\perp$. (See Exercise 53.)

55. Consider the subspace L of $\mathbb{R}^5$ spanned by the given vector. Find a basis of $L^\perp$. (See Exercise 53.)

$$\begin{bmatrix} 1 \\ 2 \\ 3 \\ 4 \\ 5 \end{bmatrix}$$

56. For which values of the constants $a, b, \ldots, m$ are the given vectors linearly independent?

$$\begin{bmatrix} a \\ b \\ c \\ d \\ 1 \\ 0 \end{bmatrix}, \begin{bmatrix} e \\ 1 \\ 0 \\ 0 \\ 0 \\ 0 \end{bmatrix}, \begin{bmatrix} f \\ g \\ h \\ i \\ j \\ 1 \end{bmatrix}, \begin{bmatrix} k \\ m \\ 1 \\ 0 \\ 0 \\ 0 \end{bmatrix}$$

57. Consider the matrix

$$A = \begin{bmatrix} 0 & 1 & 2 & 0 & 0 & 3 & 0 \\ 0 & 0 & 0 & 1 & 0 & 4 & 0 \\ 0 & 0 & 0 & 0 & 1 & 5 & 0 \\ 0 & 0 & 0 & 0 & 0 & 0 & 0 \end{bmatrix}.$$

Note that matrix A is in reduced row-echelon form.

For which positive integers $j = 1, \ldots, 7$ does there exist a vector $\vec{x}$ in the kernel of A such that the jth component x_j of $\vec{x}$ is nonzero, while all the components $x_{j+1}, \ldots, x_7$ are zero?

58. Consider an $n \times m$ matrix A. For which positive integers $j = 1, \ldots, m$ does there exist a vector $\vec{x}$ in the kernel of A such that the jth component x_j of $\vec{x}$ is nonzero, while all the components $x_{j+1}, \ldots, x_m$ are zero? Use Exercise 57 as a guide. Give your answer in terms of the redundant column vectors of A.

The Dimension of a Subspace of $\mathbb{R}^n$

Consider a plane V in $\mathbb{R}^3$. Using our geometric intuition, we observe that all bases of V consist of two vectors. (Any two nonparallel vectors in V will do: see Figure 1.) One vector is not enough to span V, and three or more vectors are linearly dependent. It turns out that, more generally, all bases of a subspace V of $\mathbb{R}^n$ consist of the same number of vectors. In order to prove this important fact, we need an auxiliary result.

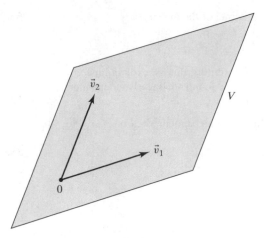

Figure 1 The vectors $\vec{v}_1$, $\vec{v}_2$ form a basis of V.

Theorem 3.3.1 Consider vectors $\vec{v}_1, \ldots, \vec{v}_p$ and $\vec{w}_1, \ldots, \vec{w}_q$ in a subspace V of $\mathbb{R}^n$. If the vectors $\vec{v}_1, \ldots, \vec{v}_p$ are linearly independent, and the vectors $\vec{w}_1, \ldots, \vec{w}_q$ span V, then $q \geq p$. ∎

For example, let V be a plane in $\mathbb{R}^3$. Our geometric intuition tells us that we can find *at most* two linearly independent vectors in V, so that $2 \geq p$, and we need *at least* two vectors to span V, so that $q \geq 2$. Therefore, the inequality $q \geq p$ does indeed hold in this case.

Proof This proof is rather technical and not very illuminating. In the next section, when we study coordinate systems, we will gain a more conceptual understanding of this matter.

Consider the matrices

$$A = \begin{bmatrix} \vec{w}_1 & \cdots & \vec{w}_q \end{bmatrix} \quad \text{and} \quad B = \begin{bmatrix} \vec{v}_1 & \cdots & \vec{v}_p \end{bmatrix}.$$

Note that $\operatorname{im}(A) = V$, since the vectors $\vec{w}_1, \ldots, \vec{w}_q$ span V. The vectors $\vec{v}_1, \ldots, \vec{v}_p$ are in the image of A, so that we can write

$$\vec{v}_1 = A\vec{u}_1, \quad \ldots, \quad \vec{v}_p = A\vec{u}_p$$

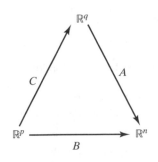

Figure 2

for some vectors $\vec{u}_1, \ldots, \vec{u}_p$ in $\mathbb{R}^q$. We can combine these equations and write

$$B = \begin{bmatrix} \vec{v}_1 & \cdots & \vec{v}_p \end{bmatrix} = A \underbrace{\begin{bmatrix} \vec{u}_1 & \cdots & \vec{u}_p \end{bmatrix}}_{C}, \quad \text{or,} \quad B = AC.$$

See Figure 2.

The kernel of C is a subset of the kernel of B (if $C\vec{x} = \vec{0}$, then $B\vec{x} = AC\vec{x} = \vec{0}$). But the kernel of B is $\{\vec{0}\}$, since the vectors $\vec{v}_1, \ldots, \vec{v}_p$ are linearly independent. Therefore, the kernel of C is $\{\vec{0}\}$ as well. Theorem 3.1.7b now tells us that the $q \times p$ matrix C has at least as many rows as it has columns, that is, $q \geq p$, as claimed. ∎

Theorem 3.3.2 Number of vectors in a basis

All bases of a subspace V of $\mathbb{R}^n$ consist of the same number of vectors.

Proof Consider two bases $\vec{v}_1, \ldots, \vec{v}_p$ and $\vec{w}_1, \ldots, \vec{w}_q$ of V. Since the vectors $\vec{v}_1, \ldots, \vec{v}_p$ are linearly independent and the vectors $\vec{w}_1, \ldots, \vec{w}_q$ span V, we have $q \geq p$, by Theorem 3.3.1. Likewise, since the vectors $\vec{w}_1, \ldots, \vec{w}_q$ are linearly independent and the vectors $\vec{v}_1, \ldots, \vec{v}_p$ span V, we have $p \geq q$. Therefore, $p = q$. ∎

Consider a line L and a plane V in $\mathbb{R}^3$. A basis of L consists of just one vector (any nonzero vector in L will do), while all bases of V consist of two vectors. A basis of $\mathbb{R}^3$ consists of three vectors. (The standard vectors $\vec{e}_1, \vec{e}_2, \vec{e}_3$ are one possible choice.) In each case, the number of vectors in a basis corresponds to what we intuitively sense to be the dimension of the subspace.

Definition 3.3.3 Dimension

Consider a subspace V of $\mathbb{R}^n$. The number of vectors in a basis of V is called the *dimension* of V, denoted by $\dim(V)$.[9]

This algebraic definition of dimension represents a major advance in the development of linear algebra, and indeed of mathematics as a whole: It allows us to conceive of spaces with more than three dimensions. This idea is often poorly understood in popular culture, where some mysticism still surrounds higher-dimensional spaces. The German mathematician Hermann Weyl (1855–1955) puts it this way: "We are by no means obliged to seek illumination from the mystic doctrines of spiritists to obtain a clearer vision of multidimensional geometry" (*Raum, Zeit, Materie*, 1918).

The first mathematician who thought about dimension from an algebraic point of view may have been the Frenchman Jean Le Rond d'Alembert (1717–1783). In the article on dimension in the *Encyclopédie*, he writes the following:

> The way of considering quantities having more than three dimensions is just as right as the other, because letters can always be viewed as representing numbers, whether rational or not. I said above that it was not possible to conceive more than three dimensions. A thoughtful gentleman [*un homme d'esprit*] with whom I am acquainted believes that nevertheless one could view duration as a fourth dimension. . . . This idea may be challenged, but it has, it seems to me, some merit, were it only that of being new [*cette idée peut être contestée, mais elle a, ce me semble, quelque mérite, quand ce serait que celui de la nouveauté*]. (*Encyclopédie*, vol. 4, 1754)

This *homme d'esprit* was no doubt d'Alembert himself, afraid of being attacked for what appeared as a risky idea at that time.

The idea of dimension was later studied much more systematically by the German mathematician Hermann Günther Grassmann (1809–1877), who introduced the concept of a subspace of $\mathbb{R}^n$. In fact, most of the concepts discussed in this chapter can be traced back to Grassmann's work. Grassmann presented his ideas in 1844 in the book *Die lineare Ausdehnungslehre, ein neuer Zweig der Mathematik* (*The Theory of Linear Extension, a New Branch of Mathematics*). Grassmann's methods were only slowly adopted, partly because of his obscure writing. He used unfamiliar authentic German terms, rather than the customary Latin, for mathematical concepts; he writes about "*Schatten*," shadows, for example, rather than projections. While his ideas have survived, most of his terminology has not.

---

[9]For this definition to make sense, we have to be sure that any subspace of $\mathbb{R}^n$ has a basis. This verification is left as Exercise 3.2.38a.

Similar work was done by the Swiss mathematician Ludwig Schläfli (1814–1895), a contemporary of Grassmann.

Today, dimension is a standard and central tool in mathematics, as well as in physics and statistics. The concept can be applied to certain nonlinear subsets of $\mathbb{R}^n$, called manifolds, generalizing the idea of curves and surfaces in $\mathbb{R}^3$.

After this brief historical digression, let us return to the more mundane: What is the dimension of $\mathbb{R}^n$? We expect this dimension to be n, of course. This is indeed the case: The vectors $\vec{e}_1, \ldots, \vec{e}_n$ form a basis, called the *standard basis* of $\mathbb{R}^n$.

A plane V in $\mathbb{R}^3$ is two dimensional. Earlier, we mentioned that we cannot find more than two linearly independent vectors in V and that we need at least two vectors to span V. If two vectors in V are linearly independent, then they form a basis of V. Likewise, if two vectors span V, then they form a basis of V.

We can generalize these observations as follows:

Theorem 3.3.4 Independent vectors and spanning vectors in a subspace of $\mathbb{R}^n$

Consider a subspace V of $\mathbb{R}^n$ with $\dim(V) = m$.

  a. We can find *at most* m linearly independent vectors in V.

  b. We need *at least* m vectors to span V.

  c. If m vectors in V are linearly independent, then they form a basis of V.

  d. If m vectors in V span V, then they form a basis of V.

Part (a) allows us to *define* the dimension of V alternatively as the maximal number of linearly independent vectors in V. Likewise, part (b) tells us that the dimension of V is the minimal number of vectors needed to span V.

In parts (c) and (d) we make the following point: By Definition 3.2.3, some vectors $\vec{v}_1, \ldots, \vec{v}_m$ in V form a basis of V if they are linearly independent *and* span V. However, if we are dealing with "the right number" of vectors (namely, m, the dimension of V), then it suffices to check only one of the two properties; the other will then follow "automatically."

Proof We prove Theorem 3.3.4, parts (a) and (c). We leave the proofs of parts (b) and (d) as Exercises 78 and 79.

  a. Consider linearly independent vectors $\vec{v}_1, \vec{v}_2, \ldots, \vec{v}_p$ in V, and let $\vec{w}_1, \vec{w}_2, \ldots, \vec{w}_m$ be a basis of V. Since the vectors $\vec{w}_1, \vec{w}_2, \ldots, \vec{w}_m$ span V, we have $p \leq m$, by Theorem 3.3.1, as claimed.

  c. Consider linearly independent vectors $\vec{v}_1, \ldots, \vec{v}_m$ in V. We have to show that the vectors $\vec{v}_1, \ldots, \vec{v}_m$ span V. If $\vec{v}$ is any vector in V, then the $m + 1$ vectors $\vec{v}_1, \ldots, \vec{v}_m, \vec{v}$ will be linearly dependent, by part (a). Since vectors $\vec{v}_1, \ldots, \vec{v}_m$ are linearly independent and therefore nonredundant, vector $\vec{v}$ must be redundant in the list $\vec{v}_1, \ldots, \vec{v}_m, \vec{v}$, meaning that $\vec{v}$ is a linear combination of $\vec{v}_1, \ldots, \vec{v}_m$. Since $\vec{v}$ is an arbitrary vector in V, we have shown that vectors $\vec{v}_1, \ldots, \vec{v}_m$ span V, as claimed. ∎

In Section 3.2, we saw that the kernel and image of a linear transformation are subspaces of the domain and the target space of the transformation, respectively. We will now examine how we can find bases of the image and kernel and thus determine their dimension.

Finding Bases of Kernel and Image

EXAMPLE I   Consider the matrix

$$A = \begin{bmatrix} 1 & 2 & 2 & -5 & 6 \\ -1 & -2 & -1 & 1 & -1 \\ 4 & 8 & 5 & -8 & 9 \\ 3 & 6 & 1 & 5 & -7 \end{bmatrix}.$$

a. Find a basis of the kernel of A, and thus determine the dimension of the kernel.

b. Find a basis of the image of A, and thus determine the dimension of the image.

Solution

a. We will solve the linear system $A\vec{x} = \vec{0}$, by Gaussian elimination. From Example 3.1.11 we know that

$$B = \text{rref}(A) = \begin{bmatrix} 1 & 2 & 0 & 3 & -4 \\ 0 & 0 & 1 & -4 & 5 \\ 0 & 0 & 0 & 0 & 0 \\ 0 & 0 & 0 & 0 & 0 \end{bmatrix}.$$

Note that $\ker(A) = \ker(B)$, by the definition of the reduced row-echelon form. In Chapter 1 we learned to solve the equation $A\vec{x} = \vec{0}$ by solving the simpler equation $B\vec{x} = \vec{0}$ instead. In Example 3.1.11 we have seen that the vectors in $\ker(A) = \ker(B)$ are of the form

$$\vec{x} = \begin{bmatrix} x_1 \\ x_2 \\ x_3 \\ x_4 \\ x_5 \end{bmatrix} = \begin{bmatrix} -2s & -3t & +4r \\ s & & \\ & 4t & -5r \\ & t & \\ & & r \end{bmatrix} = s \underbrace{\begin{bmatrix} -2 \\ 1 \\ 0 \\ 0 \\ 0 \end{bmatrix}}_{\vec{w}_1} + t \underbrace{\begin{bmatrix} -3 \\ 0 \\ 4 \\ 1 \\ 0 \end{bmatrix}}_{\vec{w}_2} + r \underbrace{\begin{bmatrix} 4 \\ 0 \\ -5 \\ 0 \\ 1 \end{bmatrix}}_{\vec{w}_3},$$

where s, t, and r are arbitrary constants.

We claim that the three vectors $\vec{w}_1, \vec{w}_2, \vec{w}_3$ form a basis of the kernel of A. The preceding equation, $\vec{x} = s\vec{w}_1 + t\vec{w}_2 + r\vec{w}_3$, shows that the vectors $\vec{w}_1, \vec{w}_2, \vec{w}_3$ span the kernel.

Theorem 3.2.5 tells us that the vectors $\vec{w}_1, \vec{w}_2, \vec{w}_3$ are linearly independent, since each has a 1 in a component where the other two vectors have a 0; these components correspond to the free variables x_2, x_4, and x_5.

Thus, a basis of the kernel of A is

$$\begin{bmatrix} -2 \\ 1 \\ 0 \\ 0 \\ 0 \end{bmatrix}, \begin{bmatrix} -3 \\ 0 \\ 4 \\ 1 \\ 0 \end{bmatrix}, \begin{bmatrix} 4 \\ 0 \\ -5 \\ 0 \\ 1 \end{bmatrix},$$

and $\dim(\ker A) = 3$.

b. To construct a basis of the image of A by means of Theorem 3.2.4, we need to find the redundant columns of A. Let's see how we can use $B = \text{rref}(A)$ to carry out this task. To keep track of the columns of A and B, we will denote the columns of A by $\vec{a}_1, \ldots, \vec{a}_5$ and those of B by $\vec{b}_1, \ldots, \vec{b}_5$.

The redundant columns of $B = \text{rref}(A)$ are easy to spot. They are the columns that do not contain a leading 1, namely, $\vec{b}_2 = 2\vec{b}_1$, $\vec{b}_4 = 3\vec{b}_1 - 4\vec{b}_3$, and $\vec{b}_5 = -4\vec{b}_1 + 5\vec{b}_3$.

And here comes the key observation: The redundant columns of A correspond to those of B, meaning that $\vec{a}_i$ is redundant if and only if $\vec{b}_i$ is redundant. We will illustrate this fact by means of an example. We know that $\vec{b}_5$ is redundant, with $\vec{b}_5 = -4\vec{b}_1 + 5\vec{b}_3$. This induces the relation $4\vec{b}_1 - 5\vec{b}_3 + \vec{b}_5 = \vec{0}$, meaning that the vector

$$\begin{bmatrix} 4 \\ 0 \\ -5 \\ 0 \\ 1 \end{bmatrix}$$

is in $\ker(B) = \ker(A)$; see part (a) of this example. But this in turn induces the relation $4\vec{a}_1 - 5\vec{a}_3 + \vec{a}_5 = \vec{0}$ among the columns of A, showing that $\vec{a}_5$ is redundant, with $\vec{a}_5 = -4\vec{a}_1 + 5\vec{a}_3$.[10]

Thus the redundant columns of A are $\vec{a}_2 = 2\vec{a}_1$, $\vec{a}_4 = 3\vec{a}_1 - 4\vec{a}_3$ and $\vec{a}_5 = -4\vec{a}_1 + 5\vec{a}_3$. By Theorem 3.2.4, the nonredundant columns $\vec{a}_1$ and $\vec{a}_3$ form a basis of the image of A.

Thus, a basis of the image of A is

$$\begin{bmatrix} 1 \\ -1 \\ 4 \\ 3 \end{bmatrix}, \begin{bmatrix} 2 \\ -1 \\ 5 \\ 1 \end{bmatrix}$$

and $\dim(\text{im } A) = 2$.   ∎

Using Example 1b as a guide, we can establish the following general rule for finding a basis of the image of a matrix.

Theorem 3.3.5   Using rref to construct a basis of the image

To construct a basis of the image of A, pick the column vectors of A that correspond to the columns of $\text{rref}(A)$ containing the leading 1's.   ∎

Again, here are the three main points that make this procedure work:

- The nonredundant column vectors of A form a basis of the image of A (Theorem 3.2.4).
- The redundant columns of A correspond to those of $\text{rref}(A)$.
- The nonredundant column vectors of $\text{rref}(A)$ are those containing the leading 1's.

Note that in Theorem 3.3.5 you need to pick columns of matrix A, not of $\text{rref}(A)$, because the matrices A and $\text{rref}(A)$ need not have the same image (see Exercise 3.1.44b).

In Theorem 3.3.5, we are constructing a basis of $\text{im}(A)$ that contains as many vectors as there are leading 1's in $\text{rref}(A)$. By Definition 1.3.2, this number is the rank of A.

---

[10] A general proof of the claim that the redundant columns of A correspond to those of B goes along similar lines. Suppose $\vec{b}_i$ is redundant, with $\vec{b}_i = c_1\vec{b}_1 + \cdots + c_{i-1}\vec{b}_{i-1}$. This induces a relation $-c_1\vec{b}_1 - \cdots - c_{i-1}\vec{b}_{i-1} + \vec{b}_i = \vec{0}$, and so forth, as above.

Theorem 3.3.6   Dimension of the image

For any matrix A,

$$\dim(\text{im } A) = \text{rank}(A).$$ ▪

Now back to the kernel. In Example 1a we are constructing a basis of the kernel of an $n \times m$ matrix A that contains as many vectors as there are free variables. Thus

$$\dim(\ker A) = \left(\begin{array}{c}\text{number of} \\ \text{free variables}\end{array}\right) = \left(\begin{array}{c}\text{total number} \\ \text{of variables}\end{array}\right) - \left(\begin{array}{c}\text{number of} \\ \text{leading variables}\end{array}\right)$$
$$= m - \text{rank}(A).$$

Adding up the equations $\dim(\ker A) = m - \text{rank}(A)$ and $\dim(\text{im } A) = \text{rank}(A)$, we find the remarkable equation $\dim(\ker A) + \dim(\text{im } A) = m$ for any $n \times m$ matrix A.

Theorem 3.3.7   Rank-Nullity Theorem

For any $n \times m$ matrix A, the equation

$$\dim(\ker A) + \dim(\text{im} A) = m$$

holds. The dimension of $\ker(A)$ is called the *nullity* of A, and in Theorem 3.3.6 we observed that $\dim(\text{im} A) = \text{rank}(A)$. Thus we can write the preceding equation alternatively as

$$(\text{nullity of } A) + (\text{rank of } A) = m.$$

Some authors go so far as to call this the *fundamental theorem of linear algebra*. ▪

We can write the rank-nullity theorem as

$$m - \dim(\ker A) = \dim(\text{im} A);$$

we can interpret this formula geometrically as follows.
    Consider the linear transformation

$$T(\vec{x}) = A\vec{x} \quad \text{from } \mathbb{R}^m \text{ to } \mathbb{R}^n.$$

Note that m is the dimension of the domain of transformation T. The quantity $\text{nullity}(A) = \dim(\ker A)$ counts the dimensions that "collapse" as we perform transformation T, and $\text{rank}(A) = \dim(\text{im} A)$ counts the dimensions that "survive" transformation T.

EXAMPLE 2   Consider the orthogonal projection T onto a plane V in $\mathbb{R}^3$ (see Figure 3). Here, the dimension of the domain is $m = 3$, one dimension collapses (the kernel of T is the

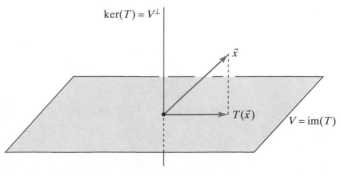

Figure 3

line $V^\perp$ orthogonal to V), and we are left with the two-dimensional $\text{im}(T) = V$. See Examples 3.1.5 and 3.1.9.

$$m \ - \ \dim(\ker T) = \dim(\text{image } T)$$
$$\uparrow \qquad \uparrow \qquad \qquad \uparrow$$
$$3 \ - \ 1 \ = \ 2 \qquad \qquad \blacksquare$$

If we can find the redundant column vectors of a matrix A by inspection; then we can construct bases for image and kernel of A without computing the reduced row-echelon form of A. This shortcut is illustrated in the following example.

EXAMPLE 3   Find bases of the image and kernel of the matrix

$$A = \begin{bmatrix} 1 & 2 & 0 & 1 & 2 \\ 1 & 2 & 0 & 2 & 3 \\ 1 & 2 & 0 & 3 & 4 \\ 1 & 2 & 0 & 4 & 5 \end{bmatrix}.$$

Solution

We can spot the redundant columns, $\vec{v}_2 = 2\vec{v}_1$, $\vec{v}_3 = \vec{0}$, and $\vec{v}_5 = \vec{v}_1 + \vec{v}_4$. Proposition 3.2.4 tells us that the nonredundant columns

$$\vec{v}_1 = \begin{bmatrix} 1 \\ 1 \\ 1 \\ 1 \end{bmatrix}, \qquad \vec{v}_4 = \begin{bmatrix} 1 \\ 2 \\ 3 \\ 4 \end{bmatrix}$$

form a basis of the image of A. Thus $\dim(\text{im } A) = 2$.

Applying the method outlined in Example 3.2.8 to the redundant vectors $\vec{v}_2$, $\vec{v}_3$, and $\vec{v}_5$, we can generate three vectors in the kernel of A. We will organize our work in a table.

Redundant Vector	Relation	Vector in Kernel of A
$\vec{v}_2 = 2\vec{v}_1$	$-2\vec{v}_1 + \vec{v}_2 = \vec{0}$	$\vec{w}_2 = \begin{bmatrix} -2 \\ 1 \\ 0 \\ 0 \\ 0 \end{bmatrix}$
$\vec{v}_3 = \vec{0}$	$\vec{v}_3 = \vec{0}$	$\vec{w}_3 = \begin{bmatrix} 0 \\ 0 \\ 1 \\ 0 \\ 0 \end{bmatrix}$
$\vec{v}_5 = \vec{v}_1 + \vec{v}_4$	$-\vec{v}_1 - \vec{v}_4 + \vec{v}_5 = \vec{0}$	$\vec{w}_5 = \begin{bmatrix} -1 \\ 0 \\ 0 \\ -1 \\ 1 \end{bmatrix}$

To facilitate the transition from the relation to the vector in the kernel, it can be useful to write the coefficients of a relation above the corresponding columns of the matrix,[11] as follows (for the last relation):

$$
\begin{array}{ccccc}
-1 & 0 & 0 & -1 & 1 \\
\end{array}
$$
$$
\begin{bmatrix}
1 & 2 & 0 & 1 & 2 \\
1 & 2 & 0 & 2 & 3 \\
1 & 2 & 0 & 3 & 4 \\
1 & 2 & 0 & 4 & 5
\end{bmatrix}.
$$

We claim that the three vectors $\vec{w}_2, \vec{w}_3, \vec{w}_5$ constructed above form a *basis of the kernel of A*. Theorem 3.2.5 tells us that these vectors are linearly independent, since vector $\vec{w}_i$ has a 1 in the ith component, while the preceding vectors have a 0 in that component.

From Theorem 3.3.7, we know that $\dim(\ker A) = 5 - \dim(\operatorname{im} A) = 3$. Because $\vec{w}_2, \vec{w}_3, \vec{w}_5$ are three linearly independent vectors in the three-dimensional space $\ker(A)$, they form a basis of $\ker(A)$, by Theorem 3.3.4c. ∎

More generally, if A is an $n \times m$ matrix, then this procedure generates as many linearly independent vectors in $\ker(A)$ as there are redundant columns vectors in A. But this number is

$$
m - \left(\begin{array}{c} \text{number of} \\ \text{nonredundant columns} \end{array} \right) = m - \dim(\operatorname{im} A) = \dim(\ker A),
$$

by Theorem 3.3.7, showing that we have enough vectors to form a basis of the kernel (again, we are invoking Theorem 3.3.4c).

Theorem 3.3.8   Finding bases of the kernel and image by inspection

Suppose you are able to spot the redundant columns of a matrix A.

Express each redundant column as a linear combination of the preceding columns, $\vec{v}_i = c_1 \vec{v}_1 + \cdots + c_{i-1} \vec{v}_{i-1}$, write a corresponding relation, $-c_1 \vec{v}_1 - \cdots - c_{i-1} \vec{v}_{i-1} + \vec{v}_i = \vec{0}$, and generate the vector

$$
\begin{bmatrix}
-c_1 \\
\vdots \\
-c_{i-1} \\
1 \\
0 \\
\vdots \\
0
\end{bmatrix}
$$

in the kernel of A. The vectors so constructed form a basis of the kernel of A.

The nonredundant columns form a basis of the image of A.

The use of Kyle Numbers can facilitate this procedure (see Example 3). ∎

Bases of $\mathbb{R}^n$

We know that any basis of $\mathbb{R}^n$ consists of n vectors, since we have the standard basis $\vec{e}_1, \ldots, \vec{e}_n$ (recall Theorem 3.3.2). Conversely, how can we tell whether n given vectors $\vec{v}_1, \ldots, \vec{v}_n$ in $\mathbb{R}^n$ form a basis?

---

[11]We will refer to these numbers above the matrix as the *Kyle Numbers*, after Kyle Burke (Colby 2003), who introduced them.

By Theorem 3.2.10, the vectors $\vec{v}_1, \ldots, \vec{v}_n$ form a basis of $\mathbb{R}^n$ if (and only if) every vector $\vec{b}$ in $\mathbb{R}^n$ can be written uniquely as a linear combination of the vectors $\vec{v}_1, \ldots, \vec{v}_n$:

$$\vec{b} = c_1\vec{v}_1 + \cdots + c_n\vec{v}_n = \begin{bmatrix} | & & | \\ \vec{v}_1 & \cdots & \vec{v}_n \\ | & & | \end{bmatrix} \begin{bmatrix} c_1 \\ \vdots \\ c_n \end{bmatrix}.$$

By definition of invertibility, the linear system

$$\begin{bmatrix} | & & | \\ \vec{v}_1 & \cdots & \vec{v}_n \\ | & & | \end{bmatrix} \begin{bmatrix} c_1 \\ \vdots \\ c_n \end{bmatrix} = \vec{b}$$

has a unique solution for all $\vec{b}$ if (and only if) the $n \times n$ matrix

$$\begin{bmatrix} | & & | \\ \vec{v}_1 & \cdots & \vec{v}_n \\ | & & | \end{bmatrix}$$

is invertible. We have shown the following result:

Theorem 3.3.9   Bases of $\mathbb{R}^n$

The vectors $\vec{v}_1, \ldots, \vec{v}_n$ in $\mathbb{R}^n$ form a basis of $\mathbb{R}^n$ if (and only if) the matrix

$$\begin{bmatrix} | & & | \\ \vec{v}_1 & \cdots & \vec{v}_n \\ | & & | \end{bmatrix}$$

is invertible.   ■

EXAMPLE 4   For which values of the constant k do the following vectors form a basis of $\mathbb{R}^3$?

$$\begin{bmatrix} 1 \\ 1 \\ 1 \end{bmatrix}, \quad \begin{bmatrix} 1 \\ -1 \\ 1 \end{bmatrix}, \quad \begin{bmatrix} 1 \\ k \\ k^2 \end{bmatrix}$$

Solution

We need to examine when the matrix

$$\begin{bmatrix} 1 & 1 & 1 \\ 1 & -1 & k \\ 1 & 1 & k^2 \end{bmatrix}$$

is invertible. This matrix reduces to

$$\begin{bmatrix} 1 & 1 & 1 \\ 0 & 1 & (1-k)/2 \\ 0 & 0 & k^2 - 1 \end{bmatrix}.$$

We can reduce this matrix all the way to I_3 if (and only if) $k^2 - 1 \neq 0$, that is, if k is neither 1 nor -1.

Thus the three given vectors form a basis of $\mathbb{R}^3$ if (and only if) k is neither 1 nor -1.   ■

Theorem 3.3.4, parts (c) and (d), applied to $V = \mathbb{R}^n$, and Theorem 3.3.9 provide us with three new characterizations of invertible matrices.

SUMMARY 3.3.10 | Various characterizations of invertible matrices

For an $n \times n$ matrix A, the following statements are equivalent.

i. A is invertible.

ii. The linear system $A\vec{x} = \vec{b}$ has a unique solution $\vec{x}$, for all $\vec{b}$ in $\mathbb{R}^n$.

iii. $\operatorname{rref}(A) = I_n$.

iv. $\operatorname{rank}(A) = n$.

v. $\operatorname{im}(A) = \mathbb{R}^n$.

vi. $\ker(A) = \{\vec{0}\}$.

vii. The column vectors of A form a basis of $\mathbb{R}^n$.

viii. The column vectors of A span $\mathbb{R}^n$.

ix. The column vectors of A are linearly independent.

EXERCISES 3.3

GOAL *Use the concept of dimension. Find a basis of the kernel and of the image of a linear transformation.*

In Exercises 1 through 20, find the redundant column vectors of the given matrix A "by inspection." Then find a basis of the image of A and a basis of the kernel of A.

1. $\begin{bmatrix} 1 & 3 \\ 2 & 6 \end{bmatrix}$

2. $\begin{bmatrix} 1 & 4 \\ 2 & 8 \end{bmatrix}$

3. $\begin{bmatrix} 1 & 2 \\ 3 & 4 \end{bmatrix}$

4. $\begin{bmatrix} 0 & 1 \\ 0 & 2 \end{bmatrix}$

5. $\begin{bmatrix} 1 & -2 & 3 \\ 2 & 4 & 6 \end{bmatrix}$

6. $\begin{bmatrix} 1 & 1 & 3 \\ 2 & 1 & 4 \end{bmatrix}$

7. $\begin{bmatrix} 1 & 2 & 3 \\ 1 & 2 & 4 \end{bmatrix}$

8. $\begin{bmatrix} 1 & -3 \\ 2 & -6 \\ 3 & -9 \end{bmatrix}$

9. $\begin{bmatrix} 1 & 2 & 1 \\ 1 & 2 & 2 \\ 1 & 2 & 3 \end{bmatrix}$

10. $\begin{bmatrix} 0 & 1 & 1 \\ 0 & 1 & 2 \\ 0 & 1 & 3 \end{bmatrix}$

11. $\begin{bmatrix} 1 & 0 & 1 \\ 0 & 1 & 0 \\ 0 & 1 & 0 \end{bmatrix}$

12. $\begin{bmatrix} 1 & 0 & 0 \\ 1 & 0 & 0 \\ 1 & 1 & 1 \end{bmatrix}$

13. $\begin{bmatrix} 1 & 2 & 3 \end{bmatrix}$

14. $\begin{bmatrix} 0 & 1 & 2 \end{bmatrix}$

15. $\begin{bmatrix} 1 & 0 & 2 & 0 \\ 0 & 1 & 2 & 0 \\ 1 & 0 & 2 & 0 \\ 0 & 1 & 2 & 0 \end{bmatrix}$

16. $\begin{bmatrix} 1 & 1 & 5 & 1 \\ 0 & 1 & 2 & 2 \\ 0 & 1 & 2 & 3 \\ 0 & 1 & 2 & 4 \end{bmatrix}$

17. $\begin{bmatrix} 0 & 1 & 2 & 0 & 3 \\ 0 & 0 & 0 & 1 & 4 \end{bmatrix}$

18. $\begin{bmatrix} 1 & -2 & 0 & -1 & 0 \\ 0 & 0 & 1 & 5 & 0 \\ 0 & 0 & 0 & 0 & 1 \end{bmatrix}$

19. $\begin{bmatrix} 1 & 0 & 5 & 3 & 0 \\ 0 & 1 & 4 & 2 & 0 \\ 0 & 0 & 0 & 0 & 1 \\ 0 & 0 & 0 & 0 & 0 \end{bmatrix}$

20. $\begin{bmatrix} 1 & 0 & 5 & 3 & -3 \\ 0 & 0 & 0 & 1 & 3 \\ 0 & 0 & 0 & 0 & 0 \\ 0 & 0 & 0 & 0 & 0 \end{bmatrix}$

In Exercises 21 through 25, find the reduced row-echelon form of the given matrix A. Then find a basis of the image of A and a basis of the kernel of A.

21. $\begin{bmatrix} 1 & 3 & 9 \\ 4 & 5 & 8 \\ 7 & 6 & 3 \end{bmatrix}$

22. $\begin{bmatrix} 2 & 4 & 8 \\ 4 & 5 & 1 \\ 7 & 9 & 3 \end{bmatrix}$

23. $\begin{bmatrix} 1 & 0 & 2 & 4 \\ 0 & 1 & -3 & -1 \\ 3 & 4 & -6 & 8 \\ 0 & -1 & 3 & 1 \end{bmatrix}$

24. $\begin{bmatrix} 4 & 8 & 1 & 1 & 6 \\ 3 & 6 & 1 & 2 & 5 \\ 2 & 4 & 1 & 9 & 10 \\ 1 & 2 & 3 & 2 & 0 \end{bmatrix}$

25. $\begin{bmatrix} 1 & 2 & 3 & 2 & 1 \\ 3 & 6 & 9 & 6 & 3 \\ 1 & 2 & 4 & 1 & 2 \\ 2 & 4 & 9 & 1 & 2 \end{bmatrix}$

26. Consider the matrices

$$C = \begin{bmatrix} 1 & 1 & 1 \\ 1 & 0 & 0 \\ 1 & 1 & 1 \end{bmatrix}, \quad H = \begin{bmatrix} 1 & 0 & 1 \\ 1 & 1 & 1 \\ 1 & 0 & 1 \end{bmatrix},$$

$$L = \begin{bmatrix} 1 & 0 & 0 \\ 1 & 0 & 0 \\ 1 & 1 & 1 \end{bmatrix}, \quad T = \begin{bmatrix} 1 & 1 & 1 \\ 0 & 1 & 0 \\ 0 & 1 & 0 \end{bmatrix},$$

$$X = \begin{bmatrix} 1 & 0 & 1 \\ 0 & 1 & 0 \\ 1 & 0 & 1 \end{bmatrix}, \quad Y = \begin{bmatrix} 1 & 0 & 1 \\ 0 & 1 & 0 \\ 0 & 1 & 0 \end{bmatrix}.$$

a. Which of the matrices in this list have the same kernel as matrix C?

b. Which of the matrices in this list have the same image as matrix C?

c. Which of these matrices has an image that is different from the images of all the other matrices in the list?

27. Determine whether the following vectors form a basis of $\mathbb{R}^4$:

$$\begin{bmatrix} 1 \\ 1 \\ 1 \\ 1 \end{bmatrix}, \quad \begin{bmatrix} 1 \\ -1 \\ 1 \\ -1 \end{bmatrix}, \quad \begin{bmatrix} 1 \\ 2 \\ 4 \\ 8 \end{bmatrix}, \quad \begin{bmatrix} 1 \\ -2 \\ 4 \\ -8 \end{bmatrix}.$$

28. For which value(s) of the constant k do the vectors below form a basis of $\mathbb{R}^4$?

$$\begin{bmatrix} 1 \\ 0 \\ 0 \\ 2 \end{bmatrix}, \quad \begin{bmatrix} 0 \\ 1 \\ 0 \\ 3 \end{bmatrix}, \quad \begin{bmatrix} 0 \\ 0 \\ 1 \\ 4 \end{bmatrix}, \quad \begin{bmatrix} 2 \\ 3 \\ 4 \\ k \end{bmatrix}.$$

29. Find a basis of the subspace of $\mathbb{R}^3$ defined by the equation

$$2x_1 + 3x_2 + x_3 = 0.$$

30. Find a basis of the subspace of $\mathbb{R}^4$ defined by the equation

$$2x_1 - x_2 + 2x_3 + 4x_4 = 0.$$

31. Let V be the subspace of $\mathbb{R}^4$ defined by the equation

$$x_1 - x_2 + 2x_3 + 4x_4 = 0.$$

Find a linear transformation T from $\mathbb{R}^3$ to $\mathbb{R}^4$ such that $\ker(T) = \{\vec{0}\}$ and $\operatorname{im}(T) = V$. Describe T by its matrix A.

32. Find a basis of the subspace of $\mathbb{R}^4$ that consists of all vectors perpendicular to both

$$\begin{bmatrix} 1 \\ 0 \\ -1 \\ 1 \end{bmatrix} \quad \text{and} \quad \begin{bmatrix} 0 \\ 1 \\ 2 \\ 3 \end{bmatrix}.$$

See Definition A.8 in the appendix.

33. A subspace V of $\mathbb{R}^n$ is called a *hyperplane* if V is defined by a homogeneous linear equation

$$c_1 x_1 + c_2 x_2 + \cdots + c_n x_n = 0,$$

where at least one of the coefficients c_i is nonzero. What is the dimension of a hyperplane in $\mathbb{R}^n$? Justify your answer carefully. What is a hyperplane in $\mathbb{R}^3$? What is it in $\mathbb{R}^2$?

34. Consider a subspace V in $\mathbb{R}^m$ that is defined by n homogeneous linear equations:

$$\begin{vmatrix} a_{11}x_1 + a_{12}x_2 + \cdots + a_{1m}x_m = 0 \\ a_{21}x_1 + a_{22}x_2 + \cdots + a_{2m}x_m = 0 \\ \vdots \qquad \vdots \qquad\qquad \vdots \quad \vdots \\ a_{n1}x_1 + a_{n2}x_2 + \cdots + a_{nm}x_m = 0 \end{vmatrix}.$$

What is the relationship between the dimension of V and the quantity $m - n$? State your answer as an inequality. Explain carefully.

35. Consider a nonzero vector $\vec{v}$ in $\mathbb{R}^n$. What is the dimension of the space of all vectors in $\mathbb{R}^n$ that are perpendicular to $\vec{v}$?

36. Can you find a 3×3 matrix A such that $\operatorname{im}(A) = \ker(A)$? Explain.

37. Give an example of a 4×5 matrix A with $\dim(\ker A) = 3$.

38. a. Consider a linear transformation T from $\mathbb{R}^5$ to $\mathbb{R}^3$. What are the possible values of $\dim(\ker T)$? Explain.

b. Consider a linear transformation T from $\mathbb{R}^4$ to $\mathbb{R}^7$. What are the possible values of $\dim(\operatorname{im}T)$? Explain.

39. We are told that a certain 5×5 matrix A can be written as

$$A = BC,$$

where B is a 5×4 matrix and C is 4×5. Explain how you know that A is not invertible.

In Exercises 40 through 43, consider the problem of fitting a conic through m given points $P_1(x_1, y_1), \ldots, P_m(x_m, y_m)$ in the plane; see Exercises 51 through 60 in Section 1.2. Recall that a conic is a curve in $\mathbb{R}^2$ that can be described by an equation of the form $f(x, y) = c_1 + c_2 x + c_3 y + c_4 x^2 + c_5 x y + c_6 y^2 = 0$.

40. Explain why fitting a conic through the points $P_1(x_1, y_1), \ldots, P_m(x_m, y_m)$ amounts to finding the kernel of an $m \times 6$ matrix A. Give the entries of the ith row of A.

Note that a one-dimensional subspace of the kernel of A defines a unique conic, since the equations $f(x, y) = 0$ and $kf(x, y) = 0$ describe the same conic.

41. How many conics can you fit through 4 distinct points $P_1(x_1, y_1), \ldots, P_4(x_4, y_4)$?

42. How many conics can you fit through 5 distinct points $P_1(x_1, y_1), \ldots, P_5(x_5, y_5)$? Describe all possible scenarios, and give an example in each case.

43. How many conics can you fit through 6 distinct points $P_1(x_1, y_1), \ldots, P_6(x_6, y_6)$? Describe all possible scenarios, and give an example in each case.

For Exercises 44 through 53, consider the problem of fitting a cubic through m given points $P_1(x_1, y_1), \ldots, P_m(x_m, y_m)$ in the plane. A cubic is a curve in $\mathbb{R}^2$ that can be described by an equation of the form $f(x, y) = c_1 + c_2x + c_3y + c_4x^2 + c_5xy + c_6y^2 + c_7x^3 + c_8x^2y + c_9xy^2 + c_{10}y^3 = 0$. If k is any nonzero constant, then the equations $f(x, y) = 0$ and $kf(x, y) = 0$ define the same cubic. Find all cubics through the given points.

44. $(0, 0), (1, 0), (2, 0), (3, 0), (0, 1), (0, 2), (0, 3), (1, 1)$

45. $(0, 0), (1, 0), (2, 0), (3, 0), (0, 1), (0, 2), (0, 3), (1, 1),$ $(2, 2)$

46. $(0, 0), (1, 0), (2, 0), (3, 0), (4, 0), (0, 1), (0, 2), (0, 3),$ $(1, 1)$

47. $(0, 0), (1, 0), (2, 0), (3, 0), (0, 1), (0, 2), (0, 3), (1, 1),$ $(2, 2), (2, 1)$

48. $(0, 0), (1, 0), (2, 0), (3, 0), (0, 1), (0, 2), (0, 3), (1, 1),$ $(2, 2), (3, 3)$

49. $(0, 0), (1, 0), (2, 0), (3, 0), (4, 0), (0, 1), (0, 2), (0, 3),$ $(0, 4), (1, 1)$

50. $(0, 0), (1, 0), (2, 0), (0, 1), (1, 1), (2, 1), (0, 2), (1, 2)$

51. $(0, 0), (1, 0), (2, 0), (0, 1), (1, 1), (2, 1), (0, 2), (1, 2),$ $(3, 2)$

52. $(0, 0), (1, 0), (2, 0), (0, 1), (1, 1), (2, 1), (0, 2), (1, 2),$ $(2, 2)$

53. $(0, 0), (1, 0), (2, 0), (0, 1), (1, 1), (2, 1), (0, 2), (1, 2),$ $(2, 2), (3, 3)$

54. Explain why fitting a cubic through the m points $P_1(x_1, y_1), \ldots, P_m(x_m, y_m)$ amounts to finding the kernel of an $m \times 10$ matrix A. Give the entries of the ith row of A.

55. How many cubics can you fit through 8 distinct points $P_1(x_1, y_1), \ldots, P_8(x_8, y_8)$?

56. How many cubics can you fit through 9 distinct points $P_1(x_1, y_1), \ldots, P_9(x_9, y_9)$? Describe all possible scenarios, and give an example in each case.

57. How many cubics can you fit through 10 distinct points $P_1(x_1, y_1), \ldots, P_{10}(x_{10}, y_{10})$? Describe all possible scenarios, and give an example in each case.

58. On September 30, 1744, the Swiss mathematician Gabriel Cramer (1704–1752) wrote a remarkable letter to his countryman Leonhard Euler, concerning the issue of fitting a cubic to given points in the plane. He states two "facts" about cubics: (1) Any 9 distinct points determine a unique cubic. (2) Two cubics can intersect in 9 points. Cramer points out that these two statements are incompatible. If we consider two specific cubics that intersect in 9 points (such as $x^3 - x = 0$ and $y^3 - y = 0$), then there is more than one cubic through these 9 points, contradicting the first "fact." Something is terribly wrong here, and Cramer asks Euler, the greatest mathematician of that time, to resolve this apparent contradiction. (This issue is now known as the Cramer–Euler Paradox.)

Euler worked on the problem for a while and put his answer into an article he submitted in 1747, "Sur one contradiction apparente dans la doctrine des lignes courbes" [*Mémoires de l'Académie des Sciences de Berlin*, 4 (1750): 219–233].

Using Examples 44 through 57 as a guide, explain which of the so-called facts stated by Cramer is wrong, thus resolving the paradox.

59. Find all points P in the plane such that you can fit infinitely many cubics through the points $(0, 0), (1, 0),$ $(2, 0), (3, 0), (0, 1), (0, 2), (0, 3), (1, 1), P$.

60. Consider two subspaces V and W of $\mathbb{R}^n$, where V is contained in W. Explain why $\dim(V) \leq \dim(W)$. (This statement seems intuitively rather obvious. Still, we cannot rely on our intuition when dealing with $\mathbb{R}^n$.)

61. Consider two subspaces V and W of $\mathbb{R}^n$, where V is contained in W. In Exercise 60 we learned that $\dim(V) \leq \dim(W)$. Show that if $\dim(V) = \dim(W)$, then $V = W$.

62. Consider a subspace V of $\mathbb{R}^n$ with $\dim(V) = n$. Explain why $V = \mathbb{R}^n$.

63. Consider two subspaces V and W of $\mathbb{R}^n$, with $V \cap W = \{\vec{0}\}$. What is the relationship between $\dim(V), \dim(W),$ and $\dim(V + W)$? (For the definition of $V + W$, see Exercise 3.2.50; Exercise 3.2.51 is helpful.)

64. Two subspaces V and W of $\mathbb{R}^n$ are called *complements* if any vector $\vec{x}$ in $\mathbb{R}^n$ can be expressed uniquely as $\vec{x} = \vec{v} + \vec{w}$, where $\vec{v}$ is in V and $\vec{w}$ is in W. Show that V and W are complements if (and only if) $V \cap W = \{\vec{0}\}$ and $\dim(V) + \dim(W) = n$.

65. Consider linearly independent vectors $\vec{v}_1, \vec{v}_2, \ldots, \vec{v}_p$ in a subspace V of $\mathbb{R}^n$ and vectors $\vec{w}_1, \vec{w}_2, \ldots, \vec{w}_q$ that span V. Show that there is a basis of V that consists of *all* the $\vec{v}_i$ and *some* of the $\vec{w}_j$. *Hint*: Find a basis of the image of the matrix

$$A = \begin{bmatrix} | & & | & | & & | \\ \vec{v}_1 & \cdots & \vec{v}_p & \vec{w}_1 & \cdots & \vec{w}_q \\ | & & | & | & & | \end{bmatrix}.$$

66. Use Exercise 65 to construct a basis of $\mathbb{R}^4$ that consists of the vectors

$$\begin{bmatrix} 1 \\ 2 \\ 3 \\ 4 \end{bmatrix}, \begin{bmatrix} 1 \\ 4 \\ 6 \\ 8 \end{bmatrix},$$

and some of the vectors $\vec{e}_1, \vec{e}_2, \vec{e}_3,$ and $\vec{e}_4$ in $\mathbb{R}^4$.

67. Consider two subspaces V and W of $\mathbb{R}^n$. Show that

$$\dim(V) + \dim(W) = \dim(V \cap W) + \dim(V + W).$$

(For the definition of $V + W$, see Exercise 3.2.50.) (*Hint*: Pick a basis $\vec{u}_1, \vec{u}_2, \ldots, \vec{u}_m$ of $V \cap W$. Using Exercise 65, construct bases $\vec{u}_1, \vec{u}_2, \ldots, \vec{u}_m, \vec{v}_1, \vec{v}_2, \ldots, \vec{v}_p$

of V and $\vec{u}_1, \vec{u}_2, \ldots, \vec{u}_m, \vec{w}_1, \vec{w}_2, \ldots, \vec{w}_q$ of W.) Show that $\vec{u}_1, \vec{u}_2, \ldots, \vec{u}_m, \vec{v}_1, \vec{v}_2, \ldots, \vec{v}_p, \vec{w}_1, \vec{w}_2, \ldots, \vec{w}_q$ is a basis of $V + W$. Demonstrating linear independence is somewhat challenging.

68. Use Exercise 67 to answer the following question: If V and W are subspaces of $\mathbb{R}^{10}$, with $\dim(V) = 6$ and $\dim(W) = 7$, what are the possible dimensions of $V \cap W$?

In Exercises 69 through 72, we will study the **row space** *of a matrix. The row space of an $n \times m$ matrix A is defined as the span of the row vectors of A (i.e., the set of their linear combinations). For example, the row space of the matrix*

$$\begin{bmatrix} 1 & 2 & 3 & 4 \\ 1 & 1 & 1 & 1 \\ 2 & 2 & 2 & 3 \end{bmatrix}$$

it is the set of all row vectors of the form

$$a \begin{bmatrix} 1 & 2 & 3 & 4 \end{bmatrix} + b \begin{bmatrix} 1 & 1 & 1 & 1 \end{bmatrix} + c \begin{bmatrix} 2 & 2 & 2 & 3 \end{bmatrix}.$$

69. Find a basis of the row space of the matrix

$$E = \begin{bmatrix} 0 & 1 & 0 & 2 & 0 \\ 0 & 0 & 1 & 3 & 0 \\ 0 & 0 & 0 & 0 & 1 \\ 0 & 0 & 0 & 0 & 0 \end{bmatrix}.$$

70. Consider an $n \times m$ matrix E in reduced row-echelon form. Using your work in Exercise 69 as a guide, explain how you can find a basis of the row space of E. What is the relationship between the dimension of the row space and the rank of E?

71. Consider an arbitrary $n \times m$ matrix A.
  a. What is the relationship between the row spaces of A and $E = \text{rref}(A)$? (*Hint*: Examine how the row space is affected by elementary row operations.)
  b. What is the relationship between the dimension of the row space of A and the rank of A?

72. Find a basis of the row space of the matrix

$$A = \begin{bmatrix} 1 & 1 & 1 & 1 \\ 2 & 2 & 2 & 2 \\ 1 & 2 & 3 & 4 \\ 1 & 3 & 5 & 7 \end{bmatrix}.$$

73. Consider an $n \times n$ matrix A. Show that there exist scalars $c_0, c_1, \ldots, c_n$ (not all zero) such that the matrix $c_0 I_n + c_1 A + c_2 A^2 + \cdots + c_n A^n$ is noninvertible. *Hint*: Pick an arbitrary nonzero vector $\vec{v}$ in $\mathbb{R}^n$. Then the $n + 1$ vectors $\vec{v}, A\vec{v}, A^2\vec{v}, \ldots, A^n\vec{v}$ will be linearly dependent. (Much more is true: There are scalars $c_0, c_1, \ldots, c_n$ such that $c_0 I_n + c_1 A + c_2 A^2 + \cdots + c_n A^n = 0$. You are not asked to demonstrate this fact here.)

74. Consider the matrix

$$A = \begin{bmatrix} 1 & -2 \\ 2 & 1 \end{bmatrix}.$$

Find scalars c_0, c_1, c_2 (not all zero) such that the matrix $c_0 I_2 + c_1 A + c_2 A^2$ is noninvertible. (See Exercise 73.)

75. Consider an $n \times m$ matrix A. Show that the rank of A is n if (and only if) A has an invertible $n \times n$ submatrix (i.e., a matrix obtained by deleting $m - n$ columns of A).

76. An $n \times n$ matrix A is called *nilpotent* if $A^m = 0$ for some positive integer m. Examples are triangular matrices whose entries on the diagonal are all 0. Consider a nilpotent $n \times n$ matrix A, and choose the smallest number m such that $A^m = 0$. Pick a vector $\vec{v}$ in $\mathbb{R}^n$ such that $A^{m-1}\vec{v} \neq \vec{0}$. Show that the vectors $\vec{v}, A\vec{v}, A^2\vec{v}, \ldots, A^{m-1}\vec{v}$ are linearly independent. (*Hint*: Consider a relation $c_0\vec{v} + c_1 A\vec{v} + c_2 A^2\vec{v} + \cdots + c_{m-1} A^{m-1}\vec{v} = \vec{0}$. Multiply both sides of the equation with A^{m-1} to show that $c_0 = 0$. Next, show that $c_1 = 0$, and so on.)

77. Consider a nilpotent $n \times n$ matrix A. Use the result demonstrated in Exercise 76 to show that $A^n = 0$.

78. Explain why you need at least m vectors to span a space of dimension m. (See Theorem 3.3.4b.)

79. Prove Theorem 3.3.4d: If m vectors span an m-dimensional space, they form a basis of the space.

80. If a 3×3 matrix A represents the projection onto a plane in $\mathbb{R}^3$, what is rank (A)?

81. Consider a 4×2 matrix A and a 2×5 matrix B.
  a. What are the possible dimensions of the *kernel* of AB?
  b. What are the possible dimensions of the *image* of AB?

82. Consider two $n \times m$ matrices A and B. What can you say about the relationship between the quantitites rank(A), rank(B), and rank$(A + B)$?

83. Consider an $n \times p$ matrix A and a $p \times m$ matrix B.
  a. What can you say about the relationship between rank(A) and rank(AB)?
  b. What can you say about the relationship between rank(B) and rank(AB)?

84. Consider the matrices

$$A = \begin{bmatrix} 1 & 0 & 2 & 0 & 4 & 0 \\ 0 & 1 & 3 & 0 & 5 & 0 \\ 0 & 0 & 0 & 1 & 6 & 0 \\ 0 & 0 & 0 & 0 & 0 & 1 \end{bmatrix}$$

and

$$B = \begin{bmatrix} 1 & 0 & 2 & 0 & 4 & 0 \\ 0 & 1 & 3 & 0 & 5 & 0 \\ 0 & 0 & 0 & 1 & 7 & 0 \\ 0 & 0 & 0 & 0 & 0 & 1 \end{bmatrix}.$$

Show that the kernels of matrices A and B are different. (*Hint*: Think about ways to write the fifth column as a linear combination of the preceding columns.)

85. Consider the matrices

$$A = \begin{bmatrix} 1 & 0 & 2 & 0 & 4 & 0 \\ 0 & 1 & 3 & 0 & 5 & 0 \\ 0 & 0 & 0 & 1 & 6 & 0 \\ 0 & 0 & 0 & 0 & 0 & 1 \end{bmatrix}$$

and

$$B = \begin{bmatrix} 1 & 0 & 2 & 0 & 0 & 4 \\ 0 & 1 & 3 & 0 & 0 & 5 \\ 0 & 0 & 0 & 1 & 0 & 6 \\ 0 & 0 & 0 & 0 & 1 & 7 \end{bmatrix}.$$

Show that the kernels of matrices A and B are different. *Hint:* Think about ways to write the fifth column as a linear combination of the preceding columns.

86. Let A and B be two different matrices of the same size, both in reduced row-echelon form. Show that the kernels of A and B are different. (*Hint:* Focus on the first column in which the two matrices differ, say, the kth columns $\vec{a}_k$ and $\vec{b}_k$ of A and B, respectively. Explain why at least one of the columns $\vec{a}_k$ and $\vec{b}_k$ fails to contain a leading 1. Thus, reversing the roles of matrices A and B if necessary, we can assume that $\vec{a}_k$ does not contain a leading 1. We can write $\vec{a}_k$ as a linear combination of preceding columns and use this representation to construct a vector in the kernel of A. Show that this vector fails to be in the kernel of B. Use Exercises 84 and 85 as a guide.)

87. Suppose a matrix A in reduced row-echelon form can be obtained from a matrix M by a sequence of elementary row operations. Show that $A = \text{rref}(M)$. (*Hint:* Both A and $\text{rref}(M)$ are in reduced row-echelon form, and they have the same kernel. Exercise 86 is helpful.)

3.4 Coordinates

Coordinates are one of the "great ideas" of mathematics. René Descartes (1596–1650) is credited with having introduced them, in an appendix to his treatise *Discours de la Méthode* (Leyden, 1637). Myth has it that the idea came to him as he was laying on his back in bed one lazy Sunday morning, watching a fly on the ceiling above him. It occurred to him that he could describe the position of the fly by giving its distance from two walls.

Descartes's countryman Pierre de Fermat (1601–1665) independently developed the basic principles of analytic geometry, at about the same time, but he did not publish his work.

We have used Cartesian coordinates in the x–y-plane and in x–y–z-space throughout Chapters 1 through 3, without much fanfare, when representing vectors in $\mathbb{R}^2$ and $\mathbb{R}^3$ geometrically. In this section and in Chapter 4, we will discuss coordinates more systematically.

EXAMPLE 1 Consider the vectors

$$\vec{v}_1 = \begin{bmatrix} 1 \\ 1 \\ 1 \end{bmatrix} \quad \text{and} \quad \vec{v}_2 = \begin{bmatrix} 1 \\ 2 \\ 3 \end{bmatrix}$$

in $\mathbb{R}^3$, and define the plane $V = \text{span}(\vec{v}_1, \vec{v}_2)$ in $\mathbb{R}^3$. Is the vector

$$\vec{x} = \begin{bmatrix} 5 \\ 7 \\ 9 \end{bmatrix}$$

on the plane V? Visualize your answer.

Solution

We have to examine whether there exist scalars c_1 and c_2 such that $\vec{x} = c_1 \vec{v}_1 + c_2 \vec{v}_2$. This problem amounts to solving the linear system with augmented matrix

$$M = \begin{bmatrix} 1 & 1 & | & 5 \\ 1 & 2 & | & 7 \\ 1 & 3 & | & 9 \end{bmatrix}, \quad \text{and} \quad \text{rref}(M) = \begin{bmatrix} 1 & 0 & | & 3 \\ 0 & 1 & | & 2 \\ 0 & 0 & | & 0 \end{bmatrix}.$$

This system is consistent, with the unique solution $c_1 = 3$ and $c_2 = 2$, so that

$$\vec{x} = c_1 \vec{v}_1 + c_2 \vec{v}_2 = 3\vec{v}_1 + 2\vec{v}_2.$$

In Figure 1, we represent this solution geometrically. It turns out that the vector is indeed on the plane V.

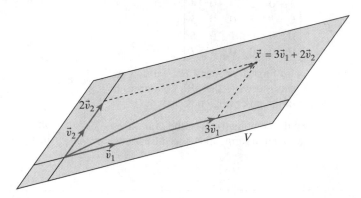

Figure 1

To visualize the coefficients 3 and 2 in the linear combination $\vec{x} = 3\vec{v}_1 + 2\vec{v}_2$, it is suggestive to introduce a *coordinate grid* on the plane V, with the axes pointing in the directions of the vectors $\vec{v}_1$ and $\vec{v}_2$, as in Figure 2, where we label the axes c_1 and c_2. In this grid, our vector $\vec{x}$ has the coordinates $c_1 = 3$ and $c_2 = 2$. The *coordinate vector* of $\vec{v} = 3\vec{v}_1 + 2\vec{v}_2$ in this coordinate system is

$$\begin{bmatrix} c_1 \\ c_2 \end{bmatrix} = \begin{bmatrix} 3 \\ 2 \end{bmatrix}.$$

We can think of $\begin{bmatrix} 3 \\ 2 \end{bmatrix}$ as the "*address*" of $\vec{x}$ in the c_1–c_2 coordinate system. By introducing c_1–c_2 coordinates in V, we transform the plane V into $\mathbb{R}^2$.

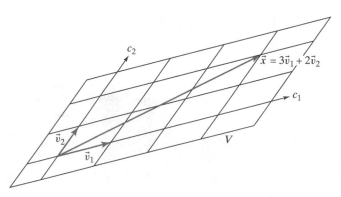

Figure 2

Don't be alarmed by the fact that the axes aren't perpendicular; Cartesian coordinates work just as well with oblique axes.

The following notation can be helpful when discussing coordinates, although it is a bit heavy. Let's denote the basis $\vec{v}_1$, $\vec{v}_2$ of V by $\mathfrak{B}$. Then the coordinate vector of $\vec{x}$ with respect to $\mathfrak{B}$ is denoted by $[\vec{x}]_{\mathfrak{B}}$.

$$\text{If } \quad \vec{x} = \begin{bmatrix} 5 \\ 7 \\ 9 \end{bmatrix} = c_1\vec{v}_1 + c_2\vec{v}_2 = 3\vec{v}_1 + 2\vec{v}_2, \quad \text{then} \quad \left[\vec{x}\right]_{\mathfrak{B}} = \begin{bmatrix} c_1 \\ c_2 \end{bmatrix} = \begin{bmatrix} 3 \\ 2 \end{bmatrix}. \quad \blacksquare$$

Let's generalize the ideas introduced in Example 1.

Definition 3.4.1   Coordinates in a subspace of $\mathbb{R}^n$

Consider a basis $\mathfrak{B} = (\vec{v}_1, \vec{v}_2, \ldots, \vec{v}_m)$ of a subspace V of $\mathbb{R}^n$. By Theorem 3.2.10, any vector $\vec{x}$ in V can be written uniquely as

$$\vec{x} = c_1\vec{v}_1 + c_2\vec{v}_2 + \cdots + c_m\vec{v}_m.$$

The scalars $c_1, c_2, \ldots, c_m$ are called the $\mathfrak{B}$-*coordinates* of $\vec{x}$, and the vector

$$\begin{bmatrix} c_1 \\ c_2 \\ \vdots \\ c_m \end{bmatrix}$$

is the $\mathfrak{B}$-*coordinate vector* of $\vec{x}$, denoted by $\left[\vec{x}\right]_{\mathfrak{B}}$. Thus

$$\left[\vec{x}\right]_{\mathfrak{B}} = \begin{bmatrix} c_1 \\ c_2 \\ \vdots \\ c_m \end{bmatrix} \quad \text{means that } \vec{x} = c_1\vec{v}_1 + c_2\vec{v}_2 + \cdots + c_m\vec{v}_m.$$

Note that

$$\vec{x} = S\left[\vec{x}\right]_{\mathfrak{B}}, \quad \text{where } S = \begin{bmatrix} \vec{v}_1 & \vec{v}_2 & \cdots & \vec{v}_m \end{bmatrix}, \text{ an } n \times m \text{ matrix.}$$

The last equation, $\vec{x} = S\left[\vec{x}\right]_{\mathfrak{B}}$, follows directly from the definition of coordinates:

$$\vec{x} = c_1\vec{v}_1 + c_2\vec{v}_2 + \cdots + c_m\vec{v}_m = \begin{bmatrix} \vec{v}_1 & \vec{v}_2 & \cdots & \vec{v}_m \end{bmatrix} \begin{bmatrix} c_1 \\ c_2 \\ \vdots \\ c_m \end{bmatrix} = S\left[\vec{x}\right]_{\mathfrak{B}}.$$

In Example 1 we considered the case where

$$\vec{x} = \begin{bmatrix} 5 \\ 7 \\ 9 \end{bmatrix}, \quad \left[\vec{x}\right]_{\mathfrak{B}} = \begin{bmatrix} 3 \\ 2 \end{bmatrix}, \quad \text{and} \quad S = \begin{bmatrix} 1 & 1 \\ 1 & 2 \\ 1 & 3 \end{bmatrix}.$$

You can verify that

$$\vec{x} = S\left[\vec{x}\right]_{\mathfrak{B}}, \quad \text{or} \quad \begin{bmatrix} 5 \\ 7 \\ 9 \end{bmatrix} = \begin{bmatrix} 1 & 1 \\ 1 & 2 \\ 1 & 3 \end{bmatrix} \begin{bmatrix} 3 \\ 2 \end{bmatrix}.$$

It turns out that coordinates have some important linearity properties:

Theorem 3.4.2   Linearity of Coordinates

If $\mathfrak{B}$ is a basis of a subspace V of $\mathbb{R}^n$, then

a. $[\vec{x} + \vec{y}]_{\mathfrak{B}} = [\vec{x}]_{\mathfrak{B}} + [\vec{y}]_{\mathfrak{B}}$,   for all vectors $\vec{x}$ and $\vec{y}$ in V, and

b. $[k\vec{x}]_{\mathfrak{B}} = k [\vec{x}]_{\mathfrak{B}}$,   for all $\vec{x}$ in V and for all scalars k.

Proof   We will prove property (b) and leave part (a) as Exercise 51. Let $\mathfrak{B} = (\vec{v}_1, \vec{v}_2, \ldots, \vec{v}_m)$. If $\vec{x} = c_1\vec{v}_1 + c_2\vec{v}_2 + \cdots + c_m\vec{v}_m$, then $k\vec{x} = kc_1\vec{v}_1 + kc_2\vec{v}_2 + \cdots + kc_m\vec{v}_m$, so that

$$[k\vec{x}]_{\mathfrak{B}} = \begin{bmatrix} kc_1 \\ kc_2 \\ \vdots \\ kc_m \end{bmatrix} = k \begin{bmatrix} c_1 \\ c_2 \\ \vdots \\ c_m \end{bmatrix} = k [\vec{x}]_{\mathfrak{B}},$$

as claimed.   ∎

As an important special case of Definition 3.4.1, consider the case when V is all of $\mathbb{R}^n$. It is often useful to work with bases of $\mathbb{R}^n$ other than the standard basis, $\vec{e}_1, \vec{e}_2, \ldots, \vec{e}_n$. When dealing with the ellipse in Figure 3, for example, the c_1–c_2 axes aligned with the principal axes may be preferable to the standard x_1–x_2 axes.

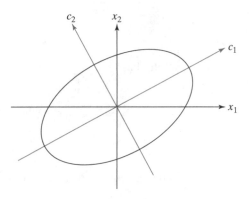

Figure 3

EXAMPLE 2   Consider the basis $\mathfrak{B}$ of $\mathbb{R}^2$ consisting of vectors $\vec{v}_1 = \begin{bmatrix} 3 \\ 1 \end{bmatrix}$ and $\vec{v}_2 = \begin{bmatrix} -1 \\ 3 \end{bmatrix}$.

a. If $\vec{x} = \begin{bmatrix} 10 \\ 10 \end{bmatrix}$, find $[\vec{x}]_{\mathfrak{B}}$.   b. If $[\vec{y}]_{\mathfrak{B}} = \begin{bmatrix} 2 \\ -1 \end{bmatrix}$, find $\vec{y}$.

Solution

a. To find the $\mathfrak{B}$-coordinates of vector $\vec{x}$, we write $\vec{x}$ as a linear combination of the basis vectors:

$$\vec{x} = c_1\vec{v}_1 + c_2\vec{v}_2 \quad \text{or} \quad \begin{bmatrix} 10 \\ 10 \end{bmatrix} = c_1 \begin{bmatrix} 3 \\ 1 \end{bmatrix} + c_2 \begin{bmatrix} -1 \\ 3 \end{bmatrix}.$$

The solution is $c_1 = 4$, $c_2 = 2$, so that $[\vec{x}]_{\mathfrak{B}} = \begin{bmatrix} 4 \\ 2 \end{bmatrix}$.

Alternatively, we can solve the equation $\vec{x} = S\left[\vec{x}\right]_{\mathfrak{B}}$ for $\left[\vec{x}\right]_{\mathfrak{B}}$:

$$\left[\vec{x}\right]_{\mathfrak{B}} = S^{-1}\vec{x} = \begin{bmatrix} 3 & -1 \\ 1 & 3 \end{bmatrix}^{-1} \begin{bmatrix} 10 \\ 10 \end{bmatrix} = \frac{1}{10}\begin{bmatrix} 3 & 1 \\ -1 & 3 \end{bmatrix} \begin{bmatrix} 10 \\ 10 \end{bmatrix} = \begin{bmatrix} 4 \\ 2 \end{bmatrix}.$$

b. By definition of coordinates, $\left[\vec{y}\right]_{\mathfrak{B}} = \begin{bmatrix} 2 \\ -1 \end{bmatrix}$ means that

$$\vec{y} = 2\vec{v}_1 + (-1)\vec{v}_2 = 2\begin{bmatrix} 3 \\ 1 \end{bmatrix} + (-1)\begin{bmatrix} -1 \\ 3 \end{bmatrix} = \begin{bmatrix} 7 \\ -1 \end{bmatrix}.$$

Alternatively, use the formula

$$\vec{y} = S\left[\vec{y}\right]_{\mathfrak{B}} = \begin{bmatrix} 3 & -1 \\ 1 & 3 \end{bmatrix}\begin{bmatrix} 2 \\ -1 \end{bmatrix} = \begin{bmatrix} 7 \\ -1 \end{bmatrix}.$$

These results are illustrated in Figure 4. ■

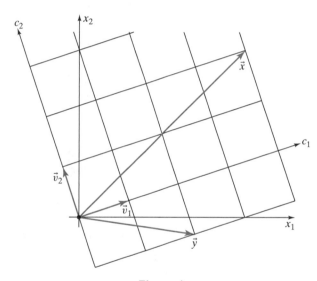

Figure 4

We will now go a step further and see how we can express a linear transformation in coordinates.

EXAMPLE 3 Let L be the line in $\mathbb{R}^2$ spanned by vector $\begin{bmatrix} 3 \\ 1 \end{bmatrix}$. Let T be the linear transformation from $\mathbb{R}^2$ to $\mathbb{R}^2$ that projects any vector $\vec{x}$ orthogonally onto line L, as shown in Figure 5. We can facilitate the study of T by introducing a coordinate system where L is one of the axes (say, the c_1-axis), with the c_2-axis perpendicular to L, as illustrated in Figure 6. If we use this coordinate system, then T transforms $\begin{bmatrix} c_1 \\ c_2 \end{bmatrix}$ into $\begin{bmatrix} c_1 \\ 0 \end{bmatrix}$. In c_1–c_2 coordinates, T is represented by the matrix $B = \begin{bmatrix} 1 & 0 \\ 0 & 0 \end{bmatrix}$, since

$$\begin{bmatrix} c_1 \\ 0 \end{bmatrix} = \begin{bmatrix} 1 & 0 \\ 0 & 0 \end{bmatrix}\begin{bmatrix} c_1 \\ c_2 \end{bmatrix}.$$

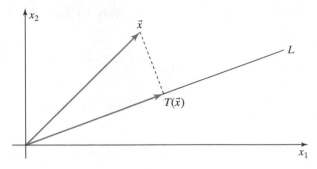

Figure 5

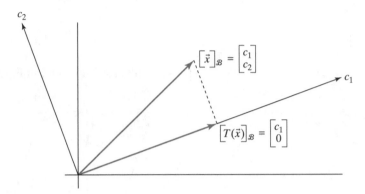

Figure 6

Let's make these ideas more precise. We start by introducing a basis $\mathfrak{B} = (\vec{v}_1, \vec{v}_2)$ of $\mathbb{R}^2$ with vector $\vec{v}_1$ on line L and vector $\vec{v}_2$ perpendicular to L, for example,

$$\vec{v}_1 = \begin{bmatrix} 3 \\ 1 \end{bmatrix} \quad \text{and} \quad \vec{v}_2 = \begin{bmatrix} -1 \\ 3 \end{bmatrix}.$$

If $\vec{x} = \underbrace{c_1 \vec{v}_1}_{\text{in } L} + \underbrace{c_2 \vec{v}_2}_{\text{in } L^\perp}$, then $T(\vec{x}) = \text{proj}_L(\vec{x}) = c_1 \vec{v}_1$. Equivalently, if

$$[\vec{x}]_{\mathfrak{B}} = \begin{bmatrix} c_1 \\ c_2 \end{bmatrix}, \quad \text{then} \quad [T(\vec{x})]_{\mathfrak{B}} = \begin{bmatrix} c_1 \\ 0 \end{bmatrix};$$

see Figure 6.

The matrix $B = \begin{bmatrix} 1 & 0 \\ 0 & 0 \end{bmatrix}$ that transforms $[\vec{x}]_{\mathfrak{B}} = \begin{bmatrix} c_1 \\ c_2 \end{bmatrix}$ into $[T(\vec{x})]_{\mathfrak{B}} = \begin{bmatrix} c_1 \\ 0 \end{bmatrix}$ is called the $\mathfrak{B}$-matrix of T:

$$[T(\vec{x})]_{\mathfrak{B}} = B [\vec{x}]_{\mathfrak{B}}.$$

We can organize our work in a diagram as follows:

$$\vec{x} = \overbrace{c_1 \vec{v}_1}^{\text{in } L} + \overbrace{c_2 \vec{v}_2}^{\text{in } L^\perp} \xrightarrow{\quad T \quad} T(\vec{x}) = c_1 \vec{v}_1$$

$$\downarrow \qquad\qquad\qquad\qquad \downarrow$$

$$[\vec{x}]_{\mathfrak{B}} = \begin{bmatrix} c_1 \\ c_2 \end{bmatrix} \xrightarrow[\ B = \begin{bmatrix} 1 & 0 \\ 0 & 0 \end{bmatrix}\ ]{} [T(\vec{x})]_{\mathfrak{B}} = \begin{bmatrix} c_1 \\ 0 \end{bmatrix}.$$

When setting up such a diagram, we begin in the top left by writing an arbitrary input vector $\vec{x}$ as a linear combination of the vectors in the given basis $\mathfrak{B}$. In the top right we have $T(\vec{x})$, again written as a linear combination of the vectors of basis $\mathfrak{B}$. The corresponding entries below are the coordinates vectors $[\vec{x}]_{\mathfrak{B}}$ and $[T(\vec{x})]_{\mathfrak{B}}$. Finding those is a routine step that requires no computational work, since $\vec{x}$ and $T(\vec{x})$ have been written as linear combinations of the basis vectors already. Finally, we find the matrix B that transforms $[\vec{x}]_{\mathfrak{B}}$ into $[T(\vec{x})]_{\mathfrak{B}}$; this is again a routine step. ∎

Let's generalize the ideas of Example 3.

Definition 3.4.3 The matrix of a linear transformation

Consider a linear transformation T from $\mathbb{R}^n$ to $\mathbb{R}^n$ and a basis $\mathfrak{B}$ of $\mathbb{R}^n$. The $n \times n$ matrix B that transforms $[\vec{x}]_{\mathfrak{B}}$ into $[T(\vec{x})]_{\mathfrak{B}}$ is called the $\mathfrak{B}$-matrix of T:

$$[T(\vec{x})]_{\mathfrak{B}} = B\,[\vec{x}]_{\mathfrak{B}},$$

for all $\vec{x}$ in $\mathbb{R}^n$. We can construct B column by column as follows: If $\mathfrak{B} = (\vec{v}_1, \dots, \vec{v}_n)$, then

$$B = \begin{bmatrix} [T(\vec{v}_1)]_{\mathfrak{B}} & \cdots & [T(\vec{v}_n)]_{\mathfrak{B}} \end{bmatrix}.$$

We need to verify that the columns of B are $[T(\vec{v}_1)]_{\mathfrak{B}}, \dots, [T(\vec{v}_n)]_{\mathfrak{B}}$. Let $\vec{x} = c_1\vec{v}_1 + \cdots + c_n\vec{v}_n$. Using first the linearity of T and then the linearity of coordinates (Theorem 3.4.2), we find that

$$T(\vec{x}) = c_1 T(\vec{v}_1) + \cdots + c_n T(\vec{v}_n)$$

and

$$[T(\vec{x})]_{\mathfrak{B}} = c_1\,[T(\vec{v}_1)]_{\mathfrak{B}} + \cdots + c_n\,[T(\vec{v}_n)]_{\mathfrak{B}}$$

$$= \underbrace{\begin{bmatrix} [T(\vec{v}_1)]_{\mathfrak{B}} & \cdots & [T(\vec{v}_n)]_{\mathfrak{B}} \end{bmatrix}}_{B} [\vec{x}]_{\mathfrak{B}},$$

as claimed. We can use this result to construct B, although it is often more enlightening to construct B by means of a diagram, as in Example 3.

EXAMPLE 4 Consider two perpendicular unit vectors $\vec{v}_1$ and $\vec{v}_2$ in $\mathbb{R}^3$. Form the basis $\mathfrak{B} = (\vec{v}_1, \vec{v}_2, \vec{v}_3)$ of $\mathbb{R}^3$, where $\vec{v}_3 = \vec{v}_1 \times \vec{v}_2$. (Take a look at Theorem A.10 in the Appendix to review the basic properties of the cross product.) Note that $\vec{v}_3$ is perpendicular to both $\vec{v}_1$ and $\vec{v}_2$, and $\vec{v}_3$ is a unit vector, since $\|\vec{v}_3\| = \|\vec{v}_1 \times \vec{v}_2\| = \|\vec{v}_1\|\|\vec{v}_2\|\sin(\pi/2) = 1 \cdot 1 \cdot 1 = 1$.

a. Draw a sketch to find $\vec{v}_1 \times \vec{v}_3$.

b. Find the $\mathfrak{B}$-matrix B of the linear transformation $T(\vec{x}) = \vec{v}_1 \times \vec{x}$.

Solution

a. Note first that $\vec{v}_1 \times \vec{v}_3$ is a unit vector.
Figure 7 illustrates that $\vec{v}_1 \times \vec{v}_3 = -\vec{v}_2$.

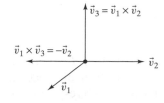

Figure 7

b. We will organize our work in a diagram, as in Example 3.

$$\vec{x} = c_1\vec{v}_1 + c_2\vec{v}_2 + c_3\vec{v}_3 \xrightarrow{\quad T \quad} \begin{aligned} T(\vec{x}) &= \vec{v}_1 \times (c_1\vec{v}_1 + c_2\vec{v}_2 + c_3\vec{v}_3) \\ &= c_1(\vec{v}_1 \times \vec{v}_1) + c_2(\vec{v}_1 \times \vec{v}_2) + c_3(\vec{v}_1 \times \vec{v}_3) \\ &= c_2\vec{v}_3 - c_3\vec{v}_2 \end{aligned}$$

$$[\vec{x}]_{\mathfrak{B}} = \begin{bmatrix} c_1 \\ c_2 \\ c_3 \end{bmatrix} \xrightarrow{\qquad\qquad} \qquad [T(\vec{x})]_{\mathfrak{B}} = \begin{bmatrix} 0 \\ -c_3 \\ c_2 \end{bmatrix}$$

$$B = \begin{bmatrix} 0 & 0 & 0 \\ 0 & 0 & -1 \\ 0 & 1 & 0 \end{bmatrix}$$

Alternatively, we can construct B column by column,

$$B = \begin{bmatrix} \Big[T(\vec{v}_1)\Big]_{\mathfrak{B}} & \Big[T(\vec{v}_2)\Big]_{\mathfrak{B}} & \Big[T(\vec{v}_3)\Big]_{\mathfrak{B}} \end{bmatrix}.$$

We have $T(\vec{v}_1) = \vec{v}_1 \times \vec{v}_1 = \vec{0}$, $T(\vec{v}_2) = \vec{v}_1 \times \vec{v}_2 = \vec{v}_3$ and $T(\vec{v}_3) = \vec{v}_1 \times \vec{v}_3 = -\vec{v}_2$, so that

$$[T(\vec{v}_1)]_{\mathfrak{B}} = \begin{bmatrix} 0 \\ 0 \\ 0 \end{bmatrix}, \qquad [T(\vec{v}_2)]_{\mathfrak{B}} = \begin{bmatrix} 0 \\ 0 \\ 1 \end{bmatrix}, \qquad [T(\vec{v}_3)]_{\mathfrak{B}} = \begin{bmatrix} 0 \\ -1 \\ 0 \end{bmatrix}.$$

$$B = \begin{bmatrix} \Big[T(\vec{v}_1)\Big]_{\mathfrak{B}} & \Big[T(\vec{v}_2)\Big]_{\mathfrak{B}} & \Big[T(\vec{v}_3)\Big]_{\mathfrak{B}} \end{bmatrix} = \begin{bmatrix} 0 & 0 & 0 \\ 0 & 0 & -1 \\ 0 & 1 & 0 \end{bmatrix}$$

Note the block $\begin{bmatrix} 0 & -1 \\ 1 & 0 \end{bmatrix}$ in matrix B, representing a rotation through $\frac{\pi}{2}$. Writing $\begin{bmatrix} 0 & 0 & 0 \\ 0 & 0 & -1 \\ 0 & 1 & 0 \end{bmatrix} = \begin{bmatrix} 1 & 0 & 0 \\ 0 & 0 & -1 \\ 0 & 1 & 0 \end{bmatrix}\begin{bmatrix} 0 & 0 & 0 \\ 0 & 1 & 0 \\ 0 & 0 & 1 \end{bmatrix}$, we can interpret transformation T geometrically. It is the orthogonal projection onto the c_2–c_3 plane followed by a rotation through $\frac{\pi}{2}$ about the c_1 axis, counterclockwise as viewed from the positive c_1 axis. ∎

EXAMPLE 5 As in Example 3, let T be the linear transformation from $\mathbb{R}^2$ to $\mathbb{R}^2$ that projects any vector orthogonally onto the line L spanned by the vector $\begin{bmatrix} 3 \\ 1 \end{bmatrix}$. In Example 3, we found that the matrix of B of T with respect to the basis $\mathfrak{B} = \left(\begin{bmatrix} 3 \\ 1 \end{bmatrix}, \begin{bmatrix} -1 \\ 3 \end{bmatrix}\right)$ is

$$B = \begin{bmatrix} 1 & 0 \\ 0 & 0 \end{bmatrix}.$$

What is the relationship between B and the *standard matrix* A of T (such that $T(\vec{x}) = A\vec{x}$)? We introduced the standard matrix of a linear transformation back in Section 2.1; alternatively, we can think of A as the matrix of T with respect to the standard basis $\mathfrak{A} = (\vec{e}_1, \vec{e}_2)$ of $\mathbb{R}^2$, in the sense of Definition 3.4.3. (Think about it!)

Solution

Recall from Definition 3.4.1 that

$$\vec{x} = S\left[\vec{x}\right]_{\mathfrak{B}}, \quad \text{where} \quad S = \begin{bmatrix} 3 & -1 \\ 1 & 3 \end{bmatrix},$$

and consider the following diagram:

Note that $T(\vec{x}) = AS\left[\vec{x}\right]_{\mathfrak{B}} = SB\left[\vec{x}\right]_{\mathfrak{B}}$ for all $\vec{x}$ in $\mathbb{R}^2$, so that

$$AS = SB, \quad B = S^{-1}AS, \quad \text{and} \quad A = SBS^{-1}.$$

We can use the last formula to find the standard matrix A of T:

$$A = SBS^{-1} = \begin{bmatrix} 3 & -1 \\ 1 & 3 \end{bmatrix} \begin{bmatrix} 1 & 0 \\ 0 & 0 \end{bmatrix} \left(\frac{1}{10} \begin{bmatrix} 3 & 1 \\ -1 & 3 \end{bmatrix} \right) = \begin{bmatrix} 0.9 & 0.3 \\ 0.3 & 0.1 \end{bmatrix}.$$

Alternatively, we could use Definition 2.2.1 to construct matrix A.   ∎

Theorem 3.4.4   **Standard matrix versus $\mathfrak{B}$-matrix**

Consider a linear transformation T from $\mathbb{R}^n$ to $\mathbb{R}^n$ and a basis $\mathfrak{B} = (\vec{v}_1, \ldots, \vec{v}_n)$ of $\mathbb{R}^n$. Let B be the $\mathfrak{B}$-matrix of T, and let A be the standard matrix of T (such that $T(\vec{x}) = A\vec{x}$ for all $\vec{x}$ in $\mathbb{R}^n$). Then

$$AS = SB, \quad B = S^{-1}AS, \quad \text{and} \quad A = SBS^{-1}, \quad \text{where} \quad S = \begin{bmatrix} \vec{v}_1 & \cdots & \vec{v}_n \end{bmatrix}. \quad ∎$$

The formulas in Theorem 3.4.4 motivate the following definition.

Definition 3.4.5   **Similar matrices**

Consider two $n \times n$ matrices A and B. We say that A is similar to B if there exists an invertible matrix S such that

$$AS = SB, \quad \text{or} \quad B = S^{-1}AS.$$

Thus two matrices are similar if they represent the same linear transformation with respect to different bases.

EXAMPLE 6   Is matrix $A = \begin{bmatrix} 1 & 2 \\ 4 & 3 \end{bmatrix}$ similar to $B = \begin{bmatrix} 5 & 0 \\ 0 & -1 \end{bmatrix}$?

Solution

At this early stage of the course, we have to tackle this problem with "brute force," using Definition 3.4.5. In Chapter 7, we will develop tools that allow a more conceptual approach.

We are looking for a matrix $S = \begin{bmatrix} x & y \\ z & t \end{bmatrix}$ such that $AS = SB$, or

$$\begin{bmatrix} x+2z & y+2t \\ 4x+3z & 4y+3t \end{bmatrix} = \begin{bmatrix} 5x & -y \\ 5z & -t \end{bmatrix}.$$

These equations simplify to

$$z = 2x, \quad t = -y,$$

so that any invertible matrix of the form

$$S = \begin{bmatrix} x & y \\ 2x & -y \end{bmatrix}$$

does the job. Note that $\det(S) = -3xy$. By Theorem 2.4.9, matrix S is invertible if $\det(S) = -3xy \neq 0$, meaning that neither x nor y is zero. For example, we can let $x = y = 1$, so that $S = \begin{bmatrix} 1 & 1 \\ 2 & -1 \end{bmatrix}$.

Matrix A turns out to be similar to B.   ■

EXAMPLE 7   Show that if matrix A is similar to B, then its power A^t is similar to B^t, for all positive integers t. (That is, A^2 is similar to B^2, A^3 is similar to B^3, and so on.)

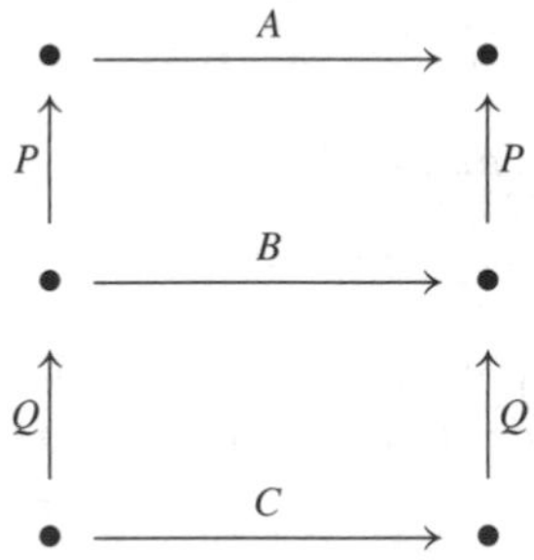

Solution

We know that $B = S^{-1}AS$ for some invertible matrix S. Now

$$B^t = \underbrace{(S^{-1}AS)(S^{-1}AS)\cdots(S^{-1}AS)(S^{-1}AS)}_{t \text{ times}} = S^{-1}A^tS,$$

proving our claim. Note the cancellation of many terms of the form SS^{-1}.   ■

We conclude this section with some noteworthy facts about similar matrices.

Figure 8

Theorem 3.4.6   Similarity is an equivalence relation

a. An $n \times n$ matrix A is similar to A itself (*reflexivity*).

b. If A is similar to B, then B is similar to A (*symmetry*).

c. If A is similar to B and B is similar to C, then A is similar to C (*transitivity*).

Proof   We will prove transitivity, leaving reflexivity and symmetry as Exercise 65.

The assumptions of part (c) mean that there exist invertible matrices P and Q such that $AP = PB$ and $BQ = QC$. Using Figure 8 as a guide, we find that $APQ = PBQ = PQC$. We see that $AS = SC$, where $S = PQ$ is invertible, proving that A is similar to C.   ■

EXERCISES 3.4

GOAL   *Use the concept of coordinates. Apply the definition of the matrix of a linear transformation with respect to a basis. Relate this matrix to the standard matrix of the transformation. Find the matrix of a linear transformation (with respect to any basis) column by column. Use the concept of similarity.*

In Exercises 1 through 18, determine whether the vector $\vec{x}$ is in the span V of the vectors $\vec{v}_1, \ldots, \vec{v}_m$ (proceed "by inspection" if possible, and use the reduced row-echelon form if necessary). If $\vec{x}$ is in V, find the coordinates of $\vec{x}$ with respect to the basis $\mathfrak{B} = (\vec{v}_1, \ldots, \vec{v}_m)$ of V, and write the coordinate vector $[\vec{x}]_{\mathfrak{B}}$.

1. $\vec{x} = \begin{bmatrix} 2 \\ 3 \end{bmatrix}$; $\vec{v}_1 = \begin{bmatrix} 1 \\ 0 \end{bmatrix}$, $\vec{v}_2 = \begin{bmatrix} 0 \\ 1 \end{bmatrix}$

2. $\vec{x} = \begin{bmatrix} 3 \\ -4 \end{bmatrix}$; $\vec{v}_1 = \begin{bmatrix} 0 \\ 1 \end{bmatrix}$, $\vec{v}_2 = \begin{bmatrix} 1 \\ 0 \end{bmatrix}$

3. $\vec{x} = \begin{bmatrix} 31 \\ 37 \end{bmatrix}$; $\vec{v}_1 = \begin{bmatrix} 23 \\ 29 \end{bmatrix}$, $\vec{v}_2 = \begin{bmatrix} 31 \\ 37 \end{bmatrix}$

4. $\vec{x} = \begin{bmatrix} 23 \\ 29 \end{bmatrix}$; $\vec{v}_1 = \begin{bmatrix} 46 \\ 58 \end{bmatrix}$, $\vec{v}_2 = \begin{bmatrix} 61 \\ 67 \end{bmatrix}$

5. $\vec{x} = \begin{bmatrix} 7 \\ 16 \end{bmatrix}$; $\vec{v}_1 = \begin{bmatrix} 2 \\ 5 \end{bmatrix}$, $\vec{v}_2 = \begin{bmatrix} 5 \\ 12 \end{bmatrix}$

6. $\vec{x} = \begin{bmatrix} -4 \\ 4 \end{bmatrix}$; $\vec{v}_1 = \begin{bmatrix} 1 \\ 2 \end{bmatrix}$, $\vec{v}_2 = \begin{bmatrix} 5 \\ 6 \end{bmatrix}$

7. $\vec{x} = \begin{bmatrix} 3 \\ 1 \\ -4 \end{bmatrix}$; $\vec{v}_1 = \begin{bmatrix} 1 \\ -1 \\ 0 \end{bmatrix}$, $\vec{v}_2 = \begin{bmatrix} 0 \\ 1 \\ -1 \end{bmatrix}$

8. $\vec{x} = \begin{bmatrix} 2 \\ 3 \\ 4 \end{bmatrix}$; $\vec{v}_1 = \begin{bmatrix} 1 \\ 1 \\ 0 \end{bmatrix}$, $\vec{v}_2 = \begin{bmatrix} 2 \\ 0 \\ 1 \end{bmatrix}$

9. $\vec{x} = \begin{bmatrix} 3 \\ 3 \\ 4 \end{bmatrix}$; $\vec{v}_1 = \begin{bmatrix} 1 \\ 1 \\ 0 \end{bmatrix}$, $\vec{v}_2 = \begin{bmatrix} 0 \\ -1 \\ 2 \end{bmatrix}$

10. $\vec{x} = \begin{bmatrix} -5 \\ 1 \\ 3 \end{bmatrix}$; $\vec{v}_1 = \begin{bmatrix} -1 \\ 0 \\ 1 \end{bmatrix}$, $\vec{v}_2 = \begin{bmatrix} -2 \\ 1 \\ 0 \end{bmatrix}$

11. $\vec{x} = \begin{bmatrix} -1 \\ 2 \\ 2 \end{bmatrix}$; $\vec{v}_1 = \begin{bmatrix} 1 \\ 2 \\ 1 \end{bmatrix}$, $\vec{v}_2 = \begin{bmatrix} -3 \\ 2 \\ 3 \end{bmatrix}$

12. $\vec{x} = \begin{bmatrix} 1 \\ -2 \\ -2 \end{bmatrix}$; $\vec{v}_1 = \begin{bmatrix} 8 \\ 4 \\ -1 \end{bmatrix}$, $\vec{v}_2 = \begin{bmatrix} 5 \\ 2 \\ -1 \end{bmatrix}$

13. $\vec{x} = \begin{bmatrix} 1 \\ 1 \\ 1 \end{bmatrix}$; $\vec{v}_1 = \begin{bmatrix} 1 \\ 2 \\ 3 \end{bmatrix}$, $\vec{v}_2 = \begin{bmatrix} 0 \\ 1 \\ 2 \end{bmatrix}$, $\vec{v}_3 = \begin{bmatrix} 0 \\ 0 \\ 1 \end{bmatrix}$

14. $\vec{x} = \begin{bmatrix} 3 \\ 7 \\ 13 \end{bmatrix}$; $\vec{v}_1 = \begin{bmatrix} 1 \\ 1 \\ 1 \end{bmatrix}$, $\vec{v}_2 = \begin{bmatrix} 0 \\ 1 \\ 1 \end{bmatrix}$, $\vec{v}_3 = \begin{bmatrix} 0 \\ 0 \\ 1 \end{bmatrix}$

15. $\vec{x} = \begin{bmatrix} 1 \\ 0 \\ 0 \end{bmatrix}$; $\vec{v}_1 = \begin{bmatrix} 1 \\ 2 \\ 1 \end{bmatrix}$, $\vec{v}_2 = \begin{bmatrix} 1 \\ 3 \\ 4 \end{bmatrix}$, $\vec{v}_3 = \begin{bmatrix} 1 \\ 4 \\ 8 \end{bmatrix}$

16. $\vec{x} = \begin{bmatrix} 7 \\ 1 \\ 3 \end{bmatrix}$; $\vec{v}_1 = \begin{bmatrix} 1 \\ 1 \\ 1 \end{bmatrix}$, $\vec{v}_2 = \begin{bmatrix} 1 \\ 2 \\ 3 \end{bmatrix}$, $\vec{v}_3 = \begin{bmatrix} 1 \\ 3 \\ 6 \end{bmatrix}$

17. $\vec{x} = \begin{bmatrix} 1 \\ 1 \\ 1 \\ -1 \end{bmatrix}$; $\vec{v}_1 = \begin{bmatrix} 1 \\ 0 \\ 2 \\ 0 \end{bmatrix}$, $\vec{v}_2 = \begin{bmatrix} 0 \\ 1 \\ 3 \\ 0 \end{bmatrix}$, $\vec{v}_3 = \begin{bmatrix} 0 \\ 0 \\ 4 \\ 1 \end{bmatrix}$

18. $\vec{x} = \begin{bmatrix} 5 \\ 4 \\ 3 \\ 2 \end{bmatrix}$; $\vec{v}_1 = \begin{bmatrix} 1 \\ 1 \\ 0 \\ 0 \end{bmatrix}$, $\vec{v}_2 = \begin{bmatrix} 0 \\ 1 \\ 1 \\ 0 \end{bmatrix}$, $\vec{v}_3 = \begin{bmatrix} 0 \\ -1 \\ 0 \\ 1 \end{bmatrix}$

In Exercises 19 through 24, find the matrix B of the linear transformation $T(\vec{x}) = A\vec{x}$ with respect to the basis $\mathfrak{B} = (\vec{v}_1, \vec{v}_2)$. For practice, solve each problem in three ways: (a) Use the formula $B = S^{-1}AS$ (b) use a commutative diagram (as in Examples 3 and 4), and (c) construct B "column by column."

19. $A = \begin{bmatrix} 0 & 1 \\ 1 & 0 \end{bmatrix}$; $\vec{v}_1 = \begin{bmatrix} 1 \\ 1 \end{bmatrix}$, $\vec{v}_2 = \begin{bmatrix} 1 \\ -1 \end{bmatrix}$

20. $A = \begin{bmatrix} 1 & 1 \\ 1 & 1 \end{bmatrix}$; $\vec{v}_1 = \begin{bmatrix} 1 \\ 1 \end{bmatrix}$, $\vec{v}_2 = \begin{bmatrix} 1 \\ -1 \end{bmatrix}$

21. $A = \begin{bmatrix} 1 & 2 \\ 3 & 6 \end{bmatrix}$; $\vec{v}_1 = \begin{bmatrix} 1 \\ 3 \end{bmatrix}$, $\vec{v}_2 = \begin{bmatrix} -2 \\ 1 \end{bmatrix}$

22. $A = \begin{bmatrix} -3 & 4 \\ 4 & 3 \end{bmatrix}$; $\vec{v}_1 = \begin{bmatrix} 1 \\ 2 \end{bmatrix}$, $\vec{v}_2 = \begin{bmatrix} -2 \\ 1 \end{bmatrix}$

23. $A = \begin{bmatrix} 5 & -3 \\ 6 & -4 \end{bmatrix}$; $\vec{v}_1 = \begin{bmatrix} 1 \\ 1 \end{bmatrix}$, $\vec{v}_2 = \begin{bmatrix} 1 \\ 2 \end{bmatrix}$

24. $A = \begin{bmatrix} 13 & -20 \\ 6 & -9 \end{bmatrix}$; $\vec{v}_1 = \begin{bmatrix} 2 \\ 1 \end{bmatrix}$, $\vec{v}_2 = \begin{bmatrix} 5 \\ 3 \end{bmatrix}$

In Exercises 25 through 30, find the matrix B of the linear transformation $T(\vec{x}) = A\vec{x}$ with respect to the basis $\mathfrak{B} = (\vec{v}_1, \ldots, \vec{v}_m)$.

25. $A = \begin{bmatrix} 1 & 2 \\ 3 & 4 \end{bmatrix}$; $\vec{v}_1 = \begin{bmatrix} 1 \\ 1 \end{bmatrix}$, $\vec{v}_2 = \begin{bmatrix} 1 \\ 2 \end{bmatrix}$

26. $A = \begin{bmatrix} 0 & 1 \\ 2 & 3 \end{bmatrix}$; $\vec{v}_1 = \begin{bmatrix} 1 \\ 2 \end{bmatrix}$, $\vec{v}_2 = \begin{bmatrix} 1 \\ 1 \end{bmatrix}$

27. $A = \begin{bmatrix} 4 & 2 & -4 \\ 2 & 1 & -2 \\ -4 & -2 & 4 \end{bmatrix}$;

$\vec{v}_1 = \begin{bmatrix} 2 \\ 1 \\ -2 \end{bmatrix}$, $\vec{v}_2 = \begin{bmatrix} 0 \\ 2 \\ 1 \end{bmatrix}$, $\vec{v}_3 = \begin{bmatrix} 1 \\ 0 \\ 1 \end{bmatrix}$

28. $A = \begin{bmatrix} 5 & -4 & -2 \\ -4 & 5 & -2 \\ -2 & -2 & 8 \end{bmatrix}$;

$\vec{v}_1 = \begin{bmatrix} 2 \\ 2 \\ 1 \end{bmatrix}$, $\vec{v}_2 = \begin{bmatrix} 1 \\ -1 \\ 0 \end{bmatrix}$, $\vec{v}_3 = \begin{bmatrix} 0 \\ 1 \\ -2 \end{bmatrix}$

29. $A = \begin{bmatrix} -1 & 1 & 0 \\ 0 & -2 & 2 \\ 3 & -9 & 6 \end{bmatrix}$;

$\vec{v}_1 = \begin{bmatrix} 1 \\ 1 \\ 1 \end{bmatrix}, \vec{v}_2 = \begin{bmatrix} 1 \\ 2 \\ 3 \end{bmatrix}, \vec{v}_3 = \begin{bmatrix} 1 \\ 3 \\ 6 \end{bmatrix}$

30. $A = \begin{bmatrix} 0 & 2 & -1 \\ 2 & -1 & 0 \\ 4 & -4 & 1 \end{bmatrix}$;

$\vec{v}_1 = \begin{bmatrix} 1 \\ 1 \\ 1 \end{bmatrix}, \vec{v}_2 = \begin{bmatrix} 0 \\ 1 \\ 2 \end{bmatrix}, \vec{v}_3 = \begin{bmatrix} 1 \\ 2 \\ 4 \end{bmatrix}$

Let $\mathfrak{B} = (\vec{v}_1, \vec{v}_2, \vec{v}_3)$ be any basis of $\mathbb{R}^3$ consisting of perpendicular unit vectors, such that $\vec{v}_3 = \vec{v}_1 \times \vec{v}_2$. In Exercises 31 through 36, find the $\mathfrak{B}$-matrix B of the given linear transformation T from $\mathbb{R}^3$ to $\mathbb{R}^3$. Interpret T geometrically.

31. $T(\vec{x}) = \vec{v}_2 \times \vec{x}$ 32. $T(\vec{x}) = \vec{x} \times \vec{v}_3$

33. $T(\vec{x}) = (\vec{v}_2 \cdot \vec{x})\vec{v}_2$ 34. $T(\vec{x}) = \vec{x} - 2(\vec{v}_3 \cdot \vec{x})\vec{v}_3$

35. $T(\vec{x}) = \vec{x} - 2(\vec{v}_1 \cdot \vec{x})\vec{v}_2$

36. $T(\vec{x}) = \vec{v}_1 \times \vec{x} + (\vec{v}_1 \cdot \vec{x})\vec{v}_1$

In Exercises 37 through 42, find a basis $\mathfrak{B}$ of $\mathbb{R}^n$ such that the $\mathfrak{B}$-matrix B of the given linear transformation T is diagonal.

37. Orthogonal projection T onto the line in $\mathbb{R}^2$ spanned by $\begin{bmatrix} 1 \\ 2 \end{bmatrix}$

38. Reflection T about the line in $\mathbb{R}^2$ spanned by $\begin{bmatrix} 2 \\ 3 \end{bmatrix}$

39. Reflection T about the line in $\mathbb{R}^3$ spanned by $\begin{bmatrix} 1 \\ 2 \\ 3 \end{bmatrix}$

40. Orthogonal projection T onto the line in $\mathbb{R}^3$ spanned by $\begin{bmatrix} 1 \\ 1 \\ 1 \end{bmatrix}$

41. Orthogonal projection T onto the plane $3x_1 + x_2 + 2x_3 = 0$ in $\mathbb{R}^3$

42. Reflection T about the plane $x_1 - 2x_2 + 2x_3 = 0$ in $\mathbb{R}^3$

43. Consider the plane $x_1 + 2x_2 + x_3 = 0$ with basis $\mathfrak{B}$ consisting of vectors $\begin{bmatrix} -1 \\ 0 \\ 1 \end{bmatrix}$ and $\begin{bmatrix} -2 \\ 1 \\ 0 \end{bmatrix}$. If $\left[\vec{x}\right]_{\mathfrak{B}} = \begin{bmatrix} 2 \\ -3 \end{bmatrix}$, find $\vec{x}$.

44. Consider the plane $2x_1 - 3x_2 + 4x_3 = 0$ with basis $\mathfrak{B}$ consisting of vectors $\begin{bmatrix} 8 \\ 4 \\ -1 \end{bmatrix}$ and $\begin{bmatrix} 5 \\ 2 \\ -1 \end{bmatrix}$. If $\left[\vec{x}\right]_{\mathfrak{B}} = \begin{bmatrix} 2 \\ -1 \end{bmatrix}$, find $\vec{x}$.

45. Consider the plane $2x_1 - 3x_2 + 4x_3 = 0$. Find a basis $\mathfrak{B}$ of this plane such that $\left[\vec{x}\right]_{\mathfrak{B}} = \begin{bmatrix} 2 \\ 3 \end{bmatrix}$ for $\vec{x} = \begin{bmatrix} 2 \\ 0 \\ -1 \end{bmatrix}$.

46. Consider the plane $x_1 + 2x_2 + x_3 = 0$. Find a basis $\mathfrak{B}$ of this plane such that $\left[\vec{x}\right]_{\mathfrak{B}} = \begin{bmatrix} 2 \\ -1 \end{bmatrix}$ for $\vec{x} = \begin{bmatrix} 1 \\ -1 \\ 1 \end{bmatrix}$.

47. Consider a linear transformation T from $\mathbb{R}^2$ to $\mathbb{R}^2$. We are told that the matrix of T with respect to the basis $\begin{bmatrix} 0 \\ 1 \end{bmatrix}, \begin{bmatrix} 1 \\ 0 \end{bmatrix}$ is $\begin{bmatrix} a & b \\ c & d \end{bmatrix}$. Find the standard matrix of T in terms of $a, b, c,$ and d.

48. In the accompanying figure, sketch the vector $\vec{x}$ with $\left[\vec{x}\right]_{\mathfrak{B}} = \begin{bmatrix} -1 \\ 2 \end{bmatrix}$, where $\mathfrak{B}$ is the basis of $\mathbb{R}^2$ consisting of the vectors $\vec{v}, \vec{w}$.

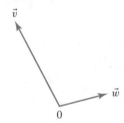

49. Consider the vectors $\vec{u}, \vec{v},$ and $\vec{w}$ sketched in the accompanying figure. Find the coordinate vector of $\vec{w}$ with respect to the basis $\vec{u}, \vec{v}$.

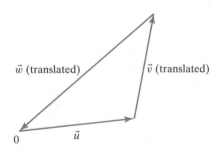

$\vec{w}$ (translated) $\vec{v}$ (translated)

50. Given a hexagonal tiling of the plane, such as you might find on a kitchen floor, consider the basis $\mathfrak{B}$ of $\mathbb{R}^2$ consisting of the vectors $\vec{v}, \vec{w}$ in the following sketch:

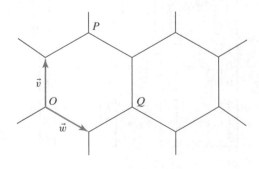

a. Find the coordinate vectors $\left[\overrightarrow{OP}\right]_{\mathfrak{B}}$ and $\left[\overrightarrow{OQ}\right]_{\mathfrak{B}}$.

Hint: Sketch the coordinate grid defined by the basis $\mathfrak{B} = (\vec{v}, \vec{w})$.

b. We are told that $\left[\overrightarrow{OR}\right]_{\mathfrak{B}} = \begin{bmatrix} 3 \\ 2 \end{bmatrix}$. Sketch the point R. Is R a vertex or a center of a tile?

c. We are told that $\left[\overrightarrow{OS}\right]_{\mathfrak{B}} = \begin{bmatrix} 17 \\ 13 \end{bmatrix}$. Is S a center or a vertex of a tile?

51. Prove part (a) of Theorem 3.4.2.

52. If $\mathfrak{B}$ is a basis of $\mathbb{R}^n$, is the transformation T from $\mathbb{R}^n$ to $\mathbb{R}^n$ given by

$$T(\vec{x}) = \left[\vec{x}\right]_{\mathfrak{B}}$$

linear? Justify your answer.

53. Consider the basis $\mathfrak{B}$ of $\mathbb{R}^2$ consisting of the vectors $\begin{bmatrix} 1 \\ 2 \end{bmatrix}$ and $\begin{bmatrix} 3 \\ 4 \end{bmatrix}$. We are told that $\left[\vec{x}\right]_{\mathfrak{B}} = \begin{bmatrix} 7 \\ 11 \end{bmatrix}$ for a certain vector $\vec{x}$ in $\mathbb{R}^2$. Find $\vec{x}$.

54. Let $\mathfrak{B}$ be the basis of $\mathbb{R}^n$ consisting of the vectors $\vec{v}_1, \vec{v}_2, \ldots, \vec{v}_n$, and let $\mathfrak{T}$ be some other basis of $\mathbb{R}^n$. Is

$$\left[\vec{v}_1\right]_{\mathfrak{T}}, \quad \left[\vec{v}_2\right]_{\mathfrak{T}}, \quad \cdots, \quad \left[\vec{v}_n\right]_{\mathfrak{T}}$$

a basis of $\mathbb{R}^n$ as well? Explain.

55. Consider the basis $\mathfrak{B}$ of $\mathbb{R}^2$ consisting of the vectors $\begin{bmatrix} 1 \\ 1 \end{bmatrix}$ and $\begin{bmatrix} 1 \\ 2 \end{bmatrix}$, and let $\mathfrak{R}$ be the basis consisting of $\begin{bmatrix} 1 \\ 2 \end{bmatrix}$, $\begin{bmatrix} 3 \\ 4 \end{bmatrix}$. Find a matrix P such that

$$\left[\vec{x}\right]_{\mathfrak{R}} = P\left[\vec{x}\right]_{\mathfrak{B}},$$

for all $\vec{x}$ in $\mathbb{R}^2$.

56. Find a basis $\mathfrak{B}$ of $\mathbb{R}^2$ such that

$$\begin{bmatrix} 1 \\ 2 \end{bmatrix}_{\mathfrak{B}} = \begin{bmatrix} 3 \\ 5 \end{bmatrix} \quad \text{and} \quad \begin{bmatrix} 3 \\ 4 \end{bmatrix}_{\mathfrak{B}} = \begin{bmatrix} 2 \\ 3 \end{bmatrix}.$$

57. Show that if a 3×3 matrix A represents the reflection about a plane, then A is similar to the matrix

$$\begin{bmatrix} 1 & 0 & 0 \\ 0 & 1 & 0 \\ 0 & 0 & -1 \end{bmatrix}.$$

58. Consider a 3×3 matrix A and a vector $\vec{v}$ in $\mathbb{R}^3$ such that $A^3\vec{v} = \vec{0}$, but $A^2\vec{v} \neq \vec{0}$.

a. Show that the vectors $A^2\vec{v}$, $A\vec{v}$, $\vec{v}$ form a basis of $\mathbb{R}^3$. (*Hint:* It suffices to show linear independence. Consider a relation $c_1 A^2\vec{v} + c_2 A\vec{v} + c_3\vec{v} = \vec{0}$ and multiply by A^2 to show that $c_3 = 0$.)

b. Find the matrix of the transformation $T(\vec{x}) = A\vec{x}$ with respect to the basis $A^2\vec{v}, A\vec{v}, \vec{v}$.

59. Is matrix $\begin{bmatrix} 2 & 0 \\ 0 & 3 \end{bmatrix}$ similar to matrix $\begin{bmatrix} 2 & 1 \\ 0 & 3 \end{bmatrix}$?

60. Is matrix $\begin{bmatrix} 1 & 0 \\ 0 & -1 \end{bmatrix}$ similar to matrix $\begin{bmatrix} 0 & 1 \\ 1 & 0 \end{bmatrix}$?

61. Find a basis $\mathfrak{B}$ of $\mathbb{R}^2$ such that the $\mathfrak{B}$-matrix of the linear transformation

$$T(\vec{x}) = \begin{bmatrix} -5 & -9 \\ 4 & 7 \end{bmatrix}\vec{x} \quad \text{is} \quad B = \begin{bmatrix} 1 & 1 \\ 0 & 1 \end{bmatrix}.$$

62. Find a basis $\mathfrak{B}$ of $\mathbb{R}^2$ such that the $\mathfrak{B}$-matrix of the linear transformation

$$T(\vec{x}) = \begin{bmatrix} 1 & 2 \\ 4 & 3 \end{bmatrix}\vec{x} \quad \text{is} \quad B = \begin{bmatrix} 5 & 0 \\ 0 & -1 \end{bmatrix}.$$

63. Is matrix $\begin{bmatrix} p & -q \\ q & p \end{bmatrix}$ similar to matrix $\begin{bmatrix} p & q \\ -q & p \end{bmatrix}$ for all p and q?

64. Is matrix $\begin{bmatrix} a & b \\ c & d \end{bmatrix}$ similar to matrix $\begin{bmatrix} a & c \\ b & d \end{bmatrix}$ for all a, b, c, d?

65. Prove parts (a) and (b) of Theorem 3.4.6.

66. Consider a matrix A of the form $A = \begin{bmatrix} a & b \\ b & -a \end{bmatrix}$, where $a^2 + b^2 = 1$ and $a \neq 1$. Find the matrix B of the linear transformation $T(\vec{x}) = A\vec{x}$ with respect to the basis $\begin{bmatrix} b \\ 1-a \end{bmatrix}$, $\begin{bmatrix} a-1 \\ b \end{bmatrix}$. Interpret the answer geometrically.

67. If $c \neq 0$, find the matrix of the linear transformation $T(\vec{x}) = \begin{bmatrix} a & b \\ c & d \end{bmatrix}\vec{x}$ with respect to basis $\begin{bmatrix} 1 \\ 0 \end{bmatrix}$, $\begin{bmatrix} a \\ c \end{bmatrix}$.

68. Find an invertible 2×2 matrix S such that

$$S^{-1}\begin{bmatrix} 1 & 2 \\ 3 & 4 \end{bmatrix}S$$

is of the form $\begin{bmatrix} 0 & b \\ 1 & d \end{bmatrix}$. See Exercise 67.

69. If A is a 2×2 matrix such that

$$A\begin{bmatrix} 1 \\ 2 \end{bmatrix} = \begin{bmatrix} 3 \\ 6 \end{bmatrix} \quad \text{and} \quad A\begin{bmatrix} 2 \\ 1 \end{bmatrix} = \begin{bmatrix} -2 \\ -1 \end{bmatrix},$$

show that A is similar to a diagonal matrix D. Find an invertible S such that $S^{-1}AS = D$.

70. Is there a basis $\mathfrak{B}$ of $\mathbb{R}^2$ such that $\mathfrak{B}$-matrix B of the linear transformation

$$T(\vec{x}) = \begin{bmatrix} 0 & -1 \\ 1 & 0 \end{bmatrix}\vec{x}$$

is upper triangular? (*Hint:* Think about the first column of B.)

71. Suppose that matrix A is similar to B, with $B = S^{-1}AS$.
a. Show that if $\vec{x}$ is in $\ker(B)$, then $S\vec{x}$ is in $\ker(A)$.

b. Show that nullity(A) = nullity(B). *Hint*: If $\vec{v}_1$, $\vec{v}_2, \ldots, \vec{v}_p$ is a basis of ker(B), then the vectors $S\vec{v}_1, S\vec{v}_2, \ldots, S\vec{v}_p$ in ker(A) are linearly independent. Now reverse the roles of A and B.

72. If A is similar to B, what is the relationship between rank(A) and rank(B)? See Exercise 71.

73. Let L be the line in $\mathbb{R}^3$ spanned by the vector

$$\vec{v} = \begin{bmatrix} 0.6 \\ 0.8 \\ 0 \end{bmatrix}.$$

Let T from $\mathbb{R}^3$ to $\mathbb{R}^3$ be the rotation about this line through an angle of $\pi/2$, in the direction indicated in the accompanying sketch. Find the matrix A such that $T(\vec{x}) = A\vec{x}$.

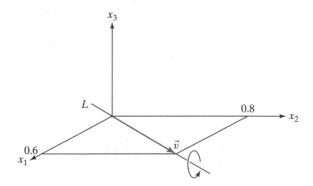

74. Consider the regular tetrahedron in the accompanying sketch whose center is at the origin. Let $\vec{v}_0, \vec{v}_1, \vec{v}_2, \vec{v}_3$ be the position vectors of the four vertices of the tetrahedron:

$$\vec{v}_0 = \overrightarrow{OP_0}, \ldots, \quad \vec{v}_3 = \overrightarrow{OP_3}.$$

a. Find the sum $\vec{v}_0 + \vec{v}_1 + \vec{v}_2 + \vec{v}_3$.
b. Find the coordinate vector of $\vec{v}_0$ with respect to the basis $\vec{v}_1, \vec{v}_2, \vec{v}_3$.

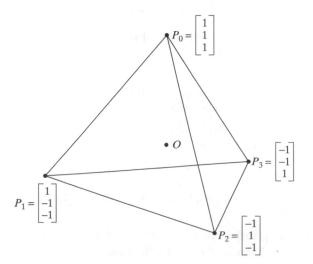

c. Let T be the linear transformation with $T(\vec{v}_0) = \vec{v}_3$, $T(\vec{v}_3) = \vec{v}_1$, and $T(\vec{v}_1) = \vec{v}_0$. What is $T(\vec{v}_2)$? Describe the transformation T geometrically (as a reflection, rotation, projection, or whatever). Find the matrix B of T with respect to the basis $\vec{v}_1, \vec{v}_2, \vec{v}_3$. What is B^3? Explain.

75. Find the matrix B of the rotation $T(\vec{x}) = \begin{bmatrix} 0 & -1 \\ 1 & 0 \end{bmatrix} \vec{x}$ with respect to the basis $\begin{bmatrix} 0 \\ 1 \end{bmatrix}$, $\begin{bmatrix} -1 \\ 0 \end{bmatrix}$. Interpret your answer geometrically.

76. If t is any real number, what is the matrix B of the linear transformation

$$T(\vec{x}) = \begin{bmatrix} \cos(t) & -\sin(t) \\ \sin(t) & \cos(t) \end{bmatrix} \vec{x}$$

with respect to basis $\begin{bmatrix} \cos(t) \\ \sin(t) \end{bmatrix}$, $\begin{bmatrix} -\sin(t) \\ \cos(t) \end{bmatrix}$? Interpret your answer geometrically.

77. Consider a linear transformation $T(\vec{x}) = A\vec{x}$ from $\mathbb{R}^n$ to $\mathbb{R}^n$. Let B be the matrix of T with respect to the basis $\vec{e}_n, \vec{e}_{n-1}, \ldots, \vec{e}_2, \vec{e}_1$ of $\mathbb{R}^n$. Describe the entries of B in terms of the entries of A.

78. This problem refers to Leontief's input–output model, first discussed in the Exercises 1.1.20 and 1.2.37. Consider three industries I_1, I_2, I_3, each of which produces only one good, with unit prices $p_1 = 2, p_2 = 5, p_3 = 10$ (in U.S. dollars), respectively. Let the three products be labeled good 1, good 2, and good 3. Let

$$A = \begin{bmatrix} a_{11} & a_{12} & a_{13} \\ a_{21} & a_{22} & a_{23} \\ a_{31} & a_{32} & a_{33} \end{bmatrix} = \begin{bmatrix} 0.3 & 0.2 & 0.1 \\ 0.1 & 0.3 & 0.3 \\ 0.2 & 0.2 & 0.1 \end{bmatrix}$$

be the matrix that lists the interindustry demand in terms of dollar amounts. The entry a_{ij} tells us how many dollars' worth of good i are required to produce one dollar's worth of good j. Alternatively, the interindustry demand can be measured in units of goods by means of the matrix

$$B = \begin{bmatrix} b_{11} & b_{12} & b_{13} \\ b_{21} & b_{22} & b_{23} \\ b_{31} & b_{32} & b_{33} \end{bmatrix},$$

where b_{ij} tells us how many units of good i are required to produce one unit of good j. Find the matrix B for the economy discussed here. Also write an equation relating the three matrices A, B, and S, where

$$S = \begin{bmatrix} 2 & 0 & 0 \\ 0 & 5 & 0 \\ 0 & 0 & 10 \end{bmatrix}$$

is the diagonal matrix listing the unit prices on the diagonal. Justify your answer carefully.

Chapter Three Exercises

TRUE OR FALSE?

1. If $\vec{v}_1, \vec{v}_2, \ldots, \vec{v}_n$ are linearly independent vectors in $\mathbb{R}^n$, then they must form a basis of $\mathbb{R}^n$.

2. There exists a 5×4 matrix whose image consists of all of $\mathbb{R}^5$.

3. The kernel of any invertible matrix consists of the zero vector only.

4. The identity matrix I_n is similar to all invertible $n \times n$ matrices.

5. If $2\vec{u} + 3\vec{v} + 4\vec{w} = 5\vec{u} + 6\vec{v} + 7\vec{w}$, then vectors $\vec{u}$, $\vec{v}$, $\vec{w}$ must be linearly dependent.

6. The column vectors of a 5×4 matrix must be linearly dependent.

7. If $\vec{v}_1, \vec{v}_2, \ldots, \vec{v}_n$ and $\vec{w}_1, \vec{w}_2, \ldots, \vec{w}_m$ are any two bases of a subspace V of $\mathbb{R}^{10}$, then n must equal m.

8. If A is a 5×6 matrix of rank 4, then the nullity of A is 1.

9. The image of a 3×4 matrix is a subspace of $\mathbb{R}^4$.

10. The span of vectors $\vec{v}_1, \vec{v}_2, \ldots, \vec{v}_n$ consists of all linear combinations of vectors $\vec{v}_1, \vec{v}_2, \ldots, \vec{v}_n$.

11. If vectors $\vec{v}_1, \vec{v}_2, \vec{v}_3, \vec{v}_4$ are linearly independent, then vectors $\vec{v}_1, \vec{v}_2, \vec{v}_3$ must be linearly independent as well.

12. The vectors of the form $\begin{bmatrix} a \\ b \\ 0 \\ a \end{bmatrix}$ (where a and b are arbitrary real numbers) form a subspace of $\mathbb{R}^4$.

13. Matrix $\begin{bmatrix} 1 & 0 \\ 0 & -1 \end{bmatrix}$ is similar to $\begin{bmatrix} 0 & 1 \\ 1 & 0 \end{bmatrix}$.

14. Vectors $\begin{bmatrix} 1 \\ 0 \\ 0 \end{bmatrix}, \begin{bmatrix} 2 \\ 1 \\ 0 \end{bmatrix}, \begin{bmatrix} 3 \\ 2 \\ 1 \end{bmatrix}$ form a basis of $\mathbb{R}^3$.

15. If the kernel of a matrix A consists of the zero vector only, then the column vectors of A must be linearly independent.

16. If the image of an $n \times n$ matrix A is all of $\mathbb{R}^n$, then A must be invertible.

17. If vectors $\vec{v}_1, \vec{v}_2, \ldots, \vec{v}_n$ span $\mathbb{R}^4$, then n must be equal to 4.

18. If vectors $\vec{u}, \vec{v},$ and $\vec{w}$ are in a subspace V of $\mathbb{R}^n$, then vector $2\vec{u} - 3\vec{v} + 4\vec{w}$ must be in V as well.

19. If matrix A is similar to matrix B, and B is similar to C, then C must be similar to A.

20. If a subspace V of $\mathbb{R}^n$ contains none of the standard vectors $\vec{e}_1, \vec{e}_2, \ldots, \vec{e}_n$, then V consists of the zero vector only.

21. If A and B are $n \times n$ matrices, and vector $\vec{v}$ is in the kernel of both A and B, then $\vec{v}$ must be in the kernel of matrix AB as well.

22. If two nonzero vectors are linearly dependent, then each of them is a scalar multiple of the other.

23. If $\vec{v}_1, \vec{v}_2, \vec{v}_3$ are any three distinct vectors in $\mathbb{R}^3$, then there must be a linear transformation T from $\mathbb{R}^3$ to $\mathbb{R}^3$ such that $T(\vec{v}_1) = \vec{e}_1$, $T(\vec{v}_2) = \vec{e}_2$, and $T(\vec{v}_3) = \vec{e}_3$.

24. If vectors $\vec{u}, \vec{v}, \vec{w}$ are linearly dependent, then vector $\vec{w}$ must be a linear combination of $\vec{u}$ and $\vec{v}$.

25. If A and B are invertible $n \times n$ matrices, then AB must be similar to BA.

26. If A is an invertible $n \times n$ matrix, then the kernels of A and A^{-1} must be equal.

27. Matrix $\begin{bmatrix} 0 & 1 \\ 0 & 0 \end{bmatrix}$ is similar to $\begin{bmatrix} 0 & 0 \\ 0 & 1 \end{bmatrix}$.

28. Vectors $\begin{bmatrix} 1 \\ 2 \\ 3 \\ 4 \end{bmatrix}, \begin{bmatrix} 5 \\ 6 \\ 7 \\ 8 \end{bmatrix}, \begin{bmatrix} 9 \\ 8 \\ 7 \\ 6 \end{bmatrix}, \begin{bmatrix} 5 \\ 4 \\ 3 \\ 2 \end{bmatrix}, \begin{bmatrix} 1 \\ 0 \\ -1 \\ -2 \end{bmatrix}$ are linearly independent.

29. If a subspace V of $\mathbb{R}^3$ contains the standard vectors $\vec{e}_1, \vec{e}_2, \vec{e}_3$, then V must be $\mathbb{R}^3$.

30. If a 2×2 matrix P represents the orthogonal projection onto a line in $\mathbb{R}^2$, then P must be similar to matrix $\begin{bmatrix} 1 & 0 \\ 0 & 0 \end{bmatrix}$.

31. $\mathbb{R}^2$ is a subspace of $\mathbb{R}^3$.

32. If an $n \times n$ matrix A is similar to matrix B, then $A + 7I_n$ must be similar to $B + 7I_n$.

33. If V is any three-dimensional subspace of $\mathbb{R}^5$, then V has infinitely many bases.

34. Matrix I_n is similar to $2I_n$.

35. If $AB = 0$ for two 2×2 matrices A and B, then BA must be the zero matrix as well.

36. If A and B are $n \times n$ matrices, and vector $\vec{v}$ is in the image of both A and B, then $\vec{v}$ must be in the image of matrix $A + B$ as well.

37. If V and W are subspaces of $\mathbb{R}^n$, then their union $V \cup W$ must be a subspace of $\mathbb{R}^n$ as well.

38. If the kernel of a 5×4 matrix A consists of the zero vector only and if $A\vec{v} = A\vec{w}$ for two vectors $\vec{v}$ and $\vec{w}$ in $\mathbb{R}^4$, then vectors $\vec{v}$ and $\vec{w}$ must be equal.

39. If $\vec{v}_1, \vec{v}_2, \ldots, \vec{v}_n$ and $\vec{w}_1, \vec{w}_2, \ldots, \vec{w}_n$ are two bases of $\mathbb{R}^n$, then there exists a linear transformation T from $\mathbb{R}^n$ to $\mathbb{R}^n$ such that $T(\vec{v}_1) = \vec{w}_1$, $T(\vec{v}_2) = \vec{w}_2, \ldots,$ $T(\vec{v}_n) = \vec{w}_n$.

40. If matrix A represents a rotation through $\pi/2$ and matrix B a rotation through $\pi/4$, then A is similar to B.

41. There exists a 2×2 matrix A such that $\text{im}(A) = \text{ker}(A)$.

42. If two $n \times n$ matrices A and B have the same rank, then they must be similar.

43. If A is similar to B, and A is invertible, then B must be invertible as well.

44. If $A^2 = 0$ for a 10×10 matrix A, then the inequality $\text{rank}(A) \leq 5$ must hold.

45. For every subspace V of $\mathbb{R}^3$ there exists a 3×3 matrix A such that $V = \text{im}(A)$.

46. There exists a nonzero 2×2 matrix A that is similar to $2A$.

47. If the 2×2 matrix R represents the reflection about a line in $\mathbb{R}^2$, then R must be similar to matrix $\begin{bmatrix} 0 & 1 \\ 1 & 0 \end{bmatrix}$.

48. If A is similar to B, then there exists one *and only one* invertible matrix S such that $S^{-1}AS = B$.

49. If the kernel of a 5×4 matrix A consists of the zero vector alone, and if $AB = AC$ for two 4×5 matrices B and C, then matrices B and C must be equal.

50. If A is any $n \times n$ matrix such that $A^2 = A$, then the image of A and the kernel of A have only the zero vector in common.

51. There exists a 2×2 matrix A such that $A^2 \neq 0$ and $A^3 = 0$.

52. If A and B are $n \times m$ matrices such that the image of A is a subset of the image of B, then there must exist an $m \times m$ matrix C such that $A = BC$.

53. Among the 3×3 matrices whose entries are all 0's and 1's, most are invertible.

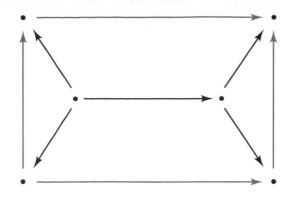

Linear Spaces

4.1 Introduction to Linear Spaces

Thus far in this text, we have applied the language of linear algebra to vectors in $\mathbb{R}^n$. Some of the key words of this language are linear combination, linear transformation, kernel, image, subspace, span, linear independence, basis, dimension, and coordinates. Note that all these concepts can be defined in terms of sums and scalar multiples of vectors. In this chapter, we will see that it can be both natural and useful to apply this language to other mathematical objects, such as functions, matrices, equations, or infinite sequences. Indeed, linear algebra provides a unifying language used throughout modern mathematics and physics.

Here is an introductory example:

EXAMPLE I   Consider the differential equation[1] (DE)

$$f''(x) + f(x) = 0, \quad \text{or} \quad f''(x) = -f(x).$$

We are asked to find all functions $f(x)$ whose second derivative is the negative of the function itself. Recalling the derivative rules from your introductory calculus class, you will (hopefully) note that

$$\sin(x) \quad \text{and} \quad \cos(x)$$

are solutions of this DE.

Can you find any other solutions?

Note that the solution set of this DE is closed under addition and under scalar multiplication. If $f_1(x)$ and $f_2(x)$ are solutions, then so is $f(x) = f_1(x) + f_2(x)$, since

$$f''(x) = f_1''(x) + f_2''(x) = -f_1(x) - f_2(x) = -f(x).$$

Likewise, if $f_1(x)$ is a solution and k is any scalar, then $f(x) = kf_1(x)$ is a solution of the DE as well. (Verify this!)

---

[1] A differential equation is an equation involving derivatives of an unknown function. No previous knowledge of DEs is expected here.

It follows that all "linear combinations"[2]

$$f(x) = c_1 \sin(x) + c_2 \cos(x)$$

are solutions of this DE. It can be shown that all solutions are of this form; we leave the proof as Exercise 58.

Let $F(\mathbb{R}, \mathbb{R})$ be the set of all functions from $\mathbb{R}$ to $\mathbb{R}$. Since the solution set V of our DE is closed under addition and scalar multiplication, we can say that V is a "subspace" of $F(\mathbb{R}, \mathbb{R})$.

How many solutions does this differential equation have? There are infinitely many solutions, of course, but we can use the language of linear algebra to give a more precise answer. The functions $\sin(x)$ and $\cos(x)$ form a "basis" of the "solution space" V, so that the "dimension" of V is 2.

In summary, the solutions of our DE form a two-dimensional subspace of $F(\mathbb{R}, \mathbb{R})$, with basis $\sin(x)$ and $\cos(x)$. ∎

We will now make the informal ideas presented in Example 1 more precise.

Note again that all the basic concepts of linear algebra can be defined in terms of sums and scalar multiples. Whenever we are dealing with a set [such as $F(\mathbb{R}, \mathbb{R})$ in Example 1] whose elements can be added and multiplied by scalars, subject to certain rules, then we can apply the language of linear algebra just as we do for vectors in $\mathbb{R}^n$. These "certain rules" are spelled out in Definition 4.1.1. (Compare this definition with the rules of vector algebra listed in Appendix A.2.)

Definition 4.1.1   Linear spaces (or vector spaces)

A *linear space*[3] V is a set endowed with a rule for addition (if f and g are in V, then so is $f + g$) and a rule for scalar multiplication (if f is in V and k in $\mathbb{R}$, then kf is in V) such that these operations satisfy the following eight rules[4] (for all f, g, h in V and all c, k in $\mathbb{R}$):

1. $(f + g) + h = f + (g + h)$.
2. $f + g = g + f$.
3. There exists a *neutral element* n in V such that $f + n = f$, for all f in V. This n is unique and denoted by 0.
4. For each f in V there exists a g in V such that $f + g = 0$. This g is unique and denoted by $(-f)$.
5. $k(f + g) = kf + kg$.
6. $(c + k)f = cf + kf$.
7. $c(kf) = (ck)f$.
8. $1f = f$.

This definition contains a lot of fine print. In brief, a linear space is a set with two reasonably defined operations, addition and scalar multiplication, that allow us to form linear combinations. All the other basic concepts of linear algebra in turn rest on the concept of a linear combination.

---

[2] We are cautious here and use quotes, since the term *linear combination* has been officially defined for vectors in $\mathbb{R}^n$ only.

[3] The term *vector space* is more commonly used in English (but it's *espace linéaire* in French). We prefer the term *linear space* to avoid the confusion that some students experience with the term *vector* in this abstract sense.

[4] These axioms were established by the Italian mathematician Giuseppe Peano (1858–1932) in his *Calcolo Geometrico* of 1888. Peano calls V a "linear system."

EXAMPLE 2   In $\mathbb{R}^n$, the prototype linear space, the neutral element is the zero vector, $\vec{0}$.   ∎

Probably the most important examples of linear spaces, besides $\mathbb{R}^n$, are *spaces of functions*.

EXAMPLE 3   Let $F(\mathbb{R}, \mathbb{R})$ be the set of all functions from $\mathbb{R}$ to $\mathbb{R}$ (see Example 1), with the operations

$$(f + g)(x) = f(x) + g(x)$$
$$\text{and}$$
$$(kf)(x) = kf(x).$$

Then $F(\mathbb{R}, \mathbb{R})$ is a linear space. The neutral element is the zero function, $f(x) = 0$ for all x.   ∎

EXAMPLE 4   If addition and scalar multiplication are given as in Definition 1.3.5, then $\mathbb{R}^{n \times m}$, the set of all $n \times m$ matrices, is a linear space. The neutral element is the zero matrix, whose entries are all zero.   ∎

EXAMPLE 5   The set of all infinite sequences of real numbers is a linear space, where addition and scalar multiplication are defined term by term:

$$(x_0, x_1, x_2, \ldots) + (y_0, y_1, y_2, \ldots) = (x_0 + y_0, x_1 + y_1, x_2 + y_2, \ldots)$$
$$k(x_0, x_1, x_2, \ldots) = (kx_0, kx_1, kx_2, \ldots).$$

The neutral element is the sequence

$$(0, 0, 0, \ldots).$$ ∎

EXAMPLE 6   The linear equations in three unknowns,

$$ax + by + cz = d,$$

where a, b, c, and d are constants, form a linear space.

The operations (addition and scalar multiplication) are familiar from the process of Gaussian elimination discussed in Chapter 1. The neutral element is the equation $0 = 0$ (with $a = b = c = d = 0$).   ∎

EXAMPLE 7   Consider the plane with a point designated as the origin, O, but without a coordinate system (the coordinate-free plane). A *geometric vector* $\vec{v}$ in this plane is an arrow (a directed line segment) with its tail at the origin, as shown in Figure 1. The sum $\vec{v} + \vec{w}$ of two vectors $\vec{v}$ and $\vec{w}$ is defined by means of a parallelogram, as illustrated in Figure 2. If k is a positive scalar, then vector $k\vec{v}$ points in the same direction as $\vec{v}$, but $k\vec{v}$ is k times as long as $\vec{v}$; see Figure 3. If k is negative, then $k\vec{v}$ points in the opposite direction, and it is $|k|$ times as long as $\vec{v}$; see Figure 4. The geometric vectors in the plane with these operations forms a linear space. The neutral element is the zero vector $\vec{0}$, with tail and head at the origin.

By introducing a *coordinate system,* we can identify the plane of geometric vectors with $\mathbb{R}^2$; this was the great idea of Descartes's *Analytic Geometry*. In Section 4.3, we will study this idea more systematically.   ∎

EXAMPLE 8   Let $\mathbb{C}$ be the set of the *complex numbers*. We trust that you have at least a fleeting acquaintance with complex numbers. Without attempting a definition, we recall that a complex number can be expressed as $z = a + bi$, where a and b are real numbers.

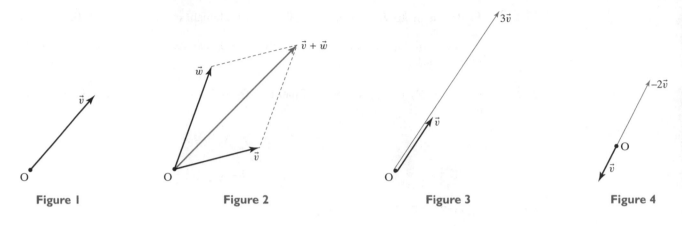

Figure 1   Figure 2   Figure 3   Figure 4

Addition of complex numbers is defined in a natural way, by the rule

$$(a + ib) + (c + id) = (a + c) + i(b + d).$$

If k is a real scalar, we define

$$k(a + ib) = ka + i(kb).$$

There is also a (less natural) rule for the multiplication of complex numbers, but we are not concerned with this operation here.

The complex numbers $\mathbb{C}$ with the two operations just given form a linear space; the neutral element is the complex number $0 = 0 + 0i$. ∎

We say that an element f of a linear space is a *linear combination* of the elements $f_1, f_2, \ldots, f_n$ if

$$f = c_1 f_1 + c_2 f_2 + \cdots + c_n f_n$$

for some scalars $c_1, c_2, \ldots, c_n$.

EXAMPLE 9   Let $A = \begin{bmatrix} 0 & 1 \\ 2 & 3 \end{bmatrix}$. Show that $A^2 = \begin{bmatrix} 2 & 3 \\ 6 & 11 \end{bmatrix}$ is a linear combination of A and I_2.

Solution

We have to find scalars c_1 and c_2 such that

$$A^2 = c_1 A + c_2 I_2,$$

or

$$\begin{bmatrix} 2 & 3 \\ 6 & 11 \end{bmatrix} = c_1 \begin{bmatrix} 0 & 1 \\ 2 & 3 \end{bmatrix} + c_2 \begin{bmatrix} 1 & 0 \\ 0 & 1 \end{bmatrix}.$$

In this simple example, we can see by inspection that $c_1 = 3$ and $c_2 = 2$. We could do this problem more systematically and solve a system of four linear equations in two unknowns. ∎

Since the basic notions of linear algebra (initially introduced for $\mathbb{R}^n$) are defined in terms of linear combinations, we can now generalize these notions without modifications. A short version of the rest of this chapter would say that the concepts of linear transformation, kernel, image, linear independence, span, subspace, basis, dimension, and coordinates can be defined for a linear space in just the same way as for $\mathbb{R}^n$. Figure 5 illustrates the logical dependencies between the key concepts of linear algebra introduced thus far.

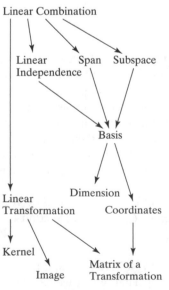

Figure 5

What follows is the long version, with many examples.

Definition 4.1.2   Subspaces

A subset W of a linear space V is called a *subspace* of V if

    a. W contains the neutral element 0 of V.

    b. W is closed under addition (if f and g are in W, then so is $f + g$).

    c. W is closed under scalar multiplication (if f is in W and k is a scalar, then kf is in W).

We can summarize parts (b) and (c) by saying that W is closed under linear combinations.

Note that a subspace W of a linear space V is a linear space in its own right. (Why do the eight rules listed in Definition 4.1.1 hold for W?)

EXAMPLE 10   Show that the polynomials of degree ≤ 2, of the form $f(x) = a + bx + cx^2$, are a subspace W of the space $F(\mathbb{R}, \mathbb{R})$ of all functions from $\mathbb{R}$ to $\mathbb{R}$.

Solution

    a. W contains the neutral element of $F(\mathbb{R}, \mathbb{R})$, the zero function $f(x) = 0$. Indeed, we can write $f(x) = 0 + 0x + 0x^2$.

    b. W is closed under addition: If two polynomials $f(x) = a + bx + cx^2$ and $g(x) = p + qx + rx^2$ are in W, then their sum $f(x) + g(x) = (a + p) + (b + q)x + (c + r)x^2$ is in W as well, since $f(x) + g(x)$ is a polynomial of degree ≤ 2.

    c. W is closed under scalar multiplication: If $f(x) = a + bx + cx^2$ is a polynomial in W and k is a constant, then $kf(x) = ka + (kb)x + (kc)x^2$ is in W as well. ∎

EXAMPLE 11   Show that the differentiable functions form a subspace W of $F(\mathbb{R}, \mathbb{R})$.

Solution

    a. The zero function $f(x) = 0$ is differentiable, with $f'(x) = 0$.

    b. W is closed under addition: You learned in your introductory calculus class that the sum of two differentiable functions $f(x)$ and $g(x)$ is differentiable, with $\big(f(x) + g(x)\big)' = f'(x) + g'(x)$.

    c. W is closed under scalar multiplication, since any scalar multiple of a differentiable function is differentiable as well. ∎

In the next example, we will build on Examples 10 and 11.

EXAMPLE 12   Here are more subspaces of $F(\mathbb{R}, \mathbb{R})$:

    a. C^∞, the smooth functions, that is, functions we can differentiate as many times as we want. This subspace contains all polynomials, exponential functions, $\sin(x)$, and $\cos(x)$, for example.

    b. P, the set of all polynomials.

    c. P_n, the set of all polynomials of degree $\leq n$. ∎

EXAMPLE 13   Show that the matrices B that commute with $A = \begin{bmatrix} 0 & 1 \\ 2 & 3 \end{bmatrix}$ form a subspace of $\mathbb{R}^{2\times 2}$.

Solution

a. The zero matrix 0 commutes with A, since $A0 = 0A = 0$.

b. If matrices B_1 and B_2 commute with A, then so does matrix $B = B_1 + B_2$, since

$$BA = (B_1 + B_2)A = B_1A + B_2A = AB_1 + AB_2 = A(B_1 + B_2) = AB.$$

c. If B commutes with A, then so does kB, since

$$(kB)A = k(BA) = k(AB) = A(kB).$$

Note that we have not used the special form of A. We have indeed shown that the $n \times n$ matrices B that commute with any given $n \times n$ matrix A form a subspace of $\mathbb{R}^{n\times n}$.   ∎

EXAMPLE 14   Consider the set W of all noninvertible 2×2 matrices. Is W a subspace of $\mathbb{R}^{2\times 2}$?

Solution

The following example shows that W isn't closed under addition:

$$\begin{bmatrix} 1 & 0 \\ 0 & 0 \end{bmatrix} + \begin{bmatrix} 0 & 0 \\ 0 & 1 \end{bmatrix} = \begin{bmatrix} 1 & 0 \\ 0 & 1 \end{bmatrix}.$$

$$\underset{\text{in } W}{\nwarrow \quad \nearrow} \qquad \underset{\text{not in } W}{\uparrow}$$

Therefore, W fails to be a subspace of $\mathbb{R}^{2\times 2}$.   ∎

Next, we will generalize the notions of span, linear independence, basis, coordinates, and dimension.

Definition 4.1.3   Span, linear independence, basis, coordinates

Consider the elements $f_1, \ldots, f_n$ in a linear space V.

a. We say that $f_1, \ldots, f_n$ *span* V if every f in V can be expressed as a linear combination of $f_1, \ldots, f_n$.

b. We say that f_i is *redundant* if it is a linear combination of $f_1, \ldots, f_{i-1}$. The elements $f_1, \ldots, f_n$ are called *linearly independent* if none of them is redundant. This is the case if the equation

$$c_1 f_1 + \cdots + c_n f_n = 0$$

has only the trivial solution

$$c_1 = \cdots = c_n = 0.$$

c. We say that elements $f_1, \ldots, f_n$ are a *basis* of V if they span V and are linearly independent. This means that every f in V can be written uniquely as a linear combination $f = c_1 f_1 + \cdots + c_n f_n$. The coefficients $c_1, \ldots, c_n$ are called the *coordinates* of f with respect to the basis $\mathcal{B} = (f_1, \ldots, f_n)$.

The vector

$$\begin{bmatrix} c_1 \\ \vdots \\ c_n \end{bmatrix}$$

in $\mathbb{R}^n$ is called the $\mathcal{B}$-*coordinate vector* of f, denoted by $[f]_{\mathcal{B}}$.

The transformation

$$L(f) = [f]_{\mathcal{B}} = \begin{bmatrix} c_1 \\ \vdots \\ c_n \end{bmatrix} \quad \text{from } V \text{ to } \mathbb{R}^n$$

is called the $\mathcal{B}$-*coordinate transformation,* sometimes denoted by $L_{\mathcal{B}}$.

The $\mathcal{B}$-coordinate transformation is invertible, with inverse

$$L^{-1} \begin{bmatrix} c_1 \\ \vdots \\ c_n \end{bmatrix} = c_1 f_1 + \cdots + c_n f_n.$$

Note in particular that $L^{-1}(\vec{e}_i) = f_i$.

We can represent the coordinate transformation in the following diagram.

$$f = c_1 f_1 + \cdots + c_n f_n \quad \text{in } V \quad \xrightarrow{\;L_{\mathcal{B}}\;} \quad [f]_{\mathcal{B}} = \begin{bmatrix} c_1 \\ \vdots \\ c_n \end{bmatrix} \quad \text{in } \mathbb{R}^n$$

As in the case of $\mathbb{R}^n$, coordinates have important linearity properties.

Theorem 4.1.4   Linearity of the coordinate transformation $L_{\mathcal{B}}$

If $\mathcal{B}$ is a basis of a linear space V, then

a. $[f + g]_{\mathcal{B}} = [f]_{\mathcal{B}} + [g]_{\mathcal{B}},$ for all elements f and g of V, and

b. $[kf]_{\mathcal{B}} = k\,[f]_{\mathcal{B}},$ for all f in V and for all scalars k.   ■

The proof is analogous to that of Theorem 3.4.2.

Now we are ready to introduce the key concept of the dimension of a linear space.

Theorem 4.1.5   Dimension

If a linear space V has a basis with n elements, then all other bases of V consist of n elements as well. We say that n is the *dimension* of V:

$$\dim(V) = n.$$ ■

To prove this important theorem, consider two bases $\mathcal{B} = (f_1, \ldots, f_n)$ and $\mathcal{C} = (g_1, \ldots, g_m)$ of V; we have to show that $n = m$.

We will show first that the m vectors $[g_1]_{\mathcal{B}}, \ldots, [g_m]_{\mathcal{B}}$ in $\mathbb{R}^n$ are linearly independent, which implies that $m \leq n$ (by Theorem 3.2.8). Consider a relation

$$c_1 [g_1]_{\mathcal{B}} + \cdots + c_m [g_m]_{\mathcal{B}} = \vec{0}.$$

By Theorem 4.1.4, we have

$$[c_1 g_1 + \cdots + c_m g_m]_{\mathcal{B}} = \vec{0}, \quad \text{so that} \quad c_1 g_1 + \cdots + c_m g_m = 0.$$

Since the elements $g_1, \ldots, g_m$ are linearly independent, it follows that $c_1 = \cdots = c_m = 0$, meaning that $c_1 \left[g_1 \right]_{\mathfrak{B}} + \cdots + c_m \left[g_m \right]_{\mathfrak{B}} = \vec{0}$ is the trivial relation, as claimed.

Reversing the roles of the two bases, we see that the n vectors $\left[f_1 \right]_{\mathfrak{C}}, \ldots, \left[f_n \right]_{\mathfrak{C}}$ in $\mathbb{R}^m$ are linearly independent, so that $n \leq m$.

We can conclude that $n = m$, as claimed.   ∎

EXAMPLE 15   Find a basis of $\mathbb{R}^{2\times 2}$, the space of all 2×2 matrices, and thus determine the dimension of $\mathbb{R}^{2\times 2}$.

Solution

We can write any 2×2 matrix $\begin{bmatrix} a & b \\ c & d \end{bmatrix}$ as

$$\begin{bmatrix} a & b \\ c & d \end{bmatrix} = a \begin{bmatrix} 1 & 0 \\ 0 & 0 \end{bmatrix} + b \begin{bmatrix} 0 & 1 \\ 0 & 0 \end{bmatrix} + c \begin{bmatrix} 0 & 0 \\ 1 & 0 \end{bmatrix} + d \begin{bmatrix} 0 & 0 \\ 0 & 1 \end{bmatrix}.$$

This shows that matrices

$$\begin{bmatrix} 1 & 0 \\ 0 & 0 \end{bmatrix}, \begin{bmatrix} 0 & 1 \\ 0 & 0 \end{bmatrix}, \begin{bmatrix} 0 & 0 \\ 1 & 0 \end{bmatrix}, \begin{bmatrix} 0 & 0 \\ 0 & 1 \end{bmatrix}$$

span $\mathbb{R}^{2\times 2}$. The four matrices are also linearly independent: None of them is a linear combination of the others, since each has a 1 in a position where the three others have a 0. This shows that

$$\mathfrak{B} = \left(\begin{bmatrix} 1 & 0 \\ 0 & 0 \end{bmatrix}, \begin{bmatrix} 0 & 1 \\ 0 & 0 \end{bmatrix}, \begin{bmatrix} 0 & 0 \\ 1 & 0 \end{bmatrix}, \begin{bmatrix} 0 & 0 \\ 0 & 1 \end{bmatrix} \right)$$

is a basis (called the *standard basis* of $\mathbb{R}^{2\times 2}$), so that $\dim(\mathbb{R}^{2\times 2}) = 4$.

The $\mathfrak{B}$-coordinate transformation $L_{\mathfrak{B}}$ is represented in the following diagram:

$$A = \begin{bmatrix} a & b \\ c & d \end{bmatrix} \quad \text{in } \mathbb{R}^{2\times 2} \quad \xrightarrow{\;L_{\mathfrak{B}}\;} \quad \left[A \right]_{\mathfrak{B}} = \begin{bmatrix} a \\ b \\ c \\ d \end{bmatrix} \quad \text{in } \mathbb{R}^4. \quad ∎$$

EXAMPLE 16   Find a basis of P_2, the space of all polynomials of degree ≤ 2, and thus determine the dimension of P_2.

Solution

We can write any polynomial $f(x)$ of degree ≤ 2 as

$$f(x) = a + bx + cx^2 = a \cdot 1 + b \cdot x + c \cdot x^2,$$

showing that the monomials $1, x, x^2$ span P_2. We leave it as an exercise to the reader to verify the linear independence of these monomials. Thus $\mathfrak{B} = (1, x, x^2)$ is a basis (called the *standard basis* of P_2), so that $\dim(P_2) = 3$.

The $\mathfrak{B}$-coordinate transformation $L_{\mathfrak{B}}$ is represented in the following diagram:

$$f(x) = a + bx + cx^2 \quad \text{in } P_2 \quad \xrightarrow{\;L_{\mathfrak{B}}\;} \quad \left[f(x) \right]_{\mathfrak{B}} = \begin{bmatrix} a \\ b \\ c \end{bmatrix} \quad \text{in } \mathbb{R}^3. \quad ∎$$

Using Examples 15 and 16 as a guide, we can present the following strategy for finding a basis of a linear space.

SUMMARY 4.1.6 | **Finding a basis of a linear space V**

a. Write down a typical element of V, in terms of some arbitrary constants.

b. Using the arbitrary constants as coefficients, express your typical element as a linear combination of some elements of V.

c. Verify that the elements of V in this linear combination are linearly independent; then they will form a basis of V.

In Examples 10 and 11 of Section 3.1, we used this method to find a basis of a kernel.

EXAMPLE 17 Find a basis of the space V of all matrices B that commute with $A = \begin{bmatrix} 0 & 1 \\ 2 & 3 \end{bmatrix}$. (See Example 13.)

Solution

We need to find all matrices $B = \begin{bmatrix} a & b \\ c & d \end{bmatrix}$ such that

$$\begin{bmatrix} a & b \\ c & d \end{bmatrix}\begin{bmatrix} 0 & 1 \\ 2 & 3 \end{bmatrix} = \begin{bmatrix} 0 & 1 \\ 2 & 3 \end{bmatrix}\begin{bmatrix} a & b \\ c & d \end{bmatrix}.$$

The entries of B must satisfy the linear equations

$$2b = c, \quad a + 3b = d, \quad 2d = 2a + 3c, \quad c + 3d = 2b + 3d.$$

The last two equations are redundant, so that the matrices B in V are of the form

$$B = \begin{bmatrix} a & b \\ 2b & a + 3b \end{bmatrix} = a\begin{bmatrix} 1 & 0 \\ 0 & 1 \end{bmatrix} + b\begin{bmatrix} 0 & 1 \\ 2 & 3 \end{bmatrix} = aI_2 + bA.$$

Since the matrices I_2 and A are linearly independent, a basis of V is

$$(I_2, A) = \left(\begin{bmatrix} 1 & 0 \\ 0 & 1 \end{bmatrix}, \begin{bmatrix} 0 & 1 \\ 2 & 3 \end{bmatrix} \right). \qquad \blacksquare$$

In the introductory example of this section, we found that the solutions of the differential equation

$$f''(x) + f(x) = 0$$

form a two-dimensional subspace of C^∞, with basis $(\cos x, \sin x)$.

We can generalize this result as follows:

Theorem 4.1.7 **Linear differential equations**

The solutions of the differential equation

$$f''(x) + af'(x) + bf(x) = 0 \quad \text{(where } a \text{ and } b \text{ are constants)}$$

form a two-dimensional subspace of the space C^∞ of smooth functions.

More generally, the solutions of the differential equation

$$f^{(n)}(x) + a_{n-1}f^{(n-1)}(x) + \cdots + a_1 f'(x) + a_0 f(x) = 0$$

(where $a_0, \ldots, a_{n-1}$ are constants) form an n-dimensional subspace of C^∞. A differential equation of this form is called an *nth-order linear differential equation with constant coefficients*.

$\blacksquare$

Second-order linear DEs are frequently used to model oscillatory phenomena in physics. Important examples are damped harmonic motion and *LC* circuits.

Consider how cumbersome it would be to state the second part of Theorem 4.1.7 without using the language of linear algebra. (Try it!) This may convince you that it can be both natural and useful to apply the language of linear algebra to functions. Theorem 4.1.7 will be proven in Section 9.3.

EXAMPLE 18   Find all solutions of the DE

$$f''(x) + f'(x) - 6f(x) = 0.$$

[*Hint*: Find all *exponential* functions $f(x) = e^{kx}$ that solve the DE.]

Solution
An exponential function $f(x) = e^{kx}$ solves the DE if

$$f''(x) + f'(x) - 6f(x) = k^2 e^{kx} + k e^{kx} - 6 e^{kx}$$
$$= (k^2 + k - 6)e^{kx} = (k+3)(k-2)e^{kx} = 0.$$

The solutions are $k = 2$ and $k = -3$. Thus, e^{2x} and e^{-3x} are solutions of the DE. (Check this!) Theorem 4.1.7 tells us that the solution space is two-dimensional. Thus the linearly independent functions e^{2x}, e^{-3x} form a basis of V, and all solutions are of the form

$$f(x) = c_1 e^{2x} + c_2 e^{-3x}. \qquad \blacksquare$$

EXAMPLE 19   Let $f_1, \ldots, f_n$ be polynomials. Explain why these polynomials will not span the space P of all polynomials.

Solution
Let N be the maximum of the degrees of the polynomials $f_1, \ldots, f_n$. Then all linear combinations of $f_1, \ldots, f_n$ are in P_N, the space of polynomials of degree $\leq N$. Any polynomial of higher degree, such as $f(x) = x^{N+1}$, will not be in the span of $f_1, \ldots, f_n$, proving our claim. $\qquad \blacksquare$

Example 19 implies that the space P of all polynomials does not have a finite basis $f_1, \ldots, f_n$.

Here we are faced with an issue that we did not encounter in Chapter 3, when studying $\mathbb{R}^n$ and its subspaces (they all have finite bases). This state of affairs calls for some new terminology.

Definition 4.1.8   Finite dimensional linear spaces

A linear space V is called *finite dimensional* if it has a (finite) basis $f_1, \ldots, f_n$, so that we can define its dimension $\dim(V) = n$. (See Definition 4.1.5.) Otherwise, the space is called *infinite dimensional*.[5]

As we have just seen, the space P of all polynomials is infinite dimensional (as was known to Peano in 1888).

Take another look at the linear spaces introduced in Examples 1 through 8 of this section and see which of them are finite dimensional.

The basic theory of infinite dimensional spaces of functions was established by David Hilbert (1862–1943) and his student Erhard Schmidt (1876–1959), in the

---

[5]More advanced texts introduce the concept of an *infinite basis*.

first decade of the twentieth century, based on their work on integral equations. A more general and axiomatic approach was presented by Stefan Banach (1892–1945) in his 1920 doctoral thesis. These topics (Hilbert spaces, Banach spaces) would be discussed in a course on *Functional Analysis* rather than linear algebra.

EXERCISES 4.1

GOAL *Find a basis of a linear space and thus determine its dimension. Examine whether a subset of a linear space is a subspace.*

Which of the subsets of P_2 given in Exercises 1 through 5 are subspaces of P_2? Find a basis for those that are subspaces.

1. $\{p(t): p(0) = 2\}$

2. $\{p(t): p(2) = 0\}$

3. $\{p(t): p'(1) = p(2)\}$ (p' is the derivative.)

4. $\{p(t): \int_0^1 p(t)\, dt = 0\}$

5. $\{p(t): p(-t) = -p(t), \text{ for all } t\}$

Which of the subsets of $\mathbb{R}^{3\times3}$ given in Exercises 6 through 11 are subspaces of $\mathbb{R}^{3\times3}$?

6. The invertible 3×3 matrices

7. The diagonal 3×3 matrices

8. The upper triangular 3×3 matrices

9. The 3×3 matrices whose entries are all greater than or equal to zero

10. The 3×3 matrices A such that vector $\begin{bmatrix} 1 \\ 2 \\ 3 \end{bmatrix}$ is in the kernel of A

11. The 3×3 matrices in reduced row-echelon form

Let V be the space of all infinite sequences of real numbers. (See Example 5.) Which of the subsets of V given in Exercises 12 through 15 are subspaces of V?

12. The arithmetic sequences [i.e., sequences of the form $(a, a + k, a + 2k, a + 3k, \ldots)$, for some constants a and k]

13. The geometric sequences [i.e., sequences of the form $(a, ar, ar^2, ar^3, \ldots)$, for some constants a and r]

14. The sequences $(x_0, x_1, \ldots)$ that converge to zero (i.e., $\lim_{n\to\infty} x_n = 0$)

15. The square-summable sequences $(x_0, x_1, \ldots)$ (i.e., those for which $\sum_{i=0}^{\infty} x_i^2$ converges)

Find a basis for each of the spaces in Exercises 16 through 36, and determine its dimension.

16. $\mathbb{R}^{3\times2}$

17. $\mathbb{R}^{n\times m}$

18. P_n

19. The real linear space $\mathbb{C}^2$

20. The space of all matrices $A = \begin{bmatrix} a & b \\ c & d \end{bmatrix}$ in $\mathbb{R}^{2\times2}$ such that $a + d = 0$

21. The space of all diagonal 2×2 matrices

22. The space of all diagonal $n \times n$ matrices

23. The space of all lower triangular 2×2 matrices

24. The space of all upper triangular 3×3 matrices

25. The space of all polynomials $f(t)$ in P_2 such that $f(1) = 0$

26. The space of all polynomials $f(t)$ in P_3 such that $f(1) = 0$ and $\int_{-1}^{1} f(t)\, dt = 0$

27. The space of all 2×2 matrices A that commute with $B = \begin{bmatrix} 1 & 0 \\ 0 & 2 \end{bmatrix}$

28. The space of all 2×2 matrices A that commute with $B = \begin{bmatrix} 1 & 1 \\ 0 & 1 \end{bmatrix}$

29. The space of all 2×2 matrices A such that $A \begin{bmatrix} 1 & 1 \\ 1 & 1 \end{bmatrix} = \begin{bmatrix} 0 & 0 \\ 0 & 0 \end{bmatrix}$

30. The space of all 2×2 matrices A such that $\begin{bmatrix} 1 & 2 \\ 3 & 6 \end{bmatrix} A = \begin{bmatrix} 0 & 0 \\ 0 & 0 \end{bmatrix}$

31. The space of all 2×2 matrices S such that $\begin{bmatrix} 0 & 1 \\ 1 & 0 \end{bmatrix} S = S \begin{bmatrix} 1 & 0 \\ 0 & -1 \end{bmatrix}$

32. The space of all 2×2 matrices S such that $\begin{bmatrix} 1 & 1 \\ 1 & 1 \end{bmatrix} S = S \begin{bmatrix} 2 & 0 \\ 0 & 0 \end{bmatrix}$

33. The space of all 2×2 matrices S such that $\begin{bmatrix} 1 & 1 \\ 1 & 1 \end{bmatrix} S = S$

34. The space of all 2×2 matrices S such that $\begin{bmatrix} 3 & 2 \\ 4 & 5 \end{bmatrix} S = S$

35. The space of all 3×3 matrices A that commute with

$$B = \begin{bmatrix} 2 & 0 & 0 \\ 0 & 3 & 0 \\ 0 & 0 & 4 \end{bmatrix}$$

36. The space of all 3×3 matrices A that commute with

$$B = \begin{bmatrix} 2 & 0 & 0 \\ 0 & 3 & 0 \\ 0 & 0 & 3 \end{bmatrix}$$

37. If B is a diagonal 3×3 matrix, what are the possible dimensions of the space of all 3×3 matrices A that commute with B? Use Exercises 35 and 36 as a guide.

38. If B is a diagonal 4×4 matrix, what are the possible dimensions of the space of all 4×4 matrices A that commute with B?

39. What is the dimension of the space of all upper triangular $n \times n$ matrices?

40. If $\vec{v}$ is any vector in $\mathbb{R}^n$, what are the possible dimensions of the space of all $n \times n$ matrices A such that $A\vec{v} = \vec{0}$?

41. If B is any 3×3 matrix, what are the possible dimensions of the space of all 3×3 matrices A such that $BA = 0$?

42. If B is any $n \times n$ matrix, what are the possible dimensions of the space of all $n \times n$ matrices A such that $BA = 0$?

43. If matrix A represents the reflection about a line L in $\mathbb{R}^2$, what is the dimension of the space of all matrices S such that

$$AS = S \begin{bmatrix} 1 & 0 \\ 0 & -1 \end{bmatrix}?$$

(*Hint*: Write $S = \begin{bmatrix} \vec{v} & \vec{w} \end{bmatrix}$, and show that $\vec{v}$ must be parallel to L, and $\vec{w}$ must be perpendicular to L.)

44. If matrix A represents the orthogonal projection onto a plane V in $\mathbb{R}^3$, what is the dimension of the space of all matrices S such that

$$AS = S \begin{bmatrix} 1 & 0 & 0 \\ 0 & 1 & 0 \\ 0 & 0 & 0 \end{bmatrix}?$$

See Exercise 43.

45. Find a basis of the space V of all 3×3 matrices A that commute with

$$B = \begin{bmatrix} 0 & 1 & 0 \\ 0 & 0 & 1 \\ 0 & 0 & 0 \end{bmatrix},$$

and thus determine the dimension of V.

46. In the linear space of infinite sequences, consider the subspace W of arithmetic sequences (see Exercise 12).

Find a basis for W, and thus determine the dimension of W.

47. A function $f(t)$ from $\mathbb{R}$ to $\mathbb{R}$ is called even if $f(-t) = f(t)$, for all t in $\mathbb{R}$, and odd if $f(-t) = -f(t)$, for all t. Are the even functions a subspace of $F(\mathbb{R}, \mathbb{R})$, the space of all functions from $\mathbb{R}$ to $\mathbb{R}$? What about the odd functions? Justify your answers carefully.

48. Find a basis of each of the following linear spaces, and thus determine their dimensions. (See Exercise 47.)
  a. $\{f$ in $P_4: f$ is even$\}$
  b. $\{f$ in $P_4: f$ is odd$\}$

49. Let $L(\mathbb{R}^m, \mathbb{R}^n)$ be the set of all linear transformations from $\mathbb{R}^m$ to $\mathbb{R}^n$. Is $L(\mathbb{R}^m, \mathbb{R}^n)$ a subspace of $F(\mathbb{R}^m, \mathbb{R}^n)$, the space of all functions from $\mathbb{R}^m$ to $\mathbb{R}^n$? Justify your answer carefully.

50. Find all the solutions of the differential equation $f''(x) + 8f'(x) - 20f(x) = 0$.

51. Find all the solutions of the differential equation $f''(x) - 7f'(x) + 12f(x) = 0$.

52. Make up a second-order linear DE whose solution space is spanned by the functions e^{-x} and e^{-5x}.

53. Show that in an n-dimensional linear space we can find at most n linearly independent elements. *Hint*: Consider the proof of Theorem 4.1.5.

54. Show that if W is a subspace of an n-dimensional linear space V, then W is finite-dimensional as well, and $\dim(W) \leq n$. Compare with Exercise 3.2.38a.

55. Show that the space $F(\mathbb{R}, \mathbb{R})$ of all functions from $\mathbb{R}$ to $\mathbb{R}$ is infinite dimensional.

56. Show that the space of infinite sequences of real numbers is infinite dimensional.

57. We say that a linear space V is *finitely generated* if it can be spanned by finitely many elements. Show that a finitely generated space is in fact finite dimensional (and vice versa, of course). Furthermore, if the elements $g_1, \ldots, g_m$ span V, then $\dim(V) \leq m$.

58. In this exercise we will show that the functions $\cos(x)$ and $\sin(x)$ span the solution space V of the differential equation $f''(x) = -f(x)$. (See Example 1 of this section.)
  a. Show that if $g(x)$ is in V, then the function $\left(g(x)\right)^2 + \left(g'(x)\right)^2$ is constant. *Hint*: Consider the derivative.
  b. Show that if $g(x)$ is in V, with $g(0) = g'(0) = 0$, then $g(x) = 0$ for all x.
  c. If $f(x)$ is in V, then $g(x) = f(x) - f(0)\cos(x) - f'(0)\sin(x)$ is in V as well (why?). Verify that $g(0) = 0$ and $g'(0) = 0$. We can conclude that $g(x) = 0$ for all x, so that $f(x) = f(0)\cos(x) + f'(0)\sin(x)$. It follows that the functions $\cos(x)$ and $\sin(x)$ span V, as claimed.

4.2 Linear Transformations and Isomorphisms

In this section, we will define the concepts of a linear transformation, image, kernel, rank, and nullity in the context of linear spaces.

Definition 4.2.1   Linear transformations, image, kernel, rank, nullity

Consider two linear spaces V and W. A function T from V to W is called a *linear transformation* if

$$T(f + g) = T(f) + T(g) \quad \text{and} \quad T(kf) = kT(f)$$

for all elements f and g of V and for all scalars k.

For a linear transformation T from V to W, we let

$$\text{im}(T) = \{T(f) : f \text{ in } V\}$$

and

$$\ker(T) = \{f \text{ in } V : T(f) = 0\}.$$

Note that $\text{im}(T)$ is a subspace of target space W and that $\ker(T)$ is a subspace of domain V.

If the image of T is finite dimensional, then $\dim(\text{im } T)$ is called the *rank* of T, and if the kernel of T is finite dimensional, then $\dim(\ker T)$ is the *nullity* of T.

If V is finite dimensional, then the rank-nullity theorem holds (see Theorem 3.3.7):

$$\dim(V) = \text{rank}(T) + \text{nullity}(T) = \dim(\text{im } T) + \dim(\ker T).$$

A proof of the rank-nullity theorem is outlined in Exercise 81.

EXAMPLE I   Consider the transformation $D(f) = f'$ from C^∞ to C^∞. It follows from the rules of calculus that D is a linear transformation:

$$D(f + g) = (f + g)' = f' + g' \quad \text{equals} \quad D(f) + D(g) = f' + g' \quad \text{and}$$
$$D(kf) = (kf)' = kf' \quad \text{equals} \quad kD(f) = kf'.$$

Here f and g are smooth functions, and k is a constant.

What is the *kernel* of D? This kernel consists of all smooth functions f such that $D(f) = f' = 0$. As you may recall from calculus, these are the constant functions $f(x) = k$. Therefore, the kernel of D is one dimensional; the function $f(x) = 1$ is a basis. The nullity of D is 1.

What about the *image* of D? The image consists of all smooth functions g such that $g = D(f) = f'$ for some function f in C^∞ (i.e., all smooth functions g that have a smooth antiderivative f). The fundamental theorem of calculus implies that all smooth functions (in fact, all continuous functions) have an antiderivative. We can conclude that

$$\text{im}(D) = C^\infty. \qquad \blacksquare$$

EXAMPLE 2   Let $C[0, 1]$ be the linear space of all continuous functions from the closed interval $[0, 1]$ to $\mathbb{R}$. We define the transformation

$$I(f) = \int_0^1 f(x)\, dx \quad \text{from } C[0, 1] \text{ to } \mathbb{R}.$$

We adopt the simplified notation $I(f) = \int_0^1 f$. To check that I is linear, we apply basic rules of integration:

$$I(f + g) = \int_0^1 (f + g) = \int_0^1 f + \int_0^1 g \quad \text{equals} \quad I(f) + I(g) = \int_0^1 f + \int_0^1 g$$

and

$$I(kf) = \int_0^1 (kf) = k \int_0^1 f \quad \text{equals} \quad kI(f) = k \int_0^1 f.$$

What is the image of I? The image of I consists of all real numbers b such that

$$b = I(f) = \int_0^1 f,$$

for some continuous function f. One of many possible choices for f is the constant function $f(x) = b$. Therefore,

$$\operatorname{im}(I) = \mathbb{R}, \quad \text{and} \quad \operatorname{rank}(I) = 1.$$

We leave it to the reader to think about the kernel of I. ∎

EXAMPLE 3 Let V be the space of all infinite sequences of real numbers. Consider the transformation

$$T(x_0, x_1, x_2, \ldots) = (x_1, x_2, x_3, \ldots)$$

from V to V. (We drop the first term, x_0, of the sequence.)

    a. Show that T is a linear transformation.

    b. Find the kernel of T.

    c. Is the sequence $(1, 2, 3, \ldots)$ in the image of T?

    d. Find the image of T.

Solution

    a. $T\big((x_0, x_1, x_2, \ldots) + (y_0, y_1, y_2, \ldots)\big) = T(x_0 + y_0, x_1 + y_1, x_2 + y_2, \ldots)$
    $= (x_1 + y_1, x_2 + y_2, x_3 + y_3, \ldots) \quad$ equals

    $T(x_0, x_1, x_2, \ldots) + T(y_0, y_1, y_2, \ldots) = (x_1, x_2, x_3, \ldots) + (y_1, y_2, y_3, \ldots)$
    $= (x_1 + y_1, x_2 + y_2, x_3 + y_3, \ldots).$

    We leave it to the reader to verify the second property of a linear transformation.

    b. The kernel consists of everything that is transformed to zero, that is, all sequences $(x_0, x_1, x_2, \ldots)$ such that

$$T(x_0, x_1, x_2, \ldots) = (x_1, x_2, x_3, \ldots) = (0, 0, 0, \ldots).$$

    This means that entries $x_1, x_2, x_3, \ldots$ all have to be zero, while x_0 is arbitrary. Thus, $\ker(T)$ consists of all sequences of the form $(x_0, 0, 0, \ldots)$, where x_0 is arbitrary. The kernel of T is one dimensional, with basis $(1, 0, 0, 0, \ldots)$. The nullity of T is 1.

    c. We need to find a sequence $(x_0, x_1, x_2, \ldots)$ such that

$$T(x_0, x_1, x_2, \ldots) = (x_1, x_2, x_3, \ldots) = (1, 2, 3, \ldots).$$

    It is required that $x_1 = 1$, $x_2 = 2$, $x_3 = 3, \ldots$, and we can choose any value for x_0, for example, $x_0 = 0$. Thus,

$$(1, 2, 3, \ldots) = T(0, 1, 2, 3, \ldots)$$

    is indeed in the image of T.

d. Mimicking our solution in part (c), we can write any sequence $(b_0, b_1, b_2, \ldots)$ as

$$(b_0, b_1, b_2, \ldots) = T(0, b_0, b_1, b_2, \ldots),$$

so that $\text{im}(T) = V$. ∎

EXAMPLE 4   Consider the transformation

$$L\left(\begin{bmatrix} a & b \\ c & d \end{bmatrix}\right) = \begin{bmatrix} a \\ b \\ c \\ d \end{bmatrix} \quad \text{from } \mathbb{R}^{2 \times 2} \text{ to } \mathbb{R}^4.$$

Note that L is the coordinate transformation $L_\mathfrak{B}$ with respect to the standard basis

$$\mathfrak{B} = \left(\begin{bmatrix} 1 & 0 \\ 0 & 0 \end{bmatrix}, \begin{bmatrix} 0 & 1 \\ 0 & 0 \end{bmatrix}, \begin{bmatrix} 0 & 0 \\ 1 & 0 \end{bmatrix}, \begin{bmatrix} 0 & 0 \\ 0 & 1 \end{bmatrix}\right)$$

of $\mathbb{R}^{2 \times 2}$; see Example 4.1.15. Being a coordinate transformation, L is both linear and invertible; see Theorem 4.1.4.

Note that the elements of both $\mathbb{R}^{2 \times 2}$ and $\mathbb{R}^4$ are described by a list of four scalars a, b, c, and d. The linear transformation L merely "rearranges" these scalars, and L^{-1} puts them back into their original places in $\mathbb{R}^{2 \times 2}$.

$$\begin{bmatrix} a & b \\ c & d \end{bmatrix} \quad \overset{L}{\underset{L^{-1}}{\rightleftarrows}} \quad \begin{bmatrix} a \\ b \\ c \\ d \end{bmatrix}$$

The linear spaces $\mathbb{R}^{2 \times 2}$ and $\mathbb{R}^4$ have essentially the *same structure*. We say that the linear spaces $\mathbb{R}^{2 \times 2}$ and $\mathbb{R}^4$ are *isomorphic*, from Greek ἴσος (isos), same, and μορφή (morphe), structure. The invertible linear transformation L is called an *isomorphism*. ∎

Definition 4.2.2   Isomorphisms and isomorphic spaces

An invertible linear transformation T is called an *isomorphism*. We say that the linear space V is isomorphic to the linear space W if there exists an isomorphism T from V to W.

We can generalize our findings in Example 4.

Theorem 4.2.3   Coordinate transformations are isomorphisms

If $\mathfrak{B} = (f_1, f_2, \ldots, f_n)$ is a basis of a linear space V, then the *coordinate transformation* $L_\mathfrak{B}(f) = [f]_\mathfrak{B}$ from V to $\mathbb{R}^n$ is an isomorphism. Thus V is isomorphic to $\mathbb{R}^n$; the linear spaces V and $\mathbb{R}^n$ have the same structure. ∎

$$f = c_1 f_1 + \cdots + c_n f_n \quad \text{in } V \quad \overset{L_\mathfrak{B}}{\underset{(L_\mathfrak{B})^{-1}}{\rightleftarrows}} \quad [f]_\mathfrak{B} = \begin{bmatrix} c_1 \\ \vdots \\ c_n \end{bmatrix} \quad \text{in } \mathbb{R}^n$$

Let's reiterate the main point: *Any n-dimensional linear space V is isomorphic to $\mathbb{R}^n$.* This means that we don't need a new theory for finite dimensional spaces. By introducing coordinates, we can transform any n-dimensional linear space into $\mathbb{R}^n$ and then apply the techniques of Chapters 1 through 3. Infinite dimensional linear spaces, on the other hand, are largely beyond the reach of the methods of elementary linear algebra.

EXAMPLE 5   Show that the transformation

$$T(A) = S^{-1}AS \quad \text{from } \mathbb{R}^{2 \times 2} \text{ to } \mathbb{R}^{2 \times 2}$$

is an isomorphism, where $S = \begin{bmatrix} 1 & 2 \\ 3 & 4 \end{bmatrix}$.

Solution

We need to show that T is a linear transformation, and that T is invertible.
Let's check the linearity first:

$$T(A_1 + A_2) = S^{-1}(A_1 + A_2)S = S^{-1}(A_1 S + A_2 S) = S^{-1}A_1 S + S^{-1}A_2 S$$

$$\text{equals}$$

$$T(A_1) + T(A_2) = S^{-1}A_1 S + S^{-1}A_2 S,$$

and

$$T(kA) = S^{-1}(kA)S = k(S^{-1}AS) \quad \text{equals} \quad kT(A) = k(S^{-1}AS).$$

The most direct way to show that a function is invertible is to exhibit the inverse. Here we need to solve the equation $B = S^{-1}AS$ for input A. We find that $A = SBS^{-1}$, so that T is indeed invertible. The inverse transformation is

$$T^{-1}(B) = SBS^{-1}. \qquad \blacksquare$$

Theorem 4.2.4   Properties of isomorphisms

 a. A linear transformation T from V to W is an isomorphism if (and only if) $\ker(T) = \{0\}$ and $\operatorname{im}(T) = W$.

In parts (b) through (d), the linear spaces V and W are assumed to be finite dimensional.

 b. If V is isomorphic to W, then $\dim(V) = \dim(W)$.

 c. Suppose T is a linear transformation from V to W with $\ker(T) = \{0\}$. If $\dim(V) = \dim(W)$, then T is an isomorphism.

 d. Suppose T is a linear transformation from V to W with $\operatorname{im}(T) = W$. If $\dim(V) = \dim(W)$, then T is an isomorphism.

Proof   **a.** Suppose first that T is an isomorphism. To find the kernel of T, we have to solve the equation $T(f) = 0$. Applying T^{-1} on both sides, we find that $f = T^{-1}(0) = 0$, so that $\ker(T) = \{0\}$, as claimed (see Exercise 76). To see that $\operatorname{im}(T) = W$, note that any g in V can be written as $g = T(T^{-1}(g))$.

 Conversely, suppose that $\ker(T) = \{0\}$ and $\operatorname{im}(T) = W$. We have to show that T is invertible, that is, the equation $T(f) = g$ has a unique solution f for every g in W (by Definition 2.4.1). There is at least one solution f, since $\operatorname{im}(T) = W$. Consider two solutions f_1 and f_2, so that $T(f_1) = T(f_2) = g$. Then

$$0 = T(f_1) - T(f_2) = T(f_1 - f_2),$$

so that $f_1 - f_2$ is in the kernel of T. Since the kernel of T is $\{0\}$, we must have $f_1 - f_2 = 0$ and $f_1 = f_2$, as claimed.

b. This claim follows from part (a) and the rank-nullity theorem (Definition 4.2.1):

$$\dim(V) = \dim(\ker T) + \dim(\operatorname{im} T) = 0 + \dim(W) = \dim(W).$$

c. By part (a), it suffices to show that $\operatorname{im}(T) = W$, or, equivalently, that $\dim(\operatorname{im} T) = \dim(W)$; compare with Exercise 3.3.61. But this claim follows from the rank-nullity theorem:

$$\dim(W) = \dim(V) = \dim(\ker T) + \dim(\operatorname{im} T) = \dim(\operatorname{im} T).$$

d. By part (a), it suffices to show that $\ker(T) = \{0\}$. The proof is analogous to part (*c*). ∎

EXAMPLE 6

a. Is the linear transformation

$$L\big(f(x)\big) = \begin{bmatrix} f(1) \\ f(2) \\ f(3) \end{bmatrix} \quad \text{from } P_3 \text{ to } \mathbb{R}^3 \text{ an isomorphism?}$$

b. Is the linear transformation

$$T\big(f(x)\big) = \begin{bmatrix} f(1) \\ f(2) \\ f(3) \end{bmatrix} \quad \text{from } P_2 \text{ to } \mathbb{R}^3 \text{ an isomorphism?}$$

Solution

a. Consider Theorem 4.2.4b. Since $\dim(P_3) = 4$ and $\dim(\mathbb{R}^3) = 3$, the spaces P_3 and $\mathbb{R}^3$ fail to be isomorphic, so that L fails to be an isomorphism.

b. In this case, domain and target space have the same dimension,

$$\dim(P_2) = \dim(\mathbb{R}^3) = 3.$$

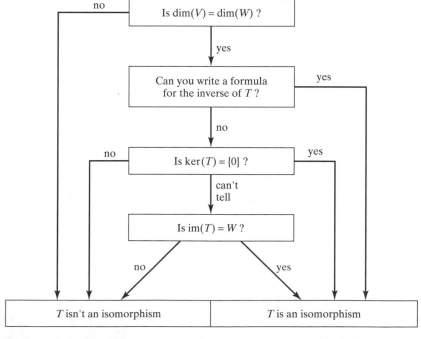

Figure 1 Is the linear transformation T from V to W an isomorphism? (V and W are finite dimensional linear spaces.)

This fact alone does not guarantee that T is an isomorphism, however. Let's find the kernel of T and use Theorem 4.2.4c. The kernel of T consists of all polynomials $f(x)$ in P_2 such that

$$T(f(x)) = \begin{bmatrix} f(1) \\ f(2) \\ f(3) \end{bmatrix} = \begin{bmatrix} 0 \\ 0 \\ 0 \end{bmatrix},$$

that is, $f(1) = 0$, $f(2) = 0$, and $f(3) = 0$. Since a *nonzero* polynomial in P_2 has at most two zeros, the zero polynomial is the only solution, so that $\ker(T) = \{0\}$. Thus T is indeed an isomorphism. ∎

The diagram in Figure 1 can be a useful guide in determining whether a given linear transformation T from V to W is an isomorphism. Here we assume that both V and W are *finite dimensional*. (We leave it as an exercise to the reader to adapt the diagram to the case of infinite dimensional spaces.)

EXERCISES 4.2

GOAL   *Examine whether a transformation is linear. Find the image and kernel of a linear transformation. Examine whether a linear transformation is an isomorphism.*

Find out which of the transformations in Exercises 1 through 50 are linear. For those that are linear, determine whether they are isomorphisms.

1. $T(M) = M + I_2$ from $\mathbb{R}^{2\times2}$ to $\mathbb{R}^{2\times2}$

2. $T(M) = 7M$ from $\mathbb{R}^{2\times2}$ to $\mathbb{R}^{2\times2}$

3. $T(M) =$ (sum of the diagonal entries of M) from $\mathbb{R}^{2\times2}$ to $\mathbb{R}$

4. $T(M) = \det(M)$ from $\mathbb{R}^{2\times2}$ to $\mathbb{R}$

5. $T(M) = M^2$ from $\mathbb{R}^{2\times2}$ to $\mathbb{R}^{2\times2}$

6. $T(M) = M\begin{bmatrix} 1 & 2 \\ 3 & 6 \end{bmatrix}$ from $\mathbb{R}^{2\times2}$ to $\mathbb{R}^{2\times2}$

7. $T(M) = \begin{bmatrix} 1 & 2 \\ 3 & 4 \end{bmatrix}M$ from $\mathbb{R}^{2\times2}$ to $\mathbb{R}^{2\times2}$

8. $T(M) = M\begin{bmatrix} 1 & 2 \\ 3 & 4 \end{bmatrix}M$ from $\mathbb{R}^{2\times2}$ to $\mathbb{R}^{2\times2}$

9. $T(M) = S^{-1}MS$, where $S = \begin{bmatrix} 3 & 4 \\ 5 & 6 \end{bmatrix}$, from $\mathbb{R}^{2\times2}$ to $\mathbb{R}^{2\times2}$

10. $T(M) = PMP^{-1}$, where $P = \begin{bmatrix} 2 & 3 \\ 5 & 7 \end{bmatrix}$, from $\mathbb{R}^{2\times2}$ to $\mathbb{R}^{2\times2}$

11. $T(M) = PMQ$, where $P = \begin{bmatrix} 2 & 3 \\ 5 & 7 \end{bmatrix}$ and $Q = \begin{bmatrix} 3 & 5 \\ 7 & 11 \end{bmatrix}$, from $\mathbb{R}^{2\times2}$ to $\mathbb{R}^{2\times2}$

12. $T(c) = c\begin{bmatrix} 2 & 3 \\ 4 & 5 \end{bmatrix}$ from $\mathbb{R}$ to $\mathbb{R}^{2\times2}$

13. $T(M) = M\begin{bmatrix} 1 & 2 \\ 0 & 1 \end{bmatrix} - \begin{bmatrix} 1 & 2 \\ 0 & 1 \end{bmatrix}M$ from $\mathbb{R}^{2\times2}$ to $\mathbb{R}^{2\times2}$

14. $T(M) = \begin{bmatrix} 2 & 3 \\ 5 & 7 \end{bmatrix}M - M\begin{bmatrix} 2 & 3 \\ 5 & 7 \end{bmatrix}$ from $\mathbb{R}^{2\times2}$ to $\mathbb{R}^{2\times2}$

15. $T(M) = \begin{bmatrix} 2 & 0 \\ 0 & 3 \end{bmatrix}M - M\begin{bmatrix} 4 & 0 \\ 0 & 5 \end{bmatrix}$ from $\mathbb{R}^{2\times2}$ to $\mathbb{R}^{2\times2}$

16. $T(M) = M\begin{bmatrix} 2 & 0 \\ 0 & 3 \end{bmatrix} - \begin{bmatrix} 3 & 0 \\ 0 & 4 \end{bmatrix}M$ from $\mathbb{R}^{2\times2}$ to $\mathbb{R}^{2\times2}$

17. $T(x + iy) = x$ from $\mathbb{C}$ to $\mathbb{C}$

18. $T(x + iy) = x^2 + y^2$ from $\mathbb{C}$ to $\mathbb{C}$

19. $T(x + iy) = i(x + iy)$ from $\mathbb{C}$ to $\mathbb{C}$

20. $T(x + iy) = x - iy$ from $\mathbb{C}$ to $\mathbb{C}$

21. $T(x + iy) = y + ix$ from $\mathbb{C}$ to $\mathbb{C}$

22. $T(f(t)) = \int_{-2}^{3} f(t)\,dt$ from P_2 to $\mathbb{R}$

23. $T(f(t)) = f(7)$ from P_2 to $\mathbb{R}$

24. $T(f(t)) = f''(t)f(t)$ from P_2 to P_2

25. $T(f(t)) = f''(t) + 4f'(t)$ from P_2 to P_2

26. $T(f(t)) = f(-t)$ from P_2 to P_2, that is, $T(a + bt + ct^2) = a - bt + ct^2$

27. $T(f(t)) = f(2t)$ from P_2 to P_2, that is, $T(a + bt + ct^2) = a + 2bt + 4ct^2$

28. $T(f(t)) = f(2t) - f(t)$ from P_2 to P_2

29. $T(f(t)) = f'(t)$ from P_2 to P_2

30. $T(f(t)) = t(f'(t))$ from P_2 to P_2

31. $T(f(t)) = \begin{bmatrix} f(0) & f(1) \\ f(2) & f(3) \end{bmatrix}$ from P_2 to $\mathbb{R}^{2\times2}$

32. $T(f(t)) = f'(t) + t^2$ from P_2 to P_2

In Exercises 33 through 36, V denotes the space of infinite sequences of real numbers.

33. $T(x_0, x_1, x_2, x_3, x_4, \ldots) = (x_0, x_2, x_4, \ldots)$ from V to V (we are dropping every other term)

34. $T(x_0, x_1, x_2, \ldots) = (0, x_0, x_1, x_2, \ldots)$ from V to V

35. $T(f(t)) = (f(0), f'(0), f''(0), f'''(0), \ldots)$ from P to V, where P denotes the space of all polynomials

36. $T(f(t)) = (f(0), f(1), f(2), f(3), \ldots)$ from P to V, where P denotes the space of all polynomials

37. $T(f) = f + f'$ from C^∞ to C^∞

38. $T(f) = f + f''$ from C^∞ to C^∞ Find kernel

39. $T(f) = f'' - 5f' + 6f$ from C^∞ to C^∞

40. $T(f) = f'' + 2f' + f$ from C^∞ to C^∞

41. $T(f(t)) = f(t) + f''(t) + \sin(t)$ from C^∞ to C^∞

42. $T(f(t)) = \begin{bmatrix} f(7) \\ f(11) \end{bmatrix}$ from P_2 to $\mathbb{R}^2$

43. $T(f(t)) = \begin{bmatrix} f(5) \\ f(7) \\ f(11) \end{bmatrix}$ from P_2 to $\mathbb{R}^3$

44. $T(f(t)) = \begin{bmatrix} f(1) \\ f'(2) \\ f(3) \end{bmatrix}$ from P_2 to $\mathbb{R}^3$

45. $T(f(t)) = t(f(t))$ from P to P

46. $T(f(t)) = (t - 1)f(t)$ from P to P

47. $T(f(t)) = \int_0^t f(x)\, dx$ from P to P

48. $T(f(t)) = f'(t)$ from P to P

49. $T(f(t)) = f(t^2)$ from P to P

50. $T(f(t)) = \dfrac{f(t + 2) - f(t)}{2}$ from P to P

51. Find kernel and nullity of the transformation in Exercise 13.

52. Find kernel and nullity of the transformation in Exercise 6.

53. Find image, rank, kernel, and nullity of the transformation in Exercise 25.

54. Find image, rank, kernel, and nullity of the transformation in Exercise 22.

55. Find image and kernel of the transformation in Exercise 33.

56. Find image, rank, kernel, and nullity of the transformation in Exercise 30.

57. Find kernel and nullity of the transformation in Exercise 39.

58. Find image and kernel of the transformation in Exercise 34.

59. Find image, rank, kernel, and nullity of the transformation in Exercise 23.

60. Find image, rank, kernel, and nullity of the transformation in Exercise 42.

61. Find image and kernel of the transformation in Exercise 45.

62. Find image and kernel of the transformation in Exercise 48.

63. Define an isomorphism from P_3 to $\mathbb{R}^3$, if you can.

64. Define an isomorphism from P_3 to $\mathbb{R}^{2 \times 2}$, if you can.

65. We will define a transformation T from $\mathbb{R}^{n \times m}$ to $F(\mathbb{R}^m, \mathbb{R}^n)$; recall that $F(\mathbb{R}^m, \mathbb{R}^n)$ is the space of all functions from $\mathbb{R}^m$ to $\mathbb{R}^n$. For a matrix A in $\mathbb{R}^{n \times m}$, the value $T(A)$ will be a function from $\mathbb{R}^m$ to $\mathbb{R}^n$; thus we need to define $(T(A))(\vec{v})$ for a vector $\vec{v}$ in $\mathbb{R}^m$. We let

$$(T(A))(\vec{v}) = A\vec{v}.$$

a. Show that T is a linear transformation.

b. Find the kernel of T.

c. Show that the image of T is the space $L(\mathbb{R}^m, \mathbb{R}^n)$ of all linear transformation from $\mathbb{R}^m$ to $\mathbb{R}^n$. (See Exercise 4.1.49.)

d. Find the dimension of $L(\mathbb{R}^m, \mathbb{R}^n)$.

66. Find the kernel and nullity of the linear transformation $T(f) = f - f'$ from C^∞ to C^∞.

67. For which constants k is the linear transformation

$$T(M) = \begin{bmatrix} 2 & 3 \\ 0 & 4 \end{bmatrix} M - M \begin{bmatrix} 3 & 0 \\ 0 & k \end{bmatrix}$$

an isomorphism from $\mathbb{R}^{2 \times 2}$ to $\mathbb{R}^{2 \times 2}$?

68. For which constants k is the linear transformation

$$T(M) = M \begin{bmatrix} 5 & 0 \\ 0 & 1 \end{bmatrix} - \begin{bmatrix} 2 & 0 \\ 0 & k \end{bmatrix} M$$

an isomorphism from $\mathbb{R}^{2 \times 2}$ to $\mathbb{R}^{2 \times 2}$?

69. If matrix A is similar to B, is $T(M) = AM - MB$ an isomorphism from $\mathbb{R}^{2 \times 2}$ to $\mathbb{R}^{2 \times 2}$?

70. For which real numbers $c_0, c_1, \ldots, c_n$ is the linear transformation

$$T(f(t)) = \begin{bmatrix} f(c_0) \\ f(c_1) \\ \vdots \\ f(c_n) \end{bmatrix}$$

an isomorphism from P_n to $\mathbb{R}^{n+1}$?

71. Does there exist a polynomial $f(t)$ of degree ≤ 4 such that $f(2) = 3$, $f(3) = 5$, $f(5) = 7$, $f(7) = 11$, and $f(11) = 2$? If so, how many such polynomials are there? *Hint:* Use Exercise 70.

In Exercises 72 through 74, let Z_n be the set of all polynomials of degree $\leq n$ such that $f(0) = 0$.

72. Show that Z_n is a subspace of P_n, and find the dimension of Z_n.

73. Is the linear transformation $T\left(f(t)\right) = \int_0^t f(x)\, dx$ an isomorphism from P_{n-1} to Z_n?

74. Define an isomorphism from Z_n to P_{n-1} (think calculus!).

75. Show that if 0 is the neutral element of a linear space V, then $0 + 0 = 0$ and $k0 = 0$, for all scalars k.

76. Show that if T is a linear transformation from V to W, then $T(0_V) = 0_W$, where 0_V and 0_W are the neutral elements of V and W, respectively.

77. If T is a linear transformation from V to W and L is a linear transformation from W to U, is the composite transformation $L \circ T$ from V to U linear? How can you tell? If T and L are isomorphisms, is $L \circ T$ an isomorphism as well?

78. Let $\mathbb{R}^+$ be the set of positive real numbers. On $\mathbb{R}^+$ we define the "exotic" operations

$$x \oplus y = xy \quad \text{(usual multiplication)}$$

and

$$k \odot x = x^k.$$

a. Show that $\mathbb{R}^+$ with these operations is a linear space; find a basis of this space.

b. Show that $T(x) = \ln(x)$ is a linear transformation from $\mathbb{R}^+$ to $\mathbb{R}$, where $\mathbb{R}$ is endowed with the ordinary operations. Is T an isomorphism?

79. Is it possible to define "exotic" operations on $\mathbb{R}^2$, so that $\dim(\mathbb{R}^2) = 1$?

80. Let X be the set of all students in your linear algebra class. Can you define operations on X that make X into a real linear space? Explain.

81. In this exercise, we will outline a proof of the rank-nullity theorem: If T is a linear transformation from V to W, where V is finite dimensional, then

$$\dim(V) = \dim(\operatorname{im} T) + \dim(\ker T)$$
$$= \operatorname{rank}(T) + \operatorname{nullity}(T).$$

a. Explain why $\ker(T)$ and image (T) are finite dimensional. *Hint:* Use Exercises 4.1.54 and 4.1.57.

Now consider a basis $v_1, \ldots, v_n$ of $\ker(T)$, where $n = \operatorname{nullity}(T)$, and a basis $w_1, \ldots, w_r$ of $\operatorname{im}(T)$, where $r = \operatorname{rank}(T)$. Consider elements $u_1, \ldots, u_r$ in V such that $T(u_i) = w_i$ for $i = 1, \ldots, r$. Our goal is to show that the $r + n$ elements $u_1, \ldots, u_r, v_1, \ldots, v_n$ form a basis of V; this will prove our claim.

b. Show that the elements $u_1, \ldots, u_r, v_1, \ldots, v_n$ are linearly independent. *Hint:* Consider a relation $c_1 u_1 + \cdots + c_r u_r + d_1 v_1 + \cdots + d_n v_n = 0$, apply transformation T to both sides, and take it from there.

c. Show that the elements $u_1, \ldots, u_r, v_1, \ldots, v_n$ span V. [*Hint:* Consider an arbitrary element v in V, and write $T(v) = d_1 w_1 + \cdots + d_r w_r$. Now show that the element $v - d_1 u_1 - \cdots - d_r u_r$ is in the kernel of T, so that $v - d_1 u_1 - \cdots - d_r u_r$ can be written as a linear combination of $v_1, \ldots, v_n$.]

82. Prove the following variant of the rank-nullity theorem: If T is a linear transformation from V to W, and if $\ker T$ and $\operatorname{im} T$ are both finite dimensional, then V is finite dimensional as well, and $\dim V = \dim(\ker T) + \dim(\operatorname{im} T)$.

83. Consider linear transformations T from V to W and L from W to U. If $\ker T$ and $\ker L$ are both finite dimensional, show that $\ker(L \circ T)$ is finite dimensional as well, and $\dim\left(\ker(L \circ T)\right) \leq \dim(\ker T) + \dim(\ker L)$. [*Hint:* Restrict T to $\ker(L \circ T)$ and apply the rank-nullity theorem, as presented in Exercise 82.]

84. Consider linear transformations T from V to W and L from W to U. If $\ker T$ and $\ker L$ are both finite dimensional, and if $\operatorname{im} T = W$, show that $\ker(L \circ T)$ is finite dimensional as well and that $\dim\left(\ker(L \circ T)\right) = \dim(\ker T) + \dim(\ker L)$.

4.3   The Matrix of a Linear Transformation

Next we will examine how we can express a linear transformation in coordinates.

EXAMPLE I   Consider the linear transformation

$$T(f) = f' + f'' \quad \text{from } P_2 \text{ to } P_2,$$

or, written more explicitly,

$$T(f(x)) = f'(x) + f''(x).$$

Since P_2 is isomorphic to $\mathbb{R}^3$, this is essentially a linear transformation from $\mathbb{R}^3$ to $\mathbb{R}^3$, given by a 3×3 matrix B. Let's see how we can find this matrix.

If we let $f(x) = a + bx + cx^2$, then we can write transformation T as

$$T(a + bx + cx^2) = (a + bx + cx^2)' + (a + bx + cx^2)''$$
$$= (b + 2cx) + 2c = (b + 2c) + 2cx.$$

Next let's write the input $f(x) = a + bx + cx^2$ and the output $T(f(x)) = (b + 2c) + 2cx$ in coordinates with respect to the standard basis $\mathfrak{B} = (1, x, x^2)$ of P_2.

$$f(x) = a + bx + cx^2 \xrightarrow{\ \ T\ \ } T(f(x)) = (b + 2c) + 2cx$$

$$L_\mathfrak{B} \downarrow \qquad\qquad\qquad\qquad \downarrow L_\mathfrak{B}$$

$$[f(x)]_\mathfrak{B} = \begin{bmatrix} a \\ b \\ c \end{bmatrix} \longrightarrow [T(f(x))]_\mathfrak{B} = \begin{bmatrix} b + 2c \\ 2c \\ 0 \end{bmatrix}$$

Written in $\mathfrak{B}$-coordinates, transformation T takes $[f(x)]_\mathfrak{B}$ to

$$[T(f(x))]_\mathfrak{B} = \begin{bmatrix} b + 2c \\ 2c \\ 0 \end{bmatrix} = \begin{bmatrix} 0 & 1 & 2 \\ 0 & 0 & 2 \\ 0 & 0 & 0 \end{bmatrix} \begin{bmatrix} a \\ b \\ c \end{bmatrix} = \begin{bmatrix} 0 & 1 & 2 \\ 0 & 0 & 2 \\ 0 & 0 & 0 \end{bmatrix} [f(x)]_\mathfrak{B}.$$

The matrix

$$B = \begin{bmatrix} 0 & 1 & 2 \\ 0 & 0 & 2 \\ 0 & 0 & 0 \end{bmatrix}$$

is called the $\mathfrak{B}$-matrix of transformation T. It describes the transformation T if input and output are written in $\mathfrak{B}$-coordinates. Let us summarize our work in two diagrams:

$$\begin{array}{ccc} P_2 & \xrightarrow{\ T\ } & P_2 \\ L_\mathfrak{B} \downarrow & & \downarrow L_\mathfrak{B} \\ \mathbb{R}^3 & \xrightarrow{\ B\ } & \mathbb{R}^3 \end{array} \qquad\qquad \begin{array}{ccc} f & \xrightarrow{\ T\ } & T(f) \\ L_\mathfrak{B} \downarrow & & \downarrow L_\mathfrak{B} \\ [f]_\mathfrak{B} & \xrightarrow{\ B\ } & [T(f)]_\mathfrak{B} \end{array}$$

Definition 4.3.1 The $\mathfrak{B}$-matrix of a linear transformation

Consider a linear transformation T from V to V, where V is an n-dimensional linear space. Let $\mathfrak{B}$ be a basis of V. Consider the linear transformation $L_\mathfrak{B} \circ T \circ L_\mathfrak{B}^{-1}$ from $\mathbb{R}^n$ to $\mathbb{R}^n$, with standard matrix B, meaning that $B\vec{x} = L_\mathfrak{B}\big(T\big(L_\mathfrak{B}^{-1}(\vec{x})\big)\big)$ for all $\vec{x}$

in $\mathbb{R}^n$. This matrix B is called the $\mathfrak{B}$-matrix of transformation T. See the diagrams below. Letting $f = L_{\mathfrak{B}}^{-1}(\vec{x})$ and $\vec{x} = [f]_{\mathfrak{B}}$, we find that

$$[T(f)]_{\mathfrak{B}} = B[f]_{\mathfrak{B}}, \qquad \text{for all } f \text{ in } V.$$

Consider the following diagrams:

$$
\begin{array}{ccc}
V & \xrightarrow{\;T\;} & V \\
L_{\mathfrak{B}} \downarrow & & \downarrow L_{\mathfrak{B}} \\
\mathbb{R}^n & \xrightarrow{\;B\;} & \mathbb{R}^n
\end{array}
\qquad\qquad
\begin{array}{ccc}
f & \xrightarrow{\;T\;} & T(f) \\
L_{\mathfrak{B}} \downarrow & & \downarrow L_{\mathfrak{B}} \\
[f]_{\mathfrak{B}} & \xrightarrow{\;B\;} & [T(f)]_{\mathfrak{B}}
\end{array}
$$

Compare this with Definition 3.4.3.

We can write B in terms of its columns. Suppose that $\mathfrak{B} = (f_1, \ldots, f_i, \ldots, f_n)$. Then

$$[T(f_i)]_{\mathfrak{B}} = B[f_i]_{\mathfrak{B}} = B\vec{e}_i = (i\text{th column of } B)$$

Theorem 4.3.2   **The columns of the $\mathfrak{B}$-matrix of a linear transformation**

Consider a linear transformation T from V to V, and let B be the matrix of T with respect to a basis $\mathfrak{B} = (f_1, \ldots, f_n)$ of V. Then

$$B = \left[\; [T(f_1)]_{\mathfrak{B}} \quad \cdots \quad [T(f_n)]_{\mathfrak{B}} \; \right].$$

The columns of B are the $\mathfrak{B}$-coordinate vectors of the transforms of the basis elements $f_1, \ldots, f_n$ of V.   ∎

EXAMPLE 2   Use Theorem 4.3.2 to find the matrix B of the linear transformation $T(f) = f' + f''$ from P_2 to P_2 with respect to the standard basis $\mathfrak{B} = (1, x, x^2)$; see Example 1.

Solution

By Theorem 4.3.2, we have

$$B = \left[\; [T(1)]_{\mathfrak{B}} \quad [T(x)]_{\mathfrak{B}} \quad [T(x^2)]_{\mathfrak{B}} \; \right].$$

Now

$$
\begin{array}{lll}
T(1) = 1' + 1'' & T(x) = x' + x'' & T(x^2) = (x^2)' + (x^2)'' \\
\quad = 0 & \quad = 1 & \quad = 2 + 2x
\end{array}
$$

$$
[T(1)]_{\mathfrak{B}} = \begin{bmatrix} 0 \\ 0 \\ 0 \end{bmatrix}
\qquad
[T(x)]_{\mathfrak{B}} = \begin{bmatrix} 1 \\ 0 \\ 0 \end{bmatrix}
\qquad
[T(x^2)]_{\mathfrak{B}} = \begin{bmatrix} 2 \\ 2 \\ 0 \end{bmatrix}
$$

$$
B = \left[\; [T(1)]_{\mathfrak{B}} \quad [T(x)]_{\mathfrak{B}} \quad [T(x^2)]_{\mathfrak{B}} \; \right] = \begin{bmatrix} 0 & 1 & 2 \\ 0 & 0 & 2 \\ 0 & 0 & 0 \end{bmatrix}.
$$ ∎

A problem concerning a linear transformation T can often be done by solving the corresponding problem for the matrix B of T with respect to some basis $\mathfrak{B}$. We can use this technique to find the image and kernel of T, to determine whether T is an isomorphism (this is the case if B is invertible), or to solve an equation $T(f) = g$ for f if g is given (this amounts to solving the linear system $B\left[f\right]_{\mathfrak{B}} = \left[g\right]_{\mathfrak{B}}$).

EXAMPLE 3   Let V be the span of $\cos(x)$ and $\sin(x)$ in C^{∞}; note that V consists of all trigonometric functions of the form $f(x) = a\cos(x) + b\sin(x)$. Consider the transformation

$$T(f) = 3f + 2f' - f'' \quad \text{from } V \text{ to } V.$$

We are told that T is a linear transformation.

   a. Using Theorem 4.3.2, find the matrix B of T with respect to the basis $\mathfrak{B} = \left(\cos(x),\ \sin(x)\right)$.

   b. Is T an isomorphism?

Solution

   a. Here

$$B = \left[\ \left[T(\cos x)\right]_{\mathfrak{B}} \quad \left[T(\sin x)\right]_{\mathfrak{B}}\ \right].$$

   Now

   $T(\cos x)$
   $= 3\cos(x) - 2\sin(x) + \cos(x)$
   $= 4\cos(x) - 2\sin(x)$

   $T(\sin x)$
   $= 3\sin(x) + 2\cos(x) + \sin(x)$
   $= 2\cos(x) + 4\sin(x)$

   $$\left[T(\cos x)\right]_{\mathfrak{B}} = \begin{bmatrix} 4 \\ -2 \end{bmatrix} \qquad \left[T(\sin x)\right]_{\mathfrak{B}} = \begin{bmatrix} 2 \\ 4 \end{bmatrix}$$

   $$B = \begin{bmatrix} 4 & 2 \\ -2 & 4 \end{bmatrix}.$$

   Matrix B represents a rotation combined with a scaling.

   b. Matrix B is invertible, since $\det(B) = ad - bc = 20 \neq 0$. Thus transformation T is invertible as well, so that T is an isomorphism (we were told that T is linear). ∎

EXAMPLE 4   Consider the linear transformation

$$T(M) = \begin{bmatrix} 0 & 1 \\ 0 & 0 \end{bmatrix} M - M \begin{bmatrix} 0 & 1 \\ 0 & 0 \end{bmatrix} \quad \text{from } \mathbb{R}^{2\times 2} \text{ to } \mathbb{R}^{2\times 2}.$$

   a. Find the matrix B of T with respect to the standard basis $\mathfrak{B}$ of $\mathbb{R}^{2\times 2}$.

   b. Find bases of the image and kernel of B.

c. Find bases of the image and kernel of T, and thus determine rank and nullity of transformation T.

Solution

a. For the sake of variety, let us find B by means of a diagram.

$$\begin{bmatrix} a & b \\ c & d \end{bmatrix} \xrightarrow{\ T\ } \begin{bmatrix} 0 & 1 \\ 0 & 0 \end{bmatrix}\begin{bmatrix} a & b \\ c & d \end{bmatrix} - \begin{bmatrix} a & b \\ c & d \end{bmatrix}\begin{bmatrix} 0 & 1 \\ 0 & 0 \end{bmatrix}$$

$$= \begin{bmatrix} c & d \\ 0 & 0 \end{bmatrix} - \begin{bmatrix} 0 & a \\ 0 & c \end{bmatrix} = \begin{bmatrix} c & d-a \\ 0 & -c \end{bmatrix}$$

$$L_{\mathfrak{B}}\downarrow \qquad\qquad\qquad \downarrow L_{\mathfrak{B}}$$

$$\begin{bmatrix} a \\ b \\ c \\ d \end{bmatrix} \xrightarrow{\ B\ } \qquad\qquad \begin{bmatrix} c \\ d-a \\ 0 \\ -c \end{bmatrix}$$

We see that

$$B = \begin{bmatrix} 0 & 0 & 1 & 0 \\ -1 & 0 & 0 & 1 \\ 0 & 0 & 0 & 0 \\ 0 & 0 & -1 & 0 \end{bmatrix}.$$

b. Note that columns $\vec{v}_2$ and $\vec{v}_4$ of B are redundant, with $\vec{v}_2 = \vec{0}$ and $\vec{v}_4 = -\vec{v}_1$, or $\vec{v}_1 + \vec{v}_4 = \vec{0}$. Thus the nonredundant columns

$$\vec{v}_1 = \begin{bmatrix} 0 \\ -1 \\ 0 \\ 0 \end{bmatrix}, \quad \vec{v}_3 = \begin{bmatrix} 1 \\ 0 \\ 0 \\ -1 \end{bmatrix} \quad \text{form a basis of im}(B),$$

and

$$\begin{bmatrix} 0 \\ 1 \\ 0 \\ 0 \end{bmatrix}, \quad \begin{bmatrix} 1 \\ 0 \\ 0 \\ 1 \end{bmatrix} \quad \text{is a basis of ker}(B).$$

c. We apply $L_{\mathfrak{B}}^{-1}$ to transform the vectors we found in part (b) back into $\mathbb{R}^{2\times 2}$, the domain and target space of transformation T:

$$\begin{bmatrix} 0 & -1 \\ 0 & 0 \end{bmatrix}, \quad \begin{bmatrix} 1 & 0 \\ 0 & -1 \end{bmatrix} \quad \text{is a basis of im}(T),$$

and

$$\begin{bmatrix} 0 & 1 \\ 0 & 0 \end{bmatrix}, \quad I_2 = \begin{bmatrix} 1 & 0 \\ 0 & 1 \end{bmatrix} \quad \text{is a basis of ker}(T).$$

Thus $\text{rank}(T) = \dim(\text{im } T) = 2$ and $\text{nullity}(T) = \dim(\text{ker } T) = 2$.   ■

Change of Basis

If $\mathfrak{A}$ and $\mathfrak{B}$ are two bases of a linear space V, what is the relationship between the coordinate vectors $[f]_{\mathfrak{A}}$ and $[f]_{\mathfrak{B}}$, for an element f of V?

Definition 4.3.3 Change of basis matrix

Consider two bases $\mathfrak{A}$ and $\mathfrak{B}$ of an n-dimensional linear space V. Consider the linear transformation $L_{\mathfrak{A}} \circ L_{\mathfrak{B}}^{-1}$ from $\mathbb{R}^n$ to $\mathbb{R}^n$, with standard matrix S, meaning that $S\vec{x} = L_{\mathfrak{A}}\left(L_{\mathfrak{B}}^{-1}(\vec{x})\right)$ for all $\vec{x}$ in $\mathbb{R}^n$. This invertible matrix S is called the *change of basis matrix* from $\mathfrak{B}$ to $\mathfrak{A}$, sometimes denoted by $S_{\mathfrak{B} \to \mathfrak{A}}$. See the accompanying diagrams. Letting $f = L_{\mathfrak{B}}^{-1}(\vec{x})$ and $\vec{x} = [f]_{\mathfrak{B}}$, we find that

$$[f]_{\mathfrak{A}} = S\,[f]_{\mathfrak{B}}, \quad \text{for all } f \text{ in } V.$$

If $\mathfrak{B} = (b_1, \ldots, b_i, \ldots, b_n)$, then

$$[b_i]_{\mathfrak{A}} = S\,[b_i]_{\mathfrak{B}} = S\vec{e}_i = (i\text{th column of } S),$$

so that

$$S_{\mathfrak{B} \to \mathfrak{A}} = \begin{bmatrix} [b_1]_{\mathfrak{A}} & \cdots & [b_n]_{\mathfrak{A}} \end{bmatrix}$$

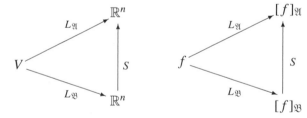

What is the relationship between the change of basis matrices $S_{\mathfrak{B} \to \mathfrak{A}}$ and $S_{\mathfrak{A} \to \mathfrak{B}}$? Solving the equation $[f]_{\mathfrak{A}} = (S_{\mathfrak{B} \to \mathfrak{A}})\,[f]_{\mathfrak{B}}$ for $[f]_{\mathfrak{B}}$, we find that $[f]_{\mathfrak{B}} = (S_{\mathfrak{B} \to \mathfrak{A}})^{-1}\,[f]_{\mathfrak{A}}$, so that

$$S_{\mathfrak{A} \to \mathfrak{B}} = (S_{\mathfrak{B} \to \mathfrak{A}})^{-1}.$$

EXAMPLE 5 Let V be the subspace of C^{∞} spanned by the functions e^x and e^{-x}, with the bases $\mathfrak{A} = (e^x, e^{-x})$ and $\mathfrak{B} = (e^x + e^{-x}, e^x - e^{-x})$. Find the change of basis matrix $S_{\mathfrak{B} \to \mathfrak{A}}$.

Solution

By Definition 4.3.3,

$$S = \begin{bmatrix} [e^x + e^{-x}]_{\mathfrak{A}} & [e^x - e^{-x}]_{\mathfrak{A}} \end{bmatrix}.$$

Now

$$e^x + e^{-x} = 1 \cdot e^x + 1 \cdot e^{-x} \qquad\qquad e^x - e^{-x} = 1 \cdot e^x + (-1) \cdot e^{-x}$$

$$\left[e^x + e^{-x} \right]_{\mathfrak{A}} = \begin{bmatrix} 1 \\ 1 \end{bmatrix} \qquad\qquad \left[e^x - e^{-x} \right]_{\mathfrak{A}} = \begin{bmatrix} 1 \\ -1 \end{bmatrix}$$

$$S_{\mathfrak{B} \to \mathfrak{A}} = \begin{bmatrix} 1 & 1 \\ 1 & -1 \end{bmatrix}$$

■

EXAMPLE 6   Let $\mathfrak{A} = (\vec{e}_1, \ldots, \vec{e}_n)$ be the standard basis of $\mathbb{R}^n$, and let $\mathfrak{B} = (\vec{b}_1, \ldots, \vec{b}_n)$ be an arbitrary basis of $\mathbb{R}^n$. Find the change of basis matrix $S_{\mathfrak{B} \to \mathfrak{A}}$.

Solution
By Definition 4.3.3,

$$S_{\mathfrak{B} \to \mathfrak{A}} = \begin{bmatrix} \left[\vec{b}_1 \right]_{\mathfrak{A}} & \cdots & \left[\vec{b}_n \right]_{\mathfrak{A}} \end{bmatrix}.$$

But note that

$$\left[\vec{x} \right]_{\mathfrak{A}} = \vec{x} \quad \text{for all } \vec{x} \text{ in } \mathbb{R}^n,$$

since

$$\vec{x} = \begin{bmatrix} x_1 \\ \vdots \\ x_n \end{bmatrix} = x_1 \vec{e}_1 + \cdots + x_n \vec{e}_n;$$

the components of $\vec{x}$ are its coordinates with respect to the standard basis. Thus

$$S_{\mathfrak{B} \to \mathfrak{A}} = \begin{bmatrix} \vec{b}_1 & \cdots & \vec{b}_n \end{bmatrix}.$$

Compare this with Definition 3.4.1.   ■

EXAMPLE 7   The equation $x_1 + x_2 + x_3 = 0$ defines a plane V in $\mathbb{R}^3$. In this plane, consider the two bases

$$\mathfrak{A} = (\vec{a}_1, \vec{a}_2) = \left(\begin{bmatrix} 0 \\ 1 \\ -1 \end{bmatrix}, \begin{bmatrix} 1 \\ 0 \\ -1 \end{bmatrix} \right) \quad \text{and} \quad \mathfrak{B} = (\vec{b}_1, \vec{b}_2) = \left(\begin{bmatrix} 1 \\ 2 \\ -3 \end{bmatrix}, \begin{bmatrix} 4 \\ -1 \\ -3 \end{bmatrix} \right).$$

a. Find the change of basis matrix S from $\mathfrak{B}$ to $\mathfrak{A}$.
b. Verify that the matrix S in part (a) satisfies the equation

$$\begin{bmatrix} \vec{b}_1 & \vec{b}_2 \end{bmatrix} = \begin{bmatrix} \vec{a}_1 & \vec{a}_2 \end{bmatrix} S.$$

Solution
a. By inspection, we find that

$$\left[\vec{b}_1 \right]_{\mathfrak{A}} = \begin{bmatrix} 2 \\ 1 \end{bmatrix}, \quad \left[\vec{b}_2 \right]_{\mathfrak{A}} = \begin{bmatrix} -1 \\ 4 \end{bmatrix}, \quad \text{so that} \quad S = \begin{bmatrix} 2 & -1 \\ 1 & 4 \end{bmatrix}.$$

b. We can verify that

$$\begin{bmatrix} \vec{b}_1 & \vec{b}_2 \end{bmatrix} = \begin{bmatrix} 1 & 4 \\ 2 & -1 \\ -3 & -3 \end{bmatrix} = \begin{bmatrix} 0 & 1 \\ 1 & 0 \\ -1 & -1 \end{bmatrix} \begin{bmatrix} 2 & -1 \\ 1 & 4 \end{bmatrix} = \begin{bmatrix} \vec{a}_1 & \vec{a}_2 \end{bmatrix} S.$$

This equation reflects the fact that the two columns of S are the coordinate vectors of $\vec{b}_1$ and $\vec{b}_2$ with respect to the basis $\mathfrak{A} = (\vec{a}_1, \vec{a}_2)$. We can illustrate this equation with a commutative diagram, where $\vec{x}$ represents a vector in V:

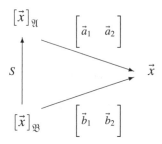

Let us remind ourselves where the equation $\vec{x} = \begin{bmatrix} \vec{b}_1 & \vec{b}_2 \end{bmatrix} \begin{bmatrix} \vec{x} \end{bmatrix}_{\mathfrak{B}}$ comes from: If c_1, c_2 are the coordinates of $\vec{x}$ with respect to $\mathfrak{B}$, then

$$\vec{x} = c_1\vec{b}_1 + c_2\vec{b}_2 = \begin{bmatrix} \vec{b}_1 & \vec{b}_2 \end{bmatrix} \begin{bmatrix} c_1 \\ c_2 \end{bmatrix} = \begin{bmatrix} \vec{b}_1 & \vec{b}_2 \end{bmatrix} \begin{bmatrix} \vec{x} \end{bmatrix}_{\mathfrak{B}}. \qquad \blacksquare$$

We can generalize.

Theorem 4.3.4 **Change of basis in a subspace of $\mathbb{R}^n$**

Consider a subspace V of $\mathbb{R}^n$ with two bases $\mathfrak{A} = (\vec{a}_1, \ldots, \vec{a}_m)$ and $\mathfrak{B} = (\vec{b}_1, \ldots, \vec{b}_m)$. Let S be the change of basis matrix from $\mathfrak{B}$ to $\mathfrak{A}$. Then the following equation holds:

$$\begin{bmatrix} \vec{b}_1 & \cdots & \vec{b}_m \end{bmatrix} = \begin{bmatrix} \vec{a}_1 & \cdots & \vec{a}_m \end{bmatrix} S \qquad \blacksquare$$

As in Example 7, this equation can be justified by means of a commutative diagram:

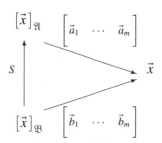

Now consider a linear transformation T from V to V, where V is a finite dimensional linear space. Let $\mathfrak{A}$ and $\mathfrak{B}$ be two bases of V, and let A and B be the $\mathfrak{A}$- and the $\mathfrak{B}$-matrix of T, respectively. What is the relationship between A, B, and the change of basis matrix S from $\mathfrak{B}$ to $\mathfrak{A}$?

Consider the following diagram.

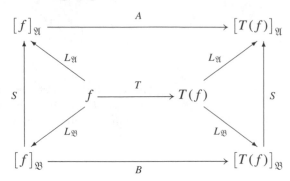

As in Section 3.4, we see that $\left[T(f)\right]_{\mathfrak{A}} = AS\left[f\right]_{\mathfrak{B}} = SB\left[f\right]_{\mathfrak{B}}$ for all f in V, so that $AS = SB$.

Theorem 4.3.5   **Change of basis for the matrix of a linear transformation**

Let V be a linear space with two given bases $\mathfrak{A}$ and $\mathfrak{B}$. Consider a linear transformation T from V to V, and let A and B be the $\mathfrak{A}$- and the $\mathfrak{B}$-matrix of T, respectively. Let S be the change of basis matrix from $\mathfrak{B}$ to $\mathfrak{A}$. Then A is similar to B, and

$$AS = SB \quad \text{or} \quad A = SBS^{-1} \quad \text{or} \quad B = S^{-1}AS. \qquad \blacksquare$$

EXAMPLE 8   As in Example 5, let V be the linear space spanned by the functions e^x and e^{-x}, with the bases $\mathfrak{A} = (e^x, e^{-x})$ and $\mathfrak{B} = (e^x + e^{-x}, e^x - e^{-x})$. Consider the linear transformation $D(f) = f'$ from V to V.

a. Find the $\mathfrak{A}$-matrix A of D.

b. Use part (a), Theorem 4.3.5, and Example 5 to find the $\mathfrak{B}$-matrix B of D.

c. Use Theorem 4.3.2 to find the $\mathfrak{B}$-matrix B of D in terms of its columns.

Solution

a. Let's use a diagram. Recall that $(e^{-x})' = -e^{-x}$, by the chain rule.

$$ae^x + be^{-x} \xrightarrow{\quad D \quad} ae^x - be^{-x}$$

$$\begin{bmatrix} a \\ b \end{bmatrix} \xrightarrow[\quad A = \begin{bmatrix} 1 & 0 \\ 0 & -1 \end{bmatrix} \quad]{} \begin{bmatrix} a \\ -b \end{bmatrix}$$

b. In Example 5 we found the change of basis matrix $S = \begin{bmatrix} 1 & 1 \\ 1 & -1 \end{bmatrix}$ from $\mathfrak{B}$ to $\mathfrak{A}$. Now

$$B = S^{-1}AS = \frac{1}{2}\begin{bmatrix} 1 & 1 \\ 1 & -1 \end{bmatrix}\begin{bmatrix} 1 & 0 \\ 0 & -1 \end{bmatrix}\begin{bmatrix} 1 & 1 \\ 1 & -1 \end{bmatrix} = \begin{bmatrix} 0 & 1 \\ 1 & 0 \end{bmatrix}.$$

c. $B = \begin{bmatrix} \left[D(e^x + e^{-x})\right]_{\mathfrak{B}} & \left[D(e^x - e^{-x})\right]_{\mathfrak{B}} \end{bmatrix}$

$$= \begin{bmatrix} \left[e^x - e^{-x}\right]_{\mathfrak{B}} & \left[e^x + e^{-x}\right]_{\mathfrak{B}} \end{bmatrix} = \begin{bmatrix} 0 & 1 \\ 1 & 0 \end{bmatrix} \qquad \blacksquare$$

EXERCISES 4.3

GOALS   *Use the concept of coordinates. Find the matrix of a linear transformation. Use this matrix to find the image and kernel of a transformation.*

1. Are the polynomials $f(t) = 7 + 3t + t^2$, $g(t) = 9 + 9t + 4t^2$, and $h(t) = 3 + 2t + t^2$ linearly independent?

2. Are the matrices
$$\begin{bmatrix} 1 & 1 \\ 1 & 1 \end{bmatrix}, \quad \begin{bmatrix} 1 & 2 \\ 3 & 4 \end{bmatrix}, \quad \begin{bmatrix} 2 & 3 \\ 5 & 7 \end{bmatrix}, \quad \begin{bmatrix} 1 & 4 \\ 6 & 8 \end{bmatrix}$$
linearly independent?

3. Do the polynomials $f(t) = 1 + 2t + 9t^2 + t^3$, $g(t) = 1 + 7t + 7t^3$, $h(t) = 1 + 8t + t^2 + 5t^3$, and $k(t) = 1 + 8t + 4t^2 + 8t^3$ form a basis of P_3?

4. Consider the polynomials $f(t) = t + 1$ and $g(t) = (t + 2)(t + k)$, where k is an arbitrary constant. For which values of the constant k are the three polynomials $f(t)$, $tf(t)$, and $g(t)$ a basis of P_2?

In Exercises 5 through 40, find the matrix of the given linear transformation T with respect to the given basis. If no basis is specified, use the standard basis: $\mathfrak{A} = (1, t, t^2)$ for P_2,
$$\mathfrak{A} = \left(\begin{bmatrix} 1 & 0 \\ 0 & 0 \end{bmatrix}, \begin{bmatrix} 0 & 1 \\ 0 & 0 \end{bmatrix}, \begin{bmatrix} 0 & 0 \\ 1 & 0 \end{bmatrix}, \begin{bmatrix} 0 & 0 \\ 0 & 1 \end{bmatrix} \right)$$
for $\mathbb{R}^{2\times2}$, and $\mathfrak{A} = (1, i)$ for $\mathbb{C}$. For the space $U^{2\times2}$ of upper triangular 2×2 matrices, use the basis
$$\mathfrak{A} = \left(\begin{bmatrix} 1 & 0 \\ 0 & 0 \end{bmatrix}, \begin{bmatrix} 0 & 1 \\ 0 & 0 \end{bmatrix}, \begin{bmatrix} 0 & 0 \\ 0 & 1 \end{bmatrix} \right)$$
unless another basis is given. In each case, determine whether T is an isomorphism. If T isn't an isomorphism, find bases of the kernel and image of T, and thus determine the rank of T.

5. $T(M) = \begin{bmatrix} 1 & 2 \\ 0 & 3 \end{bmatrix} M$ from $U^{2\times2}$ to $U^{2\times2}$

6. $T(M) = \begin{bmatrix} 1 & 2 \\ 0 & 3 \end{bmatrix} M$ from $U^{2\times2}$ to $U^{2\times2}$, with respect to the basis $\mathfrak{B} = \left(\begin{bmatrix} 1 & 0 \\ 0 & 0 \end{bmatrix}, \begin{bmatrix} 0 & 1 \\ 0 & 0 \end{bmatrix}, \begin{bmatrix} 0 & 1 \\ 0 & 1 \end{bmatrix} \right)$

7. $T(M) = M \begin{bmatrix} 1 & 2 \\ 0 & 1 \end{bmatrix} - \begin{bmatrix} 1 & 2 \\ 0 & 1 \end{bmatrix} M$ from $U^{2\times2}$ to $U^{2\times2}$, with respect to the basis
$$\mathfrak{B} = \left(\begin{bmatrix} 1 & 0 \\ 0 & 1 \end{bmatrix}, \begin{bmatrix} 0 & 1 \\ 0 & 0 \end{bmatrix}, \begin{bmatrix} 1 & 0 \\ 0 & -1 \end{bmatrix} \right)$$

8. $T(M) = M \begin{bmatrix} 1 & 2 \\ 0 & 1 \end{bmatrix} - \begin{bmatrix} 1 & 2 \\ 0 & 1 \end{bmatrix} M$ from $U^{2\times2}$ to $U^{2\times2}$

9. $T(M) = \begin{bmatrix} 1 & 0 \\ 0 & 2 \end{bmatrix}^{-1} M \begin{bmatrix} 1 & 0 \\ 0 & 2 \end{bmatrix}$ from $U^{2\times2}$ to $U^{2\times2}$

10. $T(M) = \begin{bmatrix} 1 & 2 \\ 0 & 3 \end{bmatrix}^{-1} M \begin{bmatrix} 1 & 2 \\ 0 & 3 \end{bmatrix}$ from $U^{2\times2}$ to $U^{2\times2}$

11. $T(M) = \begin{bmatrix} 1 & 2 \\ 0 & 3 \end{bmatrix}^{-1} M \begin{bmatrix} 1 & 2 \\ 0 & 3 \end{bmatrix}$ from $U^{2\times2}$ to $U^{2\times2}$, with respect to the basis
$$\mathfrak{B} = \left(\begin{bmatrix} 1 & -1 \\ 0 & 0 \end{bmatrix}, \begin{bmatrix} 0 & 1 \\ 0 & 1 \end{bmatrix}, \begin{bmatrix} 0 & 1 \\ 0 & 0 \end{bmatrix} \right)$$

12. $T(M) = M \begin{bmatrix} 2 & 0 \\ 0 & 3 \end{bmatrix}$ from $\mathbb{R}^{2\times2}$ to $\mathbb{R}^{2\times2}$

13. $T(M) = \begin{bmatrix} 1 & 1 \\ 2 & 2 \end{bmatrix} M$ from $\mathbb{R}^{2\times2}$ to $\mathbb{R}^{2\times2}$

14. $T(M) = \begin{bmatrix} 1 & 1 \\ 2 & 2 \end{bmatrix} M$ from $\mathbb{R}^{2\times2}$ to $\mathbb{R}^{2\times2}$, with respect to the basis
$$\mathfrak{B} = \left(\begin{bmatrix} 1 & 0 \\ -1 & 0 \end{bmatrix}, \begin{bmatrix} 0 & 1 \\ 0 & -1 \end{bmatrix}, \begin{bmatrix} 1 & 0 \\ 2 & 0 \end{bmatrix}, \begin{bmatrix} 0 & 1 \\ 0 & 2 \end{bmatrix} \right)$$

15. $T(x + iy) = x - iy$ from $\mathbb{C}$ to $\mathbb{C}$

16. $T(x + iy) = x - iy$ from $\mathbb{C}$ to $\mathbb{C}$, with respect to the basis $\mathfrak{B} = (1 + i, 1 - i)$

17. $T(z) = iz$ from $\mathbb{C}$ to $\mathbb{C}$

18. $T(z) = (2 + 3i)z$ from $\mathbb{C}$ to $\mathbb{C}$

19. $T(z) = (p + iq)z$ from $\mathbb{C}$ to $\mathbb{C}$, where p and q are arbitrary real numbers

20. $T(f) = f'$ from P_2 to P_2

21. $T(f) = f' - 3f$ from P_2 to P_2

22. $T(f) = f'' + 4f'$ from P_2 to P_2

23. $T(f(t)) = f(3)$ from P_2 to P_2

24. $T(f(t)) = f(3)$ from P_2 to P_2, with respect to the basis $\mathfrak{B} = (1, t - 3, (t - 3)^2)$

25. $T(f(t)) = f(-t)$ from P_2 to P_2

26. $T(f(t)) = f(2t)$ from P_2 to P_2

27. $T(f(t)) = f(2t - 1)$ from P_2 to P_2

28. $T(f(t)) = f(2t - 1)$ from P_2 to P_2, with respect to the basis $\mathfrak{B} = (1, t - 1, (t - 1)^2)$

29. $T(f(t)) = \int_0^2 f(t)\, dt$ from P_2 to P_2

30. $T\big(f(t)\big) = \dfrac{f(t+h) - f(t)}{h}$ from P_2 to P_2, where h is a nonzero constant. Interpret transformation T geometrically.

31. $T\big(f(t)\big) = \dfrac{f(t+h) - f(t-h)}{2h}$ from P_2 to P_2, where h is a nonzero constant. Interpret transformation T geometrically.

32. $T\big(f(t)\big) = f(1) + f'(1)(t-1)$ from P_2 to P_2. Interpret transformation T geometrically.

33. $T\big(f(t)\big) = f(1) + f'(1)(t-1)$ from P_2 to P_2, with respect to the basis $\mathfrak{B} = \big(1, t-1, (t-1)^2\big)$

34. $T(M) = \begin{bmatrix} 2 & 0 \\ 0 & 5 \end{bmatrix} M - M \begin{bmatrix} 2 & 0 \\ 0 & 5 \end{bmatrix}$ from $\mathbb{R}^{2\times2}$ to $\mathbb{R}^{2\times2}$

35. $T(M) = \begin{bmatrix} 2 & 1 \\ 0 & 2 \end{bmatrix} M - M \begin{bmatrix} 2 & 1 \\ 0 & 2 \end{bmatrix}$ from $\mathbb{R}^{2\times2}$ to $\mathbb{R}^{2\times2}$

36. $T(M) = \begin{bmatrix} 1 & 1 \\ 1 & 1 \end{bmatrix} M - M \begin{bmatrix} 1 & 1 \\ 1 & 1 \end{bmatrix}$ from $\mathbb{R}^{2\times2}$ to $\mathbb{R}^{2\times2}$

37. $T(M) = \begin{bmatrix} 1 & 1 \\ 1 & 1 \end{bmatrix} M - M \begin{bmatrix} 1 & 1 \\ 1 & 1 \end{bmatrix}$ from $\mathbb{R}^{2\times2}$ to $\mathbb{R}^{2\times2}$, with respect to the basis

$$\mathfrak{B} = \left(\begin{bmatrix} 1 & 1 \\ -1 & -1 \end{bmatrix}, \begin{bmatrix} 1 & -1 \\ 1 & -1 \end{bmatrix}, \begin{bmatrix} 1 & 0 \\ 0 & 1 \end{bmatrix}, \begin{bmatrix} 0 & 1 \\ 1 & 0 \end{bmatrix} \right)$$

38. $T(M) = \begin{bmatrix} 0 & 1 \\ 1 & 0 \end{bmatrix} M - M \begin{bmatrix} 1 & 0 \\ 0 & -1 \end{bmatrix}$ from $\mathbb{R}^{2\times2}$ to $\mathbb{R}^{2\times2}$, with respect to the basis

$$\mathfrak{B} = \left(\begin{bmatrix} 1 & 0 \\ 1 & 0 \end{bmatrix}, \begin{bmatrix} 1 & 0 \\ -1 & 0 \end{bmatrix}, \begin{bmatrix} 0 & 1 \\ 0 & 1 \end{bmatrix}, \begin{bmatrix} 0 & 1 \\ 0 & -1 \end{bmatrix} \right)$$

39. $T(M) = \begin{bmatrix} 0 & 1 \\ 1 & 0 \end{bmatrix} M - M \begin{bmatrix} 1 & 0 \\ 0 & -1 \end{bmatrix}$ from $\mathbb{R}^{2\times2}$ to $\mathbb{R}^{2\times2}$

40. $T(M) = \begin{bmatrix} 1 & 2 \\ 4 & 3 \end{bmatrix} M - M \begin{bmatrix} 5 & 0 \\ 0 & -1 \end{bmatrix}$ from $\mathbb{R}^{2\times2}$ to $\mathbb{R}^{2\times2}$

41. a. Find the change of basis matrix S from the basis $\mathfrak{B}$ considered in Exercise 6 to the standard basis $\mathfrak{A}$ of $U^{2\times2}$ considered in Exercise 5.

   b. Verify the formula $SB = AS$ for the matrices B and A you found in Exercises 6 and 5, respectively.

   c. Find the change of basis matrix from $\mathfrak{A}$ to $\mathfrak{B}$.

42. a. Find the change of basis matrix S from the basis $\mathfrak{B}$ considered in Exercise 7 to the standard basis $\mathfrak{A}$ of $U^{2\times2}$ considered in Exercise 8.

   b. Verify the formula $SB = AS$ for the matrices B and A you found in Exercises 7 and 8, respectively.

   c. Find the change of basis matrix from $\mathfrak{A}$ to $\mathfrak{B}$.

43. a. Find the change of basis matrix S from the basis $\mathfrak{B}$ considered in Exercise 11 to the standard basis $\mathfrak{A}$ of $U^{2\times2}$ considered in Exercise 10.

   b. Verify the formula $SB = AS$ for the matrices B and A you found in Exercises 11 and 10, respectively.

   c. Find the change of basis matrix from $\mathfrak{A}$ to $\mathfrak{B}$.

44. a. Find the change of basis matrix S from the basis $\mathfrak{B}$ considered in Exercise 14 to the standard basis $\mathfrak{A}$ of $\mathbb{R}^{2\times2}$ considered in Exercise 13.

   b. Verify the formula $SB = AS$ for the matrices B and A you found in Exercises 14 and 13, respectively.

45. a. Find the change of basis matrix S from the basis $\mathfrak{B}$ considered in Exercise 16 to the standard basis $\mathfrak{A} = (1, i)$ of $\mathbb{C}$ considered in Exercise 15.

   b. Verify the formula $SB = AS$ for the matrices B and A you found in Exercises 16 and 15, respectively.

   c. Find the change of basis matrix from $\mathfrak{A}$ to $\mathfrak{B}$.

46. a. Find the change of basis matrix S from the basis $\mathfrak{B}$ considered in Exercise 24 to the standard basis $\mathfrak{A} = (1, t, t^2)$ of P_2 considered in Exercise 23.

   b. Verify the formula $SB = AS$ for the matrices B and A you found in Exercises 24 and 23, respectively.

   c. Find the change of basis matrix from $\mathfrak{A}$ to $\mathfrak{B}$.

47. a. Find the change of basis matrix S from the basis $\mathfrak{B}$ considered in Exercise 28 to the standard basis $\mathfrak{A} = (1, t, t^2)$ of P_2 considered in Exercise 27.

   b. Verify the formula $SB = AS$ for the matrices B and A you found in Exercises 28 and 27, respectively.

   c. Find the change of basis matrix from $\mathfrak{A}$ to $\mathfrak{B}$.

In Exercises 48 through 53, let V be the space spanned by the two functions $\cos(t)$ and $\sin(t)$. In each exercise, find the matrix of the given transformation T with respect to the basis $\cos(t)$, $\sin(t)$, and determine whether T is an isomorphism.

48. $T(f) = f'$ **49.** $T(f) = f'' + 2f' + 3f$

50. $T(f) = f'' + af' + bf$, where a and b are arbitrary real numbers

51. $T\big(f(t)\big) = f(t - \pi/2)$

52. $T\big(f(t)\big) = f(t - \pi/4)$

53. $T\big(f(t)\big) = f(t - \theta)$, where θ is an arbitrary real number. (*Hint:* Use the addition theorems for sine and cosine.)

In Exercises 54 through 58, let V be the plane with equation $x_1 + 2x_2 + 3x_3 = 0$ in $\mathbb{R}^3$. In each exercise, find the matrix B of the given transformation T from V to V, with respect to the basis $\begin{bmatrix} 1 \\ 1 \\ -1 \end{bmatrix}, \begin{bmatrix} 5 \\ -4 \\ 1 \end{bmatrix}$. *Note that domain and target space of T are restricted to the plane V, so that B will be a 2×2 matrix.*

54. The orthogonal projection onto the line spanned by

vector $\begin{bmatrix} 1 \\ 1 \\ -1 \end{bmatrix}$

55. The orthogonal projection onto the line spanned by

vector $\begin{bmatrix} 1 \\ -2 \\ 1 \end{bmatrix}$

56. $T(\vec{x}) = \begin{bmatrix} 1 \\ 2 \\ 3 \end{bmatrix} \times \vec{x}$

57. $T(\vec{x}) = \begin{bmatrix} -2 & -3 & 1 \\ 1 & 0 & -2 \\ 0 & 1 & 1 \end{bmatrix} \vec{x}$

58. $T(\vec{x}) = \left(\vec{x} \cdot \begin{bmatrix} 1 \\ 1 \\ -1 \end{bmatrix} \right) \begin{bmatrix} 1 \\ 1 \\ -1 \end{bmatrix}$

59. Consider a linear transformation T from V to V with $\ker(T) = \{0\}$. If V is finite dimensional, then T is an isomorphism, since the matrix of T will be invertible. Show that this is not necessarily the case if V is infinite dimensional: Give an example of a linear transformation T from P to P with $\ker(T) = \{0\}$ that is not an isomorphism. (Recall that P is the space of all polynomials.)

60. In the plane V defined by the equation $2x_1 + x_2 - 2x_3 = 0$, consider the bases

$$\mathfrak{A} = (\vec{a}_1, \vec{a}_2) = \left(\begin{bmatrix} 1 \\ 2 \\ 2 \end{bmatrix}, \begin{bmatrix} 2 \\ -2 \\ 1 \end{bmatrix} \right)$$

and

$$\mathfrak{B} = (\vec{b}_1, \vec{b}_2) = \left(\begin{bmatrix} 1 \\ 2 \\ 2 \end{bmatrix}, \begin{bmatrix} 3 \\ 0 \\ 3 \end{bmatrix} \right).$$

a. Find the change of basis matrix S from $\mathfrak{B}$ to $\mathfrak{A}$.
b. Find the change of basis matrix from $\mathfrak{A}$ to $\mathfrak{B}$.
c. Write an equation relating the matrices $\begin{bmatrix} \vec{a}_1 & \vec{a}_2 \end{bmatrix}$, $\begin{bmatrix} \vec{b}_1 & \vec{b}_2 \end{bmatrix}$, and $S = S_{\mathfrak{B} \to \mathfrak{A}}$.

61. In $\mathbb{R}^2$, consider the bases

$$\mathfrak{A} = (\vec{a}_1, \vec{a}_2) = \left(\begin{bmatrix} 1 \\ 2 \end{bmatrix}, \begin{bmatrix} 2 \\ -1 \end{bmatrix} \right)$$

and

$$\mathfrak{B} = (\vec{b}_1, \vec{b}_2) = \left(\begin{bmatrix} 5 \\ -10 \end{bmatrix}, \begin{bmatrix} 10 \\ 5 \end{bmatrix} \right).$$

a. Find the change of basis matrix S from $\mathfrak{B}$ to $\mathfrak{A}$. Interpret the transformation defined by S geometrically.

b. Find the change of basis matrix from $\mathfrak{A}$ to $\mathfrak{B}$.
c. Write an equation relating the matrices $\begin{bmatrix} \vec{a}_1 & \vec{a}_2 \end{bmatrix}$, $\begin{bmatrix} \vec{b}_1 & \vec{b}_2 \end{bmatrix}$, and $S = S_{\mathfrak{B} \to \mathfrak{A}}$.

62. In the plane V defined by the equation $x_1 - 2x_2 + 2x_3 = 0$, consider the basis

$$\mathfrak{A} = (\vec{a}_1, \vec{a}_2) = \left(\begin{bmatrix} 2 \\ 1 \\ 0 \end{bmatrix}, \begin{bmatrix} -2 \\ 0 \\ 1 \end{bmatrix} \right).$$

a. Construct another basis $\mathfrak{B} = (\vec{b}_1, \vec{b}_2)$ of V, such that neither $\vec{b}_1$ nor $\vec{b}_2$ has any zero components.
b. Find the change of basis matrix S from $\mathfrak{B}$ to $\mathfrak{A}$.
c. Find the change of basis matrix from $\mathfrak{A}$ to $\mathfrak{B}$.
d. Write an equation relating the matrices $\begin{bmatrix} \vec{a}_1 & \vec{a}_2 \end{bmatrix}$, $\begin{bmatrix} \vec{b}_1 & \vec{b}_2 \end{bmatrix}$, and $S = S_{\mathfrak{B} \to \mathfrak{A}}$.

63. In the plane V defined by the equation $x_1 + 3x_2 - 2x_3 = 0$, consider the basis

$$\mathfrak{A} = (\vec{a}_1, \vec{a}_2) = \left(\begin{bmatrix} -3 \\ 1 \\ 0 \end{bmatrix}, \begin{bmatrix} -1 \\ 1 \\ 1 \end{bmatrix} \right).$$

a. Construct another basis $\mathfrak{B} = (\vec{b}_1, \vec{b}_2)$ of V, such that neither $\vec{b}_1$ nor $\vec{b}_2$ has any negative components.
b. Find the change of basis matrix S from $\mathfrak{B}$ to $\mathfrak{A}$.
c. Find the change of basis matrix from $\mathfrak{A}$ to $\mathfrak{B}$.
d. Write an equation relating the matrices $\begin{bmatrix} \vec{a}_1 & \vec{a}_2 \end{bmatrix}$, $\begin{bmatrix} \vec{b}_1 & \vec{b}_2 \end{bmatrix}$, and $S = S_{\mathfrak{B} \to \mathfrak{A}}$.

64. Let V be the space of all upper triangular 2×2 matrices. Consider the linear transformation

$$T \begin{bmatrix} a & b \\ 0 & c \end{bmatrix} = aI_2 + bP + cP^2$$

from V to V, where $P = \begin{bmatrix} 1 & 2 \\ 0 & 3 \end{bmatrix}$.

a. Find the matrix A of T with respect to the basis

$$\mathfrak{A} = \left(\begin{bmatrix} 1 & 0 \\ 0 & 0 \end{bmatrix}, \begin{bmatrix} 0 & 1 \\ 0 & 0 \end{bmatrix}, \begin{bmatrix} 0 & 0 \\ 0 & 1 \end{bmatrix} \right).$$

b. Find bases of the image and kernel of T, and thus determine the rank of T.

65. Let V be the subspace of $\mathbb{R}^{2 \times 2}$ spanned by the matrices I_2 and $P = \begin{bmatrix} a & b \\ c & d \end{bmatrix}$, where $b \neq 0$.

a. Compute P^2 and find the coordinate vector $\begin{bmatrix} P^2 \end{bmatrix}_{\mathfrak{B}}$, where $\mathfrak{B} = (I_2, P)$.

b. Consider the linear transformation $T(M) = MP$ from V to V. Find the $\mathfrak{B}$-matrix B of T. For which matrices P is T an isomorphism?

c. If T fails to be an isomorphism, find the image and kernel of T. What is the rank of T in that case?

66. Let V be the linear space of all functions in two variables of the form $q(x_1, x_2) = ax_1^2 + bx_1x_2 + cx_2^2$. Consider the linear transformation

$$T(f) = x_2\frac{\partial f}{\partial x_1} - x_1\frac{\partial f}{\partial x_2}$$

from V to V.

a. Find the matrix of T with respect to the basis x_1^2, x_1x_2, x_2^2 of V.

b. Find bases of the kernel and image of T.

67. Let V be the linear space of all functions of the form

$$f(t) = c_1\cos(t) + c_2\sin(t) + c_3t\cos(t) + c_4t\sin(t).$$

Consider the linear transformation T from V to V given by

$$T(f) = f'' + f.$$

a. Find the matrix of T with respect to the basis $\cos(t)$, $\sin(t)$, $t\cos(t)$, $t\sin(t)$ of V.

b. Find all solutions f in W of the differential equation

$$T(f) = f'' + f = \cos(t).$$

Graph your solution(s). [The differential equation $f'' + f = \cos(t)$ describes a forced undamped oscillator. In this example, we observe the phenomenon of *resonance*.]

68. Consider the linear space V of all infinite sequences of real numbers. We define the subset W of V consisting of all sequences $(x_0, x_1, x_2, \ldots)$ such that $x_{n+2} = x_{n+1} + 6x_n$ for all $n \geq 0$.

a. Show that W is a *subspace* of V.

b. Determine the dimension of W.

c. Does W contain any geometric sequences of the form $(1, c, c^2, c^3, \ldots)$, for some constant c? Find all such sequences in W.

d. Can you find a basis of W consisting of geometric sequences?

e. Consider the sequence in W whose first two terms are $x_0 = 0$, $x_1 = 1$. Find x_2, x_3, x_4. Find a closed formula for the nth term x_n of this sequence. *Hint*: Write this sequence as a linear combination of the sequences you found in part (d).

69. Consider a basis $f_1, \ldots, f_n$ of P_{n-1}. Let $a_1, \ldots, a_n$ be distinct real numbers. Consider the $n \times n$ matrix M whose ijth entry is $f_j(a_i)$. Show that the matrix M is invertible. [*Hint*: If the vector

$$\begin{bmatrix} c_1 \\ c_2 \\ \vdots \\ c_n \end{bmatrix}$$

is in the kernel of M, then the polynomial $f = c_1f_1 + \cdots + c_nf_n$ in P_{n-1} vanishes at $a_1, \ldots, a_n$; therefore, $f = 0$.]

70. Consider two finite dimensional linear spaces V and W. If V and W are isomorphic, then they have the same dimension (by Theorem 4.2.4b). Conversely, if V and W have the same dimension, are they necessarily isomorphic? Justify your answer carefully.

71. Let $a_1, \ldots, a_n$ be distinct real numbers. Show that there exist "weights" $w_1, \ldots, w_n$ such that

$$\int_{-1}^{1} f(t)\, dt = \sum_{i=1}^{n} w_i f(a_i),$$

for all polynomials $f(t)$ in P_{n-1}. *Hint*: It suffices to prove the claim for a basis $f_1, \ldots, f_n$ of P_{n-1}. Exercise 69 is helpful.

72. Find the weights w_1, w_2, w_3 in Exercise 71 for $a_1 = -1, a_2 = 0, a_3 = 1$. (Compare this with Simpson's rule in calculus.)

Chapter Four Exercises

TRUE OR FALSE?

1. The space $\mathbb{R}^{2\times3}$ is 5-dimensional.

2. If $f_1, \ldots, f_n$ is a basis of a linear space V, then any element of V can be written as a linear combination of $f_1, \ldots, f_n$.

3. The space P_1 is isomorphic to $\mathbb{C}$.

4. If the kernel of a linear transformation T from P_4 to P_4 is $\{0\}$, then T must be an isomorphism.

5. If W_1 and W_2 are subspaces of a linear space V, then the intersection $W_1 \cap W_2$ must be a subspace of V as well.

6. If T is a linear transformation from P_6 to $\mathbb{R}^{2\times2}$, then the kernel of T must be 3-dimensional.

7. The polynomials of degree less than 7 form a 7-dimensional subspace of the linear space of all polynomials.

8. The function $T(f) = 3f - 4f'$ from C^∞ to C^∞ is a linear transformation.

9. The lower triangular 2×2 matrices form a subspace of the space of all 2×2 matrices.

10. The kernel of a linear transformation is a subspace of the domain.

11. The linear transformation $T(f) = f + f''$ from C^∞ to C^∞ is an isomorphism.

12. All linear transformations from P_3 to $\mathbb{R}^{2\times 2}$ are isomorphisms.

13. If T is a linear transformation from V to V, then the intersection of $\operatorname{im}(T)$ and $\ker(T)$ must be $\{0\}$.

14. The space of all upper triangular 4×4 matrices is isomorphic to the space of all lower triangular 4×4 matrices.

15. Every polynomial of degree 3 can be expressed as a linear combination of the polynomial $(t-3)$, $(t-3)^2$, and $(t-3)^3$.

16. If a linear space V can be spanned by 10 elements, then the dimension of V must be ≤ 10.

17. The function $T(M) = \det(M)$ from $\mathbb{R}^{2\times 2}$ to $\mathbb{R}$ is a linear transformation.

18. There exists a 2×2 matrix A such that the space of all matrices commuting with A is 1-dimensional.

19. All bases of P_3 contain at least one polynomial of degree ≤ 2.

20. If T is an isomorphism, then T^{-1} must be an isomorphism as well.

21. If the image of a linear transformation T from P to P is all of P, then T must be an isomorphism.

22. If f_1, f_2, f_3 is a basis of a linear space V, then f_1, $f_1 + f_2$, $f_1 + f_2 + f_3$ must be a basis of V as well.

23. If a, b, and c are distinct real numbers, then the polynomials $(x-b)(x-c)$, $(x-a)(x-c)$, and $(x-a)(x-b)$ must be linearly independent.

24. The linear transformation $T\big(f(t)\big) = f(4t-3)$ from P to P is an isomorphism.

25. The linear transformation $T(M) = \begin{bmatrix} 1 & 2 \\ 3 & 6 \end{bmatrix} M$ from $\mathbb{R}^{2\times 2}$ to $\mathbb{R}^{2\times 2}$ has rank 1.

26. If the matrix of a linear transformation T (with respect to some basis) is $\begin{bmatrix} 3 & 5 \\ 0 & 4 \end{bmatrix}$, then there must exist a nonzero element f in the domain of T such that $T(f) = 3f$.

27. The kernel of the linear transformation $T\big(f(t)\big) = f(t^2)$ from P to P is $\{0\}$.

28. If S is any invertible 2×2 matrix, then the linear transformation $T(M) = SMS$ is an isomorphism from $\mathbb{R}^{2\times 2}$ to $\mathbb{R}^{2\times 2}$.

29. There exists a 2×2 matrix A such that the space of all matrices commuting with A is 2-dimensional.

30. There exists a basis of $\mathbb{R}^{2\times 2}$ that consists of four invertible matrices.

31. If W is a subspace of V, and if W is finite dimensional, then V must be finite dimensional as well.

32. There exists a linear transformation from $\mathbb{R}^{3\times 3}$ to $\mathbb{R}^{2\times 2}$ whose kernel consists of all lower triangular 3×3 matrices, while the image consists of all upper triangular 2×2 matrices.

33. Every two-dimensional subspace of $\mathbb{R}^{2\times 2}$ contains at least one invertible matrix.

34. If $\mathfrak{A} = (f, g)$ and $\mathfrak{B} = (f, f + g)$ are two bases of a linear space V, then the change of basis matrix from $\mathfrak{A}$ to $\mathfrak{B}$ is $\begin{bmatrix} 1 & 1 \\ 0 & 1 \end{bmatrix}$.

35. If the matrix of a linear transformation T with respect to a basis (f, g) is $\begin{bmatrix} 1 & 2 \\ 3 & 4 \end{bmatrix}$, then the matrix of T with respect to the basis (g, f) is $\begin{bmatrix} 2 & 1 \\ 4 & 3 \end{bmatrix}$.

36. The linear transformation $T(f) = f'$ from P_n to P_n has rank n, for all positive integers n.

37. If the matrix of a linear transformation T (with respect to some basis) is $\begin{bmatrix} 2 & 3 \\ 5 & 7 \end{bmatrix}$, then T must be an isomorphism.

38. There exists a subspace of $\mathbb{R}^{3\times 4}$ that is isomorphic to P_9.

39. There exist two distinct subspaces W_1 and W_2 of $\mathbb{R}^{2\times 2}$ whose union $W_1 \cup W_2$ is a subspace of $\mathbb{R}^{2\times 2}$ as well.

40. There exists a linear transformation from P to P_5 whose image is all of P_5.

41. If $f_1, \ldots, f_n$ are polynomials such that the degree of f_k is k (for $k = 1, \ldots, n$), then $f_1, \ldots, f_n$ must be linearly independent.

42. The transformation $D(f) = f'$ from C^∞ to C^∞ is an isomorphism.

43. If T is a linear transformation from P_4 to W with $\operatorname{im}(T) = W$, then the inequality $\dim(W) \leq 5$ must hold.

44. The kernel of the linear transformation

$$T\big(f(t)\big) = \int_0^1 f(t)\, dt$$

from P to $\mathbb{R}$ is finite dimensional.

45. If T is a linear transformation from V to V, then $\{f \text{ in } V : T(f) = f\}$ must be a subspace of V.

46. If T is a linear transformation from P_6 to P_6 that transforms t^k into a polynomial of degree k (for $k = 1, \ldots, 6$), then T must be an isomorphism.

47. There exist invertible 2×2 matrices P and Q such that the linear transformation $T(M) = PM - MQ$ is an isomorphism.

48. There exists a linear transformation from P_6 to $\mathbb{C}$ whose kernel is isomorphic to $\mathbb{R}^{2 \times 2}$.

49. If f_1, f_2, f_3 is a basis of a linear space V, and if f is any element of V, then the elements $f_1 + f$, $f_2 + f$, $f_3 + f$ must form a basis of V as well.

50. There exists a two-dimensional subspace of $\mathbb{R}^{2 \times 2}$ whose nonzero elements are all invertible.

51. The space P_{11} is isomorphic to $\mathbb{R}^{3 \times 4}$.

52. If T is a linear transformation from V to W, and if both $\operatorname{im}(T)$ and $\ker(T)$ are finite dimensional, then W must be finite dimensional.

53. If T is a linear transformation from V to $\mathbb{R}^{2 \times 2}$ with $\ker(T) = \{0\}$, then the inequality $\dim(V) \leq 4$ must hold.

54. The function

$$T\big(f(t)\big) = \frac{d}{dt} \int_2^{3t+4} f(x)\, dx$$

from P_5 to P_5 is an isomorphism.

55. Any 4-dimensional linear space has infinitely many 3-dimensional subspaces.

56. If the matrix of a linear transformation T (with respect to some basis) is $\begin{bmatrix} 3 & 5 \\ 0 & 4 \end{bmatrix}$, then there must exist a nonzero element f in the domain of T such that $T(f) = 4f$.

57. If the image of a linear transformation T is infinite dimensional, then the domain of T must be infinite dimensional.

58. There exists a 2×2 matrix A such that the space of all matrices commuting with A is 3-dimensional.

59. If A, B, C, and D are noninvertible 2×2 matrices, then the matrices AB, AC, and AD must be linearly dependent.

60. There exist two distinct 3-dimensional subspaces W_1 and W_2 of P_4 such that the union $W_1 \cup W_2$ is a subspace of P_4 as well.

61. If the elements $f_1, \ldots, f_n$ (where $f_1 \neq 0$) are linearly dependent, then one element f_k can be expressed *uniquely* as a linear combination of the preceding elements $f_1, \ldots, f_{k-1}$.

62. There exists a 3×3 matrix P such that the linear transformation $T(M) = MP - PM$ from $\mathbb{R}^{3 \times 3}$ to $\mathbb{R}^{3 \times 3}$ is an isomorphism.

63. If f_1, f_2, f_3, f_4, f_5 are elements of a linear space V, and if there are exactly two redundant elements in the list f_1, f_2, f_3, f_4, f_5, then there must be exactly two redundant elements in the list f_2, f_4, f_5, f_1, f_3 as well.

64. There exists a linear transformation T from P_6 to P_6 such that the kernel of T is isomorphic to the image of T.

65. If T is a linear transformation from V to W, and if both $\operatorname{im}(T)$ and $\ker(T)$ are finite dimensional, then V must be finite dimensional.

66. If the matrix of a linear transformation T (with respect to some basis) is $\begin{bmatrix} 3 & 5 \\ 0 & 4 \end{bmatrix}$, then there must exist a nonzero element f in the domain of T such that $T(f) = 5f$.

67. Every three-dimensional subspace of $\mathbb{R}^{2 \times 2}$ contains at least one invertible matrix.

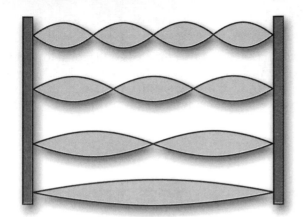

Orthogonality and Least Squares

Orthogonal Projections and Orthonormal Bases

In Section 2.2, we took a first look at some linear transformations that are important in geometry: orthogonal projections, reflections, and rotations in particular. In this chapter, we will generalize these ideas. In Sections 5.1 and 5.2, we will discuss the orthogonal projection onto a subspace V of $\mathbb{R}^n$. In Section 5.3, we will study linear transformations that preserve length, such as reflections and rotations. In Section 5.4, we will present an important application of orthogonal projections: the method of least squares in statistics. Finally, in Section 5.5, we will go a step further and generalize all these ideas from $\mathbb{R}^n$ to linear spaces.

First, we will discuss some basic concepts of geometry.

Definition 5.1.1   Orthogonality, length, unit vectors

    a. Two vectors $\vec{v}$ and $\vec{w}$ in $\mathbb{R}^n$ are called *perpendicular* or *orthogonal*[1] if $\vec{v} \cdot \vec{w} = 0$.

    b. The *length* (or magnitude or norm) of a vector $\vec{v}$ in $\mathbb{R}^n$ is $\|\vec{v}\| = \sqrt{\vec{v} \cdot \vec{v}}$.

    c. A vector $\vec{u}$ in $\mathbb{R}^n$ is called a *unit vector* if its length is 1, (i.e., $\|\vec{u}\| = 1$, or $\vec{u} \cdot \vec{u} = 1$).

If $\vec{v}$ is a nonzero vector in $\mathbb{R}^n$, then

$$\vec{u} = \frac{1}{\|\vec{v}\|}\vec{v}$$

is a unit vector. (See Exercise 25b.)

A vector $\vec{x}$ in $\mathbb{R}^n$ is said to be orthogonal to a subspace V of $\mathbb{R}^n$ if $\vec{x}$ is orthogonal to all the vectors $\vec{v}$ in V, meaning that $\vec{x} \cdot \vec{v} = 0$ for all vectors $\vec{v}$ in V.

---

[1]The two terms are synonymous: *Perpendicular* comes from Latin and *orthogonal* from Greek.

If we are given a basis $\vec{v}_1, \ldots, \vec{v}_m$ of V, then $\vec{x}$ is orthogonal to V if (and only if) $\vec{x}$ is orthogonal to all the vectors $\vec{v}_1, \ldots, \vec{v}_m$. (See Exercise 22.)

For example, a vector $\vec{x}$ in $\mathbb{R}^3$ is orthogonal to a plane V in $\mathbb{R}^3$ if (and only if) $\vec{x}$ is orthogonal to two vectors $\vec{v}_1, \vec{v}_2$ that form a basis of V. (See Figure 1).

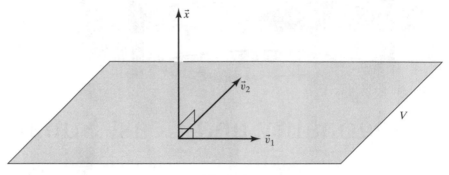

Figure 1

Definition 5.1.2   Orthonormal vectors

The vectors $\vec{u}_1, \vec{u}_2, \ldots, \vec{u}_m$ in $\mathbb{R}^n$ are called *orthonormal* if they are all unit vectors and orthogonal to one another:

$$\vec{u}_i \cdot \vec{u}_j = \begin{cases} 1 & \text{if } i = j \\ 0 & \text{if } i \neq j \end{cases}$$

EXAMPLE 1   The vectors $\vec{e}_1, \vec{e}_2, \ldots, \vec{e}_n$ in $\mathbb{R}^n$ are orthonormal.   ∎

EXAMPLE 2   For any scalar θ, the vectors $\begin{bmatrix} \cos\theta \\ \sin\theta \end{bmatrix}, \begin{bmatrix} -\sin\theta \\ \cos\theta \end{bmatrix}$ are orthonormal. (See Figure 2.)   ∎

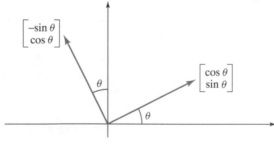

Figure 2

EXAMPLE 3   The vectors

$$\vec{u}_1 = \begin{bmatrix} 1/2 \\ 1/2 \\ 1/2 \\ 1/2 \end{bmatrix}, \quad \vec{u}_2 = \begin{bmatrix} 1/2 \\ 1/2 \\ -1/2 \\ -1/2 \end{bmatrix}, \quad \vec{u}_3 = \begin{bmatrix} 1/2 \\ -1/2 \\ 1/2 \\ -1/2 \end{bmatrix}$$

in $\mathbb{R}^4$ are orthonormal. (Verify this.) Can you find a vector $\vec{u}_4$ in $\mathbb{R}^4$ such that all the vectors $\vec{u}_1, \vec{u}_2, \vec{u}_3, \vec{u}_4$ are orthonormal? (See Exercise 16.)   ∎

The following properties of orthonormal vectors are often useful:

Theorem 5.1.3   Properties of orthonormal vectors

a. Orthonormal vectors are linearly independent.

b. Orthonormal vectors $\vec{u}_1, \ldots, \vec{u}_n$ in $\mathbb{R}^n$ form a basis of $\mathbb{R}^n$.

Proof   a. Consider a relation

$$c_1\vec{u}_1 + c_2\vec{u}_2 + \cdots + c_i\vec{u}_i + \cdots + c_m\vec{u}_m = \vec{0}$$

among the orthonormal vectors $\vec{u}_1, \vec{u}_2, \ldots, \vec{u}_m$ in $\mathbb{R}^n$. Let us form the dot product of each side of this equation with $\vec{u}_i$:

$$(c_1\vec{u}_1 + c_2\vec{u}_2 + \cdots + c_i\vec{u}_i + \cdots + c_m\vec{u}_m) \cdot \vec{u}_i = \vec{0} \cdot \vec{u}_i = 0.$$

Because the dot product is distributive (see Theorem A5 in the Appendix),

$$c_1(\vec{u}_1 \cdot \vec{u}_i) + c_2(\vec{u}_2 \cdot \vec{u}_i) + \cdots + c_i(\vec{u}_i \cdot \vec{u}_i) + \cdots + c_m(\vec{u}_m \cdot \vec{u}_i) = 0.$$

We know that $\vec{u}_i \cdot \vec{u}_i = 1$, and all other dot products $\vec{u}_j \cdot \vec{u}_i$ are zero. Therefore, $c_i = 0$. Since this holds for all $i = 1, \ldots, m$, it follows that the vectors $\vec{u}_1, \ldots, \vec{u}_m$ are linearly independent.

b. This follows from part (a) and Summary 3.3.10. (Any n linearly independent vectors in $\mathbb{R}^n$ form a basis of $\mathbb{R}^n$.)   ■

Orthogonal Projections

In Section 2.2 we discussed the basic idea behind an orthogonal projection: If $\vec{x}$ is a vector in $\mathbb{R}^2$ and L is a line in $\mathbb{R}^2$, then we can resolve the vector $\vec{x}$ into a component $\vec{x}^{\parallel}$ parallel to L and a component $\vec{x}^{\perp}$ perpendicular to L,

$$\vec{x} = \vec{x}^{\parallel} + \vec{x}^{\perp},$$

and this decomposition is unique. The vector $\vec{x}^{\parallel}$ is called the orthogonal projection of $\vec{x}$ onto L. See Figure 3.

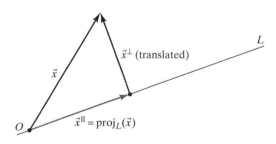

Figure 3

Let's see how we can generalize this idea to any subspace V of $\mathbb{R}^n$.

Theorem 5.1.4   Orthogonal projection

Consider a vector $\vec{x}$ in $\mathbb{R}^n$ and a subspace V of $\mathbb{R}^n$. Then we can write

$$\vec{x} = \vec{x}^{\parallel} + \vec{x}^{\perp},$$

where $\vec{x}^{\parallel}$ is in V and $\vec{x}^{\perp}$ is perpendicular to V, and this representation is unique.

The vector $\vec{x}^{\|}$ is called the *orthogonal projection* of $\vec{x}$ onto V, denoted by $\text{proj}_V(\vec{x})$. See Figure 4.

The transformation $T(\vec{x}) = \text{proj}_V(\vec{x}) = \vec{x}^{\|}$ from $\mathbb{R}^n$ to $\mathbb{R}^n$ is linear.

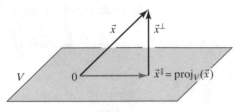

Figure 4

Proof Consider an *orthonormal* basis $\vec{u}_1, \ldots, \vec{u}_m$ of V (see Definition 5.1.2).[2] If a decomposition $\vec{x} = \vec{x}^{\|} + \vec{x}^{\perp}$ (with $\vec{x}^{\|}$ in V and $\vec{x}^{\perp}$ orthogonal to V) does exist, then we can write

$$\vec{x}^{\|} = c_1\vec{u}_1 + \cdots + c_i\vec{u}_i + \cdots + c_m\vec{u}_m,$$

for some coefficients $c_1, \ldots, c_i, \ldots, c_m$ yet to be determined (since $\vec{x}^{\|}$ is in V). We know that

$$\vec{x}^{\perp} = \vec{x} - \vec{x}^{\|} = \vec{x} - c_1\vec{u}_1 - \cdots - c_i\vec{u}_i - \cdots - c_m\vec{u}_m$$

is orthogonal to V, meaning that $\vec{x} - c_1\vec{u}_1 - \cdots - c_i\vec{u}_i - \cdots - c_m\vec{u}_m$ is orthogonal to all the vectors $\vec{u}_i$ in V:

$$\begin{aligned}
0 &= \vec{u}_i \cdot (\vec{x} - c_1\vec{u}_1 - \cdots - c_i\vec{u}_i - \cdots - c_m\vec{u}_m) \\
&= \vec{u}_i \cdot \vec{x} - c_1\underbrace{(\vec{u}_i \cdot \vec{u}_1)}_{0} - \cdots - c_i\underbrace{(\vec{u}_i \cdot \vec{u}_i)}_{1} - \cdots - c_m\underbrace{(\vec{u}_i \cdot \vec{u}_m)}_{0} = \vec{u}_i \cdot \vec{x} - c_i.
\end{aligned}$$

(See Definition 5.1.2.) It follows that $c_i = \vec{u}_i \cdot \vec{x}$, so that

$$\vec{x}^{\|} = (\vec{u}_1 \cdot \vec{x})\vec{u}_1 + \cdots + (\vec{u}_i \cdot \vec{x})\vec{u}_i + \cdots + (\vec{u}_m \cdot \vec{x})\vec{u}_m$$

and

$$\vec{x}^{\perp} = \vec{x} - \vec{x}^{\|} = \vec{x} - (\vec{u}_1 \cdot \vec{x})\vec{u}_1 - \cdots - (\vec{u}_i \cdot \vec{x})\vec{u}_i - \cdots - (\vec{u}_m \cdot \vec{x})\vec{u}_m.$$

Note that $\vec{u}_i \cdot \vec{x}^{\perp} = 0$, by construction, so that $\vec{x}^{\perp}$ is orthogonal to V, as required. (Recall the remarks preceding Figure 1.)

We leave the verification of the linearity of the transformation $T(\vec{x}) = \text{proj}_V(\vec{x}) = \vec{x}^{\|}$ as Exercise 24. ∎

---

[2]In the next section we will introduce an algorithm for constructing an orthonormal basis of any subspace V of $\mathbb{R}^n$. Here we need to convince ourselves merely that such a basis of V does indeed exist. For those who are familiar with this proof technique, we will present a proof by induction on $m = \dim(V)$. If $\dim(V) = 1$, then a unit vector $\vec{u}$ in V will form an orthonormal basis of V. Now consider a subspace V of $\mathbb{R}^n$ with $\dim(V) = m$, and let $\vec{u}$ be a unit vector in V. Consider the linear transformation $T(\vec{x}) = \vec{x} \cdot \vec{u}$ from V to $\mathbb{R}$. By the rank-nullity theorem, the kernel of T will be an $(m-1)$-dimensional subspace W of V, consisting of all vectors $\vec{x}$ in V that are orthogonal to $\vec{u}$ [since $T(\vec{x}) = \vec{x} \cdot \vec{u} = 0$]. By induction hypothesis, there exists an orthonormal basis $(\vec{u}_1, \ldots, \vec{u}_{m-1})$ of W, and $(\vec{u}_1, \ldots, \vec{u}_{m-1}, \vec{u})$ will be an orthonormal basis of V.

Theorem 5.1.5   Formula for the orthogonal projection

If V is a subspace of $\mathbb{R}^n$ with an orthonormal basis $\vec{u}_1, \ldots, \vec{u}_m$, then

$$\text{proj}_V(\vec{x}) = \vec{x}^{\parallel} = (\vec{u}_1 \cdot \vec{x})\vec{u}_1 + \cdots + (\vec{u}_m \cdot \vec{x})\vec{u}_m.$$

for all $\vec{x}$ in $\mathbb{R}^n$.   ■

Note that $\text{proj}_V(\vec{x})$ is the sum of all vectors $(\vec{u}_i \cdot \vec{x})\vec{u}_i$, for $i = 1, \ldots, m$, representing the orthogonal projections of $\vec{x}$ onto the lines spanned by the basis vectors $\vec{u}_1, \ldots, \vec{u}_m$ of V (see Definition 2.2.1).

For example, projecting a vector in $\mathbb{R}^3$ orthogonally onto the x_1–x_2-plane amounts to the same as projecting it onto the x_1-axis, then onto the x_2-axis, and then adding the resultant vectors. (See Figure 5.)

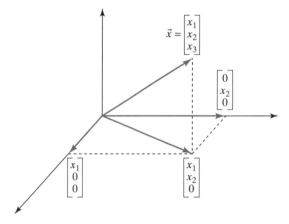

Figure 5

EXAMPLE 4   Consider the subspace $V = \text{im}(A)$ of $\mathbb{R}^4$, where

$$A = \begin{bmatrix} 1 & 1 \\ 1 & -1 \\ 1 & -1 \\ 1 & 1 \end{bmatrix}.$$

Find $\text{proj}_V \vec{x}$, for

$$\vec{x} = \begin{bmatrix} 1 \\ 3 \\ 1 \\ 7 \end{bmatrix}.$$

Solution

Since the two column vectors of A happen to be linearly independent, they form a basis of V. Since they happen to be orthogonal, we can construct an orthonormal basis of V merely by dividing each of these two vectors by its length (2 for both vectors):

$$\vec{u}_1 = \begin{bmatrix} 1/2 \\ 1/2 \\ 1/2 \\ 1/2 \end{bmatrix}, \quad \vec{u}_2 = \begin{bmatrix} 1/2 \\ -1/2 \\ -1/2 \\ 1/2 \end{bmatrix}.$$

Then

$$\text{proj}_V \vec{x} = (\vec{u}_1 \cdot \vec{x})\vec{u}_1 + (\vec{u}_2 \cdot \vec{x})\vec{u}_2 = 6\vec{u}_1 + 2\vec{u}_2 = \begin{bmatrix} 3 \\ 3 \\ 3 \\ 3 \end{bmatrix} + \begin{bmatrix} 1 \\ -1 \\ -1 \\ 1 \end{bmatrix} = \begin{bmatrix} 4 \\ 2 \\ 2 \\ 4 \end{bmatrix}.$$

To check this answer, verify that $\vec{x} - \text{proj}_V \vec{x}$ is perpendicular to both $\vec{u}_1$ and $\vec{u}_2$. ∎

What happens when we apply Theorem 5.1.5 to the subspace $V = \mathbb{R}^n$ of $\mathbb{R}^n$ with orthonormal basis $\vec{u}_1, \vec{u}_2, \ldots, \vec{u}_n$? Clearly, $\text{proj}_V \vec{x} = \vec{x}$, for all $\vec{x}$ in $\mathbb{R}^n$ by Theorem 5.1.4. Therefore,

$$\vec{x} = (\vec{u}_1 \cdot \vec{x})\vec{u}_1 + \cdots + (\vec{u}_n \cdot \vec{x})\vec{u}_n,$$

for all $\vec{x}$ in $\mathbb{R}^n$.

Theorem 5.1.6   Consider an orthonormal basis $\vec{u}_1, \ldots, \vec{u}_n$ of $\mathbb{R}^n$. Then

$$\vec{x} = (\vec{u}_1 \cdot \vec{x})\vec{u}_1 + \cdots + (\vec{u}_n \cdot \vec{x})\vec{u}_n,$$

for all $\vec{x}$ in $\mathbb{R}^n$.   ∎

This means that if you project $\vec{x}$ onto all the lines spanned by the basis vectors $\vec{u}_i$ and add the resultant vectors, you get the vector $\vec{x}$ back. Figure 6 illustrates this in the case $n = 2$.

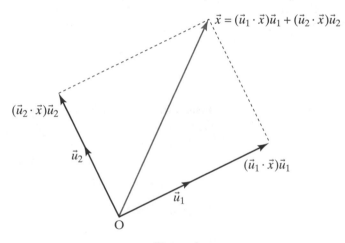

Figure 6

What is the practical significance of Theorem 5.1.6? Whenever we have a basis $\vec{v}_1, \ldots, \vec{v}_n$ of $\mathbb{R}^n$, any vector $\vec{x}$ in $\mathbb{R}^n$ can be expressed uniquely as a linear combination of the $\vec{v}_i$, by Theorem 3.2.10:

$$\vec{x} = c_1 \vec{v}_1 + c_2 \vec{v}_2 + \cdots + c_n \vec{v}_n.$$

To find the coordinates c_i, we generally need to solve a linear system, which may involve a fair amount of computation. However, if we are dealing with an *orthonormal* basis $\vec{u}_1, \ldots, \vec{u}_n$, then we can find the c_i much more easily, using the formula

$$c_i = \vec{u}_i \cdot \vec{x}.$$

EXAMPLE 5   Consider the orthogonal projection $T(\vec{x}) = \text{proj}_V(\vec{x})$ onto a subspace V of $\mathbb{R}^n$. Describe the image and kernel of T.

Solution

By definition of an orthogonal projection, the image of T is the subspace V, while the kernel of T consists of all those vectors $\vec{x}$ in $\mathbb{R}^n$ such that $T(\vec{x}) = \text{proj}_V(\vec{x}) = \vec{x}^{\parallel} = \vec{0}$, meaning that $\vec{x} = \vec{x}^{\perp}$. In other words, the kernel consists of those vectors $\vec{x}$ in $\mathbb{R}^n$ that are orthogonal to V. This space deserves a name.   ■

Definition 5.1.7   Orthogonal complement

Consider a subspace V of $\mathbb{R}^n$. The *orthogonal complement* $V^{\perp}$ of V is the set of those vectors $\vec{x}$ in $\mathbb{R}^n$ that are orthogonal to all vectors in V:

$$V^{\perp} = \{\vec{x} \text{ in } \mathbb{R}^n : \vec{v} \cdot \vec{x} = 0, \text{ for all } \vec{v} \text{ in } V\}.$$

Note that $V^{\perp}$ is the kernel of the orthogonal projection onto V.

Take another look at Figures 3 and 4, and identify the kernel of the projection in each case.

In Figure 7, we sketch the orthogonal complements of a line L and of a plane V in $\mathbb{R}^3$. Note that both $L^{\perp}$ and $V^{\perp}$ are subspaces of $\mathbb{R}^3$. Furthermore,

$$\dim(L) + \dim(L^{\perp}) = 1 + 2 = 3 = \dim(\mathbb{R}^3),$$

and

$$\dim(V) + \dim(V^{\perp}) = 2 + 1 = 3 = \dim(\mathbb{R}^3).$$

We can generalize these observations.

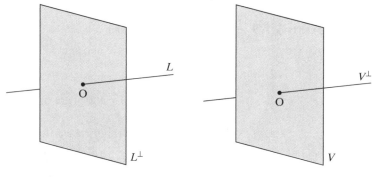

Figure 7

Theorem 5.1.8   Properties of the orthogonal complement

Consider a subspace V of $\mathbb{R}^n$.

    a. The orthogonal complement $V^{\perp}$ of V is a subspace of $\mathbb{R}^n$.
    b. The intersection of V and $V^{\perp}$ consists of the zero vector alone: $V \cap V^{\perp} = \{\vec{0}\}$.
    c. $\dim(V) + \dim(V^{\perp}) = n$
    d. $(V^{\perp})^{\perp} = V$

Proof

    a. If $T(\vec{x}) = \text{proj}_V(\vec{x})$, then $V^{\perp} = \ker(T)$, a subspace of $\mathbb{R}^n$.
    b. If a vector $\vec{x}$ is in V as well as in $V^{\perp}$, then $\vec{x}$ is orthogonal to itself: $\vec{x} \cdot \vec{x} = \|\vec{x}\|^2 = 0$, so that $\vec{x}$ must equal $\vec{0}$, as claimed.

c. We can apply the rank-nullity theorem to the linear transformation $T(\vec{x}) = \text{proj}_V(\vec{x})$:

$$n = \dim(\text{im}T) + \dim(\ker T) = \dim(V) + \dim(V^\perp)$$

d. We leave this proof as Exercise 23. ∎

From Pythagoras to Cauchy

EXAMPLE 6   Consider a line L in $\mathbb{R}^3$ and a vector $\vec{x}$ in $\mathbb{R}^3$. What can you say about the relationship between the lengths of the vectors $\vec{x}$ and $\text{proj}_L\vec{x}$?

Solution

Applying the Pythagorean theorem to the shaded right triangle in Figure 8, we find that $\|\text{proj}_L\vec{x}\| \leq \|\vec{x}\|$. The statement is an equality if (and only if) $\vec{x}$ is on L. ∎

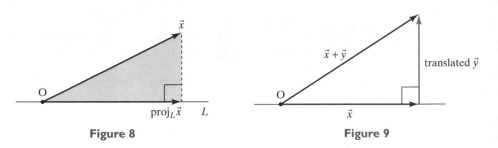

Figure 8   Figure 9

Does this inequality hold in higher dimensional cases? We have to examine whether the Pythagorean theorem holds in $\mathbb{R}^n$.

Theorem 5.1.9   **Pythagorean theorem**

Consider two vectors $\vec{x}$ and $\vec{y}$ in $\mathbb{R}^n$. The equation

$$\|\vec{x} + \vec{y}\|^2 = \|\vec{x}\|^2 + \|\vec{y}\|^2$$

holds if (and only if) $\vec{x}$ and $\vec{y}$ are orthogonal. (See Figure 9.)

Proof   The verification is straightforward:

$$\|\vec{x} + \vec{y}\|^2 = (\vec{x} + \vec{y}) \cdot (\vec{x} + \vec{y}) = \vec{x} \cdot \vec{x} + 2(\vec{x} \cdot \vec{y}) + \vec{y} \cdot \vec{y} = \|\vec{x}\|^2 + 2(\vec{x} \cdot \vec{y}) + \|\vec{y}\|^2$$
$$= \|\vec{x}\|^2 + \|\vec{y}\|^2 \quad \text{if (and only if) } \vec{x} \cdot \vec{y} = 0. \quad ∎$$

Now we can generalize Example 6.

Theorem 5.1.10   **An inequality for the magnitude of $\text{proj}_V(\vec{x})$**

Consider a subspace V of $\mathbb{R}^n$ and a vector $\vec{x}$ in $\mathbb{R}^n$. Then

$$\|\text{proj}_V\vec{x}\| \leq \|\vec{x}\|.$$

The statement is an equality if (and only if) $\vec{x}$ is in V.

Proof   We apply the Pythagorean theorem (see Figure 10):

$$\|\vec{x}\|^2 = \|\text{proj}_V\vec{x}\|^2 + \|\vec{x}^\perp\|^2.$$

It follows that $\|\text{proj}_V\vec{x}\| \leq \|\vec{x}\|$, as claimed. ∎

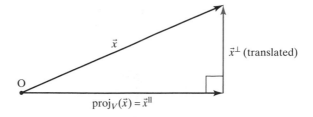

Figure 10

For example, let V be a one-dimensional subspace of $\mathbb{R}^n$ spanned by a (nonzero) vector $\vec{y}$. We introduce the unit vector

$$\vec{u} = \frac{1}{\|\vec{y}\|}\vec{y}$$

in V. (See Figure 11.)

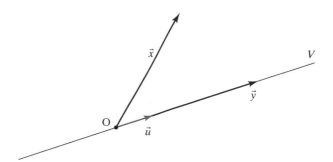

Figure 11

We know that

$$\text{proj}_V\vec{x} = (\vec{x} \cdot \vec{u})\vec{u},$$

for any $\vec{x}$ in $\mathbb{R}^n$. Theorem 5.1.10 tells us that

$$\|\vec{x}\| \geq \|\text{proj}_V(\vec{x})\| = \|(\vec{x} \cdot \vec{u})\vec{u}\| \overset{\text{Step 3}}{=} |\vec{x} \cdot \vec{u}| = \left|\vec{x} \cdot \left(\frac{1}{\|\vec{y}\|}\vec{y}\right)\right| = \frac{1}{\|\vec{y}\|}|\vec{x} \cdot \vec{y}|.$$

To justify Step 3, note that $\|k\vec{v}\| = |k|\|\vec{v}\|$, for all vectors $\vec{v}$ in $\mathbb{R}^n$ and all scalars k. (See Exercise 25a.) We conclude that

$$\frac{|\vec{x} \cdot \vec{y}|}{\|\vec{y}\|} \leq \|\vec{x}\|.$$

Multiplying both sides of this inequality by $\|\vec{y}\|$, we find the following useful result:

Theorem 5.1.11 **Cauchy–Schwarz inequality**[3]

If $\vec{x}$ and $\vec{y}$ are vectors in $\mathbb{R}^n$, then

$$|\vec{x} \cdot \vec{y}| \leq \|\vec{x}\|\|\vec{y}\|.$$

This statement is an equality if (and only if) $\vec{x}$ and $\vec{y}$ are parallel. ∎

---

[3]Named after the French mathematician Augustin-Louis Cauchy (1789–1857) and the German mathematician Hermann Amandus Schwarz (1843–1921).

Consider two nonzero vectors $\vec{x}$ and $\vec{y}$ in $\mathbb{R}^3$. You may know an expression for the dot product $\vec{x} \cdot \vec{y}$ in terms of the angle θ between the two vectors (see Figure 12):

$$\vec{x} \cdot \vec{y} = \|\vec{x}\| \|\vec{y}\| \cos\theta.$$

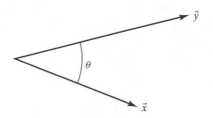

Figure 12

This formula allows us to find the angle between two nonzero vectors $\vec{x}$ and $\vec{y}$ in $\mathbb{R}^3$:

$$\cos\theta = \frac{\vec{x} \cdot \vec{y}}{\|\vec{x}\| \|\vec{y}\|} \quad \text{or} \quad \theta = \arccos \frac{\vec{x} \cdot \vec{y}}{\|\vec{x}\| \|\vec{y}\|}.$$

In $\mathbb{R}^n$, where we have no intuitive notion of an angle between two vectors, we can use this formula to *define* the angle:

Definition 5.1.12   Angle between two vectors

Consider two nonzero vectors $\vec{x}$ and $\vec{y}$ in $\mathbb{R}^n$. The angle θ between these vectors is defined as

$$\theta = \arccos \frac{\vec{x} \cdot \vec{y}}{\|\vec{x}\| \|\vec{y}\|}.$$

Note that θ is between 0 and π, by definition of the inverse cosine function.

We have to make sure that

$$\arccos \frac{\vec{x} \cdot \vec{y}}{\|\vec{x}\| \|\vec{y}\|}$$

is defined; that is,

$$\frac{\vec{x} \cdot \vec{y}}{\|\vec{x}\| \|\vec{y}\|}$$

is between -1 and 1, or, equivalently,

$$\left| \frac{\vec{x} \cdot \vec{y}}{\|\vec{x}\| \|\vec{y}\|} \right| = \frac{|\vec{x} \cdot \vec{y}|}{\|\vec{x}\| \|\vec{y}\|} \leq 1.$$

But this follows from the Cauchy–Schwarz inequality, $|\vec{x} \cdot \vec{y}| \leq \|\vec{x}\| \|\vec{y}\|$.

EXAMPLE 7   Find the angle between the vectors

$$\vec{x} = \begin{bmatrix} 1 \\ 0 \\ 0 \\ 0 \end{bmatrix} \quad \text{and} \quad \vec{y} = \begin{bmatrix} 1 \\ 1 \\ 1 \\ 1 \end{bmatrix}.$$

Solution

$$\theta = \arccos \frac{\vec{x} \cdot \vec{y}}{\|\vec{x}\| \|\vec{y}\|} = \arccos \frac{1}{1 \cdot 2} = \frac{\pi}{3} \qquad \blacksquare$$

Here is an application to statistics of some concepts introduced in this section.

Correlation (Optional)

Consider the meat consumption (in grams per day per person) and incidence of colon cancer (per 100,000 women per year) in various industrialized countries:

Country	Meat Consumption	Cancer Rate
Japan	26	7.5
Finland	101	9.8
Israel	124	16.4
Great Britain	205	23.3
United States	284	34
Mean	148	18.2

Can we detect a positive or negative *correlation*[4] between meat consumption and cancer rate? Does a country with high meat consumption have high cancer rates, and vice versa? By *high*, we mean "above average," of course. A quick look at the data shows such a *positive correlation:* In Great Britain and the United States, both meat consumption and cancer rate are above average. In the three other countries, they are below average. This positive correlation becomes more apparent when we list the preceding data as *deviations from the mean* (above or below the average):

Country	Meat Consumption (Deviation from Mean)	Cancer Rate (Deviation from Mean)
Japan	−122	−10.7
Finland	−47	−8.4
Israel	−24	−1.8
Great Britain	57	5.1
United States	136	15.8

Perhaps even more informative is a *scatter plot* of the deviation data. (See Figure 13.)

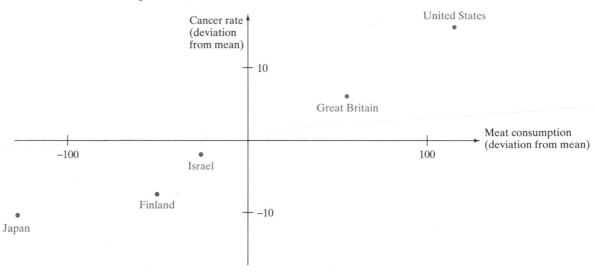

Figure 13

---

[4]We are using the term *correlation* in a colloquial, *qualitative* sense. Our goal is to *quantify* this term.

A positive correlation is indicated when most of the data points (in our case, all of them) are located in the first and third quadrant.

To process these data numerically, it is convenient to represent the deviation for both characteristics (meat consumption and cancer rate) as vectors in $\mathbb{R}^5$:

$$\vec{x} = \begin{bmatrix} -122 \\ -47 \\ -24 \\ 57 \\ 136 \end{bmatrix}, \quad \vec{y} = \begin{bmatrix} -10.7 \\ -8.4 \\ -1.8 \\ 5.1 \\ 15.8 \end{bmatrix}$$

We will call these two vectors the *deviation vectors* of the two characteristics.

In the case of a positive correlation, most of the corresponding entries x_i, y_i of the deviation vectors have the same sign (both positive or both negative). In our example, this is the case for all entries. This means that the product $x_i y_i$ will be positive most of the time; hence, the sum of all these products will be positive. But this sum is simply the dot product of the two deviation vectors.

Still using the term *correlation* in a colloquial sense, we conclude the following:

> Consider two characteristics of a population, with deviation vectors $\vec{x}$ and $\vec{y}$. There is a *positive correlation* between the two characteristics if (and only if) $\vec{x} \cdot \vec{y} > 0$.

A positive correlation between the characteristics means that the angle θ between the deviation vectors is less than $90°$. (See Figure 14.)

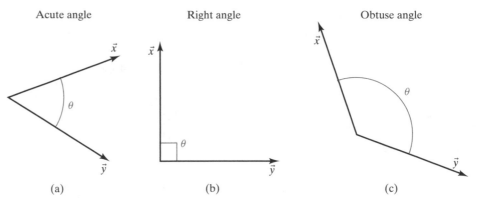

Figure 14 (a) Positive correlation: $\vec{x} \cdot \vec{y} > 0$. (b) No correlation: $\vec{x} \cdot \vec{y} = 0$. (c) Negative correlation: $\vec{x} \cdot \vec{y} < 0$.

We can use the cosine of the angle θ between $\vec{x}$ and $\vec{y}$ as a quantitative measure for the correlation between the two characteristics.

Definition 5.1.13   Correlation coefficient

The *correlation coefficient r* between two characteristics of a population is the cosine of the angle θ between the deviation vectors $\vec{x}$ and $\vec{y}$ for the two characteristics:

$$r = \cos(\theta) = \frac{\vec{x} \cdot \vec{y}}{\|\vec{x}\| \|\vec{y}\|}$$

In the case of meat consumption and cancer, we find that

$$r \approx \frac{4182.9}{198.53 \cdot 21.539} \approx 0.9782.$$

The angle between the two deviation vectors is $\arccos(r) \approx 0.21$ (radians) $\approx 12°$.

Note that the length of the deviation vectors is irrelevant for the correlation: If we had measured the cancer rate per 1,000,000 women (instead of 100,000), the vector $\vec{y}$ would be 10 times longer, but the correlation would be the same.

The correlation coefficient r is always between -1 and 1; the cases when $r = 1$ (representing a perfect positive correlation) and $r = -1$ (perfect negative correlation) are of particular interest. (See Figure 15.) In both cases, the data points (x_i, y_i) will be on the straight line $y = mx$. (See Figure 16.)

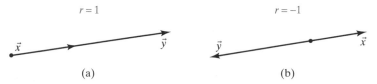

Figure 15 (a) $\vec{y} = m\vec{x}$, for positive m. (b) $\vec{y} = m\vec{x}$, for negative m.

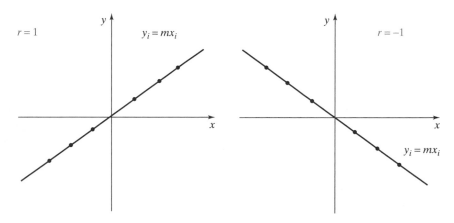

Figure 16

Note that even a strong positive correlation (an r close to 1) does not necessarily imply a causal relationship. Based only on the work we did above, we cannot conclude that high meat consumption causes colon cancer. Take a statistics course to learn more about these important issues!

EXERCISES 5.1

GOAL *Apply the basic concepts of geometry in $\mathbb{R}^n$: length, angles, orthogonality. Use the idea of an orthogonal projection onto a subspace. Find this projection if an orthonormal basis of the subspace is given.*

Find the length of each of the vectors $\vec{v}$ in Exercises 1 through 3.

1. $\vec{v} = \begin{bmatrix} 7 \\ 11 \end{bmatrix}$ 2. $\vec{v} = \begin{bmatrix} 2 \\ 3 \\ 4 \end{bmatrix}$ 3. $\vec{v} = \begin{bmatrix} 2 \\ 3 \\ 4 \\ 5 \end{bmatrix}$

Find the angle θ between each of the pairs of vectors $\vec{u}$ and $\vec{v}$ in Exercises 4 through 6.

4. $\vec{u} = \begin{bmatrix} 1 \\ 1 \end{bmatrix}$, $\vec{v} = \begin{bmatrix} 7 \\ 11 \end{bmatrix}$ **5.** $\vec{u} = \begin{bmatrix} 1 \\ 2 \\ 3 \end{bmatrix}$, $\vec{v} = \begin{bmatrix} 2 \\ 3 \\ 4 \end{bmatrix}$

6. $\vec{u} = \begin{bmatrix} 1 \\ -1 \\ 2 \\ -2 \end{bmatrix}$, $\vec{v} = \begin{bmatrix} 2 \\ 3 \\ 4 \\ 5 \end{bmatrix}$

For each pair of vectors $\vec{u}$, $\vec{v}$ listed in Exercises 7 through 9, determine whether the angle θ between $\vec{u}$ and $\vec{v}$ is acute, obtuse, or right.

7. $\vec{u} = \begin{bmatrix} 2 \\ -3 \end{bmatrix}$, $\vec{v} = \begin{bmatrix} 5 \\ 4 \end{bmatrix}$ **8.** $\vec{u} = \begin{bmatrix} 2 \\ 3 \\ 4 \end{bmatrix}$, $\vec{v} = \begin{bmatrix} 2 \\ -8 \\ 5 \end{bmatrix}$

9. $\vec{u} = \begin{bmatrix} 1 \\ -1 \\ 1 \\ -1 \end{bmatrix}$, $\vec{v} = \begin{bmatrix} 3 \\ 4 \\ 5 \\ 3 \end{bmatrix}$

10. For which value(s) of the constant k are the vectors

$$\vec{u} = \begin{bmatrix} 2 \\ 3 \\ 4 \end{bmatrix} \quad \text{and} \quad \vec{v} = \begin{bmatrix} 1 \\ k \\ 1 \end{bmatrix}$$

   perpendicular?

11. Consider the vectors

$$\vec{u} = \begin{bmatrix} 1 \\ 1 \\ \vdots \\ 1 \end{bmatrix} \quad \text{and} \quad \vec{v} = \begin{bmatrix} 1 \\ 0 \\ \vdots \\ 0 \end{bmatrix} \quad \text{in } \mathbb{R}^n.$$

   a. For $n = 2, 3, 4$, find the angle θ between $\vec{u}$ and $\vec{v}$. For $n = 2$ and 3, represent the vectors graphically.

   b. Find the limit of θ as n approaches infinity.

12. Give an algebraic proof for the *triangle inequality*

$$\|\vec{v} + \vec{w}\| \le \|\vec{v}\| + \|\vec{w}\|.$$

   Draw a sketch. [*Hint*: Expand $\|\vec{v} + \vec{w}\|^2 = (\vec{v} + \vec{w}) \cdot (\vec{v} + \vec{w})$. Then use the Cauchy–Schwarz inequality.]

13. *Leg traction.* The accompanying figure shows how a leg may be stretched by a pulley line for therapeutic purposes. We denote by $\vec{F}_1$ the vertical force of the weight. The string of the pulley line has the same tension everywhere; hence, the forces $\vec{F}_2$ and $\vec{F}_3$ have the same magnitude as $\vec{F}_1$. Assume that the magnitude of each force is 10 pounds. Find the angle θ so that the magnitude of the force exerted on the leg is 16 pounds. Round your answer to the nearest degree. (Adapted from E. Batschelet, *Introduction to Mathematics for Life Scientists*, Springer, 1979.)

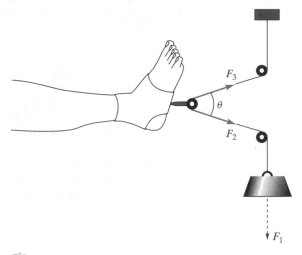

14. *Leonardo da Vinci and the resolution of forces.* Leonardo (1452–1519) asked himself how the weight of a body, supported by two strings of different length, is apportioned between the two strings.

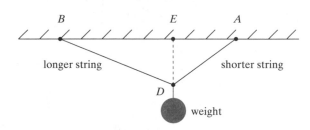

Three forces are acting at the point D: the tensions $\vec{F}_1$ and $\vec{F}_2$ in the strings and the weight $\vec{W}$. Leonardo believed that

$$\frac{\|\vec{F}_1\|}{\|\vec{F}_2\|} = \frac{EA}{EB}.$$

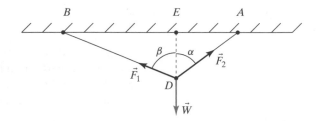

Was he right? (Source: *Les Manuscrits de Léonard de Vinci*, published by Ravaisson-Mollien, Paris, 1890.) (*Hint*: Resolve $\vec{F}_1$ into a horizontal and a vertical component; do the same for $\vec{F}_2$. Since the system is at rest, the equation $\vec{F}_1 + \vec{F}_2 + \vec{W} = \vec{0}$ holds. Express the ratios

$$\frac{\|\vec{F}_1\|}{\|\vec{F}_2\|} \quad \text{and} \quad \frac{EA}{EB}$$

in terms of α and β, using trigonometric functions, and compare the results.)

15. Consider the vector

$$\vec{v} = \begin{bmatrix} 1 \\ 2 \\ 3 \\ 4 \end{bmatrix} \quad \text{in } \mathbb{R}^4.$$

Find a basis of the subspace of $\mathbb{R}^4$ consisting of all vectors perpendicular to $\vec{v}$.

16. Consider the vectors

$$\vec{u}_1 = \begin{bmatrix} 1/2 \\ 1/2 \\ 1/2 \\ 1/2 \end{bmatrix}, \quad \vec{u}_2 = \begin{bmatrix} 1/2 \\ 1/2 \\ -1/2 \\ -1/2 \end{bmatrix}, \quad \vec{u}_3 = \begin{bmatrix} 1/2 \\ -1/2 \\ 1/2 \\ -1/2 \end{bmatrix}$$

in $\mathbb{R}^4$. Can you find a vector $\vec{u}_4$ in $\mathbb{R}^4$ such that the vectors $\vec{u}_1, \vec{u}_2, \vec{u}_3, \vec{u}_4$ are orthonormal? If so, how many such vectors are there?

17. Find a basis for $W^\perp$, where

$$W = \text{span} \left(\begin{bmatrix} 1 \\ 2 \\ 3 \\ 4 \end{bmatrix}, \begin{bmatrix} 5 \\ 6 \\ 7 \\ 8 \end{bmatrix} \right).$$

18. Here is an infinite-dimensional version of Euclidean space: In the space of all infinite sequences, consider the subspace ℓ_2 of square-summable sequences [i.e., those sequences $(x_1, x_2, \ldots)$ for which the infinite series $x_1^2 + x_2^2 + \cdots$ converges]. For $\vec{x}$ and $\vec{y}$ in ℓ_2, we define

$$\|\vec{x}\| = \sqrt{x_1^2 + x_2^2 + \cdots}, \quad \vec{x} \cdot \vec{y} = x_1 y_1 + x_2 y_2 + \cdots.$$

(Why does the series $x_1 y_1 + x_2 y_2 + \cdots$ converge?)

a. Check that $\vec{x} = (1, \frac{1}{2}, \frac{1}{4}, \frac{1}{8}, \frac{1}{16}, \ldots)$ is in ℓ_2, and find $\|\vec{x}\|$. Recall the formula for the geometric series: $1 + a + a^2 + a^3 + \cdots = 1/(1-a)$, if $-1 < a < 1$.

b. Find the angle between $(1, 0, 0, \ldots)$ and $(1, \frac{1}{2}, \frac{1}{4}, \frac{1}{8}, \ldots)$.

c. Give an example of a sequence $(x_1, x_2, \ldots)$ that converges to 0 (i.e., $\lim_{n \to \infty} x_n = 0$) but does not belong to ℓ_2.

d. Let L be the subspace of ℓ_2 spanned by $(1, \frac{1}{2}, \frac{1}{4}, \frac{1}{8}, \ldots)$. Find the orthogonal projection of $(1, 0, 0, \ldots)$ onto L.

The Hilbert space ℓ_2 was initially used mostly in physics: Werner Heisenberg's formulation of quantum mechanics is in terms of ℓ_2. Today, this space is used in many other applications, including economics. (See, for example, the work of the economist Andreu Mas-Colell of the University of Barcelona.)

19. For a line L in $\mathbb{R}^2$, draw a sketch to interpret the following transformations geometrically:

a. $T(\vec{x}) = \vec{x} - \text{proj}_L \vec{x}$
b. $T(\vec{x}) = \vec{x} - 2\text{proj}_L \vec{x}$
c. $T(\vec{x}) = 2\text{proj}_L \vec{x} - \vec{x}$

20. Refer to Figure 13 of this section. The *least-squares line* for these data is the line $y = mx$ that fits the data best, in that the sum of the squares of the vertical distances between the line and the data points is minimal. We want to minimize the sum

$$(mx_1 - y_1)^2 + (mx_2 - y_2)^2 + \cdots + (mx_5 - y_5)^2.$$

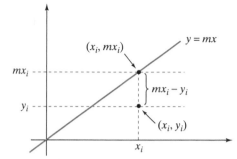

In vector notation, to minimize the sum means to find the scalar m such that

$$\|m\vec{x} - \vec{y}\|^2$$

is minimal. Arguing geometrically, explain how you can find m. Use the accompanying sketch, which is not drawn to scale.

Find m numerically, and explain the relationship between m and the correlation coefficient r. You may find the following information helpful:

$$\vec{x} \cdot \vec{y} = 4182.9, \quad \|\vec{x}\| \approx 198.53, \quad \|\vec{y}\| \approx 21.539.$$

To check whether your solution m is reasonable, draw the line $y = mx$ in Figure 13. (A more thorough discussion of least-squares approximations will follow in Section 5.4.)

21. Find scalars a, b, c, d, e, f, g such that the vectors

$$\begin{bmatrix} a \\ d \\ f \end{bmatrix}, \begin{bmatrix} b \\ 1 \\ g \end{bmatrix}, \begin{bmatrix} c \\ e \\ 1/2 \end{bmatrix}$$

are orthonormal.

22. Consider a basis $\vec{v}_1, \vec{v}_2, \ldots, \vec{v}_m$ of a subspace V of $\mathbb{R}^n$. Show that a vector $\vec{x}$ in $\mathbb{R}^n$ is orthogonal to V if (and only if) $\vec{x}$ is orthogonal to all the vectors $\vec{v}_1, \ldots, \vec{v}_m$.

23. Prove Theorem 5.1.8d. $(V^\perp)^\perp = V$ for any subspace V of $\mathbb{R}^n$. *Hint:* Show that $V \subseteq (V^\perp)^\perp$, by the definition of $V^\perp$; then show that $\dim(V) = \dim(V^\perp)^\perp$, by Theorem 5.1.8c.

24. Complete the proof of Theorem 5.1.4: Orthogonal projections are linear transformations.

25. a. Consider a vector $\vec{v}$ in $\mathbb{R}^n$, and a scalar k. Show that

$$\|k\vec{v}\| = |k|\|\vec{v}\|.$$

b. Show that if $\vec{v}$ is a nonzero vector in $\mathbb{R}^n$, then $\vec{u} = \dfrac{1}{\|\vec{v}\|}\vec{v}$ is a unit vector.

26. Find the orthogonal projection of $\begin{bmatrix} 49 \\ 49 \\ 49 \end{bmatrix}$ onto the subspace of $\mathbb{R}^3$ spanned by

$$\begin{bmatrix} 2 \\ 3 \\ 6 \end{bmatrix} \quad \text{and} \quad \begin{bmatrix} 3 \\ -6 \\ 2 \end{bmatrix}.$$

27. Find the orthogonal projection of $9\vec{e}_1$ onto the subspace of $\mathbb{R}^4$ spanned by

$$\begin{bmatrix} 2 \\ 2 \\ 1 \\ 0 \end{bmatrix} \quad \text{and} \quad \begin{bmatrix} -2 \\ 2 \\ 0 \\ 1 \end{bmatrix}.$$

28. Find the orthogonal projection of

$$\begin{bmatrix} 1 \\ 0 \\ 0 \\ 0 \end{bmatrix}$$

onto the subspace of $\mathbb{R}^4$ spanned by

$$\begin{bmatrix} 1 \\ 1 \\ 1 \\ 1 \end{bmatrix}, \quad \begin{bmatrix} 1 \\ 1 \\ -1 \\ -1 \end{bmatrix}, \quad \begin{bmatrix} 1 \\ -1 \\ -1 \\ 1 \end{bmatrix}.$$

29. Consider the orthonormal vectors $\vec{u}_1, \vec{u}_2, \vec{u}_3, \vec{u}_4, \vec{u}_5$ in $\mathbb{R}^{10}$. Find the length of the vector

$$\vec{x} = 7\vec{u}_1 - 3\vec{u}_2 + 2\vec{u}_3 + \vec{u}_4 - \vec{u}_5.$$

30. Consider a subspace V of $\mathbb{R}^n$ and a vector $\vec{x}$ in $\mathbb{R}^n$. Let $\vec{y} = \text{proj}_V\vec{x}$. What is the relationship between the following quantities?

$$\|\vec{y}\|^2 \quad \text{and} \quad \vec{y} \cdot \vec{x}$$

31. Consider the orthonormal vectors $\vec{u}_1, \vec{u}_2, \ldots, \vec{u}_m$ in $\mathbb{R}^n$, and an arbitrary vector $\vec{x}$ in $\mathbb{R}^n$. What is the relationship between the following two quantities?

$$p = (\vec{u}_1 \cdot \vec{x})^2 + (\vec{u}_2 \cdot \vec{x})^2 + \cdots + (\vec{u}_m \cdot \vec{x})^2 \quad \text{and} \quad \|\vec{x}\|^2$$

When are the two quantities equal?

32. Consider two vectors $\vec{v}_1$ and $\vec{v}_2$ in $\mathbb{R}^n$. Form the matrix

$$G = \begin{bmatrix} \vec{v}_1 \cdot \vec{v}_1 & \vec{v}_1 \cdot \vec{v}_2 \\ \vec{v}_2 \cdot \vec{v}_1 & \vec{v}_2 \cdot \vec{v}_2 \end{bmatrix}.$$

For which choices of $\vec{v}_1$ and $\vec{v}_2$ is the matrix G invertible?

33. Among all the vectors in $\mathbb{R}^n$ whose components add up to 1, find the vector of minimal length. In the case $n = 2$, explain your solution geometrically.

34. Among all the unit vectors in $\mathbb{R}^n$, find the one for which the sum of the components is maximal. In the case $n = 2$, explain your answer geometrically, in terms of the unit circle and the level curves of the function $x_1 + x_2$.

35. Among all the unit vectors $\vec{u} = \begin{bmatrix} x \\ y \\ z \end{bmatrix}$ in $\mathbb{R}^3$, find the one for which the sum $x + 2y + 3z$ is minimal.

36. There are three exams in your linear algebra class, and you theorize that your score in each exam (out of 100) will be numerically equal to the number of hours you study for that exam. The three exams count 20%, 30%, and 50%, respectively, toward the final grade. If your (modest) goal is to score 76% in the course, how many hours a, b, and c should you study for each of the three exams to minimize quantity $a^2 + b^2 + c^2$? This quadratic model reflects the fact that it may be four times as painful to study for 10 hours than for just 5 hours.

37. Consider a plane V in $\mathbb{R}^3$ with orthonormal basis $\vec{u}_1, \vec{u}_2$. Let $\vec{x}$ be a vector in $\mathbb{R}^3$. Find a formula for the reflection $R(\vec{x})$ of $\vec{x}$ about the plane V.

38. Consider three unit vectors $\vec{v}_1, \vec{v}_2$, and $\vec{v}_3$ in $\mathbb{R}^n$. We are told that $\vec{v}_1 \cdot \vec{v}_2 = \vec{v}_1 \cdot \vec{v}_3 = 1/2$. What are the possible values of $\vec{v}_2 \cdot \vec{v}_3$? What could the angle between the vectors $\vec{v}_2$ and $\vec{v}_3$ be? Give examples; draw sketches for the cases $n = 2$ and $n = 3$.

39. Can you find a line L in $\mathbb{R}^n$ and a vector $\vec{x}$ in $\mathbb{R}^n$ such that

$$\vec{x} \cdot \text{proj}_L\vec{x}$$

is negative? Explain, arguing algebraically.

In Exercises 40 through 46, consider vectors $\vec{v}_1, \vec{v}_2, \vec{v}_3$ in $\mathbb{R}^4$; we are told that $\vec{v}_i \cdot \vec{v}_j$ is the entry a_{ij} of matrix A.

$$A = \begin{bmatrix} 3 & 5 & 11 \\ 5 & 9 & 20 \\ 11 & 20 & 49 \end{bmatrix}$$

40. Find $\|\vec{v}_2\|$.

41. Find the angle enclosed by vectors $\vec{v}_2$ and $\vec{v}_3$.

42. Find $\|\vec{v}_1 + \vec{v}_2\|$.

43. Find $\text{proj}_{\vec{v}_2}(\vec{v}_1)$, expressed as a scalar multiple of $\vec{v}_2$.

44. Find a nonzero vector $\vec{v}$ in $\text{span}(\vec{v}_2, \vec{v}_3)$ such that $\vec{v}$ is orthogonal to $\vec{v}_3$. Express $\vec{v}$ as a linear combination of $\vec{v}_2$ and $\vec{v}_3$.

45. Find $\text{proj}_V(\vec{v}_1)$, where $V = \text{span}(\vec{v}_2, \vec{v}_3)$. Express your answer as a linear combination of $\vec{v}_2$ and $\vec{v}_3$.

46. Find $\text{proj}_V(\vec{v}_3)$, where $V = \text{span}(\vec{v}_1, \vec{v}_2)$. Express your answer as a linear combination of $\vec{v}_1$ and $\vec{v}_2$.

5.2 Gram–Schmidt Process and *QR* Factorization

One of the main themes of this chapter is the study of the orthogonal projection onto a subspace V of $\mathbb{R}^n$. In Theorem 5.1.5, we gave a formula for this projection,

$$\operatorname{proj}_V(\vec{x}) = (\vec{u}_1 \cdot \vec{x})\vec{u}_1 + \cdots + (\vec{u}_m \cdot \vec{x})\vec{u}_m,$$

where $\vec{u}_1, \ldots, \vec{u}_m$ is an orthonormal basis of V. Now we will show how to construct such an orthonormal basis. We will present an algorithm that allows us to convert any basis $\vec{v}_1, \ldots, \vec{v}_m$ of a subspace V of $\mathbb{R}^n$ into an orthonormal basis $\vec{u}_1, \ldots, \vec{u}_m$ of V.

Let us first think about low-dimensional cases. If V is a line with basis $\vec{v}_1$, we can find an orthonormal basis $\vec{u}_1$ simply by dividing $\vec{v}_1$ by its length:

$$\vec{u}_1 = \frac{1}{\|\vec{v}_1\|} \vec{v}_1.$$

When V is a plane with basis $\vec{v}_1, \vec{v}_2$, we first construct

$$\vec{u}_1 = \frac{1}{\|\vec{v}_1\|} \vec{v}_1$$

as before (see Figure 1).

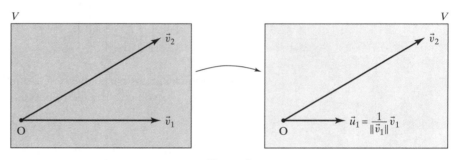

Figure 1

Now comes the crucial step: We have to find a vector in V orthogonal to $\vec{u}_1$. (Initially, we will not insist that this vector be a unit vector.) Let's resolve the vector $\vec{v}_2$ into its components parallel and perpendicular to the line L spanned by $\vec{u}_1$:

$$\vec{v}_2 = \vec{v}_2^{\parallel} + \vec{v}_2^{\perp}.$$

See Figure 2. Then the vector

$$\vec{v}_2^{\perp} = \vec{v}_2 - \vec{v}_2^{\parallel} = \vec{v}_2 - \operatorname{proj}_L(\vec{v}_2) = \vec{v}_2 - (\vec{u}_1 \cdot \vec{v}_2)\vec{u}_1$$

is orthogonal to $\vec{u}_1$.

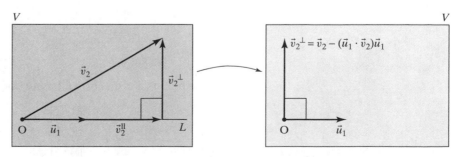

Figure 2

The last step is straightforward: We divide the vector $\vec{v_2}^{\perp}$ by its length to get the second vector $\vec{u_2}$ of an orthonormal basis. (See Figure 3.)

$$\vec{u_2} = \frac{1}{\|\vec{v_2}^{\perp}\|} \vec{v_2}^{\perp}$$

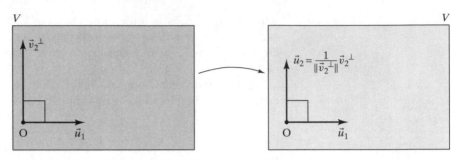

Figure 3

EXAMPLE 1

a. Find an orthonormal basis $\vec{u_1}, \vec{u_2}$ of the subspace

$$V = \text{span}\left(\begin{bmatrix} 1 \\ 1 \\ 1 \\ 1 \end{bmatrix}, \begin{bmatrix} 1 \\ 9 \\ 9 \\ 1 \end{bmatrix} \right) \quad \text{of } \mathbb{R}^4, \text{ with basis } \vec{v_1} = \begin{bmatrix} 1 \\ 1 \\ 1 \\ 1 \end{bmatrix}, \vec{v_2} = \begin{bmatrix} 1 \\ 9 \\ 9 \\ 1 \end{bmatrix}.$$

b. Find the change of basis matrix R from the basis $\mathfrak{B} = (\vec{v_1}, \vec{v_2})$ to the basis $\mathfrak{A} = (\vec{u_1}, \vec{u_2})$ you constructed in part a.

Solution

a. Following the three steps illustrated in Figures 1, 2, and 3, we will compute first $\vec{u_1}$, then $\vec{v_2}^{\perp}$, and finally $\vec{u_2}$.

$$\vec{u_1} = \frac{1}{\|\vec{v_1}\|} \vec{v_1} = \frac{1}{2} \begin{bmatrix} 1 \\ 1 \\ 1 \\ 1 \end{bmatrix} = \begin{bmatrix} 1/2 \\ 1/2 \\ 1/2 \\ 1/2 \end{bmatrix},$$

$$\vec{v_2}^{\perp} = \vec{v_2} - \vec{v_2}^{\|} = \vec{v_2} - (\vec{u_1} \cdot \vec{v_2})\vec{u_1} = \begin{bmatrix} 1 \\ 9 \\ 9 \\ 1 \end{bmatrix} - 10 \begin{bmatrix} 1/2 \\ 1/2 \\ 1/2 \\ 1/2 \end{bmatrix} = \begin{bmatrix} -4 \\ 4 \\ 4 \\ -4 \end{bmatrix},$$

$$\vec{u_2} = \frac{1}{\|\vec{v_2}^{\perp}\|} \vec{v_2}^{\perp} = \frac{1}{8} \begin{bmatrix} -4 \\ 4 \\ 4 \\ -4 \end{bmatrix} = \begin{bmatrix} -1/2 \\ 1/2 \\ 1/2 \\ -1/2 \end{bmatrix}.$$

We have found an orthonormal basis $\mathfrak{A}$ of V:

$$\vec{u_1} = \begin{bmatrix} 1/2 \\ 1/2 \\ 1/2 \\ 1/2 \end{bmatrix}, \quad \vec{u_2} = \begin{bmatrix} -1/2 \\ 1/2 \\ 1/2 \\ -1/2 \end{bmatrix}.$$

b. Recall that the columns of the change of basis matrix R from $\mathfrak{B}$ to $\mathfrak{A}$ are the coordinate vectors of $\vec{v}_1$ and $\vec{v}_2$ with respect to basis $\mathfrak{A}$ (see Definition 4.3.3):

$$R = \left[\ [\vec{v}_1]_{\mathfrak{A}} \quad [\vec{v}_2]_{\mathfrak{A}} \ \right].$$

A straightforward computation shows that

$$[\vec{v}_1]_{\mathfrak{A}} = \begin{bmatrix} 2 \\ 0 \end{bmatrix} \quad \text{and} \quad [\vec{v}_2]_{\mathfrak{A}} = \begin{bmatrix} 10 \\ 8 \end{bmatrix}, \quad \text{so that} \quad R = \begin{bmatrix} 2 & 10 \\ 0 & 8 \end{bmatrix}.$$

(Later in this section we will develop more efficient methods for finding the entries of R.)

To express the relationship between the bases $\mathfrak{A}$ and $\mathfrak{B}$ in matrix form, we can use Theorem 4.3.4 and write

$$\begin{bmatrix} \vec{v}_1 & \vec{v}_2 \end{bmatrix} = \begin{bmatrix} \vec{u}_1 & \vec{u}_2 \end{bmatrix} R, \quad \text{or,} \quad \underbrace{\begin{bmatrix} 1 & 1 \\ 1 & 9 \\ 1 & 9 \\ 1 & 1 \end{bmatrix}}_{M} = \underbrace{\begin{bmatrix} 1/2 & -1/2 \\ 1/2 & 1/2 \\ 1/2 & 1/2 \\ 1/2 & -1/2 \end{bmatrix}}_{Q} \underbrace{\begin{bmatrix} 2 & 10 \\ 0 & 8 \end{bmatrix}}_{R}.$$

In this context, it is customary to denote the matrices on the right-hand side by Q and R. Note that we have written the 4×2 matrix M with columns $\vec{v}_1$ and $\vec{v}_2$ as the product of the 4×2 matrix Q with orthonormal columns and the upper triangular 2×2 matrix R with positive entries on the diagonal. This is referred to as the *QR factorization* of matrix M. Matrix Q stores the orthonormal basis $\vec{u}_1, \vec{u}_2$ we constructed, and matrix R gives the relationship between the "old" basis $\vec{v}_1, \vec{v}_2$, and the "new" basis $\vec{u}_1, \vec{u}_2$ of V. ∎

Now that we know how to find an orthonormal basis of a plane, how would we proceed in the case of a three-dimensional subspace V of $\mathbb{R}^n$ with basis $\vec{v}_1, \vec{v}_2, \vec{v}_3$? We can first find an orthonormal basis $\vec{u}_1, \vec{u}_2$ of the plane $E = \text{span}(\vec{v}_1, \vec{v}_2)$, as illustrated in Example 1. Next we resolve the vector $\vec{v}_3$ into its components parallel and perpendicular to the plane E:

$$\vec{v}_3 = \vec{v}_3^{\parallel} + \vec{v}_3^{\perp},$$

so that

$$\vec{v}_3^{\perp} = \vec{v}_3 - \vec{v}_3^{\parallel} = \vec{v}_3 - \text{proj}_E(\vec{v}_3) = \vec{v}_3 - (\vec{u}_1 \cdot \vec{v}_3)\vec{u}_1 - (\vec{u}_2 \cdot \vec{v}_3)\vec{u}_2.$$

Finally, we let

$$\vec{u}_3 = \frac{1}{\|\vec{v}_3^{\perp}\|}\vec{v}_3^{\perp}.$$

See Figure 4.

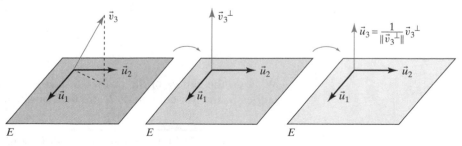

Figure 4

Generalizing this method, we can construct an orthonormal basis of any subspace V of $\mathbb{R}^n$. Unfortunately, the terminology gets a bit heavy in the general case. Conceptually, the method is pretty straightforward, however: We keep computing perpendicular components of vectors, using the formula for projection, and we keep dividing vectors by their length to generate unit vectors.

Theorem 5.2.1   **The Gram–Schmidt process[5]**

Consider a basis $\vec{v}_1, \ldots, \vec{v}_m$ of a subspace V of $\mathbb{R}^n$. For $j = 2, \ldots, m$, we resolve the vector $\vec{v}_j$ into its components parallel and perpendicular to the span of the preceding vectors, $\vec{v}_1, \ldots, \vec{v}_{j-1}$:

$$\vec{v}_j = \vec{v}_j^{\parallel} + \vec{v}_j^{\perp}, \quad \text{with respect to span}(\vec{v}_1, \ldots, \vec{v}_{j-1}).$$

Then

$$\vec{u}_1 = \frac{1}{\|\vec{v}_1\|}\vec{v}_1, \quad \vec{u}_2 = \frac{1}{\|\vec{v}_2^{\perp}\|}\vec{v}_2^{\perp}, \ldots, \quad \vec{u}_j = \frac{1}{\|\vec{v}_j^{\perp}\|}\vec{v}_j^{\perp}, \ldots, \quad \vec{u}_m = \frac{1}{\|\vec{v}_m^{\perp}\|}\vec{v}_m^{\perp}$$

is an orthonormal basis of V. By Theorem 5.1.7 we have

$$\vec{v}_j^{\perp} = \vec{v}_j - \vec{v}_j^{\parallel} = \vec{v}_j - (\vec{u}_1 \cdot \vec{v}_j)\vec{u}_1 - \cdots - (\vec{u}_{j-1} \cdot \vec{v}_j)\vec{u}_{j-1}. \qquad \blacksquare$$

If you are puzzled by these formulas, go back to the cases where V is a two- or three-dimensional space; take another good look at Figures 1 through 4.

The QR Factorization

The Gram–Schmidt process represents a change of basis from the "old" basis $\mathfrak{B} = (\vec{v}_1, \ldots, \vec{v}_m)$ to a "new", orthonormal basis $\mathfrak{A} = (\vec{u}_1, \ldots, \vec{u}_m)$ of V; it is most succinctly described in terms of the change of basis matrix R from $\mathfrak{B}$ to $\mathfrak{A}$, as discussed in Example 1. Using Theorem 4.3.4, we can write

$$\underbrace{\begin{bmatrix} \vec{v}_1 & \cdots & \vec{v}_m \end{bmatrix}}_{M} = \underbrace{\begin{bmatrix} \vec{u}_1 & \cdots & \vec{u}_m \end{bmatrix}}_{Q} R.$$

Again, it is customary to denote the matrices on the right-hand side by Q and R; the preceding equation is called the QR *factorization* of M.

We can represent the relationship among the matrices M, Q, and R in a commutative diagram, where $\vec{x}$ is a vector in V.

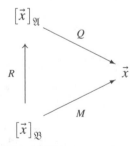

What do the entries of the change of basis matrix R look like? We know that the entries in the jth column of R are the coordinates of $\vec{v}_j$ with respect to the basis

---

[5]Named after the Danish actuary Jörgen Gram (1850–1916) and the German mathematician Erhardt Schmidt (1876–1959).

$\mathfrak{A} = (\vec{u}_1, \ldots, \vec{u}_m)$. Using the equations in Theorem 5.2.1, we can write

$$\vec{v}_j = \vec{v}_j^{\parallel} + \vec{v}_j^{\perp}$$

$$= \underbrace{\overbrace{(\vec{u}_1 \cdot \vec{v}_j)}^{r_{1j}} \vec{u}_1 + \cdots + \overbrace{(\vec{u}_i \cdot \vec{v}_j)}^{r_{ij}} \vec{u}_i + \cdots + \overbrace{(\vec{u}_{j-1} \cdot \vec{v}_j)}^{r_{j-1,j}} \vec{u}_{j-1}}_{\vec{v}_j^{\parallel}} + \underbrace{\overbrace{\left\| \vec{v}_j^{\perp} \right\|}^{r_{jj}} \vec{u}_j}_{\vec{v}_j^{\perp}}.$$

It follows that $r_{ij} = \vec{u}_i \cdot \vec{v}_j$ if $i < j$; $r_{jj} = \left\| \vec{v}_j^{\perp} \right\|$; and $r_{ij} = 0$ if $i > j$. The last equation implies that R is upper triangular. (The first diagonal entry is $r_{11} = \|\vec{v}_1\|$, since $\vec{v}_1 = \|\vec{v}_1\| \vec{u}_1$.)

Theorem 5.2.2   QR factorization

Consider an $n \times m$ matrix M with linearly independent columns $\vec{v}_1, \ldots, \vec{v}_m$. Then there exists an $n \times m$ matrix Q whose columns $\vec{u}_1, \ldots, \vec{u}_m$ are orthonormal and an upper triangular matrix R with positive diagonal entries such that

$$M = QR.$$

This representation is unique. Furthermore, $r_{11} = \|\vec{v}_1\|$, $r_{jj} = \left\| \vec{v}_j^{\perp} \right\|$ (for $j = 2, \ldots, m$), and $r_{ij} = \vec{u}_i \cdot \vec{v}_j$ (for $i < j$).   ■

Take another look at Example 1, where $L = V_1 = \text{span}(\vec{v}_1)$.

The verification of the uniqueness of the QR factorization is left as Exercise 5.3.51. To find the QR factorization of a matrix M, we perform the Gram–Schmidt process on the columns of M, constructing R and Q column by column. No extra computations are required: All the information necessary to build R and Q is provided by the Gram–Schmidt process. QR factorization is an effective way to organize and record the work performed in the Gram–Schmidt process; it is useful for many computational and theoretical purposes.

EXAMPLE 2   Find the QR factorization of the matrix $M = \begin{bmatrix} 2 & 2 \\ 1 & 7 \\ -2 & -8 \end{bmatrix}$.

Solution
Here

$$\vec{v}_1 = \begin{bmatrix} 2 \\ 1 \\ -2 \end{bmatrix}, \quad \text{and} \quad \vec{v}_2 = \begin{bmatrix} 2 \\ 7 \\ -8 \end{bmatrix}.$$

As in Example 1, the QR factorization of M will have the form

$$M = \begin{bmatrix} \vec{v}_1 & \vec{v}_2 \end{bmatrix} = \underbrace{\begin{bmatrix} \vec{u}_1 & \vec{u}_2 \end{bmatrix}}_{Q} \underbrace{\begin{bmatrix} \|\vec{v}_1\| & \vec{u}_1 \cdot \vec{v}_2 \\ 0 & \left\| \vec{v}_2^{\perp} \right\| \end{bmatrix}}_{R}.$$

We will compute the entries of R and the columns of Q step by step.

$$r_{11} = \|\vec{v}_1\| = 3, \quad \vec{u}_1 = \frac{1}{r_{11}}\vec{v}_1 = \frac{1}{3}\begin{bmatrix} 2 \\ 1 \\ -2 \end{bmatrix},$$

$$r_{12} = \vec{u}_1 \cdot \vec{v}_2 = \frac{1}{3}\begin{bmatrix} 2 \\ 1 \\ -2 \end{bmatrix} \cdot \begin{bmatrix} 2 \\ 7 \\ -8 \end{bmatrix} = 9, \quad \vec{v}_2^{\perp} = \vec{v}_2 - r_{12}\vec{u}_1 = \begin{bmatrix} -4 \\ 4 \\ -2 \end{bmatrix},$$

$$r_{22} = \left\|\vec{v}_2^{\perp}\right\| = \sqrt{36} = 6, \quad \vec{u}_2 = \frac{1}{r_{22}}\vec{v}_2^{\perp} = \frac{1}{3}\begin{bmatrix} -2 \\ 2 \\ -1 \end{bmatrix}$$

Now

$$\begin{bmatrix} 2 & 2 \\ 1 & 7 \\ -2 & -8 \end{bmatrix} = M = QR = \begin{bmatrix} \vec{u}_1 & \vec{u}_2 \end{bmatrix}\begin{bmatrix} r_{11} & r_{12} \\ 0 & r_{22} \end{bmatrix}$$

$$= \underbrace{\left(\frac{1}{3}\begin{bmatrix} 2 & -2 \\ 1 & 2 \\ -2 & -1 \end{bmatrix}\right)}_{Q} \underbrace{\left(\begin{bmatrix} 3 & 9 \\ 0 & 6 \end{bmatrix}\right)}_{R}.$$

Draw pictures analogous to Figures 1 through 3 to illustrate these computations! ■

Let us outline the algorithm we used in Example 2.

Theorem 5.2.3 **QR factorization**

Consider an $n \times m$ matrix M with linearly independent columns $\vec{v}_1, \ldots, \vec{v}_m$. Then the columns $\vec{u}_1, \ldots, \vec{u}_m$ of Q and the entries r_{ij} of R can be computed in the following order:

First column of R, first column of Q;

second column of R, second column of Q;

third column of R, third column of Q;

and so on.

More specifically,

$$r_{11} = \|\vec{v}_1\|, \quad \vec{u}_1 = \frac{1}{r_{11}}\vec{v}_1$$

$$r_{12} = \vec{u}_1 \cdot \vec{v}_2, \quad \vec{v}_2^{\perp} = \vec{v}_2 - r_{12}\vec{u}_1, \quad r_{22} = \left\|\vec{v}_2^{\perp}\right\|, \quad \vec{u}_2 = \frac{1}{r_{22}}\vec{v}_2^{\perp}$$

$$r_{13} = \vec{u}_1 \cdot \vec{v}_3, \quad r_{23} = \vec{u}_2 \cdot \vec{v}_3, \quad \vec{v}_3^{\perp} = \begin{cases} \vec{v}_3 - r_{13}\vec{u}_1 \\ -r_{23}\vec{u}_2 \end{cases}$$

$$r_{33} = \left\|\vec{v}_3^{\perp}\right\|, \quad \vec{u}_3 = \frac{1}{r_{33}}\vec{v}_3^{\perp}$$

and so on. ■

For matrices M with more than three columns, the computation of the QR factorization is tedious, and may best be left to a machine (unless M is of a particularly simple form).

EXERCISES 5.2

GOAL *Perform the Gram–Schmidt process, and thus find the QR factorization of a matrix.*

Using paper and pencil, perform the Gram–Schmidt process on the sequences of vectors given in Exercises 1 through 14.

1. $\begin{bmatrix} 2 \\ 1 \\ -2 \end{bmatrix}$

2. $\begin{bmatrix} 6 \\ 3 \\ 2 \end{bmatrix}, \begin{bmatrix} 2 \\ -6 \\ 3 \end{bmatrix}$

3. $\begin{bmatrix} 4 \\ 0 \\ 3 \end{bmatrix}, \begin{bmatrix} 25 \\ 0 \\ -25 \end{bmatrix}$

4. $\begin{bmatrix} 4 \\ 0 \\ 3 \end{bmatrix}, \begin{bmatrix} 25 \\ 0 \\ -25 \end{bmatrix}, \begin{bmatrix} 0 \\ -2 \\ 0 \end{bmatrix}$

5. $\begin{bmatrix} 2 \\ 2 \\ 1 \end{bmatrix}, \begin{bmatrix} 1 \\ 1 \\ 5 \end{bmatrix}$

6. $\begin{bmatrix} 2 \\ 0 \\ 0 \end{bmatrix}, \begin{bmatrix} 3 \\ 4 \\ 0 \end{bmatrix}, \begin{bmatrix} 5 \\ 6 \\ 7 \end{bmatrix}$

7. $\begin{bmatrix} 2 \\ 2 \\ 1 \end{bmatrix}, \begin{bmatrix} -2 \\ 1 \\ 2 \end{bmatrix}, \begin{bmatrix} 18 \\ 0 \\ 0 \end{bmatrix}$

8. $\begin{bmatrix} 5 \\ 4 \\ 2 \\ 2 \end{bmatrix}, \begin{bmatrix} 3 \\ 6 \\ 7 \\ -2 \end{bmatrix}$

9. $\begin{bmatrix} 1 \\ 1 \\ 1 \\ 1 \end{bmatrix}, \begin{bmatrix} 1 \\ 9 \\ -5 \\ 3 \end{bmatrix}$

10. $\begin{bmatrix} 1 \\ 1 \\ 1 \\ 1 \end{bmatrix}, \begin{bmatrix} 6 \\ 4 \\ 6 \\ 4 \end{bmatrix}$

11. $\begin{bmatrix} 4 \\ 0 \\ 0 \\ 3 \end{bmatrix}, \begin{bmatrix} 5 \\ 2 \\ 14 \\ 10 \end{bmatrix}$

12. $\begin{bmatrix} 2 \\ 3 \\ 0 \\ 6 \end{bmatrix}, \begin{bmatrix} 4 \\ 4 \\ 2 \\ 13 \end{bmatrix}$

13. $\begin{bmatrix} 1 \\ 1 \\ 1 \\ 1 \end{bmatrix}, \begin{bmatrix} 1 \\ 0 \\ 0 \\ 1 \end{bmatrix}, \begin{bmatrix} 0 \\ 2 \\ 1 \\ -1 \end{bmatrix}$

14. $\begin{bmatrix} 1 \\ 7 \\ 1 \\ 7 \end{bmatrix}, \begin{bmatrix} 0 \\ 7 \\ 2 \\ 7 \end{bmatrix}, \begin{bmatrix} 1 \\ 8 \\ 1 \\ 6 \end{bmatrix}$

Using paper and pencil, find the QR factorizations of the matrices in Exercises 15 through 28. (Compare with Exercises 1 through 14.)

15. $\begin{bmatrix} 2 \\ 1 \\ -2 \end{bmatrix}$

16. $\begin{bmatrix} 6 & 2 \\ 3 & -6 \\ 2 & 3 \end{bmatrix}$

17. $\begin{bmatrix} 4 & 25 \\ 0 & 0 \\ 3 & -25 \end{bmatrix}$

18. $\begin{bmatrix} 4 & 25 & 0 \\ 0 & 0 & -2 \\ 3 & -25 & 0 \end{bmatrix}$

19. $\begin{bmatrix} 2 & 1 \\ 2 & 1 \\ 1 & 5 \end{bmatrix}$

20. $\begin{bmatrix} 2 & 3 & 5 \\ 0 & 4 & 6 \\ 0 & 0 & 7 \end{bmatrix}$

21. $\begin{bmatrix} 2 & -2 & 18 \\ 2 & 1 & 0 \\ 1 & 2 & 0 \end{bmatrix}$

22. $\begin{bmatrix} 5 & 3 \\ 4 & 6 \\ 2 & 7 \\ 2 & -2 \end{bmatrix}$

23. $\begin{bmatrix} 1 & 1 \\ 1 & 9 \\ 1 & -5 \\ 1 & 3 \end{bmatrix}$

24. $\begin{bmatrix} 1 & 6 \\ 1 & 4 \\ 1 & 6 \\ 1 & 4 \end{bmatrix}$

25. $\begin{bmatrix} 4 & 5 \\ 0 & 2 \\ 0 & 14 \\ 3 & 10 \end{bmatrix}$

26. $\begin{bmatrix} 2 & 4 \\ 3 & 4 \\ 0 & 2 \\ 6 & 13 \end{bmatrix}$

27. $\begin{bmatrix} 1 & 1 & 0 \\ 1 & 0 & 2 \\ 1 & 0 & 1 \\ 1 & 1 & -1 \end{bmatrix}$

28. $\begin{bmatrix} 1 & 0 & 1 \\ 7 & 7 & 8 \\ 1 & 2 & 1 \\ 7 & 7 & 6 \end{bmatrix}$

29. Perform the Gram–Schmidt process on the following basis of $\mathbb{R}^2$:
$$\vec{v}_1 = \begin{bmatrix} -3 \\ 4 \end{bmatrix}, \quad \vec{v}_2 = \begin{bmatrix} 1 \\ 7 \end{bmatrix}.$$
Illustrate your work with sketches, as in Figures 1 through 3 of this section.

30. Consider two linearly independent vectors $\vec{v}_1 = \begin{bmatrix} a \\ b \end{bmatrix}$ and $\vec{v}_2 = \begin{bmatrix} c \\ d \end{bmatrix}$ in $\mathbb{R}^2$. Draw sketches (as in Figures 1 through 3 of this section) to illustrate the Gram–Schmidt process for $\vec{v}_1, \vec{v}_2$. You need not perform the process algebraically.

31. Perform the Gram–Schmidt process on the following basis of $\mathbb{R}^3$:
$$\vec{v}_1 = \begin{bmatrix} a \\ 0 \\ 0 \end{bmatrix}, \quad \vec{v}_2 = \begin{bmatrix} b \\ c \\ 0 \end{bmatrix}, \quad \vec{v}_3 = \begin{bmatrix} d \\ e \\ f \end{bmatrix}.$$
Here, a, c, and f are positive constants, and the other constants are arbitrary. Illustrate your work with a sketch, as in Figure 4 of this section.

32. Find an orthonormal basis of the plane
$$x_1 + x_2 + x_3 = 0.$$

33. Find an orthonormal basis of the kernel of the matrix
$$A = \begin{bmatrix} 1 & 1 & 1 & 1 \\ 1 & -1 & -1 & 1 \end{bmatrix}.$$

34. Find an orthonormal basis of the kernel of the matrix
$$A = \begin{bmatrix} 1 & 1 & 1 & 1 \\ 1 & 2 & 3 & 4 \end{bmatrix}.$$

35. Find an orthonormal basis of the image of the matrix

$$A = \begin{bmatrix} 1 & 2 & 1 \\ 2 & 1 & 1 \\ 2 & -2 & 0 \end{bmatrix}.$$

36. Consider the matrix

$$M = \frac{1}{2} \begin{bmatrix} 1 & 1 & 1 \\ 1 & -1 & -1 \\ 1 & -1 & 1 \\ 1 & 1 & -1 \end{bmatrix} \begin{bmatrix} 2 & 3 & 5 \\ 0 & -4 & 6 \\ 0 & 0 & 7 \end{bmatrix}.$$

Find the QR factorization of M.

37. Consider the matrix

$$M = \frac{1}{2} \begin{bmatrix} 1 & 1 & 1 & 1 \\ 1 & -1 & -1 & 1 \\ 1 & -1 & 1 & -1 \\ 1 & 1 & -1 & -1 \end{bmatrix} \begin{bmatrix} 3 & 4 \\ 0 & 5 \\ 0 & 0 \\ 0 & 0 \end{bmatrix}.$$

Find the QR factorization of M.

38. Find the QR factorization of

$$A = \begin{bmatrix} 0 & -3 & 0 \\ 0 & 0 & 0 \\ 2 & 0 & 0 \\ 0 & 0 & 4 \end{bmatrix}.$$

39. Find an orthonormal basis $\vec{u}_1, \vec{u}_2, \vec{u}_3$ of $\mathbb{R}^3$ such that

$$\text{span}(\vec{u}_1) = \text{span} \left(\begin{bmatrix} 1 \\ 2 \\ 3 \end{bmatrix} \right)$$

and

$$\text{span}(\vec{u}_1, \vec{u}_2) = \text{span} \left(\begin{bmatrix} 1 \\ 2 \\ 3 \end{bmatrix}, \begin{bmatrix} 1 \\ 1 \\ -1 \end{bmatrix} \right).$$

40. Consider an invertible $n \times n$ matrix A whose columns are orthogonal, but not necessarily orthonormal. What does the QR factorization of A look like?

41. Consider an invertible upper triangular $n \times n$ matrix A. What does the QR factorization of A look like?

42. The two column vectors $\vec{v}_1$ and $\vec{v}_2$ of a 2×2 matrix A are shown in the accompanying figure. Let $A = QR$ be the QR factorization of A. Represent the diagonal entries r_{11} and r_{22} of R as lengths in the figure. Interpret the product $r_{11}r_{22}$ as an area.

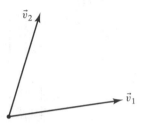

43. Consider a block matrix

$$A = \begin{bmatrix} A_1 & A_2 \end{bmatrix}$$

with linearly independent columns. (A_1 is an $n \times m_1$ matrix, and A_2 is $n \times m_2$.) Suppose you know the QR factorization of A. Explain how this allows you to find the QR factorization of A_1.

44. Consider an $n \times m$ matrix A with rank$(A) < m$. Is it always possible to write

$$A = QR,$$

where Q is an $n \times m$ matrix with orthonormal columns and R is upper triangular? Explain.

45. Consider an $n \times m$ matrix A with rank$(A) = m$. Is it always possible to write A as

$$A = QL,$$

where Q is an $n \times m$ matrix with orthonormal columns and L is a lower triangular $m \times m$ matrix with positive diagonal entries? Explain.

5.3 Orthogonal Transformations and Orthogonal Matrices

In geometry, we are particularly interested in those linear transformations that preserve the length of vectors.

Definition 5.3.1 Orthogonal transformations and orthogonal matrices

A linear transformation T from $\mathbb{R}^n$ to $\mathbb{R}^n$ is called *orthogonal* if it preserves the length of vectors:

$$\|T(\vec{x})\| = \|\vec{x}\|, \quad \text{for all } \vec{x} \text{ in } \mathbb{R}^n.$$

If $T(\vec{x}) = A\vec{x}$ is an orthogonal transformation, we say that A is an *orthogonal matrix*.[6]

EXAMPLE 1 The rotation

$$T(\vec{x}) = \begin{bmatrix} \cos\theta & -\sin\theta \\ \sin\theta & \cos\theta \end{bmatrix} \vec{x}$$

is an orthogonal transformation from $\mathbb{R}^2$ to $\mathbb{R}^2$, and

$$A = \begin{bmatrix} \cos\theta & -\sin\theta \\ \sin\theta & \cos\theta \end{bmatrix}$$

is an orthogonal matrix, for all angles θ. ■

EXAMPLE 2 Consider a subspace V of $\mathbb{R}^n$. For a vector $\vec{x}$ in $\mathbb{R}^n$, the vector $\text{ref}_V(\vec{x}) = \vec{x}^{\parallel} - \vec{x}^{\perp}$ is called the reflection of $\vec{x}$ about V. (Compare this with Definition 2.2.2; see Figure 1). Show that reflections are orthogonal transformations.

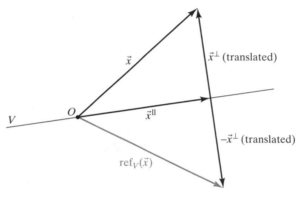

Figure 1

Solution

By the Pythagorean theorem, we have

$$\|\text{ref}_V(\vec{x})\|^2 = \left\|\vec{x}^{\parallel}\right\|^2 + \left\|-\vec{x}^{\perp}\right\|^2 = \left\|\vec{x}^{\parallel}\right\|^2 + \left\|\vec{x}^{\perp}\right\|^2 = \|\vec{x}\|^2.$$

■

As the name suggests, orthogonal transformations preserve right angles. In fact, orthogonal transformations preserve all angles. (See Exercise 29.)

Theorem 5.3.2 Orthogonal transformations preserve orthogonality

Consider an orthogonal transformation T from $\mathbb{R}^n$ to $\mathbb{R}^n$. If the vectors $\vec{v}$ and $\vec{w}$ in $\mathbb{R}^n$ are orthogonal, then so are $T(\vec{v})$ and $T(\vec{w})$.

Proof By the theorem of Pythagoras, we have to show that

$$\|T(\vec{v}) + T(\vec{w})\|^2 = \|T(\vec{v})\|^2 + \|T(\vec{w})\|^2.$$

---

[6]A list of alternative characterizations of an orthogonal matrix will be presented in Summary 5.3.8.

Let's see:

$$\|T(\vec{v}) + T(\vec{w})\|^2 = \|T(\vec{v} + \vec{w})\|^2 \qquad (T \text{ is linear})$$
$$= \|\vec{v} + \vec{w}\|^2 \qquad (T \text{ is orthogonal})$$
$$= \|\vec{v}\|^2 + \|\vec{w}\|^2 \qquad (\vec{v} \text{ and } \vec{w} \text{ are orthogonal})$$
$$= \|T(\vec{v})\|^2 + \|T(\vec{w})\|^2 \qquad (T \text{ is orthogonal}). \qquad \blacksquare$$

Theorem 5.3.2 is perhaps better explained with a sketch. (See Figure 2.)

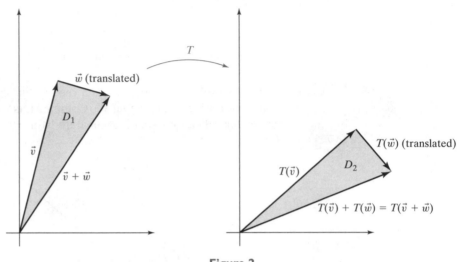

Figure 2

The two shaded triangles are congruent, because corresponding sides are the same length (since T preserves length). Since D_1 is a right triangle, so is D_2.

Here is an alternative characterization of orthogonal transformations:

Theorem 5.3.3   Orthogonal transformations and orthonormal bases

a. A linear transformation T from $\mathbb{R}^n$ to $\mathbb{R}^n$ is orthogonal if (and only if) the vectors $T(\vec{e}_1), T(\vec{e}_2), \ldots, T(\vec{e}_n)$ form an orthonormal basis of $\mathbb{R}^n$.

b. An $n \times n$ matrix A is orthogonal if (and only if) its columns form an orthonormal basis of $\mathbb{R}^n$.

Figure 3 illustrates part (a) for a linear transformation from $\mathbb{R}^2$ to $\mathbb{R}^2$.

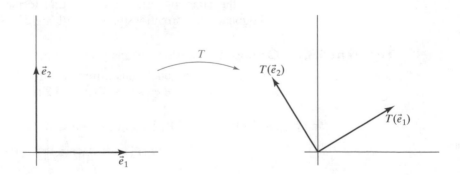

Figure 3

Proof We prove part (a); part (b) then follows from Theorem 2.1.2. If T is an orthogonal transformation, then, by definition, the $T(\vec{e}_i)$ are unit vectors, and, by Theorem 5.3.2, they are orthogonal. Conversely, suppose the $T(\vec{e}_i)$ form an orthonormal basis. Consider a vector $\vec{x} = x_1\vec{e}_1 + x_2\vec{e}_2 + \cdots + x_n\vec{e}_n$ in $\mathbb{R}^n$. Then

$$
\begin{aligned}
\|T(\vec{x})\|^2 &= \|x_1 T(\vec{e}_1) + x_2 T(\vec{e}_2) + \cdots + x_n T(\vec{e}_n)\|^2 \\
&= \|x_1 T(\vec{e}_1)\|^2 + \|x_2 T(\vec{e}_2)\|^2 + \cdots + \|x_n T(\vec{e}_n)\|^2 \quad \text{(by Pythagoras)} \\
&= x_1^2 + x_2^2 + \cdots + x_n^2 \\
&= \|\vec{x}\|^2.
\end{aligned}
$$

∎

Warning: A matrix with orthogonal columns need not be an orthogonal matrix. As an example, consider the matrix $A = \begin{bmatrix} 4 & -3 \\ 3 & 4 \end{bmatrix}$.

EXAMPLE 3 Show that the matrix A is orthogonal:

$$
A = \frac{1}{2} \begin{bmatrix} 1 & -1 & -1 & -1 \\ 1 & -1 & 1 & 1 \\ 1 & 1 & -1 & 1 \\ 1 & 1 & 1 & -1 \end{bmatrix}.
$$

Solution

Check that the columns of A form an orthonormal basis of $\mathbb{R}^4$. ∎

Here are some algebraic properties of orthogonal matrices.

Theorem 5.3.4 **Products and inverses of orthogonal matrices**

a. The product AB of two orthogonal $n \times n$ matrices A and B is orthogonal.

b. The inverse A^{-1} of an orthogonal $n \times n$ matrix A is orthogonal.

Proof In part (a), the linear transformation $T(\vec{x}) = AB\vec{x}$ preserves length, because $\|T(\vec{x})\| = \|A(B\vec{x})\| = \|B\vec{x}\| = \|\vec{x}\|$. In part (b), the linear transformation $T(\vec{x}) = A^{-1}\vec{x}$ preserves length, because $\|A^{-1}\vec{x}\| = \|A(A^{-1}\vec{x})\| = \|\vec{x}\|$. Figure 4 illustrates property (a). ∎

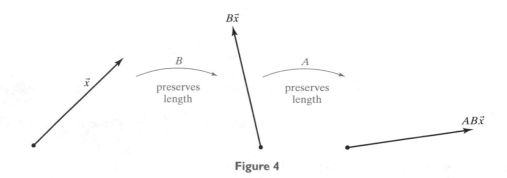

Figure 4

The Transpose of a Matrix

EXAMPLE 4   Consider the orthogonal matrix

$$A = \frac{1}{7} \begin{bmatrix} 2 & 6 & 3 \\ 3 & 2 & -6 \\ 6 & -3 & 2 \end{bmatrix}.$$

Form another 3×3 matrix B whose ijth entry is the jith entry of A:

$$B = \frac{1}{7} \begin{bmatrix} 2 & 3 & 6 \\ 6 & 2 & -3 \\ 3 & -6 & 2 \end{bmatrix}.$$

Note that the rows of B correspond to the columns of A.
    Compute BA, and explain the result.

Solution

$$BA = \frac{1}{49} \begin{bmatrix} 2 & 3 & 6 \\ 6 & 2 & -3 \\ 3 & -6 & 2 \end{bmatrix} \begin{bmatrix} 2 & 6 & 3 \\ 3 & 2 & -6 \\ 6 & -3 & 2 \end{bmatrix} = \frac{1}{49} \begin{bmatrix} 49 & 0 & 0 \\ 0 & 49 & 0 \\ 0 & 0 & 49 \end{bmatrix} = I_3$$

This result is no coincidence: The ijth entry of BA is the dot product of the ith row of B and the jth column of A. By definition of B, this is just the dot product of the ith column of A and the jth column of A. Since A is orthogonal, this product is 1 if $i = j$ and 0 otherwise.   ∎

Before we can generalize the findings of Example 4, we introduce some new terminology.

Definition 5.3.5   The transpose of a matrix; symmetric and skew-symmetric matrices

Consider an $m \times n$ matrix A.
    The *transpose* A^T of A is the $n \times m$ matrix whose ijth entry is the jith entry of A: The roles of rows and columns are reversed.
    We say that a square matrix A is *symmetric* if $A^T = A$, and A is called *skew-symmetric* if $A^T = -A$.

EXAMPLE 5   If $A = \begin{bmatrix} 1 & 2 & 3 \\ 9 & 7 & 5 \end{bmatrix}$, then $A^T = \begin{bmatrix} 1 & 9 \\ 2 & 7 \\ 3 & 5 \end{bmatrix}$.   ∎

EXAMPLE 6   The *symmetric* 2×2 matrices are those of the form $A = \begin{bmatrix} a & b \\ b & c \end{bmatrix}$, for example, $A = \begin{bmatrix} 1 & 2 \\ 2 & 3 \end{bmatrix}$. The symmetric 2×2 matrices form a three-dimensional subspace of $\mathbb{R}^{2 \times 2}$, with basis $\begin{bmatrix} 1 & 0 \\ 0 & 0 \end{bmatrix}, \begin{bmatrix} 0 & 1 \\ 1 & 0 \end{bmatrix}, \begin{bmatrix} 0 & 0 \\ 0 & 1 \end{bmatrix}$.

The *skew-symmetric* 2×2 matrices are those of the form $A = \begin{bmatrix} 0 & b \\ -b & 0 \end{bmatrix}$, for example, $A = \begin{bmatrix} 0 & 2 \\ -2 & 0 \end{bmatrix}$. These form a one-dimensional space with basis $\begin{bmatrix} 0 & 1 \\ -1 & 0 \end{bmatrix}$.   ∎

Note that the transpose of a (column) vector $\vec{v}$ is a row vector: If

$$\vec{v} = \begin{bmatrix} 1 \\ 2 \\ 3 \end{bmatrix}, \quad \text{then} \quad \vec{v}^T = \begin{bmatrix} 1 & 2 & 3 \end{bmatrix}.$$

The transpose gives us a convenient way to express the dot product of two (column) vectors as a matrix product.

Theorem 5.3.6   If $\vec{v}$ and $\vec{w}$ are two (column) vectors in $\mathbb{R}^n$, then

$$\vec{v} \cdot \vec{w} = \vec{v}^T \vec{w}.$$

$$\underset{\substack{\text{Dot} \\ \text{product}}}{\uparrow} \qquad \underset{\substack{\text{Matrix} \\ \text{product}}}{\uparrow}$$

∎

Here we are identifying the 1×1 matrix $\vec{v}^T \vec{w}$ with its sole entry, the scalar $\vec{v} \cdot \vec{w}$. Purists may prefer to write $\vec{v}^T \vec{w} = \begin{bmatrix} \vec{v} \cdot \vec{w} \end{bmatrix}$.

For example,

$$\begin{bmatrix} 1 \\ 2 \\ 3 \end{bmatrix} \cdot \begin{bmatrix} 1 \\ -1 \\ 1 \end{bmatrix} = \begin{bmatrix} 1 & 2 & 3 \end{bmatrix} \begin{bmatrix} 1 \\ -1 \\ 1 \end{bmatrix} = 2.$$

Now we can succinctly state the observation made in Example 4.

Theorem 5.3.7   Consider an $n \times n$ matrix A. The matrix A is orthogonal if (and only if) $A^T A = I_n$ or, equivalently, if $A^{-1} = A^T$.

Proof   To justify this fact, write A in terms of its columns:

$$A = \begin{bmatrix} | & | & & | \\ \vec{v}_1 & \vec{v}_2 & \cdots & \vec{v}_n \\ | & | & & | \end{bmatrix}.$$

Then

$$A^T A = \begin{bmatrix} - & \vec{v}_1^T & - \\ - & \vec{v}_2^T & - \\ & \vdots & \\ - & \vec{v}_n^T & - \end{bmatrix} \begin{bmatrix} | & | & & | \\ \vec{v}_1 & \vec{v}_2 & \cdots & \vec{v}_n \\ | & | & & | \end{bmatrix} = \begin{bmatrix} \vec{v}_1 \cdot \vec{v}_1 & \vec{v}_1 \cdot \vec{v}_2 & \cdots & \vec{v}_1 \cdot \vec{v}_n \\ \vec{v}_2 \cdot \vec{v}_1 & \vec{v}_2 \cdot \vec{v}_2 & \cdots & \vec{v}_2 \cdot \vec{v}_n \\ \vdots & \vdots & \ddots & \vdots \\ \vec{v}_n \cdot \vec{v}_1 & \vec{v}_n \cdot \vec{v}_2 & \cdots & \vec{v}_n \cdot \vec{v}_n \end{bmatrix}.$$

By Theorem 5.3.3b this product is I_n if (and only if) A is orthogonal. ∎

Later in this text, we will frequently work with matrices of the form $A^T A$. It is helpful to think of $A^T A$ as a table displaying the dot products $\vec{v}_i \cdot \vec{v}_j$ among the columns of A, as shown above.

We summarize the various characterizations we have found of orthogonal matrices.

SUMMARY 5.3.8 | **Orthogonal matrices**

Consider an $n \times n$ matrix A. Then the following statements are equivalent:

  i. A is an orthogonal matrix.
  ii. The transformation $L(\vec{x}) = A\vec{x}$ preserves length, that is, $\|A\vec{x}\| = \|\vec{x}\|$ for all $\vec{x}$ in $\mathbb{R}^n$.
  iii. The columns of A form an orthonormal basis of $\mathbb{R}^n$.
  iv. $A^T A = I_n$.
  v. $A^{-1} = A^T$.

Here are some algebraic properties of transposes:

Theorem 5.3.9   **Properties of the transpose**

  a. If A is an $n \times p$ matrix and B a $p \times m$ matrix, then

$$(AB)^T = B^T A^T.$$

Note the order of the factors.

  b. If an $n \times n$ matrix A is invertible, then so is A^T, and

$$(A^T)^{-1} = (A^{-1})^T.$$

  c. For any matrix A,

$$\operatorname{rank}(A) = \operatorname{rank}(A^T).$$

Proof   **a.** Compare entries:

$$ij\text{th entry of } (AB)^T = ji\text{th entry of } AB$$
$$= (j\text{th row of } A) \cdot (i\text{th column of } B)$$
$$ij\text{th entry of } B^T A^T = (i\text{th row of } B^T) \cdot (j\text{th column of } A^T)$$
$$= (i\text{th column of } B) \cdot (j\text{th row of } A).$$

  b. We know that

$$AA^{-1} = I_n.$$

Transposing both sides and using part (a), we find that

$$(AA^{-1})^T = (A^{-1})^T A^T = I_n.$$

By Theorem 2.4.8, it follows that

$$(A^{-1})^T = (A^T)^{-1}.$$

  c. Consider the row space of A (i.e., the span of the rows of A). It is not hard to show that the dimension of this space is $\operatorname{rank}(A)$ (see Exercises 69–72 in Section 3.3):

$$\operatorname{rank}(A^T) = \text{dimension of the span of the columns of } A^T$$
$$= \text{dimension of the span of the rows of } A$$
$$= \operatorname{rank}(A). \qquad \blacksquare$$

The Matrix of an Orthogonal Projection

The transpose allows us to write a formula for the matrix of an orthogonal projection. Consider first the orthogonal projection

$$\text{proj}_L \vec{x} = (\vec{u}_1 \cdot \vec{x})\vec{u}_1$$

onto a line L in $\mathbb{R}^n$, where $\vec{u}_1$ is a unit vector in L. If we view the vector $\vec{u}_1$ as an $n \times 1$ matrix and the scalar $\vec{u}_1 \cdot \vec{x}$ as a 1×1 matrix, we can write

$$\begin{aligned} \text{proj}_L \vec{x} &= \vec{u}_1(\vec{u}_1 \cdot \vec{x}) \\ &= \vec{u}_1 \vec{u}_1^T \vec{x} \\ &= M\vec{x}, \end{aligned}$$

where $M = \vec{u}_1 \vec{u}_1^T$. Note that $\vec{u}_1$ is an $n \times 1$ matrix and $\vec{u}_1^T$ is $1 \times n$, so that M is $n \times n$, as expected.

More generally, consider the projection

$$\text{proj}_V \vec{x} = (\vec{u}_1 \cdot \vec{x})\vec{u}_1 + \cdots + (\vec{u}_m \cdot \vec{x})\vec{u}_m$$

onto a subspace V of $\mathbb{R}^n$ with orthonormal basis $\vec{u}_1, \ldots, \vec{u}_m$. We can write

$$\begin{aligned} \text{proj}_V \vec{x} &= \vec{u}_1 \vec{u}_1^T \vec{x} + \cdots + \vec{u}_m \vec{u}_m^T \vec{x} \\ &= (\vec{u}_1 \vec{u}_1^T + \cdots + \vec{u}_m \vec{u}_m^T)\vec{x} \\ &= \begin{bmatrix} | & & | \\ \vec{u}_1 & \cdots & \vec{u}_m \\ | & & | \end{bmatrix} \begin{bmatrix} - & \vec{u}_1^T & - \\ & \vdots & \\ - & \vec{u}_m^T & - \end{bmatrix} \vec{x}. \end{aligned}$$

We have shown the following result:

Theorem 5.3.10   **The matrix of an orthogonal projection**

Consider a subspace V of $\mathbb{R}^n$ with orthonormal basis $\vec{u}_1, \vec{u}_2, \ldots, \vec{u}_m$. The matrix of the orthogonal projection onto V is

$$QQ^T, \quad \text{where} \quad Q = \begin{bmatrix} | & | & & | \\ \vec{u}_1 & \vec{u}_2 & \cdots & \vec{u}_m \\ | & | & & | \end{bmatrix}.$$

Pay attention to the order of the factors (QQ^T as opposed to $Q^T Q$). ∎

EXAMPLE 7   Find the matrix of the orthogonal projection onto the subspace of $\mathbb{R}^4$ spanned by

$$\vec{u}_1 = \frac{1}{2} \begin{bmatrix} 1 \\ 1 \\ 1 \\ 1 \end{bmatrix}, \quad \vec{u}_2 = \frac{1}{2} \begin{bmatrix} 1 \\ -1 \\ -1 \\ 1 \end{bmatrix}.$$

Solution

Note that the vectors $\vec{u}_1$ and $\vec{u}_2$ are orthonormal. Therefore, the matrix is

$$QQ^T = \frac{1}{4} \begin{bmatrix} 1 & 1 \\ 1 & -1 \\ 1 & -1 \\ 1 & 1 \end{bmatrix} \begin{bmatrix} 1 & 1 & 1 & 1 \\ 1 & -1 & -1 & 1 \end{bmatrix} = \frac{1}{2} \begin{bmatrix} 1 & 0 & 0 & 1 \\ 0 & 1 & 1 & 0 \\ 0 & 1 & 1 & 0 \\ 1 & 0 & 0 & 1 \end{bmatrix}. \quad \blacksquare$$

EXERCISES 5.3

GOAL *Use the various characterizations of orthogonal transformations and orthogonal matrices. Find the matrix of an orthogonal projection. Use the properties of the transpose.*

Which of the matrices in Exercises 1 through 4 orthogonal?

1. $\begin{bmatrix} 0.6 & 0.8 \\ 0.8 & 0.6 \end{bmatrix}$

2. $\begin{bmatrix} -0.8 & 0.6 \\ 0.6 & 0.8 \end{bmatrix}$

3. $\frac{1}{3}\begin{bmatrix} 2 & -2 & 1 \\ 1 & 2 & 2 \\ 2 & 1 & -2 \end{bmatrix}$

4. $\frac{1}{7}\begin{bmatrix} 2 & 6 & -3 \\ 6 & -3 & 2 \\ 3 & 2 & 6 \end{bmatrix}$

If the $n \times n$ matrices A and B are orthogonal, which of the matrices in Exercises 5 through 11 must be orthogonal as well?

5. $3A$ 6. $-B$ 7. AB 8. $A + B$

9. B^{-1} 10. $B^{-1}AB$ 11. A^T

If the $n \times n$ matrices A and B are symmetric and B is invertible, which of the matrices in Exercises 13 through 20 must be symmetric as well?

13. $3A$ 14. $-B$ 15. AB 16. $A + B$

17. B^{-1} 18. A^{10}

19. $2I_n + 3A - 4A^2$ 20. AB^2A

If A and B are arbitrary $n \times n$ matrices, which of the matrices in Exercises 21 through 26 must be symmetric?

21. $A^T A$ 22. BB^T 23. $A - A^T$

24. $A^T BA$ 25. $A^T B^T BA$ 26. $B(A + A^T)B^T$

27. Consider an $n \times m$ matrix A, a vector $\vec{v}$ in $\mathbb{R}^m$, and a vector $\vec{w}$ in $\mathbb{R}^n$. Show that
$$(A\vec{v}) \cdot \vec{w} = \vec{v} \cdot (A^T \vec{w}).$$

28. Consider an orthogonal transformation L from $\mathbb{R}^n$ to $\mathbb{R}^n$. Show that L preserves the dot product:
$$\vec{v} \cdot \vec{w} = L(\vec{v}) \cdot L(\vec{w}),$$
for all $\vec{v}$ and $\vec{w}$ in $\mathbb{R}^n$.

29. Show that an orthogonal transformation L from $\mathbb{R}^n$ to $\mathbb{R}^n$ preserves angles: The angle between two nonzero vectors $\vec{v}$ and $\vec{w}$ in $\mathbb{R}^n$ equals the angle between $L(\vec{v})$ and $L(\vec{w})$. Conversely, is any linear transformation that preserves angles orthogonal?

30. Consider a linear transformation L from $\mathbb{R}^m$ to $\mathbb{R}^n$ that preserves length. What can you say about the kernel of L? What is the dimension of the image? What can you say about the relationship between n and m? If A is the matrix of L, what can you say about the columns of A? What is $A^T A$? What about AA^T? Illustrate your answers with an example where $m = 2$ and $n = 3$.

31. Are the *rows* of an orthogonal matrix A necessarily orthonormal?

32. a. Consider an $n \times m$ matrix A such that $A^T A = I_m$. Is it necessarily true that $AA^T = I_n$? Explain.
    b. Consider an $n \times n$ matrix A such that $A^T A = I_n$. Is it necessarily true that $AA^T = I_n$? Explain.

33. Find all orthogonal 2×2 matrices.

34. Find all orthogonal 3×3 matrices of the form
$$\begin{bmatrix} a & b & 0 \\ c & d & 1 \\ e & f & 0 \end{bmatrix}.$$

35. Find an orthogonal transformation T from $\mathbb{R}^3$ to $\mathbb{R}^3$ such that
$$T\begin{bmatrix} 2/3 \\ 2/3 \\ 1/3 \end{bmatrix} = \begin{bmatrix} 0 \\ 0 \\ 1 \end{bmatrix}.$$

36. Find an orthogonal matrix of the form
$$\begin{bmatrix} 2/3 & 1/\sqrt{2} & a \\ 2/3 & -1/\sqrt{2} & b \\ 1/3 & 0 & c \end{bmatrix}.$$

37. Is there an orthogonal transformation T from $\mathbb{R}^3$ to $\mathbb{R}^3$ such that
$$T\begin{bmatrix} 2 \\ 3 \\ 0 \end{bmatrix} = \begin{bmatrix} 3 \\ 0 \\ 2 \end{bmatrix} \quad \text{and} \quad T\begin{bmatrix} -3 \\ 2 \\ 0 \end{bmatrix} = \begin{bmatrix} 2 \\ -3 \\ 0 \end{bmatrix} ?$$

38. a. Give an example of a (nonzero) skew-symmetric 3×3 matrix A, and compute A^2.
    b. If an $n \times n$ matrix A is skew-symmetric, is matrix A^2 necessarily skew-symmetric as well? Or is A^2 necessarily symmetric?

39. Consider a line L in $\mathbb{R}^n$, spanned by a unit vector
$$\vec{u} = \begin{bmatrix} u_1 \\ u_2 \\ \vdots \\ u_n \end{bmatrix}.$$

Consider the matrix A of the orthogonal projection onto L. Describe the ijth entry of A, in terms of the components u_i of $\vec{u}$.

40. Consider the subspace W of $\mathbb{R}^4$ spanned by the vectors

$$\vec{v}_1 = \begin{bmatrix} 1 \\ 1 \\ 1 \\ 1 \end{bmatrix} \quad \text{and} \quad \vec{v}_2 = \begin{bmatrix} 1 \\ 9 \\ -5 \\ 3 \end{bmatrix}.$$

Find the matrix of the orthogonal projection onto W.

41. Find the matrix A of the orthogonal projection onto the line in $\mathbb{R}^n$ spanned by the vector

$$\left.\begin{bmatrix} 1 \\ 1 \\ \vdots \\ 1 \end{bmatrix}\right\} \quad \text{all } n \text{ components are 1.}$$

42. Let A be the matrix of an orthogonal projection. Find A^2 in two ways:

a. Geometrically. (Consider what happens when you apply an orthogonal projection twice.)

b. By computation, using the formula given in Theorem 5.3.10.

43. Consider a unit vector $\vec{u}$ in $\mathbb{R}^3$. We define the matrices

$$A = 2\vec{u}\vec{u}^T - I_3 \quad \text{and} \quad B = I_3 - 2\vec{u}\vec{u}^T.$$

Describe the linear transformations defined by these matrices geometrically.

44. Consider an $n \times m$ matrix A. Find

$$\dim\big(\text{im}(A)\big) + \dim\big(\text{ker}(A^T)\big),$$

in terms of m and n.

45. For which $n \times m$ matrices A does the equation

$$\dim\big(\text{ker}(A)\big) = \dim\big(\text{ker}(A^T)\big)$$

hold? Explain.

46. Consider a QR factorization

$$M = QR.$$

Show that

$$R = Q^T M.$$

47. If $A = QR$ is a QR factorization, what is the relationship between $A^T A$ and $R^T R$?

48. Consider an invertible $n \times n$ matrix A. Can you write A as $A = LQ$, where L is a *lower* triangular matrix and Q is orthogonal? *Hint:* Consider the QR factorization of A^T.

49. Consider an invertible $n \times n$ matrix A. Can you write $A = RQ$, where R is an upper triangular matrix and Q is orthogonal?

50. a. Find all $n \times n$ matrices that are both orthogonal and upper triangular, with positive diagonal entries.

b. Show that the QR factorization of an invertible $n \times n$ matrix is unique. *Hint:* If $A = Q_1 R_1 = Q_2 R_2$, then the matrix $Q_2^{-1} Q_1 = R_2 R_1^{-1}$ is both orthogonal and upper triangular, with positive diagonal entries.

51. a. Consider the matrix product $Q_1 = Q_2 S$, where both Q_1 and Q_2 are $n \times m$ matrices with orthonormal columns. Show that S is an orthogonal matrix. *Hint:* Compute $Q_1^T Q_1 = (Q_2 S)^T Q_2 S$. Note that $Q_1^T Q_1 = Q_2^T Q_2 = I_m$.

b. Show that the QR factorization of an $n \times m$ matrix M is unique. *Hint:* If $M = Q_1 R_1 = Q_2 R_2$, then $Q_1 = Q_2 R_2 R_1^{-1}$. Now use part (a) and Exercise 50a.

52. Find a basis of the space V of all symmetric 3×3 matrices, and thus determine the dimension of V.

53. Find a basis of the space V of all skew-symmetric 3×3 matrices, and thus determine the dimension of V.

54. Find the dimension of the space of all skew-symmetric $n \times n$ matrices.

55. Find the dimension of the space of all symmetric $n \times n$ matrices.

56. Is the transformation $L(A) = A^T$ from $\mathbb{R}^{2\times 3}$ to $\mathbb{R}^{3\times 2}$ linear? Is L an isomorphism?

57. Is the transformation $L(A) = A^T$ from $\mathbb{R}^{m\times n}$ to $\mathbb{R}^{n\times m}$ linear? Is L an isomorphism?

58. Find image and kernel of the linear transformation $L(A) = \frac{1}{2}(A + A^T)$ from $\mathbb{R}^{n\times n}$ to $\mathbb{R}^{n\times n}$. *Hint:* Think about symmetric and skew-symmetric matrices.

59. Find the image and kernel of the linear transformation $L(A) = \frac{1}{2}(A - A^T)$ from $\mathbb{R}^{n\times n}$ to $\mathbb{R}^{n\times n}$. *Hint:* Think about symmetric and skew-symmetric matrices.

60. Find the matrix of the linear transformation $L(A) = A^T$ from $\mathbb{R}^{2\times 2}$ to $\mathbb{R}^{2\times 2}$ with respect to the basis

$$\begin{bmatrix} 1 & 0 \\ 0 & 0 \end{bmatrix}, \begin{bmatrix} 0 & 0 \\ 0 & 1 \end{bmatrix}, \begin{bmatrix} 0 & 1 \\ 1 & 0 \end{bmatrix}, \begin{bmatrix} 0 & 1 \\ -1 & 0 \end{bmatrix}.$$

61. Find the matrix of the linear transformation $L(A) = A - A^T$ from $\mathbb{R}^{2\times 2}$ to $\mathbb{R}^{2\times 2}$ with respect to the basis

$$\begin{bmatrix} 1 & 0 \\ 0 & 0 \end{bmatrix}, \begin{bmatrix} 0 & 0 \\ 0 & 1 \end{bmatrix}, \begin{bmatrix} 0 & 1 \\ 1 & 0 \end{bmatrix}, \begin{bmatrix} 0 & 1 \\ -1 & 0 \end{bmatrix}.$$

62. Consider the matrix

$$A = \begin{bmatrix} 1 & 1 & -1 \\ 3 & 2 & -5 \\ 2 & 2 & 0 \end{bmatrix}$$

with *LDU*-factorization

$$A = \begin{bmatrix} 1 & 0 & 0 \\ 3 & 1 & 0 \\ 2 & 0 & 1 \end{bmatrix} \begin{bmatrix} 1 & 0 & 0 \\ 0 & -1 & 0 \\ 0 & 0 & 2 \end{bmatrix} \begin{bmatrix} 1 & 1 & -1 \\ 0 & 1 & 2 \\ 0 & 0 & 1 \end{bmatrix}.$$

Find the *LDU*-factorization of A^T. (See Exercise 2.4.90d.)

63. Consider a symmetric invertible $n \times n$ matrix A which admits an *LDU*-factorization $A = LDU$. (See Exercises 90, 93, and 94 of Section 2.4.) Recall that this factorization is unique. (See Exercise 2.4.94.) Show that $U = L^T$. (This is sometimes called the LDL^T-factorization of a symmetric matrix A.)

64. This exercise shows one way to define the *quaternions*, discovered in 1843 by the Irish mathematician Sir W. R. Hamilton (1805–1865). Consider the set H of all 4×4 matrices M of the form

$$M = \begin{bmatrix} p & -q & -r & -s \\ q & p & s & -r \\ r & -s & p & q \\ s & r & -q & p \end{bmatrix},$$

where p, q, r, s are arbitrary real numbers. We can write M more succinctly in partitioned form as

$$M = \begin{bmatrix} A & -B^T \\ B & A^T \end{bmatrix},$$

where A and B are rotation–scaling matrices.

a. Show that H is closed under addition: If M and N are in H, then so is $M + N$.

b. Show that H is closed under scalar multiplication: If M is in H and k is an arbitrary scalar, then kM is in H.

c. Parts (a) and (b) show that H is a subspace of the linear space $\mathbb{R}^{4\times4}$. Find a basis of H, and thus determine the dimension of H.

d. Show that H is closed under multiplication: If M and N are in H, then so is MN.

e. Show that if M is in H, then so is M^T.

f. For a matrix M in H, compute $M^T M$.

g. Which matrices M in H are invertible? If a matrix M in H is invertible, is M^{-1} necessarily in H as well?

h. If M and N are in H, does the equation $MN = NM$ always hold?

65. Find all orthogonal 2×2 matrices A such that all the entries of $10A$ are integers and such that both entries in the first column are positive.

66. Find an orthogonal 2×2 matrices A such that all the entries of $100A$ are integers while all the entries of $10A$ fail to be integers.

67. Consider a subspace V of $\mathbb{R}^n$ with a basis $\vec{v}_1, \ldots, \vec{v}_m$; suppose we wish to find a formula for the orthogonal projection onto V. Using the methods we have developed thus far, we can proceed in two steps: We use the Gram–Schmidt process to construct an *orthonormal* basis $\vec{u}_1, \ldots, \vec{u}_m$ of V, and then we use Theorem 5.3.10: The matrix of the orthogonal projection is QQ^T, where

$$Q = \begin{bmatrix} \vec{u}_1 & \cdots & \vec{u}_m \end{bmatrix}.$$

In this exercise we will see how we can write the matrix of the projection directly in terms of the basis $\vec{v}_1, \ldots, \vec{v}_m$ and the matrix

$$A = \begin{bmatrix} \vec{v}_1 & \cdots & \vec{v}_m \end{bmatrix}.$$

(This issue will be discussed more thoroughly in Section 5.4; see Theorem 5.4.7.)

Since $\text{proj}_V \vec{x}$ is in V, we can write

$$\text{proj}_V \vec{x} = c_1 \vec{v}_1 + \cdots + c_m \vec{v}_m$$

for some scalars $c_1, \ldots, c_m$ yet to be determined. Now $\vec{x} - \text{proj}_V(\vec{x}) = \vec{x} - c_1 \vec{v}_1 - \cdots - c_m \vec{v}_m$ is orthogonal to V, meaning that $\vec{v}_i \cdot (\vec{x} - c_1 \vec{v}_1 - \cdots - c_m \vec{v}_m) = 0$ for $i = 1, \ldots, m$.

a. Use the equation $\vec{v}_i \cdot (\vec{x} - c_1 \vec{v}_1 - \cdots - c_m \vec{v}_m) = 0$ to show that $A^T A \vec{c} = A^T \vec{x}$, where $\vec{c} = \begin{bmatrix} c_1 \\ \vdots \\ c_m \end{bmatrix}$.

b. Conclude that $\vec{c} = (A^T A)^{-1} A^T \vec{x}$ and $\text{proj}_V \vec{x} = A\vec{c} = A(A^T A)^{-1} A^T \vec{x}$.

68. The formula $A(A^T A)^{-1} A^T$ for the matrix of an orthogonal projection is derived in Exercise 67. Now consider the QR factorization of A, and express the matrix $A(A^T A)^{-1} A^T$ in terms of Q.

5.4 Least Squares and Data Fitting

In this section, we will present an important application of the ideas introduced in this chapter. First, we take another look at orthogonal complements and orthogonal projections.

Another Characterization of Orthogonal Complements

Consider a subspace $V = \text{image}(A)$ of $\mathbb{R}^n$, where $A = \begin{bmatrix} \vec{v}_1 & \vec{v}_2 & \cdots & \vec{v}_m \end{bmatrix}$. Then

$$
\begin{aligned}
V^\perp &= \{\vec{x} \text{ in } \mathbb{R}^n : \vec{v} \cdot \vec{x} = 0, \text{ for all } \vec{v} \text{ in } V\} \\
&= \{\vec{x} \text{ in } \mathbb{R}^n : \vec{v}_i \cdot \vec{x} = 0, \text{ for } i = 1, \ldots, m\} \\
&= \{\vec{x} \text{ in } \mathbb{R}^n : \vec{v}_i^T \vec{x} = 0, \text{ for } i = 1, \ldots, m\}.
\end{aligned}
$$

In other words, $V^\perp = (\text{im}A)^\perp$ is the kernel of the matrix

$$
A^T = \begin{bmatrix} - & \vec{v}_1^T & - \\ - & \vec{v}_2^T & - \\ & \vdots & \\ - & \vec{v}_m^T & - \end{bmatrix}.
$$

Theorem 5.4.1 For any matrix A,

$$
(\text{im}A)^\perp = \ker(A^T). \qquad \blacksquare
$$

Here is a very simple example: Consider the line

$$
V = \text{im} \begin{bmatrix} 1 \\ 2 \\ 3 \end{bmatrix}.
$$

Then

$$
V^\perp = \ker \begin{bmatrix} 1 & 2 & 3 \end{bmatrix}
$$

is the plane with equation $x_1 + 2x_2 + 3x_3 = 0$. (See Figure 1.)

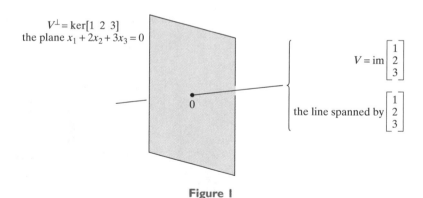

$V^\perp = \ker[1\ 2\ 3]$
the plane $x_1 + 2x_2 + 3x_3 = 0$

$V = \text{im} \begin{bmatrix} 1 \\ 2 \\ 3 \end{bmatrix}$

the line spanned by $\begin{bmatrix} 1 \\ 2 \\ 3 \end{bmatrix}$

Figure I

The following somewhat technical result will be useful later:

Theorem 5.4.2 **a.** If A is an $n \times m$ matrix, then

$$
\ker(A) = \ker(A^T A).
$$

b. If A is an $n \times m$ matrix with $\ker(A) = \{\vec{0}\}$, then $A^T A$ is invertible.

Proof **a.** Clearly, the kernel of A is contained in the kernel of $A^T A$. Conversely, consider a vector $\vec{x}$ in the kernel of $A^T A$, so that $A^T A \vec{x} = \vec{0}$. Then $A\vec{x}$ is in the image of A and in the kernel of A^T. Since $\ker(A^T)$ is the orthogonal complement of $\operatorname{im}(A)$ by Theorem 5.4.1, the vector $A\vec{x}$ is $\vec{0}$ by Theorem 5.1.8b, that is, $\vec{x}$ is in the kernel of A.

b. Note that $A^T A$ is an $m \times m$ matrix. By part (a), $\ker(A^T A) = \{\vec{0}\}$, and the square matrix $A^T A$ is therefore invertible. (See Theorem 3.3.10.)   ∎

An Alternative Characterization of Orthogonal Projections

Theorem 5.4.3   Consider a vector $\vec{x}$ in $\mathbb{R}^n$ and a subspace V of $\mathbb{R}^n$. Then the orthogonal projection $\operatorname{proj}_V \vec{x}$ is the vector in V *closest* to $\vec{x}$, in that

$$\|\vec{x} - \operatorname{proj}_V \vec{x}\| < \|\vec{x} - \vec{v}\|,$$

for all $\vec{v}$ in V different from $\operatorname{proj}_V \vec{x}$.   ∎

To justify this fact, apply the Pythagorean theorem to the shaded right triangle in Figure 2.

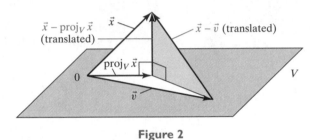

Figure 2

Least-Squares Approximations

Consider an *inconsistent* linear system $A\vec{x} = \vec{b}$. The fact that this system is inconsistent means that the vector $\vec{b}$ is not in the image of A. (See Figure 3.)

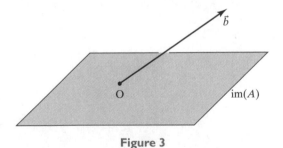

Figure 3

Although this system cannot be solved, we might be interested in finding a good approximate solution. We can try to find a vector $\vec{x}^*$ such that $A\vec{x}^*$ is "as close as possible" to $\vec{b}$. In other words, we try to minimize the *error* $\|\vec{b} - A\vec{x}\|$.

Definition 5.4.4   Least-squares solution

Consider a linear system

$$A\vec{x} = \vec{b},$$

where A is an $n \times m$ matrix. A vector $\vec{x}^*$ in $\mathbb{R}^m$ is called a *least-squares solution* of this system if $\|\vec{b} - A\vec{x}^*\| \leq \|\vec{b} - A\vec{x}\|$ for all $\vec{x}$ in $\mathbb{R}^m$.

See Figure 4.

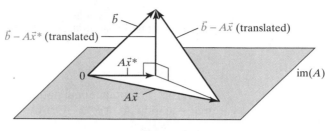

Figure 4

The term *least-squares solution* reflects the fact that we are minimizing the sum of the squares of the components of the vector $\vec{b} - A\vec{x}$.

If the system $A\vec{x} = \vec{b}$ happens to be consistent, then the least-squares solutions are its exact solutions: The error $\|\vec{b} - A\vec{x}\|$ is zero.

How can we find the least-squares solutions of a linear system $A\vec{x} = \vec{b}$? Consider the following string of equivalent statements:

The vector $\vec{x}^*$ is a least-squares solution
of the system $A\vec{x} = \vec{b}$.

$\Updownarrow$ Def. 5.4.4

$\|\vec{b} - A\vec{x}^*\| \le \|\vec{b} - A\vec{x}\|$ for all $\vec{x}$ in $\mathbb{R}^m$.

$\Updownarrow$ Theorem 5.4.3

$A\vec{x}^* = \mathrm{proj}_V \vec{b}$,   where $V = \mathrm{im}(A)$

$\Updownarrow$ Theorems 5.1.4 and 5.4.1

$\vec{b} - A\vec{x}^*$ is in $V^\perp = (\mathrm{im}A)^\perp = \ker(A^T)$

$\Updownarrow$

$A^T(\vec{b} - A\vec{x}^*) = \vec{0}$

$\Updownarrow$

$A^T A\vec{x}^* = A^T \vec{b}$

Take another look at Figures 2 and 4.

Theorem 5.4.5   The normal equation

The least-squares solutions of the system

$$A\vec{x} = \vec{b}$$

are the exact solutions of the (consistent) system

$$A^T A\vec{x} = A^T \vec{b}.$$

The system $A^T A\vec{x} = A^T \vec{b}$ is called the *normal equation* of $A\vec{x} = \vec{b}$. ■

The case when $\ker(A) = \{\vec{0}\}$ is of particular importance. Then, the matrix $A^T A$ is invertible (by Theorem 5.4.2b), and we can give a closed formula for the least-squares solution.

Theorem 5.4.6   If $\ker(A) = \{\vec{0}\}$, then the linear system

$$A\vec{x} = \vec{b}$$

has the unique least-squares solution

$$\vec{x}^* = (A^T A)^{-1} A^T \vec{b}.$$ ∎

From a computational point of view, it may be more efficient to solve the normal equation $A^T A\vec{x} = A^T \vec{b}$ by Gauss–Jordan elimination, rather than by using Theorem 5.4.6.

EXAMPLE I   Use Theorem 5.4.6 to find the least-squares solution $\vec{x}^*$ of the system

$$A\vec{x} = \vec{b}, \quad \text{where} \quad A = \begin{bmatrix} 1 & 1 \\ 1 & 2 \\ 1 & 3 \end{bmatrix} \quad \text{and} \quad \vec{b} = \begin{bmatrix} 0 \\ 0 \\ 6 \end{bmatrix}.$$

What is the geometric relationship between $A\vec{x}^*$ and $\vec{b}$?

Solution
We compute

$$\vec{x}^* = (A^T A)^{-1} A^T \vec{b} = \begin{bmatrix} -4 \\ 3 \end{bmatrix} \quad \text{and} \quad A\vec{x}^* = \begin{bmatrix} -1 \\ 2 \\ 5 \end{bmatrix}.$$

Recall that $A\vec{x}^*$ is the orthogonal projection of $\vec{b}$ onto the image of A. Check that

$$\vec{b} - A\vec{x}^* = \begin{bmatrix} 1 \\ -2 \\ 1 \end{bmatrix}$$

is indeed perpendicular to the two column vectors of A. (See Figure 5.)   ∎

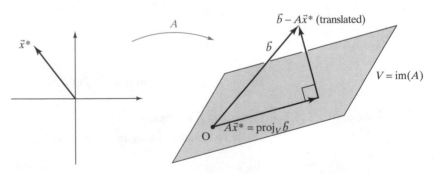

Figure 5

If $\vec{x}^*$ is a least-squares solution of the system $A\vec{x} = \vec{b}$, then $A\vec{x}^*$ is the orthogonal projection of $\vec{b}$ onto $\text{im}(A)$. We can use this fact to find a new formula for orthogonal projections. (Compare this with Theorem 5.1.5 and 5.3.10.) Consider a subspace V of $\mathbb{R}^n$ and a vector $\vec{b}$ in $\mathbb{R}^n$. Choose a basis $\vec{v}_1, \ldots, \vec{v}_m$ of V, and form the matrix $A = [\vec{v}_1 \ \cdots \ \vec{v}_m]$. Note that $\ker(A) = \{\vec{0}\}$, since the columns of A are linearly independent. The least-squares solution of the system $A\vec{x} = \vec{b}$ is $\vec{x}^* = (A^T A)^{-1} A^T \vec{b}$. Thus, the orthogonal projection of $\vec{b}$ onto V is $\text{proj}_V \vec{b} = A\vec{x}^* = A(A^T A)^{-1} A^T \vec{b}$.

Theorem 5.4.7   **The matrix of an orthogonal projection**

Consider a subspace V of $\mathbb{R}^n$ with basis $\vec{v}_1, \vec{v}_2, \ldots, \vec{v}_m$. Let

$$A = \begin{bmatrix} \vec{v}_1 & \vec{v}_2 & \cdots & \vec{v}_m \end{bmatrix}.$$

Then the matrix of the orthogonal projection onto V is

$$A(A^T A)^{-1} A^T.$$

We are not required to find an *orthonormal* basis of V here. If the vectors $\vec{v}_1, \ldots, \vec{v}_m$ happen to be orthonormal, then $A^T A = I_m$ and the formula simplifies to AA^T. (See Theorem 5.3.10.)

EXAMPLE 2   Find the matrix of the orthogonal projection onto the subspace of $\mathbb{R}^4$ spanned by the vectors

$$\begin{bmatrix} 1 \\ 1 \\ 1 \\ 1 \end{bmatrix} \quad \text{and} \quad \begin{bmatrix} 1 \\ 2 \\ 3 \\ 4 \end{bmatrix}.$$

Solution

Let

$$A = \begin{bmatrix} 1 & 1 \\ 1 & 2 \\ 1 & 3 \\ 1 & 4 \end{bmatrix},$$

and compute

$$A(A^T A)^{-1} A^T = \frac{1}{10} \begin{bmatrix} 7 & 4 & 1 & -2 \\ 4 & 3 & 2 & 1 \\ 1 & 2 & 3 & 4 \\ -2 & 1 & 4 & 7 \end{bmatrix}.$$

Data Fitting

Scientists are often interested in fitting a function of a certain type to data they have gathered. The functions considered could be linear, polynomial, rational, trigonometric, or exponential. The equations we have to solve as we fit data are frequently linear. (See Exercises 29 through 36 of Section 1.1, and Exercises 30 through 33 of Section 1.2.)

EXAMPLE 3   Find a cubic polynomial whose graph passes through the points $(1, 3)$, $(-1, 13)$, $(2, 1)$, $(-2, 33)$.

Solution

We are looking for a function

$$f(t) = c_0 + c_1 t + c_2 t^2 + c_3 t^3$$

such that $f(1) = 3$, $f(-1) = 13$, $f(2) = 1$, $f(-2) = 33$; that is, we have to solve the linear system

$$\begin{vmatrix} c_0 + & c_1 + & c_2 + & c_3 = & 3 \\ c_0 - & c_1 + & c_2 - & c_3 = & 13 \\ c_0 + & 2c_1 + & 4c_2 + & 8c_3 = & 1 \\ c_0 - & 2c_1 + & 4c_2 - & 8c_3 = & 33 \end{vmatrix}.$$

This linear system has the unique solution

$$\begin{bmatrix} c_0 \\ c_1 \\ c_2 \\ c_3 \end{bmatrix} = \begin{bmatrix} 5 \\ -4 \\ 3 \\ -1 \end{bmatrix}.$$

Thus, the cubic polynomial whose graph passes through the four given data points is $f(t) = 5 - 4t + 3t^2 - t^3$, as shown in Figure 6. ∎

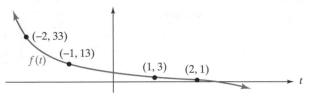

(-2, 33)

(-1, 13)

$f(t)$

(1, 3)   (2, 1)

t

Figure 6

Frequently, a data-fitting problem leads to a linear system with more equations than variables. (This happens when the number of data points exceeds the number of parameters in the function we seek.) Such a system is usually inconsistent, and we will look for the least-squares solution(s).

EXAMPLE 4   Fit a quadratic function to the four data points $(a_1, b_1) = (-1, 8)$, $(a_2, b_2) = (0, 8)$, $(a_3, b_3) = (1, 4)$, and $(a_4, b_4) = (2, 16)$.

Solution

We are looking for a function $f(t) = c_0 + c_1 t + c_2 t^2$ such that

$$\begin{vmatrix} f(a_1) = b_1 \\ f(a_2) = b_2 \\ f(a_3) = b_3 \\ f(a_4) = b_4 \end{vmatrix} \quad \text{or} \quad \begin{vmatrix} c_0 - & c_1 + & c_2 = & 8 \\ c_0 & & = & 8 \\ c_0 + & c_1 + & c_2 = & 4 \\ c_0 + & 2c_1 + & 4c_2 = & 16 \end{vmatrix} \quad \text{or} \quad A \begin{bmatrix} c_0 \\ c_1 \\ c_2 \end{bmatrix} = \vec{b},$$

where

$$A = \begin{bmatrix} 1 & -1 & 1 \\ 1 & 0 & 0 \\ 1 & 1 & 1 \\ 1 & 2 & 4 \end{bmatrix} \quad \text{and} \quad \vec{b} = \begin{bmatrix} 8 \\ 8 \\ 4 \\ 16 \end{bmatrix}.$$

We have four equations, corresponding to the four data points, but only three unknowns, the three coefficients of a quadratic polynomial. Check that this system is indeed inconsistent. The least-squares solution is

$$\vec{c}^* = \begin{bmatrix} c_0^* \\ c_1^* \\ c_2^* \end{bmatrix} = (A^T A)^{-1} A^T \vec{b} = \begin{bmatrix} 5 \\ -1 \\ 3 \end{bmatrix}.$$

The least-squares approximation is $f^*(t) = 5 - t + 3t^2$, as shown in Figure 7.

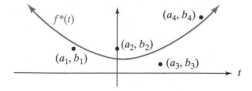

Figure 7

This quadratic function $f^*(t)$ fits the data points best, in that the vector

$$A\vec{c}^* = \begin{bmatrix} f^*(a_1) \\ f^*(a_2) \\ f^*(a_3) \\ f^*(a_4) \end{bmatrix}$$

is as close as possible to

$$\vec{b} = \begin{bmatrix} b_1 \\ b_2 \\ b_3 \\ b_4 \end{bmatrix}.$$

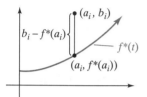

Figure 8

This means that

$$\|\vec{b} - A\vec{c}^*\|^2 = \left(b_1 - f^*(a_1)\right)^2 + \left(b_2 - f^*(a_2)\right)^2 \\ + \left(b_3 - f^*(a_3)\right)^2 + \left(b_4 - f^*(a_4)\right)^2$$

is minimal: The sum of the squares of the vertical distances between graph and data points is minimal. (See Figure 8.) ∎

EXAMPLE 5 Find the linear function $c_0 + c_1 t$ that best fits the data points (a_1, b_1), (a_2, b_2), …, (a_n, b_n), using least squares. Assume that $a_1 \neq a_2$.

Solution

We attempt to solve the system

$$\begin{vmatrix} c_0 + c_1 a_1 = b_1 \\ c_0 + c_1 a_2 = b_2 \\ \vdots \quad \vdots \quad \vdots \\ c_0 + c_1 a_n = b_n \end{vmatrix},$$

or

$$\begin{bmatrix} 1 & a_1 \\ 1 & a_2 \\ \vdots & \vdots \\ 1 & a_n \end{bmatrix} \begin{bmatrix} c_0 \\ c_1 \end{bmatrix} = \begin{bmatrix} b_1 \\ b_2 \\ \vdots \\ b_n \end{bmatrix},$$

or

$$A \begin{bmatrix} c_0 \\ c_1 \end{bmatrix} = \vec{b}.$$

Note that $\text{rank}(A) = 2$, since $a_1 \neq a_2$.

The least-squares solution is

$$
\begin{bmatrix} c_0^* \\ c_1^* \end{bmatrix} = (A^T A)^{-1} A^T \vec{b} = \left(\begin{bmatrix} 1 & \cdots & 1 \\ a_1 & \cdots & a_n \end{bmatrix} \begin{bmatrix} 1 & a_1 \\ \vdots & \vdots \\ 1 & a_n \end{bmatrix} \right)^{-1} \begin{bmatrix} 1 & \cdots & 1 \\ a_1 & \cdots & a_n \end{bmatrix} \begin{bmatrix} b_1 \\ \vdots \\ b_n \end{bmatrix}
$$

$$
= \begin{bmatrix} n & \sum_i a_i \\ \sum_i a_i & \sum_i a_i^2 \end{bmatrix}^{-1} \begin{bmatrix} \sum_i b_i \\ \sum_i a_i b_i \end{bmatrix} \qquad \text{(where } \sum_i \text{ refers to the sum for } i = 1, \ldots, n)
$$

$$
= \frac{1}{n(\sum_i a_i^2) - (\sum_i a_i)^2} \begin{bmatrix} \sum_i a_i^2 & -\sum_i a_i \\ -\sum_i a_i & n \end{bmatrix} \begin{bmatrix} \sum_i b_i \\ \sum_i a_i b_i \end{bmatrix}.
$$

We have found that

$$
c_0^* = \frac{\left(\sum_i a_i^2\right)\left(\sum_i b_i\right) - \left(\sum_i a_i\right)\left(\sum_i a_i b_i\right)}{n\left(\sum_i a_i^2\right) - \left(\sum_i a_i\right)^2},
$$

$$
c_1^* = \frac{n\left(\sum_i a_i b_i\right) - \left(\sum_i a_i\right)\left(\sum_i b_i\right)}{n\left(\sum_i a_i^2\right) - \left(\sum_i a_i\right)^2}.
$$

These formulas are well known to statisticians. There is no need to memorize them. ∎

We conclude this section with an example for multivariate data fitting.

EXAMPLE 6   In the accompanying table, we list the scores of five students in the three exams given in a class.

	h: Hour Exam	m: Midterm Exam	f: Final Exam
Gabriel	76	48	43
Kyle	92	92	90
Faruk	68	82	64
Yasmine	86	68	69
Alec	54	70	50

Find the function of the form $f = c_0 + c_1 h + c_2 m$ that best fits these data, using least squares. What score f does your formula predict for Marilyn, another student, whose scores in the first two exams were $h = 92$ and $m = 72$?

Solution

We attempt to solve the system

$$
\begin{vmatrix} c_0 + 76c_1 + 48c_2 = 43 \\ c_0 + 92c_1 + 92c_2 = 90 \\ c_0 + 68c_1 + 82c_2 = 64 \\ c_0 + 86c_1 + 68c_2 = 69 \\ c_0 + 54c_1 + 70c_2 = 50 \end{vmatrix}.
$$

The least-squares solution is

$$\begin{bmatrix} c_0^* \\ c_1^* \\ c_2^* \end{bmatrix} = (A^T A)^{-1} A^T \vec{b} \approx \begin{bmatrix} -42.4 \\ 0.639 \\ 0.799 \end{bmatrix}.$$

The function which gives the best fit is approximately

$$f = -42.4 + 0.639h + 0.799m.$$

This formula predicts the score

$$f = -42.4 + 0.639 \cdot 92 + 0.799 \cdot 72 \approx 74$$

for Marilyn. ∎

EXERCISES 5.4

GOAL *Use the formula* $(\text{im } A)^\perp = \ker(A^T)$. *Apply the characterization of* $\text{proj}_V \vec{x}$ *as the vector in* V *"closest to* $\vec{x}$." *Find the least-squares solutions of a linear system* $A\vec{x} = \vec{b}$ *using the normal equation* $A^T A\vec{x} = A^T \vec{b}$.

1. Consider the subspace $\text{im}(A)$ of $\mathbb{R}^2$, where

$$A = \begin{bmatrix} 2 & 4 \\ 3 & 6 \end{bmatrix}.$$

Find a basis of $\ker(A^T)$, and draw a sketch illustrating the formula

$$(\text{im } A)^\perp = \ker(A^T)$$

in this case.

2. Consider the subspace $\text{im}(A)$ of $\mathbb{R}^3$, where

$$A = \begin{bmatrix} 1 & 1 \\ 1 & 2 \\ 1 & 3 \end{bmatrix}.$$

Find a basis of $\ker(A^T)$, and draw a sketch illustrating the formula $(\text{im } A)^\perp = \ker(A^T)$ in this case.

3. Consider a subspace V of $\mathbb{R}^n$. Let $\vec{v}_1, \ldots, \vec{v}_p$ be a basis of V and $\vec{w}_1, \ldots, \vec{w}_q$ a basis of $V^\perp$. Is $\vec{v}_1, \ldots, \vec{v}_p, \vec{w}_1, \ldots, \vec{w}_q$ a basis of $\mathbb{R}^n$? Explain.

4. Let A be an $n \times m$ matrix. Is the formula

$$(\ker A)^\perp = \text{im}(A^T)$$

necessarily true? Explain.

5. Let V be the solution space of the linear system

$$\begin{vmatrix} x_1 + x_2 + x_3 + x_4 = 0 \\ x_1 + 2x_2 + 5x_3 + 4x_4 = 0 \end{vmatrix}.$$

Find a basis of $V^\perp$.

6. If A is an $n \times m$ matrix, is the formula

$$\text{im}(A) = \text{im}(AA^T)$$

necessarily true? Explain.

7. Consider a symmetric $n \times n$ matrix A. What is the relationship between $\text{im}(A)$ and $\ker(A)$?

8. Consider a linear transformation $L(\vec{x}) = A\vec{x}$ from $\mathbb{R}^n$ to $\mathbb{R}^m$, with $\ker(L) = \{\vec{0}\}$. The *pseudoinverse* L^+ of L is the transformation from $\mathbb{R}^m$ to $\mathbb{R}^n$ given by

$$L^+(\vec{y}) = \left(\text{the least-squares solution of } L(\vec{x}) = \vec{y}\right).$$

a. Show that the transformation L^+ is linear. Find the matrix A^+ of L^+, in terms of the matrix A of L.
b. If L is invertible, what is the relationship between L^+ and L^{-1}?
c. What is $L^+\left(L(\vec{x})\right)$, for $\vec{x}$ in $\mathbb{R}^n$?
d. What is $L\left(L^+(\vec{y})\right)$, for $\vec{y}$ in $\mathbb{R}^m$?
e. Find L^+ for the linear transformation

$$L(\vec{x}) = \begin{bmatrix} 1 & 0 \\ 0 & 1 \\ 0 & 0 \end{bmatrix} \vec{x}.$$

9. Consider the linear system $A\vec{x} = \vec{b}$, where

$$A = \begin{bmatrix} 1 & 3 \\ 2 & 6 \end{bmatrix} \quad \text{and} \quad \vec{b} = \begin{bmatrix} 10 \\ 20 \end{bmatrix}.$$

a. Draw a sketch showing the following subsets of $\mathbb{R}^2$:
   - the kernel of A, and $(\ker A)^\perp$
   - the image of A^T
   - the solution set S of the system $A\vec{x} = \vec{b}$
b. What relationship do you observe between $\ker(A)$ and $\text{im}(A^T)$? Explain.
c. What relationship do you observe between $\ker(A)$ and S? Explain.
d. Find the unique vector $\vec{x}_0$ in the intersection of S and $(\ker A)^\perp$. Show $\vec{x}_0$ on your sketch.
e. What can you say about the length of $\vec{x}_0$ compared with the length of all other vectors in S?

10. Consider a consistent system $A\vec{x} = \vec{b}$.

  a. Show that this system has a solution $\vec{x}_0$ in $(\ker A)^\perp$. [*Hint*: An arbitrary solution $\vec{x}$ of the system can be written as $\vec{x} = \vec{x}_h + \vec{x}_0$, where $\vec{x}_h$ is in $\ker(A)$ and $\vec{x}_0$ is in $(\ker A)^\perp$.]

  b. Show that the system $A\vec{x} = \vec{b}$ has only one solution in $(\ker A)^\perp$. [*Hint*: If $\vec{x}_0$ and $\vec{x}_1$ are two solutions in $(\ker A)^\perp$, think about $\vec{x}_1 - \vec{x}_0$.]

  c. If $\vec{x}_0$ is the solution in $(\ker A)^\perp$ and $\vec{x}_1$ is another solution of the system $A\vec{x} = \vec{b}$, show that $\|\vec{x}_0\| < \|\vec{x}_1\|$. The vector $\vec{x}_0$ is called the *minimal solution* of the linear system $A\vec{x} = \vec{b}$.

11. Consider a linear transformation $L(\vec{x}) = A\vec{x}$ from $\mathbb{R}^n$ to $\mathbb{R}^m$, where $\mathrm{rank}(A) = m$. The *pseudoinverse* L^+ of L is the transformation from $\mathbb{R}^m$ to $\mathbb{R}^n$ given by

$$L^+(\vec{y}) = \big(\text{the minimal solution of the system } L(\vec{x}) = \vec{y}\big).$$

(See Exercise 10.)

  a. Show that the transformation L^+ is linear.
  b. What is $L\big(L^+(\vec{y})\big)$, for $\vec{y}$ in $\mathbb{R}^m$?
  c. What is $L^+\big(L(\vec{x})\big)$, for $\vec{x}$ in $\mathbb{R}^n$?
  d. Determine the image and kernel of L^+.
  e. Find L^+ for the linear transformation

$$L(\vec{x}) = \begin{bmatrix} 1 & 0 & 0 \\ 0 & 1 & 0 \end{bmatrix} \vec{x}.$$

12. Using Exercise 10 as a guide, define the term *minimal least-squares solution* of a linear system. Explain why the minimal least-squares solution $\vec{x}^*$ of a linear system $A\vec{x} = \vec{b}$ is in $(\ker A)^\perp$.

13. Consider a linear transformation $L(\vec{x}) = A\vec{x}$ from $\mathbb{R}^n$ to $\mathbb{R}^m$. The pseudoinverse L^+ of L is the transformation from $\mathbb{R}^m$ to $\mathbb{R}^n$ given by

$$L^+(\vec{y}) = \big(\text{the minimal least-squares solution} \\ \text{of the system } L(\vec{x}) = \vec{y}\big).$$

(See Exercises 8, 11, and 12 for special cases.)

  a. Show that the transformation L^+ is linear.
  b. What is $L^+\big(L(\vec{x})\big)$, for $\vec{x}$ in $\mathbb{R}^n$?
  c. What is $L\big(L^+(\vec{y})\big)$, for $\vec{y}$ in $\mathbb{R}^m$?
  d. Determine the image and kernel of L^+ $\big($in terms of $\mathrm{im}(A^T)$ and $\ker(A^T)\big)$.
  e. Find L^+ for the linear transformation

$$L(\vec{x}) = \begin{bmatrix} 2 & 0 & 0 \\ 0 & 0 & 0 \end{bmatrix} \vec{x}.$$

14. In the accompanying figure, we show the kernel and the image of a linear transformation L from $\mathbb{R}^2$ to $\mathbb{R}^2$, together with some vectors $\vec{v}_1, \vec{w}_1, \vec{w}_2, \vec{w}_3$. We are told that $L(\vec{v}_1) = \vec{w}_1$. For $i = 1, 2, 3$, find the vectors $L^+(\vec{w}_i)$, where L^+ is the pseudoinverse of L defined in

Exercise 13. Show your solutions in the figure, and explain how you found them.

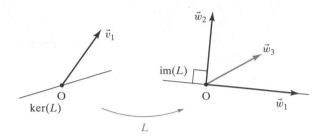

15. Consider an $m \times n$ matrix A with $\ker(A) = \{\vec{0}\}$. Show that there exists an $n \times m$ matrix B such that $BA = I_n$. (*Hint*: $A^T A$ is invertible.)

16. Use the formula $(\mathrm{im}\, A)^\perp = \ker(A^T)$ to prove the equation

$$\mathrm{rank}(A) = \mathrm{rank}(A^T).$$

17. Does the equation

$$\mathrm{rank}(A) = \mathrm{rank}(A^T A)$$

hold for all $n \times m$ matrices A? Explain.

18. Does the equation

$$\mathrm{rank}(A^T A) = \mathrm{rank}(A A^T)$$

hold for all $n \times m$ matrices A? Explain. *Hint*: Exercise 17 is useful.

19. Find the least-squares solution $\vec{x}^*$ of the system

$$A\vec{x} = \vec{b}, \quad \text{where} \quad A = \begin{bmatrix} 1 & 0 \\ 0 & 1 \\ 0 & 0 \end{bmatrix} \quad \text{and} \quad \vec{b} = \begin{bmatrix} 1 \\ 1 \\ 1 \end{bmatrix}.$$

Use paper and pencil. Draw a sketch showing the vector $\vec{b}$, the image of A, the vector $A\vec{x}^*$, and the vector $\vec{b} - A\vec{x}^*$.

20. By using paper and pencil, find the least-squares solution $\vec{x}^*$ of the system

$$A\vec{x} = \vec{b}, \quad \text{where} \quad A = \begin{bmatrix} 1 & 1 \\ 1 & 0 \\ 0 & 1 \end{bmatrix} \quad \text{and} \quad \vec{b} = \begin{bmatrix} 3 \\ 3 \\ 3 \end{bmatrix}.$$

Verify that the vector $\vec{b} - A\vec{x}^*$ is perpendicular to the image of A.

21. Find the least-squares solution $\vec{x}^*$ of the system

$$A\vec{x} = \vec{b}, \quad \text{where} \quad A = \begin{bmatrix} 6 & 9 \\ 3 & 8 \\ 2 & 10 \end{bmatrix} \quad \text{and} \quad \vec{b} = \begin{bmatrix} 0 \\ 49 \\ 0 \end{bmatrix}.$$

Determine the error $\|\vec{b} - A\vec{x}^*\|$.

22. Find the least-squares solution $\vec{x}^*$ of the system

$$A\vec{x} = \vec{b}, \quad \text{where} \quad A = \begin{bmatrix} 3 & 2 \\ 5 & 3 \\ 4 & 5 \end{bmatrix} \quad \text{and} \quad \vec{b} = \begin{bmatrix} 5 \\ 9 \\ 2 \end{bmatrix}.$$

Determine the error $\|\vec{b} - A\vec{x}^*\|$.

23. Find the least-squares solution $\vec{x}^*$ of the system

$$A\vec{x} = \vec{b}, \quad \text{where} \quad A = \begin{bmatrix} 1 & 1 \\ 2 & 8 \\ 1 & 5 \end{bmatrix} \quad \text{and} \quad \vec{b} = \begin{bmatrix} 1 \\ -2 \\ 3 \end{bmatrix}.$$

Explain.

24. Find the least-squares solution $\vec{x}^*$ of the system

$$A\vec{x} = \vec{b}, \quad \text{where} \quad A = \begin{bmatrix} 1 \\ 2 \\ 3 \end{bmatrix} \quad \text{and} \quad \vec{b} = \begin{bmatrix} 3 \\ 2 \\ 7 \end{bmatrix}.$$

Draw a sketch showing the vector $\vec{b}$, the image of A, the vector $A\vec{x}^*$, and the vector $\vec{b} - A\vec{x}^*$.

25. Find the least-squares solutions $\vec{x}^*$ of the system $A\vec{x} = \vec{b}$, where

$$A = \begin{bmatrix} 1 & 3 \\ 2 & 6 \end{bmatrix} \quad \text{and} \quad \vec{b} = \begin{bmatrix} 5 \\ 0 \end{bmatrix}.$$

Use only paper and pencil. Draw a sketch.

26. Find the least-squares solutions $\vec{x}^*$ of the system $A\vec{x} = \vec{b}$, where

$$A = \begin{bmatrix} 1 & 2 & 3 \\ 4 & 5 & 6 \\ 7 & 8 & 9 \end{bmatrix} \quad \text{and} \quad \vec{b} = \begin{bmatrix} 1 \\ 0 \\ 0 \end{bmatrix}.$$

27. Consider an inconsistent linear system $A\vec{x} = \vec{b}$, where A is a 3×2 matrix. We are told that the least-squares solution of this system is $\vec{x}^* = \begin{bmatrix} 7 \\ 11 \end{bmatrix}$. Consider an orthogonal 3×3 matrix S. Find the least-squares solution(s) of the system $SA\vec{x} = S\vec{b}$.

28. Consider an orthonormal basis $\vec{u}_1, \vec{u}_2, \ldots, \vec{u}_n$ in $\mathbb{R}^n$. Find the least-squares solution(s) of the system

$$A\vec{x} = \vec{u}_n,$$

where

$$A = \begin{bmatrix} | & | & & | \\ \vec{u}_1 & \vec{u}_2 & \cdots & \vec{u}_{n-1} \\ | & | & & | \end{bmatrix}.$$

29. Find the least-squares solution of the system

$$A\vec{x} = \vec{b}, \text{ where } A = \begin{bmatrix} 1 & 1 \\ 10^{-10} & 0 \\ 0 & 10^{-10} \end{bmatrix} \text{ and}$$

$$\vec{b} = \begin{bmatrix} 1 \\ 10^{-10} \\ 10^{-10} \end{bmatrix}.$$

Describe and explain the difficulties you may encounter if you use technology. Then find the solution using paper and pencil.

30. Fit a linear function of the form $f(t) = c_0 + c_1 t$ to the data points $(0, 0)$, $(0, 1)$, $(1, 1)$, using least squares. Use only paper and pencil. Sketch your solution, and explain why it makes sense.

31. Fit a linear function of the form $f(t) = c_0 + c_1 t$ to the data points $(0, 3)$, $(1, 3)$, $(1, 6)$, using least squares. Sketch the solution.

32. Fit a quadratic polynomial to the data points $(0, 27)$, $(1, 0)$, $(2, 0)$, $(3, 0)$, using least squares. Sketch the solution.

33. Find the trigonometric function of the form $f(t) = c_0 + c_1 \sin(t) + c_2 \cos(t)$ that best fits the data points $(0, 0)$, $(1, 1)$, $(2, 2)$, $(3, 3)$, using least squares. Sketch the solution together with the function $g(t) = t$.

34. Find the function of the form

$$f(t) = c_0 + c_1 \sin(t) + c_2 \cos(t) + c_3 \sin(2t) + c_4 \cos(2t)$$

that best fits the data points $(0, 0)$, $(0.5, 0.5)$, $(1, 1)$, $(1.5, 1.5)$, $(2, 2)$, $(2.5, 2.5)$, $(3, 3)$, using least squares. Sketch the solution, together with the function $g(t) = t$.

35. Suppose you wish to fit a function of the form

$$f(t) = c + p \sin(t) + q \cos(t)$$

to a given continuous function $g(t)$ on the closed interval from 0 to 2π. One approach is to choose n equally spaced points a_i between 0 and 2π [$a_i = i \cdot (2\pi/n)$, for $i = 1, \ldots, n$, say]. We can fit a function

$$f_n(t) = c_n + p_n \sin(t) + q_n \cos(t)$$

to the data points $(a_i, g(a_i))$, for $i = 1, \ldots, n$. Now examine what happens to the coefficients c_n, p_n, q_n of $f_n(t)$ as n approaches infinity:

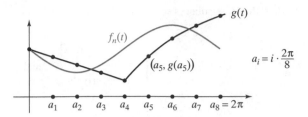

To find $f_n(t)$, we make an attempt to solve the equations

$$f_n(a_i) = g(a_i), \quad \text{for} \quad i = 1, \ldots, n,$$

or

$$\begin{vmatrix} c_n + p_n \sin(a_1) + q_n \cos(a_1) = g(a_1) \\ c_n + p_n \sin(a_2) + q_n \cos(a_2) = g(a_2) \\ \vdots \qquad \qquad \vdots \\ c_n + p_n \sin(a_n) + q_n \cos(a_n) = g(a_n) \end{vmatrix},$$

or

$$A_n \begin{bmatrix} c_n \\ p_n \\ q_n \end{bmatrix} = \vec{b}_n,$$

where

$$A_n = \begin{bmatrix} 1 & \sin(a_1) & \cos(a_1) \\ 1 & \sin(a_2) & \cos(a_2) \\ \vdots & \vdots & \vdots \\ 1 & \sin(a_n) & \cos(a_n) \end{bmatrix}, \quad \vec{b}_n = \begin{bmatrix} g(a_1) \\ g(a_2) \\ \vdots \\ g(a_n) \end{bmatrix}.$$

a. Find the entries of the matrix $A_n^T A_n$ and the components of the vector $A_n^T \vec{b}_n$.

b. Find

$$\lim_{n \to \infty} \left(\frac{2\pi}{n} A_n^T A_n \right) \quad \text{and} \quad \lim_{n \to \infty} \left(\frac{2\pi}{n} A_n^T \vec{b} \right).$$

[*Hint*: Interpret the entries of the matrix $(2\pi/n)A_n^T A_n$ and the components of the vector $(2\pi/n)A^T \vec{b}$ as Riemann sums. Then the limits are the corresponding Riemann integrals. Evaluate as many integrals as you can. Note that

$$\lim_{n \to \infty} \left(\frac{2\pi}{n} A_n^T A_n \right)$$

is a diagonal matrix.]

c. Find

$$\lim_{n \to \infty} \begin{bmatrix} c_n \\ p_n \\ q_n \end{bmatrix} = \lim_{n \to \infty} (A_n^T A_n)^{-1} A_n^T \vec{b}_n$$

$$= \lim_{n \to \infty} \left[\left(\frac{2\pi}{n} A_n^T A_n \right)^{-1} \left(\frac{2\pi}{n} A_n^T \vec{b}_n \right) \right]$$

$$= \left[\lim_{n \to \infty} \left(\frac{2\pi}{n} A_n^T A_n \right) \right]^{-1} \lim_{n \to \infty} \left(\frac{2\pi}{n} A_n^T \vec{b}_n \right).$$

The resulting vector $\begin{bmatrix} c \\ p \\ q \end{bmatrix}$ gives you the coefficient of the desired function

$$f(t) = \lim_{n \to \infty} f_n(t).$$

Write $f(t)$. The function $f(t)$ is called the first *Fourier approximation* of $g(t)$. The Fourier approximation satisfies a "continuous" least-squares condition, an idea we will make more precise in the next section.

36. Let $S(t)$ be the number of daylight hours on the tth day of the year 2008 in Rome, Italy. We are given the following data for $S(t)$:

Day	t	$S(t)$
February 1	32	10
March 17	77	12
April 30	121	14
May 31	152	15

We wish to fit a trigonometric function of the form

$$f(t) = a + b \sin\left(\frac{2\pi}{366}t\right) + c \cos\left(\frac{2\pi}{366}t\right)$$

to these data. Find the best approximation of this form, using least squares.

How many daylight hours does your model predict for the longest day of the year 2008? (The actual value is 15 hours, 13 minutes, 39 seconds.)

37. The accompanying table lists several commercial airplanes, the year they were introduced, and the number of displays in the cockpit.

Plane	Year t	Displays d
Douglas DC-3	'35	35
Lockheed Constellation	'46	46
Boeing 707	'59	77
Concorde	'69	133

a. Fit a linear function of the form $\log(d) = c_0 + c_1 t$ to the data points $(t_i, \log(d_i))$, using least squares.

b. Use your answer in part (a) to fit an exponential function $d = ka^t$ to the data points (t_i, d_i).

c. The Airbus A320 was introduced in 1988. Based on your answer in part (b), how many displays do you expect in the cockpit of this plane? (There are 93 displays in the cockpit of an Airbus A320. Explain.)

38. In the accompanying table, we list the height h, the gender g, and the weight w of some young adults.

Height h (in Inches above 5 Feet)	Gender g ($1 =$ "Female," $0 =$ "Male")	Weight w (in Pounds)
2	1	110
12	0	180
5	1	120
11	1	160
6	0	160

Fit a function of the form

$$w = c_0 + c_1 h + c_2 g$$

to these data, using least squares. Before you do the computations, think about the signs of c_1 and c_2. What signs would you expect if these data were representative of the general population? Why? What is the sign of c_0? What is the practical significance of c_0?

39. In the accompanying table, we list the estimated number g of genes and the estimated number z of cell types for various organisms.

Organism	Number of Genes, g	Number of Cell Types, z
Humans	600,000	250
Annelid worms	200,000	60
Jellyfish	60,000	25
Sponges	10,000	12
Yeasts	2,500	5

a. Fit a function of the form $\log(z) = c_0 + c_1 \log(g)$ to the data points $\big(\log(g_i), \log(z_i)\big)$, using least squares.

b. Use your answer in part (a) to fit a power function $z = kg^n$ to the data points (g_i, z_i).

c. Using the theory of self-regulatory systems, scientists developed a model that predicts that z is a square-root

function of g (i.e., $a = k\sqrt{g}$, for some constant k). Is your answer in part (b) reasonably close to this form?

40. Consider the data in the following table:

Planet	a Mean Distance from the Sun (in Astronomical Units)	D Period of Revolution (in Earth Years)
Mercury	0.387	0.241
Earth	1	1
Jupiter	5.20	11.86
Uranus	19.18	84.0
Pluto	39.53	248.5

Use the methods discussed in Exercise 39 to fit a power function of the form $D = ka^n$ to these data. Explain, in terms of Kepler's laws of planetary motion. Explain why the constant k is close to 1.

41. In the accompanying table, we list the public debt D of the United States (in billions of dollars), in various years t (as of September 30).

Year	1975	1985	1995	2005
D	533	1,823	4,974	7,933

a. Letting $t = 0$ in 1975, fit a linear function of the form $\log(D) = c_0 + c_1 t$ to the data points $(t_i, \log(D_i))$, using least squares. Use the result to fit an exponential function to the data points (t_i, D_i).

b. What debt does your formula in part (a) predict for 2015?

42. If A is any matrix, show that the linear transformation $L(\vec{x}) = A\vec{x}$ from $\operatorname{im}(A^T)$ to $\operatorname{im}(A)$ is an isomorphism. This provides yet another proof of the formula $\operatorname{rank}(A) = \operatorname{rank}(A^T)$.

5.5 Inner Product Spaces

Let's take a look back at what we have done thus far in this text. In Chapters 1 through 3, we studied the basic concepts of linear algebra in the concrete context of $\mathbb{R}^n$. Recall that these concepts are all defined in terms of two operations: addition and scalar multiplication. In Chapter 4, we saw that it can be both natural and useful to apply the language of linear algebra to objects other than vectors in $\mathbb{R}^n$, for example, to functions. We introduced the term *linear space* (or *vector space*) for a set that behaves like $\mathbb{R}^n$ as far as addition and scalar multiplication are concerned.

In this chapter, a new operation for vectors in $\mathbb{R}^n$ takes center stage: the *dot product*. In Sections 5.1 through 5.4, we studied concepts that are defined in terms of the dot product, the most important of them being the *length* of vectors and *orthogonality* of vectors. In this section, we will see that it can be useful to define a product analogous to the dot product in a linear space other than $\mathbb{R}^n$. These generalized dot products are called *inner products*. Once we have an inner product in a linear space, we can define length and orthogonality in that space just as in $\mathbb{R}^n$, and we can generalize all the key ideas and theorems of Sections 5.1 through 5.4.

Definition 5.5.1   Inner products and inner product spaces

An *inner product* in a linear space V is a rule that assigns a real scalar (denoted by $\langle f, g \rangle$) to any pair f, g of elements of V, such that the following properties hold for all f, g, h in V, and all c in $\mathbb{R}$:

a. $\langle f, g \rangle = \langle g, f \rangle$ (symmetry)
b. $\langle f + h, g \rangle = \langle f, g \rangle + \langle h, g \rangle$
c. $\langle cf, g \rangle = c \langle f, g \rangle$
d. $\langle f, f \rangle > 0$, for all nonzero f in V (positive definiteness)

A linear space endowed with an inner product is called an *inner product space*.

Properties b and c express the fact that $T(f) = \langle f, g \rangle$ is a linear transformation from V to $\mathbb{R}$, for a fixed g in V.

Compare these rules with those for the dot product in $\mathbb{R}^n$, listed in the Appendix, Theorem A.5. Roughly speaking, an inner product space behaves like $\mathbb{R}^n$ as far as addition, scalar multiplication, and the dot product are concerned.

EXAMPLE 1   Consider the linear space $C[a, b]$ consisting of all continuous functions whose domain is the closed interval $[a, b]$, where $a < b$. See Figure 1.

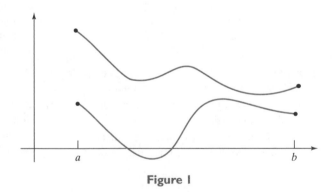

Figure 1

For functions f and g in $C[a, b]$, we define

$$\langle f, g \rangle = \int_a^b f(t)g(t)\, dt.$$

The verification of the first three axioms for an inner product is straightforward. For example,

$$\langle f, g \rangle = \int_a^b f(t)g(t)\, dt = \int_a^b g(t)f(t)\, dt = \langle g, f \rangle.$$

The verification of the last axiom requires a bit of calculus. We leave it as Exercise 1.

Recall that the Riemann integral $\int_a^b f(t)g(t)\, dt$ is the limit of the Riemann sum $\sum_{i=1}^m f(t_k)g(t_k)\Delta t$, where the t_k can be chosen as equally spaced points on the interval $[a, b]$. See Figure 2.

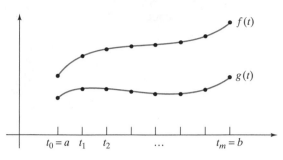

Figure 2

Then

$$\langle f, g \rangle = \int_a^b f(t)g(t)\,dt \approx \sum_{k=1}^m f(t_k)g(t_k)\Delta t = \left(\begin{bmatrix} f(t_1) \\ f(t_2) \\ \vdots \\ f(t_m) \end{bmatrix} \cdot \begin{bmatrix} g(t_1) \\ g(t_2) \\ \vdots \\ g(t_m) \end{bmatrix} \right) \Delta t$$

for large m.

This approximation shows that the inner product $\langle f, g \rangle = \int_a^b f(t)g(t)\,dt$ for functions is a continuous version of the dot product: The more subdivisions you choose, the better the dot product on the right will approximate the inner product $\langle f, g \rangle$. ∎

EXAMPLE 2 Let ℓ_2 be the space of all "square-summable" infinite sequences, that is, sequences

$$\vec{x} = (x_0, x_1, x_2, \ldots, x_n, \ldots)$$

such that $\sum_{i=0}^\infty x_i^2 = x_0^2 + x_1^2 + \cdots$ converges. In this space we can define the inner product

$$\langle \vec{x}, \vec{y} \rangle = \sum_{i=0}^\infty x_i y_i = x_0 y_0 + x_1 y_1 + \cdots.$$

(Show that this series converges.) The verification of the axioms is straightforward. Compare this with Exercises 4.1.15 and 5.1.18. ∎

EXAMPLE 3 The *trace* of a square matrix is the sum of its diagonal entries. For example,

$$\mathrm{trace} \begin{bmatrix} 1 & 2 \\ 3 & 4 \end{bmatrix} = 1 + 4 = 5.$$

In $\mathbb{R}^{n \times m}$, the space of all $n \times m$ matrices, we can define the inner product

$$\langle A, B \rangle = \mathrm{trace}(A^T B).$$

We will verify the first and the fourth axioms.

$$\langle A, B \rangle = \mathrm{trace}(A^T B) = \mathrm{trace}\left((A^T B)^T\right) = \mathrm{trace}(B^T A) = \langle B, A \rangle$$

To check that $\langle A, A \rangle > 0$ for nonzero A, write A in terms of its columns:

$$A = \begin{bmatrix} | & | & & | \\ \vec{v}_1 & \vec{v}_2 & \dots & \vec{v}_m \\ | & | & & | \end{bmatrix}$$

$$\langle A, A \rangle = \operatorname{trace}(A^T A) = \operatorname{trace}\left(\begin{bmatrix} - & \vec{v}_1^T & - \\ - & \vec{v}_2^T & - \\ & \vdots & \\ - & \vec{v}_m^T & - \end{bmatrix} \begin{bmatrix} | & | & & | \\ \vec{v}_1 & \vec{v}_2 & \dots & \vec{v}_m \\ | & | & & | \end{bmatrix} \right)$$

$$= \operatorname{trace}\left(\begin{bmatrix} \|\vec{v}_1\|^2 & \dots & & \dots \\ \dots & \|\vec{v}_2\|^2 & & \dots \\ \vdots & \vdots & \ddots & \vdots \\ & & \dots & \|\vec{v}_m\|^2 \end{bmatrix} \right)$$

$$= \|\vec{v}_1\|^2 + \|\vec{v}_2\|^2 + \dots + \|\vec{v}_m\|^2.$$

If A is nonzero, then at least one of the column vectors $\vec{v}_i$ is nonzero, so that the sum $\|\vec{v}_1\|^2 + \|\vec{v}_2\|^2 + \dots + \|\vec{v}_m\|^2$ is positive, as desired. ∎

We can introduce the basic concepts of geometry for an inner product space exactly as we did in $\mathbb{R}^n$ for the dot product.

Definition 5.5.2 Norm, orthogonality

The *norm* (or magnitude) of an element f of an inner product space is

$$\|f\| = \sqrt{\langle f, f \rangle}.$$

Two elements f and g of an inner product space are called *orthogonal* (or perpendicular) if

$$\langle f, g \rangle = 0.$$

We can define the *distance* between two elements of an inner product space as the norm of their difference:

$$\operatorname{dist}(f, g) = \|f - g\|.$$

Consider the space $C[a, b]$, with the inner product defined in Example 1.

In physics, the quantity $\|f\|^2$ can often be interpreted as *energy*. For example, it describes the acoustic energy of a periodic sound wave $f(t)$ and the elastic potential energy of a uniform string with vertical displacement $f(x)$. (See Figure 3.) The quantity $\|f\|^2$ may also measure thermal or electric energy.

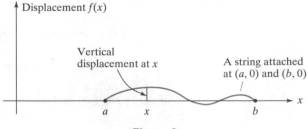

Figure 3

EXAMPLE 4   In the inner product space $C[0, 1]$ with $\langle f, g \rangle = \int_0^1 f(t)g(t)\,dt$, find $\|f\|$ for $f(t) = t^2$.

Solution

$$\|f\| = \sqrt{\langle f, f \rangle} = \sqrt{\int_0^1 t^4\,dt} = \sqrt{\frac{1}{5}}$$ ∎

EXAMPLE 5   Show that $f(t) = \sin(t)$ and $g(t) = \cos(t)$ are orthogonal in the inner product space $C[0, 2\pi]$ with $\langle f, g \rangle = \int_0^{2\pi} f(t)g(t)\,dt$.

Solution

$$\langle f, g \rangle = \int_0^{2\pi} \sin(t)\cos(t)\,dt = \left[\frac{1}{2}\sin^2(t)\right]\Big|_0^{2\pi} = 0$$ ∎

EXAMPLE 6   Find the distance between $f(t) = t$ and $g(t) = 1$ in $C[0, 1]$.

Solution

$$\operatorname{dist}\langle f, g \rangle = \sqrt{\int_0^1 (t-1)^2\,dt} = \sqrt{\left[\frac{1}{3}(t-1)^3\right]\Big|_0^1} = \frac{1}{\sqrt{3}}$$ ∎

The results and procedures discussed for the dot product generalize to arbitrary inner product spaces. For example, the Pythagorean theorem holds; the Gram–Schmidt process can be used to construct an orthonormal basis of a (finite dimensional) inner product space; and the Cauchy–Schwarz inequality tells us that $|\langle f, g \rangle| \le \|f\|\,\|g\|$, for two elements f and g of an inner product space.

Orthogonal Projections

In an inner product space V, consider a finite dimensional subspace W with orthonormal basis $g_1, \ldots, g_m$. The orthogonal projection $\operatorname{proj}_W f$ of an element f of V onto W is defined as the unique element of W such that $f - \operatorname{proj}_W f$ is orthogonal to W. As in the case of the dot product in $\mathbb{R}^n$, the orthogonal projection is given by the following formula.

Theorem 5.5.3   Orthogonal projection

If $g_1, \ldots, g_m$ is an orthonormal basis of a subspace W of an inner product space V, then

$$\operatorname{proj}_W f = \langle g_1, f \rangle g_1 + \cdots + \langle g_m, f \rangle g_m,$$

for all f in V.   ∎

(Verify this by checking that $\langle f - \operatorname{proj}_W f, g_i \rangle = 0$ for $i = 1, \ldots, m$.)

We may think of $\operatorname{proj}_W f$ as the element of W closest to f. In other words, if we choose another element h of W, then the distance between f and h will exceed the distance between f and $\operatorname{proj}_W f$.

As an example, consider a subspace W of $C[a, b]$, with the inner product introduced in Example 1. Then $\operatorname{proj}_W f$ is the function g in W that is closest to f, in the sense that

$$\text{dist}(f, g) = \|f - g\| = \sqrt{\int_a^b \left(f(t) - g(t)\right)^2 dt}$$

is least.

The requirement that

$$\int_a^b \left(f(t) - g(t)\right)^2 dt$$

be minimal is a *continuous least-squares condition,* as opposed to the discrete least-squares conditions we discussed in Section 5.4. We can use the discrete least-squares condition to fit a function g of a certain type to some data points (a_k, b_k), while the continuous least-squares condition can be used to fit a function g of a certain type to a given function f. (Functions of a certain type are frequently polynomials of a certain degree or trigonometric functions of a certain form.) See Figures 4(a) and 4(b).

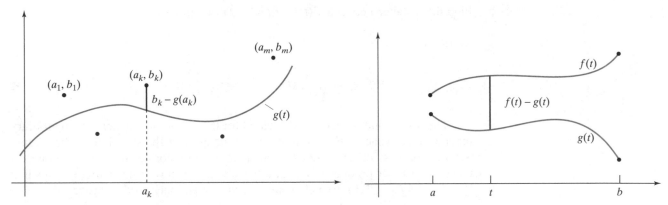

Figure 4 (a) Discrete least-squares condition: $\sum_{k=1}^m \left(b_k - g(a_k)\right)^2$ is minimal. (b) Continuous least-squares condition: $\int_a^b \left(f(t) - g(t)\right)^2 dt$ is minimal.

We can think of the continuous least-squares condition as a limiting case of a discrete least-squares condition by writing

$$\int_a^b \left(f(t) - g(t)\right)^2 dt = \lim_{m \to \infty} \sum_{k=1}^m \left(f(t_k) - g(t_k)\right)^2 \Delta t.$$

EXAMPLE 7   Find the linear function of the form $g(t) = a + bt$ that best fits the function $f(t) = e^t$ over the interval from -1 to 1, in a continuous least-squares sense.

Solution

We need to find $\text{proj}_{P_1} f$. We first find an orthonormal basis of P_1 for the given inner product; then, we will use Theorem 5.5.3. In general, we have to use the Gram–Schmidt process to find an orthonormal basis of an inner product space. Because the two functions $1, t$ in the standard basis of P_1 are orthogonal already, or

$$\langle 1, t \rangle = \int_{-1}^1 t \, dt = 0,$$

we merely need to divide each function by its norm:

$$\|1\| = \sqrt{\int_{-1}^{1} 1 \, dt} = \sqrt{2} \quad \text{and} \quad \|t\| = \sqrt{\int_{-1}^{1} t^2 \, dt} = \sqrt{\frac{2}{3}}.$$

An orthonormal basis of P_1 is

$$\frac{1}{\sqrt{2}} 1 \quad \text{and} \quad \sqrt{\frac{3}{2}} t.$$

Now,

$$\text{proj}_{P_1} f = \frac{1}{2} \langle 1, f \rangle 1 + \frac{3}{2} \langle t, f \rangle t$$

$$= \frac{1}{2} (e - e^{-1}) + 3e^{-1} t. \quad \text{(We omit the straightforward computations.)}$$

See Figure 5. ∎

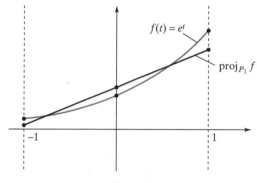

Figure 5

What follows is one of the major applications of this theory.

Fourier Analysis[7]

In the space $C[-\pi, \pi]$, we introduce an inner product that is a slight modification of the definition given in Example 1:

$$\langle f, g \rangle = \frac{1}{\pi} \int_{-\pi}^{\pi} f(t) g(t) \, dt.$$

The factor $1/\pi$ is introduced to facilitate the computations. Convince yourself that this is indeed an inner product. (Compare with Exercise 7.)

More generally, we can consider this inner product in the space of all *piecewise continuous functions* defined in the interval $[-\pi, \pi]$. These are functions $f(t)$ that are continuous except for a finite number of *jump-discontinuities* [i.e., points c where the one-sided limits $\lim_{t \to c^-} f(t)$ and $\lim_{t \to c^+} f(t)$ both exist, but are not equal]. Also, it is required that $f(c)$ equal one of the two one-sided limits. Let us consider the piecewise continuous functions with $f(c) = \lim_{t \to c^-} f(t)$. See Figure 6.

For a positive integer n, consider the subspace T_n of $C[-\pi, \pi]$ that is defined as the span of the functions $1, \sin(t), \cos(t), \sin(2t), \cos(2t), \ldots, \sin(nt), \cos(nt)$.

---

[7]Named after the French mathematician Jean-Baptiste-Joseph Fourier (1768–1830), who developed the subject in his *Théorie analytique de la chaleur* (1822), where he investigated the conduction of heat in very thin sheets of metal. Baron Fourier was also an Egyptologist and government administrator; he accompanied Napoléon on his expedition to Egypt in 1798.

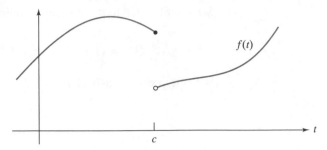

Figure 6 $f(t)$ has a jump-discontinuity at $t = c$.

The space T_n consists of all functions of the form

$$f(t) = a + b_1 \sin(t) + c_1 \cos(t) + \cdots + b_n \sin(nt) + c_n \cos(nt),$$

called *trigonometric polynomials* of order $\leq n$.

From calculus, you may recall the *Euler identities*:

$$\int_{-\pi}^{\pi} \sin(pt) \cos(mt) \, dt = 0, \quad \text{for integers } p, m,$$

$$\int_{-\pi}^{\pi} \sin(pt) \sin(mt) \, dt = 0, \quad \text{for distinct integers } p, m,$$

$$\int_{-\pi}^{\pi} \cos(pt) \cos(mt) \, dt = 0, \quad \text{for distinct integers } p, m.$$

These equations tell us that the functions $1, \sin(t), \cos(t), \ldots, \sin(nt), \cos(nt)$ are orthogonal to one another (and therefore linearly independent).

Another of Euler's identities tells us that

$$\int_{-\pi}^{\pi} \sin^2(mt) \, dt = \int_{-\pi}^{\pi} \cos^2(mt) \, dt = \pi,$$

for positive integers m. This means that the functions $\sin(t), \cos(t), \ldots, \sin(nt)$, $\cos(nt)$ all have norm 1 with respect to the given inner product. This is why we chose the inner product as we did, with the factor $\frac{1}{\pi}$.

The norm of the function $f(t) = 1$ is

$$\|f\| = \sqrt{\frac{1}{\pi} \int_{-\pi}^{\pi} 1 \, dt} = \sqrt{2};$$

therefore,

$$g(t) = \frac{f(t)}{\|f(t)\|} = \frac{1}{\sqrt{2}}$$

is a function of norm 1.

Theorem 5.5.4 An orthonormal basis of T_n

Let T_n be the space of all trigonometric polynomials of order $\leq n$, with the inner product

$$\langle f, g \rangle = \frac{1}{\pi} \int_{-\pi}^{\pi} f(t) g(t) \, dt.$$

Then the functions

$$\frac{1}{\sqrt{2}}, \sin(t), \cos(t), \sin(2t), \cos(2t), \ldots, \sin(nt), \cos(nt)$$

form an orthonormal basis of T_n. ∎

For a piecewise continuous function f, we can consider

$$f_n = \text{proj}_{T_n} f.$$

As discussed after Theorem 5.5.3, f_n is the trigonometric polynomial in T_n that best approximates f, in the sense that

$$\text{dist}(f, f_n) < \text{dist}(f, g),$$

for all other g in T_n.

We can use Theorems 5.5.3 and 5.5.4 to find a formula for $f_n = \text{proj}_{T_n} f$.

Theorem 5.5.5   Fourier coefficients

If f is a piecewise continuous function defined on the interval $[-\pi, \pi]$, then its best approximation f_n in T_n is

$$f_n(t) = \text{proj}_{T_n} f(t)$$
$$= a_0 \frac{1}{\sqrt{2}} + b_1 \sin(t) + c_1 \cos(t) + \cdots + b_n \sin(nt) + c_n \cos(nt),$$

where

$$b_k = \langle f(t), \sin(kt) \rangle = \frac{1}{\pi} \int_{-\pi}^{\pi} f(t) \sin(kt)\, dt,$$

$$c_k = \langle f(t), \cos(kt) \rangle = \frac{1}{\pi} \int_{-\pi}^{\pi} f(t) \cos(kt)\, dt,$$

$$a_0 = \left\langle f(t), \frac{1}{\sqrt{2}} \right\rangle = \frac{1}{\sqrt{2\pi}} \int_{-\pi}^{\pi} f(t)\, dt.$$

The b_k, the c_k, and a_0 are called the *Fourier coefficients* of the function f. The function

$$f_n(t) = a_0 \frac{1}{\sqrt{2}} + b_1 \sin(t) + c_1 \cos(t) + \cdots + b_n \sin(nt) + c_n \cos(nt)$$

is called the nth-order *Fourier approximation* of f.   ∎

Note that the constant term, written somewhat awkwardly, is

$$a_0 \frac{1}{\sqrt{2}} = \frac{1}{2\pi} \int_{-\pi}^{\pi} f(t)\, dt,$$

which is the average value of the function f between $-\pi$ and π. It makes sense that the best way to approximate $f(t)$ by a constant function is to take the average value of $f(t)$.

The function $b_k \sin(kt) + c_k \cos(kt)$ is called the kth *harmonic* of $f(t)$. Using elementary trigonometry, we can write the harmonics alternatively as

$$b_k \sin(kt) + c_k \cos(kt) = A_k \sin\big(k(t - \delta_k)\big),$$

where $A_k = \sqrt{b_k^2 + c_k^2}$ is the *amplitude* of the harmonic and δ_k is the *phase shift*.

Consider the sound generated by a vibrating string, such as in a piano or on a violin. Let $f(t)$ be the air pressure at your eardrum as a function of time t. [The function $f(t)$ is measured as a deviation from the normal atmospheric pressure.] In this case, the harmonics have a simple physical interpretation: They correspond to the various sinusoidal modes at which the string can vibrate. See Figure 7.

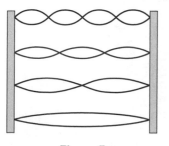

Figure 7

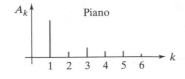

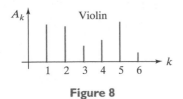

Figure 8

The *fundamental frequency* (corresponding to the vibration shown at the bottom of Figure 7) gives us the *first harmonic* of $f(t)$, while the *overtones* (with frequencies that are integer multiples of the fundamental frequency) give us the other terms of the harmonic series. The quality of a tone is in part determined by the relative amplitudes of the harmonics. When you play concert A (440 Hertz) on a piano, the first harmonic is much more prominent than the higher ones, but the same tone played on a violin gives prominence to higher harmonics (especially the fifth). See Figure 8. Similar considerations apply to wind instruments; they have a vibrating column of air instead of a vibrating string.

The human ear cannot hear tones whose frequencies exceed 20,000 Hertz. We pick up only finitely many harmonics of a tone. What we hear is the projection of $f(t)$ onto a certain T_n.

EXAMPLE 8   Find the Fourier coefficients for the function $f(t) = t$ on the interval $-\pi \leq t \leq \pi$:

$$b_k = \langle f, \sin(kt) \rangle = \frac{1}{\pi} \int_{-\pi}^{\pi} \sin(kt) t \, dt$$

$$= \frac{1}{\pi} \left\{ - \left[\frac{1}{k} \cos(kt) t \right] \Big|_{-\pi}^{\pi} + \frac{1}{k} \int_{-\pi}^{\pi} \cos(kt) \, dt \right\} \quad \text{(integration by parts)}$$

$$= \begin{cases} -\dfrac{2}{k} & \text{if } k \text{ is even} \\[2mm] \dfrac{2}{k} & \text{if } k \text{ is odd} \end{cases}$$

$$= (-1)^{k+1} \frac{2}{k}.$$

All c_k and a_0 are zero, since the integrands are odd functions.

The first few Fourier polynomials are

$$f_1 = 2\sin(t),$$
$$f_2 = 2\sin(t) - \sin(2t),$$
$$f_3 = 2\sin(t) - \sin(2t) + \frac{2}{3}\sin(3t),$$
$$f_4 = 2\sin(t) - \sin(2t) + \frac{2}{3}\sin(3t) - \frac{1}{2}\sin(4t).$$

See Figure 9.   ∎

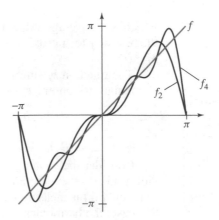

Figure 9

How do the errors $\|f - f_n\|$ and $\|f - f_{n+1}\|$ of the nth and the $(n+1)$st Fourier approximation compare? We hope that f_{n+1} will be a better approximation than f_n, or at least no worse:

$$\|f - f_{n+1}\| \leq \|f - f_n\|.$$

This is indeed the case, by definition: f_n is a polynomial in T_{n+1}, since T_n is contained in T_{n+1}, and

$$\|f - f_{n+1}\| \leq \|f - g\|,$$

for all g in T_{n+1}, in particular for $g = f_n$. In other words, as n goes to infinity, the error $\|f - f_n\|$ becomes smaller and smaller (or at least not larger). Using somewhat advanced calculus, we can show that this error approaches zero:

$$\lim_{n \to \infty} \|f - f_n\| = 0.$$

What does this tell us about $\lim_{n \to \infty} \|f_n\|$? By the theorem of Pythagoras, we have

$$\|f - f_n\|^2 + \|f_n\|^2 = \|f\|^2.$$

As n goes to infinity, the first summand, $\|f - f_n\|^2$, approaches 0, so that

$$\lim_{n \to \infty} \|f_n\| = \|f\|.$$

We have an expansion of f_n in terms of an orthonormal basis

$$f_n = a_0 \frac{1}{\sqrt{2}} + b_1 \sin(t) + c_1 \cos(t) + \cdots + b_n \sin(nt) + c_n \cos(nt),$$

where the b_k, the c_k, and a_0 are the Fourier coefficients. We can express $\|f_n\|$ in terms of these Fourier coefficients, using the Pythagorean theorem:

$$\|f_n\|^2 = a_0^2 + b_1^2 + c_1^2 + \cdots + b_n^2 + c_n^2.$$

Combining the last two "shaded" equations, we get the following identity:

Theorem 5.5.6

$$a_0^2 + b_1^2 + c_1^2 + \cdots + b_n^2 + c_n^2 + \cdots = \|f\|^2$$

The infinite series of the squares of the Fourier coefficients of a piecewise continuous function f converges to $\|f\|^2$. ∎

For the function $f(t)$ studied in Example 8, this means that

$$4 + \frac{4}{4} + \frac{4}{9} + \cdots + \frac{4}{n^2} + \cdots = \frac{1}{\pi} \int_{-\pi}^{\pi} t^2 \, dt = \frac{2}{3}\pi^2,$$

or

$$\sum_{n=1}^{\infty} \frac{1}{n^2} = 1 + \frac{1}{4} + \frac{1}{9} + \frac{1}{16} + \cdots = \frac{\pi^2}{6},$$

an equation discovered by Euler.

Theorem 5.5.6 has a physical interpretation when $\|f\|^2$ represents energy. For example, if $f(x)$ is the displacement of a vibrating string, then $b_k^2 + c_k^2$ represents the energy of the kth harmonic, and Theorem 5.5.6 tells us that the total energy $\|f\|^2$ is the sum of the energies of the harmonics.

There is an interesting application of Fourier analysis in quantum mechanics. In the 1920s, quantum mechanics was presented in two distinct forms:

Werner Heisenberg's matrix mechanics and Erwin Schrödinger's wave mechanics. Schrödinger (1887–1961) later showed that the two theories are mathematically equivalent: They use isomorphic inner product spaces. Heisenberg works with the space ℓ_2 introduced in Example 2, while Schrödinger works with a function space related to $C[-\pi, \pi]$. The isomorphism from Schrödinger's space to ℓ_2 is established by taking Fourier coefficients. (See Exercise 13.)

EXERCISES 5.5

GOAL *Use the idea of an inner product, and apply the basic results derived earlier for the dot product in $\mathbb{R}^n$ to inner product spaces.*

1. In $C[a, b]$, define the product
$$\langle f, g \rangle = \int_a^b f(t)g(t)\, dt.$$
   Show that this product satisfies the property
$$\langle f, f \rangle > 0$$
   for all nonzero f.

2. Does the equation
$$\langle f, g + h \rangle = \langle f, g \rangle + \langle f, h \rangle$$
   hold for all elements f, g, h of an inner product space? Explain.

3. Consider a matrix S in $\mathbb{R}^{n \times n}$. In $\mathbb{R}^n$, define the product
$$\langle \vec{x}, \vec{y} \rangle = (S\vec{x})^T S\vec{y}.$$
   a. For matrices S is this an inner product?
   b. For matrices S is $\langle \vec{x}, \vec{y} \rangle = \vec{x} \cdot \vec{y}$ (the dot product)?

4. In $\mathbb{R}^{n \times m}$, consider the inner product
$$\langle A, B \rangle = \text{trace}(A^T B)$$
   defined in Example 3.
   a. Find a formula for this inner product in $\mathbb{R}^{n \times 1} = \mathbb{R}^n$.
   b. Find a formula for this inner product in $\mathbb{R}^{1 \times m}$ (i.e., the space of row vectors with m components).

5. Is $\langle\!\langle A, B \rangle\!\rangle = \text{trace}(AB^T)$ an inner product in $\mathbb{R}^{n \times m}$? (The notation $\langle\!\langle A, B \rangle\!\rangle$ is chosen to distinguish this product from the one considered in Example 3 and Exercise 4.)

6. **a.** Consider an $n \times m$ matrix P and an $m \times n$ matrix Q. Show that
$$\text{trace}(PQ) = \text{trace}(QP).$$
   b. Compare the following two inner products in $\mathbb{R}^{n \times m}$:
$$\langle A, B \rangle = \text{trace}(A^T B),$$
   and
$$\langle\!\langle A, B \rangle\!\rangle = \text{trace}(AB^T).$$
   (See Example 3 and Exercises 4 and 5.)

7. Consider an inner product $\langle v, w \rangle$ in a space V, and a scalar k. For which choices of k is
$$\langle\!\langle v, w \rangle\!\rangle = k \langle v, w \rangle$$
   an inner product?

8. Consider an inner product $\langle v, w \rangle$ in a space V. Let w be a fixed element of V. Is the transformation $T(v) = \langle v, w \rangle$ from V to $\mathbb{R}$ linear? What is its image? Give a geometric interpretation of its kernel.

9. Recall that a function $f(t)$ from $\mathbb{R}$ to $\mathbb{R}$ is called
$$\text{even if } f(-t) = f(t), \quad \text{for all } t,$$
   and
$$\text{odd if } f(-t) = -f(t), \quad \text{for all } t.$$
   Show that if $f(x)$ is an odd continuous function and $g(x)$ is an even continuous function, then functions $f(x)$ and $g(x)$ are orthogonal in the space $C[-1, 1]$ with the inner product defined in Example 1.

10. Consider the space P_2 with inner product
$$\langle f, g \rangle = \frac{1}{2} \int_{-1}^{1} f(t)g(t)\, dt.$$
    Find an *orthonormal* basis of the space of all functions in P_2 that are orthogonal to $f(t) = t$.

11. The angle between two nonzero elements v and w of an inner product space is defined as
$$\angle(v, w) = \arccos \frac{\langle v, w \rangle}{\|v\| \|w\|}.$$
    In the space $C[-\pi, \pi]$ with inner product
$$\langle f, g \rangle = \frac{1}{\pi} \int_{-\pi}^{\pi} f(t)g(t)\, dt,$$
    find the angle between $f(t) = \cos(t)$ and $g(t) = \cos(t + \delta)$, where $0 \leq \delta \leq \pi$. *Hint:* Use the formula $\cos(t + \delta) = \cos(t)\cos(\delta) - \sin(t)\sin(\delta)$.

12. Find all Fourier coefficients of the absolute value function
$$f(t) = |t|.$$

13. For a function f in $C[-\pi, \pi]$ (with the inner product defined on page 239), consider the sequence of all its

Fourier coefficients,

$$(a_0, b_1, c_1, b_2, c_2, \ldots, b_n, c_n, \ldots).$$

Is this infinite sequence in ℓ_2? If so, what is the relationship between

$$\|f\| \quad \text{(the norm in } C[-\pi, \pi])$$

and

$$\|(a_0, b_1, c_1, b_2, c_2, \ldots)\| \quad \text{(the norm in } \ell_2)?$$

The inner product space ℓ_2 was introduced in Example 2.

14. Which of the following is an inner product in P_2? Explain.
  a. $\langle f, g \rangle = f(1)g(1) + f(2)g(2)$
  b. $\langle\!\langle f, g \rangle\!\rangle = f(1)g(1) + f(2)g(2) + f(3)g(3)$

15. For which values of the constants b, c, and d is the following an inner product in $\mathbb{R}^2$?

$$\left\langle \begin{bmatrix} x_1 \\ x_2 \end{bmatrix}, \begin{bmatrix} y_1 \\ y_2 \end{bmatrix} \right\rangle = x_1 y_1 + b x_1 y_2 + c x_2 y_1 + d x_2 y_2$$

(*Hint*: Be prepared to complete a square.)

16. a. Find an *orthonormal* basis of the space P_1 with inner product

$$\langle f, g \rangle = \int_0^1 f(t)g(t)\, dt.$$

  b. Find the linear polynomial $g(t) = a + bt$ that best approximates the function $f(t) = t^2$ on the interval $[0, 1]$ in the (continuous) least-squares sense. Draw a sketch.

17. Consider a linear space V. For which linear transformations T from V to $\mathbb{R}^n$ is

$$\langle v, w \rangle = \underset{\underset{\text{Dot product}}{\uparrow}}{T(v) \cdot T(w)}$$

an inner product in V?

18. Consider an orthonormal basis $\mathfrak{B}$ of the inner product space V. For an element f of V, what is the relationship between $\|f\|$ and $\|[f]_\mathfrak{B}\|$ (the norm in $\mathbb{R}^n$ defined by the dot product)?

19. For which 2×2 matrices A is

$$\langle \vec{v}, \vec{w} \rangle = \vec{v}^T A \vec{w}$$

an inner product in $\mathbb{R}^2$? *Hint*: Be prepared to complete a square.

20. Consider the inner product

$$\langle \vec{v}, \vec{w} \rangle = \vec{v}^T \begin{bmatrix} 1 & 2 \\ 2 & 8 \end{bmatrix} \vec{w}$$

in $\mathbb{R}^2$. (See Exercise 19.)

  a. Find all vectors in $\mathbb{R}^2$ that are perpendicular to $\begin{bmatrix} 1 \\ 0 \end{bmatrix}$ with respect to this inner product.

  b. Find an orthonormal basis of $\mathbb{R}^2$ with respect to this inner product.

21. If $\|\vec{v}\|$ denotes the standard norm in $\mathbb{R}^n$, does the formula

$$\langle \vec{v}, \vec{w} \rangle = \|\vec{v} + \vec{w}\|^2 - \|\vec{v}\|^2 - \|\vec{w}\|^2$$

define an inner product in $\mathbb{R}^n$?

22. If $f(t)$ is a continuous function, what is the relationship between

$$\int_0^1 \left(f(t)\right)^2 dt \quad \text{and} \quad \left(\int_0^1 f(t)\, dt\right)^2 ?$$

(*Hint*: Use the Cauchy–Schwarz inequality.)

23. In the space P_1 of the polynomials of degree ≤ 1, we define the inner product

$$\langle f, g \rangle = \frac{1}{2}\left(f(0)g(0) + f(1)g(1)\right).$$

Find an orthonormal basis for this inner product space.

24. Consider the linear space P of all polynomials, with inner product

$$\langle f, g \rangle = \int_0^1 f(t)g(t)\, dt.$$

For three polynomials f, g, and h we are given the following inner products:

$\langle \cdot \rangle$	f	g	h
f	4	0	8
g	0	1	3
h	8	3	50

For example, $\langle f, f \rangle = 4$ and $\langle g, h \rangle = \langle h, g \rangle = 3$.
  a. Find $\langle f, g + h \rangle$.
  b. Find $\|g + h\|$.
  c. Find $\text{proj}_E h$, where $E = \text{span}(f, g)$. Express your solution as linear combinations of f and g.
  d. Find an orthonormal basis of $\text{span}(f, g, h)$. Express the functions in your basis as linear combinations of f, g, and h.

25. Find the norm $\|\vec{x}\|$ of

$$\vec{x} = \left(1, \frac{1}{2}, \frac{1}{3}, \ldots, \frac{1}{n}, \ldots\right) \quad \text{in } \ell_2.$$

(ℓ_2 is defined in Example 2.)

26. Find the Fourier coefficients of the piecewise continuous function

$$f(t) = \begin{cases} -1 & \text{if } t \leq 0 \\ 1 & \text{if } t > 0. \end{cases}$$

Sketch the graphs of the first few Fourier polynomials.

27. Find the Fourier coefficients of the piecewise continuous function

$$f(t) = \begin{cases} 0 & \text{if } t \leq 0 \\ 1 & \text{if } t > 0. \end{cases}$$

28. Apply Theorem 5.5.6 to your answer in Exercise 26.

29. Apply Theorem 5.5.6 to your answer in Exercise 27.

30. Consider an ellipse E in $\mathbb{R}^2$ centered at the origin. Show that there is an inner product $\langle \cdot, \cdot \rangle$ in $\mathbb{R}^2$ such that E consists of all vectors $\vec{x}$ with $\|\vec{x}\| = 1$, where the norm is taken with respect to the inner product $\langle \cdot, \cdot \rangle$.

31. *Gaussian integration.* In an introductory calculus course, you may have seen approximation formulas for integrals of the form

$$\int_a^b f(t)\, dt \approx \sum_{i=i}^n w_i f(a_i),$$

where the a_i are equally spaced points on the interval (a, b), and the w_i are certain "weights" (Riemann sums, trapezoidal sums, and Simpson's rule). Gauss has shown that, with the same computational effort, we can get better approximations if we drop the requirement that the a_i be equally spaced. Next, we outline his approach.

Consider the space P_n with the inner product

$$\langle f, g \rangle = \int_{-1}^1 f(t)g(t)\, dt.$$

Let $f_0, f_1, \ldots, f_n$ be an orthonormal basis of this space, with $degree(f_k) = k$. (To construct such a basis, apply the Gram–Schmidt process to the standard basis $1, t, \ldots, t^n$.) It can be shown that f_n has n distinct roots $a_1, a_2, \ldots, a_n$ on the interval $(-1, 1)$. We can find "weights" $w_1, w_2, \ldots, w_n$ such that

$$\int_{-1}^1 f(t)\, dt = \sum_{i=1}^n w_i f(a_i),$$

for all polynomials of degree less than n. (See Exercise 4.3.71.) In fact, much more is true: This formula holds for all polynomials $f(t)$ of degree less than $2n$.

You are not asked to prove the foregoing assertions for arbitrary n, but work out the case $n = 2$: Find a_1, a_2 and w_1, w_2, and show that the formula

$$\int_{-1}^1 f(t)\, dt = w_1 f(a_1) + w_2 f(a_2)$$

holds for all cubic polynomials.

32. In the space $C[-1, 1]$, we introduce the inner product

$$\langle f, g \rangle = \frac{1}{2}\int_{-1}^1 f(t)g(t)dt.$$

a. Find $\langle t^n, t^m \rangle$, where n and m are positive integers.

b. Find the norm of $f(t) = t^n$, where n is a positive integer.

c. Applying the Gram-Schmidt process to the standard basis $1, t, t^2, t^3$ of P_3, construct an orthonormal basis $g_0(t), \ldots, g_3(t)$ of P_3 for the given inner product.

d. Find the polynomials $\dfrac{g_0(t)}{g_0(1)}, \ldots, \dfrac{g_3(t)}{g_3(1)}$. (Those are the first few *Legendre Polynomials,* named after the great French mathematician Adrien-Marie Legendre, 1752–1833. These polynomials have a wide range of applications in math, physics, and engineering. Note that the Legendre polynomials are normalized so that their value at 1 is 1.)

e. Find the polynomial $g(t)$ in P_3 that best approximates the function $f(t) = \dfrac{1}{1 + t^2}$ on the interval $[-1, 1]$, for the inner product introduced in this exercise. Draw a sketch.

33. a. Let $w(t)$ be a positive-valued function in $C[a, b]$, where $b > a$. Verify that the rule $\langle f, g \rangle = \int_a^b w(t)f(t)g(t)dt$ defines an inner product on $C[a, b]$.

b. If we chose the weight function $w(t)$ so that $\int_a^b w(t)dt = 1$, what is the norm of the constant function $f(t) = 1$ in this inner product space?

34. In the space $C[-1, 1]$, we define the inner product $\langle f, g \rangle = \int_{-1}^1 \frac{2}{\pi}\sqrt{1 - t^2}f(t)g(t)\, dt = \frac{2}{\pi}\int_{-1}^1 \sqrt{1 - t^2}f(t)g(t)\, dt$. See Exercise 33; here we let $w(t) = \frac{2}{\pi}\sqrt{1 - t^2}$. [This function $w(t)$ is called a *Wigner Semicircle Distribution,* after the Hungarian physicist and mathematician E. P. Wigner (1902–1995), who won the 1963 Nobel Prize in Physics.] Since this is not a course in calculus, here are some inner products that will turn out to be useful: $\langle 1, t^2 \rangle = 1/4$, $\langle t, t^3 \rangle = 1/8$, and $\langle t^3, t^3 \rangle = 5/64$.

a. Find $\int_{-1}^1 w(t)dt$. Sketch a rough graph of the weight function $w(t)$.

b. Find the norm of the constant function $f(t) = 1$.

c. Find $\langle t^2, t^3 \rangle$; explain. More generally, find $\langle t^n, t^m \rangle$ for positive integers n and m whose sum is odd.

d. Find $\langle t, t \rangle$ and $\langle t^2, t^2 \rangle$. Also, find the norms of the functions t and t^2.

e. Applying the Gram–Schmidt process to the standard basis $1, t, t^2, t^3$ of P_3, construct an orthonormal basis $g_0(t), \ldots, g_3(t)$ of P_3 for the given inner product. [The polynomials $g_0(t), \ldots, g_3(t)$ are the first few *Chebyshev Polynomials of the Second Kind,* named after the Russian mathematician Pafnuty Chebyshev (1821–1894). They have a wide range of applications in math, physics, and engineering.]

f. Find the polynomial $g(t)$ in P_3 that best approximates the function $f(t) = t^4$ on the interval $[-1, 1]$, for the inner product introduced in this exercise.

35. In this exercise, we compare the inner products and norms introduced in Problems 32 and 34. Let's denote the two norms by $\|f\|_{32}$ and $\|f\|_{34}$, respectively.

    a. Compute $\|t\|_{32}$ and $\|t\|_{34}$. Which is larger? Explain the answer conceptually. Graph the weight functions $w_{32}(t) = \frac{1}{2}$ and $w_{34}(t) = \frac{2}{\pi}\sqrt{1 - t^2}$ on the same axes. Then graph the functions $w_{32}(t)t^2$ and $w_{34}(t)t^2$ on the same axes.

    b. Give an example of a continuous function $f(t)$ such that $\|f\|_{34} > \|f\|_{32}$.

Chapter Five Exercises

TRUE OR FALSE?

1. If A and B are symmetric $n \times n$ matrices, then $A + B$ must be symmetric as well.

2. If matrices A and S are orthogonal, then $S^{-1}AS$ is orthogonal as well.

3. All nonzero symmetric matrices are invertible.

4. If A is an $n \times n$ matrix such that $AA^T = I_n$, then A must be an orthogonal matrix.

5. If $\vec{u}$ is a unit vector in $\mathbb{R}^n$, and $L = \text{span}(\vec{u})$, then $\text{proj}_L(\vec{x}) = (\vec{x} \cdot \vec{u})\vec{x}$ for all vectors $\vec{x}$ in $\mathbb{R}^n$.

6. If A is a symmetric matrix, then $7A$ must be symmetric as well.

7. If T is a linear transformation from $\mathbb{R}^n$ to $\mathbb{R}^n$ such that $T(\vec{e}_1), T(\vec{e}_2), \ldots, T(\vec{e}_n)$ are all unit vectors, then T must be an orthogonal transformation.

8. If A is an invertible matrix, then the equation $(A^T)^{-1} = (A^{-1})^T$ must hold.

9. If matrix A is orthogonal, then matrix A^2 must be orthogonal as well.

10. The equation $(AB)^T = A^T B^T$ holds for all $n \times n$ matrices A and B.

11. If matrix A is orthogonal, then A^T must be orthogonal as well.

12. If A and B are symmetric $n \times n$ matrices, then AB must be symmetric as well.

13. If matrices A and B commute, then A must commute with B^T as well.

14. If A is any matrix with $\ker(A) = \{\vec{0}\}$, then the matrix AA^T represents the orthogonal projection onto the image of A.

15. If A and B are symmetric $n \times n$ matrices, then $ABBA$ must be symmetric as well.

16. If matrices A and B commute, then matrices A^T and B^T must commute as well.

17. There exists a subspace V of $\mathbb{R}^5$ such that $\dim(V) = \dim(V^\perp)$, where $V^\perp$ denotes the orthogonal complement of V.

18. Every invertible matrix A can be expressed as the product of an orthogonal matrix and an upper triangular matrix.

19. If $\vec{x}$ and $\vec{y}$ are two vectors in $\mathbb{R}^n$, then the equation $\|\vec{x} + \vec{y}\|^2 = \|\vec{x}\|^2 + \|\vec{y}\|^2$ must hold.

20. The equation $\det(A^T) = \det(A)$ holds for all 2×2 matrices A.

21. If A and B are orthogonal 2×2 matrices, then $AB = BA$.

22. If A is a symmetric matrix, vector $\vec{v}$ is in the image of A, and $\vec{w}$ is in the kernel of A, then the equation $\vec{v} \cdot \vec{w} = 0$ must hold.

23. The formula $\ker(A) = \ker(A^T A)$ holds for all matrices A.

24. If $A^T A = AA^T$ for an $n \times n$ matrix A, then A must be orthogonal.

25. The determinant of all orthogonal 2×2 matrices is 1.

26. If A is any square matrix, then matrix $\frac{1}{2}(A - A^T)$ is skew-symmetric.

27. The entries of an orthogonal matrix are all less than or equal to 1.

28. Every nonzero subspace of $\mathbb{R}^n$ has an orthonormal basis.

29. $\begin{bmatrix} 3 & -4 \\ 4 & 3 \end{bmatrix}$ is an orthogonal matrix.

30. If V is a subspace of $\mathbb{R}^n$ and $\vec{x}$ is a vector in $\mathbb{R}^n$, then vector $\text{proj}_V \vec{x}$ must be orthogonal to vector $\vec{x} - \text{proj}_V \vec{x}$.

31. There exist orthogonal 2×2 matrices A and B such that $A + B$ is orthogonal as well.

32. If $\|A\vec{x}\| \le \|\vec{x}\|$ for all $\vec{x}$ in $\mathbb{R}^n$, then A must represent the orthogonal projection onto a subspace V of $\mathbb{R}^n$.

33. If A is an invertible matrix such that $A^{-1} = A$, then A must be orthogonal.

34. If the entries of two vectors $\vec{v}$ and $\vec{w}$ in $\mathbb{R}^n$ are all positive, then $\vec{v}$ and $\vec{w}$ must enclose an acute angle.

35. The formula $(\ker B)^\perp = \text{im}(B^T)$ holds for all matrices B.

36. The matrix $A^T A$ is symmetric for all matrices A.

37. If matrix A is similar to B and A is orthogonal, then B must be orthogonal as well.

38. The formula $\text{im}(B) = \text{im}(B^T B)$ holds for all square matrices B.

39. If matrix A is symmetric and matrix S is orthogonal, then matrix $S^{-1} A S$ must be symmetric.

40. If A is a square matrix such that $A^T A = A A^T$, then $\ker(A) = \ker(A^T)$.

41. Any square matrix can be written as the sum of a symmetric and a skew-symmetric matrix.

42. If $x_1, x_2, \ldots, x_n$ are any real numbers, then the inequality

$$\left(\sum_{k=1}^{n} x_k \right)^2 \leq n \sum_{k=1}^{n} (x_k^2)$$

must hold.

43. If $A A^T = A^2$ for a 2×2 matrix A, then A must be symmetric.

44. If V is a subspace of $\mathbb{R}^n$ and $\vec{x}$ is a vector in $\mathbb{R}^n$, then the inequality $\vec{x} \cdot (\text{proj}_V \vec{x}) \geq 0$ must hold.

45. If A is an $n \times n$ matrix such that $\|A\vec{u}\| = 1$ for all unit vectors $\vec{u}$, then A must be an orthogonal matrix.

46. If A is any symmetric 2×2 matrix, then there must exist a real number x such that matrix $A - x I_2$ fails to be invertible.

47. There exists a basis of $\mathbb{R}^{2 \times 2}$ that consists of orthogonal matrices.

48. If $A = \begin{bmatrix} 1 & 2 \\ 2 & 1 \end{bmatrix}$, then the matrix Q in the QR factorization of A is a rotation matrix.

49. There exists a linear transformation L from $\mathbb{R}^{3 \times 3}$ to $\mathbb{R}^{2 \times 2}$ whose kernel is the space of all skew-symmetric 3×3 matrices.

50. If a 3×3 matrix A represents the orthogonal projection onto a plane V in $\mathbb{R}^3$, then there must exist an orthogonal 3×3 matrix S such that $S^T A S$ is diagonal.

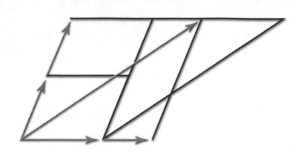

Determinants

6.1 Introduction to Determinants

In Chapter 2 we found a criterion for the invertibility of a 2×2 matrix: The matrix

$$A = \begin{bmatrix} a & b \\ c & d \end{bmatrix}$$

is invertible if (and only if)

$$\det A = ad - bc \neq 0,$$

by Theorem 2.4.9a.

You may wonder whether the concept of a determinant can be generalized to square matrices of arbitrary size. Can we assign a number $\det A$ to a square matrix A, expressed in terms of the entries of A, such that A is invertible if (and only if) $\det A \neq 0$?

The Determinant of a 3×3 Matrix

Let

$$A = \begin{bmatrix} a_{11} & a_{12} & a_{13} \\ a_{21} & a_{22} & a_{23} \\ a_{31} & a_{32} & a_{33} \end{bmatrix} = \begin{bmatrix} | & | & | \\ \vec{u} & \vec{v} & \vec{w} \\ | & | & | \end{bmatrix}$$

(we denote the three column vectors $\vec{u}$, $\vec{v}$, and $\vec{w}$). (See Figure 1.)

The matrix A fails to be invertible if the image of A isn't all of $\mathbb{R}^3$, meaning that the three column vectors $\vec{u}$, $\vec{v}$, and $\vec{w}$ are contained in some plane V. In this case, the cross product[1] $\vec{v} \times \vec{w}$, being perpendicular to V, is perpendicular to vector $\vec{u}$, so that

$$\vec{u} \cdot (\vec{v} \times \vec{w}) = 0.$$

If A is invertible, on the other hand, then $\vec{v} \times \vec{w}$ fails to be perpendicular to $\vec{u}$, so that $\vec{u} \cdot (\vec{v} \times \vec{w}) \neq 0$.

---

[1]To review the definition of the cross product, see Definition A.9 in the Appendix.

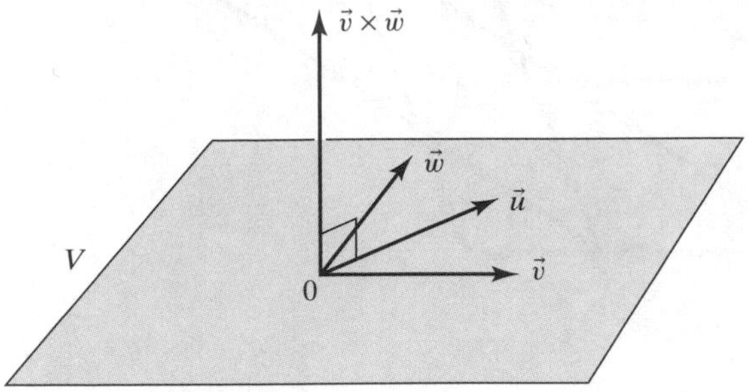

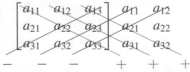

$$\vec{v} \times \vec{w}$$

Figure 1

Thus the quantity $\vec{u} \cdot (\vec{v} \times \vec{w})$ has the property we expect from the determinant: It is nonzero if (and only if) matrix

$$A = \begin{bmatrix} \vec{u} & \vec{v} & \vec{w} \end{bmatrix}$$

is invertible. This motivates the following definition.

Definition 6.1.1   Determinant of a 3×3 matrix, in terms of the columns
If $A = \begin{bmatrix} \vec{u} & \vec{v} & \vec{w} \end{bmatrix}$, then

$$\det A = \vec{u} \cdot (\vec{v} \times \vec{w}).$$

A 3×3 matrix A is invertible if (and only if) $\det A \neq 0$.

Let's express the determinant $\det A = \vec{u} \cdot (\vec{v} \times \vec{w})$ in terms of the entries of matrix A:

$$\det A = \vec{u} \cdot (\vec{v} \times \vec{w})$$
$$= \begin{bmatrix} a_{11} \\ a_{21} \\ a_{31} \end{bmatrix} \cdot \left(\begin{bmatrix} a_{12} \\ a_{22} \\ a_{32} \end{bmatrix} \times \begin{bmatrix} a_{13} \\ a_{23} \\ a_{33} \end{bmatrix} \right) = \begin{bmatrix} a_{11} \\ a_{21} \\ a_{31} \end{bmatrix} \cdot \begin{bmatrix} a_{22}a_{33} - a_{32}a_{23} \\ a_{32}a_{13} - a_{12}a_{33} \\ a_{12}a_{23} - a_{22}a_{13} \end{bmatrix}$$
$$= a_{11}(a_{22}a_{33} - a_{32}a_{23}) + a_{21}(a_{32}a_{13} - a_{12}a_{33}) + a_{31}(a_{12}a_{23} - a_{22}a_{13})$$
$$= a_{11}a_{22}a_{33} - a_{11}a_{32}a_{23} + a_{21}a_{32}a_{13} - a_{21}a_{12}a_{33} + a_{31}a_{12}a_{23} - a_{31}a_{22}a_{13}.$$

Here is a memory aid for the determinant of a 3×3 matrix.

Theorem 6.1.2   Sarrus's rule[2]
To find the determinant of a 3×3 matrix A, write the first two columns of A to the right of A. Then multiply the entries along the six diagonals shown below.

$$\begin{bmatrix} a_{11} & a_{12} & a_{13} \\ a_{21} & a_{22} & a_{23} \\ a_{31} & a_{32} & a_{33} \end{bmatrix} \begin{matrix} a_{11} & a_{12} \\ a_{21} & a_{22} \\ a_{31} & a_{32} \end{matrix}$$

$$-\quad -\quad -\quad +\quad +\quad +$$

Add or subtract these diagonal products, as shown in the diagram:

$$\det A = a_{11}a_{22}a_{33} + a_{12}a_{23}a_{31} + a_{13}a_{21}a_{32} - a_{13}a_{22}a_{31} - a_{11}a_{23}a_{32} - a_{12}a_{21}a_{33}$$

∎

---

[2]Stated by Pierre Frédéric Sarrus (1798–1861) of Strasbourg, c. 1820.

EXAMPLE I   Find the determinant of

$$A = \begin{bmatrix} 1 & 2 & 3 \\ 4 & 5 & 6 \\ 7 & 8 & 10 \end{bmatrix}.$$

Solution

By Sarrus's rule, $\det A = 1 \cdot 5 \cdot 10 + 2 \cdot 6 \cdot 7 + 3 \cdot 4 \cdot 8 - 3 \cdot 5 \cdot 7 - 1 \cdot 6 \cdot 8 - 2 \cdot 4 \cdot 10 = -3$. Matrix A is invertible. ∎

EXAMPLE 2   Find the determinant of the upper triangular matrix

$$A = \begin{bmatrix} a & b & c \\ 0 & d & e \\ 0 & 0 & f \end{bmatrix}.$$

Solution

We find that $\det A = adf$, since all other terms in Sarrus's formula are zero. The determinant of an upper (or lower) triangular 3×3 matrix is the product of the diagonal entries. Thus a triangular 3×3 matrix is invertible if (and only if) all its diagonal entries are nonzero. ∎

EXAMPLE 3   For which values of the scalar λ is the matrix

$$A = \begin{bmatrix} \lambda & 1 & 1 \\ 1 & \lambda & -1 \\ 1 & 1 & \lambda \end{bmatrix}$$

invertible?

Solution

$$\det A = \lambda^3 - 1 + 1 + \lambda - \lambda - \lambda = \lambda^3 - \lambda$$
$$= \lambda(\lambda^2 - 1) = \lambda(\lambda - 1)(\lambda + 1).$$

The determinant is 0 if $\lambda = 0$, $\lambda = 1$, or $\lambda = -1$. Matrix A is invertible if λ is any real number other than 0, 1, and -1. ∎

EXAMPLE 4   For three column vectors $\vec{u}, \vec{v}, \vec{w}$ in $\mathbb{R}^3$, what is the relationship between the determinants of $A = \begin{bmatrix} \vec{u} & \vec{v} & \vec{w} \end{bmatrix}$ and $B = \begin{bmatrix} \vec{u} & \vec{w} & \vec{v} \end{bmatrix}$? Note that matrix B is obtained by swapping the last two columns of A.

Solution

$$\det B = \det \begin{bmatrix} \vec{u} & \vec{w} & \vec{v} \end{bmatrix} = \vec{u} \cdot (\vec{w} \times \vec{v}) = -\vec{u} \cdot (\vec{v} \times \vec{w}) = -\det \begin{bmatrix} \vec{u} & \vec{v} & \vec{w} \end{bmatrix} = -\det A.$$

We have used the fact that the cross product is anticommutative: $\vec{w} \times \vec{v} = -(\vec{v} \times \vec{w})$. See Theorem A.10 in the Appendix. ∎

It turns out that $\det B = -\det A$ if B is obtained by swapping any two columns or any two rows of a 3×3 matrix A; we can verify this by direct computation. This is referred to as the *alternating* property of the determinant on the columns and on the rows. The 2×2 determinant is alternating on the rows and columns as well (verify this!), and we will see that this property generalizes to determinants of square matrices of any size.

Linearity Properties of the Determinant

EXAMPLE 5   Is the function $F(A) = \det A$ from the linear space $\mathbb{R}^{3 \times 3}$ to $\mathbb{R}$ a linear transformation?

Solution

The answer is negative. For example, $F(I_3 + I_3) = F(2I_3) = 8$ while $F(I_3) + F(I_3) = 1 + 1 = 2$. ∎

However, the determinant does have some noteworthy linearity properties.

EXAMPLE 6   Is the function

$$T \begin{bmatrix} x_1 \\ x_2 \\ x_3 \end{bmatrix} = \det \begin{bmatrix} 2 & x_1 & 5 \\ 3 & x_2 & 6 \\ 4 & x_3 & 7 \end{bmatrix}$$

from $\mathbb{R}^3$ to $\mathbb{R}$ a linear transformation? Here we are placing the input variables x_1, x_2, x_3 in the second column, choosing arbitrary constants for all the other entries.

Solution

Note that

$$T \begin{bmatrix} x_1 \\ x_2 \\ x_3 \end{bmatrix} = \det \begin{bmatrix} 2 & x_1 & 5 \\ 3 & x_2 & 6 \\ 4 & x_3 & 7 \end{bmatrix} = (6 \cdot 4 - 3 \cdot 7)x_1 + (2 \cdot 7 - 5 \cdot 4)x_2 + (5 \cdot 3 - 2 \cdot 6)x_3$$

$$= 3x_1 - 6x_2 + 3x_3.$$

Therefore, T is a linear transformation, by Definition 2.1.1, since the output is a linear combination of the input variables. ∎

We say that the 3×3 determinant is *linear in the second column*. Likewise, the determinant is linear in the two other columns and in all three rows. For example, linearity in the third row means that

$$L(\vec{x}) = \det \begin{bmatrix} - & \vec{v}_1 & - \\ - & \vec{v}_2 & - \\ - & \vec{x} & - \end{bmatrix}$$

is linear on row vectors $\vec{x}$ with three components, for any two fixed row vectors $\vec{v}_1$ and $\vec{v}_2$.

Alternatively, we can express the linearity of L by the equations

$$L(\vec{x} + \vec{y}) = L(\vec{x}) + L(\vec{y}) \qquad \text{and} \qquad L(k\vec{x}) = kL(\vec{x})$$

or

$$\det \begin{bmatrix} - & \vec{v}_1 & - \\ - & \vec{v}_2 & - \\ - & \vec{x} + y & - \end{bmatrix} = \det \begin{bmatrix} - & \vec{v}_1 & - \\ - & \vec{v}_2 & - \\ - & \vec{x} & - \end{bmatrix} + \det \begin{bmatrix} - & \vec{v}_1 & - \\ - & \vec{v}_2 & - \\ - & \vec{y} & - \end{bmatrix} \qquad \text{and}$$

$$\det \begin{bmatrix} - & \vec{v}_1 & - \\ - & \vec{v}_2 & - \\ - & k\vec{x} & - \end{bmatrix} = k \det \begin{bmatrix} - & \vec{v}_1 & - \\ - & \vec{v}_2 & - \\ - & \vec{x} & - \end{bmatrix}$$

The Determinant of an $n \times n$ Matrix

We may be tempted to define the determinant of an $n \times n$ matrix by generalizing Sarrus's rule. (See Theorem 6.1.2.) For a 4×4 matrix, a naive generalization of

Sarrus's rule produces the expression

$$a_{11}a_{22}a_{33}a_{44} + \cdots + a_{14}a_{21}a_{32}a_{43} - a_{14}a_{23}a_{32}a_{41} - \cdots - a_{13}a_{22}a_{31}a_{44}.$$

$$\begin{bmatrix} a_{11} & a_{12} & a_{13} & a_{14} \\ a_{21} & a_{22} & a_{23} & a_{24} \\ a_{31} & a_{32} & a_{33} & a_{34} \\ a_{41} & a_{42} & a_{43} & a_{44} \end{bmatrix} \begin{matrix} a_{11} & a_{12} & a_{13} \\ a_{21} & a_{22} & a_{23} \\ a_{31} & a_{32} & a_{33} \\ a_{41} & a_{42} & a_{43} \end{matrix}$$

$$- \quad - \quad - \quad - \quad + \quad + \quad + \quad +$$

For example, for the *invertible* matrix

$$A = \begin{bmatrix} 1 & 0 & 0 & 0 \\ 0 & 1 & 0 & 0 \\ 0 & 0 & 0 & 1 \\ 0 & 0 & 1 & 0 \end{bmatrix},$$

the expression given by this generalization of Sarrus's rule is 0. This shows that we cannot define the determinant by generalizing Sarrus's rule in this way: recall that we want the determinant of an invertible matrix to be nonzero.

We have to look for a more subtle structure in the formula

$$\det A = a_{11}a_{22}a_{33} + a_{12}a_{23}a_{31} + a_{13}a_{21}a_{32} - a_{13}a_{22}a_{31} - a_{11}a_{23}a_{32} - a_{12}a_{21}a_{33}$$

for the determinant of a 3×3 matrix. Note that each of the six terms in this expression is a product of three factors involving exactly one entry from each row and each column of the matrix:

$$\begin{bmatrix} \textcircled{a_{11}} & a_{12} & a_{13} \\ a_{21} & \textcircled{a_{22}} & a_{23} \\ a_{31} & a_{32} & \textcircled{a_{33}} \end{bmatrix} \begin{bmatrix} a_{11} & \textcircled{a_{12}} & a_{13} \\ a_{21} & a_{22} & \textcircled{a_{23}} \\ \textcircled{a_{31}} & a_{32} & a_{33} \end{bmatrix} \begin{bmatrix} a_{11} & a_{12} & \textcircled{a_{13}} \\ \textcircled{a_{21}} & a_{22} & a_{23} \\ a_{31} & \textcircled{a_{32}} & a_{33} \end{bmatrix}$$

$$\begin{bmatrix} a_{11} & a_{12} & \textcircled{a_{13}} \\ a_{21} & \textcircled{a_{22}} & a_{23} \\ \textcircled{a_{31}} & a_{32} & a_{33} \end{bmatrix} \begin{bmatrix} \textcircled{a_{11}} & a_{12} & a_{13} \\ a_{21} & a_{22} & \textcircled{a_{23}} \\ a_{31} & \textcircled{a_{32}} & a_{33} \end{bmatrix} \begin{bmatrix} a_{11} & \textcircled{a_{12}} & a_{13} \\ \textcircled{a_{21}} & a_{22} & a_{23} \\ a_{31} & a_{32} & \textcircled{a_{33}} \end{bmatrix}.$$

For lack of a better word, we call such a choice of a number in each row and column of a square matrix a *pattern* in the matrix.[3] The simplest pattern is the *diagonal pattern,* where we choose all numbers a_{ii} on the main diagonal. For you chess players, a pattern in an 8×8 matrix corresponds to placing 8 rooks on a chessboard so that none of them can attack another.

How many patterns are there in an $n \times n$ matrix? Let us see how we can construct a pattern column by column. In the first column we have n choices. For each of these, we then have $n - 1$ choices left in the second column. Therefore, we have $n(n - 1)$ choices for the numbers in the first two columns. For each of these, there are $n - 2$ choices in the third column, and so on. When we come to the last column, we have no choice, because there is only one row left. We conclude that there are $n(n - 1)(n - 2) \cdots 3 \cdot 2 \cdot 1$ patterns in an $n \times n$ matrix. The quantity $1 \cdot 2 \cdot 3 \cdots (n - 2) \cdot (n - 1) \cdot n$ is written $n!$ (read "n factorial").

For a pattern P in a 3×3 matrix we consider the product of all the entries in the pattern, denoted prod P. For example, for the pattern $P = (a_{12}, a_{23}, a_{31})$, we have

---

[3]This theory is usually phrased in the language of *permutations*. Here we attempt a less technical presentation, without sacrificing content.

prod $P = a_{12}a_{23}a_{31}$. Then we can write

$$\det A = \sum \pm \text{prod } P,$$

where the sum is taken over all six patterns P in a 3×3 matrix A. Next we need to examine how the signs of the six summands are chosen.

It turns out that these signs are related to the alternating property of the determinant we discussed after Example 4:

$$\det \begin{bmatrix} 0 & a_{12} & 0 \\ 0 & 0 & a_{23} \\ a_{31} & 0 & 0 \end{bmatrix} = -\det \begin{bmatrix} 0 & a_{12} & 0 \\ a_{31} & 0 & 0 \\ 0 & 0 & a_{23} \end{bmatrix}$$

$$= \det \begin{bmatrix} a_{31} & 0 & 0 \\ 0 & a_{12} & 0 \\ 0 & 0 & a_{23} \end{bmatrix} = a_{31}a_{12}a_{23} = a_{12}a_{23}a_{31},$$

since we perform two row swaps to bring the matrix into diagonal form, while

$$\det \begin{bmatrix} 0 & 0 & a_{13} \\ 0 & a_{22} & 0 \\ a_{31} & 0 & 0 \end{bmatrix} = -\det \begin{bmatrix} a_{31} & 0 & 0 \\ 0 & a_{22} & 0 \\ 0 & 0 & a_{13} \end{bmatrix} = -a_{31}a_{22}a_{13} = -a_{13}a_{22}a_{31}.$$

There is an equivalent way to predict this sign without actually counting row swaps. We say that two numbers in a pattern are *inverted* if one of them is to the right and above the other. Let's indicate the number of inversions for each of the six patterns in a 3×3 matrix.

$$\det A = a_{11}a_{22}a_{33} + a_{12}a_{23}a_{31} + a_{13}a_{21}a_{32} - a_{13}a_{22}a_{31} - a_{11}a_{23}a_{32} - a_{12}a_{21}a_{33}$$

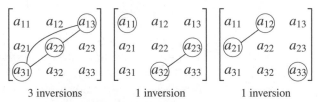

We see that the sign of prod P in the formula $\det A = \sum \pm \text{prod } P$ depends on the number of inversions in P. We get the plus sign if the number of inversions is even and the minus sign if that number is odd. We can write

$$\det A = \sum (-1)^{(\text{number of inversions in } P)} \text{prod } P$$

If we define the *signature* of a pattern P as $\text{sgn } P = (-1)^{(\text{number of inversions in } P)}$, then we can write more succinctly

$$\det A = \sum (\text{sgn } P)(\text{prod } P),$$

where the sum is taken over all six patterns P in the matrix A.

Alternatively, we can describe the signature in terms of row swaps: If we can bring a pattern P into diagonal form by means of p row swaps, then $\text{sgn } P = (-1)^p$. (See Theorem 6.2.3b.)

Using these definitions and observations as a guide, we are now ready to define the determinant of an $n \times n$ matrix.

Definition 6.1.3 Patterns, inversions, and determinants[4]

A *pattern* in an $n \times n$ matrix A is a way to choose n entries of the matrix so that there is one chosen entry in each row and in each column of A.

With a pattern P we associate the product of all its entries, denoted prod P.

Two entries in a pattern are said to be *inverted* if one of them is located to the right and above the other in the matrix.

The *signature* of a pattern P is defined as sgn $P = (-1)^{\text{(number of inversions in } P)}$.

The determinant of A is defined as

$$\det A = \sum (\text{sgn } P)(\text{prod } P),$$

where the sum is taken over all $n!$ patterns P in the matrix A. Thus we are summing up the products associated with all patterns with an even number of inversions, and we are subtracting the products associated with the patterns with an odd number of inversions.

EXAMPLE 7 Apply Definition 6.1.3 to a 2×2 matrix, and verify that the result agrees with the formula given in Theorem 2.4.9a.

Solution

There are two patterns in the 2×2 matrix $A = \begin{bmatrix} a & b \\ c & d \end{bmatrix}$:

$$\begin{bmatrix} \textcircled{a} & b \\ c & \textcircled{d} \end{bmatrix} \qquad \begin{bmatrix} a & \textcircled{b} \\ \textcircled{c} & d \end{bmatrix}.$$

No inversions   One inversion

Therefore, $\det A = (-1)^0 ad + (-1)^1 bc = ad - bc$. ∎

EXAMPLE 8 Find det A for

$$A = \begin{bmatrix} 0 & 2 & 0 & 0 & 0 & 0 \\ 0 & 0 & 0 & 8 & 0 & 0 \\ 0 & 0 & 0 & 0 & 0 & 2 \\ 3 & 0 & 0 & 0 & 0 & 0 \\ 0 & 0 & 0 & 0 & 5 & 0 \\ 0 & 0 & 1 & 0 & 0 & 0 \end{bmatrix}.$$

---

[4]It appears that determinants were first considered by the Japanese mathematician Seki Kowa (1642–1708). Seki may have known that the determinant of an $n \times n$ matrix has $n!$ terms and that rows and columns are interchangeable. (See Theorem 6.2.1.) The French mathematician Alexandre-Théophile Vandermonde (1735–1796) was the first to give a coherent and systematic exposition of the theory of determinants. Throughout the 19th century, determinants were considered the ultimate tool in linear algebra, used extensively by Cauchy, Jacobi, Kronecker, and others. Recently, determinants have gone somewhat out of fashion, and some people would like to see them eliminated altogether from linear algebra. (See, for example, Sheldon Axler's article "Down with Determinants" in *The American Mathematical Monthly,* February 1995, where we read: "This paper will show how linear algebra can be done better without determinants." Read it and see what you think.)

Solution

Only one pattern P makes a nonzero contribution toward the determinant:

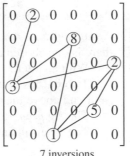

7 inversions

Thus $\det A = (\operatorname{sgn} P)(\operatorname{prod} P) = (-1)^7 2 \cdot 8 \cdot 2 \cdot 3 \cdot 5 \cdot 1 = -480.$ ∎

EXAMPLE 9   Find $\det A$ for

$$A = \begin{bmatrix} 6 & 0 & 1 & 0 & 0 \\ 9 & 3 & 2 & 3 & 7 \\ 8 & 0 & 3 & 2 & 9 \\ 0 & 0 & 4 & 0 & 0 \\ 5 & 0 & 5 & 0 & 1 \end{bmatrix}.$$

Solution

Again, let's look for patterns with a nonzero product. We pick the entries column by column this time. In the second column, we must choose the second component, 3. Then, in the fourth column, we must choose the third component, 2. Next, think about the last column, and so on. It turns out that there is only one pattern P with a nonzero product.

$$\begin{bmatrix} ⑥ & 0 & 1 & 0 & 0 \\ 9 & ③ & 2 & 3 & 7 \\ 8 & 0 & 3 & ② & 9 \\ 0 & 0 & ④ & 0 & 0 \\ 5 & 0 & 5 & 0 & ① \end{bmatrix}$$

1 inversion

$$\det A = (\operatorname{sgn} P)(\operatorname{prod} P) = (-1)^1 6 \cdot 3 \cdot 2 \cdot 4 \cdot 1 = -144.$$ ∎

EXAMPLE 10   Find $\det A$ for

$$A = \begin{bmatrix} 1 & 2 & 3 & 4 & 5 \\ 0 & 2 & 3 & 4 & 5 \\ 0 & 0 & 3 & 4 & 5 \\ 0 & 0 & 0 & 4 & 5 \\ 0 & 0 & 0 & 0 & 5 \end{bmatrix}.$$

Solution

Note that A is an upper triangular matrix. To have a nonzero product, a pattern must contain the first component of the first column, then the second component of the second column, and so on. Thus, only the diagonal pattern P makes a nonzero contribution. We conclude that

$$\det A = (\operatorname{sgn} P)(\operatorname{prod} P) = (-1)^0 1 \cdot 2 \cdot 3 \cdot 4 \cdot 5 = 5! = 120.$$ ∎

We can generalize this result:

Theorem 6.1.4   **Determinant of a triangular matrix**

The determinant of an (upper or lower) triangular matrix is the product of the diagonal entries of the matrix.

In particular, the determinant of a diagonal matrix is the product of its diagonal entries. ∎

The Determinant of a Block Matrix (optional)

EXAMPLE 11   Find $\det M$ for

$$M = \begin{bmatrix} a_{11} & a_{12} & b_{11} & b_{12} \\ a_{21} & a_{22} & b_{21} & b_{22} \\ 0 & 0 & c_{11} & c_{12} \\ 0 & 0 & c_{21} & c_{22} \end{bmatrix}.$$

Solution

It is natural to partition the 4×4 matrix M into four 2×2 blocks, one of which is zero:

$$M = \begin{bmatrix} A & B \\ 0 & C \end{bmatrix}.$$

Let's see whether we can express $\det \begin{bmatrix} A & B \\ 0 & C \end{bmatrix}$ in terms of $\det A$, $\det B$, and $\det C$. Let's find the patterns in M that may have a nonzero product.

$$\begin{bmatrix} \textcircled{a_{11}} & a_{12} & b_{11} & b_{12} \\ a_{21} & \textcircled{a_{22}} & b_{21} & b_{22} \\ 0 & 0 & \textcircled{c_{11}} & c_{12} \\ 0 & 0 & c_{21} & \textcircled{c_{22}} \end{bmatrix}, \quad \begin{bmatrix} \textcircled{a_{11}} & a_{12} & b_{11} & b_{12} \\ a_{21} & \textcircled{a_{22}} & b_{21} & b_{22} \\ 0 & 0 & c_{11} & \textcircled{c_{12}} \\ 0 & 0 & \textcircled{c_{21}} & c_{22} \end{bmatrix}, \quad \begin{bmatrix} a_{11} & \textcircled{a_{12}} & b_{11} & b_{12} \\ \textcircled{a_{21}} & a_{22} & b_{21} & b_{22} \\ 0 & 0 & \textcircled{c_{11}} & c_{12} \\ 0 & 0 & c_{21} & \textcircled{c_{22}} \end{bmatrix},$$

$$\begin{bmatrix} a_{11} & \textcircled{a_{12}} & b_{11} & b_{12} \\ \textcircled{a_{21}} & a_{22} & b_{21} & b_{22} \\ 0 & 0 & c_{11} & \textcircled{c_{12}} \\ 0 & 0 & \textcircled{c_{21}} & c_{22} \end{bmatrix},$$

Thus

$$\det M = a_{11}a_{22}c_{11}c_{22} - a_{11}a_{22}c_{12}c_{21} - a_{12}a_{21}c_{11}c_{22} + a_{12}a_{21}c_{12}c_{21}$$
$$= a_{11}a_{22}(c_{11}c_{22} - c_{12}c_{21}) - a_{12}a_{21}(c_{11}c_{22} - c_{12}c_{21})$$
$$= (a_{11}a_{22} - a_{12}a_{21})(c_{11}c_{22} - c_{12}c_{21}) = (\det A)(\det C)$$

In summary,

$$\det M = \det \begin{bmatrix} A & B \\ 0 & C \end{bmatrix} = (\det A)(\det C)$$ ∎

In turns out that the formula we derived in Example 11 holds for block matrices of any size.

Theorem 6.1.5   Determinant of a block matrix

If $M = \begin{bmatrix} A & B \\ 0 & C \end{bmatrix}$, where A and C are square matrices (not necessarily of the same size), then

$$\det \begin{bmatrix} A & B \\ 0 & C \end{bmatrix} = (\det A)(\det C).$$

Likewise,

$$\det \begin{bmatrix} A & 0 \\ B & C \end{bmatrix} = (\det A)(\det C).$$

However, the formula

$$\det \begin{bmatrix} A & B \\ C & D \end{bmatrix} = (\det A)(\det D) - (\det B)(\det C)$$

does not always hold (see Exercise 48).

Proof   Let's outline a proof for Theorem 6.1.5. As you follow this somewhat technical presentation, use Example 11 as a guide.

If P_A is a pattern in A and P_C is a pattern in C, then their concatenation, $P_M = (P_A, P_C)$, will be a pattern in M, with prod $P_M = (\text{prod } P_A)(\text{prod } P_C)$ and sgn $P_M = (\text{sgn } P_A)(\text{sgn } P_C)$, since the number of inversions in P_M will be the sum of those in P_A and P_C. Conversely, any pattern P_M in M with a nonzero product will be of this form, $P_M = (P_A, P_C)$, since the pattern entries cannot be taken from the zero block in matrix M. Now

$$(\det A)(\det C) = \left(\sum_{P_A} (\text{sgn } P_A)(\text{prod } P_A) \right) \left(\sum_{P_C} (\text{sgn } P_C)(\text{prod } P_C) \right)$$

$$= \sum_{(P_A, P_C)} (\text{sgn } P_A)(\text{sgn } P_C)(\text{prod } P_A)(\text{prod } P_C)$$

$$= \sum_{P_M} (\text{sgn } P_M)(\text{prod } P_M) = \det M$$

Here is another example illustrating this proof:

$$M = \begin{bmatrix} A & B \\ 0 & C \end{bmatrix} = \left[\begin{array}{ccc|ccc} 1 & ② & 3 & 0 & 0 & 0 \\ ④ & 5 & 6 & 0 & 0 & 0 \\ 7 & 8 & ⑦ & 0 & 0 & 0 \\ \hline 6 & 5 & 4 & 3 & 2 & ① \\ 2 & 3 & 4 & ⑤ & 6 & 7 \\ 8 & 7 & 6 & 5 & ④ & 3 \end{array} \right]$$

Here, prod $P_M = 2 \cdot 4 \cdot 7 \cdot 1 \cdot 5 \cdot 4 = (2 \cdot 4 \cdot 7)(1 \cdot 5 \cdot 4) = (\text{prod } P_A)(\text{prod } P_C)$. There is one inversion in P_A and there are two inversions in P_C, for a total of three inversions in P_M. Thus sgn $P_M = (-1)^3 = (-1)^1(-1)^2 = (\text{sgn } P_A)(\text{sgn } P_C)$. ∎

EXAMPLE 12   Find

$$\det \begin{bmatrix} 1 & 0 & 0 & 0 & 0 & 0 \\ 2 & 7 & 0 & 0 & 0 & 0 \\ 3 & 8 & 6 & 0 & 0 & 0 \\ 4 & 9 & 5 & 2 & 1 & 4 \\ 5 & 8 & 4 & 0 & 2 & 5 \\ 6 & 7 & 3 & 0 & 3 & 6 \end{bmatrix}.$$

Solution

$$\det \begin{bmatrix} 1 & 0 & 0 & 0 & 0 & 0 \\ 2 & 7 & 0 & 0 & 0 & 0 \\ 3 & 8 & 6 & 0 & 0 & 0 \\ \hline 4 & 9 & 5 & 2 & 1 & 4 \\ 5 & 8 & 4 & 0 & 2 & 5 \\ 6 & 7 & 3 & 0 & 3 & 6 \end{bmatrix} = \det \begin{bmatrix} 1 & 0 & 0 \\ 2 & 7 & 0 \\ 3 & 8 & 6 \end{bmatrix} \det \begin{bmatrix} 2 & 1 & 4 \\ 0 & 2 & 5 \\ 0 & 3 & 6 \end{bmatrix}$$

$$= (1 \cdot 7 \cdot 6)(2 \cdot 2 \cdot 6 - 2 \cdot 5 \cdot 3)$$
$$= 42(-6) = -252 \qquad \blacksquare$$

EXERCISES 6.1

Find the determinants of the matrices in Exercises 1 through 10, and find out which of these matrices are invertible.

1. $\begin{bmatrix} 1 & 2 \\ 3 & 6 \end{bmatrix}$

2. $\begin{bmatrix} 2 & 3 \\ 4 & 5 \end{bmatrix}$

3. $\begin{bmatrix} 3 & 5 \\ 7 & 11 \end{bmatrix}$

4. $\begin{bmatrix} 1 & 4 \\ 2 & 8 \end{bmatrix}$

5. $\begin{bmatrix} 2 & 5 & 7 \\ 0 & 11 & 7 \\ 0 & 0 & 5 \end{bmatrix}$

6. $\begin{bmatrix} 6 & 0 & 0 \\ 5 & 4 & 0 \\ 3 & 2 & 1 \end{bmatrix}$

7. $\begin{bmatrix} 1 & 1 & 1 \\ 2 & 2 & 2 \\ 3 & 3 & 3 \end{bmatrix}$

8. $\begin{bmatrix} 1 & 2 & 3 \\ 1 & 1 & 1 \\ 3 & 2 & 1 \end{bmatrix}$

9. $\begin{bmatrix} 0 & 1 & 2 \\ 7 & 8 & 3 \\ 6 & 5 & 4 \end{bmatrix}$

10. $\begin{bmatrix} 1 & 1 & 1 \\ 1 & 2 & 3 \\ 1 & 3 & 6 \end{bmatrix}$

In Exercises 11 through 22, use the determinant to find out for which values of the constant k the given matrix is invertible.

11. $\begin{bmatrix} k & 2 \\ 3 & 4 \end{bmatrix}$

12. $\begin{bmatrix} 1 & k \\ k & 4 \end{bmatrix}$

13. $\begin{bmatrix} k & 3 & 5 \\ 0 & 2 & 6 \\ 0 & 0 & 4 \end{bmatrix}$

14. $\begin{bmatrix} 4 & 0 & 0 \\ 3 & k & 0 \\ 2 & 1 & 0 \end{bmatrix}$

15. $\begin{bmatrix} 0 & k & 1 \\ 2 & 3 & 4 \\ 5 & 6 & 7 \end{bmatrix}$

16. $\begin{bmatrix} 1 & 2 & 3 \\ 4 & k & 5 \\ 6 & 7 & 8 \end{bmatrix}$

17. $\begin{bmatrix} 1 & 1 & 1 \\ 1 & k & -1 \\ 1 & k^2 & 1 \end{bmatrix}$

18. $\begin{bmatrix} 0 & 1 & k \\ 3 & 2k & 5 \\ 9 & 7 & 5 \end{bmatrix}$

19. $\begin{bmatrix} 1 & 1 & k \\ 1 & k & k \\ k & k & k \end{bmatrix}$

20. $\begin{bmatrix} 1 & k & 1 \\ 1 & k+1 & k+2 \\ 1 & k+2 & 2k+4 \end{bmatrix}$

21. $\begin{bmatrix} k & 1 & 1 \\ 1 & k & 1 \\ 1 & 1 & k \end{bmatrix}$

22. $\begin{bmatrix} \cos k & 1 & -\sin k \\ 0 & 2 & 0 \\ \sin k & 0 & \cos k \end{bmatrix}$

In Exercises 23 through 30, use the determinant to find out for which values of the constant λ the matrix $A - \lambda I_n$ fails to be invertible.

23. $\begin{bmatrix} 1 & 2 \\ 0 & 4 \end{bmatrix}$

24. $\begin{bmatrix} 2 & 0 \\ 1 & 0 \end{bmatrix}$

25. $\begin{bmatrix} 4 & 2 \\ 4 & 6 \end{bmatrix}$

26. $\begin{bmatrix} 4 & 2 \\ 2 & 7 \end{bmatrix}$

27. $\begin{bmatrix} 2 & 0 & 0 \\ 5 & 3 & 0 \\ 7 & 6 & 4 \end{bmatrix}$

28. $\begin{bmatrix} 5 & 7 & 11 \\ 0 & 3 & 13 \\ 0 & 0 & 2 \end{bmatrix}$

29. $\begin{bmatrix} 3 & 5 & 6 \\ 0 & 4 & 2 \\ 0 & 2 & 7 \end{bmatrix}$

30. $\begin{bmatrix} 4 & 2 & 0 \\ 4 & 6 & 0 \\ 5 & 2 & 3 \end{bmatrix}$

Find the determinants of the matrices in Exercises 31 through 42.

31. $\begin{bmatrix} 1 & 9 & 8 & 7 \\ 0 & 2 & 9 & 6 \\ 0 & 0 & 3 & 5 \\ 0 & 0 & 0 & 4 \end{bmatrix}$

32. $\begin{bmatrix} 2 & 5 & 7 & 11 \\ 0 & 3 & 5 & 13 \\ 0 & 0 & 5 & 11 \\ 0 & 0 & 0 & 7 \end{bmatrix}$

33. $\begin{bmatrix} 1 & 2 & 3 & 4 \\ 8 & 7 & 6 & 5 \\ 0 & 0 & 2 & 3 \\ 0 & 0 & 7 & 5 \end{bmatrix}$

34. $\begin{bmatrix} 4 & 5 & 0 & 0 \\ 3 & 6 & 0 & 0 \\ 2 & 7 & 1 & 4 \\ 1 & 8 & 2 & 3 \end{bmatrix}$

35. $\begin{bmatrix} 2 & 3 & 0 & 2 \\ 4 & 3 & 2 & 1 \\ 6 & 0 & 0 & 3 \\ 7 & 0 & 0 & 4 \end{bmatrix}$

36. $\begin{bmatrix} 0 & 0 & 0 & 1 \\ 0 & 0 & 1 & 0 \\ 0 & 1 & 0 & 0 \\ 1 & 0 & 0 & 0 \end{bmatrix}$

37. $\begin{bmatrix} 5 & 4 & 0 & 0 & 0 \\ 6 & 7 & 0 & 0 & 0 \\ 3 & 4 & 5 & 6 & 7 \\ 2 & 1 & 0 & 1 & 2 \\ 2 & 1 & 0 & 0 & 1 \end{bmatrix}$

38. $\begin{bmatrix} 1 & 2 & 3 & 4 & 5 \\ 3 & 0 & 4 & 5 & 6 \\ 2 & 1 & 2 & 3 & 4 \\ 0 & 0 & 0 & 6 & 5 \\ 0 & 0 & 0 & 5 & 6 \end{bmatrix}$

39. $\begin{bmatrix} 0 & 0 & 0 & 0 & 1 \\ 0 & 0 & 2 & 0 & 0 \\ 0 & 4 & 0 & 0 & 0 \\ 0 & 0 & 0 & 3 & 0 \\ 5 & 0 & 0 & 0 & 0 \end{bmatrix}$

40. $\begin{bmatrix} 0 & 0 & 3 & 0 & 0 \\ 0 & 0 & 0 & 0 & 2 \\ 0 & 4 & 0 & 0 & 0 \\ 0 & 0 & 0 & 1 & 0 \\ 5 & 0 & 0 & 0 & 0 \end{bmatrix}$

41. $\begin{bmatrix} 0 & 0 & 1 & 0 & 2 \\ 5 & 4 & 3 & 2 & 1 \\ 1 & 3 & 5 & 0 & 7 \\ 2 & 0 & 4 & 0 & 6 \\ 0 & 0 & 3 & 0 & 4 \end{bmatrix}$

42. $\begin{bmatrix} 0 & 0 & 2 & 3 & 1 \\ 0 & 0 & 0 & 2 & 2 \\ 0 & 9 & 7 & 9 & 3 \\ 0 & 0 & 0 & 0 & 5 \\ 3 & 4 & 5 & 8 & 5 \end{bmatrix}$

43. If A is an $n \times n$ matrix, what is the relationship between $\det A$ and $\det(-A)$?

44. If A is an $n \times n$ matrix and k is an arbitrary constant, what is the relationship between $\det A$ and $\det(kA)$?

45. If A is a 2×2 matrix, what is the relationship between $\det A$ and $\det(A^T)$?

46. If A is an invertible 2×2 matrix, what is the relationship between $\det A$ and $\det(A^{-1})$?

47. Find nonzero numbers a, b, c, d, e, f, g, h such that the matrix $\begin{bmatrix} a & b & c \\ d & k & e \\ f & g & h \end{bmatrix}$ is invertible for all real numbers k, or explain why no such matrix exists.

48. Find 2×2 matrices A, B, C, D such that

$$\det \begin{bmatrix} A & B \\ C & D \end{bmatrix} \neq (\det A)(\det D) - (\det B)(\det C).$$

49. For two nonparallel vectors $\vec{v}$ and $\vec{w}$ in $\mathbb{R}^3$, consider the linear transformation

$$T(\vec{x}) = \det \begin{bmatrix} \vec{x} & \vec{v} & \vec{w} \end{bmatrix}$$

from $\mathbb{R}^3$ to $\mathbb{R}$. Describe the kernel of T geometrically. What is the image of T?

50. If $\vec{u}, \vec{v}, \vec{w}$ are three unit vectors in $\mathbb{R}^3$, what are the possible values of $\det \begin{bmatrix} \vec{u} & \vec{v} & \vec{w} \end{bmatrix}$?

51. Explain why any pattern P in a matrix A, other than the diagonal pattern, contains at least one entry below the diagonal and at least one entry above the diagonal.

52. Consider two vectors $\vec{v}$ and $\vec{w}$ in $\mathbb{R}^3$. Form the matrix

$$A = \begin{bmatrix} \vec{v} \times \vec{w} & \vec{v} & \vec{w} \end{bmatrix}.$$

Express $\det A$ in terms of $\|\vec{v} \times \vec{w}\|$. For which choices of $\vec{v}$ and $\vec{w}$ is A invertible?

53. Find the determinant of the $(2n) \times (2n)$ matrix

$$A = \begin{bmatrix} 0 & I_n \\ I_n & 0 \end{bmatrix}.$$

54. Is the determinant of the matrix

$$A = \begin{bmatrix} 1 & 1000 & 2 & 3 & 4 \\ 5 & 6 & 7 & 1000 & 8 \\ 1000 & 9 & 8 & 7 & 6 \\ 5 & 4 & 3 & 2 & 1000 \\ 1 & 2 & 1000 & 3 & 4 \end{bmatrix}$$

positive or negative? How can you tell? Do not use technology.

55. Does the following matrix have an LU factorization? (See Exercises 2.4.90 and 2.4.93.)

$$A = \begin{bmatrix} 7 & 4 & 2 \\ 5 & 3 & 1 \\ 3 & 1 & 4 \end{bmatrix}$$

56. Let M_n be the $n \times n$ matrix with all 1's along "the other diagonal," and 0's everywhere else. For example,

$$M_4 = \begin{bmatrix} 0 & 0 & 0 & 1 \\ 0 & 0 & 1 & 0 \\ 0 & 1 & 0 & 0 \\ 1 & 0 & 0 & 0 \end{bmatrix}.$$

a. Find $\det(M_n)$ for $n = 2, 3, 4, 5, 6, 7$.
b. Find a formula for $\det(M_n)$, in terms of n.

57. A square matrix is called a *permutation matrix* if each row and each column contains exactly one entry 1, with all other entries being 0. Examples are I_n, $\begin{bmatrix} 0 & 1 & 0 \\ 0 & 0 & 1 \\ 1 & 0 & 0 \end{bmatrix}$, and the matrices considered in Exercises 53 and 56. What are the possible values of the determinant of a permutation matrix?

58. a. Find a noninvertible 2×2 matrix whose entries are four distinct prime numbers, or explain why no such matrix exists.

b. Find a noninvertible 3×3 matrix whose entries are nine distinct prime numbers, or explain why no such matrix exists.

59. Consider the function $F(A) = F \begin{bmatrix} \vec{v} & \vec{w} \end{bmatrix} = \vec{v} \cdot \vec{w}$ from $\mathbb{R}^{2 \times 2}$ to $\mathbb{R}$, the dot product of the column vectors of A.

a. Is F linear in both columns of A? (See Example 6.)

b. Is F linear in both rows of A?

c. Is F alternating on the columns of A? (See Example 4.)

60. Which of the following functions F of $A = \begin{bmatrix} a & b \\ c & d \end{bmatrix}$ are linear in both columns? Which are linear in both rows? Which are alternating on the columns?

a. $F(A) = bc$ **b.** $F(A) = cd$ **c.** $F(A) = ac$

d. $F(A) = bc - ad$ **e.** $F(A) = c$

61. Show that the function

$$F \begin{bmatrix} a & b & c \\ d & e & f \\ g & h & j \end{bmatrix} = bfg$$

is linear in all three columns and in all three rows. (See Example 6.) Is F alternating on the columns? (See Example 4.)

In Exercises 62 through 64, consider a function D from $\mathbb{R}^{2 \times 2}$ to $\mathbb{R}$ that is linear in both columns and alternating on the columns. (See Examples 4 and 6 and the subsequent discussions.) Assume that $D(I_2) = 1$.

62. Show that $D(A) = 0$ for any 2×2 matrix A whose two columns are equal.

63. Show that $D \begin{bmatrix} a & b \\ 0 & d \end{bmatrix} = ad$. {*Hint:* Write $\begin{bmatrix} b \\ d \end{bmatrix} = \begin{bmatrix} b \\ 0 \end{bmatrix} + \begin{bmatrix} 0 \\ d \end{bmatrix}$ and use linearity in the second column: $D \begin{bmatrix} a & b \\ 0 & d \end{bmatrix} = D \begin{bmatrix} a & b \\ 0 & 0 \end{bmatrix} + D \begin{bmatrix} a & 0 \\ 0 & d \end{bmatrix} = ab D \begin{bmatrix} 1 & 1 \\ 0 & 0 \end{bmatrix} + \dots$. Use Exercise 62.}

64. Using Exercises 62 and 63 as a guide, show that $D(A) = ad - bc = \det A$ for all 2×2 matrices A.

65. Consider a function D from $\mathbb{R}^{3 \times 3}$ to $\mathbb{R}$ that is linear in all three columns and alternating on the columns. Assume that $D(I_3) = 1$. Using Exercises 62 through 64 as a guide, show that $D(A) = \det A$ for all 3×3 matrices A.

66. a. Let V be the linear space of all functions F from $\mathbb{R}^{2 \times 2}$ to $\mathbb{R}$ that are linear in both columns. Find a basis of V, and thus determine the dimension of V.

b. Let W be the linear space of all functions D from $\mathbb{R}^{2 \times 2}$ to $\mathbb{R}$ that are linear in both columns and alternating on the columns. Find a basis of W, and thus determine the dimension of W.

6.2 Properties of the Determinant

The main goal of this section is to show that a square matrix of any size is invertible if (and only if) its determinant is nonzero. As we work toward this goal, we will discuss a number of other remarkable properties of the determinant.

The Determinant of the Transpose[5]

EXAMPLE I   Let

$$A = \begin{bmatrix} 1 & 2 & 3 & 4 & 5 \\ 6 & 7 & 8 & 9 & 8 \\ 7 & 6 & 5 & 4 & 3 \\ 2 & 1 & 2 & 3 & 4 \\ 5 & 6 & 7 & 8 & 9 \end{bmatrix}.$$

Express $\det(A^T)$ in terms of $\det A$. You need not compute $\det A$.

---

[5]If you skipped Chapter 5, read Definition 5.3.5.

Solution

For each pattern P in A, we can consider the corresponding (transposed) pattern P^T in A^T; for example,

$$A = \begin{bmatrix} 1 & 2 & ③ & 4 & 5 \\ ⑥ & 7 & 8 & 9 & 8 \\ 7 & 6 & 5 & 4 & ③ \\ 2 & ① & 2 & 3 & 4 \\ 5 & 6 & 7 & ⑧ & 9 \end{bmatrix}, \qquad A^T = \begin{bmatrix} 1 & ⑥ & 7 & 2 & 5 \\ 2 & 7 & 6 & ① & 6 \\ ③ & 8 & 5 & 2 & 7 \\ 4 & 9 & 4 & 3 & ⑧ \\ 5 & 8 & ③ & 4 & 9 \end{bmatrix}.$$

$$\qquad\quad P \qquad\qquad\qquad\qquad\qquad P^T$$

The two patterns P and P^T involve the same numbers, and they contain the same number of inversions, but the role of the two numbers in each inversion is reversed. Therefore, the two patterns make the same contributions to the respective determinants $(\text{sgn } P)(\text{prod } P) = (\text{sgn } P^T)(\text{prod } P^T)$. Since these observations apply to all patterns of A, we can conclude that $\det(A^T) = \det A$. ■

Since we have not used any special properties of the matrix A in Example 1, we can state more generally:

Theorem 6.2.1   **Determinant of the transpose**

If A is a square matrix, then

$$\det(A^T) = \det A.$$

■

This symmetry property will prove very useful. Any property of the determinant expressed in terms of the rows holds for the columns as well, and vice versa.

Linearity Properties of the Determinant

In Section 6.1 we observed that the 3×3 determinant is linear in the rows and in the columns; take another look at Example 6 of Section 6.1. It turns out that these linearity properties generalize to the determinant of $n \times n$ matrices.

Theorem 6.2.2   **Linearity of the determinant in the rows and columns**

Consider fixed row vectors $\vec{v}_1, \ldots, \vec{v}_{i-1}, \vec{v}_{i+1}, \ldots, \vec{v}_n$ with n components. Then the function

$$T(\vec{x}) = \det \begin{bmatrix} — & \vec{v}_1 & — \\ & \ldots & \\ — & \vec{v}_{i-1} & — \\ — & \vec{x} & — \\ — & \vec{v}_{i+1} & — \\ & \ldots & \\ & \vec{v}_n & \end{bmatrix} \qquad \text{from } \mathbb{R}^{1 \times n} \text{ to } \mathbb{R}$$

is a linear transformation. This property is referred to as *linearity of the determinant in the ith row*. Likewise, the determinant is *linear in all the columns*. ■

To prove Theorem 6.2.2, observe that the product prod P associated with a pattern P is linear in all the rows and columns, since this product contains exactly one factor from each row and one from each column. Thus the determinant itself is linear in all the rows and columns, being a linear combination of pattern products.

We can express the linearity of the transformation T in Theorem 6.2.2 in terms of the equations $T(\vec{x} + \vec{y}) = T(\vec{x}) + T(\vec{y})$ and $T(k\vec{x}) = kT(\vec{x})$, or

$$\det \begin{bmatrix} - & \vec{v}_1 & - \\ & \cdots & \\ - & \vec{x} + \vec{y} & - \\ & \cdots & \\ - & \vec{v}_n & - \end{bmatrix} = \det \begin{bmatrix} - & \vec{v}_1 & - \\ & \cdots & \\ - & \vec{x} & - \\ & \cdots & \\ - & \vec{v}_n & - \end{bmatrix} + \begin{bmatrix} - & \vec{v}_1 & - \\ & \cdots & \\ - & \vec{y} & - \\ & \cdots & \\ - & \vec{v}_n & - \end{bmatrix} \quad \text{and}$$

$$\det \begin{bmatrix} - & \vec{v}_1 & - \\ & \cdots & \\ - & k\vec{x} & - \\ & \cdots & \\ - & \vec{v}_n & - \end{bmatrix} = k \det \begin{bmatrix} - & \vec{v}_1 & - \\ & \cdots & \\ - & \vec{x} & - \\ & \cdots & \\ - & \vec{v}_n & - \end{bmatrix}$$

In these equations, all rows except the ith are fixed, $\vec{x}$ and $\vec{y}$ are arbitrary row vectors with n components, and k is an arbitrary real number.

Determinants and Gauss–Jordan Elimination

Consider a 30×30 matrix A, a rather small matrix by the standards of contemporary scientific and engineering applications. Then $29 \cdot 30! \approx 7 \cdot 10^{33}$ multiplications are required to compute the determinant of A by Definition 6.1.3, using patterns. If a super computer performs 10 trillion (10^{13}) multiplications a second, it will take over a trillion years to carry out these computations; our universe might long be gone by then. Clearly, we have to look for more efficient ways to compute the determinant.

To use the language of computer science, is there an algorithm for the determinant that runs on polynomial rather than on exponential time?

So far in this text, Gauss–Jordan elimination has served us well as a powerful tool for solving numerical problems in linear algebra. If we could understand what happens to the determinant of a matrix as we row-reduce it, we could use Gauss–Jordan elimination to compute determinants as well. We have to understand how the three elementary row operations affect the determinant:

a. Row division: dividing a row by a nonzero scalar k,

b. Row swap: swapping two rows, and

c. Row addition: adding a multiple of a row to another row.

Let's look at the case of a 2×2 matrix $A = \begin{bmatrix} a & b \\ c & d \end{bmatrix}$ first, with $\det A = ad - bc$.

a. If $B = \begin{bmatrix} a/k & b/k \\ c & d \end{bmatrix}$, then $\det B = \dfrac{a}{k}d - \dfrac{b}{k}c = \dfrac{1}{k}\det A$.

Verify that $\det B = \dfrac{1}{k}\det A$ if B is obtained from A by dividing the *second* row by k.

b. If $B = \begin{bmatrix} c & d \\ a & b \end{bmatrix}$, then $\det B = cb - da = -\det A$.

c. If $B = \begin{bmatrix} a + kc & b + kd \\ c & d \end{bmatrix}$, then $\det B = (a + kc)d - (b + kd)c = ad + kcd - bc - kdc = \det A$. Verify that $\det B = \det A$ if B is obtained from A by adding k times the *first* row to the *second*.

Next, we will examine the effect of the elementary row operations on the determinant of square matrices of arbitrary size.

a. Row division: If

$$A = \begin{bmatrix} - & \vec{v}_1 & - \\ & \vdots & \\ - & \vec{v}_i & - \\ & \vdots & \\ - & \vec{v}_n & - \end{bmatrix} \quad \text{and} \quad B = \begin{bmatrix} - & \vec{v}_1 & - \\ & \vdots & \\ - & \vec{v}_i/k & - \\ & \vdots & \\ - & \vec{v}_n & - \end{bmatrix},$$

then $\det B = (1/k) \det A$, by linearity in the ith row, Theorem 6.2.2.

b. Row swap: Refer to Example 2.

EXAMPLE 2   Consider the matrices

$$A = \begin{bmatrix} 1 & 2 & 3 & 4 & 5 \\ 6 & 7 & 8 & 9 & 8 \\ 7 & 6 & 5 & 4 & 3 \\ 2 & 1 & 2 & 3 & 4 \\ 5 & 6 & 7 & 8 & 9 \end{bmatrix} \quad \text{and} \quad B = \begin{bmatrix} 6 & 7 & 8 & 9 & 8 \\ 1 & 2 & 3 & 4 & 5 \\ 7 & 6 & 5 & 4 & 3 \\ 2 & 1 & 2 & 3 & 4 \\ 5 & 6 & 7 & 8 & 9 \end{bmatrix}.$$

Note that B is obtained from A by swapping the first two rows. Express $\det B$ in terms of $\det A$.

Solution

For each pattern P in A, we can consider the corresponding pattern P_{swap} in B; for example,

$$A = \begin{bmatrix} 1 & 2 & ③ & 4 & 5 \\ ⑥ & 7 & 8 & 9 & 8 \\ 7 & 6 & 5 & 4 & ③ \\ 2 & ① & 2 & 3 & 4 \\ 5 & 6 & 7 & ⑧ & 9 \end{bmatrix} \quad \text{and} \quad B = \begin{bmatrix} ⑥ & 7 & 8 & 9 & 8 \\ 1 & 2 & ③ & 4 & 5 \\ 7 & 6 & 5 & 4 & ③ \\ 2 & ① & 2 & 3 & 4 \\ 5 & 6 & 7 & ⑧ & 9 \end{bmatrix}.$$
$$\qquad\qquad P \qquad\qquad\qquad\qquad\qquad\qquad P_{swap}$$

These two patterns P and P_{swap} involve the same numbers, but the number of inversions in P_{swap} is one less than in P, since we are losing the inversion formed by the entries in the first two rows of A. Thus prod $P_{swap} = $ prod P, but sgn $P_{swap} = -$sgn P, so that the two patterns make opposite contributions to the respective determinants. Since these remarks apply to all patterns in A, we can conclude that

$$\det B = -\det A.$$

(If P is a pattern in A such that the entries in the first two rows do not form an inversion, then an additional inversion is created in P_{swap}; again, sgn $P_{swap} = -$sgn$P$.) ∎

What if B is obtained from A by swapping any two rows, rather than the first two? If we swap two *adjacent* rows, then everything works the same way as in Example 2, and $\det B = -\det A$. But what if B is obtained from A by swapping two arbitrary rows? Observe that swapping any two rows amounts to an odd number of swaps of adjacent rows. (See Exercise 60.) Since the determinant changes its sign with each swap of adjacent rows, the equation $\det B = -\det A$ still holds.

EXAMPLE 3   If a matrix A has two equal rows, what can you say about $\det A$?

Solution

Swap the two equal rows and call the resulting matrix B. Since we have swapped two *equal* rows, we have $A = B$. Now

$$\det A = \det B = -\det A,$$

so that

$$\det A = 0. \qquad \blacksquare$$

c. Row addition: Finally, what happens to the determinant if we add k times the ith row to the jth row?

$$A = \begin{bmatrix} \vdots \\ - & \vec{v}_i & - \\ \vdots \\ - & \vec{v}_j & - \\ \vdots \end{bmatrix} \longrightarrow \quad B = \begin{bmatrix} \vdots \\ - & \vec{v}_i & - \\ \vdots \\ - & \vec{v}_j + k\vec{v}_i & - \\ \vdots \end{bmatrix}$$

By linearity in the jth row, we find that

$$\det B = \det \begin{bmatrix} \vdots \\ - & \vec{v}_i & - \\ \vdots \\ - & \vec{v}_j & - \\ \vdots \end{bmatrix} + k \det \begin{bmatrix} \vdots \\ - & \vec{v}_i & - \\ \vdots \\ - & \vec{v}_i & - \\ \vdots \end{bmatrix} = \det A,$$

by Example 3.

Theorem 6.2.3 **Elementary row operations and determinants**

a. If B is obtained from A by dividing a row of A by a scalar k, then

$$\det B = (1/k) \det A.$$

b. If B is obtained from A by a row swap, then

$$\det B = - \det A.$$

We say that the determinant is *alternating* on the rows.

c. If B is obtained from A by adding a multiple of a row of A to another row, then

$$\det B = \det A.$$

Analogous results hold for elementary column operations.   ■

Now that we understand how elementary row operations affect determinants, we can analyze the relationship between the determinant of a square matrix A and that of rref A. Suppose that in the course of Gauss–Jordan elimination we swap rows s times and divide various rows by the scalars $k_1, k_2, \ldots, k_r$. Then

$$\det(\text{rref } A) = (-1)^s \frac{1}{k_1 k_2 \cdots k_r} (\det A),$$

or

$$\det A = (-1)^s k_1 k_2 \cdots k_r \det(\text{rref } A),$$

by Theorem 6.2.3.

Let us examine the cases when A is invertible and when it is not.

If A is invertible, then rref $A = I_n$, so that $\det(\text{rref } A) = \det(I_n) = 1$, and

$$\det A = (-1)^s k_1 k_2 \cdots k_r \neq 0.$$

Note that $\det A$ is not zero since all the scalars k_i are nonzero.

If A is noninvertible, then the last row of rref A contains all zeros, so that $\det(\text{rref } A) = 0$ (by linearity in the last row). It follows that $\det A = 0$.

We have established the following fundamental result.

Theorem 6.2.4   Invertibility and determinant

A square matrix A is invertible if and only if $\det A \neq 0$.   ∎

The foregoing discussion provides us with an efficient method for computing the determinant, using Gauss–Jordan elimination.

Algorithm 6.2.5 Using Gauss-Jordan elimination to compute the determinant

a. Consider an invertible $n \times n$ matrix A. Suppose you swap rows s times as you compute rref $A = I_n$, and you divide various rows by the scalars $k_1, k_2, \ldots, k_r$. Then

$$\det A = (-1)^s k_1 k_2 \cdots k_r.$$

b. In fact, it is not always sensible to reduce A all the way to rref A. Suppose you can use elementary row operations to transform A into some matrix B whose determinant is easy to compute (B might be a triangular matrix, for example). Suppose you swap rows s times as you transform A into B, and you divide various rows by the scalars $k_1, k_2, \ldots, k_r$. Then

$$\det A = (-1)^s k_1 k_2 \cdots k_r \det B.$$

EXAMPLE 4   Find

$$\det \begin{bmatrix} 0 & 7 & 5 & 3 \\ 1 & 1 & 2 & 1 \\ 1 & 1 & 2 & -1 \\ 1 & 1 & 1 & 2 \end{bmatrix}.$$

Solution

We go through the elimination process, keeping a note of all the row swaps and row divisions we perform (if any). In view of part (b) of Algorithm 6.2.5, we realize that it suffices to reduce A to an upper triangular matrix: There is no need to eliminate entries above the diagonal, or to make the diagonal entries equal to 1.

$$A = \begin{bmatrix} 0 & 7 & 5 & 3 \\ 1 & 1 & 2 & 1 \\ 1 & 1 & 2 & -1 \\ 1 & 1 & 1 & 2 \end{bmatrix} \longrightarrow \begin{bmatrix} 1 & 1 & 2 & 1 \\ 0 & 7 & 5 & 3 \\ 1 & 1 & 2 & -1 \\ 1 & 1 & 1 & 2 \end{bmatrix} \begin{matrix} \\ \\ -(I) \\ -(I) \end{matrix} \longrightarrow$$

$$\begin{bmatrix} 1 & 1 & 2 & 1 \\ 0 & 7 & 5 & 3 \\ 0 & 0 & 0 & -2 \\ 0 & 0 & -1 & 1 \end{bmatrix} \longrightarrow B = \begin{bmatrix} 1 & 1 & 2 & 1 \\ 0 & 7 & 5 & 3 \\ 0 & 0 & -1 & 1 \\ 0 & 0 & 0 & -2 \end{bmatrix}$$

We have performed two row swaps, so that $\det A = (-1)^2(\det B) = 7(-1)(-2) = 14$. We have used Theorem 6.1.4: The determinant of the triangular matrix B is the product of its diagonal entries. ∎

Determinant of a Product

If A and B are two $n \times n$ matrices, what is the relationship between $\det A$, $\det B$, and $\det(AB)$? The answer is as simple as could be:

Theorem 6.2.6 **Determinant of a product of matrices**

If A and B are $n \times n$ matrices, then

$$\det(AB) = (\det A)(\det B).$$

Proof   Let's first consider the case when A is invertible. In Exercise 34 the reader is asked to show that

$$\text{rref}\left[\, A \mid AB \,\right] = \left[\, I_n \mid B \,\right].$$

Suppose we swap rows s times, and we divide various rows by $k_1, k_2, \ldots, k_r$ as we perform this elimination.

Considering the left and the right halves of the matrices $\left[\, A \mid AB \,\right]$ and $\left[\, I_n \mid B \,\right]$ separately, and using Algorithm 6.2.5, we conclude that

$$\det(A) = (-1)^s k_1 k_2 \cdots k_r$$

and

$$\det(AB) = (-1)^s k_1 k_2 \cdots k_r (\det B) = (\det A)(\det B),$$

as claimed. If A is not invertible, then neither is AB (think about the image), so that $(\det A)(\det B) = 0(\det B) = 0 = \det(AB)$, as claimed. ∎

EXAMPLE 5   If matrix A is similar to B, what is the relationship between $\det A$ and $\det B$?

Solution

By Definition 3.4.5, there exists an invertible matrix S such that $AS = SB$. By Theorem 6.2.6 we have

$$(\det A)(\det S) = (\det S)(\det B).$$

Dividing both sides by the nonzero scalar $\det S$, we find that

$$\det A = \det B.$$ ∎

Theorem 6.2.7 **Determinants of similar matrices**

If matrix A is similar to B, then $\det A = \det B$. ∎

Conversely, if $\det A = \det B$, are the matrices A and B necessarily similar? See Exercise 59.

The Determinant of an Inverse

If A is an invertible $n \times n$ matrix, what is the relationship between $\det A$ and $\det(A^{-1})$? By definition of an inverse, the equation $I_n = AA^{-1}$ holds. By taking determinants of both sides and using Theorem 6.2.6, we find that

$$1 = \det(I_n) = \det(AA^{-1}) = \det(A)\det(A^{-1}),$$

so that

$$\det(A^{-1}) = \frac{1}{\det A}.$$

It turns out that $\det(A^{-1})$ is the reciprocal of $\det A$.

Theorem 6.2.8   Determinant of an inverse

If A is an invertible matrix, then

$$\det(A^{-1}) = \frac{1}{\det A} = (\det A)^{-1}. \qquad \blacksquare$$

Minors and Laplace Expansion[6] (Optional)

Recall the formula

$$\det A = a_{11}a_{22}a_{33} + a_{12}a_{23}a_{31} + a_{13}a_{21}a_{32} - a_{13}a_{22}a_{31} - a_{11}a_{23}a_{32} - a_{12}a_{21}a_{33}$$

for the determinant of a 3×3 matrix. (See Theorem 6.1.2.) Collecting the two terms involving a_{11} and then those involving a_{21} and a_{31}, we can write

$$\det A = a_{11}(a_{22}a_{33} - a_{32}a_{23})$$
$$+a_{21}(a_{32}a_{13} - a_{12}a_{33})$$
$$+a_{31}(a_{12}a_{23} - a_{22}a_{13}).$$

(Where have we seen this formula before?)

Note that computing the determinant this way requires only 9 multiplications, compared with the 12 for Sarrus's formula. Let us analyze the structure of this formula more closely. The terms $a_{22}a_{33} - a_{32}a_{23}$, $a_{32}a_{13} - a_{12}a_{33}$, and $a_{12}a_{23} - a_{22}a_{13}$ are the determinants of submatrices of A, up to the signs. The expression $a_{22}a_{33} - a_{32}a_{23}$ is the determinant of the matrix we get when we omit the first row and the first column of A:

$$\begin{bmatrix} a_{11} & a_{12} & a_{13} \\ a_{21} & a_{22} & a_{23} \\ a_{31} & a_{32} & a_{33} \end{bmatrix}.$$

Likewise for the other summands:

$$\det A = a_{11} \det \begin{bmatrix} a_{11} & a_{12} & a_{13} \\ a_{21} & a_{22} & a_{23} \\ a_{31} & a_{32} & a_{33} \end{bmatrix} - a_{21} \det \begin{bmatrix} a_{11} & a_{12} & a_{13} \\ a_{21} & a_{22} & a_{23} \\ a_{31} & a_{32} & a_{33} \end{bmatrix}$$

$$+ a_{31} \det \begin{bmatrix} a_{11} & a_{12} & a_{13} \\ a_{21} & a_{22} & a_{23} \\ a_{31} & a_{32} & a_{33} \end{bmatrix}.$$

To state these observations more succinctly, we introduce some terminology.

---

[6]Named after the French mathematician Pierre-Simon Marquis de Laplace (1749–1827). Laplace is perhaps best known for his investigation into the stability of the solar system. He was also a prominent member of the committee that aided in the organization of the metric system.

Definition 6.2.9   Minors

For an $n \times n$ matrix A, let A_{ij} be the matrix obtained by omitting the ith row and the jth column of A. The determinant of the $(n-1) \times (n-1)$ matrix A_{ij} is called a *minor* of A.

$$A = \begin{bmatrix} a_{11} & a_{12} & \cdots & a_{1j} & \cdots & a_{1n} \\ a_{21} & a_{22} & \cdots & a_{2j} & \cdots & a_{2n} \\ \vdots & \vdots & & \vdots & & \vdots \\ a_{i1} & a_{i2} & \cdots & a_{ij} & \cdots & a_{in} \\ \vdots & \vdots & & \vdots & & \vdots \\ a_{n1} & a_{n2} & \cdots & a_{nj} & \cdots & a_{nn} \end{bmatrix},$$

$$A_{ij} = \begin{bmatrix} a_{11} & a_{12} & \cdots & a_{1j} & \cdots & a_{1n} \\ a_{21} & a_{22} & \cdots & a_{2j} & \cdots & a_{2n} \\ \vdots & \vdots & & & & \vdots \\ a_{i1} & a_{i2} & \cdots & a_{ij} & \cdots & a_{in} \\ \vdots & \vdots & & & & \vdots \\ a_{n1} & a_{n2} & \cdots & a_{nj} & \cdots & a_{nn} \end{bmatrix}$$

We can now represent the determinant of a 3×3 matrix more succinctly:

$$\det A = a_{11} \det(A_{11}) - a_{21} \det(A_{21}) + a_{31} \det(A_{31}).$$

This representation of the determinant is called the *Laplace expansion* (or *cofactor expansion*) of $\det A$ *down the first column*. Likewise, we can expand along the first *row* $\left(\text{since } \det(A^T) = \det A \right)$:

$$\det A = a_{11} \det(A_{11}) - a_{12} \det(A_{12}) + a_{13} \det(A_{13})$$

In fact, we can expand along any row or down any column. (We can verify this directly or argue in terms of row or column swaps.) For example, the Laplace expansion down the second column is

$$\det A = -a_{12} \det(A_{12}) + a_{22} \det(A_{22}) - a_{32} \det(A_{32}),$$

and the Laplace expansion along the third row is

$$\det A = a_{31} \det(A_{31}) - a_{32} \det(A_{32}) + a_{33} \det(A_{33}).$$

The rule for the signs is as follows: The summand $a_{ij} \det(A_{ij})$ has a negative sign if the sum of the two indices, $i + j$, is odd. The signs follow a checkerboard pattern:

$$\begin{bmatrix} + & - & + \\ - & + & - \\ + & - & + \end{bmatrix}$$

We can generalize the Laplace expansion to $n \times n$ matrices.

We will focus on the expansion down the jth column of A. The formula for the expansion along the ith row then follows from the fact that $\det A = \det(A^T)$, Theorem 6.2.1.

Consider a pattern P in an $n \times n$ matrix A. For a fixed j, the pattern P will contain exactly one entry a_{ij} in the jth column of A. Let P_{ij} be the pattern in A_{ij}

that contains the same entries as P, except for the omitted entry a_{ij}. See the example below, where $j = 4$ and $i = 3$.

Note that prod $P = a_{ij} \text{prod}(P_{ij})$. In Exercise 68 we see that sgn $P = (-1)^{i+j} \text{sgn}(P_{ij})$, so that $(\text{sgn} P)(\text{prod } P) = (-1)^{i+j} a_{ij}(\text{sgn} P_{ij})(\text{prod } P_{ij})$. Verify this formula in the example above, where $(\text{sgn} P)(\text{prod } P) = 7! = 5{,}040$. Now we can compute the determinant of A, collecting the patterns containing a_{1j}, then those containing a_{2j}, and so forth, just as we did on page 268 in the case of a 3×3 matrix, with $j = 1$.

$$\det A = \sum (\text{sgn} P)(\text{prod } P) = \sum_{i=1}^{n} \sum_{P \text{ contains } a_{ij}} (\text{sgn} P)(\text{prod } P)$$

$$\sum_{i=1}^{n} \sum_{P \text{ contains } a_{ij}} (-1)^{i+j} a_{ij} (\text{sgn} P_{ij})(\text{prod } P_{ij})$$

$$= \sum_{i=1}^{n} (-1)^{i+j} a_{ij} \sum_{P \text{ contains } a_{ij}} (\text{sgn} P_{ij})(\text{prod } P_{ij}) = \sum_{i=1}^{n} (-1)^{i+j} a_{ij} \det(A_{ij})$$

Theorem 6.2.10   **Laplace expansion (or cofactor expansion)**

We can compute the determinant of an $n \times n$ matrix A by Laplace expansion down any column or along any row.

Expansion down the jth column:

$$\det A = \sum_{i=1}^{n} (-1)^{i+j} a_{ij} \det(A_{ij}).$$

Expansion along the ith row:

$$\det A = \sum_{j=1}^{n} (-1)^{i+j} a_{ij} \det(A_{ij}).$$

Again, the signs follow a checkerboard pattern:

$$\begin{bmatrix} + & - & + & - & \cdots \\ - & + & - & + & \cdots \\ + & - & + & - & \cdots \\ - & + & - & + & \cdots \\ \vdots & \vdots & \vdots & \vdots & \ddots \end{bmatrix}.$$

EXAMPLE 6   Use Laplace expansion to compute det A for

$$A = \begin{bmatrix} 1 & 0 & 1 & 2 \\ 9 & 1 & 3 & 0 \\ 9 & 2 & 2 & 0 \\ 5 & 0 & 0 & 3 \end{bmatrix}.$$

Solution

Looking for rows or columns with as many zeros as possible, we choose the second column:

$$\det A = -a_{12}\det(A_{12}) + a_{22}\det(A_{22}) - a_{32}\det(A_{32}) + a_{42}\det(A_{42})$$

$$= 1\,\det\begin{bmatrix} 1 & \cancel{0} & 1 & 2 \\ \cancel{9} & \cancel{1} & \cancel{3} & \cancel{0} \\ 9 & \cancel{2} & 2 & 0 \\ 5 & \cancel{0} & 0 & 3 \end{bmatrix} - 2\,\det\begin{bmatrix} 1 & \cancel{0} & 1 & 2 \\ 9 & \cancel{1} & 3 & 0 \\ \cancel{9} & \cancel{2} & \cancel{2} & \cancel{0} \\ 5 & \cancel{0} & 0 & 3 \end{bmatrix}$$

$$= \det\begin{bmatrix} 1 & 1 & 2 \\ 9 & 2 & 0 \\ 5 & 0 & 3 \end{bmatrix} - 2\,\det\begin{bmatrix} 1 & 1 & 2 \\ 9 & 3 & 0 \\ 5 & 0 & 3 \end{bmatrix}$$

$$= 2\,\det\begin{bmatrix} 9 & 2 \\ 5 & 0 \end{bmatrix} + 3\,\det\begin{bmatrix} 1 & 1 \\ 9 & 2 \end{bmatrix} - 2\left(2\,\det\begin{bmatrix} 9 & 3 \\ 5 & 0 \end{bmatrix} + 3\,\det\begin{bmatrix} 1 & 1 \\ 9 & 3 \end{bmatrix}\right)$$

↗

Expand down
the last
column

$$= -20 - 21 - 2(-30 - 18) = 55. \qquad \blacksquare$$

Computing the determinant using Laplace expansion is a bit more efficient than using the definition of the determinant, but a lot less efficient than Gauss–Jordan elimination.

The Determinant of a Linear Transformation (Optional)

(for those who have studied Chapter 4)

   If $T(\vec{x}) = A\vec{x}$ is a linear transformation from $\mathbb{R}^n$ to $\mathbb{R}^n$, then it is natural to define the determinant of T as the determinant of matrix A:

$$\det T = \det A.$$

This definition makes sense in view of the fact that an $n \times n$ matrix is essentially the same thing as a linear transformation from $\mathbb{R}^n$ to $\mathbb{R}^n$.

   If T is a linear transformation from V to V, where V is a finite-dimensional linear space, then we can introduce coordinates to define the determinant of T. If $\mathfrak{B}$ is a basis of V and B is the $\mathfrak{B}$-matrix of T, then we define

$$\det T = \det B.$$

We need to think about one issue though. If you pick another basis, $\mathfrak{A}$, of V and consider the $\mathfrak{A}$-matrix A of T, will you end up with the same determinant; that is, will det A equal det B?

Fortunately, there is no reason to worry. We know that matrix A is similar to B (by Theorem 4.3.5), so that determinants $\det A$ and $\det B$ are indeed equal, by Theorem 6.2.7.

Definition 6.2.11 The determinant of a linear transformation

Consider a linear transformation T from V to V, where V is a finite-dimensional linear space. If $\mathfrak{B}$ is a basis of V and B is the $\mathfrak{B}$-matrix of T, then we define

$$\det T = \det B.$$

This determinant is independent of the basis $\mathfrak{B}$ we choose.

EXAMPLE 7 Let V be the space spanned by functions $\cos(2x)$ and $\sin(2x)$. Find the determinant of the linear transformation $D(f) = f'$ from V to V.

Solution

The matrix B of D with respect to the basis $\cos(2x)$, $\sin(2x)$ is

$$B = \begin{bmatrix} 0 & 2 \\ -2 & 0 \end{bmatrix},$$

so that

$$\det D = \det B = 4. \qquad \blacksquare$$

Determinants: Focus on History

Most of the results of this and the preceding section (with the notable exception of the product rule, Theorem 6.2.6) were known to Gottfried Wilhelm von Leibniz (1646–1716). In 1678, while studying the solutions of systems of three equations in three unknowns, he used a method that amounts to expanding the determinant of a 3×3 matrix down the third column. Later that year he attempted the same for 4×4 matrices but made a sign error in his computations. In a manuscript of 1684, however, Leibniz states the sign rule for determinants in the correct, general form. His work remained unpublished and was discovered only after 1850, through careful examination of his manuscripts.

Meanwhile, the greatest mathematician of ancient Japan, Seki Kowa (1642–1708), came up with remarkably similar results, in his manuscript *Kai Fukudai no Ho*. It appears that he found the correct sign rule for determinants of 4×4 matrices. However, it is hard to assess his work, as he was an extremely secretive fellow. Florian Cajori, the eminent Swiss historian of mathematics, puts it this way: "Seki was a great teacher who attracted many gifted pupils. Like Pythagoras, he discouraged divulgence of mathematical discoveries made by himself and his school. For that reason it is difficult to determine with certainty the exact origin and nature of some of the discoveries attributed to him. He is said to have left hundreds of manuscripts; the translations of only a few of them still remain" (Cajori, *A History of Mathematics*, 1919).

Apparently without detailed knowledge of Leibniz's work, the Swiss mathematician Gabriel Cramer (1704–1752) developed the general theory of determinants (still without the product rule, though) and published his results in the *Introduction à l'analyse des lignes courbes algébriques* (1750). The mathematical community quickly discovered the power of this new technique, and during the next 100 years many mathematicians made important advances: Bézout, Vandermonde, Laplace, Binet, and Cayley, to name just a few. In 1812, Augustin Louis Cauchy (1789–1857) contributed the product rule. In the 1880's, Karl Weierstrass (1817–1897) offered an axiomatic definition of the determinant that allows a more elegant exposition of the theory (see Exercise 6.2.55).

EXERCISES 6.2

Use Gaussian elimination to find the determinant of the matrices in Exercises 1 through 10.

1. $\begin{bmatrix} 1 & 1 & 1 \\ 1 & 3 & 3 \\ 2 & 2 & 5 \end{bmatrix}$

2. $\begin{bmatrix} 1 & 2 & 3 \\ 1 & 6 & 8 \\ -2 & -4 & 0 \end{bmatrix}$

3. $\begin{bmatrix} 1 & 3 & 2 & 4 \\ 1 & 6 & 4 & 8 \\ 1 & 3 & 0 & 0 \\ 2 & 6 & 4 & 12 \end{bmatrix}$

4. $\begin{bmatrix} 1 & -1 & 2 & -2 \\ -1 & 2 & 1 & 6 \\ 2 & 1 & 14 & 10 \\ -2 & 6 & 10 & 33 \end{bmatrix}$

5. $\begin{bmatrix} 0 & 2 & 3 & 4 \\ 0 & 0 & 0 & 4 \\ 1 & 2 & 3 & 4 \\ 0 & 0 & 3 & 4 \end{bmatrix}$

6. $\begin{bmatrix} 1 & 1 & 1 & 1 \\ 1 & 1 & 4 & 4 \\ 1 & -1 & 2 & -2 \\ 1 & -1 & 8 & -8 \end{bmatrix}$

7. $\begin{bmatrix} 0 & 0 & 0 & 0 & 1 \\ 0 & 0 & 0 & 1 & 2 \\ 0 & 0 & 1 & 2 & 3 \\ 0 & 1 & 2 & 3 & 4 \\ 1 & 2 & 3 & 4 & 5 \end{bmatrix}$

8. $\begin{bmatrix} 0 & 0 & 0 & 0 & 2 \\ 1 & 0 & 0 & 0 & 3 \\ 0 & 1 & 0 & 0 & 4 \\ 0 & 0 & 1 & 0 & 5 \\ 0 & 0 & 0 & 1 & 6 \end{bmatrix}$

9. $\begin{bmatrix} 1 & 1 & 1 & 1 & 1 \\ 1 & 2 & 2 & 2 & 2 \\ 1 & 1 & 3 & 3 & 3 \\ 1 & 1 & 1 & 4 & 4 \\ 1 & 1 & 1 & 1 & 5 \end{bmatrix}$

10. $\begin{bmatrix} 1 & 1 & 1 & 1 & 1 \\ 1 & 2 & 3 & 4 & 5 \\ 1 & 3 & 6 & 10 & 15 \\ 1 & 4 & 10 & 20 & 35 \\ 1 & 5 & 15 & 35 & 70 \end{bmatrix}$

Consider a 4×4 matrix A with rows $\vec{v}_1, \vec{v}_2, \vec{v}_3, \vec{v}_4$. If $\det(A) = 8$, find the determinants in Exercises 11 through 16.

11. $\det \begin{bmatrix} \vec{v}_1 \\ \vec{v}_2 \\ -9\vec{v}_3 \\ \vec{v}_4 \end{bmatrix}$

12. $\det \begin{bmatrix} \vec{v}_4 \\ \vec{v}_2 \\ \vec{v}_3 \\ \vec{v}_1 \end{bmatrix}$

13. $\det \begin{bmatrix} \vec{v}_2 \\ \vec{v}_3 \\ \vec{v}_1 \\ \vec{v}_4 \end{bmatrix}$

14. $\det \begin{bmatrix} \vec{v}_1 \\ \vec{v}_2 + 9\vec{v}_4 \\ \vec{v}_3 \\ \vec{v}_4 \end{bmatrix}$

15. $\det \begin{bmatrix} \vec{v}_1 \\ \vec{v}_1 + \vec{v}_2 \\ \vec{v}_1 + \vec{v}_2 + \vec{v}_3 \\ \vec{v}_1 + \vec{v}_2 + \vec{v}_3 + \vec{v}_4 \end{bmatrix}$

16. $\det \begin{bmatrix} 6\vec{v}_1 + 2\vec{v}_4 \\ \vec{v}_2 \\ \vec{v}_3 \\ 3\vec{v}_1 + \vec{v}_4 \end{bmatrix}$

Find the determinants of the linear transformations in Exercises 17 through 28.

17. $T(f) = 2f + 3f'$ from P_2 to P_2

18. $T(f(t)) = f(3t - 2)$ from P_2 to P_2

19. $T(f(t)) = f(-t)$ from P_2 to P_2

20. $L(A) = A^T$ from $\mathbb{R}^{2 \times 2}$ to $\mathbb{R}^{2 \times 2}$

21. $T(f(t)) = f(-t)$ from P_3 to P_3

22. $T(f(t)) = f(-t)$ from P_n to P_n

23. $L(A) = A^T$ from $\mathbb{R}^{n \times n}$ to $\mathbb{R}^{n \times n}$

24. $T(z) = (2 + 3i)z$ from $\mathbb{C}$ to $\mathbb{C}$

25. $T(M) = \begin{bmatrix} 2 & 3 \\ 0 & 4 \end{bmatrix} M$ from the space V of upper triangular 2×2 matrices to V

26. $T(M) = \begin{bmatrix} 1 & 2 \\ 2 & 3 \end{bmatrix} M + M \begin{bmatrix} 1 & 2 \\ 2 & 3 \end{bmatrix}$ from the space V of symmetric 2×2 matrices to V

27. $T(f) = af' + bf''$, where a and b are arbitrary constants, from the space V spanned by $\cos(x)$ and $\sin(x)$ to V

28. $T(\vec{v}) = \begin{bmatrix} 1 \\ 2 \\ 3 \end{bmatrix} \times \vec{v}$ from the plane V given by $x_1 + 2x_2 + 3x_3 = 0$ to V

29. Let P_n be the $n \times n$ matrix whose entries are all ones, except for zeros directly below the main diagonal; for example,

$$P_5 = \begin{bmatrix} 1 & 1 & 1 & 1 & 1 \\ 0 & 1 & 1 & 1 & 1 \\ 1 & 0 & 1 & 1 & 1 \\ 1 & 1 & 0 & 1 & 1 \\ 1 & 1 & 1 & 0 & 1 \end{bmatrix}.$$

Find the determinant of P_n.

30. Consider two distinct numbers, a and b. We define the function

$$f(t) = \det \begin{bmatrix} 1 & 1 & 1 \\ a & b & t \\ a^2 & b^2 & t^2 \end{bmatrix}.$$

a. Show that $f(t)$ is a quadratic function. What is the coefficient of t^2?

b. Explain why $f(a) = f(b) = 0$. Conclude that $f(t) = k(t - a)(t - b)$, for some constant k. Find k, using your work in part (a).

c. For which values of t is the matrix invertible?

31. *Vandermonde determinants* (introduced by Alexandre Théophile Vandermonde). Consider distinct scalars a_0,

$a_1, \ldots, a_n$. We define the $(n+1) \times (n+1)$ matrix

$$A = \begin{bmatrix} 1 & 1 & \cdots & 1 \\ a_0 & a_1 & \cdots & a_n \\ a_0^2 & a_1^2 & \cdots & a_n^2 \\ \vdots & \vdots & & \vdots \\ a_0^n & a_1^n & \cdots & a_n^n \end{bmatrix}.$$

Vandermonde showed that

$$\det(A) = \prod_{i>j}(a_i - a_j),$$

the product of all differences $(a_i - a_j)$, where i exceeds j.

a. Verify this formula in the case of $n = 1$.

b. Suppose the Vandermonde formula holds for $n - 1$. You are asked to demonstrate it for n. Consider the function

$$f(t) = \det \begin{bmatrix} 1 & 1 & \cdots & 1 & 1 \\ a_0 & a_1 & \cdots & a_{n-1} & t \\ a_0^2 & a_1^2 & \cdots & a_{n-1}^2 & t^2 \\ \vdots & \vdots & & \vdots & \vdots \\ a_0^n & a_1^n & \cdots & a_{n-1}^n & t^n \end{bmatrix}.$$

Explain why $f(t)$ is a polynomial of nth degree. Find the coefficient k of t^n using Vandermonde's formula for $a_0, \ldots, a_{n-1}$. Explain why

$$f(a_0) = f(a_1) = \cdots = f(a_{n-1}) = 0.$$

Conclude that

$$f(t) = k(t - a_0)(t - a_1)\cdots(t - a_{n-1})$$

for the scalar k you found above. Substitute $t = a_n$ to demonstrate Vandermonde's formula.

32. Use Exercise 31 to find

$$\det \begin{bmatrix} 1 & 1 & 1 & 1 & 1 \\ 1 & 2 & 3 & 4 & 5 \\ 1 & 4 & 9 & 16 & 25 \\ 1 & 8 & 27 & 64 & 125 \\ 1 & 16 & 81 & 256 & 625 \end{bmatrix}.$$

Do not use technology.

33. For n distinct scalars $a_1, a_2, \ldots, a_n$, find

$$\det \begin{bmatrix} a_1 & a_2 & \cdots & a_n \\ a_1^2 & a_2^2 & \cdots & a_n^2 \\ \vdots & \vdots & & \vdots \\ a_1^n & a_2^n & \cdots & a_n^n \end{bmatrix}.$$

34. a. For an invertible $n \times n$ matrix A and an arbitrary $n \times n$ matrix B, show that

$$\text{rref}\begin{bmatrix} A & | & AB \end{bmatrix} = \begin{bmatrix} I_n & | & B \end{bmatrix}.$$

{*Hint*: The left part of rref $\begin{bmatrix} A & | & AB \end{bmatrix}$ is rref$(A) = I_n$. Write rref $\begin{bmatrix} A & | & AB \end{bmatrix} = \begin{bmatrix} I_n & | & M \end{bmatrix}$; we have to show that $M = B$. To demonstrate this, note that the columns of matrix

$$\begin{bmatrix} B \\ -I_n \end{bmatrix}$$

are in the kernel of $\begin{bmatrix} A & | & AB \end{bmatrix}$ and therefore in the kernel of $\begin{bmatrix} I_n & | & M \end{bmatrix}$.}

b. What does the formula

$$\text{rref}\begin{bmatrix} A & | & AB \end{bmatrix} = \begin{bmatrix} I_n & | & B \end{bmatrix}$$

tell you if $B = A^{-1}$?

35. Consider two distinct points $\begin{bmatrix} a_1 \\ a_2 \end{bmatrix}$ and $\begin{bmatrix} b_1 \\ b_2 \end{bmatrix}$ in the plane. Explain why the solutions $\begin{bmatrix} x_1 \\ x_2 \end{bmatrix}$ of the equation

$$\det \begin{bmatrix} 1 & 1 & 1 \\ x_1 & a_1 & b_1 \\ x_2 & a_2 & b_2 \end{bmatrix} = 0$$

form a line and why this line goes through the two points $\begin{bmatrix} a_1 \\ a_2 \end{bmatrix}$ and $\begin{bmatrix} b_1 \\ b_2 \end{bmatrix}$.

36. Consider three distinct points $\begin{bmatrix} a_1 \\ a_2 \end{bmatrix}$, $\begin{bmatrix} b_1 \\ b_2 \end{bmatrix}$, $\begin{bmatrix} c_1 \\ c_2 \end{bmatrix}$ in the plane. Describe the set of all points $\begin{bmatrix} x_1 \\ x_2 \end{bmatrix}$ satisfying the equation

$$\det \begin{bmatrix} 1 & 1 & 1 & 1 \\ x_1 & a_1 & b_1 & c_1 \\ x_2 & a_2 & b_2 & c_2 \\ x_1^2 + x_2^2 & a_1^2 + a_2^2 & b_1^2 + b_2^2 & c_1^2 + c_2^2 \end{bmatrix} = 0.$$

37. Consider an $n \times n$ matrix A such that both A and A^{-1} have integer entries. What are the possible values of $\det A$?

38. If $\det A = 3$ for some $n \times n$ matrix, what is $\det(A^T A)$?

39. If A is an invertible matrix, what can you say about the sign of $\det(A^T A)$?

40. If A is an orthogonal matrix, what are the possible values of $\det A$?

41. Consider a skew-symmetric $n \times n$ matrix A, where n is odd. Show that A is noninvertible, by showing that $\det A = 0$.

42. Consider an $n \times m$ matrix

$$A = QR,$$

where Q is an $n \times m$ matrix with orthonormal columns and R is an upper triangular $m \times m$ matrix with positive diagonal entries $r_{11}, \ldots, r_{mm}$. Express $\det(A^T A)$ in terms of the scalars r_{ii}. What can you say about the sign of $\det(A^T A)$?

43. Consider two vectors $\vec{v}$ and $\vec{w}$ in $\mathbb{R}^n$. Form the matrix $A = \begin{bmatrix} \vec{v} & \vec{w} \end{bmatrix}$. Express $\det(A^T A)$ in terms of $\|\vec{v}\|$, $\|\vec{w}\|$, and $\vec{v} \cdot \vec{w}$. What can you say about the sign of the result?

44. *The cross product in $\mathbb{R}^n$.* Consider the vectors $\vec{v}_2$, $\vec{v}_3, \ldots, \vec{v}_n$ in $\mathbb{R}^n$. The transformation

$$T(\vec{x}) = \det \begin{bmatrix} | & | & | & & | \\ \vec{x} & \vec{v}_2 & \vec{v}_3 & \cdots & \vec{v}_n \\ | & | & | & & | \end{bmatrix}$$

is linear. Therefore, there exists a unique vector $\vec{u}$ in $\mathbb{R}^n$ such that

$$T(\vec{x}) = \vec{x} \cdot \vec{u}$$

for all $\vec{x}$ in $\mathbb{R}^n$. (Compare this with Exercise 2.1.43c.) This vector $\vec{u}$ is called the *cross product* of $\vec{v}_2, \vec{v}_3, \ldots, \vec{v}_n$, written as

$$\vec{u} = \vec{v}_2 \times \vec{v}_3 \times \cdots \times \vec{v}_n.$$

In other words, the cross product is defined by the fact that

$$\vec{x} \cdot (\vec{v}_2 \times \vec{v}_3 \times \cdots \times \vec{v}_n) = \det \begin{bmatrix} | & | & | & & | \\ \vec{x} & \vec{v}_2 & \vec{v}_3 & \cdots & \vec{v}_n \\ | & | & | & & | \end{bmatrix},$$

for all $\vec{x}$ in $\mathbb{R}^n$. Note that the cross product in $\mathbb{R}^n$ is defined for $n-1$ vectors only. (For example, you cannot form the cross product of just 2 vectors in $\mathbb{R}^4$.) Since the ith component of a vector $\vec{w}$ is $\vec{e}_i \cdot \vec{w}$, we can find the cross product by components as follows:

ith component of $\vec{v}_2 \times \vec{v}_3 \times \cdots \times \vec{v}_n$
$$= \vec{e}_i \cdot (\vec{v}_2 \times \cdots \times \vec{v}_n)$$
$$= \det \begin{bmatrix} | & | & | & & | \\ \vec{e}_i & \vec{v}_2 & \vec{v}_3 & \cdots & \vec{v}_n \\ | & | & | & & | \end{bmatrix}.$$

a. When is $\vec{v}_2 \times \vec{v}_3 \times \cdots \times \vec{v}_n = \vec{0}$? Give your answer in terms of linear independence.
b. Find $\vec{e}_2 \times \vec{e}_3 \times \cdots \times \vec{e}_n$.
c. Show that $\vec{v}_2 \times \vec{v}_3 \times \cdots \times \vec{v}_n$ is orthogonal to all the vectors $\vec{v}_i$, for $i = 2, \ldots, n$.
d. What is the relationship between $\vec{v}_2 \times \vec{v}_3 \times \cdots \times \vec{v}_n$ and $\vec{v}_3 \times \vec{v}_2 \times \cdots \times \vec{v}_n$? (We swap the first two factors.)
e. Express $\det \begin{bmatrix} \vec{v}_2 \times \vec{v}_3 \times \cdots \times \vec{v}_n & \vec{v}_2 & \vec{v}_3 & \cdots & \vec{v}_n \end{bmatrix}$ in terms of $\|\vec{v}_2 \times \vec{v}_3 \times \cdots \times \vec{v}_n\|$.
f. How do we know that the cross product of two vectors in $\mathbb{R}^3$, as defined here, is the same as the standard cross product in $\mathbb{R}^3$? (See Definition A.9 of the Appendix.)

45. Find the derivative of the function

$$f(x) = \det \begin{bmatrix} 1 & 1 & 2 & 3 & 4 \\ 9 & 0 & 2 & 3 & 4 \\ 9 & 0 & 0 & 3 & 4 \\ x & 1 & 2 & 9 & 1 \\ 7 & 0 & 0 & 0 & 4 \end{bmatrix}.$$

46. Given some numbers a, b, c, d, e, and f such that

$$\det \begin{bmatrix} a & 1 & d \\ b & 1 & e \\ c & 1 & f \end{bmatrix} = 7 \quad \text{and} \quad \det \begin{bmatrix} a & 1 & d \\ b & 2 & e \\ c & 3 & f \end{bmatrix} = 11,$$

a. Find
$$\det \begin{bmatrix} a & 3 & d \\ b & 3 & e \\ c & 3 & f \end{bmatrix}.$$

b. Find
$$\det \begin{bmatrix} a & 3 & d \\ b & 5 & e \\ c & 7 & f \end{bmatrix}.$$

47. Is the function

$$T \begin{bmatrix} a & b \\ c & d \end{bmatrix} = ad + bc$$

linear in the rows and columns of the matrix?

48. Consider the linear transformation

$$T(\vec{x}) = \det \begin{bmatrix} | & | & | & & | \\ \vec{x} & \vec{v}_2 & \vec{v}_3 & \cdots & \vec{v}_n \\ | & | & | & & | \end{bmatrix}$$

from $\mathbb{R}^n$ to $\mathbb{R}$, where $\vec{v}_2, \vec{v}_3, \ldots, \vec{v}_n$ are linearly independent vectors in $\mathbb{R}^n$. Describe image and kernel of this transformation, and determine their dimensions.

49. Give an example of a 3×3 matrix A with all nonzero entries such that $\det A = 13$.

50. Find the determinant of the matrix

$$M_n = \begin{bmatrix} 1 & 1 & 1 & \cdots & 1 \\ 1 & 2 & 2 & \cdots & 2 \\ 1 & 2 & 3 & \cdots & 3 \\ \vdots & \vdots & \vdots & & \vdots \\ 1 & 2 & 3 & \cdots & n \end{bmatrix}$$

for arbitrary n. (The ijth entry of M_n is the minimum of i and j.)

51. Find the determinant of the $(2n) \times (2n)$ matrix

$$A = \begin{bmatrix} 0 & I_n \\ I_n & 0 \end{bmatrix}.$$

52. Consider a 2×2 matrix

$$A = \begin{bmatrix} a & b \\ c & d \end{bmatrix}$$

with column vectors

$$\vec{v} = \begin{bmatrix} a \\ c \end{bmatrix} \quad \text{and} \quad \vec{w} = \begin{bmatrix} b \\ d \end{bmatrix}.$$

We define the linear transformation

$$T(\vec{x}) = \begin{bmatrix} \det \begin{bmatrix} \vec{x} & \vec{w} \end{bmatrix} \\ \det \begin{bmatrix} \vec{v} & \vec{x} \end{bmatrix} \end{bmatrix}$$

from $\mathbb{R}^2$ to $\mathbb{R}^2$.

a. Find the standard matrix B of T. (Write the entries of B in terms of the entries a, b, c, d of A.)

b. What is the relationship between the determinants of A and B?

c. Show that BA is a scalar multiple of I_2. What about AB?

d. If A is noninvertible (but nonzero), what is the relationship between the image of A and the kernel of B? What about the kernel of A and the image of B?

e. If A is invertible, what is the relationship between B and A^{-1}?

53. Consider an invertible 2×2 matrix A with integer entries.

a. Show that if the entries of A^{-1} are integers, then $\det A = 1$ or $\det A = -1$.

b. Show the converse: If $\det A = 1$ or $\det A = -1$, then the entries of A^{-1} are integers.

54. Let A and B be 2×2 matrices with integer entries such that $A, A+B, A+2B, A+3B$, and $A+4B$ are all invertible matrices whose inverses have integer entries. Show that $A + 5B$ is invertible and that its inverse has integer entries. This question was in the William Lowell Putnam Mathematical Competition in 1994. *Hint:* Consider the function $f(t) = \left(\det(A + tB) \right)^2 - 1$. Show that this is a polynomial; what can you say about its degree? Find the values $f(0), f(1), f(2), f(3), f(4)$, using Exercise 53. Now you can determine $f(t)$ by using a familiar result: If a polynomial $f(t)$ of degree $\leq m$ has more than m zeros, then $f(t) = 0$ for all t.

55. For a fixed positive integer n, let D be a function which assigns to any $n \times n$ matrix A a number $D(A)$ such that

a. D is linear in the rows (see Theorem 6.2.2).

b. $D(B) = -D(A)$ if B is obtained from A by a row swap, and

c. $D(I_n) = 1$.

Show that $D(A) = \det(A)$ for all $n \times n$ matrices A. *Hint:* Consider $E = \text{rref } A$. Think about the relationship between $D(A)$ and $D(E)$, mimicking Algorithm 6.2.5.

The point of this exercise is that $\det(A)$ can be characterized by the three properties a, b, and c; the determinant can in fact be *defined* in terms of these properties. Ever since this approach was first presented in the 1880s by the German mathematician Karl Weierstrass (1817–1897), this definition has been generally used in advanced linear algebra courses because it allows a more elegant presentation of the theory of determinants.

56. Use the characterization of the determinant given in Exercise 55 to show that

$$\det(AM) = (\det A)(\det M).$$

{*Hint:* For a fixed invertible matrix M consider the function

$$D(A) = \frac{\det(AM)}{\det M}.$$

Show that this function has the three properties a, b, and c listed in Exercise 55, and therefore $D(A) = \det A$.}

57. Consider a linear transformation T from $\mathbb{R}^{m+n}$ to $\mathbb{R}^m$. The matrix A of T can be written in block form as $A = \begin{bmatrix} A_1 & A_2 \end{bmatrix}$, where A_1 is $m \times m$ and A_2 is $m \times n$. Suppose that $\det(A_1) \neq 0$. Show that for every vector $\vec{x}$ in $\mathbb{R}^n$ there exists a unique $\vec{y}$ in $\mathbb{R}^m$ such that

$$T \begin{bmatrix} \vec{y} \\ \vec{x} \end{bmatrix} = \vec{0}.$$

Show that the transformation

$$\vec{x} \rightarrow \vec{y}$$

from $\mathbb{R}^n$ to $\mathbb{R}^m$ is linear, and find its matrix M (in terms of A_1 and A_2). (This is the linear version of the *implicit function theorem* of multivariable calculus.)

58. Find the matrix M introduced in Exercise 57 for the linear transformation

$$T(\vec{v}) = \begin{bmatrix} 1 & 2 & 1 & 2 \\ 3 & 7 & 4 & 3 \end{bmatrix} \vec{v}.$$

You can either follow the approach outlined in Exercise 57 or use Gaussian elimination, expressing the leading variables y_1, y_2 in terms of the free variables x_1, x_2, where

$$\vec{v} = \begin{bmatrix} y_1 \\ y_2 \\ x_1 \\ x_2 \end{bmatrix}.$$

Note that this procedure amounts to finding the kernel of T, in the familiar way; we just interpret the result somewhat differently.

59. If the equation $\det A = \det B$ holds for two $n \times n$ matrices A and B, is A necessarily similar to B?

60. Consider an $n \times n$ matrix A. Show that swapping the ith and the jth rows of A (where $i < j$) amounts to performing $2(j - i) - 1$ swaps of *adjacent* rows.

61. Consider $n \times n$ matrices A, B , C, and D, where A is invertible and commutes with C. Show that

$$\det \begin{bmatrix} A & B \\ C & D \end{bmatrix} = \det(AD - CB).$$

{*Hint:* Consider the product

$$\begin{bmatrix} I_n & 0 \\ -C & A \end{bmatrix} \begin{bmatrix} A & B \\ C & D \end{bmatrix} .\}$$

62. Consider $n \times n$ matrices A, B, C, and D such that

$$\text{rank}(A) = \text{rank} \begin{bmatrix} A & B \\ C & D \end{bmatrix} = n.$$

Show that

a. $D = CA^{-1}B$, and

b. The 2×2 matrix $\begin{bmatrix} \det(A) & \det(B) \\ \det(C) & \det(D) \end{bmatrix}$ is noninvertible. {*Hint:* Consider the product

$$\begin{bmatrix} I_n & 0 \\ -CA^{-1} & I_n \end{bmatrix} \begin{bmatrix} A & B \\ C & D \end{bmatrix} .\}$$

63. Show that more than $n! = 1 \cdot 2 \cdot 3 \cdot \cdots \cdot n$ multiplications are required to compute the determinant of an $n \times n$ matrix by Laplace expansion (for $n > 2$).

64. Show that fewer than $e \cdot n!$ algebraic operations (additions and multiplications) are required to compute the determinant of an $n \times n$ matrix by Laplace expansion. *Hint:* Let L_n be the number of operations required to compute the determinant of a "general" $n \times n$ matrix by Laplace expansion. Find a formula expressing L_n in terms of L_{n-1}. Use this formula to show, by induction, that

$$\frac{L_n}{n!} = 1 + 1 + \frac{1}{2!} + \frac{1}{3!} + \cdots + \frac{1}{(n-1)!} - \frac{1}{n!}.$$

Use the Taylor series of e^x, $e^x = \sum_{n=0}^{\infty} \frac{x^n}{n!}$, to show that the right-hand side of this equation is less than e.

65. Let M_n be the $n \times n$ matrix with 1's on the main diagonal and directly above the main diagonal, -1's directly

below the main diagonal, and 0's elsewhere. For example,

$$M_4 = \begin{bmatrix} 1 & 1 & 0 & 0 \\ -1 & 1 & 1 & 0 \\ 0 & -1 & 1 & 1 \\ 0 & 0 & -1 & 1 \end{bmatrix} .$$

Let $d_n = \det(M_n)$.

a. For $n \geq 3$, find a formula expressing d_n in terms of d_{n-1} and d_{n-2}.

b. Find d_1, d_2, d_3, d_4, and d_{10}.

c. For which positive integers n is the matrix M_n invertible?

66. Let M_n be the matrix with all 1's along the main diagonal, directly above the main diagonal, and directly below the diagonal, and 0's everywhere else. For example,

$$M_4 = \begin{bmatrix} 1 & 1 & 0 & 0 \\ 1 & 1 & 1 & 0 \\ 0 & 1 & 1 & 1 \\ 0 & 0 & 1 & 1 \end{bmatrix} .$$

Let $d_n = \det(M_n)$.

a. Find a formula expressing d_n in terms of d_{n-1} and d_{n-2}, for positive integers $n \geq 3$.

b. Find d_1, d_2, $\ldots$, d_8.

c. What is the relationship between d_n and d_{n+3}? What about d_n and d_{n+6}?

d. Find d_{100}.

67. Consider a pattern P in an $n \times n$ matrix, and choose an entry a_{ij} in this pattern. Show that the number of inversions involving a_{ij} is even if $(i + j)$ is even and odd if $(i + j)$ is odd. *Hint:* Suppose there are k entries in the pattern to the left and above a_{ij}. Express the number of inversions involving a_{ij} in terms of k.

68. Using the terminology introduced in the proof of Theorem 6.2.10, show that $\text{sgn} P = (-1)^{i+j} \text{sgn}(P_{ij})$. See Exercise 67.

6.3 Geometrical Interpretations of the Determinant; Cramer's Rule

We now present several ways to think about the determinant in geometrical terms. Here is a preliminary exercise.

EXAMPLE 1   What are the possible values of the determinant of an orthogonal matrix A?

Solution

We know that

$$A^T A = I_n$$

(by Theorem 5.3.7). Taking the determinants of both sides and using Theorems 6.2.1 and 6.2.6, we find that

$$\det(A^T A) = \det(A^T)\det A = (\det A)^2 = 1.$$

Therefore, $\det A$ is either 1 or -1. ∎

Theorem 6.3.1   The determinant of an orthogonal matrix is either 1 or -1. ∎

For example,

$$\det \begin{bmatrix} 0.6 & -0.8 \\ 0.8 & 0.6 \end{bmatrix} = 1,$$

representing a rotation, and

$$\det \begin{bmatrix} 0.6 & 0.8 \\ 0.8 & -0.6 \end{bmatrix} = -1,$$

representing a reflection about a line.

Definition 6.3.2   Rotation matrices

An orthogonal $n \times n$ matrix A with $\det A = 1$ is called a *rotation matrix*, and the linear transformation $T(\vec{x}) = A\vec{x}$ is called a *rotation*.

The Determinant as Area and Volume

In Theorem 2.4.10 we give a geometrical interpretation of the determinant of a 2×2 matrix A, based on the formula

$$\det A = \det \begin{bmatrix} \vec{v}_1 & \vec{v}_2 \end{bmatrix} = \|\vec{v}_1\| \sin\theta \|\vec{v}_2\|,$$

where θ is the oriented angle from $\vec{v}_1$ to $\vec{v}_2$.

Figure 1a illustrates the fact that

$$|\det A| = \left|\det \begin{bmatrix} \vec{v}_1 & \vec{v}_2 \end{bmatrix}\right| = \|\vec{v}_1\| |\sin\theta| \|\vec{v}_2\|$$

is the area of the parallelogram spanned by the vectors $\vec{v}_1$ and $\vec{v}_2$. In Theorem 2.4.10, we provide a geometrical interpretation of the *sign* of $\det A$ as well; here we will focus on interpreting the absolute value.

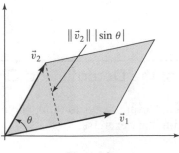

Figure 1a

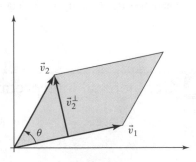

Figure 1b

Alternatively, we can write $|\det A|$ in terms of the Gram-Schmidt process, Theorem 5.2.1. Observe that $|\sin\theta| \|\vec{v}_2\| = \|\vec{v}_2^{\perp}\|$, where $\vec{v}_2^{\perp}$ denotes the component of $\vec{v}_2$ perpendicular to $\vec{v}_1$. (See Figure 1b.) Thus

$$|\det A| = \|\vec{v}_1\| \left\|\vec{v}_2^{\perp}\right\|.$$

More generally, consider an invertible $n \times n$ matrix

$$A = \begin{bmatrix} \vec{v}_1 & \vec{v}_2 & \cdots & \vec{v}_n \end{bmatrix}.$$

By Theorem 5.2.2, we can write $A = QR$, where Q is an orthogonal matrix and R is an upper triangular matrix whose diagonal entries are

$$r_{11} = \|\vec{v}_1\| \quad \text{and} \quad r_{jj} = \left\|\vec{v}_j^\perp\right\|, \quad \text{for } j \geq 2.$$

We conclude that

$$|\det A| = |\det Q||\det R| = \|\vec{v}_1\| \left\|\vec{v}_2^\perp\right\| \cdots \left\|\vec{v}_n^\perp\right\|.$$

Indeed, $|\det Q| = 1$ by Theorem 6.3.1, and the determinant of R is the product of its diagonal entries, by Theorem 6.1.4.

Theorem 6.3.3   **The determinant in terms of the columns**

If A is an $n \times n$ matrix with columns $\vec{v}_1, \vec{v}_2, \ldots, \vec{v}_n$, then

$$\cdot \quad |\det A| = \|\vec{v}_1\| \left\|\vec{v}_2^\perp\right\| \cdots \left\|\vec{v}_n^\perp\right\|,$$

where $\vec{v}_k^\perp$ is the component of $\vec{v}_k$ perpendicular to span $(\vec{v}_1, \ldots, \vec{v}_{k-1})$. (See Theorem 5.2.1.) ∎

The proof of Theorem 6.3.3 in the case of a noninvertible matrix A is left as Exercise 8.

As an example, consider the 3×3 matrix

$$A = \begin{bmatrix} | & | & | \\ \vec{v}_1 & \vec{v}_2 & \vec{v}_3 \\ | & | & | \end{bmatrix},$$

with

$$|\det A| = \|\vec{v}_1\| \left\|\vec{v}_2^\perp\right\| \left\|\vec{v}_3^\perp\right\|.$$

As in Figure 1b, $\|\vec{v}_1\|\left\|\vec{v}_2^\perp\right\|$ is the area of the parallelogram defined by $\vec{v}_1$ and $\vec{v}_2$. Now consider the parallelepiped defined by $\vec{v}_1, \vec{v}_2$, and $\vec{v}_3$ (i.e., the set of all vectors of the form $c_1\vec{v}_1 + c_2\vec{v}_2 + c_3\vec{v}_3$, where the c_i are between 0 and 1, as shown in Figure 2).

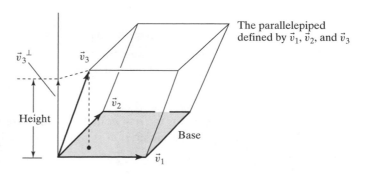

Figure 2

The volume of this parallelepiped is

$$\text{Volume} = \underbrace{\|\vec{v}_1\|\|\vec{v}_2^{\perp}\|}_{\text{Base area}} \underbrace{\|\vec{v}_3^{\perp}\|}_{\text{Height}} = |\det A|$$

(by Theorem 6.3.3).

Theorem 6.3.4   **Volume of a parallelepiped in $\mathbb{R}^3$**

Consider a 3×3 matrix $A = \begin{bmatrix} \vec{v}_1 & \vec{v}_2 & \vec{v}_3 \end{bmatrix}$. Then the volume of the parallelepiped defined by $\vec{v}_1, \vec{v}_2,$ and $\vec{v}_3$ is $|\det A|$. ∎

For a geometrical interpretation of the *sign* of det A, see Exercises 19 through 21.
Let us generalize these observations to higher dimensions.

Definition 6.3.5   **Parallelepipeds in $\mathbb{R}^n$**

Consider the vectors $\vec{v}_1, \vec{v}_2, \ldots, \vec{v}_m$ in $\mathbb{R}^n$. The m-parallelepiped defined by the vectors $\vec{v}_1, \ldots, \vec{v}_m$ is the set of all vectors in $\mathbb{R}^n$ of the form $c_1\vec{v}_1 + c_2\vec{v}_2 + \cdots + c_m\vec{v}_m$, where $0 \le c_i \le 1$. The m-volume $V(\vec{v}_1, \ldots, \vec{v}_m)$ of this m-parallelepiped is defined recursively by $V(\vec{v}_1) = \|\vec{v}_1\|$ and

$$V(\vec{v}_1, \ldots, \vec{v}_m) = V(\vec{v}_1, \ldots, \vec{v}_{m-1}) \left\| \vec{v}_m^{\perp} \right\|.$$

Note that this formula for the m-volume generalizes the formula

$$(\text{base})(\text{height})$$

we used to compute the area of a parallelogram (i.e., a 2-parallelepiped) and the volume of a 3-parallelepiped in $\mathbb{R}^3$. Take another look at Figures 1 and 2.

Alternatively, we can write the formula for the m-volume as

$$V(\vec{v}_1, \ldots, \vec{v}_m) = \|\vec{v}_1\| \left\| \vec{v}_2^{\perp} \right\| \cdots \left\| \vec{v}_m^{\perp} \right\|.$$

Let A be the $n \times m$ matrix whose columns are $\vec{v}_1, \ldots, \vec{v}_m$. If the columns of A are linearly independent, we can consider the QR factorization $A = QR$. Then, $A^T A = R^T Q^T Q R = R^T R$, because $Q^T Q = I_m$ (since the columns of Q are orthonormal). Therefore,

$$\det(A^T A) = \det(R^T R) = (\det R)^2 = (r_{11}r_{22}\cdots r_{mm})^2$$
$$= \left(\|\vec{v}_1\| \left\| \vec{v}_2^{\perp} \right\| \cdots \left\| \vec{v}_m^{\perp} \right\| \right)^2 = (V(\vec{v}_1, \ldots, \vec{v}_m))^2.$$

We can conclude

Theorem 6.3.6   **Volume of a parallelepiped in $\mathbb{R}^n$**

Consider the vectors $\vec{v}_1, \vec{v}_2, \ldots, \vec{v}_m$ in $\mathbb{R}^n$. Then the m-volume of the m-parallelepiped defined by the vectors $\vec{v}_1, \ldots, \vec{v}_m$ is

$$\sqrt{\det(A^T A)},$$

where A is the $n \times m$ matrix with columns $\vec{v}_1, \vec{v}_2, \ldots, \vec{v}_m$.
In particular, if $m = n$, this volume is

$$|\det A|.$$

(Compare this with Theorem 6.3.3.) ∎

We leave it to the reader to verify Theorem 6.3.6 for linearly dependent vectors $\vec{v}_1, \ldots, \vec{v}_m$. (See Exercise 15.)

As a simple example, consider the 2-volume (i.e., area) of the 2-parallelepiped (i.e., parallelogram) defined by the vectors

$$\vec{v}_1 = \begin{bmatrix} 1 \\ 1 \\ 1 \end{bmatrix} \quad \text{and} \quad \vec{v}_2 = \begin{bmatrix} 1 \\ 2 \\ 3 \end{bmatrix}$$

in $\mathbb{R}^3$. By Theorem 6.3.6, this area is

$$\sqrt{\det\left(\begin{bmatrix} 1 & 1 & 1 \\ 1 & 2 & 3 \end{bmatrix} \begin{bmatrix} 1 & 1 \\ 1 & 2 \\ 1 & 3 \end{bmatrix}\right)} = \sqrt{\det \begin{bmatrix} 3 & 6 \\ 6 & 14 \end{bmatrix}} = \sqrt{6}.$$

In this special case, we can also determine the area as the norm $\|\vec{v}_1 \times \vec{v}_2\|$ of the cross product of the two vectors.

The Determinant as Expansion Factor

Consider a linear transformation T from $\mathbb{R}^2$ to $\mathbb{R}^2$. In Chapter 5, we examined how a linear transformation T affects various geometric quantities such as lengths and angles. For example, we observed that a rotation preserves both the length of vectors and the angle between vectors. Similarly, we can ask how a linear transformation T affects the *area* of a region Ω in the plane. (See Figure 3.)

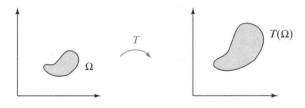

Figure 3

We might be interested in finding the *expansion factor,* the ratio

$$\frac{\text{area of } T(\Omega)}{\text{area of } \Omega}.$$

The simplest example is the unit square Ω shown in Figure 4.

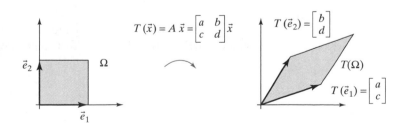

Figure 4

Since the area of Ω is 1 here, the expansion factor is simply the area of the parallelogram $T(\Omega)$, which is $|\det A|$, by Theorem 2.4.10.

More generally, let Ω be the parallelogram defined by $\vec{v}_1$ and $\vec{v}_2$, as shown in Figure 5.

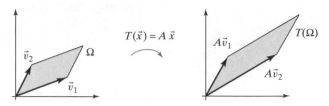

Figure 5

Let $B = \begin{bmatrix} \vec{v}_1 & \vec{v}_2 \end{bmatrix}$. Then

$$\text{area of } \Omega = |\det B|,$$

and

$$\text{area of } T(\Omega) = \left|\det \begin{bmatrix} A\vec{v}_1 & A\vec{v}_2 \end{bmatrix}\right| = |\det(AB)| = |\det A||\det B|,$$

and the expansion factor is

$$\frac{\text{area of } T(\Omega)}{\text{area of } \Omega} = \frac{|\det A||\det B|}{|\det B|} = |\det A|.$$

It is remarkable that the linear transformation $T(\vec{x}) = A\vec{x}$ expands the area of *all* parallelograms by the same factor, namely, $|\det A|$.

Theorem 6.3.7 **Expansion factor**

Consider a linear transformation $T(\vec{x}) = A\vec{x}$ from $\mathbb{R}^2$ to $\mathbb{R}^2$. Then $|\det A|$ is the *expansion factor*

$$\frac{\text{area of } T(\Omega)}{\text{area of } \Omega}$$

of T on parallelograms Ω.

Likewise, for a linear transformation $T(\vec{x}) = A\vec{x}$ from $\mathbb{R}^n$ to $\mathbb{R}^n$, $|\det A|$ is the expansion factor of T on n-parallelepipeds:

$$V(A\vec{v}_1, \ldots, A\vec{v}_n) = |\det A| V(\vec{v}_1, \ldots, \vec{v}_n),$$

for all vectors $\vec{v}_1, \ldots, \vec{v}_n$ in $\mathbb{R}^n$. ∎

This interpretation allows us to think about the formulas $\det(A^{-1}) = 1/\det A$ and $\det(AB) = (\det A)(\det B)$ from a geometric point of view. See Figures 6 and 7.

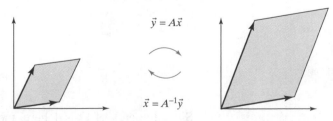

Figure 6

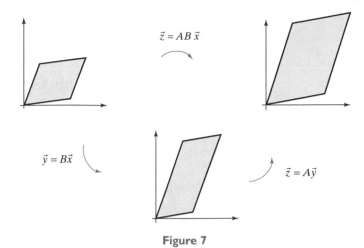

Figure 7

The expansion factor $|\det(A^{-1})|$ is the reciprocal of the expansion factor $|\det A|$:

$$|\det(A^{-1})| = \frac{1}{|\det A|}.$$

The expansion factor $|\det(AB)|$ of the composite transformation is the product of the expansion factors $|\det A|$ and $|\det B|$:

$$|\det(AB)| = |\det A||\det B|.$$

Using techniques of calculus, you can verify that $|\det A|$ gives us the expansion factor of the transformation $T(\vec{x}) = A\vec{x}$ on any region Ω in the plane. The approach uses inscribed parallelograms (or even squares) to approximate the area of the region, as shown in Figure 8. Note that the expansion factor of T on each of these squares is $|\det A|$. Choosing smaller and smaller squares and applying calculus, you can conclude that the expansion factor of T on Ω itself is $|\det A|$.

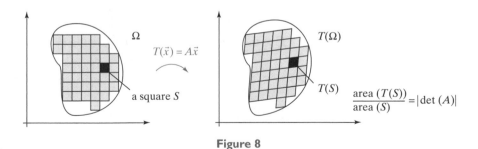

Figure 8

We will conclude this chapter with the discussion of a *closed-form solution* for the linear system $A\vec{x} = \vec{b}$ in the case when the coefficient matrix A is invertible.

Cramer's Rule

If a matrix

$$A = \begin{bmatrix} a_{11} & a_{12} \\ a_{21} & a_{22} \end{bmatrix}$$

is invertible, we can express its inverse in terms of its determinant:

$$A^{-1} = \frac{1}{\det(A)} \begin{bmatrix} a_{22} & -a_{12} \\ -a_{21} & a_{11} \end{bmatrix}.$$

This formula can be used to find a *closed-formula solution* for a linear system

$$\begin{vmatrix} a_{11}x_1 + a_{12}x_2 = b_1 \\ a_{21}x_1 + a_{22}x_2 = b_2 \end{vmatrix}$$

when the coefficient matrix is invertible. We write the system as $A\vec{x} = \vec{b}$, where

$$A = \begin{bmatrix} a_{11} & a_{12} \\ a_{21} & a_{22} \end{bmatrix}, \quad \vec{x} = \begin{bmatrix} x_1 \\ x_2 \end{bmatrix}, \quad \vec{b} = \begin{bmatrix} b_1 \\ b_2 \end{bmatrix}.$$

Then

$$\begin{bmatrix} x_1 \\ x_2 \end{bmatrix} = \vec{x} = A^{-1}\vec{b} = \frac{1}{\det A} \begin{bmatrix} a_{22} & -a_{12} \\ -a_{21} & a_{11} \end{bmatrix} \begin{bmatrix} b_1 \\ b_2 \end{bmatrix}$$

$$= \frac{1}{\det A} \begin{bmatrix} a_{22}b_1 - a_{12}b_2 \\ a_{11}b_2 - a_{21}b_1 \end{bmatrix}.$$

To write this formula more succinctly, we observe that

$$a_{22}b_1 - a_{12}b_2 = \det \begin{bmatrix} b_1 & a_{12} \\ b_2 & a_{22} \end{bmatrix} \quad \longleftarrow \quad \text{replace the first column of } A \text{ by } \vec{b}.$$

$$a_{11}b_2 - a_{21}b_1 = \det \begin{bmatrix} a_{11} & b_1 \\ a_{21} & b_2 \end{bmatrix} \quad \longleftarrow \quad \text{replace the second column of } A \text{ by } \vec{b}.$$

Let $A_{\vec{b},i}$ be the matrix obtained by replacing the ith column of A by $\vec{b}$:

$$A_{\vec{b},1} = \begin{bmatrix} b_1 & a_{12} \\ b_2 & a_{22} \end{bmatrix}, \quad A_{\vec{b},2} = \begin{bmatrix} a_{11} & b_1 \\ a_{21} & b_2 \end{bmatrix}.$$

The solution of the system $A\vec{x} = \vec{b}$ can now be written as

$$x_1 = \frac{\det(A_{\vec{b},1})}{\det A}, \quad x_2 = \frac{\det(A_{\vec{b},2})}{\det A}.$$

EXAMPLE 2 Use the preceding formula to solve the system

$$\begin{vmatrix} 2x_1 + 3x_2 = 7 \\ 4x_1 + 5x_2 = 13 \end{vmatrix}.$$

Solution

$$x_1 = \frac{\det \begin{bmatrix} 7 & 3 \\ 13 & 5 \end{bmatrix}}{\det \begin{bmatrix} 2 & 3 \\ 4 & 5 \end{bmatrix}} = 2, \quad x_2 = \frac{\det \begin{bmatrix} 2 & 7 \\ 4 & 13 \end{bmatrix}}{\det \begin{bmatrix} 2 & 3 \\ 4 & 5 \end{bmatrix}} = 1 \qquad \blacksquare$$

This method is not particularly helpful for solving numerically given linear systems; Gauss–Jordan elimination is preferable in this case. However, in many applications we have to deal with systems whose coefficients contain parameters. Often we want to know how the solution changes as we change the parameters. The closed-formula solution given before is well suited to deal with questions of this kind.

EXAMPLE 3   Solve the system

$$\begin{vmatrix} (b-1)x_1 + & ax_2 = 0 \\ -ax_1 + (b-1)x_2 = C \end{vmatrix},$$

where a, b, C are arbitrary positive constants.

Solution

$$x_1 = \frac{\det \begin{bmatrix} 0 & a \\ C & b-1 \end{bmatrix}}{\det \begin{bmatrix} b-1 & a \\ -a & b-1 \end{bmatrix}} = \frac{-aC}{(b-1)^2 + a^2}$$

$$x_2 = \frac{\det \begin{bmatrix} b-1 & 0 \\ -a & C \end{bmatrix}}{\det \begin{bmatrix} b-1 & a \\ -a & b-1 \end{bmatrix}} = \frac{(b-1)C}{(b-1)^2 + a^2} \qquad \blacksquare$$

EXAMPLE 4   Consider the linear system

$$\begin{vmatrix} ax + by = 1 \\ cx + dy = 1 \end{vmatrix}, \qquad \text{where } d > b > 0 \text{ and } a > c > 0.$$

This system always has a unique solution, since the determinant $ad - bc$ is positive (note that $ad > bc$). Thus we can think of the solution vector $\begin{bmatrix} x \\ y \end{bmatrix}$ as a (nonlinear) function of the vector

$$\begin{bmatrix} a \\ b \\ c \\ d \end{bmatrix}$$

of the parameters. How does x change as we change the parameters a and c? More precisely, find $\partial x / \partial a$ and $\partial x / \partial c$, and determine the signs of these quantities.

Solution

$$x = \frac{\det \begin{bmatrix} 1 & b \\ 1 & d \end{bmatrix}}{\det \begin{bmatrix} a & b \\ c & d \end{bmatrix}} = \frac{d - b}{ad - bc} > 0, \qquad \frac{\partial x}{\partial a} = \frac{-d(d-b)}{(ad-bc)^2} < 0,$$

$$\frac{\partial x}{\partial c} = \frac{b(d-b)}{(ad-bc)^2} > 0$$

See Figure 9. $\blacksquare$

An interesting application of these simple results in biology is to the study of castes.[7]

The closed formula for solving linear systems of two equations with two unknowns generalizes easily to larger systems.

---

[7]See E. O. Wilson, "The Ergonomics of Caste in the Social Insects," *American Naturalist,* 102, 923 (1968): 41–66.

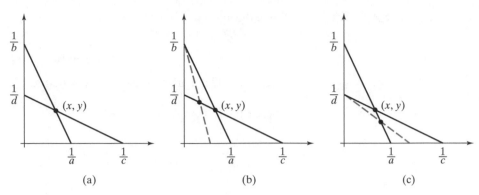

Figure 9 (a) Both components x and y of the solution are positive. (b) $\partial x / \partial a < 0$: as a increases, the component x of the solution decreases. (c) $\partial x / \partial c > 0$: as c increases, the component x of the solution increases.

Theorem 6.3.8 **Cramer's rule**

Consider the linear system

$$A\vec{x} = \vec{b},$$

where A is an invertible $n \times n$ matrix. The components x_i of the solution vector $\vec{x}$ are

$$x_i = \frac{\det(A_{\vec{b},i})}{\det A},$$

where $A_{\vec{b},i}$ is the matrix obtained by replacing the ith column of A by $\vec{b}$.

This result is due to the Swiss mathematician Gabriel Cramer (1704–1752). The rule appeared in an appendix to his 1750 book, *Introduction à l'analyse des lignes courbes algébriques.*

Proof Write $A = \begin{bmatrix} \vec{w}_1 & \vec{w}_2 & \cdots & \vec{w}_i & \cdots & \vec{w}_n \end{bmatrix}$. If $\vec{x}$ is the solution of the system $A\vec{x} = \vec{b}$, then

$$\begin{aligned}
\det(A_{\vec{b},i}) &= \det \begin{bmatrix} \vec{w}_1 & \vec{w}_2 & \cdots & \vec{b} & \cdots & \vec{w}_n \end{bmatrix} \\
&= \det \begin{bmatrix} \vec{w}_1 & \vec{w}_2 & \cdots & A\vec{x} & \cdots & \vec{w}_n \end{bmatrix} \\
&= \det \begin{bmatrix} \vec{w}_1 & \vec{w}_2 & \cdots & (x_1\vec{w}_1 + x_2\vec{w}_2 + \cdots + x_i\vec{w}_i + \cdots + x_n\vec{w}_n) & \cdots & \vec{w}_n \end{bmatrix} \\
&= \det \begin{bmatrix} \vec{w}_1 & \vec{w}_2 & \cdots & x_i\vec{w}_i & \cdots & \vec{w}_n \end{bmatrix} \\
&= x_i \det \begin{bmatrix} \vec{w}_1 & \vec{w}_2 & \cdots & \vec{w}_i & \cdots & \vec{w}_n \end{bmatrix} \\
&= x_i \det A.
\end{aligned}$$

Note that we have used the linearity of the determinant in the ith column (Theorem 6.2.2).

Therefore,

$$x_i = \frac{\det(A_{\vec{b},i})}{\det A}. \qquad \blacksquare$$

Cramer's rule allows us to find a closed formula for A^{-1}, generalizing the result

$$\begin{bmatrix} a & b \\ c & d \end{bmatrix}^{-1} = \frac{1}{\det A} \begin{bmatrix} d & -b \\ -c & a \end{bmatrix}$$

for 2×2 matrices.

Consider an invertible $n \times n$ matrix A and write

$$A^{-1} = \begin{bmatrix} m_{11} & m_{12} & \cdots & m_{1j} & \cdots & m_{1n} \\ m_{21} & m_{22} & \cdots & m_{2j} & \cdots & m_{2n} \\ \vdots & \vdots & & \vdots & & \vdots \\ m_{n1} & m_{n2} & \cdots & m_{nj} & \cdots & m_{nn} \end{bmatrix}.$$

We know that $AA^{-1} = I_n$. Picking out the jth column of A^{-1}, we find that

$$A \begin{bmatrix} m_{1j} \\ m_{2j} \\ \vdots \\ m_{nj} \end{bmatrix} = \vec{e}_j.$$

By Cramer's rule, $m_{ij} = \det(A_{\vec{e}_j, i}) / \det A$.

$$A_{\vec{e}_j, i} = \begin{bmatrix} a_{11} & a_{12} & \cdots & 0 & \cdots & a_{1n} \\ a_{21} & a_{22} & \cdots & 0 & \cdots & a_{2n} \\ \vdots & \vdots & & \vdots & & \vdots \\ a_{j1} & a_{j2} & \cdots & 1 & \cdots & a_{jn} \\ \vdots & \vdots & & \vdots & & \vdots \\ a_{n1} & a_{n2} & \cdots & 0 & \cdots & a_{nn} \end{bmatrix} \leftarrow j\text{th row}$$

$$\uparrow$$
$$i\text{th column}$$

Now $\det(A_{\vec{e}_j, i}) = (-1)^{i+j} \det(A_{ji})$ by Laplace expansion down the ith column, so that

$$m_{ij} = (-1)^{i+j} \frac{\det(A_{ji})}{\det A}.$$

We have shown the following result:

Theorem 6.3.9 Adjoint and inverse of a matrix

Consider an invertible $n \times n$ matrix A. The *classical adjoint* adj(A) is the $n \times n$ matrix whose ijth entry is $(-1)^{i+j} \det(A_{ji})$. Then

$$A^{-1} = \frac{1}{\det A} \text{adj}(A). \qquad \blacksquare$$

For an invertible 2×2 matrix $A = \begin{bmatrix} a & b \\ c & d \end{bmatrix}$, we find

$$\text{adj}(A) = \begin{bmatrix} d & -b \\ -c & a \end{bmatrix} \quad \text{and} \quad A^{-1} = \frac{1}{ad - bc} \begin{bmatrix} d & -b \\ -c & a \end{bmatrix}.$$

(Compare this with Theorem 2.4.9.)

For an invertible 3×3 matrix

$$A = \begin{bmatrix} a & b & c \\ d & e & f \\ g & h & k \end{bmatrix},$$

the formula is

$$A^{-1} = \frac{1}{\det A} \begin{bmatrix} ek - fh & ch - bk & bf - ce \\ fg - dk & ak - cg & cd - af \\ dh - eg & bg - ah & ae - bd \end{bmatrix}.$$

We can interpret Cramer's rule geometrically.

EXAMPLE 5 For the vectors $\vec{w}_1$, $\vec{w}_2$, and $\vec{b}$ shown in Figure 10, consider the linear system $A\vec{x} = \vec{b}$, where $A = \begin{bmatrix} \vec{w}_1 & \vec{w}_2 \end{bmatrix}$.

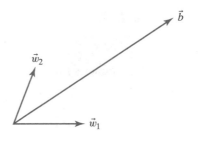

Figure 10

Using the terminology introduced in Cramer's rule, let $A_{\vec{b},2} = \begin{bmatrix} \vec{w}_1 & \vec{b} \end{bmatrix}$. Note that $\det(A)$ and $\det(A_{\vec{b},2})$ are both positive, according to Theorem 2.4.10. Cramer's rule tells us that

$$x_2 = \frac{\det(A_{\vec{b},2})}{\det A} \quad \text{or} \quad \det(A_{\vec{b},2}) = x_2 \det A.$$

Explain this last equation geometrically, in terms of areas of parallelograms.

Solution

We can write the system $A\vec{x} = \vec{b}$ as $x_1\vec{w}_1 + x_2\vec{w}_2 = \vec{b}$. The geometrical solution is shown in Figure 11.

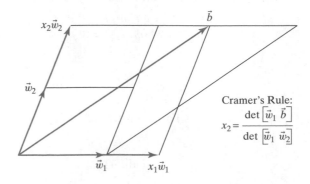

Figure 11

Now,

$$\det(A_{\vec{b},2}) = \det \begin{bmatrix} \vec{w}_1 & \vec{b} \end{bmatrix} = \text{area of the parallelogram defined by } \vec{w}_1 \text{ and } \vec{b}$$

$$= \text{area of the parallelogram}[8] \text{ defined by } \vec{w}_1 \text{ and } x_2\vec{w}_2$$

$$= x_2(\text{area of the parallelogram}[9] \text{ defined by } \vec{w}_1 \text{ and } \vec{w}_2)$$

$$= x_2 \det A,$$

as claimed. Note that this geometrical proof mimics the algebraic proof of Cramer's rule, Theorem 6.3.8. ∎

The ambitious and artistically inclined reader is encouraged to draw an analogous figure illustrating Cramer's rule for a system of three linear equations with three unknowns.

---

[8] The two parallelograms have the same base and the same height.

[9] Again, think about base and height.

EXERCISES 6.3

GOAL *Interpret the determinant as an area or volume and as an expansion factor. Use Cramer's rule.*

1. Find the area of the parallelogram defined by $\begin{bmatrix} 3 \\ 7 \end{bmatrix}$ and $\begin{bmatrix} 8 \\ 2 \end{bmatrix}$.

2. Find the area of the triangle defined by $\begin{bmatrix} 3 \\ 7 \end{bmatrix}$ and $\begin{bmatrix} 8 \\ 2 \end{bmatrix}$.

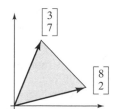

3. Find the area of the following triangle:

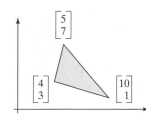

4. Consider the area A of the triangle with vertices $\begin{bmatrix} a_1 \\ a_2 \end{bmatrix}$, $\begin{bmatrix} b_1 \\ b_2 \end{bmatrix}$, $\begin{bmatrix} c_1 \\ c_2 \end{bmatrix}$. Express A in terms of

$$\det \begin{bmatrix} a_1 & b_1 & c_1 \\ a_2 & b_2 & c_2 \\ 1 & 1 & 1 \end{bmatrix}.$$

5. The tetrahedron defined by three vectors $\vec{v}_1, \vec{v}_2, \vec{v}_3$ in $\mathbb{R}^3$ is the set of all vectors of the form $c_1\vec{v}_1 + c_2\vec{v}_2 + c_3\vec{v}_3$, where $c_i \geq 0$ and $c_1 + c_2 + c_3 \leq 1$. Explain why the volume of this tetrahedron is one-sixth of the volume of the parallelepiped defined by $\vec{v}_1, \vec{v}_2, \vec{v}_3$.

6. What is the relationship between the volume of the tetrahedron defined by the vectors

$$\begin{bmatrix} a_1 \\ a_2 \\ 1 \end{bmatrix}, \quad \begin{bmatrix} b_1 \\ b_2 \\ 1 \end{bmatrix}, \quad \begin{bmatrix} c_1 \\ c_2 \\ 1 \end{bmatrix}$$

and the area of the triangle with vertices

$$\begin{bmatrix} a_1 \\ a_2 \end{bmatrix}, \quad \begin{bmatrix} b_1 \\ b_2 \end{bmatrix}, \quad \begin{bmatrix} c_1 \\ c_2 \end{bmatrix}?$$

(See Exercises 4 and 5.) Explain this relationship geometrically. *Hint:* Consider the top face of the tetrahedron.

7. Find the area of the following region:

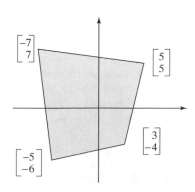

8. Demonstrate the equation

$$| \det A| = \|\vec{v}_1\| \, \|\vec{v}_2^{\perp}\| \cdots \|\vec{v}_n^{\perp}\|$$

for a noninvertible $n \times n$ matrix $A = \begin{bmatrix} \vec{v}_1 & \vec{v}_2 & \cdots & \vec{v}_n \end{bmatrix}$ (Theorem 6.3.3).

9. If $\vec{v}_1$ and $\vec{v}_2$ are linearly independent vectors in $\mathbb{R}^2$, what is the relationship between $\det \begin{bmatrix} \vec{v}_1 & \vec{v}_2 \end{bmatrix}$ and $\det \begin{bmatrix} \vec{v}_1 & \vec{v}_2^{\perp} \end{bmatrix}$, where $\vec{v}_2^{\perp}$ is the component of $\vec{v}_2$ orthogonal to $\vec{v}_1$?

10. Consider an $n \times n$ matrix $A = \begin{bmatrix} \vec{v}_1 & \vec{v}_2 & \cdots & \vec{v}_n \end{bmatrix}$. What is the relationship between the product $\|\vec{v}_1\| \, \|\vec{v}_2\| \cdots \|\vec{v}_n\|$ and $|\det A|$? When is $|\det A| = \|\vec{v}_1\| \, \|\vec{v}_2\| \cdots \|\vec{v}_n\|$?

11. Consider a linear transformation $T(\vec{x}) = A\vec{x}$ from $\mathbb{R}^2$ to $\mathbb{R}^2$. Suppose for two vectors $\vec{v}_1$ and $\vec{v}_2$ in $\mathbb{R}^2$ we have $T(\vec{v}_1) = 3\vec{v}_1$ and $T(\vec{v}_2) = 4\vec{v}_2$. What can you say about $\det A$? Justify your answer carefully.

12. Consider those 4×4 matrices whose entries are all 1, -1, or 0. What is the maximal value of the determinant of a matrix of this type? Give an example of a matrix whose determinant has this maximal value.

13. Find the area (or 2-volume) of the parallelogram (or 2-parallelepiped) defined by the vectors

$$\begin{bmatrix} 1 \\ 1 \\ 1 \\ 1 \end{bmatrix} \quad \text{and} \quad \begin{bmatrix} 1 \\ 2 \\ 3 \\ 4 \end{bmatrix}.$$

14. Find the 3-volume of the 3-parallelepiped defined by the vectors

$$\begin{bmatrix} 1 \\ 0 \\ 0 \\ 0 \end{bmatrix}, \quad \begin{bmatrix} 1 \\ 1 \\ 1 \\ 1 \end{bmatrix}, \quad \begin{bmatrix} 1 \\ 2 \\ 3 \\ 4 \end{bmatrix}.$$

15. Demonstrate Theorem 6.3.6 for linearly dependent vectors $\vec{v}_1, \ldots, \vec{v}_m$.

16. *True or false?* If Ω is a parallelogram in $\mathbb{R}^3$ and $T(\vec{x}) = A\vec{x}$ is a linear transformation from $\mathbb{R}^3$ to $\mathbb{R}^3$, then

$$\text{area of } T(\Omega) = |\det A|(\text{area of } \Omega).$$

17. (For some background on the cross product in $\mathbb{R}^n$, see Exercise 6.2.44.) Consider three linearly independent vectors $\vec{v}_1, \vec{v}_2, \vec{v}_3$ in $\mathbb{R}^4$.

   a. What is the relationship between $V(\vec{v}_1, \vec{v}_2, \vec{v}_3)$ and $V(\vec{v}_1, \vec{v}_2, \vec{v}_3, \vec{v}_1 \times \vec{v}_2 \times \vec{v}_3)$? See Definition 6.3.5. Exercise 6.2.44c is helpful.

   b. Express $V(\vec{v}_1, \vec{v}_2, \vec{v}_3, \vec{v}_1 \times \vec{v}_2 \times \vec{v}_3)$ in terms of $\|\vec{v}_1 \times \vec{v}_2 \times \vec{v}_3\|$.

   c. Use parts (a) and (b) to express $V(\vec{v}_1, \vec{v}_2, \vec{v}_3)$ in terms of $\|\vec{v}_1 \times \vec{v}_2 \times \vec{v}_3\|$. Is your result still true when the $\vec{v}_i$ are linearly dependent?

(Note the analogy to the fact that for two vectors $\vec{v}_1$ and $\vec{v}_2$ in $\mathbb{R}^3$, $\|\vec{v}_1 \times \vec{v}_2\|$ is the area of the parallelogram defined by $\vec{v}_1$ and $\vec{v}_2$.)

18. If $T(\vec{x}) = A\vec{x}$ is an invertible linear transformation from $\mathbb{R}^2$ to $\mathbb{R}^2$, then the image $T(\Omega)$ of the unit circle Ω is an ellipse. (See Exercise 2.2.50.)

   a. Sketch this ellipse when $A = \begin{bmatrix} p & 0 \\ 0 & q \end{bmatrix}$, where p and q are positive. What is its area?

   b. For an arbitrary invertible transformation $T(\vec{x}) = A\vec{x}$, denote the lengths of the semimajor and the semiminor axes of $T(\Omega)$ by a and b, respectively. What is the relationship between a, b, and $\det(A)$?

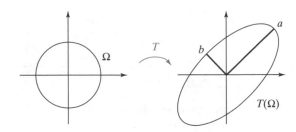

   c. For the transformation $T(\vec{x}) = \begin{bmatrix} 3 & 1 \\ 1 & 3 \end{bmatrix} \vec{x}$, sketch this ellipse and determine its axes. {*Hint*: Consider $T \begin{bmatrix} 1 \\ 1 \end{bmatrix}$ and $T \begin{bmatrix} 1 \\ -1 \end{bmatrix}$.}

19. A basis $\vec{v}_1, \vec{v}_2, \vec{v}_3$ of $\mathbb{R}^3$ is called *positively oriented* if $\vec{v}_1$ encloses an acute angle with $\vec{v}_2 \times \vec{v}_3$. Illustrate this definition with a sketch. Show that the basis is positively oriented if (and only if) $\det \begin{bmatrix} \vec{v}_1 & \vec{v}_2 & \vec{v}_3 \end{bmatrix}$ is positive.

20. We say that a linear transformation T from $\mathbb{R}^3$ to $\mathbb{R}^3$ *preserves orientation* if it transforms any positively oriented basis into another positively oriented basis. (See Exercise 19.) Explain why a linear transformation $T(\vec{x}) = A\vec{x}$ preserves orientation if (and only if) $\det A$ is positive.

21. Arguing geometrically, determine whether the following orthogonal transformations from $\mathbb{R}^3$ to $\mathbb{R}^3$ preserve or reverse orientation. (See Exercise 20.)

   a. Reflection about a plane   b. Reflection about a line
   c. Reflection about the origin

Use Cramer's rule to solve the systems in Exercises 22 through 24.

22. $\begin{vmatrix} 3x + 7y = 1 \\ 4x + 11y = 3 \end{vmatrix}$

23. $\begin{vmatrix} 5x_1 - 3x_2 = 1 \\ -6x_1 + 7x_2 = 0 \end{vmatrix}$

24. $\begin{vmatrix} 2x + 3y & = 8 \\ 4y + 5z = 3 \\ 6x & + 7z = -1 \end{vmatrix}$

25. Find the classical adjoint of the matrix

$$A = \begin{bmatrix} 1 & 0 & 1 \\ 0 & 1 & 0 \\ 2 & 0 & 1 \end{bmatrix},$$

and use the result to find A^{-1}.

26. Consider an $n \times n$ matrix A with integer entries such that $\det A = 1$. Are the entries of A^{-1} necessarily integers? Explain.

27. Consider two positive numbers a and b. Solve the following system:

$$\begin{vmatrix} ax - by = 1 \\ bx + ay = 0 \end{vmatrix}.$$

What are the signs of the solutions x and y? How does x change as b increases?

28. In an economics text,[10] we find the following system:

$$sY + ar = I° + G$$
$$mY - hr = M_s - M°.$$

Solve for Y and r.

29. In an economics text[11] we find the following system:

$$\begin{bmatrix} -R_1 & R_1 & -(1-\alpha) \\ \alpha & 1-\alpha & -(1-\alpha)^2 \\ R_2 & -R_2 & \dfrac{-(1-\alpha)^2}{\alpha} \end{bmatrix} \begin{bmatrix} dx_1 \\ dy_1 \\ dp \end{bmatrix} = \begin{bmatrix} 0 \\ 0 \\ -R_2 de_2 \end{bmatrix}.$$

Solve for dx_1, dy_1, and dp. In your answer, you may refer to the determinant of the coefficient matrix as D. (You need not compute D.) The quantities R_1, R_2, and D are positive, and α is between zero and one. If de_2 is positive, what can you say about the signs of dy_1 and dp?

30. Find the classical adjoint of $A = \begin{bmatrix} 1 & 0 & 0 \\ 2 & 3 & 0 \\ 4 & 5 & 6 \end{bmatrix}$.

31. Find the classical adjoint of $A = \begin{bmatrix} 1 & 1 & 1 \\ 1 & 2 & 3 \\ 1 & 6 & 6 \end{bmatrix}$.

32. Find the classical adjoint of $A = \begin{bmatrix} 0 & 0 & 0 & 1 \\ 0 & 1 & 0 & 0 \\ 0 & 0 & 1 & 0 \\ 1 & 0 & 0 & 0 \end{bmatrix}$.

33. Find the classical adjoint of $A = \begin{bmatrix} 1 & 0 & 0 & 0 \\ 0 & 2 & 0 & 0 \\ 0 & 0 & 3 & 0 \\ 0 & 0 & 0 & 4 \end{bmatrix}$.

34. For an invertible $n \times n$ matrix A, find the product $A(\text{adj} A)$. What about $(\text{adj } A)(A)$?

35. For an invertible $n \times n$ matrix A, what is the relationship between $\det(A)$ and $\det(\text{adj } A)$?

36. For an invertible $n \times n$ matrix A, what is $\text{adj}(\text{adj } A)$?

37. For an invertible $n \times n$ matrix A, what is the relationship between $\text{adj}(A)$ and $\text{adj}(A^{-1})$?

38. For two invertible $n \times n$ matrices A and B, what is the relationship between $\text{adj}(A)$, $\text{adj}(B)$, and $\text{adj}(AB)$?

39. If A and B are invertible $n \times n$ matrices, and if A is similar to B, is $\text{adj}(A)$ necessarily similar to $\text{adj}(B)$?

40. For an invertible $n \times n$ matrix A, consider the linear transformation

$$T(\vec{x}) = \begin{bmatrix} \det(A_{\vec{x},1}) \\ \det(A_{\vec{x},2}) \\ \vdots \\ \det(A_{\vec{x},n}) \end{bmatrix} \quad \text{from } \mathbb{R}^n \text{ to } \mathbb{R}^n.$$

Express the standard matrix of T in terms of $\text{adj}(A)$.

41. Show that an $n \times n$ matrix A has at least one nonzero minor if (and only if) $\text{rank}(A) \geq n - 1$.

42. Even if an $n \times n$ matrix A fails to be invertible, we can define the adjoint $\text{adj}(A)$ as in Theorem 6.3.9. The ijth entry of $\text{adj}(A)$ is $(-1)^{i+j} \det(A_{ji})$. For which $n \times n$ matrices A is $\text{adj}(A) = 0$? Give your answer in terms of the rank of A. See Exercise 41.

43. Show that $A(\text{adj} A) = 0 = (\text{adj} A)A$ for all noninvertible $n \times n$ matrices A. See Exercise 42.

44. If A is an $n \times n$ matrix of rank $n - 1$, what is the rank of $\text{adj}(A)$? See Exercises 42 and 43.

45. Find all 2×2 matrices A such that $\text{adj}(A) = A^T$.

46. (For those who have studied multivariable calculus.) Let T be an invertible linear transformation from $\mathbb{R}^2$ to $\mathbb{R}^2$, represented by the matrix M. Let Ω_1 be the unit square in $\mathbb{R}^2$ and Ω_2 its image under T. Consider a continuous function $f(x, y)$ from $\mathbb{R}^2$ to $\mathbb{R}$, and define the function $g(u, v) = f(T(u, v))$. What is the relationship between

---

[10] Simon and Blume, *Mathematics for Economists*, Norton, 1994.

[11] Simon and Blume, *op. cit.*

the following two double integrals?

$$\iint\limits_{\Omega_2} f(x, y)\, dA \quad \text{and} \quad \iint\limits_{\Omega_1} g(u, v)\, dA$$

Your answer will involve the matrix M. *Hint*: What happens when $f(x, y) = 1$, for all x, y?

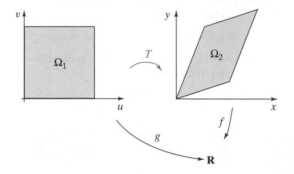

47. Consider the quadrilateral in the accompanying figure, with vertices $P_i = (x_i, y_i)$, for $i = 1, 2, 3, 4$. Show that the area of this quadrilateral is

$$\frac{1}{2}\left(\det\begin{bmatrix} x_1 & x_2 \\ y_1 & y_2 \end{bmatrix} + \det\begin{bmatrix} x_2 & x_3 \\ y_2 & y_3 \end{bmatrix} \right.$$
$$\left. + \det\begin{bmatrix} x_3 & x_4 \\ y_3 & y_4 \end{bmatrix} + \det\begin{bmatrix} x_4 & x_1 \\ y_4 & y_1 \end{bmatrix}\right).$$

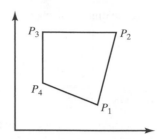

48. What is the area of the largest ellipse you can inscribe into a triangle with side lengths 3, 4, and 5? (*Hint*: The largest ellipse you can inscribe into an equilateral triangle is a circle.)

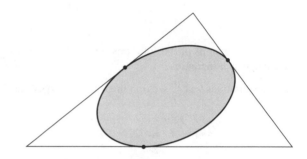

49. What are the lengths of the semiaxes of the largest ellipse you can inscribe into a triangle with sides 3, 4, and 5? See Exercise 48.

Chapter Six Exercises

TRUE OR FALSE?

1. If $A = \begin{bmatrix} \vec{u} & \vec{v} & \vec{w} \end{bmatrix}$ is any 3×3 matrix, then $\det A = \vec{u} \cdot (\vec{v} \times \vec{w})$.

2. $\det(4A) = 4 \det A$ for all 4×4 matrices A.

3. $\det(A + B) = \det A + \det B$ for all 5×5 matrices A and B.

4. The equation $\det(-A) = \det A$ holds for all 6×6 matrices.

5. If all the entries of a 7×7 matrix A are 7, then $\det A$ must be 7^7.

6. An 8×8 matrix fails to be invertible if (and only if) its determinant is nonzero.

7. If B is obtained be multiplying a column of A by 9, then the equation $\det B = 9 \det A$ must hold.

8. $\det(A^{10}) = (\det A)^{10}$ for all 10×10 matrices A.

9. The determinant of any diagonal $n \times n$ matrix is the product of its diagonal entries.

10. If matrix B is obtained by swapping two rows of an $n \times n$ matrix A, then the equation $\det B = -\det A$ must hold.

11. Matrix $\begin{bmatrix} 9 & 100 & 3 & 7 \\ 5 & 4 & 100 & 8 \\ 100 & 9 & 8 & 7 \\ 6 & 5 & 4 & 100 \end{bmatrix}$ is invertible.

12. If A is an invertible $n \times n$ matrix, then $\det(A^T)$ must equal $\det(A^{-1})$.

13. If the determinant of a 4×4 matrix A is 4, then its rank must be 4.

14. There exists a nonzero 4×4 matrix A such that $\det A = \det(4A)$.

15. If two $n \times n$ matrices A and B are similar, then the equation $\det A = \det B$ must hold.

16. The determinant of all orthogonal matrices is 1.

17. If A is any $n \times n$ matrix, then $\det(AA^T) = \det(A^T A)$.

18. There exists an invertible matrix of the form
$$\begin{bmatrix} a & e & f & j \\ b & 0 & g & 0 \\ c & 0 & h & 0 \\ d & 0 & i & 0 \end{bmatrix}.$$

19. The matrix $\begin{bmatrix} k^2 & 1 & 4 \\ k & -1 & -2 \\ 1 & 1 & 1 \end{bmatrix}$ is invertible for all positive constants k.

20. $\det \begin{bmatrix} 0 & 1 & 0 & 0 \\ 0 & 0 & 1 & 0 \\ 0 & 0 & 0 & 1 \\ 1 & 0 & 0 & 0 \end{bmatrix} = 1.$

21. There exists a 4×4 matrix A whose entries are all 1 or -1, and such that $\det A = 16$.

22. If the determinant of a 2×2 matrix A is 4, then the inequality $\|A\vec{v}\| \le 4\|\vec{v}\|$ must hold for all vectors $\vec{v}$ in $\mathbb{R}^2$.

23. If $A = \begin{bmatrix} \vec{u} & \vec{v} & \vec{w} \end{bmatrix}$ is a 3×3 matrix, then the formula $\det(A) = \vec{v} \cdot (\vec{u} \times \vec{w})$ must hold.

24. There exist invertible 2×2 matrices A and B such that $\det(A + B) = \det A + \det B$.

25. If all the entries of a square matrix are 1 or 0, then $\det A$ must be 1, 0, or -1.

26. If all the entries of a square matrix A are integers and $\det A = 1$, then the entries of matrix A^{-1} must be integers as well.

27. If all the columns of a square matrix A are unit vectors, then the determinant of A must be less than or equal to 1.

28. If A is any noninvertible square matrix, then $\det A = \det(\text{rref } A)$.

29. If the determinant of a square matrix is -1, then A must be an orthogonal matrix.

30. If all the entries of an invertible matrix A are integers, then the entries of A^{-1} must be integers as well.

31. There exist invertible 3×3 matrices A and S such that $S^{-1}AS = 2A$.

32. There exist invertible 3×3 matrices A and S such that $S^T A S = -A$.

33. If A is any symmetric matrix, then $\det A = 1$ or $\det A = -1$.

34. If A is any skew-symmetric 4×4 matrix, then $\det A = 0$.

35. If $\det A = \det B$ for two $n \times n$ matrices A and B, then A must be similar to B.

36. Suppose A is an $n \times n$ matrix and B is obtained from A by swapping two rows of A. If $\det B < \det A$, then A must be invertible.

37. If an $n \times n$ matrix A is invertible, then there must be an $(n-1) \times (n-1)$ submatrix of A (obtained by deleting a row and a column of A) that is invertible as well.

38. If all the entries of matrices A and A^{-1} are integers, then the equation $\det A = \det(A^{-1})$ must hold.

39. If a square matrix A is invertible, then its classical adjoint $\text{adj}(A)$ is invertible as well.

40. There exists a 3×3 matrix A such that $A^2 = -I_3$.

41. If all the diagonal entries of an $n \times n$ matrix A are odd integers and all the other entries are even integers, then A must be an invertible matrix.

42. If all the diagonal entries of an $n \times n$ matrix A are even integers and all the other entries are odd integers, then A must be an invertible matrix.

43. For every nonzero 2×2 matrix A there exists a 2×2 matrix B such that $\det(A + B) \ne \det A + \det B$.

44. If A is a 4×4 matrix whose entries are all 1 or -1, then $\det A$ must be divisible by 8 [i.e., $\det A = 8k$ for some integer k].

45. If A is an invertible $n \times n$ matrix, then A must commute with its adjoint, $\text{adj}(A)$.

46. There exists a real number k such that the matrix
$$\begin{bmatrix} 1 & 2 & 3 & 4 \\ 5 & 6 & k & 7 \\ 8 & 9 & 8 & 7 \\ 0 & 0 & 6 & 5 \end{bmatrix}$$
is invertible.

47. If A and B are orthogonal $n \times n$ matrices such that $\det A = \det B = 1$, then matrices A and B must commute.

Eigenvalues and Eigenvectors

7.1 Dynamical Systems and Eigenvectors: An Introductory Example

A stretch of desert in northwestern Mexico is populated mainly by two species of animals: coyotes and roadrunners. We wish to model the populations $c(t)$ and $r(t)$ of coyotes and roadrunners t years from now if the current populations c_0 and r_0 are known.[1, 2]

---

[1]The point of this lighthearted story is to present an introductory example where neither messy data nor a complicated scenario distracts us from the mathematical ideas we wish to develop.

[2]For those who don't have the time or inclination to work through this long introductory example, we propose an alternative approach:

Example A: Give a geometrical interpretation of the linear transformation $T(\vec{x}) = A\vec{x}$, where $A = \begin{bmatrix} 4 & 3 \\ 3 & -4 \end{bmatrix}$.

Solution: Write $A = 5\begin{bmatrix} 0.8 & 0.6 \\ 0.6 & -0.8 \end{bmatrix}$ to see that T is a refection about a line combined with a scaling by a factor of 5; see Definition 2.2.2. Let's find the line L of reflection. If $\vec{v} = \begin{bmatrix} x \\ y \end{bmatrix}$ is any vector parallel to L, then

$$A\vec{v} = 5\vec{v}, \quad \text{or} \quad \begin{bmatrix} 4 & 3 \\ 3 & -4 \end{bmatrix}\begin{bmatrix} x \\ y \end{bmatrix} = \begin{bmatrix} 5x \\ 5y \end{bmatrix}, \quad \text{or} \quad \begin{vmatrix} 4x + 3y = 5x \\ 3x - 4y = 5y \end{vmatrix}, \quad \text{or} \quad x = 3y.$$

Thus $L = \text{span}\begin{bmatrix} 3 \\ 1 \end{bmatrix}$.

In summary, T is the reflection about the line $L = \text{span}\begin{bmatrix} 3 \\ 1 \end{bmatrix}$, combined with a scaling by a factor of 5.

Here is a follow-up question: Are there any other vectors $\vec{v}$, besides those parallel to the line L, such that $A\vec{v}$ is a scalar multiple of $\vec{v}$? ∎

Example B: Give a geometrical interpretation of the linear transformation $T(\vec{x}) = B\vec{x}$, where $B = \begin{bmatrix} 1 & 4 \\ 4 & 7 \end{bmatrix}$.

For this habitat, the following equations model the transformation of this system from one year to the next, from time t to time $(t + 1)$:

$$\begin{vmatrix} c(t + 1) = & 0.86c(t) + 0.08r(t) \\ r(t + 1) = & -0.12c(t) + 1.14r(t) \end{vmatrix}.$$

Why is the coefficient of $c(t)$ in the first equation less than 1, while the coefficient of $r(t)$ in the second equation exceeds 1? What is the practical significance of the signs of the other two coefficients, 0.08 and -0.12?

The two equations can be written in matrix form, as

$$\begin{bmatrix} c(t + 1) \\ r(t + 1) \end{bmatrix} = \begin{bmatrix} 0.86c(t) + 0.08r(t) \\ -0.12c(t) + 1.14r(t) \end{bmatrix} = \begin{bmatrix} 0.86 & 0.08 \\ -0.12 & 1.14 \end{bmatrix} \begin{bmatrix} c(t) \\ r(t) \end{bmatrix}.$$

The vector

$$\vec{x}(t) = \begin{bmatrix} c(t) \\ r(t) \end{bmatrix}$$

is called the *state vector* of the system at time t, because it completely describes this system at time t. If we let

$$A = \begin{bmatrix} 0.86 & 0.08 \\ -0.12 & 1.14 \end{bmatrix},$$

we can write the preceding matrix equation more succinctly as

$$\vec{x}(t + 1) = A\vec{x}(t).$$

The transformation the system undergoes over the period of one year is linear, represented by the matrix A.

$$\vec{x}(t) \xrightarrow{A} \vec{x}(t + 1)$$

Suppose we know the initial state

$$\vec{x}(0) = \vec{x}_0 = \begin{bmatrix} c_0 \\ r_0 \end{bmatrix}.$$

We wish to find $\vec{x}(t)$, for an arbitrary positive integer t:

$$\vec{x}(0) \xrightarrow{A} \vec{x}(1) \xrightarrow{A} \vec{x}(2) \xrightarrow{A} \vec{x}(3) \xrightarrow{A} \cdots \xrightarrow{A} \vec{x}(t) \xrightarrow{A} \cdots.$$

Solution: This problem is harder than Example A, since matrix B is not directly associated with any of the geometrical transformations we discussed in Section 2.2: Reflections, Rotations, Scalings, Projections, and Shears.

Using Example A as a guide, we can look for vectors $\vec{v}$ such that $B\vec{v}$ is a scalar multiple of $\vec{v}$:

$$B\vec{v} = \lambda\vec{v}, \text{ for some scalar } \lambda \text{ (lambda).}$$

In Sections 7.2 and 7.3, we will learn how to find such vectors $\vec{v}$. At this early stage, I will just pull two such vectors out of a hat and tell you that

$$\begin{bmatrix} 1 & 4 \\ 4 & 7 \end{bmatrix} \begin{bmatrix} 1 \\ 2 \end{bmatrix} = \begin{bmatrix} 9 \\ 18 \end{bmatrix} = 9 \begin{bmatrix} 1 \\ 2 \end{bmatrix} \quad \text{and} \quad \begin{bmatrix} 1 & 4 \\ 4 & 7 \end{bmatrix} \begin{bmatrix} -2 \\ 1 \end{bmatrix} = \begin{bmatrix} 2 \\ -1 \end{bmatrix} = (-1) \begin{bmatrix} -2 \\ 1 \end{bmatrix}.$$

These results allow us to interpret T as a scaling by a factor of 9 in the direction of the vector $\begin{bmatrix} 1 \\ 2 \end{bmatrix}$ combined with a the reflection about the line spanned by $\begin{bmatrix} 1 \\ 2 \end{bmatrix}$. Draw a sketch! Imagine performing those transformations on a rubber sheet. ∎

Having recognized that it can be useful to know the solutions of the equation $A\vec{x} = \lambda\vec{x}$ for a square matrix A, the reader may now proceed to Definition 7.1.1.

We can find $\vec{x}(t)$ by applying the transformation t times to $\vec{x}(0)$:

$$\vec{x}(t) = A^t \vec{x}(0) = A^t \vec{x}_0.$$

Although it is extremely tedious to find $\vec{x}(t)$ with paper and pencil for large t, we can easily compute $\vec{x}(t)$ using technology. For example, given

$$\vec{x}_0 = \begin{bmatrix} 100 \\ 100 \end{bmatrix},$$

we find that

$$\vec{x}(10) = A^{10}\vec{x}_0 \approx \begin{bmatrix} 80 \\ 170 \end{bmatrix}.$$

To understand the long-term behavior of this system and how it depends on the initial values, we must go beyond numerical experimentation. It would be useful to have *closed formulas* for $c(t)$ and $r(t)$, expressing these quantities as functions of t. We will first do this for certain (carefully chosen) initial state vectors.

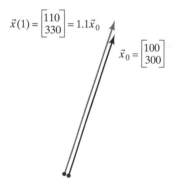

$$\vec{x}(1) = \begin{bmatrix} 110 \\ 330 \end{bmatrix} = 1.1\vec{x}_0$$

$$\vec{x}_0 = \begin{bmatrix} 100 \\ 300 \end{bmatrix}$$

Figure I

Case 1 ■ Suppose we have $c_0 = 100$ and $r_0 = 300$. Initially, there are 100 coyotes and 300 roadrunners, so that $\vec{x}_0 = \begin{bmatrix} 100 \\ 300 \end{bmatrix}$. Then

$$\vec{x}(1) = A\vec{x}_0 = \begin{bmatrix} 0.86 & 0.08 \\ -0.12 & 1.14 \end{bmatrix} \begin{bmatrix} 100 \\ 300 \end{bmatrix} = \begin{bmatrix} 110 \\ 330 \end{bmatrix}.$$

Note that each population has grown by 10%. This means that the state vector $\vec{x}(1)$ is a scalar multiple of $\vec{x}_0$ (see Figure 1):

$$\vec{x}(1) = A\vec{x}_0 = 1.1\vec{x}_0.$$

It is now easy to compute $\vec{x}(t)$ for arbitrary t, using linearity:

$$\vec{x}(2) = A\vec{x}(1) = A(1.1\vec{x}_0) = 1.1A\vec{x}_0 = 1.1^2\vec{x}_0$$
$$\vec{x}(3) = A\vec{x}(2) = A(1.1^2\vec{x}_0) = 1.1^2A\vec{x}_0 = 1.1^3\vec{x}_0$$
$$\vdots$$
$$\vec{x}(t) = 1.1^t\vec{x}_0.$$

We keep multiplying the state vector by 1.1 each time we apply the transformation A. Recall that our goal is to find closed formulas for $c(t)$ and $r(t)$. We have

$$\vec{x}(t) = \begin{bmatrix} c(t) \\ r(t) \end{bmatrix} = 1.1^t\vec{x}_0 = 1.1^t \begin{bmatrix} 100 \\ 300 \end{bmatrix},$$

so that

$$c(t) = 100(1.1)^t \quad \text{and} \quad r(t) = 300(1.1)^t.$$

Both populations will grow exponentially, by 10% each year.

Case 2 ■ Suppose we have $c_0 = 200$ and $r_0 = 100$. Then

$$\vec{x}(1) = A\vec{x}_0 = \begin{bmatrix} 0.86 & 0.08 \\ -0.12 & 1.14 \end{bmatrix} \begin{bmatrix} 200 \\ 100 \end{bmatrix} = \begin{bmatrix} 180 \\ 90 \end{bmatrix} = 0.9\vec{x}_0.$$

Both populations decline by 10% in the first year and will therefore decline another 10% each subsequent year. Thus

$$\vec{x}(t) = 0.9^t\vec{x}_0,$$

so that

$$c(t) = 200(0.9)^t \qquad \text{and} \qquad r(t) = 100(0.9)^t.$$

The initial populations are mismatched: Too many coyotes are chasing too few roadrunners, a bad state of affairs for both species.

Case 3 ■ Suppose we have $c_0 = r_0 = 1,000$. Then

$$\vec{x}(1) = A\vec{x}_0 = \begin{bmatrix} 0.86 & 0.08 \\ -0.12 & 1.14 \end{bmatrix} \begin{bmatrix} 1,000 \\ 1,000 \end{bmatrix} = \begin{bmatrix} 940 \\ 1,020 \end{bmatrix}.$$

Things are not working out as nicely as in the first two cases we considered: The state vector $\vec{x}(1)$ fails to be a scalar multiple of the initial state $\vec{x}_0$. Just by computing $\vec{x}(2), \vec{x}(3), \ldots$, we could not easily detect a trend that would allow us to generate closed formulas for $c(t)$ and $r(t)$. We have to look for another approach.

The idea is to work with the two vectors

$$\vec{v}_1 = \begin{bmatrix} 100 \\ 300 \end{bmatrix} \quad \text{and} \quad \vec{v}_2 = \begin{bmatrix} 200 \\ 100 \end{bmatrix}$$

considered in the first two cases, for which $A^t \vec{v}_i$ was easy to compute. Since the vectors $\vec{v}_1$ and $\vec{v}_2$ form a basis of $\mathbb{R}^2$, any vector in $\mathbb{R}^2$ can be written uniquely as a linear combination of $\vec{v}_1$ and $\vec{v}_2$. This holds in particular for the initial state vector

$$\vec{x}_0 = \begin{bmatrix} 1,000 \\ 1,000 \end{bmatrix}$$

of the coyote–roadrunner system:

$$\vec{x}_0 = c_1 \vec{v}_1 + c_2 \vec{v}_2.$$

A straightforward computation shows that the coordinates are $c_1 = 2$ and $c_2 = 4$:

$$\vec{x}_0 = 2\vec{v}_1 + 4\vec{v}_2.$$

Recall that $A^t \vec{v}_1 = (1.1)^t \vec{v}_1$ and $A^t \vec{v}_2 = (0.9)^t \vec{v}_2$. Therefore,

$$\begin{aligned} \vec{x}(t) = A^t \vec{x}_0 = A^t (2\vec{v}_1 + 4\vec{v}_2) &= 2A^t \vec{v}_1 + 4A^t \vec{v}_2 \\ &= 2(1.1)^t \vec{v}_1 + 4(0.9)^t \vec{v}_2 \\ &= 2(1.1)^t \begin{bmatrix} 100 \\ 300 \end{bmatrix} + 4(0.9)^t \begin{bmatrix} 200 \\ 100 \end{bmatrix}. \end{aligned}$$

Considering the components of this equation, we can find formulas for $c(t)$ and $r(t)$:

$$c(t) = 200(1.1)^t + 800(0.9)^t$$
$$r(t) = 600(1.1)^t + 400(0.9)^t.$$

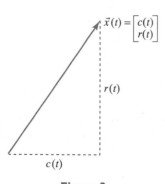

Since the terms involving 0.9^t approach zero as t increases, both populations eventually grow by about 10% a year, and their ratio $r(t)/c(t)$ approaches $600/200 = 3$.

Note that the ratio $r(t)/c(t)$ can be interpreted as the slope of the state vector $\vec{x}(t)$, as shown in Figure 2.

Figure 2

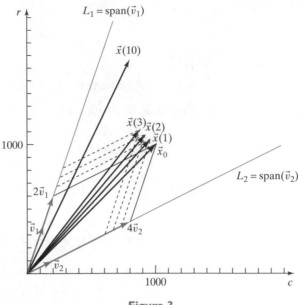

Figure 3

How can we represent the preceding computations graphically?

Figure 3 shows the representation $\vec{x}_0 = 2\vec{v}_1 + 4\vec{v}_2$ of $\vec{x}_0$ as the sum of a vector on $L_1 = \text{span}(\vec{v}_1)$ and a vector on $L_2 = \text{span}(\vec{v}_2)$. The formula

$$\vec{x}(t) = (1.1)^t 2\vec{v}_1 + (0.9)^t 4\vec{v}_2$$

now tells us that the component in L_1 grows by 10% each year, while the component in L_2 shrinks by 10%. The component $(0.9)^t 4\vec{v}_2$ in L_2 approaches $\vec{0}$, which means that the tip of the state vector $\vec{x}(t)$ approaches the line L_1, so that the slope of the state vector will approach 3, the slope of L_1.

To show the evolution of the system more clearly, we can sketch just the endpoints of the state vectors $\vec{x}(t)$. Then the changing state of the system will be traced out as a sequence of points in the c-r plane.

It is natural to connect the dots to create the illusion of a continuous *trajectory*. (Although, of course, we do not know what really happens between times t and $t + 1$.)

Sometimes we are interested in the state of the system in the past, at times $-1, -2, \ldots$. Note that $\vec{x}(0) = A\vec{x}(-1)$, so that $\vec{x}(-1) = A^{-1}\vec{x}_0$ if A is invertible (as in our example). Likewise, $\vec{x}(-t) = (A^t)^{-1}\vec{x}_0$, for $t = 2, 3, \ldots$. The trajectory (future and past) for our coyote–roadrunner system is shown in Figure 4.

To get a feeling for the long-term behavior of this system and how it depends on the initial state, we can draw a rough sketch that shows a number of different trajectories, representing the various qualitative types of behavior. Such a sketch is called a *phase portrait* of the system. In our example, a phase portrait might show the foregoing three cases, as well as a trajectory that starts above L_1 and one that starts below L_2. See Figure 5.

To sketch these trajectories, express the initial state vector $\vec{x}_0$ as the sum of a vector $\vec{w}_1$ on L_1 and a vector $\vec{w}_2$ on L_2. Then see how these two vectors change over time. If $\vec{x}_0 = \vec{w}_1 + \vec{w}_2$, then

$$\vec{x}(t) = (1.1)^t \vec{w}_1 + (0.9)^t \vec{w}_2.$$

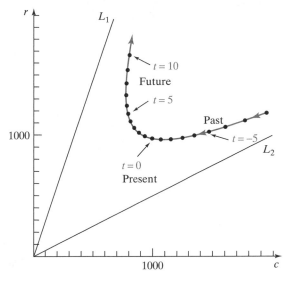

Figure 4

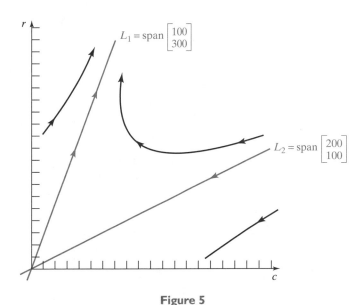

Figure 5

We see that the two populations will prosper over the long term if the ratio r_0/c_0 of the initial populations exceeds $1/2$; otherwise, both populations will die out.

From a mathematical point of view, it is informative to sketch a phase portrait of this system in the whole c–r-plane, even though the trajectories outside the first quadrant are meaningless in terms of our population study. (See Figure 6.)

Eigenvectors and Eigenvalues

In the example just discussed, it turned out to be very useful to have some nonzero vectors $\vec{v}$ such that $A\vec{v}$ is a scalar multiple of $\vec{v}$, or

$$A\vec{v} = \lambda\vec{v},$$

for some scalar λ.

Such vectors are important in many other contexts as well.

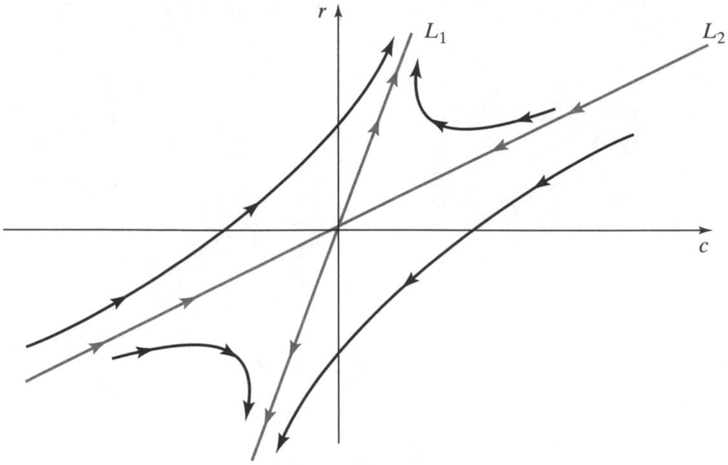

Figure 6

Definition 7.1.1   Eigenvectors[3] and eigenvalues

Consider an $n \times n$ matrix A. A nonzero vector $\vec{v}$ in $\mathbb{R}^n$ is called an *eigenvector* of A if $A\vec{v}$ is a scalar multiple of $\vec{v}$, that is, if

$$A\vec{v} = \lambda\vec{v},$$

for some scalar λ. Note that this scalar λ may be zero.

The scalar λ is called the *eigenvalue* associated with the eigenvector $\vec{v}$.

A nonzero vector $\vec{v}$ is an eigenvector of A if the vectors $\vec{v}$ and $A\vec{v}$ are parallel, as shown in Figure 7. (See Definition A.3 in the Appendix.)

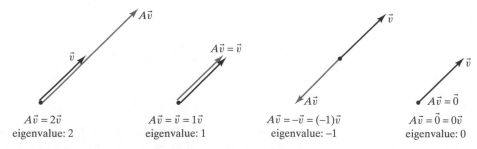

| $A\vec{v} = 2\vec{v}$ | $A\vec{v} = \vec{v} = 1\vec{v}$ | $A\vec{v} = -\vec{v} = (-1)\vec{v}$ | $A\vec{v} = \vec{0} = 0\vec{v}$ |
| eigenvalue: 2 | eigenvalue: 1 | eigenvalue: -1 | eigenvalue: 0 |

Figure 7

If $\vec{v}$ is an eigenvector of matrix A, then $\vec{v}$ is an eigenvector of matrices A^2, $A^3, \ldots$, as well, with

$$A^2\vec{v} = \lambda^2\vec{v}, \quad A^3\vec{v} = \lambda^3\vec{v}, \ldots, \quad A^t\vec{v} = \lambda^t\vec{v},$$

for all positive integers t.

---

[3]From German *eigen*: proper, characteristic.

EXAMPLE 1   Find all eigenvectors and eigenvalues of the identity matrix I_n.

Solution
Since $I_n \vec{v} = \vec{v} = 1\vec{v}$ for all $\vec{v}$ in $\mathbb{R}^n$, all nonzero vectors in $\mathbb{R}^n$ are eigenvectors, with eigenvalue 1. ∎

EXAMPLE 2   Let T be the orthogonal projection onto a line L in $\mathbb{R}^2$. Describe the eigenvectors of T geometrically and find all eigenvalues of T.

Solution
We find the eigenvectors by inspection: can you think of any nonzero vectors $\vec{v}$ in the plane such that $T(\vec{v})$ is a scalar multiple of $\vec{v}$? Clearly, any vector $\vec{v}$ on L will do [with $T(\vec{v}) = 1\vec{v}$], as well as any vector $\vec{w}$ perpendicular to L [with $T(\vec{w}) = \vec{0} = 0\vec{w}$]. See Figure 8.

    The eigenvalues are 1 and 0. ∎

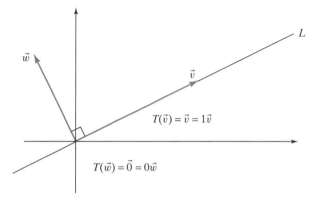

Figure 8

EXAMPLE 3   Let T from $\mathbb{R}^2$ to $\mathbb{R}^2$ be the rotation in the plane through an angle of $90°$ in the counterclockwise direction. Find all eigenvalues and eigenvectors of T.

Solution
If $\vec{v}$ is any nonzero vector in $\mathbb{R}^2$, then $T(\vec{v})$ fails to be parallel to $\vec{v}$ (it's perpendicular). See Figure 9. There are no real eigenvectors and eigenvalues here.[4] ∎

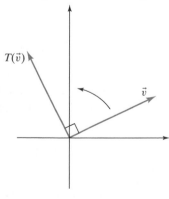

Figure 9

[4]In Section 7.5, we will consider *complex* eigenvalues.

EXAMPLE 4   What are the possible real eigenvalues of an orthogonal[5] matrix A?

Solution

Recall that the linear transformation $T(\vec{x}) = A\vec{x}$ preserves length: $\|T(\vec{x})\| = \|A\vec{x}\| = \|\vec{x}\|$, for all vectors $\vec{x}$. (See Definition 5.3.1.) Consider an eigenvector $\vec{v}$ of A, with eigenvalue λ:

$$A\vec{v} = \lambda\vec{v}.$$

Then

$$\|\vec{v}\| = \|A\vec{v}\| = \|\lambda\vec{v}\| = |\lambda|\|\vec{v}\|,$$

so that $\lambda = 1$ or $\lambda = -1$. The two possibilities are illustrated in Figure 10.   ∎

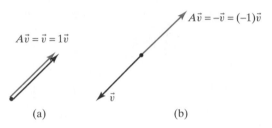

Figure 10 (a) $\lambda = 1$. (b) $\lambda = -1$.

Theorem 7.1.2   The possible real eigenvalues of an orthogonal[5] matrix are 1 and -1.   ∎

As an example, consider the reflection about a line in $\mathbb{R}^2$. (See Exercise 15.)

Dynamical Systems and Eigenvectors

Consider a physical system whose state at any given time t is described by some quantities $x_1(t), x_2(t), \ldots, x_n(t)$. [In our introductory example, there were two such quantities, the populations $c(t)$ and $r(t)$.] We can represent the quantities $x_1(t), x_2(t), \ldots, x_n(t)$ by the state vector

$$\vec{x}(t) = \begin{bmatrix} x_1(t) \\ x_2(t) \\ \vdots \\ x_n(t) \end{bmatrix}.$$

Suppose that the state of the system at time $t + 1$ is determined by the state at time t and that the transformation of the system from time t to time $t + 1$ is linear, represented by an $n \times n$ matrix A:

$$\vec{x}(t + 1) = A\vec{x}(t).$$

Then

$$\vec{x}(t) = A^t\vec{x}_0.$$

---

[5]Example 4 and Theorem 7.1.2 are for those who have studied Chapter 5.

Such a system is called a *discrete linear dynamical system*. (*Discrete* indicates that we model the change of the system from time t to time $t + 1$, rather than modeling the *continuous* rate of change, which is described by differential equations.)

For an initial state $\vec{x}_0$, it is often our goal to find *closed formulas* for $x_1(t)$, $x_2(t)$, ..., $x_n(t)$ [i.e., formulas expressing $x_i(t)$ as a function of t alone, as opposed to a recursive formula, for example, which would merely express $x_i(t + 1)$ in terms of $x_1(t), x_2(t), \ldots, x_n(t)$].

Theorem 7.1.3   Discrete dynamical systems

Consider the dynamical system

$$\vec{x}(t + 1) = A\vec{x}(t) \quad \text{with} \quad \vec{x}(0) = \vec{x}_0.$$

Then

$$\vec{x}(t) = A^t \vec{x}_0.$$

Suppose we can find a basis

$$\vec{v}_1, \vec{v}_2, \ldots, \vec{v}_n \quad \text{of } \mathbb{R}^n$$

consisting of eigenvectors of A, with

$$A\vec{v}_1 = \lambda_1\vec{v}_1, A\vec{v}_2 = \lambda_2\vec{v}_2, \ldots, A\vec{v}_n = \lambda_n\vec{v}_n.$$

Find the coordinates $c_1, c_2, \ldots, c_n$ of vector $\vec{x}_0$ with respect to basis $\vec{v}_1, \vec{v}_2, \ldots, \vec{v}_n$:

$$\vec{x}_0 = c_1\vec{v}_1 + c_2\vec{v}_2 + \cdots + c_n\vec{v}_n.$$

Then

$$\vec{x}(t) = c_1\lambda_1^t\vec{v}_1 + c_2\lambda_2^t\vec{v}_2 + \cdots + c_n\lambda_n^t\vec{v}_n.$$

We can write this equation in matrix form as

$$\vec{x}(t) = \begin{bmatrix} | & | & & | \\ \vec{v}_1 & \vec{v}_2 & \cdots & \vec{v}_n \\ | & | & & | \end{bmatrix} \begin{bmatrix} \lambda_1^t & 0 & \cdots & 0 \\ 0 & \lambda_2^t & \cdots & 0 \\ \vdots & \vdots & \ddots & \vdots \\ 0 & 0 & \cdots & \lambda_n^t \end{bmatrix} \begin{bmatrix} c_1 \\ c_2 \\ \vdots \\ c_n \end{bmatrix}$$

$$= S \begin{bmatrix} \lambda_1 & 0 & \cdots & 0 \\ 0 & \lambda_2 & \cdots & 0 \\ \vdots & \vdots & \ddots & \vdots \\ 0 & 0 & \cdots & \lambda_n \end{bmatrix}^t S^{-1}\vec{x}_0,$$

where $S = \begin{bmatrix} | & | & & | \\ \vec{v}_1 & \vec{v}_2 & \cdots & \vec{v}_n \\ | & | & & | \end{bmatrix}$. Note that $\vec{x}_0 = S \begin{bmatrix} c_1 \\ c_2 \\ \vdots \\ c_n \end{bmatrix}$, so that $\begin{bmatrix} c_1 \\ c_2 \\ \vdots \\ c_n \end{bmatrix} = S^{-1}\vec{x}_0.$ ■

We are left with two questions: How can we find the eigenvalues and eigenvectors of an $n \times n$ matrix A? When is there a basis of $\mathbb{R}^n$ consisting of eigenvectors of A? These issues are central to linear algebra; they will keep us busy for the rest of this long chapter.

Definition 7.1.4   Discrete trajectories and phase portraits

Consider a discrete dynamical system

$$\vec{x}(t+1) = A\vec{x}(t) \quad \text{with initial value } \vec{x}(0) = \vec{x}_0,$$

where A is a 2×2 matrix. In this case, the state vector $\vec{x}(t) = \begin{bmatrix} x_1(t) \\ x_2(t) \end{bmatrix}$ can be represented geometrically in the x_1-x_2-plane.

The endpoints of state vectors $\vec{x}(0) = \vec{x}_0, \vec{x}(1) = A\vec{x}_0, \vec{x}(2) = A^2\vec{x}_0, \ldots$ form the (discrete) *trajectory* of this system, representing its evolution in the future. Sometimes we are interested in the past states $\vec{x}(-1) = A^{-1}\vec{x}_0, \vec{x}(-2) = (A^2)^{-1}\vec{x}_0, \ldots$ as well. It is suggestive to "connect the dots" to create the illusion of a continuous trajectory. Take another look at Figure 4.

A (discrete) *phase portrait* of the system $\vec{x}(t+1) = A\vec{x}(t)$ shows trajectories for various initial states, capturing all the qualitatively different scenarios (as in Figure 6).

In Figure 11, we sketch phase portraits for the case when A has two eigenvalues $\lambda_1 > \lambda_2 > 0$ with associated eigenvectors $\vec{v}_1$ and $\vec{v}_2$. We leave out the special case when one of the eigenvalues is 1. Start by sketching the trajectories along the lines $L_1 = \text{span}(\vec{v}_1)$ and $L_2 = \text{span}(\vec{v}_2)$. As you sketch the other trajectories

$$\vec{x}(t) = c_1 \lambda_1^t \vec{v}_1 + c_2 \lambda_2^t \vec{v}_2,$$

think about the summands $c_1 \lambda_1^t \vec{v}_1$ and $c_2 \lambda_2^t \vec{v}_2$. Note that for a large positive t the vector $\vec{x}(t)$ will be almost parallel to L_1, since λ_1^t will be much larger than λ_2^t. Likewise, for large negative t the vector $\vec{x}(t)$ will be almost parallel to L_2.

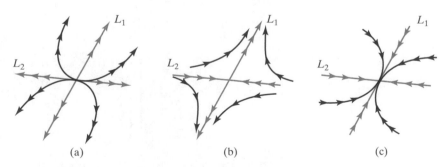

(a)   (b)   (c)

Figure 11   (a) $\lambda_1 > \lambda_2 > 1$. (b) $\lambda_1 > 1 > \lambda_2 > 0$. (c) $1 > \lambda_1 > \lambda_2 > 0$.

We conclude this section with a third (and final) version of the summing-up theorem. (Compare this with Summary 3.3.10.)

SUMMARY 7.1.5 | **Various characterizations of invertible matrices**

For an $n \times n$ matrix A, the following statements are equivalent.

    i. A is invertible.

    ii. The linear system $A\vec{x} = \vec{b}$ has a unique solution $\vec{x}$, for all $\vec{b}$ in $\mathbb{R}^n$.

    iii. $\text{rref}(A) = I_n$.

    iv. $\text{rank}(A) = n$.

    v. $\text{im}(A) = \mathbb{R}^n$.

    vi. $\ker(A) = \{\vec{0}\}$.

    vii. The column vectors of A form a basis of $\mathbb{R}^n$.

    viii. The column vectors of A span $\mathbb{R}^n$.

    ix. The column vectors of A are linearly independent.

    x. $\det(A) \neq 0$.

    xi. 0 fails to be an eigenvalue of A.

Characterization (x) was given in Theorem 6.2.4. The equivalence of (vi) and (xi) follows from the definition of an eigenvalue. (Note that an eigenvector of A with eigenvalue 0 is a nonzero vector in ker A.)

EXERCISES 7.1

GOAL *Apply the concept of eigenvalues and eigenvectors. Use eigenvectors to analyze discrete dynamical systems.*

In Exercises 1 through 4, let A be an invertible $n \times n$ matrix and $\vec{v}$ an eigenvector of A with associated eigenvalue λ.

1. Is $\vec{v}$ an eigenvector of A^3? If so, what is the eigenvalue?

2. Is $\vec{v}$ an eigenvector of A^{-1}? If so, what is the eigenvalue?

3. Is $\vec{v}$ an eigenvector of $A + 2I_n$? If so, what is the eigenvalue?

4. Is $\vec{v}$ an eigenvector of $7A$? If so, what is the eigenvalue?

5. If a vector $\vec{v}$ is an eigenvector of both A and B, is $\vec{v}$ necessarily an eigenvector of $A + B$?

6. If a vector $\vec{v}$ is an eigenvector of both A and B, is $\vec{v}$ necessarily an eigenvector of AB?

7. If $\vec{v}$ is an eigenvector of the $n \times n$ matrix A with associated eigenvalue λ, what can you say about

$$\ker(A - \lambda I_n)?$$

Is the matrix $A - \lambda I_n$ invertible?

8. Find all 2×2 matrices for which $\vec{e}_1 = \begin{bmatrix} 1 \\ 0 \end{bmatrix}$ is an eigenvector with associated eigenvalue 5.

9. Find all 2×2 matrices for which $\vec{e}_1$ is an eigenvector.

10. Find all 2×2 matrices for which $\begin{bmatrix} 1 \\ 2 \end{bmatrix}$ is an eigenvector with associated eigenvalue 5.

11. Find all 2×2 matrices for which $\begin{bmatrix} 2 \\ 3 \end{bmatrix}$ is an eigenvector with associated eigenvalue -1.

12. Consider the matrix $A = \begin{bmatrix} 2 & 0 \\ 3 & 4 \end{bmatrix}$. Show that 2 and 4 are eigenvalues of A and find all corresponding eigenvectors.

13. Show that 4 is an eigenvalue of $A = \begin{bmatrix} -6 & 6 \\ -15 & 13 \end{bmatrix}$ and find all corresponding eigenvectors.

14. Find all 4×4 matrices for which $\vec{e}_2$ is an eigenvector.

Arguing geometrically, find all eigenvectors and eigenvalues of the linear transformations in Exercises 15 through 22. Find a basis consisting of eigenvectors if possible.

15. Reflection about a line L in $\mathbb{R}^2$

16. Rotation through an angle of $180°$ in $\mathbb{R}^2$

17. Counterclockwise rotation through an angle of $45°$ followed by a scaling by 2 in $\mathbb{R}^2$

18. Reflection about a plane V in $\mathbb{R}^3$

19. Orthogonal projection onto a line L in $\mathbb{R}^3$

20. Rotation about the $\vec{e}_3$-axis through an angle of $90°$, counterclockwise as viewed from the positive $\vec{e}_3$-axis in $\mathbb{R}^3$

21. Scaling by 5 in $\mathbb{R}^3$

22. The shear with $T(\vec{v}) = \vec{v}$ and $T(\vec{w}) = \vec{v} + \vec{w}$ for the vectors $\vec{v}$ and $\vec{w}$ in $\mathbb{R}^2$ sketched below

23. **a.** Consider an invertible matrix $S = \begin{bmatrix} \vec{v}_1 & \vec{v}_2 & \ldots & \vec{v}_n \end{bmatrix}$. Find $S^{-1}\vec{v}_i$. *Hint*: What is $S\vec{e}_i$?

    b. Consider an $n \times n$ matrix A and an invertible $n \times n$ matrix $S = \begin{bmatrix} \vec{v}_1 & \vec{v}_2 & \ldots & \vec{v}_n \end{bmatrix}$ whose columns are eigenvectors of A with $A\vec{v}_i = \lambda_i \vec{v}_i$. Find $S^{-1}AS$. *Hint*: Think about the product $S^{-1}AS$ column by column.

In Exercises 24 through 29, consider a dynamical system

$$\vec{x}(t + 1) = A\vec{x}(t)$$

with two components. The accompanying sketch shows the initial state vector $\vec{x}_0$ and two eigenvectors, $\vec{v}_1$ and $\vec{v}_2$, of A (with eigenvalues λ_1 and λ_2, respectively). For the given values of λ_1 and λ_2, sketch a rough trajectory. Think about the future and the past of the system.

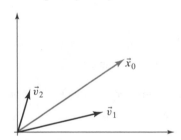

24. $\lambda_1 = 1.1, \lambda_2 = 0.9$ 25. $\lambda_1 = 1, \lambda_2 = 0.9$

26. $\lambda_1 = 1.1, \lambda_2 = 1$ 27. $\lambda_1 = 0.9, \lambda_2 = 0.8$

28. $\lambda_1 = 1.2, \lambda_2 = 1.1$ 29. $\lambda_1 = 0.9, \lambda_2 = 0.9$

In Exercises 30 through 32, consider the dynamical system

$$\vec{x}(t + 1) = \begin{bmatrix} 1.1 & 0 \\ 0 & \lambda \end{bmatrix} \vec{x}(t).$$

Sketch a phase portrait of this system for the given values of λ:

30. $\lambda = 1.2$ 31. $\lambda = 1$ 32. $\lambda = 0.9$

33. Find a 2×2 matrix A such that

$$\vec{x}(t) = \begin{bmatrix} 2^t - 6^t \\ 2^t + 6^t \end{bmatrix}$$

is a trajectory of the dynamical system

$$\vec{x}(t + 1) = A\vec{x}(t).$$

34. Suppose $\vec{v}$ is an eigenvector of the $n \times n$ matrix A, with eigenvalue 4. Explain why $\vec{v}$ is an eigenvector of $A^2 + 2A + 3I_n$. What is the associated eigenvalue?

35. Show that similar matrices have the same eigenvalues. (*Hint*: If $\vec{v}$ is an eigenvector of $S^{-1}AS$, then $S\vec{v}$ is an eigenvector of A.)

36. Find a 2×2 matrix A such that $\begin{bmatrix} 3 \\ 1 \end{bmatrix}$ and $\begin{bmatrix} 1 \\ 2 \end{bmatrix}$ are eigenvectors of A, with eigenvalues 5 and 10, respectively.

37. Consider the matrix

$$A = \begin{bmatrix} 3 & 4 \\ 4 & -3 \end{bmatrix}.$$

    a. Use the geometric interpretation of this transformation as a reflection combined with a scaling to find the eigenvalues of A.

    b. Find two linearly independent eigenvectors for A.

38. We are told that $\begin{bmatrix} 1 \\ -1 \\ -1 \end{bmatrix}$ is an eigenvector of the matrix $\begin{bmatrix} 4 & 1 & 1 \\ -5 & 0 & -3 \\ -1 & -1 & 2 \end{bmatrix}$; what is the associated eigenvalue?

39. Find a basis of the linear space V of all 2×2 matrices A for which $\begin{bmatrix} 0 \\ 1 \end{bmatrix}$ is an eigenvector, and thus determine the dimension of V.

40. Find a basis of the linear space V of all 2×2 matrices A for which $\begin{bmatrix} 1 \\ -3 \end{bmatrix}$ is an eigenvector, and thus determine the dimension of V.

41. Find a basis of the linear space V of all 2×2 matrices A for which both $\begin{bmatrix} 1 \\ 1 \end{bmatrix}$ and $\begin{bmatrix} 1 \\ 2 \end{bmatrix}$ are eigenvectors, and thus determine the dimension of V.

42. Find a basis of the linear space V of all 3×3 matrices A for which both $\begin{bmatrix} 1 \\ 0 \\ 0 \end{bmatrix}$ and $\begin{bmatrix} 0 \\ 0 \\ 1 \end{bmatrix}$ are eigenvectors, and thus determine the dimension of V.

43. Consider the linear space V of all $n \times n$ matrices for which all the vectors $\vec{e}_1, \ldots, \vec{e}_n$ are eigenvectors. Describe the space V (the matrices in V "have a name"), and determine the dimension of V.

44. For $m \leq n$, find the dimension of the space of all $n \times n$ matrices A for which all the vectors $\vec{e}_1, \ldots, \vec{e}_m$ are eigenvectors.

45. If $\vec{v}$ is any nonzero vector in $\mathbb{R}^2$, what is the dimension of the space V of all 2×2 matrices for which $\vec{v}$ is an eigenvector?

46. In all parts of this problem, let V be the linear space of all 2×2 matrices for which $\begin{bmatrix} 1 \\ 2 \end{bmatrix}$ is an eigenvector.

a. Find a basis of V and thus determine the dimension of V.

b. Consider the linear transformation $T(A) = A \begin{bmatrix} 1 \\ 2 \end{bmatrix}$ from V to $\mathbb{R}^2$. Find a basis of the image of T and a basis of the kernel of T. Determine the rank of T.

c. Consider the linear transformation $L(A) = A \begin{bmatrix} 1 \\ 3 \end{bmatrix}$ from V to $\mathbb{R}^2$. Find a basis of the image of L and a basis of the kernel of L. Determine the rank of L.

47. Consider an $n \times n$ matrix A. A subspace V of $\mathbb{R}^n$ is said to be A-*invariant* if $A\vec{v}$ is in V for all $\vec{v}$ in V. Describe all the one-dimensional A-invariant subspaces of $\mathbb{R}^n$, in terms of the eigenvectors of A.

48. a. Give an example of a 3×3 matrix A with as many nonzero entries as possible such that both $\text{span}(\vec{e}_1)$ and $\text{span}(\vec{e}_1, \vec{e}_2)$ are A-invariant subspaces of $\mathbb{R}^3$ (see Exercise 47).

b. Consider the linear space V of all 3×3 matrices A such that both $\text{span}(\vec{e}_1)$ and $\text{span}(\vec{e}_1, \vec{e}_2)$ are A-invariant subspaces of $\mathbb{R}^3$. Describe the space V (the matrices in V "have a name"), and determine the dimension of V.

49. Consider the coyotes–roadrunner system discussed in this section. Find closed formulas for $c(t)$ and $r(t)$, for the initial populations $c_0 = 100$, $r_0 = 800$.

50. Two interacting populations of hares and foxes can be modeled by the recursive equations

$$h(t + 1) = 4h(t) - 2f(t)$$
$$f(t + 1) = h(t) + f(t).$$

For each of the initial populations given in parts (a) through (c), find closed formulas for $h(t)$ and $f(t)$.

a. $h(0) = f(0) = 100$

b. $h(0) = 200$, $f(0) = 100$

c. $h(0) = 600$, $f(0) = 500$

51. Two interacting populations of coyots and roadrunners can be modeled by the recursive equations

$$c(t + 1) = \qquad\quad 0.75r(t)$$
$$r(t + 1) = -1.5c(t) + 2.25r(t).$$

For each of the initial populations given in parts (a) through (c), find closed formulas for $c(t)$ and $r(t)$.

a. $c(0) = 100$, $r(0) = 200$

b. $c(0) = r(0) = 100$

c. $c(0) = 500$, $r(0) = 700$

52. Imagine that you are diabetic and have to pay close attention to how your body metabolizes glucose. Let $g(t)$ represent the *excess glucose concentration* in your blood, usually measured in milligrams of glucose per 100 milliliters of blood. (*Excess* means that we measure how much the glucose concentration deviates from your fasting level, i.e., the level your system approaches after many hours of fasting.) A negative value of $g(t)$ indicates that the glucose concentration is below fasting level at time t. Shortly after you eat a heavy meal, the function $g(t)$ will reach a peak, and then it will slowly return to 0. Certain hormones help regulate glucose, especially the hormone *insulin*. Let $h(t)$ represent the excess hormone concentration in your blood. Researchers have developed mathematical models for the glucose regulatory system. The following is one such model, in slightly simplified form (these formulas apply between meals; obviously, the system is disturbed during and right after a meal):

$$\begin{vmatrix} g(t + 1) = ag(t) - bh(t) \\ h(t + 1) = cg(t) + dh(t) \end{vmatrix},$$

where time t is measured in minutes; a and d are constants slightly less than 1; and b and c are small positive constants. For your system, the equations might be

$$\begin{vmatrix} g(t + 1) = 0.978g(t) - 0.006h(t) \\ h(t + 1) = 0.004g(t) + 0.992h(t) \end{vmatrix}.$$

The term $-0.006h(t)$ in the first equation is negative, because insulin helps your body absorb glucose. The term $0.004g(t)$ is positive, because glucose in your blood stimulates the cells of the pancreas to secrete insulin. (For a more thorough discussion of this model, read E. Ackerman et al., "Blood glucose regulation and diabetes," Chapter 4 in *Concepts and Models of Biomathematics,* Marcel Dekker, 1969.)

Consider the coefficient matrix

$$A = \begin{bmatrix} 0.978 & -0.006 \\ 0.004 & 0.992 \end{bmatrix}$$

of this dynamical system.

a. We are told that $\begin{bmatrix} -1 \\ 2 \end{bmatrix}$ and $\begin{bmatrix} 3 \\ -1 \end{bmatrix}$ are eigenvectors of A. Find the associated eigenvalues.

b. After you have consumed a heavy meal, the concentrations in your blood are $g_0 = 100$ and $h_0 = 0$. Find closed formulas for $g(t)$ and $h(t)$. Sketch the trajectory. Briefly describe the evolution of this system in practical terms.

c. For the case discussed in part (b), how long does it take for the glucose concentration to fall below fasting level? (This quantity is useful in diagnosing diabetes: a period of more than four hours may indicate mild diabetes.)

53. Three holy men (let's call them Abraham, Benjamin, and Chaim) put little stock in material things; their only earthly possession is a small purse with a bit of gold dust. Each day they get together for the following bizarre bonding ritual: Each of them takes his purse and gives his gold away to the two others, in equal parts. For example, if Abraham has 4 ounces one day, he will give 2 ounces each to Benjamin and Chaim.

a. If Abraham starts out with 6 ounces, Benjamin with 1 ounce, and Chaim with 2 ounces, find formulas for the amounts $a(t)$, $b(t)$, and $c(t)$ each will have after t distributions.

{*Hint*: The vectors $\begin{bmatrix} 1 \\ 1 \\ 1 \end{bmatrix}$, $\begin{bmatrix} 1 \\ -1 \\ 0 \end{bmatrix}$, and $\begin{bmatrix} 1 \\ 0 \\ -1 \end{bmatrix}$ will be useful.}

b. Who will have the most gold after one year, that is, after 365 distributions?

54. Consider the growth of a lilac bush. The state of this lilac bush for several years (at year's end) is shown in the accompanying sketch. Let $n(t)$ be the number of new branches (grown in the year t) and $a(t)$ the number of old branches. In the sketch, the new branches are represented by shorter lines. Each old branch will grow two new branches in the following year. We assume that no branches ever die.

year 0	year 1	year 2	year 3	year 4
$n(0) = 1$	$n(1) = 0$	$n(2) = 2$	$n(3) = 2$	$n(4) = 6$
$a(0) = 0$	$a(1) = 1$	$a(2) = 1$	$a(3) = 3$	$a(4) = 5$

a. Find the matrix A such that
$$\begin{bmatrix} n(t+1) \\ a(t+1) \end{bmatrix} = A \begin{bmatrix} n(t) \\ a(t) \end{bmatrix}.$$

b. Verify that $\begin{bmatrix} 1 \\ 1 \end{bmatrix}$ and $\begin{bmatrix} 2 \\ -1 \end{bmatrix}$ are eigenvectors of A. Find the associated eigenvalues.

c. Find closed formulas for $n(t)$ and $a(t)$.

7.2 Finding the Eigenvalues of a Matrix

In the previous section, we used eigenvalues to analyze a dynamical system

$$\vec{x}(t+1) = A\vec{x}(t).$$

Now we will see how we can actually find those eigenvalues.

Consider an $n \times n$ matrix A and a scalar λ. By Definition 7.1.1, λ is an eigenvalue of A if there exists a nonzero vector $\vec{v}$ in $\mathbb{R}^n$ such that

$$A\vec{v} = \lambda\vec{v}, \quad \text{or,} \quad A\vec{v} - \lambda\vec{v} = \vec{0}, \quad \text{or,} \quad A\vec{v} - \lambda I_n \vec{v} = \vec{0} \quad \text{or,} \quad (A - \lambda I_n)\vec{v} = \vec{0}.$$

This means, by definition of the kernel, that

$$\ker(A - \lambda I_n) \neq \{\vec{0}\}.$$

(That is, there are other vectors in the kernel besides the zero vector.) This is the case if (and only if) the matrix $A - \lambda I_n$ fails to be invertible (by Theorem 3.1.7c), that is, if $\det(A - \lambda I_n) = 0$ (by Theorem 6.2.4).

Theorem 7.2.1 **Eigenvalues and determinants; characteristic equation**

Consider an $n \times n$ matrix A and a scalar λ. Then λ is an eigenvalue[6] of A if (and only if)

$$\det(A - \lambda I_n) = 0.$$

This is called the *characteristic equation* (or the *secular equation*) of matrix A. ∎

---

[6] Alternatively, the eigenvalues are the solutions of the equation $\det(\lambda I_n - A) = 0$. Our formula $\det(A - \lambda I_n) = 0$ is usually more convenient for numerical work.

Let's write the observations we made previously as a string of equivalent statements.

λ is an eigenvalue of A.

$\Updownarrow$

There exists a nonzero vector $\vec{v}$ such that
$$A\vec{v} = \lambda\vec{v} \text{ or } (A - \lambda I_n)\vec{v} = \vec{0}.$$

$\Updownarrow$

$$\ker(A - \lambda I_n) \neq \{\vec{0}\}.$$

$\Updownarrow$

Matrix $A - \lambda I_n$ fails to be invertible.

$\Updownarrow$

$$\det(A - \lambda I_n) = 0.$$

The idea of the characteristic equation is implicit in the work of Jean d'Alembert (1717–1783), in his *Traité de Dynamique* of 1743. Joseph Louis Lagrange (1736–1813) was the first to study the equation systematically (naming it *équation séculaire*), in his works on gravitational attraction between heavenly bodies. Augustin-Louis Cauchy (1789–1857) wrote the equation in its modern form, involving a determinant. It appears that the term *eigenvalue* (*Eigenwert* in German) was first used by David Hilbert in 1904, based perhaps on Helmholtz's notion of an *Eigenton* in acoustics.

EXAMPLE 1   Find the eigenvalues of the matrix
$$A = \begin{bmatrix} 1 & 2 \\ 4 & 3 \end{bmatrix}.$$

Solution

By Theorem 7.2.1, we have to solve the characteristic equation $\det(A - \lambda I_2) = 0$. Now

$$\det(A - \lambda I_2) = \det\left(\begin{bmatrix} 1 & 2 \\ 4 & 3 \end{bmatrix} - \begin{bmatrix} \lambda & 0 \\ 0 & \lambda \end{bmatrix}\right) = \det\begin{bmatrix} 1-\lambda & 2 \\ 4 & 3-\lambda \end{bmatrix}$$
$$= (1-\lambda)(3-\lambda) - 2\cdot 4 = \lambda^2 - 4\lambda - 5 = (\lambda - 5)(\lambda + 1) = 0.$$

The equation $\det(A - \lambda I_2) = (\lambda - 5)(\lambda + 1) = 0$ holds for $\lambda_1 = 5$ and $\lambda_2 = -1$. These two scalars are the eigenvalues of A. In Section 7.3 we will find the corresponding eigenvectors. ∎

EXAMPLE 2   Find the eigenvalues of
$$A = \begin{bmatrix} 2 & 3 & 4 \\ 0 & 3 & 4 \\ 0 & 0 & 4 \end{bmatrix}.$$

Solution

We have to solve the characteristic equation $\det(A - \lambda I_3) = 0$.

$$\det(A - \lambda I_3) = \det\begin{bmatrix} 2-\lambda & 3 & 4 \\ 0 & 3-\lambda & 4 \\ 0 & 0 & 4-\lambda \end{bmatrix} \overset{\text{Step 2}}{=} (2-\lambda)(3-\lambda)(4-\lambda) = 0$$

In Step 2 we use the fact that the determinant of a triangular matrix is the product of its diagonal entries (Theorem 6.1.4). The solutions of the characteristic equation are 2, 3, and 4; these are the eigenvalues of A.   ∎

Theorem 7.2.2   Eigenvalues of a triangular matrix

The eigenvalues of a triangular matrix are its diagonal entries.   ∎

EXAMPLE 3   Find the characteristic equation for an arbitrary 2×2 matrix $A = \begin{bmatrix} a & b \\ c & d \end{bmatrix}$.

Solution

$$\det(A - \lambda I_2) = \det \begin{bmatrix} a - \lambda & b \\ c & d - \lambda \end{bmatrix}$$

$$= (a - \lambda)(d - \lambda) - bc = \lambda^2 - (a + d)\lambda + (ad - bc) = 0$$

This is a quadratic equation. The constant term of $\det(A - \lambda I_2)$ is $ad - bc = \det A$, the value of $\det(A - \lambda I_2)$ at $\lambda = 0$. The coefficient of λ is $-(a + d)$, the opposite of the sum of the diagonal entries a and d of A. Since this sum is important in many other contexts as well, we introduce a name for it.   ∎

Definition 7.2.3   Trace

The sum of the diagonal entries of a square matrix A is called the *trace* of A, denoted by tr A.

Let us highlight the result of Example 3.

Theorem 7.2.4   Characteristic equation of a 2×2 matrix A

$$\det(A - \lambda I_2) = \lambda^2 - (\text{tr } A)\lambda + \det A = 0$$ ∎

For the matrix $A = \begin{bmatrix} 1 & 2 \\ 4 & 3 \end{bmatrix}$, we have tr $A = 1 + 3 = 4$ and det $A = 3 - 8 = -5$, so that the characteristic equation is

$$\lambda^2 - (\text{tr}A)\lambda + \det A = \lambda^2 - 4\lambda - 5 = 0,$$

as we found in Example 1.

If A is a 3×3 matrix, what does the characteristic equation $\det(A - \lambda I_3) = 0$ look like?

$$\det \begin{bmatrix} a_{11} - \lambda & a_{12} & a_{13} \\ a_{21} & a_{22} - \lambda & a_{23} \\ a_{31} & a_{32} & a_{33} - \lambda \end{bmatrix}$$

$$= (a_{11} - \lambda)(a_{22} - \lambda)(a_{33} - \lambda) + (\text{a polynomial of degree} \leq 1)$$

$$= (\lambda^2 - (a_{11} + a_{22})\lambda + a_{11}a_{22})(a_{33} - \lambda) + (\text{a polynomial of degree} \leq 1)$$

$$= -\lambda^3 + (a_{11} + a_{22} + a_{33})\lambda^2 + (\text{a polynomial of degree} \leq 1)$$

$$= -\lambda^3 + (\text{tr } A)\lambda^2 + (\text{a polynomial of degree} \leq 1)$$

$$= 0,$$

a cubic equation. Again, the constant term is det A, so that the characteristic equation has the form

$$\det(A - \lambda I_3) = -\lambda^3 + \text{tr}(A)\lambda^2 - c\lambda + \det(A) = 0, \quad \text{for some scalar } c.$$

It is possible to give a formula for c, in terms of the entries of A, but this formula is complicated, and we will not need it in this introductory text.

Based on the quadratic and the cubic case, we might conjecture that the characteristic equation of any $n \times n$ matrix A is a polynomial equation of degree n, of the form

$$\det(A - \lambda I_n) = (-1)^n \lambda^n + (-1)^{n-1}(\operatorname{tr} A)\lambda^{n-1} + \cdots + \det A = 0.$$

It makes sense to write this equation in terms of $-\lambda$ rather than λ:

$$\det(A - \lambda I_n) = (-\lambda)^n + (\operatorname{tr} A)(-\lambda)^{n-1} + \cdots + \det A = 0.$$

Let us state and then prove that the characteristic equation is indeed of this form.

Theorem 7.2.5   Characteristic polynomial

If A is an $n \times n$ matrix, then $\det(A - \lambda I_n)$ is a polynomial of degree n, of the form

$$(-\lambda)^n + (\operatorname{tr} A)(-\lambda)^{n-1} + \cdots + \det A$$
$$= (-1)^n \lambda^n + (-1)^{n-1}(\operatorname{tr} A)\lambda^{n-1} + \cdots + \det A.$$

This is called the *characteristic polynomial* of A, denoted by $f_A(\lambda)$.

Proof

$$f_A(\lambda) = \det(A - \lambda I_n) = \det \begin{bmatrix} a_{11} - \lambda & a_{12} & \cdots & a_{1n} \\ a_{21} & a_{22} - \lambda & \cdots & a_{2n} \\ \vdots & \vdots & \ddots & \vdots \\ a_{n1} & a_{n2} & \cdots & a_{nn} - \lambda \end{bmatrix}.$$

The product associated with any pattern in the matrix $A - \lambda I_n$ is a polynomial of degree less than or equal to n. This implies that $\det(A - \lambda I_n)$ is a polynomial of degree less than or equal to n.

We can be more precise: The diagonal pattern gives the product

$$(a_{11} - \lambda)(a_{22} - \lambda) \cdots (a_{nn} - \lambda)$$
$$= (-\lambda)^n + (a_{11} + a_{22} + \cdots + a_{nn})(-\lambda)^{n-1} + (\text{a polynomial of degree } \leq n - 2)$$
$$= (-\lambda)^n + (\operatorname{tr} A)(-\lambda)^{n-1} + (\text{a polynomial of degree } \leq n - 2).$$

Any other pattern involves at least two scalars off the diagonal (see Exercise 6.1.51), and its product is therefore a polynomial of degree less than or equal to $n - 2$. This implies that

$$f_A(\lambda) = (-\lambda)^n + (\operatorname{tr} A)(-\lambda)^{n-1} + (\text{a polynomial of degree } \leq n - 2).$$

The constant term is $f_A(0) = \det(A)$. ∎

Note that Theorem 7.2.4 represents a special case of Theorem 7.2.5, when $n = 2$.

What does Theorem 7.2.5 tell us about the number of eigenvalues of an $n \times n$ matrix A? We know from elementary algebra that a polynomial of degree n has at most n zeros. Therefore, an $n \times n$ matrix has at most n eigenvalues. If n is odd, then $f_A(\lambda)$ has at least one zero, by the intermediate value theorem (see Exercise 2.2.47c), since

$$\lim_{\lambda \to \infty} f_A(\lambda) = -\infty \quad \text{and} \quad \lim_{\lambda \to -\infty} f_A(\lambda) = \infty.$$

See Figure 1.

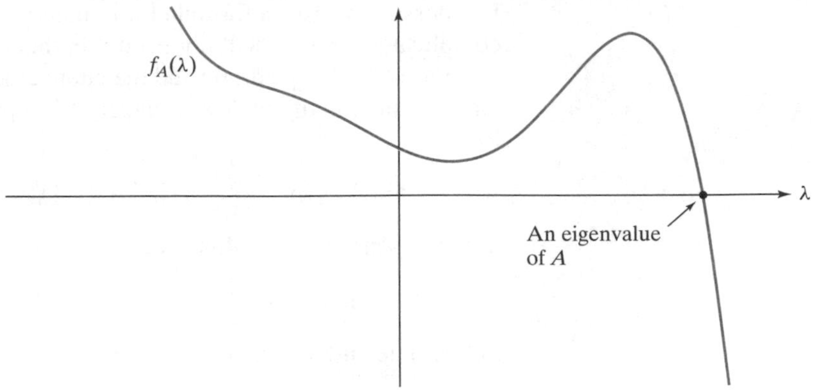

$f_A(\lambda)$

An eigenvalue
of A

λ

Figure 1

EXAMPLE 4   Find all eigenvalues of

$$A = \begin{bmatrix} 1 & 2 & 3 & 4 & 5 \\ 0 & 2 & 3 & 4 & 5 \\ 0 & 0 & 1 & 2 & 3 \\ 0 & 0 & 0 & 2 & 3 \\ 0 & 0 & 0 & 0 & 1 \end{bmatrix}.$$

Solution

The characteristic polynomial is $f_A(\lambda) = (1-\lambda)^3(2-\lambda)^2$, so that the eigenvalues are 1 and 2. Since 1 is a root of multiplicity 3 of the characteristic polynomial, we say that the eigenvalue 1 has algebraic multiplicity 3. Likewise, the eigenvalue 2 has algebraic multiplicity 2. ∎

Definition 7.2.6   Algebraic multiplicity of an eigenvalue

We say that an eigenvalue λ_0 of a square matrix A has *algebraic multiplicity* k if λ_0 is a root of multiplicity k of the characteristic polynomial $f_A(\lambda)$, meaning that we can write

$$f_A(\lambda) = (\lambda_0 - \lambda)^k g(\lambda)$$

for some polynomial $g(\lambda)$ with $g(\lambda_0) \neq 0$.

In Example 4, the algebraic multiplicity of the eigenvalue $\lambda_0 = 1$ is $k = 3$, since

$$f_A(\lambda) = (1 - \lambda)^3 \underbrace{(2 - \lambda)^2}_{g(\lambda)}.$$

k above $(1-\lambda)^3$; λ_0 below $(1-\lambda)$.

EXAMPLE 5   Find the eigenvalues of

$$A = \begin{bmatrix} 1 & 1 & 1 \\ 1 & 1 & 1 \\ 1 & 1 & 1 \end{bmatrix},$$

with their algebraic multiplicities.

Solution

We leave it to the reader to verify that

$$f_A(\lambda) = \lambda^2(3 - \lambda) = (0 - \lambda)^2(3 - \lambda).$$

We have two distinct eigenvalues, 0 and 3, with algebraic multiplicities 2 and 1, respectively. We can write, more succinctly, that the eigenvalues are 0, 0, 3. ∎

Let us summarize.

Theorem 7.2.7   Number of eigenvalues

An $n \times n$ matrix has *at most* n real eigenvalues, even if they are counted with their algebraic multiplicities.

If n is odd, then an $n \times n$ matrix has *at least* one real eigenvalue. ∎

If n is even, an $n \times n$ matrix A need not have any real eigenvalues. Consider

$$A = \begin{bmatrix} 0 & -1 \\ 1 & 0 \end{bmatrix},$$

with $f_A(\lambda) = \det \begin{bmatrix} -\lambda & -1 \\ 1 & -\lambda \end{bmatrix} = \lambda^2 + 1$. See Figure 2.

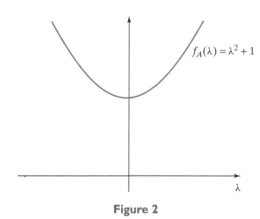

Figure 2

Recall that the transformation $T(\vec{x}) = A\vec{x}$ is a counterclockwise rotation through an angle of 90°. Geometrically, it makes sense that A has no real eigenvalues: Compare with Example 7.1.3.

EXAMPLE 6   Describe all possible cases for the number of real eigenvalues (with their algebraic multiplicities) of a 3×3 matrix A.

Solution

The characteristic polynomial either factors completely,

$$f_A(\lambda) = (\lambda_1 - \lambda)(\lambda_2 - \lambda)(\lambda_3 - \lambda),$$

or it has a quadratic factor without real zeros:

$$f_A(\lambda) = (\lambda_1 - \lambda)p(\lambda), \quad \text{where } p(\lambda) \neq 0 \text{ for all real } \lambda.$$

In the first case, the λ_i could all be distinct, two of them could be equal, or they could all be equal. This leads to the following possibilities:

Case	No. of Distinct Eigenvalues	Algebraic Multiplicities
1	3	1 each
2	2	2 and 1
3	1	3
4	1	1

Examples for each case follow:

Case 1 ■ (see Figure 3)

$$A = \begin{bmatrix} 1 & 0 & 0 \\ 0 & 2 & 0 \\ 0 & 0 & 3 \end{bmatrix}, \quad f_A(\lambda) = (1 - \lambda)(2 - \lambda)(3 - \lambda), \quad \text{Eigenvalues 1, 2, 3}$$

Case 2 ■ (see Figure 4)

$$A = \begin{bmatrix} 1 & 0 & 0 \\ 0 & 1 & 0 \\ 0 & 0 & 2 \end{bmatrix}, \quad f_A(\lambda) = (1 - \lambda)^2(2 - \lambda), \quad \text{Eigenvalues 1, 1, 2}$$

Case 3 ■ (see Figure 5)

$$A = I_3, \quad f_A(\lambda) = (1 - \lambda)^3, \quad \text{Eigenvalues 1, 1, 1}$$

Case 4 ■ (see Figure 6)

$$A = \begin{bmatrix} 1 & 0 & 0 \\ 0 & 0 & -1 \\ 0 & 1 & 0 \end{bmatrix}, \quad f_A(\lambda) = (1 - \lambda)(\lambda^2 + 1), \quad \text{Eigenvalue 1}$$

You can recognize an eigenvalue λ_0 whose algebraic multiplicity exceeds 1 on the graph of $f_A(\lambda)$ by the fact that $f_A(\lambda_0) = f_A'(\lambda_0) = 0$. (The derivative is zero, so that the tangent is horizontal.) The verification of this observation is left as Exercise 37.   ■

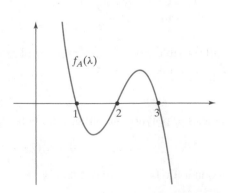

Figure 3

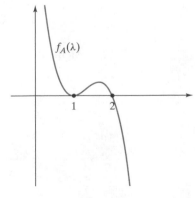

Figure 4

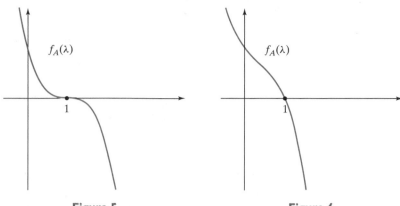

Figure 5   Figure 6

EXAMPLE 7   Suppose A is a 2×2 matrix with eigenvalues λ_1 and λ_2 (we allow $\lambda_1 = \lambda_2$ if λ_1 has algebraic multiplicity 2). Explore the relationship between the sum $\lambda_1 + \lambda_2$, the product $\lambda_1 \lambda_2$, the determinant $\det A$, and the trace $\operatorname{tr} A$. You may want to consider a numerical example first, such as $A = \begin{bmatrix} 1 & 2 \\ 4 & 3 \end{bmatrix}$, with eigenvalues $\lambda_1 = 5$ and $\lambda_2 = -1$ (see Example 1).

Solution

In the case of $A = \begin{bmatrix} 1 & 2 \\ 4 & 3 \end{bmatrix}$, we observe that $\det A = \lambda_1 \lambda_2 = -5$ and $\operatorname{tr} A = \lambda_1 + \lambda_2 = 4$. To see that these results hold in general, write the characteristic polynomial in two ways, as

$$f_A(\lambda) = \lambda^2 - (\operatorname{tr} A)\lambda + \det A$$

and as

$$f_A(\lambda) = (\lambda_1 - \lambda)(\lambda_2 - \lambda) = \lambda^2 - (\lambda_1 + \lambda_2)\lambda + \lambda_1 \lambda_2.$$

Comparing coefficients, we conclude that $\det A = \lambda_1 \lambda_2$ and $\operatorname{tr} A = \lambda_1 + \lambda_2$. ∎

It turns out that the observations we made in Example 7 generalize to $n \times n$ matrices.

Theorem 7.2.8   Eigenvalues, determinant, and trace

If an $n \times n$ matrix A has the eigenvalues $\lambda_1, \lambda_2, \ldots, \lambda_n$, listed with their algebraic multiplicities, then

$$\det A = \lambda_1 \lambda_2 \cdots \lambda_n, \quad \text{the product of the eigenvalues}$$

and

$$\operatorname{tr} A = \lambda_1 + \lambda_2 + \cdots + \lambda_n, \quad \text{the sum of the eigenvalues.} ∎$$

We will prove the claim concerning the determinant and leave the case of the trace as Exercise 21 to the reader.

Since the characteristic polynomial factors completely in this case, we can write

$$f_A(\lambda) = \det(A - \lambda I_n) = (\lambda_1 - \lambda)(\lambda_2 - \lambda) \cdots (\lambda_n - \lambda).$$

Now

$$f_A(0) = \det A = \lambda_1\lambda_2 \cdots \lambda_n,$$

as claimed.

Note that the claims of Theorem 7.2.8 are trivial in the case of a triangular matrix, since the eigenvalues are the diagonal entries in this case.

Finding the Eigenvalues of a Matrix in Practice

To find the eigenvalues of an $n \times n$ matrix A with the method developed in this section, we have to find the zeros of $f_A(\lambda)$, a polynomial of degree n. For $n = 2$, this is a trivial matter: We can either factor the polynomial by inspection or use the quadratic formula (this formula was known over 3500 years ago in Mesopotamia, the present-day Iraq). The problem of finding the zeros of a polynomial of higher degree is nontrivial; it has been of considerable interest throughout the history of mathematics. In the early 1500s, Italian mathematicians found formulas in the cases $n = 3$ and $n = 4$, published in the *Ars Magna* by Gerolamo Cardano.[7] (See Exercise 50 for the case $n = 3$.) During the next 300 years, people tried hard to find a general formula to solve the quintic (a polynomial equation of fifth degree). In 1824, the Norwegian mathematician Niels Henrik Abel (1802–1829) showed that no such general solution is possible, putting an end to the long search. The French mathematician Evariste Galois (1811–1832) was the first to give a numerical example of a quintic that cannot be solved by radicals. (Note the short life spans of these two brilliant mathematicians. Abel died from tuberculosis, and Galois died in a duel.)

When finding the eigenvalues of a matrix by means of the characteristic polynomial, it may be worth trying out a few small integers, such as ± 1 and ± 2. The matrices considered in introductory linear algebra texts often just happen to have such eigenvalues.

In light of the preceding discussion, it is usually impossible to find the exact eigenvalues of a matrix. To find approximations for the eigenvalues, you could graph the characteristic polynomial, using technology. The graph may give you an idea of the number of eigenvalues and their approximate values. Numerical analysts tell us that this is not a very efficient way to go about finding the eigenvalues of large matrices; other techniques are used in practice. See Exercise 7.5.33 for an example; another approach uses the QR factorization (Theorem 5.2.2). There is a lot of ongoing research in this area. A text in numerical linear algebra[8] characterizes the eigenvalue problem as "the third major problem area in matrix computations," after linear equations and least squares, dedicating some 200 pages to this topic.

---

[7]Cardano (1501–1576) was a Renaissance man with a wide range of interests. In his book *Liber de ludo aleae* he presents the first systematic computations of probabilities. Trained as a physician, he gave the first clinical description of typhoid fever. In his book *Somniorum Synesiorum* (Basel, 1562) he explores the meaning of dreams. He was also a leading astrologer of his day whose predictions won him access to some of the most powerful people in sixteenth-century Europe. Still, he is best known today as the most outstanding mathematician of his time and the author of the *Ars Magna*. In 1570, he was arrested on accusation of heresy; he lost his academic position and the right to publish.

To learn more about this fascinating fellow, read the award-winning biography, *Cardano's Cosmos*, by Anthony Grafton (Harvard University Press, 2000), focusing on Cardano's work as an astrologer.

For an English translation of part XI of the *Ars Magna* (dealing with cubic equations) see D. J. Struik (editor), *A Source Book in Mathematics 1200–1800*, Princeton University Press, 1986.

[8]G. H. Golub and C. F. van Loan, *Matrix Computations,* 3rd ed., Johns Hopkins University Press, 1996.

EXERCISES 7.2

GOAL *Use the characteristic polynomial $f_A(\lambda)$*
$= \det(A - \lambda I_n)$ to find the eigenvalues of a matrix A, with
their algebraic multiplicities.

For each of the matrices in Exercises 1 through 13, find all
real eigenvalues, with their algebraic multiplicities. Show
your work. Do not use technology.

1. $\begin{bmatrix} 1 & 2 \\ 0 & 3 \end{bmatrix}$

2. $\begin{bmatrix} 2 & 0 & 0 & 0 \\ 2 & 1 & 0 & 0 \\ 2 & 1 & 2 & 0 \\ 2 & 1 & 2 & 1 \end{bmatrix}$

3. $\begin{bmatrix} 5 & -4 \\ 2 & -1 \end{bmatrix}$

4. $\begin{bmatrix} 0 & 4 \\ -1 & 4 \end{bmatrix}$

5. $\begin{bmatrix} 11 & -15 \\ 6 & -7 \end{bmatrix}$

6. $\begin{bmatrix} 1 & 2 \\ 3 & 4 \end{bmatrix}$

7. I_3

8. $\begin{bmatrix} -1 & -1 & -1 \\ -1 & -1 & -1 \\ -1 & -1 & -1 \end{bmatrix}$

9. $\begin{bmatrix} 3 & -2 & 5 \\ 1 & 0 & 7 \\ 0 & 0 & 2 \end{bmatrix}$

10. $\begin{bmatrix} -3 & 0 & 4 \\ 0 & -1 & 0 \\ -2 & 7 & 3 \end{bmatrix}$

$\lambda^3 + \lambda^2 - \lambda - 1$

$(\lambda+1)(\lambda^2-1)$

11. $\begin{bmatrix} 5 & 1 & -5 \\ 2 & 1 & 0 \\ 8 & 2 & -7 \end{bmatrix}$

12. $\begin{bmatrix} 2 & -2 & 0 & 0 \\ 1 & -1 & 0 & 0 \\ 0 & 0 & 3 & -4 \\ 0 & 0 & 2 & -3 \end{bmatrix}$

13. $\begin{bmatrix} 0 & 1 & 0 \\ 0 & 0 & 1 \\ 1 & 0 & 0 \end{bmatrix}$

14. Consider a 4×4 matrix $A = \begin{bmatrix} B & C \\ 0 & D \end{bmatrix}$, where B, C, and D are 2×2 matrices. What is the relationship between the eigenvalues of A, B, C, and D?

15. Consider the matrix $A = \begin{bmatrix} 1 & k \\ 1 & 1 \end{bmatrix}$, where k is an arbitrary constant. For which values of k does A have two distinct real eigenvalues? When is there no real eigenvalue?

16. Consider the matrix $A = \begin{bmatrix} a & b \\ b & c \end{bmatrix}$, where a, b, and c are nonzero constants. For which values of a, b, and c does A have two distinct eigenvalues?

17. Consider the matrix $A = \begin{bmatrix} a & b \\ b & -a \end{bmatrix}$, where a and b are arbitrary constants. Find all eigenvalues of A. Explain in terms of the geometric interpretation of the linear transformation $T(\vec{x}) = A\vec{x}$.

18. Consider the matrix $A = \begin{bmatrix} a & b \\ b & a \end{bmatrix}$, where a and b are arbitrary constants. Find all eigenvalues of A.

19. *True or false?* If the determinant of a 2×2 matrix A is negative, then A has two distinct real eigenvalues.

20. If a 2×2 matrix A has two distinct eigenvalues λ_1 and λ_2, show that tr $A = \lambda_1 + \lambda_2$.

21. Prove the part of Theorem 7.2.8 that concerns the trace: If an $n \times n$ matrix A has n eigenvalues $\lambda_1, \ldots, \lambda_n$, listed with their algebraic multiplicities, then tr $A = \lambda_1 + \cdots + \lambda_n$.

22. Consider an arbitrary $n \times n$ matrix A. What is the relationship between the characteristic polynomials of A and A^T? What does your answer tell you about the eigenvalues of A and A^T?

23. Suppose matrix A is similar to B. What is the relationship between the characteristic polynomials of A and B? What does your answer tell you about the eigenvalues of A and B?

24. Find all eigenvalues of the matrix

$$A = \begin{bmatrix} 0.5 & 0.25 \\ 0.5 & 0.75 \end{bmatrix}.$$

25. Consider a 2×2 matrix of the form

$$A = \begin{bmatrix} a & b \\ c & d \end{bmatrix},$$

where a, b, c, d are positive numbers such that $a + c = b + d = 1$. (The matrix in Exercise 24 has this form.) Such a matrix is called a *regular transition matrix*. Verify that $\begin{bmatrix} b \\ c \end{bmatrix}$ and $\begin{bmatrix} 1 \\ -1 \end{bmatrix}$ are eigenvectors of A. What are the associated eigenvalues? Is the absolute value of these eigenvalues more or less than 1? Sketch a phase portrait.

26. Based on your answer to Exercise 25, sketch a phase portrait of the dynamical system

$$\vec{x}(t+1) = \begin{bmatrix} 0.5 & 0.25 \\ 0.5 & 0.75 \end{bmatrix} \vec{x}(t).$$

27. a. Based on your answer to Exercise 25, find closed formulas for the components of the dynamical system

$$\vec{x}(t+1) = \begin{bmatrix} 0.5 & 0.25 \\ 0.5 & 0.75 \end{bmatrix} \vec{x}(t),$$

with initial value $\vec{x}_0 = \vec{e}_1$. Then do the same for the initial value $\vec{x}_0 = \vec{e}_2$. Sketch the two trajectories.

b. Consider the matrix

$$A = \begin{bmatrix} 0.5 & 0.25 \\ 0.5 & 0.75 \end{bmatrix}.$$

Using technology, compute some powers of the matrix A, say, A^2, A^5, A^{10}, ... What do you observe? Explain your answer carefully.

c. If $A = \begin{bmatrix} a & b \\ c & d \end{bmatrix}$ is an arbitrary regular transition matrix, what can you say about the powers A^t as t goes to infinity?

28. Consider the isolated Swiss town of Andelfingen, inhabited by 1,200 families. Each family takes a weekly shopping trip to the only grocery store in town, run by Mr. and Mrs. Wipf, until the day when a new, fancier (and cheaper) chain store, Migros, opens its doors. It is not expected that everybody will immediately run to the new store, but we do anticipate that 20% of those shopping at Wipf's each week switch to Migros the following week. Some people who do switch miss the personal service (and the gossip) and switch back: We expect that 10% of those shopping at Migros each week go to Wipf's the following week. The state of this town (as far as grocery shopping is concerned) can be represented by the vector

$$\vec{x}(t) = \begin{bmatrix} w(t) \\ m(t) \end{bmatrix},$$

where $w(t)$ and $m(t)$ are the numbers of families shopping at Wipf's and at Migros, respectively, t weeks after Migros opens. Suppose $w(0) = 1,200$ and $m(0) = 0$.

a. Find a 2×2 matrix A such that $\vec{x}(t+1) = A\vec{x}(t)$. Verify that A is a regular transition matrix. (See Exercise 25.)

b. How many families will shop at each store after t weeks? Give closed formulas.

c. The Wipfs expect that they must close down when they have less than 250 customers a week. When does that happen?

29. Consider an $n \times n$ matrix A such that the sum of the entries in each *row* is 1. Show that the vector

$$\vec{e} = \begin{bmatrix} 1 \\ 1 \\ \vdots \\ 1 \end{bmatrix}$$

in $\mathbb{R}^n$ is an eigenvector of A. What is the corresponding eigenvalue?

30. a. Consider an $n \times n$ matrix A such that the sum of the entries in each row is 1 and that all entries are positive. Consider an eigenvector $\vec{v}$ of A with positive components. Show that the associated eigenvalue is less than or equal to 1. (*Hint*: Consider the largest entry of $\vec{v}$. What can you say about the corresponding entry of $A\vec{v}$?)

b. If we drop the requirement that the components of the eigenvector $\vec{v}$ be positive, is it still true that the associated eigenvalue is less than or equal to 1 in absolute value? Justify your answer.

31. Consider a matrix A with positive entries such that the entries in each column add up to 1. Explain why 1 is an eigenvalue of A. What can you say about the other eigenvalues? Is

$$\vec{e} = \begin{bmatrix} 1 \\ 1 \\ \vdots \\ 1 \end{bmatrix}$$

necessarily an eigenvector? *Hint*: Consider Exercises 22, 29, and 30.

32. Consider the matrix

$$A = \begin{bmatrix} 0 & 1 & 0 \\ 0 & 0 & 1 \\ k & 3 & 0 \end{bmatrix},$$

where k is an arbitrary constant. For which values of k does A have three distinct real eigenvalues? What happens in the other cases? (*Hint*: Graph the function $g(\lambda) = \lambda^3 - 3\lambda$. Find its local maxima and minima.)

33. a. Find the characteristic polynomial of the matrix

$$A = \begin{bmatrix} 0 & 1 & 0 \\ 0 & 0 & 1 \\ a & b & c \end{bmatrix}.$$

b. Can you find a 3×3 matrix M whose characteristic polynomial is

$$-\lambda^3 + 17\lambda^2 - 5\lambda + \pi?$$

34. Suppose a certain 4×4 matrix A has two distinct real eigenvalues. What could the algebraic multiplicities of these eigenvalues be? Give an example for each possible case and sketch the characteristic polynomial.

35. Give an example of a 4×4 matrix A without real eigenvalues.

36. For an arbitrary positive integer n, give a $2n \times 2n$ matrix A without real eigenvalues.

37. Consider an eigenvalue λ_0 of an $n \times n$ matrix A. We are told that the algebraic multiplicity of λ_0 exceeds 1. Show that $f'_A(\lambda_0) = 0$ (i.e., the *derivative* of the characteristic polynomial of A vanishes at λ_0).

38. If A is a 2×2 matrix with tr $A = 5$ and det $A = -14$, what are the eigenvalues of A?

39. If A and B are 2×2 matrices, show that tr$(AB) =$ tr(BA).

40. If A and B are $n \times n$ matrices, show that tr$(AB) =$ tr(BA).

41. If matrix A is similar to B, show that tr $B = $ tr A. (*Hint:* Exercise 40 is helpful.)

42. Consider two $n \times n$ matrices A and B such that $BA = 0$. Show that tr$\big((A + B)^2\big) = $ tr$(A^2) + $ tr(B^2). (*Hint:* Exercise 40 is helpful.)

43. Do there exist $n \times n$ matrices A and B such that $AB - BA = I_n$? Explain. (*Hint:* Exercise 40 is helpful.)

44. Do there exist invertible $n \times n$ matrices A and B such that $AB - BA = A$? Explain.

45. For which value of the constant k does the matrix $A = \begin{bmatrix} -1 & k \\ 4 & 3 \end{bmatrix}$ have 5 as an eigenvalue?

46. In all the parts of this problem, consider a matrix $A = \begin{bmatrix} a & b \\ c & d \end{bmatrix}$ with the eigenvalues λ_1 and λ_2.

  a. Show that $\lambda_1^2 + \lambda_2^2 = a^2 + d^2 + 2bc$.
  b. Show that $\lambda_1^2 + \lambda_2^2 \le a^2 + b^2 + c^2 + d^2$.
  c. For which matrices A does the equality $\lambda_1^2 + \lambda_2^2 = a^2 + b^2 + c^2 + d^2$ hold?

47. For which 2×2 matrices A does there exist a nonzero matrix M such that $AM = MD$, where $D = \begin{bmatrix} 2 & 0 \\ 0 & 3 \end{bmatrix}$? Give your answer in terms of the eigenvalues of A.

48. For which 2×2 matrices A does there exist an invertible matrix S such that $AS = SD$, where $D = \begin{bmatrix} 2 & 0 \\ 0 & 3 \end{bmatrix}$? Give your answer in terms of the eigenvalues of A.

49. For which 3×3 matrices A does there exist a nonzero matrix M such that $AM = MD$, where $D = \begin{bmatrix} 2 & 0 & 0 \\ 0 & 3 & 0 \\ 0 & 0 & 4 \end{bmatrix}$? Give your answer in terms of the eigenvalues of A.

50. In his groundbreaking text *Ars Magna* (Nuremberg, 1545), the Italian mathematician Gerolamo Cardano explains how to solve cubic equations. In Chapter XI, he considers the following example:

$$x^3 + 6x = 20.$$

a. Explain why this equation has exactly one (real) solution. Here, this solution is easy to find by inspection. The point of the exercise is to show a systematic way to find it.

b. Cardano explains his method as follows (we are using modern notation for the variables): "I take two cubes v^3 and u^3 whose difference shall be 20, so that the product vu shall be 2, that is, a third of the coefficient of the unknown x. Then, I say that $v - u$ is the value of the unknown x." Show that if v and u are chosen as stated by Cardano, then $x = v - u$ is indeed the solution of the equation $x^3 + 6x = 20$.

c. Solve the system

$$\left| \begin{array}{l} v^3 - u^3 = 20 \\ vu = 2 \end{array} \right|$$

to find u and v.

d. Consider the equation

$$x^3 + px = q,$$

where p is positive. Using your work in parts (a), (b), and (c) as a guide, show that the unique solution of this equation is

$$x = \sqrt[3]{\frac{q}{2} + \sqrt{\left(\frac{q}{2}\right)^2 + \left(\frac{p}{3}\right)^3}} - \sqrt[3]{-\frac{q}{2} + \sqrt{\left(\frac{q}{2}\right)^2 + \left(\frac{p}{3}\right)^3}}.$$

This solution can also be written as

$$x = \sqrt[3]{\frac{q}{2} + \sqrt{\left(\frac{q}{2}\right)^2 + \left(\frac{p}{3}\right)^3}} + \sqrt[3]{\frac{q}{2} - \sqrt{\left(\frac{q}{2}\right)^2 + \left(\frac{p}{3}\right)^3}}.$$

What can go wrong when p is negative?

e. Consider an arbitrary cubic equation

$$x^3 + ax^2 + bx + c = 0.$$

Show that the substitution $x = t - (a/3)$ allows you to write this equation as

$$t^3 + pt = q.$$

7.3 Finding the Eigenvectors of a Matrix

Having found an eigenvalue λ of an $n \times n$ matrix A, we will now turn our attention to the corresponding eigenvectors. We have to find the vectors $\vec{v}$ in $\mathbb{R}^n$ such that

$$A\vec{v} = \lambda\vec{v}, \quad \text{or} \quad (A - \lambda I_n)\vec{v} = \vec{0}.$$

In other words, we have to find the kernel of the matrix $A - \lambda I_n$. In this context, the following definition is useful.

Definition 7.3.1 Eigenspaces

Consider an eigenvalue λ of an $n \times n$ matrix A. Then the kernel of the matrix $A - \lambda I_n$ is called the *eigenspace* associated with λ, denoted by E_λ:

$$E_\lambda = \ker(A - \lambda I_n) = \{\vec{v} \text{ in } \mathbb{R}^n : A\vec{v} = \lambda\vec{v}\}.$$

Note that the eigenvectors with eigenvalue λ are the *nonzero* vectors in the eigenspace E_λ.

EXAMPLE 1 Let $T(\vec{x}) = A\vec{x}$ be the orthogonal projection onto a plane V in $\mathbb{R}^3$. Describe the eigenspaces E_1 and E_0 geometrically.

Solution

Eigenspace E_1 consists of the solutions of the equation $A\vec{v} = 1\vec{v} = \vec{v}$; those are the vectors on plane V. Thus $E_1 = V$.

Eigenspace $E_0 = \ker A$ consists of the solutions of the equation $A\vec{v} = 0\vec{v} = \vec{0}$; those are the vectors on the line $V^\perp$ perpendicular to plane V. See Figure 1. ■

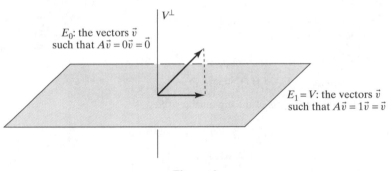

E_0: the vectors $\vec{v}$ such that $A\vec{v} = 0\vec{v} = \vec{0}$

$V^\perp$

$E_1 = V$: the vectors $\vec{v}$ such that $A\vec{v} = 1\vec{v} = \vec{v}$

Figure 1

To find the eigenvectors associated with a known eigenvalue λ algebraically, we seek a basis of the eigenspace $E_\lambda = \ker(A - \lambda I_n)$, a problem we can handle. (See Section 3.3.)

EXAMPLE 2 Find the eigenvectors of the matrix $A = \begin{bmatrix} 1 & 2 \\ 4 & 3 \end{bmatrix}$.

Solution

In Example 1 of Section 7.2 we saw that the eigenvalues are 5 and -1. Now

$$E_5 = \ker(A - 5I_2) = \ker \begin{bmatrix} -4 & 2 \\ 4 & -2 \end{bmatrix}.$$

Finding the kernel amounts to finding the relations between the columns. In the case of a 2×2 matrix, this can be done by inspection. Consider the Kyle numbers,

$$\begin{matrix} 1 & 2 \\ \begin{bmatrix} -4 & 2 \\ 4 & -2 \end{bmatrix} \end{matrix}$$

so that

$$E_5 = \operatorname{span} \begin{bmatrix} 1 \\ 2 \end{bmatrix}.$$

Similarly,

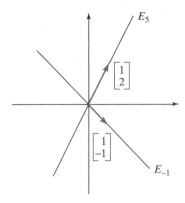

Figure 2

$$E_{-1} = \ker(A + I_2) = \ker \begin{bmatrix} \overset{1}{2} & \overset{-1}{2} \\ 4 & 4 \end{bmatrix} = \text{span} \begin{bmatrix} 1 \\ -1 \end{bmatrix}.$$

We can (and should) check that the vectors we found are indeed eigenvectors of A, with the eigenvalues we claim:

$$\begin{bmatrix} 1 & 2 \\ 4 & 3 \end{bmatrix} \begin{bmatrix} 1 \\ 2 \end{bmatrix} = \begin{bmatrix} 5 \\ 10 \end{bmatrix} = 5 \begin{bmatrix} 1 \\ 2 \end{bmatrix}$$

and

$$\begin{bmatrix} 1 & 2 \\ 4 & 3 \end{bmatrix} \begin{bmatrix} 1 \\ -1 \end{bmatrix} = \begin{bmatrix} -1 \\ 1 \end{bmatrix} = (-1) \begin{bmatrix} 1 \\ -1 \end{bmatrix}. \quad \checkmark$$

Both eigenspaces are lines, as shown in Figure 2.   ∎

Geometrically, the matrix A represents a scaling by a factor of 5 along the line spanned by vector $\begin{bmatrix} 1 \\ 2 \end{bmatrix}$, while the line spanned by $\begin{bmatrix} 1 \\ -1 \end{bmatrix}$ is flipped over the origin.

EXAMPLE 3 Find the eigenvectors of

$$A = \begin{bmatrix} 1 & 1 & 1 \\ 0 & 0 & 1 \\ 0 & 0 & 1 \end{bmatrix}.$$

Solution

The eigenvalues are 1 and 0, the diagonal entries of the upper triangular matrix A, with algebraic multiplicities 2 and 1, respectively. Now

$$E_1 = \ker(A - I_2) = \ker \begin{bmatrix} \overset{1}{0} & \overset{0}{1} & \overset{0}{1} \\ 0 & -1 & 1 \\ 0 & 0 & 0 \end{bmatrix} = \text{span} \begin{bmatrix} 1 \\ 0 \\ 0 \end{bmatrix}$$

and

$$E_0 = \ker A = \ker \begin{bmatrix} \overset{-1}{1} & \overset{1}{1} & \overset{0}{1} \\ 0 & 0 & 1 \\ 0 & 0 & 1 \end{bmatrix} = \text{span} \begin{bmatrix} -1 \\ 1 \\ 0 \end{bmatrix}.$$

Both eigenspaces are lines in the x_1-x_2 plane, as shown in Figure 3.   ∎

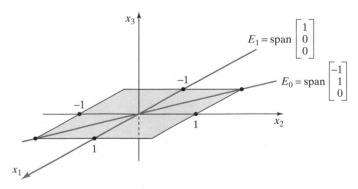

Figure 3

Note that Example 3 is qualitatively different from Example 1, where we studied the orthogonal projection onto a plane in $\mathbb{R}^3$. There, too, we had two eigenvalues, 1 and 0, but one of the eigenspaces, E_1, was a plane, while both eigenspaces in Example 3 turn out to be lines.

To discuss these different cases, it is useful to introduce the following terminology.

Definition 7.3.2   Geometric multiplicity

Consider an eigenvalue λ of an $n \times n$ matrix A. The dimension of eigenspace $E_\lambda = \ker(A - \lambda I_n)$ is called the *geometric multiplicity* of eigenvalue λ. Thus, the geometric multiplicity is the nullity of matrix $A - \lambda I_n$, or $n - \operatorname{rank}(A - \lambda I_n)$.

Example 3 shows that the geometric multiplicity of an eigenvalue may be different from the algebraic multiplicity (but see Theorem 7.3.7 at the end of this section). We have

$$(\text{algebraic multiplicity of } 1) = 2$$

but

$$(\text{geometric multiplicity of } 1) = \dim(E_1) = 1.$$

When we analyze a dynamical system

$$\vec{x}(t + 1) = A\vec{x}(t),$$

where A is an $n \times n$ matrix, we are often interested in finding a basis of $\mathbb{R}^n$ that consists of eigenvectors of A. (Recall the introductory example of Section 7.1, and Theorem 7.1.3.) Such a basis deserves a name.

Definition 7.3.3   Eigenbasis

Consider an $n \times n$ matrix A. A basis of $\mathbb{R}^n$ consisting of eigenvectors of A is called an *eigenbasis* for A.

Recall Examples 1 through 3.

Example 1 Revisited ■ *Orthogonal projection onto a plane V in $\mathbb{R}^3$.* Pick a basis $\vec{v}_1, \vec{v}_2$ of $V = E_1$ and a nonzero $\vec{v}_3$ in $V^\perp = E_0$. Then the vectors $\vec{v}_1, \vec{v}_2, \vec{v}_3$ form an eigenbasis for the projection. See Figure 4. ■

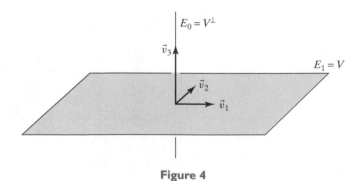

Figure 4

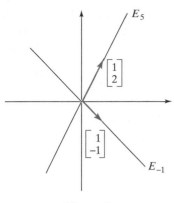

Figure 5

Example 2 Revisited ■ $A = \begin{bmatrix} 1 & 2 \\ 4 & 3 \end{bmatrix}$.

The vectors $\begin{bmatrix} 1 \\ 2 \end{bmatrix}$ in E_5 and $\begin{bmatrix} 1 \\ -1 \end{bmatrix}$ in E_{-1} form an eigenbasis for A. See Figure 5. ■

Example 3 Revisited ■ $A = \begin{bmatrix} 1 & 1 & 1 \\ 0 & 0 & 1 \\ 0 & 0 & 1 \end{bmatrix}$.

Here we can find only two linearly independent eigenvectors, namely, one in each of the two one-dimensional eigenspaces, E_1 and E_0. There is no eigenbasis for A. See Figure 6. ■

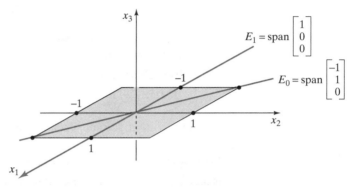

Figure 6

When analyzing dynamical systems, as well as in many other applications of eigenvectors, we need to think about the following two key questions.

---

a. For which square matrices does there exist an eigenbasis?

b. If eigenbases exist, how can we find one?

Consider an $n \times n$ matrix A. If the sum s of the geometric multiplicities of the eigenvalues is less than n (as in Example 3, where $s = 2$ and $n = 3$), then there are not enough linearly independent eigenvectors to form an eigenbasis. In fact, we can find no more than s linearly independent eigenvectors.

Conversely, suppose that the geometric multiplicities of the eigenvalues do add up to n, as in Examples 1 and 2. Can we construct an eigenbasis for A by finding a basis of each eigenspace and *concatenating*[9] these bases? This method does work in Examples 1 and 2. Next we will state and prove that this approach works in general.

Theorem 7.3.4   Eigenbases and geometric multiplicities

a. Consider an $n \times n$ matrix A. If we find a basis of each eigenspace of A and concatenate all these bases, then the resulting eigenvectors $\vec{v}_1, \ldots, \vec{v}_s$ will be linearly independent. (Note that s is the sum of the geometric multiplicities of the eigenvalues of A.)

---

[9]The *concatenation* of two lists $(a_1, a_2, \ldots, a_p)$ and $(b_1, b_2, \ldots, b_q)$ is the list $(a_1, a_2, \ldots, a_p,$ $b_1, b_2, \ldots, b_q)$.

b. There exists an eigenbasis for an $n \times n$ matrix A if (and only if) the geometric multiplicities of the eigenvalues add up to n (meaning that $s = n$ in part a).

Proof   **a.** We will argue indirectly, assuming that the eigenvectors $\vec{v}_1, \ldots, \vec{v}_s$ are linearly dependent. Let $\vec{v}_m$ be the *first* redundant vector in this list, with $\vec{v}_m = c_1 \vec{v}_1 + \cdots + c_{m-1} \vec{v}_{m-1}$. Suppose that $A\vec{v}_i = \lambda_i \vec{v}_i$. There must be at least one *nonzero* coefficient c_k such that $\lambda_k \neq \lambda_m$, since $\vec{v}_m$ cannot be expressed as a linear combination of vectors $\vec{v}_i$ that are all in the same eigenspace E_{λ_m}. Premultiplying the equation $\vec{v}_m = c_1 \vec{v}_1 + \cdots + c_k \vec{v}_k + \cdots + c_{m-1} \vec{v}_{m-1}$ by $A - \lambda_m I_n$, and realizing that $(A - \lambda_m I_n)\vec{v}_i = (\lambda_i - \lambda_m)\vec{v}_i$, we find that

$$(\lambda_m - \lambda_m)\vec{v}_m = \vec{0}$$
$$= (\lambda_1 - \lambda_m)c_1 \vec{v}_1 + \cdots + \underbrace{(\lambda_k - \lambda_m)c_k}_{\neq 0} \vec{v}_k + \cdots + (\lambda_{m-1} - \lambda_m)c_{m-1}\vec{v}_{m-1}.$$

This is a nontrivial relation among vectors $\vec{v}_1, \ldots, \vec{v}_{m-1}$, contradicting our assumption that $\vec{v}_m$ is the *first* redundant vector in the list.

b. This claim follows directly from (a). There exists an eigenbasis if (and only if) $s = n$ in part (a). ∎

Here is an important special case of Theorem 7.3.4.

Theorem 7.3.5   **An $n \times n$ matrix with n distinct eigenvalues**

If an $n \times n$ matrix A has n distinct eigenvalues, then there exists an eigenbasis for A. We can construct an eigenbasis by finding an eigenvector for each eigenvalue. ∎

EXAMPLE 4   Does there exist an eigenbasis for the following matrix A?

$$A = \begin{bmatrix} 1 & 2 & 3 & 4 & 5 & 6 \\ 0 & 2 & 3 & 4 & 5 & 6 \\ 0 & 0 & 3 & 4 & 5 & 6 \\ 0 & 0 & 0 & 4 & 5 & 6 \\ 0 & 0 & 0 & 0 & 5 & 6 \\ 0 & 0 & 0 & 0 & 0 & 6 \end{bmatrix}$$

Solution

Yes, since the 6×6 matrix A has 6 distinct eigenvalues, namely, the diagonal entries 1, 2, 3, 4, 5, 6. ∎

Let us work another example of a dynamical system.

EXAMPLE 5   Consider an Anatolian mountain farmer who raises goats. This particular breed of goats has a maximum life span of three years. At the end of each year t, the farmer conducts a census of his goats. He counts the number of young goats $j(t)$, born in the year t; the middle-aged goats $m(t)$, born the year before; and the old ones $a(t)$, born in the year $t - 2$. The state of the herd can be described by the vector

$$\vec{x}(t) = \begin{bmatrix} j(t) \\ m(t) \\ a(t) \end{bmatrix}.$$

Suppose that for this breed and this environment the evolution of the system can be modeled by the equation

$$\vec{x}(t+1) = A\vec{x}(t), \quad \text{where} \quad A = \begin{bmatrix} 0 & 0.95 & 0.6 \\ 0.8 & 0 & 0 \\ 0 & 0.5 & 0 \end{bmatrix}.$$

For example, $m(t+1) = 0.8j(t)$, meaning that 80% of the young goats will survive to the next census. We leave it as an exercise to the reader to interpret the other 3 nonzero entries of A as reproduction and survival rates.

Suppose the initial populations are $j_0 = 750$ and $m_0 = a_0 = 200$. What will the populations be after t years, according to this model? What will happen in the long term?

We are told that the eigenvalues of A are $\lambda_1 = 1$, $\lambda_2 = -0.6$, and $\lambda_3 = -0.4$.

Solution

To answer these questions, we have to find an eigenbasis for A. (See Theorem 7.1.3.) By Theorem 7.3.5, we know that such an eigenbasis does exist, since the 3×3 matrix A has three distinct eigenvalues. Let's find an eigenvector for each of the eigenvalues. With a little experimentation, we can find the Kyle numbers in each case (or, alternatively, use rref):

$$\begin{matrix} 5 & 4 & 2 \end{matrix}$$
$$E_1 = \ker \begin{bmatrix} -1 & 0.95 & 0.6 \\ 0.8 & -1 & 0 \\ 0 & 0.5 & -1 \end{bmatrix} = \text{span} \begin{bmatrix} 5 \\ 4 \\ 2 \end{bmatrix},$$

$$E_{-0.6} = \text{span} \begin{bmatrix} 9 \\ -12 \\ 10 \end{bmatrix}, \quad E_{-0.4} = \text{span} \begin{bmatrix} -2 \\ 4 \\ -5 \end{bmatrix}.$$

We have constructed an eigenbasis: $\vec{v}_1 = \begin{bmatrix} 5 \\ 4 \\ 2 \end{bmatrix}, \vec{v}_2 = \begin{bmatrix} 9 \\ -12 \\ 10 \end{bmatrix}, \vec{v}_3 = \begin{bmatrix} -2 \\ 4 \\ -5 \end{bmatrix}.$

Next, we need to express the initial state vector

$$\vec{x}_0 = \begin{bmatrix} j_0 \\ m_0 \\ a_0 \end{bmatrix} = \begin{bmatrix} 750 \\ 200 \\ 200 \end{bmatrix}$$

as a linear combination of the eigenvectors: $\vec{x}_0 = c_1\vec{v}_1 + c_2\vec{v}_2 + c_3\vec{v}_3$. A somewhat tedious computation reveals that $c_1 = 100$, $c_2 = 50$, $c_3 = 100$.

Now that we know the eigenvalues λ_i, the eigenvectors $\vec{v}_i$, and the coefficients c_i, we are ready to write down the solution:

$$\vec{x}(t) = A^t\vec{x}_0 = c_1 A^t\vec{v}_1 + c_2 A^t\vec{v}_2 + c_3 A^t\vec{v}_3 = c_1\lambda_1^t\vec{v}_1 + c_2\lambda_2^t\vec{v}_2 + c_3\lambda_3^t\vec{v}_3$$

$$= 100 \begin{bmatrix} 5 \\ 4 \\ 2 \end{bmatrix} + 50(-0.6)^t \begin{bmatrix} 9 \\ -12 \\ 10 \end{bmatrix} + 100(-0.4)^t \begin{bmatrix} -2 \\ 4 \\ -5 \end{bmatrix}.$$

The individual populations are

$$j(t) = 500 + 450(-0.6)^t - 200(-0.4)^t,$$
$$m(t) = 400 - 600(-0.6)^t + 400(-0.4)^t,$$
$$a(t) = 200 + 500(-0.6)^t - 500(-0.4)^t.$$

In the long run, the populations approach the equilibrium values

$$j = 500, \quad m = 400, \quad a = 200.$$ ■

Eigenvalues and Similarity

If matrix A is similar to B, what is the relationship between the eigenvalues of A and B? The following theorem shows that this relationship is very close indeed.

Theorem 7.3.6   **The eigenvalues of similar matrices**

Suppose matrix A is similar to B. Then

a. Matrices A and B have the same characteristic polynomial, that is, $f_A(\lambda) = f_B(\lambda)$.

b. $\text{rank}(A) = \text{rank}(B)$ and $\text{nullity}(A) = \text{nullity}(B)$.

c. Matrices A and B have the same eigenvalues, with the same algebraic and geometric multiplicities. (However, the eigenvectors need not be the same.)

d. Matrices A and B have the same determinant and the same trace: $\det A = \det B$ and $\text{tr } A = \text{tr } B$.

Proof

a. If $B = S^{-1}AS$, then $f_B(\lambda) = \det(B - \lambda I_n) = \det(S^{-1}AS - \lambda I_n) = \det(S^{-1}(A - \lambda I_n)S) = (\det S)^{-1}\det(A - \lambda I_n)\det(S) = (\det S)^{-1}(\det S)\det(A - \lambda I_n) = \det(A - \lambda I_n) = f_A(\lambda)$ for all scalars λ.

b. See Exercises 71 and 72 of Section 3.4. An alternative proof is suggested in Exercise 34 of this section.

c. It follows from part (a) that matrices A and B have the same eigenvalues, with the same algebraic multiplicities. (See Theorem 7.2.1 and Definition 7.2.6.) As for the geometric multiplicity, note that $A - \lambda I_n$ is similar to $B - \lambda I_n$ for all λ (see Exercise 33), so that $\text{nullity}(A - \lambda I_n) = \text{nullity}(B - \lambda I_n)$ for all eigenvalues λ, by part (b). (See Definition 7.3.2.)

d. These equations follow from part (a) and Theorem 7.2.5: Trace and determinant are coefficients of the characteristic polynomial, up to signs.   ■

EXAMPLE 6   Is the matrix $A = \begin{bmatrix} 2 & 3 \\ 5 & 7 \end{bmatrix}$ similar to $B = \begin{bmatrix} 3 & 2 \\ 8 & 5 \end{bmatrix}$?

Solution

No, since $\text{tr } A = 9$ and $\text{tr } B = 8$. See Theorem 7.3.6d.   ■

Earlier in this section we observed that the algebraic and the geometric multiplicity of an eigenvalue are not necessarily the same. However, the following inequality always holds.

Theorem 7.3.7   **Algebraic versus geometric multiplicity**

If λ is an eigenvalue of a square matrix A, then

(geometric multiplicity of λ) $\leq$ (algebraic multiplicity of λ).

Proof   Suppose λ_0 is an eigenvalue of an $n \times n$ matrix A, with geometric multiplicity m, meaning that the dimension of eigenspace E_{λ_0} is m. Let $\vec{v}_1, \ldots, \vec{v}_m$ be a basis of E_{λ_0}, and consider an invertible $n \times n$ matrix S whose first m columns are $\vec{v}_1, \ldots, \vec{v}_m$. (How would you find such an S?) Let $B = S^{-1}AS$, a matrix similar to A. Now compute

$B\vec{e}_i$, for $i = 1, \ldots, m$, keeping in mind that $S\vec{e}_i = \vec{v}_i$, and therefore $S^{-1}\vec{v}_i = \vec{e}_i$:

$$(i\text{th column of } B) = B\vec{e}_i = S^{-1}AS\vec{e}_i = S^{-1}A\vec{v}_i = S^{-1}(\lambda_0\vec{v}_i) = \lambda_0(S^{-1}\vec{v}_i) = \lambda_0\vec{e}_i.$$

This computation shows that the first m columns of B look like those of $\lambda_0 I_n$.

$$B = \begin{bmatrix} \lambda_0 & 0 & \cdots & 0 & * & \cdots & * \\ 0 & \lambda_0 & \cdots & 0 & * & \cdots & * \\ \vdots & \vdots & \ddots & \vdots & \vdots & \vdots & \vdots \\ 0 & 0 & \cdots & \lambda_0 & * & \cdots & * \\ 0 & 0 & \cdots & 0 & * & \cdots & * \\ \vdots & \vdots & \vdots & \vdots & \vdots & \ddots & \vdots \\ 0 & 0 & \cdots & 0 & * & \cdots & * \end{bmatrix} = \begin{bmatrix} \lambda_0 I_m & P \\ 0 & Q \end{bmatrix}$$

Since B is similar to A, we have

$$f_A(\lambda) \overset{\text{Step 1}}{=} f_B(\lambda) = \det(B - \lambda I_n) \overset{\text{Step 3}}{=} (\lambda_0 - \lambda)^m f_Q(\lambda),$$

showing that the algebraic multiplicity of eigenvalue λ_0 is at least m, as claimed. In Step 1 we use Theorem 7.3.6a, and in Step 3 we use Theorem 6.1.5. ∎

EXERCISES 7.3

GOAL *For a given eigenvalue, find a basis of the associated eigenspace. Use the geometric multiplicities of the eigenvalues to determine whether there is an eigenbasis for a matrix.*

For each of the matrices in Exercises 1 through 18, find all (real) eigenvalues. Then find a basis of each eigenspace, and find an eigenbasis, if you can. Do not use technology.

1. $\begin{bmatrix} 7 & 8 \\ 0 & 9 \end{bmatrix}$ **2.** $\begin{bmatrix} 1 & 1 \\ 1 & 1 \end{bmatrix}$

3. $\begin{bmatrix} 6 & 3 \\ 2 & 7 \end{bmatrix}$ **4.** $\begin{bmatrix} 0 & -1 \\ 1 & 2 \end{bmatrix}$

5. $\begin{bmatrix} 4 & 5 \\ -2 & -2 \end{bmatrix}$ **6.** $\begin{bmatrix} 2 & 3 \\ 4 & 5 \end{bmatrix}$

7. $\begin{bmatrix} 1 & 0 & 0 \\ 0 & 2 & 0 \\ 0 & 0 & 3 \end{bmatrix}$ **8.** $\begin{bmatrix} 1 & 1 & 0 \\ 0 & 2 & 2 \\ 0 & 0 & 3 \end{bmatrix}$

9. $\begin{bmatrix} 1 & 0 & 1 \\ 0 & 1 & 0 \\ 0 & 0 & 0 \end{bmatrix}$ **10.** $\begin{bmatrix} 1 & 1 & 0 \\ 0 & 1 & 0 \\ 0 & 0 & 0 \end{bmatrix}$

11. $\begin{bmatrix} 1 & 1 & 1 \\ 1 & 1 & 1 \\ 1 & 1 & 1 \end{bmatrix}$ **12.** $\begin{bmatrix} 1 & 1 & 0 \\ 0 & 1 & 1 \\ 0 & 0 & 1 \end{bmatrix}$

13. $\begin{bmatrix} 3 & 0 & -2 \\ -7 & 0 & 4 \\ 4 & 0 & -3 \end{bmatrix}$ **14.** $\begin{bmatrix} 1 & 0 & 0 \\ -5 & 0 & 2 \\ 0 & 0 & 1 \end{bmatrix}$

15. $\begin{bmatrix} -1 & 0 & 1 \\ -3 & 0 & 1 \\ -4 & 0 & 3 \end{bmatrix}$ **16.** $\begin{bmatrix} 1 & 1 & 0 \\ 0 & -1 & -1 \\ 2 & 2 & 0 \end{bmatrix}$

17. $\begin{bmatrix} 0 & 0 & 0 & 0 \\ 0 & 1 & 1 & 0 \\ 0 & 0 & 0 & 0 \\ 0 & 0 & 0 & 1 \end{bmatrix}$ **18.** $\begin{bmatrix} 0 & 0 & 0 & 0 \\ 0 & 1 & 0 & 1 \\ 0 & 0 & 0 & 0 \\ 0 & 0 & 0 & 1 \end{bmatrix}$

19. Consider the matrix

$$A = \begin{bmatrix} 1 & a & b \\ 0 & 1 & c \\ 0 & 0 & 1 \end{bmatrix},$$

where a, b, c are arbitrary constants. How does the geometric multiplicity of the eigenvalue 1 depend on the constants a, b, c? When is there an eigenbasis for A?

20. Consider the matrix

$$A = \begin{bmatrix} 1 & a & b \\ 0 & 1 & c \\ 0 & 0 & 2 \end{bmatrix},$$

where a, b, c are arbitrary constants. How do the geometric multiplicities of the eigenvalues 1 and 2 depend on the constants a, b, c? When is there an eigenbasis for A?

21. Find a 2×2 matrix A for which

$$E_1 = \text{span}\begin{bmatrix} 1 \\ 2 \end{bmatrix} \quad \text{and} \quad E_2 = \text{span}\begin{bmatrix} 2 \\ 3 \end{bmatrix}.$$

How many such matrices are there?

22. Find all 2×2 matrices A for which

$$E_7 = \mathbb{R}^2.$$

23. Find all eigenvalues and eigenvectors of $A = \begin{bmatrix} 1 & 1 \\ 0 & 1 \end{bmatrix}$.

Is there an eigenbasis? Interpret your result geometrically.

24. Find a 2×2 matrix A for which

$$E_1 = \operatorname{span} \begin{bmatrix} 2 \\ 1 \end{bmatrix}$$

is the only eigenspace.

25. What can you say about the geometric multiplicity of the eigenvalues of a matrix of the form

$$A = \begin{bmatrix} 0 & 1 & 0 \\ 0 & 0 & 1 \\ a & b & c \end{bmatrix},$$

where a, b, c are arbitrary constants?

26. Show that if a 6×6 matrix A has a negative determinant, then A has at least one positive eigenvalue. (*Hint:* Sketch the graph of the characteristic polynomial.)

27. Consider a 2×2 matrix A. Suppose that tr $A = 5$ and det $A = 6$. Find the eigenvalues of A.

28. Consider the matrix

$$J_n(k) = \begin{bmatrix} k & 1 & 0 & \cdots & 0 & 0 \\ 0 & k & 1 & \cdots & 0 & 0 \\ 0 & 0 & k & \cdots & 0 & 0 \\ \vdots & \vdots & \vdots & \ddots & \vdots & \vdots \\ 0 & 0 & 0 & \cdots & k & 1 \\ 0 & 0 & 0 & \cdots & 0 & k \end{bmatrix}$$

(with all k's on the diagonal and 1's directly above), where k is an arbitrary constant. Find the eigenvalue(s) of $J_n(k)$, and determine their algebraic and geometric multiplicities.

29. Consider a diagonal $n \times n$ matrix A with rank$(A) = r < n$. Find the algebraic and the geometric multiplicity of the eigenvalue 0 of A in terms of r and n.

30. Consider an upper triangular $n \times n$ matrix A with $a_{ii} \neq 0$ for $i = 1, 2, \ldots, m$ and $a_{ii} = 0$ for $i = m + 1, \ldots, n$. Find the algebraic multiplicity of the eigenvalue 0 of A. Without using Theorem 7.3.7, what can you say about the geometric multiplicity?

31. Suppose there is an eigenbasis for a matrix A. What is the relationship between the algebraic and geometric multiplicities of its eigenvalues?

32. Consider an eigenvalue λ of an $n \times n$ matrix A. We know that λ is an eigenvalue of A^T as well (since A and A^T have the same characteristic polynomial). Compare the geometric multiplicities of λ as an eigenvalue of A and A^T.

33. Show that if matrix A is similar to B, then $A - \lambda I_n$ is similar to $B - \lambda I_n$, for all scalars λ.

34. Suppose that $B = S^{-1}AS$ for some $n \times n$ matrices $A, B,$ and S.

a. Show that if $\vec{x}$ is in ker B, then $S\vec{x}$ is in ker A.

b. Show that the linear transformation $T(\vec{x}) = S\vec{x}$ from ker B to ker A is an isomorphism.

c. Show that nullity$(A) = $ nullity(B) and rank$(A) = $ rank(B).

35. Is matrix $\begin{bmatrix} 1 & 2 \\ 0 & 3 \end{bmatrix}$ similar to $\begin{bmatrix} 3 & 0 \\ 1 & 2 \end{bmatrix}$?

36. Is matrix $\begin{bmatrix} 0 & 1 \\ 5 & 3 \end{bmatrix}$ similar to $\begin{bmatrix} 1 & 2 \\ 4 & 3 \end{bmatrix}$?

37. Consider a symmetric $n \times n$ matrix A.

a. Show that if $\vec{v}$ and $\vec{w}$ are two vectors in $\mathbb{R}^n$, then

$$A\vec{v} \cdot \vec{w} = \vec{v} \cdot A\vec{w}.$$

b. Show that if $\vec{v}$ and $\vec{w}$ are two eigenvectors of A, with distinct eigenvalues, then $\vec{w}$ is orthogonal to $\vec{v}$.

38. Consider a rotation $T(\vec{x}) = A\vec{x}$ in $\mathbb{R}^3$. (That is, A is an orthogonal 3×3 matrix with determinant 1.) Show that T has a nonzero fixed point [i.e., a vector $\vec{v}$ with $T(\vec{v}) = \vec{v}$]. This result is known as *Euler's theorem,* after the great Swiss mathematician Leonhard Euler (1707–1783). (*Hint:* Consider the characteristic polynomial $f_A(\lambda)$. Pay attention to the intercepts with both axes. Use Theorem 7.1.2.)

39. Consider a subspace V of $\mathbb{R}^n$ with $\dim(V) = m$.

a. Suppose the $n \times n$ matrix A represents the orthogonal projection onto V. What can you say about the eigenvalues of A and their algebraic and geometric multiplicities?

b. Suppose the $n \times n$ matrix B represents the reflection about V. What can you say about the eigenvalues of B and their algebraic and geometric multiplicities?

40. Let $x(t)$ and $y(t)$ be the annual defense budgets of two antagonistic nations (expressed in billions of U.S. dollars). The change of these budgets is modeled by the equations

$$x(t + 1) = ax(t) + by(t),$$
$$y(t + 1) = bx(t) + ay(t),$$

where a is a constant slightly less than 1, expressing the fact that defense budgets tend to decline when there is no perceived threat. The constant b is a small positive number. You may assume that a exceeds b.

Suppose $x(0) = 3$ and $y(0) = 0.5$. What will happen in the long term? There are three possible cases, depending on the numerical values of a and b. Sketch a trajectory for each case, and discuss the outcome in practical terms. Include the eigenspaces in all your sketches.

41. Consider a modification of Example 5: Suppose the transformation matrix is

$$A = \begin{bmatrix} 0 & 1.4 & 1.2 \\ 0.8 & 0 & 0 \\ 0 & 0.4 & 0 \end{bmatrix}.$$

The initial populations are $j(0) = 600$, $m(0) = 100$, and $a(0) = 250$. Find closed formulas for $j(t)$, $m(t)$, and $a(t)$. Describe the long-term behavior. What can you say about the proportion $j(t) : m(t) : a(t)$ in the long term?

42. A street magician at Montmartre begins to perform at 11:00 P.M. on Saturday night. He starts out with no on-lookers, but he attracts passersby at a rate of 10 per minute. Some get bored and wander off: Of the people present t minutes after 11:00 P.M., 20% will have left a minute later (but everybody stays for at least a minute). Let $C(t)$ be the size of the crowd t minutes after 11:00 P.M. Find a 2×2 matrix A such that

$$\begin{bmatrix} C(t+1) \\ 1 \end{bmatrix} = A \begin{bmatrix} C(t) \\ 1 \end{bmatrix}.$$

Find a closed formula for $C(t)$, and graph this function. What is the long-term behavior of $C(t)$?

43. Three friends, Alberich, Brunnhilde, and Carl, play a number game together: Each thinks of a (real) number and announces it to the others. In the first round, each player finds the average of the numbers chosen by the two others; that is his or her new score. In the second round, the corresponding averages of the scores in the first round are taken, and so on. Here is an example:

	A	**B**	**C**
Initial choice	7	11	5
After 1st round	8	6	9
After 2nd round	7.5	8.5	7

Whoever is ahead after 1,001 rounds wins.

a. The state of the game after t rounds can be represented as a vector:

$$\vec{x}(t) = \begin{bmatrix} a(t) \\ b(t) \\ c(t) \end{bmatrix} \begin{array}{l} \text{Alberich's score} \\ \text{Brunnhilde's score.} \\ \text{Carl's score} \end{array}$$

Find the matrix A such that $\vec{x}(t+1) = A\vec{x}(t)$.

b. With the initial values mentioned earlier ($a_0 = 7$, $b_0 = 11$, $c_0 = 5$), what is the score after 10 rounds? After 50 rounds? Use technology.

c. Now suppose that Alberich and Brunnhilde initially pick the numbers 1 and 2, respectively. If Carl picks the number c_0, what is the state of the game after t rounds? [Find closed formulas for $a(t)$, $b(t)$, $c(t)$, in terms of c_0.] For which choices of c_0 does Carl win the game?

44. In an unfortunate accident involving an Austrian truck, 100 kg of a highly toxic substance are spilled into Lake Sils, in the Swiss Engadine Valley. The river Inn carries the pollutant down to Lake Silvaplana and later to Lake St. Moritz.

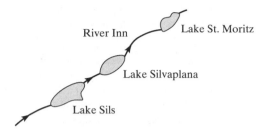

This sorry state, t weeks after the accident, can be described by the vector

$$\vec{x}(t) = \begin{bmatrix} x_1(t) \\ x_2(t) \\ x_3(t) \end{bmatrix} \begin{array}{l} \text{pollutant in Lake Sils} \\ \text{pollutant in Lake Silvaplana} \\ \text{pollutant in Lake St. Moritz} \end{array} \Big\} \text{ (in kg).}$$

Suppose that

$$\vec{x}(t+1) = \begin{bmatrix} 0.7 & 0 & 0 \\ 0.1 & 0.6 & 0 \\ 0 & 0.2 & 0.8 \end{bmatrix} \vec{x}(t).$$

a. Explain the significance of the entries of the transformation matrix in practical terms.

b. Find closed formulas for the amount of pollutant in each of the three lakes t weeks after the accident. Graph the three functions against time (on the same axes). When does the pollution in Lake Silvaplana reach a maximum?

45. Consider a dynamical system

$$\vec{x}(t) = \begin{bmatrix} x_1(t) \\ x_2(t) \end{bmatrix}$$

whose transformation from time t to time $t + 1$ is given by the following equations:

$$x_1(t+1) = 0.1x_1(t) + 0.2x_2(t) + 1,$$
$$x_2(t+1) = 0.4x_1(t) + 0.3x_2(t) + 2.$$

Such a system, with constant terms in the equations, is not linear, but *affine*.

a. Find a 2×2 matrix A and a vector $\vec{b}$ in $\mathbb{R}^2$ such that

$$\vec{x}(t+1) = A\vec{x}(t) + \vec{b}.$$

b. Introduce a new state vector

$$\vec{y}(t) = \begin{bmatrix} x_1(t) \\ x_2(t) \\ 1 \end{bmatrix},$$

with a "dummy" 1 in the last component. Find a 3×3 matrix B such that

$$\vec{y}(t+1) = B\vec{y}(t).$$

How is B related to the matrix A and the vector $\vec{b}$ in part (a)? Can you write B as a block matrix involving A and $\vec{b}$?

c. What is the relationship between the eigenvalues of A and B? What about eigenvectors?

d. For arbitrary values of $x_1(0)$ and $x_2(0)$, what can you say about the long-term behavior of $x_1(t)$ and $x_2(t)$?

46. A machine contains the grid of wires shown in the accompanying sketch. At the seven indicated points, the temperature is kept fixed at the given values (in °C). Consider the temperatures $T_1(t)$, $T_2(t)$, and $T_3(t)$ at the other three mesh points. Because of heat flow along the wires, the temperatures $T_i(t)$ changes according to the formula

$$T_i(t+1) = T_i(t) - \frac{1}{10}\sum\left(T_i(t) - T_{\text{adj}}(t)\right),$$

where the sum is taken over the four adjacent points in the grid and time is measured in minutes. For example,

$$T_2(t+1) = T_2(t) - \frac{1}{10}\left(T_2(t) - T_1(t)\right) - \frac{1}{10}\left(T_2(t) - 200\right)$$
$$- \frac{1}{10}\left(T_2(t) - 0\right) - \frac{1}{10}\left(T_2(t) - T_3(t)\right).$$

Note that each of the four terms we subtract represents the cooling caused by heat flowing along one of the wires. Let

$$\vec{x}(t) = \begin{bmatrix} T_1(t) \\ T_2(t) \\ T_3(t) \end{bmatrix}.$$

a. Find a 3×3 matrix A and a vector $\vec{b}$ in $\mathbb{R}^3$ such that

$$\vec{x}(t+1) = A\vec{x}(t) + \vec{b}.$$

b. Introduce the state vector

$$\vec{y}(t) = \begin{bmatrix} T_1(t) \\ T_2(t) \\ T_3(t) \\ 1 \end{bmatrix},$$

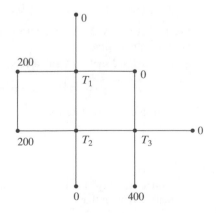

with a "dummy" 1 as the last component. Find a 4×4 matrix B such that

$$\vec{y}(t+1) = B\vec{y}(t).$$

(This technique for converting an affine system into a linear system is introduced in Exercise 45; see also Exercise 42.)

c. Suppose the initial temperatures are $T_1(0) = T_2(0) = T_3(0) = 0$. Using technology, find the temperatures at the three points at $t = 10$ and $t = 30$. What long-term behavior do you expect?

d. Using technology, find numerical approximations for the eigenvalues of the matrix B. Find an eigenvector for the largest eigenvalue. Use the results to confirm your conjecture in part (c).

47. The color of snapdragons is determined by a pair of genes, which we designate by the letters A and a. The pair of genes is called the flower's *genotype*. Genotype AA produces red flowers, genotype Aa pink ones, and genotype aa white ones. A biologist undertakes a breeding program, starting with a large population of flowers of genotype AA. Each flower is fertilized with pollen from a plant of genotype Aa (taken from another population), and one offspring is produced. Since it is a matter of chance which of the genes a parent passes on, we expect half of the flowers in the next generation to be red (genotype AA) and the other half pink (genotype Aa). All the flowers in this generation are now fertilized with pollen from plants of genotype Aa (taken from another population), and so on.

a. Find closed formulas for the fractions of red, pink, and white flowers in the tth generation. We know that $r(0) = 1$ and $p(0) = w(0) = 0$, and we found that $r(1) = p(1) = \frac{1}{2}$ and $w(1) = 0$.

b. What is the proportion $r(t) : p(t) : w(t)$ in the long run?

48. *Leonardo of Pisa: The rabbit problem.* Leonardo of Pisa (c. 1170–1240), also known as Fibonacci, was the first outstanding European mathematician after the ancient Greeks. He traveled widely in the Islamic world and studied Arabic mathematical writing. His work is in the spirit of the Arabic mathematics of his day. Fibonacci brought the decimal-position system to Europe. In his book *Liber abaci* (1202),[10] Fibonacci discusses the following problem:

> How many pairs of rabbits can be bred from one pair in one year? A man has one pair of rabbits at a certain place entirely surrounded by a wall. We wish to know how many pairs can be bred from it in one year, if the nature of these rabbits is such that they breed every month one other

---

[10] For a translation into modern English, see Laurence E. Sigler, *Fibonacci's Liber Abaci*, Springer-Verlag, 2002.

pair and begin to breed in the second month after their birth. Let the first pair breed a pair in the first month, then duplicate it and there will be 2 pairs in a month. From these pairs one, namely, the first, breeds a pair in the second month, and thus there are 3 pairs in the second month. From these, in one month, two will become pregnant, so that in the third month 2 pairs of rabbits will be born. Thus, there are 5 pairs in this month. From these, in the same month, 3 will be pregnant, so that in the fourth month there will be 8 pairs. From these pairs, 5 will breed 5 other pairs, which, added to the 8 pairs, gives 13 pairs in the fifth month, from which 5 pairs (which were bred in that same month) will not conceive in that month, but the other 8 will be pregnant. Thus, there will be 21 pairs in the sixth month. When we add to these the 13 pairs that are bred in the seventh month, then there will be in that month 34 pairs [and so on, 55, 89, 144, 233, 377, ...]. Finally, there will be 377, and this number of pairs has been born from the first-mentioned pair at the given place in one year.

Let $j(t)$ be the number of juvenile pairs and $a(t)$ the number of adult pairs after t months. Fibonacci starts his thought experiment in rabbit breeding with one adult pair, so $j(0) = 0$ and $a(0) = 1$. At $t = 1$, the adult pair will have bred a (juvenile) pair, so $a(1) = 1$ and $j(1) = 1$. At $t = 2$, the initial adult pair will have bred another (juvenile) pair, and last month's juvenile pair will have grown up, so $a(2) = 2$ and $j(2) = 1$.

a. Find formulas expressing $a(t + 1)$ and $j(t + 1)$ in terms of $a(t)$ and $j(t)$. Find the matrix A such that

$$\vec{x}(t + 1) = A\vec{x}(t),$$

where

$$\vec{x}(t) = \begin{bmatrix} a(t) \\ j(t) \end{bmatrix}.$$

b. Find closed formulas for $a(t)$ and $j(t)$. (*Note*: You will have to deal with irrational quantities here.)

c. Find the limit of the ratio $a(t)/j(t)$ as t approaches infinity. The result is known as the *golden section*. The golden section of a line segment AB is given by the point P such that

$$\frac{AB}{AP} = \frac{AP}{PB}.$$

A ———————————— P ———— B

49. Consider an $n \times n$ matrix A with zeros on the diagonal and below the diagonal, and "random" entries above the diagonal. What is the geometric multiplicity of the eigenvalue 0 likely to be?

50. a. Sketch a phase portrait for the dynamical system $\vec{x}(t + 1) = A\vec{x}(t)$, where

$$A = \begin{bmatrix} 2 & 1 \\ 3 & 2 \end{bmatrix}.$$

b. In his paper "On the Measurement of the Circle," the great Greek mathematician Archimedes (c. 280–210 B.C.) uses the approximation

$$\frac{265}{153} < \sqrt{3} < \frac{1351}{780}$$

to estimate $\cos(30°)$. He does not explain how he arrived at these estimates. Explain how we can obtain these approximations from the dynamical system in part (a). {*Hint*:

$$A^4 = \begin{bmatrix} 97 & 56 \\ 168 & 97 \end{bmatrix}, \quad A^6 = \begin{bmatrix} 1351 & 780 \\ 2340 & 1351 \end{bmatrix}.\}$$

c. Without using technology, explain why

$$\frac{1351}{780} - \sqrt{3} < 10^{-6}.$$

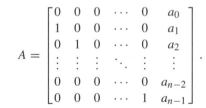

[*Hint*: Consider $\det(A^6)$.]

d. Based on the data in part (b), give an underestimate of the form p/q of $\sqrt{3}$ that is better than the one given by Archimedes.

51. Find the characteristic polynomial of the matrix

$$A = \begin{bmatrix} 0 & 0 & a \\ 1 & 0 & b \\ 0 & 1 & c \end{bmatrix},$$ where a, b, and c are arbitrary constants.

52. Find the characteristic polynomial of the $n \times n$ matrix

$$A = \begin{bmatrix} 0 & 0 & 0 & \cdots & 0 & a_0 \\ 1 & 0 & 0 & \cdots & 0 & a_1 \\ 0 & 1 & 0 & \cdots & 0 & a_2 \\ \vdots & \vdots & \vdots & \ddots & \vdots & \vdots \\ 0 & 0 & 0 & \cdots & 0 & a_{n-2} \\ 0 & 0 & 0 & \cdots & 1 & a_{n-1} \end{bmatrix}.$$

Note that the ith column of A is $\vec{e}_{i+1}$, for $i = 1, \ldots, n - 1$, while the last column has the arbitrary entries $a_0, \ldots, a_{n-1}$. (See Exercise 51 for the special case $n = 3$.)

53. Consider a 5×5 matrix A and a vector $\vec{v}$ in $\mathbb{R}^5$. Suppose the vectors $\vec{v}$, $A\vec{v}$, $A^2\vec{v}$ are linearly independent, while $A^3\vec{v} = a\vec{v} + bA\vec{v} + cA^2\vec{v}$ for some scalars a, b, c. We can take the linearly independent vectors $\vec{v}$, $A\vec{v}$, $A^2\vec{v}$ and expand them to a basis $\mathfrak{B} = (\vec{v}, A\vec{v}, A^2\vec{v}, \vec{w}_4, \vec{w}_5)$ of $\mathbb{R}^5$.

a. Consider the matrix B of the linear transformation $T(\vec{x}) = A\vec{x}$ with respect to the basis $\mathfrak{B}$. Write the entries of the first three columns of B. (Note that we do not know anything about the entries of the last two columns of B.)

b. Explain why $f_A(\lambda) = f_B(\lambda)$
$= h(\lambda)(-\lambda^3 + c\lambda^2 + b\lambda + a)$, for some quadratic polynomial $h(\lambda)$. See Exercise 51.

c. Explain why $f_A(A)\vec{v} = \vec{0}$. Here, $f_A(A)$ is the characteristic polynomial evaluated at A, that is, if $f_A(\lambda) = c_n\lambda^n + \cdots + c_1\lambda + c_0$, then $f_A(A) = c_n A^n + \cdots + c_1 A + c_0 I_n$.

54. Consider an $n \times n$ matrix A and a vector $\vec{v}$ in $\mathbb{R}^n$. Form the vectors $\vec{v}, A\vec{v}, A^2\vec{v}, A^3\vec{v}, \ldots$, and let $A^m\vec{v}$ be the first redundant vector in this list. Then the m vectors $\vec{v}, A\vec{v}, A^2\vec{v}, \ldots, A^{m-1}\vec{v}$ are linearly independent; note that $m \le n$. Since $A^m\vec{v}$ is redundant, we can write $A^m\vec{v} = a_0\vec{v} + a_1 A\vec{v} + a_2 A^2\vec{v} + \cdots + a_{m-1}A^{m-1}\vec{v}$ for some scalars $a_0, \ldots, a_{m-1}$. Form a basis $\mathfrak{B} = (\vec{v}, A\vec{v}, A^2\vec{v}, \ldots, A^{m-1}\vec{v}, \vec{w}_{m+1}, \ldots, \vec{w}_n)$ of $\mathbb{R}^n$.

a. Consider the matrix B of the linear transformation $T(\vec{x}) = A\vec{x}$ with respect to the basis $\mathfrak{B}$. Write B in block form, $B = \begin{bmatrix} B_{11} & B_{12} \\ B_{21} & B_{22} \end{bmatrix}$, where B_{11} is an

$m \times m$ matrix. Describe B_{11} column by column, paying particular attention to the mth column. What can you say about B_{21}? (Note that we do not know anything about the entries of B_{12} and B_{22}.)

b. Explain why $f_A(\lambda) = f_B(\lambda) = f_{B_{22}}(\lambda)f_{B_{11}}(\lambda) = (-1)^m f_{B_{22}}(\lambda)(\lambda^m - a_{m-1}\lambda^{m-1} - \cdots - a_1\lambda - a_0)$. See Exercise 52.

c. Explain why $f_A(A)\vec{v} = \vec{0}$. See Exercise 53.

d. Explain why $f_A(A) = 0$.

The equation $f_A(A) = 0$ is referred to as the *Cayley-Hamilton theorem:* A square matrix satisfies its characteristic polynomial. The English mathematician Arthur Cayley (1821–1895) played a leading role in the development of the algebra of matrices, and the Irish mathematician Sir William Rowan Hamilton (1805–1865) is best remembered today for his discovery of the quaternions. See Exercises 5.3.64 and 7.5.37.

7.4 Diagonalization

EXAMPLE 1 In Example 2 of Section 7.3, we found that vectors $\vec{v}_1 = \begin{bmatrix} 1 \\ -1 \end{bmatrix}$ and $\vec{v}_2 = \begin{bmatrix} 1 \\ 2 \end{bmatrix}$ form an eigenbasis for matrix $A = \begin{bmatrix} 1 & 2 \\ 4 & 3 \end{bmatrix}$, with associated eigenvalues $\lambda_1 = -1$ and $\lambda_2 = 5$. Find the matrix B of the linear transformation $T(\vec{x}) = A\vec{x}$ with respect to this eigenbasis $\mathfrak{B} = (\vec{v}_1, \vec{v}_2)$.

Solution

We are looking for the matrix B such that $\left[T(\vec{x})\right]_{\mathfrak{B}} = B\left[\vec{x}\right]_{\mathfrak{B}}$ for all $\vec{x}$ in $\mathbb{R}^2$, by Definition 3.4.3. We can use a commutative diagram to find B.

$$\vec{x} = c_1\vec{v}_1 + c_2\vec{v}_2 \xrightarrow{\quad A \quad} \begin{aligned} T(\vec{x}) &= c_1 A\vec{v}_1 + c_2 A\vec{v}_2 \\ &= c_1\lambda_1\vec{v}_1 + c_2\lambda_2\vec{v}_2 \\ &= -c_1\vec{v}_1 + 5c_2\vec{v}_2 \end{aligned}$$

$$\left[\vec{x}\right]_{\mathfrak{B}} = \begin{bmatrix} c_1 \\ c_2 \end{bmatrix} \xrightarrow[B = \begin{bmatrix} -1 & 0 \\ 0 & 5 \end{bmatrix}]{\quad} \left[T(\vec{x})\right]_{\mathfrak{B}} = \begin{bmatrix} \lambda_1 c_1 \\ \lambda_2 c_2 \end{bmatrix} = \begin{bmatrix} -c_1 \\ 5c_2 \end{bmatrix}$$

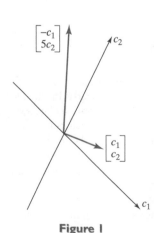

Figure 1

Note that B is a diagonal matrix. If we use the c_1-c_2 coordinate system defined by the eigenbasis $\mathfrak{B}$, then $\begin{bmatrix} c_1 \\ c_2 \end{bmatrix}$ is transformed into $\begin{bmatrix} -c_1 \\ 5c_2 \end{bmatrix}$, meaning that the c_1-component is reversed and the c_2-component is magnified by a factor of 5, as shown in Figure 1. ■

It turns out that the matrix of a linear transformation T with respect to an eigenbasis is diagonal. In light of this fact, we will often denote an eigenbasis by $\mathfrak{D}$; the matrix of T with respect to $\mathfrak{D}$ is then denoted D.

Theorem 7.4.1   The matrix of a linear transformation with respect to an eigenbasis

Consider a linear transformation $T(\vec{x}) = A\vec{x}$, where A is a square matrix. Suppose $\mathfrak{D} = (\vec{v}_1, \vec{v}_2, \ldots, \vec{v}_n)$ is an eigenbasis for T, with $A\vec{v}_i = \lambda_i \vec{v}_i$. Then the $\mathfrak{D}$-matrix D of T is

$$D = S^{-1}AS = \begin{bmatrix} \lambda_1 & 0 & \cdots & 0 \\ 0 & \lambda_2 & \cdots & 0 \\ \vdots & \vdots & \ddots & \vdots \\ 0 & 0 & \cdots & \lambda_n \end{bmatrix}, \quad \text{where} \quad S = \begin{bmatrix} \vec{v}_1 & \vec{v}_2 & \ldots & \vec{v}_n \end{bmatrix}.$$

Matrix D is diagonal, and its diagonal entries are the eigenvalues $\lambda_1, \lambda_2, \ldots, \lambda_n$ of T. ∎

Recall that the formula $D = S^{-1}AS$ was discussed in Theorem 3.4.4.

To justify Theorem 7.4.1, use a commutative diagram as in Example 1; we leave the details to the reader.

The converse of Theorem 7.4.1 holds as well: If the matrix D of a linear transformation T with respect to a basis $\mathfrak{D}$ is diagonal, then $\mathfrak{D}$ must be an eigenbasis for T (see Exercise 53).

Theorem 7.4.1 motivates the following definition.

Definition 7.4.2   Diagonalizable matrices

An $n \times n$ matrix A is called *diagonalizable* if A is similar to some diagonal matrix D, that is, if there exists an invertible $n \times n$ matrix S such that $S^{-1}AS$ is diagonal.

As we just observed, matrix $S^{-1}AS$ is diagonal if (and only if) the column vectors of S form an eigenbasis for A. This implies the following result.

Theorem 7.4.3   Eigenbases and diagonalization

a. Matrix A is diagonalizable if (and only if) there exists an eigenbasis for A.

b. If an $n \times n$ matrix A has n distinct eigenvalues, then A is diagonalizable. ∎

If an $n \times n$ matrix A has fewer than n distinct eigenvalues, then A may or may not be diagonalizable. Consider the diagonalizable matrix

$$\begin{bmatrix} 1 & 0 \\ 0 & 1 \end{bmatrix}$$

and the nondiagonalizable

$$\begin{bmatrix} 1 & 1 \\ 0 & 1 \end{bmatrix}.$$

Asking whether a matrix A is diagonalizable amounts to asking whether there exists an eigenbasis for A. In the past three sections, we outlined a method for answering this question. Let us summarize this process.

Theorem 7.4.4   Diagonalization

Suppose we are asked to determine whether a given $n \times n$ matrix A is diagonalizable. If so, we wish to find an invertible matrix S such that $S^{-1}AS$ is diagonal.

We can proceed as follows.

a. Find the eigenvalues of A, that is, solve the characteristic equation $f_A(\lambda) = \det(A - \lambda I_n) = 0$.

   b. For each eigenvalue λ, find a basis of the eigenspace $E_\lambda = \ker(A - \lambda I_n)$.

   c. Matrix A is diagonalizable if (and only if) the dimensions of the eigenspaces add up to n. In this case, we find an eigenbasis $\vec{v}_1, \vec{v}_2, \ldots, \vec{v}_n$ by concatenating the bases of the eigenspaces we found in step (b). Let $S = \begin{bmatrix} \vec{v}_1 & \vec{v}_2 & \cdots & \vec{v}_n \end{bmatrix}$. Then matrix $S^{-1}AS = D$ is diagonal, and the ith diagonal entry of D is the eigenvalue λ_i associated with $\vec{v}_i$. ∎

EXAMPLE 2   Diagonalize the matrix

$$A = \begin{bmatrix} 1 & 1 & 1 \\ 0 & 0 & 0 \\ 0 & 0 & 0 \end{bmatrix},$$

if you can.

Solution

We will follow the steps outlined in Theorem 7.4.4:

   a. The eigenvalues of the triangular matrix A are 0 and 1, the diagonal entries; eigenvalue 0 has algebraic multiplicity 2. We cannot tell yet whether A is diagonalizable; the answer will depend on the dimension of eigenspace E_0 (the kernel of matrix A).

   b. $E_0 = \ker A = \ker \begin{bmatrix} 1 & 1 & 1 \\ 0 & 0 & 0 \\ 0 & 0 & 0 \end{bmatrix} = \text{span}\left(\begin{bmatrix} -1 \\ 1 \\ 0 \end{bmatrix}, \begin{bmatrix} -1 \\ 0 \\ 1 \end{bmatrix} \right)$ and

   $E_1 = \ker(A - I_3) = \begin{bmatrix} 0 & 1 & 1 \\ 0 & -1 & 0 \\ 0 & 0 & -1 \end{bmatrix} = \text{span} \begin{bmatrix} 1 \\ 0 \\ 0 \end{bmatrix}$.

   c. Matrix A is diagonalizable, since the sum of the dimensions of the eigenspaces E_0 and E_1 is $2 + 1 = 3$. We have the eigenbasis

$$\vec{v}_1 = \begin{bmatrix} -1 \\ 1 \\ 0 \end{bmatrix}, \qquad \vec{v}_2 = \begin{bmatrix} -1 \\ 0 \\ 1 \end{bmatrix}, \qquad \vec{v}_3 = \begin{bmatrix} 1 \\ 0 \\ 0 \end{bmatrix}.$$

$$\updownarrow \qquad\qquad \updownarrow \qquad\qquad \updownarrow$$
$$\lambda_1 = 0 \qquad\quad \lambda_2 = 0 \qquad\quad \lambda_3 = 1$$

   If we let

$$S = \begin{bmatrix} \vec{v}_1 & \vec{v}_2 & \vec{v}_3 \end{bmatrix} = \begin{bmatrix} -1 & -1 & 1 \\ 1 & 0 & 0 \\ 0 & 1 & 0 \end{bmatrix},$$

   then

$$D = S^{-1}AS = \begin{bmatrix} \lambda_1 & 0 & 0 \\ 0 & \lambda_2 & 0 \\ 0 & 0 & \lambda_3 \end{bmatrix} = \begin{bmatrix} 0 & 0 & 0 \\ 0 & 0 & 0 \\ 0 & 0 & 1 \end{bmatrix}.$$

Note that you need not actually compute $S^{-1}AS$. Theorem 7.4.1 guarantees that $S^{-1}AS$ is diagonal, with the eigenvalues lined up along the diagonal. To check your work, you may wish to verify that $S^{-1}AS = D$, or, equivalently, that $AS = SD$

(this way you need not find the inverse of S):

$$AS = \begin{bmatrix} 1 & 1 & 1 \\ 0 & 0 & 0 \\ 0 & 0 & 0 \end{bmatrix} \begin{bmatrix} -1 & -1 & 1 \\ 1 & 0 & 0 \\ 0 & 1 & 0 \end{bmatrix} = \begin{bmatrix} 0 & 0 & 1 \\ 0 & 0 & 0 \\ 0 & 0 & 0 \end{bmatrix}$$

$$SD = \begin{bmatrix} -1 & -1 & 1 \\ 1 & 0 & 0 \\ 0 & 1 & 0 \end{bmatrix} \begin{bmatrix} 0 & 0 & 0 \\ 0 & 0 & 0 \\ 0 & 0 & 1 \end{bmatrix} = \begin{bmatrix} 0 & 0 & 1 \\ 0 & 0 & 0 \\ 0 & 0 & 0 \end{bmatrix} \quad \checkmark \qquad \blacksquare$$

EXAMPLE 3   For which values of the constants a, b, and c is the matrix

$$A = \begin{bmatrix} 1 & a & b \\ 0 & 0 & c \\ 0 & 0 & 1 \end{bmatrix}$$

diagonalizable?

Solution

The eigenvalues are 0, 1, 1. Theorem 7.3.7 tells us that $E_0 = \ker A$ is one dimensional. Now

$$E_1 = \ker(A - I_3) = \ker \begin{bmatrix} 0 & a & b \\ 0 & -1 & c \\ 0 & 0 & 0 \end{bmatrix} \overset{\text{Step 3}}{=} \ker \begin{bmatrix} 0 & 1 & -c \\ 0 & 0 & b + ac \\ 0 & 0 & 0 \end{bmatrix}.$$

In Step 3 we perform some elementary row operations. Matrix A is diagonalizable if (and only if) the eigenspace E_1 is two dimensional [so that $\dim(E_0) + \dim(E_1) = 3$]. This is the case if (and only if) $b + ac = 0$, or $b = -ac$. $\qquad \blacksquare$

Powers of a Matrix

EXAMPLE 4   Consider the matrix $A = \begin{bmatrix} \frac{1}{2} & \frac{3}{4} \\ \frac{1}{2} & \frac{1}{4} \end{bmatrix}$, a regular transition matrix (see Exercise 7.2.25).

a. Find formulas for the entries of A^t, for any positive integer t.
b. Find $\lim\limits_{t \to \infty} A^t$.

(*Hint*: Diagonalize A.)

Solution

It may be informative to compute A^t for a few values of t first, using a calculator or computer. For example,

$$A^{10} \approx \begin{bmatrix} 0.6000003815 & 0.5999994278 \\ 0.399996183 & 0.4000005722 \end{bmatrix}.$$

Further numerical experimentation suggests that

$$\lim_{t \to \infty} A^t = \begin{bmatrix} 0.6 & 0.6 \\ 0.4 & 0.4 \end{bmatrix}.$$

We cannot be certain of this result, of course, for (at least) two reasons: We can only check finitely many values of t, and the results the calculator produces are afflicted with round-off errors.

To prove our conjecture, $\lim\limits_{t \to \infty} A^t = \begin{bmatrix} 0.6 & 0.6 \\ 0.4 & 0.4 \end{bmatrix}$, we follow the hint and diagonalize matrix A:

- Eigenvalues: $f_A(\lambda) = \lambda^2 - (\text{tr} A)\lambda + (\det A) = \lambda^2 - \frac{3}{4}\lambda - \frac{1}{4}$
  $= (\lambda - 1)\left(\lambda + \frac{1}{4}\right) = 0$. The eigenvalues are 1 and $-\frac{1}{4}$.

- Eigenvectors: $E_1 = \ker \begin{bmatrix} -\frac{1}{2} & \frac{3}{4} \\ \frac{1}{2} & -\frac{3}{4} \end{bmatrix} = \text{span} \begin{bmatrix} 3 \\ 2 \end{bmatrix}$,

  $E_{-1/4} = \ker \begin{bmatrix} \frac{3}{4} & \frac{3}{4} \\ \frac{1}{2} & \frac{1}{2} \end{bmatrix} = \text{span} \begin{bmatrix} 1 \\ -1 \end{bmatrix}$.

- Thus $S^{-1} A S = D = \begin{bmatrix} 1 & 0 \\ 0 & -\frac{1}{4} \end{bmatrix}$, where $S = \begin{bmatrix} 3 & 1 \\ 2 & -1 \end{bmatrix}$.

To compute the powers of matrix A, we solve the equation $S^{-1} A S = D$ for A and find that $A = SDS^{-1}$. Now

$$A^t = (SDS^{-1})^t = \overbrace{(SDS^{-1})(SDS^{-1}) \cdots (SDS^{-1})}^{t \text{ times}} = SD^t S^{-1};$$

note the cancellation of the terms of the form $S^{-1} S$. (Compare this with Example 7 of Section 3.4.)

Matrices D^t and S^{-1} are easy to compute: To find D^t, raise the diagonal entries of D to the tth power. Thus, the problem of finding A^t is essentially solved.

$$S^{-1} = \frac{1}{5} \begin{bmatrix} 1 & 1 \\ 2 & -3 \end{bmatrix}, \quad D^t = \begin{bmatrix} 1 & 0 \\ 0 & \left(-\frac{1}{4}\right)^t \end{bmatrix}$$

so that

$$A^t = SD^t S^{-1} = \frac{1}{5} \begin{bmatrix} 3 & 1 \\ 2 & -1 \end{bmatrix} \begin{bmatrix} 1 & 0 \\ 0 & \left(-\frac{1}{4}\right)^t \end{bmatrix} \begin{bmatrix} 1 & 1 \\ 2 & -3 \end{bmatrix}$$

$$= \frac{1}{5} \begin{bmatrix} 3 + 2\left(-\frac{1}{4}\right)^t & 3 - 3\left(-\frac{1}{4}\right)^t \\ 2 - 2\left(-\frac{1}{4}\right)^t & 2 + 3\left(-\frac{1}{4}\right)^t \end{bmatrix}$$

Since the term $\left(-\frac{1}{4}\right)^t$ approaches 0 as t goes to infinity, we have

$$\lim_{t \to \infty} A^t = \frac{1}{5} \begin{bmatrix} 3 & 3 \\ 2 & 2 \end{bmatrix} = \begin{bmatrix} 0.6 & 0.6 \\ 0.4 & 0.4 \end{bmatrix},$$

confirming our earlier conjecture. ∎

Theorem 7.4.5 **Powers of a diagonalizable matrix**

To compute the powers A^t of a diagonalizable matrix A (where t is a positive integer), proceed as follows:

Use Theorem 7.4.4 to diagonalize A, that is, find an invertible S and a diagonal D such that $S^{-1} A S = D$.

Then

$$A = SDS^{-1} \quad \text{and} \quad A^t = SD^t S^{-1}.$$

To compute D^t, raise the diagonal entries of D to the tth power. ∎

EXAMPLE 5   Consider the dynamical system

$$\vec{x}(t+1) = \begin{bmatrix} \frac{1}{2} & \frac{3}{4} \\ \frac{1}{2} & \frac{1}{4} \end{bmatrix} \vec{x}(t) \quad \text{with initial value} \quad \vec{x}_0 = \begin{bmatrix} 100 \\ 0 \end{bmatrix}.$$

Use your work in Example 4 to give closed formulas for the entries of $\vec{x}(t)$, and find $\lim_{t \to \infty} \vec{x}(t)$.

Solution

We know that

$$\vec{x}(t) = A^t \vec{x}_0, \quad \text{where} \quad A = \begin{bmatrix} \frac{1}{2} & \frac{3}{4} \\ \frac{1}{2} & \frac{1}{4} \end{bmatrix} \quad \text{and} \quad \vec{x}_0 = \begin{bmatrix} 100 \\ 0 \end{bmatrix}.$$

Now we can use our work in Example 4:

$$\vec{x}(t) = A^t \vec{x}_0 = SD^t S^{-1} \vec{x}_0 = \frac{1}{5} \begin{bmatrix} 3 + 2\left(-\frac{1}{4}\right)^t & 3 - 3\left(-\frac{1}{4}\right)^t \\ 2 - 2\left(-\frac{1}{4}\right)^t & 2 + 3\left(-\frac{1}{4}\right)^t \end{bmatrix} \begin{bmatrix} 100 \\ 0 \end{bmatrix}$$

$$= \begin{bmatrix} 60 + 40\left(-\frac{1}{4}\right)^t \\ 40 - 40\left(-\frac{1}{4}\right)^t \end{bmatrix},$$

and

$$\lim_{t \to \infty} \vec{x}(t) = \begin{bmatrix} 60 \\ 40 \end{bmatrix}.$$

Note that we have seen the formula $\vec{x}(t) = SD^t S^{-1} \vec{x}_0$ in Theorem 7.1.3 already, although it was derived in a different way. ∎

The Eigenvalues of a Linear Transformation (for those who have studied Chapter 4)

In the preceding three sections, we developed the theory of eigenvalues and eigenvectors for $n \times n$ matrices, or, equivalently, for linear transformations $T(\vec{x}) = A\vec{x}$ from $\mathbb{R}^n$ to $\mathbb{R}^n$. These concepts can be generalized to linear transformations from V to V, where V is any linear space. In the case of a finite dimensional space V, we can generalize the idea of diagonalization as well.

Definition 7.4.6   The eigenvalues of a linear transformation

Consider a linear transformation T from V to V, where V is a linear space. A scalar λ is called an *eigenvalue* of T if there exists a nonzero element f of V such that

$$T(f) = \lambda f.$$

Such an f is called an *eigenfunction* if V consists of functions, an *eigenmatrix* if V consists of matrices, and so on. In theoretical work, the inclusive term *eigenvector* is often used for f.

Now suppose that V is finite dimensional. Then a basis $\mathfrak{D}$ of V consisting of eigenvectors of T is called an *eigenbasis* for T. We say that transformation T is *diagonalizable* if the matrix of T with respect to some basis is diagonal. Transformation T is diagonalizable if (and only if) there exists an eigenbasis for T (see Theorem 7.4.3a).

EXAMPLE 6   Consider the linear transformation $D(f) = f'$ (the derivative) from C^∞ to C^∞. Show that all real numbers are eigenvalues of D. *Hint*: Apply D to exponential functions.

Solution

Following the hint, we observe that $D(e^x) = (e^x)' = e^x = 1e^x$. This shows that e^x is an eigenfunction of D, with associated eigenvalue 1. More generally,

$$D(e^{kx}) = (e^{kx})' = k(e^{kx}) \quad \text{(use the chain rule)},$$

showing that e^{kx} is an eigenfunction of D with associated eigenvalue k. Here k can be any real number, proving our claim.   ■

EXAMPLE 7   Consider the linear transformation $L(A) = A^T$ (the transpose[11]) from $\mathbb{R}^{2\times 2}$ to $\mathbb{R}^{2\times 2}$. Is transformation L diagonalizable? If so, find an eigenbasis for L. *Hint*: Consider symmetric and skew-symmetric matrices.

Solution

If A is symmetric, then $L(A) = A^T = A = 1A$, so that A is an eigenmatrix with eigenvalue 1. The symmetric 2×2 matrices $\begin{bmatrix} a & b \\ b & c \end{bmatrix}$ form a three-dimensional space, with basis

$$\begin{bmatrix} 1 & 0 \\ 0 & 0 \end{bmatrix}, \begin{bmatrix} 0 & 1 \\ 1 & 0 \end{bmatrix}, \begin{bmatrix} 0 & 0 \\ 0 & 1 \end{bmatrix}.$$

We need only one more matrix to form an eigenbasis for L, since $\mathbb{R}^{2\times 2}$ is four-dimensional.

If A is skew symmetric, then $L(A) = A^T = -A = (-1)A$, so that A is an eigenmatrix with eigenvalue -1. The skew-symmetric 2×2 matrices $\begin{bmatrix} 0 & a \\ -a & 0 \end{bmatrix}$ form a one-dimensional space, with basis $\begin{bmatrix} 0 & 1 \\ -1 & 0 \end{bmatrix}$.

We have found enough eigenmatrices to form an eigenbasis for L:

$$\begin{bmatrix} 1 & 0 \\ 0 & 0 \end{bmatrix}, \begin{bmatrix} 0 & 1 \\ 1 & 0 \end{bmatrix}, \begin{bmatrix} 0 & 0 \\ 0 & 1 \end{bmatrix}, \begin{bmatrix} 0 & 1 \\ -1 & 0 \end{bmatrix}.$$

Thus, L is diagonalizable.   ■

EXAMPLE 8   Consider the linear transformation $T(f(x)) = f(2x - 1)$ from P_2 to P_2. Is transformation T diagonalizable? If so, find an eigenbasis $\mathfrak{D}$ and the $\mathfrak{D}$-matrix D of T.

Solution

Here it would be hard to find eigenvalues and eigenfunctions "by inspection;" we need a systematic approach. The idea is to find the matrix A of T with respect to some convenient basis $\mathfrak{A}$. Then we can use Theorem 7.4.4 to determine whether A is diagonalizable, and, if so, to find an eigenbasis for A. Finally we can transform this basis back into P_2 to find an eigenbasis $\mathfrak{D}$ for T.

---

[11]If you have skipped Chapter 5, read Definition 5.3.5 and Examples 5 and 6 following that definition.

We will use a commutative diagram to find the matrix A of T with respect to the standard basis $\mathfrak{A} = (1, x, x^2)$.

$$a + bx + cx^2 \xrightarrow{\quad T \quad} \begin{array}{l} T(a + bx + cx^2) \\ = a + b(2x - 1) + c(2x - 1)^2 \\ = a - b + c + (2b - 4c)x + 4cx^2 \end{array}$$

$$\Big\downarrow L_{\mathfrak{A}} \qquad\qquad\qquad\qquad\qquad \Big\downarrow L_{\mathfrak{A}}$$

$$\begin{bmatrix} a \\ b \\ c \end{bmatrix} \xrightarrow{\qquad\qquad} \qquad A = \begin{bmatrix} 1 & -1 & 1 \\ 0 & 2 & -4 \\ 0 & 0 & 4 \end{bmatrix} \qquad \begin{bmatrix} a - b + c \\ 2b - 4c \\ 4c \end{bmatrix}$$

The upper triangular matrix A has the three distinct eigenvalues, 1, 2, and 4, so that A is diagonalizable, by Theorem 7.4.3b. A straightforward computation produces the eigenbasis

$$\begin{bmatrix} 1 \\ 0 \\ 0 \end{bmatrix}, \quad \begin{bmatrix} -1 \\ 1 \\ 0 \end{bmatrix}, \quad \begin{bmatrix} 1 \\ -2 \\ 1 \end{bmatrix}$$

for A. Transforming these vectors back into P_2, we find the eigenbasis $\mathfrak{D}$ for T consisting of

$$1, \quad x - 1, \quad x^2 - 2x + 1 = (x - 1)^2.$$

To check our work, we can verify that these are indeed eigenfunctions of T:

$$T(1) = 1,$$
$$T(x - 1) = (2x - 1) - 1 = 2x - 2 = 2(x - 1),$$
$$T\big((x - 1)^2\big) = \big((2x - 1) - 1\big)^2 = (2x - 2)^2 = 4(x - 1)^2. \quad \checkmark$$

The $\mathfrak{D}$-matrix of T is

$$D = \begin{bmatrix} 1 & 0 & 0 \\ 0 & 2 & 0 \\ 0 & 0 & 4 \end{bmatrix}.$$

Consider Figure 2, where

$$S = \begin{bmatrix} 1 & -1 & 1 \\ 0 & 1 & -2 \\ 0 & 0 & 1 \end{bmatrix}$$

is the change of basis matrix from $\mathfrak{D}$ to $\mathfrak{A}$. ∎

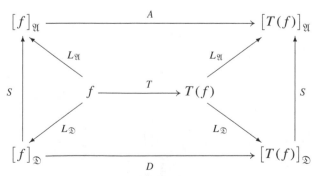

Figure 2

EXAMPLE 9 Let V be the space of all infinite sequences of real numbers. We define the linear transformation

$$T(x_0, x_1, x_2, x_3, x_4, \dots) = (x_1, x_2, x_3, x_4, \dots)$$

from V to V (we omit the first term, x_0). Find all the eigenvalues and eigensequences of T.

Solution

Since V is infinite dimensional, we cannot use the matrix techniques of Example 8 here. We have to go back to the definition of an eigenvalue: For a fixed scalar λ, we are looking for the infinite sequences $(x_0, x_1, x_2, x_3, \dots)$ such that

$$T(x_0, x_1, x_2, x_3, \dots) = \lambda(x_0, x_1, x_2, x_3, \dots)$$

or

$$(x_1, x_2, x_3, \dots) = \lambda(x_0, x_1, x_2, x_3, \dots)$$

or

$$x_1 = \lambda x_0, \quad x_2 = \lambda x_1 = \lambda^2 x_0, \quad x_3 = \lambda x_2 = \lambda^3 x_0, \dots.$$

The solutions are the *geometric sequences* of the form

$$(x_0, \lambda x_0, \lambda^2 x_0, \lambda^3 x_0, \dots) = x_0(1, \lambda, \lambda^2, \lambda^3, \dots).$$

Thus all real numbers λ are eigenvalues of T, and the eigenspace E_λ is one dimensional for all λ, with the geometric sequence $(1, \lambda, \lambda^2, \lambda^3, \dots)$ as a basis.

For example, when $\lambda = 3$, we have

$$T(1, 3, 9, 27, 81, \dots) = (3, 9, 27, 81, \dots) = 3(1, 3, 9, 27, \dots),$$

demonstrating that $(1, 3, 9, 27, 81, \dots)$ is an eigensequence of T with eigenvalue 3. ∎

EXERCISES 7.4

GOAL *Use the concept of a diagonalizable matrix. Find the eigenvalues of a linear transformation.*

Decide which of the matrices A in Exercises 1 through 20 are diagonalizable. If possible, find an invertible S and a diagonal D such that $S^{-1}AS = D$. Do not use technology.

1. $A = \begin{bmatrix} 2 & 0 \\ 0 & 3 \end{bmatrix}$

2. $A = \begin{bmatrix} 2 & 1 \\ 0 & 3 \end{bmatrix}$

3. $A = \begin{bmatrix} 1 & 1 \\ 2 & 2 \end{bmatrix}$

4. $A = \begin{bmatrix} 1 & 2 \\ 3 & 6 \end{bmatrix}$

5. $A = \begin{bmatrix} 1 & 1 \\ 0 & 1 \end{bmatrix}$

6. $A = \begin{bmatrix} 2 & 0 \\ -1 & 2 \end{bmatrix}$

7. $\begin{bmatrix} 1 & 4 \\ 1 & -2 \end{bmatrix}$

8. $A = \begin{bmatrix} 1 & 3 \\ 3 & 1 \end{bmatrix}$

9. $A = \begin{bmatrix} 4 & 9 \\ -1 & -2 \end{bmatrix}$

10. $A = \begin{bmatrix} 3 & 4 \\ -1 & -1 \end{bmatrix}$

11. $A = \begin{bmatrix} 1 & 0 & 1 \\ 0 & 2 & 0 \\ 0 & 0 & 1 \end{bmatrix}$

12. $A = \begin{bmatrix} 2 & 0 & 1 \\ 0 & 1 & 0 \\ 0 & 0 & 1 \end{bmatrix}$

13. $A = \begin{bmatrix} 1 & 1 & 0 \\ 0 & 2 & 2 \\ 0 & 0 & 3 \end{bmatrix}$

14. $A = \begin{bmatrix} 3 & 1 & 1 \\ 0 & 2 & 1 \\ 0 & 0 & 1 \end{bmatrix}$

15. $A = \begin{bmatrix} 3 & -4 & 0 \\ 2 & -3 & 0 \\ 0 & 0 & 1 \end{bmatrix}$

16. $A = \begin{bmatrix} 4 & 0 & -2 \\ 0 & 1 & 0 \\ 1 & 0 & 1 \end{bmatrix}$

17. $A = \begin{bmatrix} 1 & 1 & 1 \\ 1 & 1 & 1 \\ 1 & 1 & 1 \end{bmatrix}$

18. $A = \begin{bmatrix} 1 & 0 & 1 \\ 0 & 1 & 0 \\ 1 & 0 & 1 \end{bmatrix}$

19. $A = \begin{bmatrix} 1 & 1 & 1 \\ 0 & 1 & 0 \\ 0 & 1 & 0 \end{bmatrix}$

20. $A = \begin{bmatrix} 1 & 0 & 1 \\ 1 & 1 & 1 \\ 1 & 0 & 1 \end{bmatrix}$

For which values of constants a, b, and c are the matrices in Exercises 21 through 30 diagonalizable?

21. $\begin{bmatrix} 1 & a \\ 0 & 2 \end{bmatrix}$

22. $\begin{bmatrix} 1 & a \\ 0 & b \end{bmatrix}$

23. $\begin{bmatrix} 1 & 1 \\ a & 1 \end{bmatrix}$

24. $\begin{bmatrix} a & b \\ b & c \end{bmatrix}$

25. $\begin{bmatrix} 1 & a & b \\ 0 & 2 & c \\ 0 & 0 & 3 \end{bmatrix}$

26. $\begin{bmatrix} 1 & a & b \\ 0 & 2 & c \\ 0 & 0 & 1 \end{bmatrix}$

27. $\begin{bmatrix} 1 & a & b \\ 0 & 1 & c \\ 0 & 0 & 1 \end{bmatrix}$

28. $\begin{bmatrix} 0 & 0 & 0 \\ 1 & 0 & a \\ 0 & 1 & 0 \end{bmatrix}$

29. $\begin{bmatrix} 0 & 0 & a \\ 1 & 0 & 0 \\ 0 & 1 & 0 \end{bmatrix}$

30. $\begin{bmatrix} 0 & 0 & a \\ 1 & 0 & 3 \\ 0 & 1 & 0 \end{bmatrix}$

For the matrices A in Exercises 31 through 34, find formulas for the entries of A^t, where t is a positive integer. Also, find the vector $A^t \begin{bmatrix} 1 \\ 2 \end{bmatrix}$.

31. $A = \begin{bmatrix} 1 & 2 \\ 4 & 3 \end{bmatrix}$

32. $A = \begin{bmatrix} 4 & -2 \\ 1 & 1 \end{bmatrix}$

33. $A = \begin{bmatrix} 1 & 2 \\ 3 & 6 \end{bmatrix}$

34. $A = \begin{bmatrix} 1/2 & 1/4 \\ 1/2 & 3/4 \end{bmatrix}$

35. Is matrix $\begin{bmatrix} -1 & 6 \\ -2 & 6 \end{bmatrix}$ similar to $\begin{bmatrix} 3 & 0 \\ 0 & 2 \end{bmatrix}$?

36. Is matrix $\begin{bmatrix} -1 & 6 \\ -2 & 6 \end{bmatrix}$ similar to $\begin{bmatrix} 1 & 2 \\ -1 & 4 \end{bmatrix}$?

37. If A and B are two 2×2 matrices with $\det A = \det B = \operatorname{tr} A = \operatorname{tr} B = 7$, is A necessarily similar to B?

38. If A and B are two 2×2 matrices with $\det A = \det B = \operatorname{tr} A = \operatorname{tr} B = 4$, is A necessarily similar to B?

Find all the eigenvalues and "eigenvectors" of the linear transformations in Exercises 39 through 52.

39. $T(f) = f' - f$ from C^∞ to C^∞

40. $T(f) = 5f' - 3f$ from C^∞ to C^∞

41. $L(A) = A + A^T$ from $\mathbb{R}^{2\times2}$ to $\mathbb{R}^{2\times2}$. Is L diagonalizable?

42. $L(A) = A - A^T$ from $\mathbb{R}^{2\times2}$ to $\mathbb{R}^{2\times2}$. Is L diagonalizable?

43. $T(x + iy) = x - iy$ from $\mathbb{C}$ to $\mathbb{C}$. Is T diagonalizable?

44. $T(x_0, x_1, x_2, \ldots) = (x_2, x_3, \ldots)$ from the space V of infinite sequences into V. (We drop the first two terms of the sequence.)

45. $T(x_0, x_1, x_2, \ldots) = (0, x_0, x_1, x_2, \ldots)$ from the space V of infinite sequences into V. (We insert a zero at the beginning.)

46. $T(x_0, x_1, x_2, x_3, x_4, \ldots) = (x_0, x_2, x_4, \ldots)$ from the space V of infinite sequences into V. (We drop every other term.)

47. $T(f(x)) = f(-x)$ from P_2 to P_2. Is T diagonalizable?

48. $T(f(x)) = f(2x)$ from P_2 to P_2. Is T diagonalizable?

49. $T(f(x)) = f(3x - 1)$ from P_2 to P_2. Is T diagonalizable?

50. $T(f(x)) = f(x - 3)$ from P_2 to P_2. Is T diagonalizable?

51. $T(f) = f'$ from P to P.

52. $T(f(x)) = x(f'(x))$ from P to P.

53. Show that if the matrix D of a linear transformation T with respect to a basis $\mathfrak{D}$ is diagonal, then $\mathfrak{D}$ is an eigenbasis for T.

54. Are the following matrices similar?

$$A = \begin{bmatrix} 0 & 1 & 0 & 0 \\ 0 & 0 & 0 & 0 \\ 0 & 0 & 0 & 1 \\ 0 & 0 & 0 & 0 \end{bmatrix}, \quad B = \begin{bmatrix} 0 & 1 & 0 & 0 \\ 0 & 0 & 1 & 0 \\ 0 & 0 & 0 & 0 \\ 0 & 0 & 0 & 0 \end{bmatrix}$$

(*Hint:* Compute A^2 and B^2.)

55. Find two 2×2 matrices A and B such that AB fails to be similar to BA. (*Hint:* It can be arranged that AB is zero, but BA isn't.)

56. Show that if A and B are two $n \times n$ matrices, then the matrices AB and BA have the same characteristic polynomial, and thus the same eigenvalues (matrices AB and BA need not be similar though; see Exercise 55). *Hint:*

$$\begin{bmatrix} AB & 0 \\ B & 0 \end{bmatrix} \begin{bmatrix} I_n & A \\ 0 & I_n \end{bmatrix} = \begin{bmatrix} I_n & A \\ 0 & I_n \end{bmatrix} \begin{bmatrix} 0 & 0 \\ B & BA \end{bmatrix}.$$

57. Consider an $m \times n$ matrix A and an $n \times m$ matrix B. Using Exercise 56 as a guide, show that matrices AB and BA have the same *nonzero* eigenvalues, with the same algebraic multiplicities. What about eigenvalue 0?

58. Consider a nonzero 3×3 matrix A such that $A^2 = 0$.
  a. Show that the image of A is a subspace of the kernel of A.
  b. Find the dimensions of the image and kernel of A.
  c. Pick a nonzero vector $\vec{v}_1$ in the image of A, and write $\vec{v}_1 = A\vec{v}_2$ for some $\vec{v}_2$ in $\mathbb{R}^3$. Let $\vec{v}_3$ be a vector in the kernel of A that fails to be a scalar multiple of $\vec{v}_1$. Show that $\mathfrak{B} = (\vec{v}_1, \vec{v}_2, \vec{v}_3)$ is a basis of $\mathbb{R}^3$.
  d. Find the matrix B of the linear transformation $T(\vec{x}) = A\vec{x}$ with respect to basis $\mathfrak{B}$.

59. If A and B are two nonzero 3×3 matrices such that $A^2 = B^2 = 0$, is A necessarily similar to B? (*Hint:* Exercise 58 is useful.)

60. For the matrix $A = \begin{bmatrix} 1 & -2 & 1 \\ 2 & -4 & 2 \\ 3 & -6 & 3 \end{bmatrix}$, find an invertible matrix S such that $S^{-1}AS = \begin{bmatrix} 0 & 1 & 0 \\ 0 & 0 & 0 \\ 0 & 0 & 0 \end{bmatrix}$. (*Hint:* Exercise 58 is useful.)

61. Consider an $n \times n$ matrix A such that $A^2 = 0$, with rank $A = r$. (In Example 58 we consider the case when $n = 3$ and $r = 1$.) Show that A is similar to the block matrix

$$B = \begin{bmatrix} J & 0 & \cdots & 0 & \cdots & 0 \\ 0 & J & \cdots & 0 & \cdots & 0 \\ \vdots & \vdots & \ddots & \vdots & \vdots & \vdots \\ 0 & 0 & \cdots & J & \cdots & 0 \\ \vdots & \vdots & \vdots & \vdots & \ddots & \vdots \\ 0 & 0 & \cdots & 0 & \cdots & 0 \end{bmatrix}, \quad \text{where } J = \begin{bmatrix} 0 & 1 \\ 0 & 0 \end{bmatrix}$$

Matrix B has r blocks of the form J along the diagonal, with all other entries being 0. (*Hint:* Mimic the approach outlined in Exercise 58. Pick a basis $\vec{v}_1, \ldots, \vec{v}_r$ of the image if A, write $\vec{v}_i = A\vec{w}_i$ for $i = 1, \ldots, r$, and expand $\vec{v}_1, \ldots, \vec{v}_r$ to a basis $\vec{v}_1, \ldots, \vec{v}_r, \vec{u}_1, \ldots, \vec{u}_m$ of the kernel of A. Show that $\vec{v}_1, \vec{w}_1, \vec{v}_2, \vec{w}_2, \ldots, \vec{v}_r, \vec{w}_r, \vec{u}_1, \ldots, \vec{u}_m$ is a basis of $\mathbb{R}^n$, and show that B is the matrix of $T(\vec{x}) = A\vec{x}$ with respect to this basis.)

62. Consider the linear transformation $T(f) = f'' + af' + bf$ from C^∞ to C^∞, where a and b are arbitrary constants. What does Theorem 4.1.7 tell you about the eigenvalues of T? What about the dimension of the eigenspaces of T?

63. Consider the linear transformation $T(f) = f''$ from C^∞ to C^∞. For each of the following eigenvalues, find a basis of the associated eigenspace. See Exercise 62.
a. $\lambda = 1$ **b.** $\lambda = 0$ **c.** $\lambda = -1$ **d.** $\lambda = -4$

64. If $A = \begin{bmatrix} 1 & 2 \\ 0 & 3 \end{bmatrix}$, find a basis of the linear space V of all 2×2 matrices S such that $AS = SD$, where $D = \begin{bmatrix} 1 & 0 \\ 0 & 3 \end{bmatrix}$. Find the dimension of V.

65. If $A = \begin{bmatrix} 1 & 2 \\ 4 & 3 \end{bmatrix}$, find a basis of the linear space V of all 2×2 matrices S such that $AS = SD$, where $D = \begin{bmatrix} 5 & 0 \\ 0 & -1 \end{bmatrix}$. Find the dimension of V.

66. If $A = \begin{bmatrix} 1 & 1 & 1 \\ 0 & 2 & 1 \\ 0 & 0 & 1 \end{bmatrix}$, find a basis of the linear space V of all 3×3 matrices S such that $AS = SD$, where $D = \begin{bmatrix} 1 & 0 & 0 \\ 0 & 1 & 0 \\ 0 & 0 & 2 \end{bmatrix}$. Find the dimension of V.

67. Consider a 5×5 matrix A with two distinct eigenvalues, λ_1 and λ_2, with geometric multiplicities 3 and 2, respectively. What is the dimension of the linear space of all 5×5 matrices S such that $AS = SD$, where D is the diagonal matrix with the diagonal entries $\lambda_1, \lambda_1, \lambda_1, \lambda_2, \lambda_2$?

68. If A is an $n \times n$ matrix with n distinct eigenvalues $\lambda_1, \ldots, \lambda_n$, what is the dimension of the linear space of all $n \times n$ matrices S such that $AS = SD$, where D is the diagonal matrix with the diagonal entries $\lambda_1, \ldots, \lambda_n$? Use Exercises 64 and 65 as a guide.

69. We say that two $n \times n$ matrices A and B are *simultaneously diagonalizable* if there exists an invertible $n \times n$ matrix S such that $S^{-1}AS$ and $S^{-1}BS$ are both diagonal.
a. Are the matrices

$$A = \begin{bmatrix} 1 & 0 & 0 \\ 0 & 1 & 0 \\ 0 & 0 & 1 \end{bmatrix} \quad \text{and} \quad B = \begin{bmatrix} 1 & 2 & 3 \\ 0 & 2 & 3 \\ 0 & 0 & 3 \end{bmatrix}$$

simultaneously diagonalizable? Explain.
b. Show that if A and B are simultaneously diagonalizable then $AB = BA$.
c. Give an example of two $n \times n$ matrices such that $AB = BA$, but A and B are not simultaneously diagonalizable.
d. Let D be a diagonal $n \times n$ matrix with n distinct entries on the diagonal. Find all $n \times n$ matrices B that commute with D.
e. Show that if $AB = BA$ and A has n distinct eigenvalues, then A and B are simultaneously diagonalizable. (*Hint:* Part (d) is useful.)

70. Consider an $n \times n$ matrix A with m distinct eigenvalues $\lambda_1, \ldots, \lambda_m$. Show that matrix A is diagonalizable if (and only if) $(A - \lambda_1 I_n)(A - \lambda_2 I_n) \cdots (A - \lambda_m I_n) = 0$. (*Hint:* If $(A - \lambda_1 I_n)(A - \lambda_2 I_n) \cdots (A - \lambda_m I_n) = 0$, use Exercise 4.2.83 to show that the sum of the dimensions of the eigenspaces is n.)

71. Use the method outlined in Exercise 70 to check whether the matrix $A = \begin{bmatrix} 2 & 0 & 1 \\ -1 & 1 & -1 \\ 0 & 0 & 1 \end{bmatrix}$ is diagonalizable.

72. Use the method outlined in Exercise 70 to check for which values of the constants a, b, and c the matrix

$$A = \begin{bmatrix} 1 & a & b \\ 0 & 0 & c \\ 0 & 0 & 1 \end{bmatrix} \text{ is diagonalizable.}$$

73. Prove the Cayley-Hamilton theorem, $f_A(A) = 0$, for diagonalizable matrices A. See Exercise 7.3.54.

74. In both parts of this problem, consider the matrix

$$A = \begin{bmatrix} 1 & 2 \\ 4 & 3 \end{bmatrix},$$

with eigenvalues $\lambda_1 = 5$ and $\lambda_2 = -1$ (see Example 1).

a. Are the column vectors of the matrices $A - \lambda_1 I_2$ and $A - \lambda_2 I_2$ eigenvectors of A? Explain. Does this work for other 2×2 matrices? What about diagonalizable $n \times n$ matrices with two distinct eigenvalues, such as projections or reflections? (*Hint:* Exercise 70 is helpful.)

b. Are the column vectors of

$$A - \begin{bmatrix} \lambda_1 & 0 \\ 0 & \lambda_2 \end{bmatrix}$$

eigenvectors of A? Explain.

7.5 Complex Eigenvalues

Imagine that you are diabetic and have to pay close attention to how your body metabolizes glucose. After you eat a heavy meal, the glucose concentration will reach a peak, and then it will slowly return to the fasting level. Certain hormones help regulate the glucose metabolism, especially the hormone insulin. (Compare with Exercise 7.1.52.) Let $g(t)$ be the excess glucose concentration in your blood, usually measured in milligrams of glucose per 100 milliliters of blood. (*Excess* means that we measure how much the glucose concentration deviates from the fasting level.) A negative value of $g(t)$ indicates that the glucose concentration is below fasting level at time t. Let $h(t)$ be the excess insulin concentration in your blood. Researchers have developed mathematical models for the glucose regulatory system. The following is one such model, in slightly simplified (linearized) form.

$$g(t + 1) = ag(t) - bh(t)$$
$$h(t + 1) = cg(t) + dh(t)$$

(These formulas apply between meals; obviously, the system is disturbed during and right after a meal.)

In these formulas, a, b, c, and d are positive constants; constants a and d will be less than 1. The term $-bh(t)$ expresses the fact that insulin helps your body absorb glucose, and the term $cg(t)$ represents the fact that the glucose in your blood stimulates the pancreas to secrete insulin.

For your system, the equations might be

$$g(t + 1) = 0.9g(t) - 0.4h(t)$$
$$h(t + 1) = 0.1g(t) + 0.9h(t),$$

with initial values $g(0) = 100$ and $h(0) = 0$, after a heavy meal. Here, time t is measured in hours.

After one hour, the values will be $g(1) = 90$ and $h(1) = 10$. Some of the glucose has been absorbed, and the excess glucose has stimulated the pancreas to produce 10 extra units of insulin.

The rounded values of $g(t)$ and $h(t)$ in the following table give you some sense for the evolution of this dynamical system.

t	0	1	2	3	4	5	6	7	8	15	22	29
$g(t)$	100	90	77	62.1	46.3	30.6	15.7	2.3	−9.3	−29	1.6	9.5
$h(t)$	0	10	18	23.9	27.7	29.6	29.7	28.3	25.7	−2	−8.3	0.3

We can "connect the dots" to sketch a rough trajectory, visualizing the long-term behavior. See Figure 1.

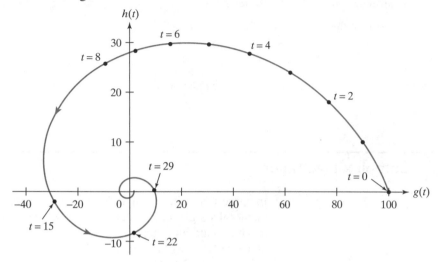

Figure 1

We see that after 7 hours the excess glucose is almost gone, but now there are about 30 units of excess insulin in the system. Since this excess insulin helps to reduce glucose further, the glucose concentration will now fall below fasting level, reaching about −30 after 15 hours. (You will feel awfully hungry by now.) Under normal circumstances, you would have taken another meal in the meantime, of course, but let's consider the case of (voluntary or involuntary) fasting.

We leave it to the reader to explain the concentrations after 22 and 29 hours, in terms of how glucose and insulin concentrations influence each other, according to our model. The *spiraling trajectory* indicates an *oscillatory behavior* of the system: Both glucose and insulin levels will swing back and forth around the fasting level, like a damped pendulum. Both concentrations will approach the fasting level (thus the name).

Another way to visualize this oscillatory behavior is to graph the functions $g(t)$ and $h(t)$ against time, using the values from our table. See Figure 2.

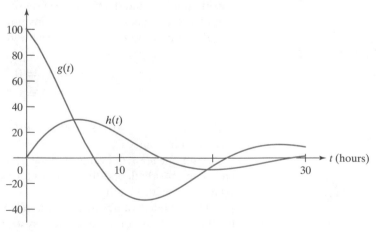

Figure 2

Next, we try to use the tools developed in the last four sections to analyze this system. We can introduce the transformation matrix

$$A = \begin{bmatrix} 0.9 & -0.4 \\ 0.1 & 0.9 \end{bmatrix}$$

and the state vector

$$\vec{x}(t) = \begin{bmatrix} g(t) \\ h(t) \end{bmatrix}.$$

Then

$$\vec{x}(t+1) = A\vec{x}(t) \quad \text{and thus} \quad \vec{x}(t) = A^t \vec{x}(0) = A^t \begin{bmatrix} 100 \\ 0 \end{bmatrix}.$$

To find formulas for $g(t)$ and $h(t)$, we need to know the eigenvalues and eigenvectors of matrix A. The characteristic polynomial of A is

$$f_A(\lambda) = \lambda^2 - 1.8\lambda + 0.85,$$

so that

$$\lambda_{1,2} = \frac{1.8 \pm \sqrt{3.24 - 3.4}}{2} = \frac{1.8 \pm \sqrt{-0.16}}{2}.$$

Since the square of a real number cannot be negative, there are no *real* eigenvalues here. However, if we allow complex solutions, then we have the eigenvalues

$$\lambda_{1,2} = \frac{1.8 \pm \sqrt{-0.16}}{2} = \frac{1.8 \pm i\sqrt{0.16}}{2} = 0.9 \pm 0.2i.$$

In this section, we will first review some basic facts on complex numbers. Then we will examine how the theory of eigenvalues and eigenvectors developed in Sections 7.1 through 7.4 can be adapted to the complex case. In Section 7.6 we will apply this work to dynamical systems. A great many dynamical systems, in physics, chemistry, biology, and economics, show oscillatory behavior; we will see that we can expect complex eigenvalues in this case.

These tools will enable you to find formulas for $g(t)$ and $h(t)$. (See Exercise 7.6.32.)

Complex Numbers: A Brief Review

Let us review some basic facts about complex numbers. We trust that you have at least a fleeting acquaintance with complex numbers. Without attempting a formal definition, we recall that a complex number can be expressed as

$$z = a + ib,$$

where a and b are real numbers.[12] Addition of complex numbers is defined in a natural way, by the rule

$$(a + ib) + (c + id) = (a + c) + i(b + d),$$

and multiplication is defined by the rule

$$(a + ib)(c + id) = (ac - bd) + i(ad + bc);$$

that is, we let $i \cdot i = -1$ and distribute.

---

[12]The letter i for the imaginary unit was introduced by Leonhard Euler, the most prolific mathematician in history. For a fascinating glimpse at the history of the complex numbers see Tobias Dantzig, *Number: The Language of Science,* Macmillan, 1954. For another intriguing introduction, full of poetry, history, and philosophy, see Barry Mazur: *Imagining Numbers (particularly the square root of minus fifteen),* Farrar, Straus, and Giroux, 2003.

If $z = a + ib$ is a complex number, we call a its *real part* [denoted by $\text{Re}(z)$] and b its *imaginary part* [denoted by $\text{Im}(z)$]. A complex number of the form ib (with $a = 0$) is called *imaginary*.

The set of all complex numbers is denoted by $\mathbb{C}$. The real numbers, $\mathbb{R}$, form a subset of $\mathbb{C}$ (namely, those complex numbers with imaginary part 0).

Complex numbers can be represented as vectors (or points) in the complex plane,[13] as shown in Figure 3. This is a graphical representation of the isomorphism

$$T \begin{bmatrix} a \\ b \end{bmatrix} = a + ib \quad \text{from } \mathbb{R}^2 \text{ to } \mathbb{C}.$$

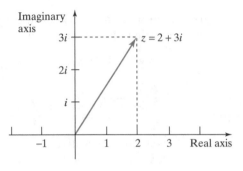

Figure 3

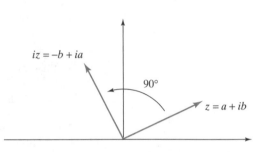

Figure 4

EXAMPLE 1   Consider a nonzero complex number z. What is the geometric relationship between z and iz in the complex plane?

Solution

If $z = a + ib$, then $iz = -b + ia$. We obtain the vector $\begin{bmatrix} -b \\ a \end{bmatrix}$ (representing iz) by rotating the vector $\begin{bmatrix} a \\ b \end{bmatrix}$ (representing z) through an angle of $90°$ in the counterclockwise direction. (See Figure 4.)   ■

The *conjugate* of a complex number $z = a + ib$ is defined by

$$\bar{z} = a - ib.$$

(The sign of the imaginary part is reversed.) We say that z and $\bar{z}$ form a *conjugate pair* of complex numbers. Geometrically, the conjugate $\bar{z}$ is the reflection of z about the real axis, as shown in Figure 5.

Sometimes it is useful to describe a complex number in *polar coordinates,* as shown in Figure 6.

The length r of the vector is called the *modulus* of z, denoted by $|z|$. The angle θ is called an *argument* of z; note that the argument is determined only up to a multiple of 2π. (Mathematicians say "modulo 2π.") For example, for $z = -1$, we can choose the argument π, $-\pi$, or 3π.

---

[13] Also called "Argand plane," after the Swiss mathematician Jean Robert Argand (1768–1822). The representation of complex numbers in the plane was introduced independently by Argand, by Gauss, and by the Norwegian mathematician Caspar Wessel (1745–1818).

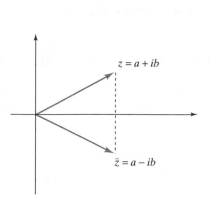

Figure 5

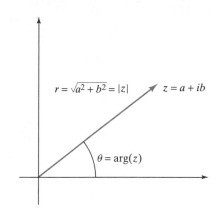

Figure 6

EXAMPLE 2   Find the modulus and an argument of $z = -2 + 2i$.

Solution

$|z| = \sqrt{2^2 + 2^2} = \sqrt{8}$. Representing z in the complex plane, we see that $\frac{3}{4}\pi$ is an argument of z. (See Figure 7.)   ∎

If z is a complex number with modulus r and argument θ, we can write z as

$$z = r(\cos\theta) + ir(\sin\theta) = r(\cos\theta + i\sin\theta),$$

as shown in Figure 8.

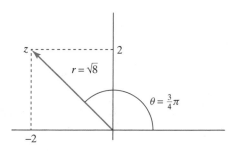

Figure 7

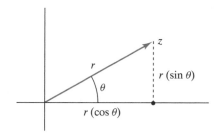

Figure 8

The representation

$$z = r(\cos\theta + i\sin\theta)$$

is called the *polar form* of the complex number z.

EXAMPLE 3   Consider the complex numbers $z = \cos\alpha + i\sin\alpha$ and $w = \cos\beta + i\sin\beta$. Find the polar form of the product zw.

Solution

Apply the addition formulas from trigonometry (see Exercise 2.2.32):

$$\begin{aligned}
zw &= (\cos\alpha + i\sin\alpha)(\cos\beta + i\sin\beta) \\
&= (\cos\alpha\cos\beta - \sin\alpha\sin\beta) + i(\sin\alpha\cos\beta + \cos\alpha\sin\beta) \\
&= \cos(\alpha + \beta) + i\sin(\alpha + \beta).
\end{aligned}$$

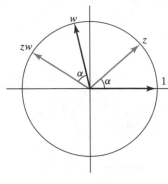

Figure 9

We conclude that the modulus of zw is 1, and $\alpha + \beta$ is an argument of zw. (See Figure 9.) ■

In general, if $z = r(\cos\alpha + i\sin\alpha)$ and $w = s(\cos\beta + i\sin\beta)$, then

$$zw = rs\big(\cos(\alpha + \beta) + i\sin(\alpha + \beta)\big).$$

When we multiply two complex numbers, we multiply the moduli, and we add the arguments:

$$|zw| = |z||w|$$
$$\arg(zw) = \arg z + \arg w \quad \text{(modulo } 2\pi\text{)}.$$

EXAMPLE 4   Describe the transformation $T(z) = (3 + 4i)z$ from $\mathbb{C}$ to $\mathbb{C}$ geometrically.

Solution

$$|T(z)| = |3 + 4i||z| = 5|z|,$$
$$\arg\big(T(z)\big) = \arg(3 + 4i) + \arg(z) = \arctan\left(\frac{4}{3}\right) + \arg(z) \approx 53° + \arg(z)$$

The transformation T is a rotation combined with a scaling in the complex plane. (See Figure 10.)

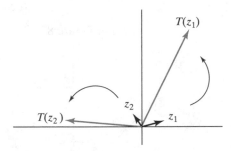

Figure 10 Rotate through about 53° and stretch the vector by a factor of 5.

Alternatively, we observe that the matrix of the linear transformation T with respect to the basis $1, i$ is $\begin{bmatrix} 3 & -4 \\ 4 & 3 \end{bmatrix}$, representing a rotation combined with a scaling.

The polar form is convenient for finding *powers* of a complex number z: If

$$z = r(\cos\theta + i\sin\theta),$$

then

$$z^2 = r^2\big(\cos(2\theta) + i\sin(2\theta)\big),$$
$$\vdots$$
$$z^n = r^n\big(\cos(n\theta) + i\sin(n\theta)\big),$$

for any positive integer n. Each time we multiply by z, the modulus is multiplied by r and the argument increases by θ. The preceding formula was found by the French mathematician Abraham de Moivre (1667–1754). ■

Theorem 7.5.1   **De Moivre's formula**

$$(\cos\theta + i\sin\theta)^n = \cos(n\theta) + i\sin(n\theta)$$ ■

EXAMPLE 5   Consider the complex number $z = 0.5 + 0.8i$. Represent the powers $z^2, z^3, \ldots$ in the complex plane. What is $\lim_{n \to \infty} z^n$?

Solution

To study the powers, write z in polar form:

$$z = r(\cos \theta + i \sin \theta).$$

Here

$$r = \sqrt{0.5^2 + 0.8^2} \approx 0.943$$

and

$$\theta = \arctan \frac{0.8}{0.5} \approx 58°.$$

We have

$$z^n = r^n \big(\cos(n\theta) + i \sin(n\theta)\big).$$

The vector representation of z^{n+1} is a little shorter than that of z^n (by about 5.7%), and z^{n+1} makes an angle $\theta \approx 58°$ with z^n. If we connect the tips of consecutive vectors, we see a trajectory that spirals in toward the origin, as shown in Figure 11. Note that $\lim_{n \to \infty} z^n = 0$, since $r = |z| < 1$.   ■

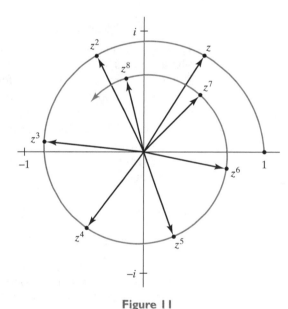

Figure 11

Perhaps the most remarkable property of the complex numbers is expressed in the fundamental theorem of algebra, first demonstrated by Carl Friedrich Gauss (in his thesis, at age 22).

Theorem 7.5.2   Fundamental theorem of algebra

Any polynomial $p(\lambda)$ with complex coefficients splits, that is, it can be written as a product of linear factors

$$p(\lambda) = k(\lambda - \lambda_1)(\lambda - \lambda_2) \ldots (\lambda - \lambda_n),$$

for some complex numbers $\lambda_1, \lambda_2, \ldots, \lambda_n$, and k. (The λ_i need not be distinct.)

Therefore, a polynomial $p(\lambda)$ of degree n has precisely n complex roots if they are properly counted with their multiplicities. ∎

For example, the polynomial

$$p(\lambda) = \lambda^2 + 1,$$

which does not have any real zeros, splits over $\mathbb{C}$:

$$p(\lambda) = (\lambda + i)(\lambda - i).$$

More generally, for a quadratic polynomial

$$q(\lambda) = \lambda^2 + b\lambda + c,$$

where b and c are real, we can find the complex roots

$$\lambda_{1,2} = \frac{-b \pm \sqrt{b^2 - 4c}}{2}$$

and

$$q(\lambda) = (\lambda - \lambda_1)(\lambda - \lambda_2).$$

Proving the fundamental theorem of algebra would lead us too far afield. Read any introduction to complex analysis or check Gauss's original proof.[14]

Complex Eigenvalues and Eigenvectors

The complex numbers share some basic algebraic properties with the real numbers.[15] Mathematicians summarize these properties by saying that both the real numbers $\mathbb{R}$ and the complex numbers $\mathbb{C}$ form a *field*. The rational numbers $\mathbb{Q}$ are another important example of a field; the integers $\mathbb{Z}$ on the other hand don't form a field. (Which of the 10 properties listed in the footnote fail to hold in this case?)

Which of the results and techniques derived in this text thus far still apply when we work with complex numbers throughout, that is, when we consider complex scalars, vectors with complex components, and matrices with complex entries? We observe that everything works the same way except for those geometrical concepts that are defined in terms of the dot product (length, angles, orthogonality, and so on, discussed in Chapter 5 and Section 6.3). The dot product in $\mathbb{C}^n$ is defined in a way that we will not discuss in this introductory text. The whole body of "core linear algebra" can be generalized without difficulty, however: echelon form, linear

---

[14]C. F. Gauss, *Werke,* III, 3–56. For an English translation see D. J. Struik (editor), *A Source Book in Mathematics 1200–1800*, Princeton University Press, 1986.

[15]Here is a list of these properties:
  1. Addition is commutative.
  2. Addition is associative.
  3. There exists a unique number n such that $a + n = a$, for all numbers a. This number n is denoted by 0.
  4. For each number a there exists a unique number b such that $a + b = 0$. This number b is denoted by $-a$. (*Comment*: This property says that we can subtract in this number system.)
  5. Multiplication is commutative.
  6. Multiplication is associative.
  7. There is a unique number e such that $ea = a$, for all numbers a. This number e is denoted by 1.
  8. For each nonzero number a there exists a unique number b such that $ab = 1$. This number b is denoted by a^{-1}. (*Comment*: This property says that we can divide by a nonzero number.)
  9. Multiplication distributes over addition: $a(b + c) = ab + ac$.
  10. The numbers 0 and 1 are not equal.

transformation, kernel, image, linear independence, basis, dimension, coordinates, linear spaces, determinant, eigenvalues, eigenvectors, and diagonalization.

EXAMPLE 6   Diagonalize the rotation-scaling matrix $A = \begin{bmatrix} a & -b \\ b & a \end{bmatrix}$ "over $\mathbb{C}$." Here, a and b are real numbers, and b is nonzero.

Solution

Following Theorem 7.4.4, we will find the eigenvalues of A first:

$$f_A(\lambda) = \det \begin{bmatrix} a - \lambda & -b \\ b & a - \lambda \end{bmatrix} = (a - \lambda)^2 + b^2 = 0$$

when

$$(a - \lambda)^2 = -b^2 \quad \text{or} \quad a - \lambda = \pm ib \quad \text{or} \quad \lambda = a \pm ib.$$

Now we find the eigenvectors:

$$E_{a+ib} = \ker \begin{bmatrix} -ib & -b \\ b & -ib \end{bmatrix} = \operatorname{span} \begin{bmatrix} i \\ 1 \end{bmatrix},$$

$$E_{a-ib} = \ker \begin{bmatrix} ib & -b \\ b & ib \end{bmatrix} = \operatorname{span} \begin{bmatrix} -i \\ 1 \end{bmatrix}.$$

Thus,

$$R^{-1} \begin{bmatrix} a & -b \\ b & a \end{bmatrix} R = \begin{bmatrix} a + ib & 0 \\ 0 & a - ib \end{bmatrix}, \quad \text{where} \quad R = \begin{bmatrix} i & -i \\ 1 & 1 \end{bmatrix}. \quad ∎$$

EXAMPLE 7   Let A be a real 2×2 matrix with eigenvalues $a \pm ib$ (where $b \neq 0$). Show that A is similar (over $\mathbb{R}$) to the matrix $\begin{bmatrix} a & -b \\ b & a \end{bmatrix}$, representing a rotation combined with a scaling.

Solution

Let $\vec{v} \pm i\vec{w}$ be eigenvectors of A with eigenvalues $a \pm ib$. (See Exercise 42.) By Theorem 7.4.1, matrix A is similar to $\begin{bmatrix} a + ib & 0 \\ 0 & a - ib \end{bmatrix}$; more precisely,

$$\begin{bmatrix} a + ib & 0 \\ 0 & a - ib \end{bmatrix} = P^{-1} A P,$$

where $P = \begin{bmatrix} \vec{v} + i\vec{w} & \vec{v} - i\vec{w} \end{bmatrix}$. By Example 6, matrix

$$\begin{bmatrix} a & -b \\ b & a \end{bmatrix}$$

is similar to

$$\begin{bmatrix} a + ib & 0 \\ 0 & a - ib \end{bmatrix}$$

as well, with

$$\begin{bmatrix} a + ib & 0 \\ 0 & a - ib \end{bmatrix} = R^{-1} \begin{bmatrix} a & -b \\ b & a \end{bmatrix} R, \quad \text{where } R = \begin{bmatrix} i & -i \\ 1 & 1 \end{bmatrix}.$$

Thus,

$$P^{-1} A P = R^{-1} \begin{bmatrix} a & -b \\ b & a \end{bmatrix} R,$$

and

$$\begin{bmatrix} a & -b \\ b & a \end{bmatrix} = RP^{-1}APR^{-1} = S^{-1}AS,$$

where $S = PR^{-1}$ and $S^{-1} = (PR^{-1})^{-1} = RP^{-1}$.

A straightforward computation shows that

$$S = PR^{-1} = \frac{1}{2i} \begin{bmatrix} v + i\vec{w} & \vec{v} - i\vec{w} \end{bmatrix} \begin{bmatrix} 1 & i \\ -1 & i \end{bmatrix} = \begin{bmatrix} \vec{w} & \vec{v} \end{bmatrix};$$

note that S has real entries, as claimed. ∎

Theorem 7.5.3   **Complex eigenvalues and rotation-scaling matrices**

If A is a real 2×2 matrix with eigenvalues $a \pm ib$ (where $b \neq 0$), and if $\vec{v} + i\vec{w}$ is an eigenvector of A with eigenvalue $a + ib$, then

$$S^{-1}AS = \begin{bmatrix} a & -b \\ b & a \end{bmatrix}, \quad \text{where} \quad S = \begin{bmatrix} \vec{w} & \vec{v} \end{bmatrix}. \quad ∎$$

We see that matrix A is similar to a rotation-scaling matrix. Those who have studied Section 5.5 can go a step further: If we introduce the inner product $\langle \vec{x}, \vec{y} \rangle = (S^{-1}\vec{x}) \cdot (S^{-1}\vec{y})$ in $\mathbb{R}^2$ and define the length of vectors and the angle between vectors with respect to this inner product, then the transformation $T(\vec{x}) = A\vec{x}$ is a rotation combined with a scaling in that inner product space. (Think about it!)

EXAMPLE 8   For $A = \begin{bmatrix} 3 & -5 \\ 1 & -1 \end{bmatrix}$, find an invertible 2×2 matrix S such that $S^{-1}AS$ is a rotation-scaling matrix.

Solution

We will use the method outlined in Theorem 7.5.3:

$$f_A(\lambda) = \lambda^2 - 2\lambda + 2, \quad \text{so that} \quad \lambda_{1,2} = \frac{2 \pm \sqrt{4 - 8}}{2} = 1 \pm i.$$

Now

$$E_{1+i} = \ker \begin{bmatrix} 2 - i & -5 \\ 1 & -2 - i \end{bmatrix} = \operatorname{span} \begin{bmatrix} -5 \\ -2 + i \end{bmatrix},$$

and

$$\begin{bmatrix} -5 \\ -2 + i \end{bmatrix} = \begin{bmatrix} -5 \\ -2 \end{bmatrix} + i \begin{bmatrix} 0 \\ 1 \end{bmatrix}, \quad \text{so that} \quad \vec{w} = \begin{bmatrix} 0 \\ 1 \end{bmatrix}, \quad \vec{v} = \begin{bmatrix} -5 \\ -2 \end{bmatrix}.$$

Therefore,

$$S^{-1}AS = \begin{bmatrix} 1 & -1 \\ 1 & 1 \end{bmatrix}, \quad \text{where} \quad S = \begin{bmatrix} 0 & -5 \\ 1 & -2 \end{bmatrix}. \quad ∎$$

The great advantage of complex eigenvalues is that there are so many of them. By the fundamental theorem of algebra, Theorem 7.5.2, the characteristic polynomial always splits:

$$f_A(\lambda) = (\lambda_1 - \lambda)(\lambda_2 - \lambda) \cdots (\lambda_n - \lambda).$$

Theorem 7.5.4   A complex $n \times n$ matrix has n complex eigenvalues if they are counted with their algebraic multiplicities. ∎

Although a complex $n \times n$ matrix may have fewer than n *distinct* complex eigenvalues (examples are I_n or $\begin{bmatrix} 0 & 1 \\ 0 & 0 \end{bmatrix}$), this is literally a coincidence. (Some of the λ_i in the factorization of the characteristic polynomial $f_A(\lambda)$ coincide.) "Most" complex $n \times n$ matrices do have n distinct eigenvalues, so that *most complex $n \times n$ matrices are diagonalizable* (by Theorem 7.4.3b). An example of a matrix that fails to be diagonalizable over $\mathbb{C}$ is $\begin{bmatrix} 0 & 1 \\ 0 & 0 \end{bmatrix}$.

In this text, we focus on diagonalizable matrices and often dismiss others as rare aberrations. Some theorems will be proven in the diagonalizable case only, with the nondiagonalizable case being left as an exercise. Much attention is given to nondiagonalizable matrices in more advanced linear algebra courses.

EXAMPLE 9 Consider an $n \times n$ matrix A with complex eigenvalues $\lambda_1, \lambda_2, \ldots, \lambda_n$, listed with their algebraic multiplicities. What is the relationship between the λ_i and the determinant of A? Compare with Theorem 7.2.8.

Solution

$$f_A(\lambda) = \det(A - \lambda I_n) = (\lambda_1 - \lambda)(\lambda_2 - \lambda) \cdots (\lambda_n - \lambda),$$
$$f_A(0) = \det A = \lambda_1 \lambda_2 \cdots \lambda_n$$

so that

$$\det(A) = \lambda_1 \lambda_2 \cdots \lambda_n \qquad \blacksquare$$

Can you interpret this result geometrically when A is a 3×3 matrix with a real eigenbasis? (*Hint:* Think about the expansion factor. See Exercise 18.)

In Example 9, we found that the determinant of a matrix is the product of its complex eigenvalues. Likewise, the trace is the *sum* of the eigenvalues. The verification is left as Exercise 35.

Theorem 7.5.5 **Trace, determinant, and eigenvalues**

Consider an $n \times n$ matrix A with complex eigenvalues $\lambda_1, \lambda_2, \ldots, \lambda_n$, listed with their algebraic multiplicities. Then

$$\operatorname{tr} A = \lambda_1 + \lambda_2 + \cdots + \lambda_n,$$

and

$$\det A = \lambda_1 \lambda_2 \cdots \lambda_n. \qquad \blacksquare$$

Note that this result is obvious for a triangular matrix: In this case, the eigenvalues are the diagonal entries.

EXERCISES 7.5

GOAL *Use the basic properties of complex numbers. Write products and powers of complex numbers in polar form. Apply the fundamental theorem of algebra.*

1. Write the complex number $z = 3 - 3i$ in polar form.

2. Find all complex numbers z such that $z^4 = 1$. Represent your answers graphically in the complex plane.

3. For an arbitrary positive integer n, find all complex numbers z such that $z^n = 1$ (in polar form). Represent your answers graphically.

4. Show that if z is a nonzero complex number, then there are exactly two complex numbers w such that $w^2 = z$. If z is in polar form, describe w in polar form.

5. Show that if z is a nonzero complex number, then there exist exactly n complex numbers w such that $w^n = z$. If z is in polar form, write w in polar form. Represent the vectors w in the complex plane.

6. If z is a nonzero complex number in polar form, describe $1/z$ in polar form. What is the relationship between the complex conjugate $\bar{z}$ and $1/z$? Represent the numbers z, $\bar{z}$, and $1/z$ in the complex plane.

7. Describe the transformation $T(z) = (1 - i)z$ from $\mathbb{C}$ to $\mathbb{C}$ geometrically.

8. Use de Moivre's formula to express $\cos(3\theta)$ and $\sin(3\theta)$ in terms of $\cos\theta$ and $\sin\theta$.

9. Consider the complex number $z = 0.8 - 0.7i$. Represent the powers $z^2, z^3, \dots$ in the complex plane and explain their long-term behavior.

10. Prove the fundamental theorem of algebra for cubic polynomials with real coefficients.

11. Express the polynomial $f(\lambda) = \lambda^3 - 3\lambda^2 + 7\lambda - 5$ as a product of linear factors over $\mathbb{C}$.

12. Consider a polynomial $f(\lambda)$ with real coefficients. Show that if a complex number λ_0 is a root of $f(\lambda)$, then so is its complex conjugate, $\bar{\lambda}_0$.

For the matrices A listed in Exercises 13 through 17, find an invertible matrix S such that $S^{-1}AS = \begin{bmatrix} a & -b \\ b & a \end{bmatrix}$, where a and b are real numbers.

13. $\begin{bmatrix} 0 & -4 \\ 1 & 0 \end{bmatrix}$

14. $\begin{bmatrix} 1 & -2 \\ 1 & -1 \end{bmatrix}$

15. $\begin{bmatrix} 0 & 1 \\ -5 & 4 \end{bmatrix}$

16. $\begin{bmatrix} 3 & 1 \\ -2 & 5 \end{bmatrix}$

17. $\begin{bmatrix} 5 & 4 \\ -5 & 1 \end{bmatrix}$ $\quad \lambda^2 - 8\lambda + 17$

18. Consider a real 2×2 matrix A with two distinct real eigenvalues, λ_1 and λ_2. Explain the formula $\det A = \lambda_1 \lambda_2$ geometrically, thinking of $|\det A|$ as an expansion factor. Illustrate your explanation with a sketch. Is there a similar geometric interpretation for a 3×3 matrix?

19. Consider a subspace V of $\mathbb{R}^n$, with $\dim(V) = m < n$.
   a. If the $n \times n$ matrix A represents the orthogonal projection onto V, what is tr A? What is $\det A$?
   b. If the $n \times n$ matrix B represents the reflection about V, what is tr B? What is $\det B$?

Find all complex eigenvalues of the matrices in Exercises 20 through 26 (including the real ones, of course). Do not use technology. Show all your work.

20. $\begin{bmatrix} 3 & -5 \\ 2 & -3 \end{bmatrix}$

21. $\begin{bmatrix} 11 & -15 \\ 6 & -7 \end{bmatrix}$

22. $\begin{bmatrix} 1 & 3 \\ -4 & 10 \end{bmatrix}$ $\quad \lambda^2$

23. $\begin{bmatrix} 0 & 0 & 1 \\ 1 & 0 & 0 \\ 0 & 1 & 0 \end{bmatrix}$

24. $\begin{bmatrix} 0 & 1 & 0 \\ 0 & 0 & 1 \\ 5 & -7 & 3 \end{bmatrix}$

25. $\begin{bmatrix} 0 & 0 & 0 & 1 \\ 1 & 0 & 0 & 0 \\ 0 & 1 & 0 & 0 \\ 0 & 0 & 1 & 0 \end{bmatrix}$

26. $\begin{bmatrix} 1 & -1 & 1 & -1 \\ 1 & 1 & 1 & 1 \\ 0 & 0 & 1 & 1 \\ 0 & 0 & 1 & 1 \end{bmatrix}$

$\lambda^3 - 3\lambda^2 + 7\lambda - 5$
$(\lambda - 1)(\lambda^2 - 2\lambda + 5)$

27. Suppose a real 3×3 matrix A has only two distinct eigenvalues. Suppose that tr $A = 1$ and $\det A = 3$. Find the eigenvalues of A with their algebraic multiplicities.

28. Suppose a 3×3 matrix A has the real eigenvalue 2 and two complex conjugate eigenvalues. Also, suppose that $\det A = 50$ and tr $A = 8$. Find the complex eigenvalues.

29. Consider a matrix of the form
$$A = \begin{bmatrix} 0 & a & b \\ c & 0 & 0 \\ 0 & d & 0 \end{bmatrix},$$
where a, b, c, and d are positive real numbers. Suppose the matrix A has three distinct real eigenvalues. What can you say about the signs of the eigenvalues? (How many of them are positive, negative, zero?) Is the eigenvalue with the largest absolute value positive or negative?

30. A real $n \times n$ matrix A is called a regular transition matrix if all entries of A are positive, and the entries in each column add up to 1. (See Exercises 24 through 31 of Section 7.2.) An example is
$$A = \begin{bmatrix} 0.4 & 0.3 & 0.1 \\ 0.5 & 0.1 & 0.2 \\ 0.1 & 0.6 & 0.7 \end{bmatrix}.$$

You may take the following properties of a regular transition matrix for granted (a partial proof is outlined in Exercise 7.2.31):
- 1 is an eigenvalue of A, with $\dim(E_1) = 1$.
- If λ is a complex eigenvalue of A other than 1, then $|\lambda| < 1$.

   a. Consider a regular $n \times n$ transition matrix A and a vector $\vec{x}$ in $\mathbb{R}^n$ whose entries add up to 1. Show that the entries of $A\vec{x}$ will also add up to 1.
   b. Pick a regular transition matrix A, and compute some powers of A (using technology): $A^2, \dots, A^{10}, \dots, A^{100}, \dots$. What do you observe? Explain your observation. Here, you may assume that there is a complex eigenbasis for A.

31. Form a 5×5 matrix by writing the integers 1, 2, 3, 4, 5 into each column in any order you like. An example is

$$A = \begin{bmatrix} 5 & 1 & 2 & 2 & 1 \\ 1 & 4 & 3 & 3 & 2 \\ 3 & 3 & 5 & 4 & 3 \\ 2 & 5 & 4 & 1 & 4 \\ 4 & 2 & 1 & 5 & 5 \end{bmatrix}.$$

(Optional question for combinatorics aficionados: How many such matrices are there?) Take higher and higher powers of the matrix you have chosen (using technology), and compare the columns of the matrices you get. What do you observe? Explain the result. (*Hint:* Exercise 30 is helpful.)

32. Most long-distance telephone service in the United States is provided by three companies: AT&T, MCI, and Sprint. The three companies are in fierce competition, offering discounts or even cash to those who switch. If the figures advertised by the companies are to be believed, people are switching their long distance provider from one month to the next according to the following diagram:

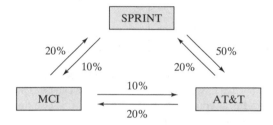

For example, 20% of the people who use AT&T go to Sprint one month later.

a. We introduce the state vector

$$\vec{x}(t) = \begin{bmatrix} a(t) \\ m(t) \\ s(t) \end{bmatrix} \begin{array}{l} \text{fraction using AT\&T} \\ \text{fraction using MCI} \\ \text{fraction using Sprint.} \end{array}$$

Find the matrix A such that $\vec{x}(t+1) = A\vec{x}(t)$, assuming that the customer base remains unchanged. Note that A is a regular transition matrix.

b. Which fraction of the customers will be with each company in the long term? Do you have to know the current market shares to answer this question? Use the power method introduced in Exercise 30.

33. *The power method for finding eigenvalues.* Consider Exercises 30 and 31 for some background. Using technology, generate a random 5×5 matrix A with nonnegative entries. (Depending on the technology you are using, the entries could be integers between zero and nine, or numbers between zero and one.) Using technology, compute $B = A^{20}$ (or another high power of A). We wish to compare the columns of B. This is hard to do by inspection, particularly because the entries of B may get rather large.

To get a better hold on B, form the diagonal 5×5 matrix D whose ith diagonal element is b_{1i}, the ith entry of the first row of B. Compute $C = BD^{-1}$.

a. How is C obtained from B? Give your answer in terms of elementary row or column operations.

b. Take a look at the columns of the matrix C you get. What do you observe? What does your answer tell you about the columns of $B = A^{20}$?

c. Explain the observations you made in part (b). You may assume that A has five distinct (complex) eigenvalues and that the eigenvalue with maximal modulus is real and positive. (We cannot explain here why this will usually be the case.)

d. Compute AC. What is the significance of the entries in the top row of this matrix in terms of the eigenvalues of A? What is the significance of the columns of C (or B) in terms of the eigenvectors of A?

34. Exercise 33 illustrates how you can use the powers of a matrix to find its dominant eigenvalue (i.e., the eigenvalue with maximal modulus), at least when this eigenvalue is real. But what about the other eigenvalues?

a. Consider an $n \times n$ matrix A with n distinct complex eigenvalues $\lambda_1, \lambda_2, \ldots, \lambda_n$, where λ_1 is real. Suppose you have a good (real) approximation λ of λ_1 (good in that $|\lambda - \lambda_1| < |\lambda - \lambda_i|$, for $i = 2, \ldots, n$). Consider the matrix $A - \lambda I_n$. What are its eigenvalues? Which has the smallest modulus? Now consider the matrix $(A - \lambda I_n)^{-1}$. What are its eigenvalues? Which has the largest modulus? What is the relationship between the eigenvectors of A and those of $(A - \lambda I_n)^{-1}$? Consider higher and higher powers of $(A - \lambda I_n)^{-1}$. How does this help you to find an eigenvector of A with eigenvalue λ_1, and λ_1 itself? Use the results of Exercise 33.

b. As an example of part (a), consider the matrix

$$A = \begin{bmatrix} 1 & 2 & 3 \\ 4 & 5 & 6 \\ 7 & 8 & 10 \end{bmatrix}.$$

We wish to find the eigenvectors and eigenvalues of A without using the corresponding commands on the computer (which is, after all, a "black box"). First, we find approximations for the eigenvalues by graphing the characteristic polynomial (use technology). Approximate the three real eigenvalues of A to the nearest integer. One of the three eigenvalues of A is negative. Find a good approximation for this eigenvalue and a corresponding eigenvector by using the procedure outlined in part (a). You are not asked to do the same for the two other eigenvalues.

35. Demonstrate the formula

$$\text{tr } A = \lambda_1 + \lambda_2 + \cdots + \lambda_n,$$

where the λ_i are the complex eigenvalues of the matrix A, counted with their algebraic multiplicities.

Hint: Consider the coefficient of λ^{n-1} in $f_A(\lambda)$ $= (\lambda_1 - \lambda)(\lambda_2 - \lambda) \cdots (\lambda_n - \lambda)$, and compare the result with Fact 7.2.5.

36. In 1990, the population of the African country Benin was about 4.6 million people. Its composition by age was as follows:

Age Bracket	0–15	15–30	30–45	45–60	60–75	75–90
Percent of Population	46.6	25.7	14.7	8.4	3.8	0.8

We represent these data in a state vector whose components are the populations in the various age brackets, in millions:

$$\vec{x}(0) = 4.6 \begin{bmatrix} 0.466 \\ 0.257 \\ 0.147 \\ 0.084 \\ 0.038 \\ 0.008 \end{bmatrix} \approx \begin{bmatrix} 2.14 \\ 1.18 \\ 0.68 \\ 0.39 \\ 0.17 \\ 0.04 \end{bmatrix}.$$

We measure time in increments of 15 years, with $t = 0$ in 1990. For example, $\vec{x}(3)$ gives the age composition in the year 2035 ($1990 + 3 \cdot 15$). If current age-dependent birth and death rates are extrapolated, we have the following model:

$$\vec{x}(t+1) = \begin{bmatrix} 1.1 & 1.6 & 0.6 & 0 & 0 & 0 \\ 0.82 & 0 & 0 & 0 & 0 & 0 \\ 0 & 0.89 & 0 & 0 & 0 & 0 \\ 0 & 0 & 0.81 & 0 & 0 & 0 \\ 0 & 0 & 0 & 0.53 & 0 & 0 \\ 0 & 0 & 0 & 0 & 0.29 & 0 \end{bmatrix} \vec{x}(t)$$

$$= A\vec{x}(t).$$

a. Explain the significance of all the entries in the matrix A in terms of population dynamics.

b. Find the eigenvalue of A with largest modulus and an associated eigenvector. (Use technology.) What is the significance of these quantities in terms of population dynamics? (For a summary on matrix techniques used in the study of age-structured populations, see Dmitrii O. Logofet, *Matrices and Graphs: Stability Problems in Mathematical Ecology*, Chapters 2 and 3, CRC Press, 1993.)

37. Consider the set $\mathbb{H}$ of all complex 2×2 matrices of the form

$$A = \begin{bmatrix} w & -\bar{z} \\ z & \bar{w} \end{bmatrix},$$

where w and z are arbitrary complex numbers.

a. Show that $\mathbb{H}$ is closed under addition and multiplication. (That is, show that the sum and the product of two matrices in $\mathbb{H}$ are again in $\mathbb{H}$.)

b. Which matrices in $\mathbb{H}$ are invertible?

c. If a matrix in $\mathbb{H}$ is invertible, is the inverse in $\mathbb{H}$ as well?

d. Find two matrices A and B in $\mathbb{H}$ such that $AB \neq BA$.

$\mathbb{H}$ is an example of a *skew field*: It satisfies all axioms for a field, except for the commutativity of multiplication. [The skew field $\mathbb{H}$ was introduced by the Irish mathematician Sir William Hamilton (1805–1865); its elements are called the *quaternions*. Another way to define the quaternions is discussed in Exercise 5.3.64.]

38. Consider the matrix

$$C_4 = \begin{bmatrix} 0 & 0 & 0 & 1 \\ 1 & 0 & 0 & 0 \\ 0 & 1 & 0 & 0 \\ 0 & 0 & 1 & 0 \end{bmatrix}.$$

a. Find the powers $C_4^2, C_4^3, C_4^4, \ldots$.

b. Find all complex eigenvalues of C_4, and construct a complex eigenbasis.

c. A 4×4 matrix M is called *circulant* if it is of the form

$$M = \begin{bmatrix} a & d & c & b \\ b & a & d & c \\ c & b & a & d \\ d & c & b & a \end{bmatrix}.$$

Circulant matrices play an important role in statistics. Show that any circulant 4×4 matrix M can be expressed as a linear combination of I_4, C_4, C_4^2, C_4^3. Use this representation to find an eigenbasis for M. What are the eigenvalues (in terms of a, b, c, d)?

39. Consider the $n \times n$ matrix C_n which has ones directly below the main diagonal and in the right upper corner, and zeros everywhere else. (See Exercise 38 for a discussion of C_4.)

a. Describe the powers of C_n.

b. Find all complex eigenvalues of C_n, and construct a complex eigenbasis.

c. Generalize part (c) of Exercise 38.

40. Consider a cubic equation

$$x^3 + px = q,$$

where $(p/3)^3 + (q/2)^2$ is negative. Show that this equation has three real solutions; write the solutions in the form $x_j = A \cos(\theta_j)$ for $j = 1, 2, 3$, expressing A and θ_j in terms of p and q. How many of the solutions are in the interval $(\sqrt{-p/3}, 2\sqrt{-p/3})$? Can there be solutions larger than $2\sqrt{-p/3}$? (*Hint*: Cardano's formula derived in Exercise 7.2.50 is useful.)

41. In his high school final examination (Aarau, Switzerland, 1896), young Albert Einstein (1879–1955) was given the following problem: In a triangle ABC, let P be the center of the inscribed circle. We are told that

$\overline{AP} = 1$, $\overline{BP} = \frac{1}{2}$, and $\overline{CP} = \frac{1}{3}$. Find the radius ρ of the inscribed circle. Einstein worked through this problem as follows:

$$\sin\left(\frac{\alpha}{2}\right) = \rho,$$

$$\sin\left(\frac{\beta}{2}\right) = 2\rho,$$

$$\sin\left(\frac{\gamma}{2}\right) = 3\rho.$$

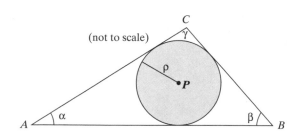

(not to scale)

For every triangle the following equation holds:

$$\sin^2\left(\frac{\alpha}{2}\right) + \sin^2\left(\frac{\beta}{2}\right) + \sin^2\left(\frac{\gamma}{2}\right)$$
$$+ 2\sin\left(\frac{\alpha}{2}\right)\sin\left(\frac{\beta}{2}\right)\sin\left(\frac{\gamma}{2}\right) = 1.$$

In our case

$$14\rho^2 + 12\rho^3 - 1 = 0.$$

Now let

$$\rho = \frac{1}{x}.$$

At this point we interrupt Einstein's work and ask you to finish the job. [*Hint:* Exercise 40 is helpful. Find the exact solution (in terms of trigonometric and inverse trigonometric functions), and give a numerical approximation as well.] (By the way, Einstein, who was allowed to use a logarithm table, solved the problem correctly.) (*Source: The Collected Papers of Albert Einstein*, Vol. 1, Princeton University Press, 1987.)

42. Consider a complex $n \times m$ matrix A. The conjugate $\overline{A}$ is defined by taking the conjugate of each entry of A. For example, if

$$A = \begin{bmatrix} 2 + 3i & 5 \\ 2i & 9 \end{bmatrix}, \quad \text{then} \quad \overline{A} = \begin{bmatrix} 2 - 3i & 5 \\ -2i & 9 \end{bmatrix}.$$

a. Show that if A and B are complex $n \times p$ and $p \times m$ matrices, respectively, then

$$\overline{AB} = \overline{A}\,\overline{B}.$$

b. Let A be a real $n \times n$ matrix and $\vec{v} + i\vec{w}$ an eigenvector of A with eigenvalue $p + iq$. Show that the vector $\vec{v} - i\vec{w}$ is an eigenvector of A with eigenvalue $p - iq$.

43. Consider two real $n \times n$ matrices A and B that are "similar over $\mathbb{C}$": That is, there is a complex invertible $n \times n$ matrix S such that $B = S^{-1}AS$. Show that A and B are in fact "similar over $\mathbb{R}$": That is, there is a real R such that $B = R^{-1}AR$. (*Hint:* Write $S = S_1 + iS_2$, where S_1 and S_2 are real. Consider the function $f(z) = \det(S_1 + zS_2)$, where z is a complex variable. Show that $f(z)$ is a nonzero polynomial. Conclude that there is a real number x such that $f(x) \neq 0$. Show that $R = S_1 + xS_2$ does the job.)

44. Show that every complex 2×2 matrix is similar to an upper triangular 2×2 matrix. Can you generalize this result to square matrices of larger size? (*Hint:* Argue by induction.)

For which values of the real constant a are the matrices in Exercises 45 through 50 diagonalizable over $\mathbb{C}$?

45. $\begin{bmatrix} 1 & 1 \\ a & 1 \end{bmatrix}$ 46. $\begin{bmatrix} 0 & -a \\ a & 0 \end{bmatrix}$ 47. $\begin{bmatrix} 0 & 0 & 0 \\ 1 & 0 & a \\ 0 & 1 & 0 \end{bmatrix}$

48. $\begin{bmatrix} 0 & 0 & a \\ 1 & 0 & 3 \\ 0 & 1 & 0 \end{bmatrix}$ 49. $\begin{bmatrix} 0 & 1 & 0 \\ 0 & 0 & 1 \\ 0 & 1-a & a \end{bmatrix}$

50. $\begin{bmatrix} -a & a & -a \\ -a-1 & a+1 & -a-1 \\ 0 & 0 & 0 \end{bmatrix}$

For Exercises 51 through 55 state whether the given set is a field (with the customary addition and multiplication).

51. The rational numbers $\mathbb{Q}$

52. The integers $\mathbb{Z}$

53. The binary digits (introduced in Exercises 3.1.53 and 3.1.54)

54. The rotation-scaling matrices of the form $\begin{bmatrix} p & -q \\ q & p \end{bmatrix}$, where p and q are real numbers

55. The set H considered in Exercise 5.3.64

7.6 Stability

In applications, the long-term behavior is often the most important qualitative feature of a dynamical system. We are frequently faced with the following situation: The state $\vec{0}$ represents an equilibrium of the system (in physics, ecology, or economics, for example). If the system is disturbed (moved into another state, away from the equilibrium $\vec{0}$) and then left to its own devices, will it always return to the equilibrium state $\vec{0}$?

EXAMPLE 1   Consider a dynamical system $\vec{x}(t + 1) = A\vec{x}(t)$, where A is an $n \times n$ matrix. Suppose an initial state vector $\vec{x}_0$ is given. We are told that A has n distinct complex eigenvalues and that the modulus of each eigenvalue is less than 1. What can you say about the long-term behavior of the system, that is, about $\lim_{t \to \infty} \vec{x}(t)$?

Solution

For each complex eigenvalue λ_i, we can choose a complex eigenvector $\vec{v}_i$. Then the $\vec{v}_i$ form a complex eigenbasis for A (by Theorem 7.3.5). We can write $\vec{x}_0$ as a complex linear combination of the $\vec{v}_i$:

$$\vec{x}_0 = c_1 \vec{v}_1 + \cdots + c_n \vec{v}_n$$

Then

$$\vec{x}(t) = A^t \vec{x}_0 = c_1 \lambda_1^t \vec{v}_1 + \cdots + c_n \lambda_n^t \vec{v}_n.$$

By Example 5 of Section 7.5,

$$\lim_{t \to \infty} \lambda_i^t = 0, \quad \text{since} \quad |\lambda_i| < 1.$$

Therefore,

$$\lim_{t \to \infty} \vec{x}(t) = \vec{0}. \qquad \blacksquare$$

For the discussion of the long-term behavior of a dynamical system, the following definition is useful:

Definition 7.6.1   Stable equilibrium

Consider a dynamical system

$$\vec{x}(t + 1) = A\vec{x}(t).$$

We say that $\vec{0}$ is an (asymptotically) *stable equilibrium* for this system if

$$\lim_{t \to \infty} \vec{x}(t) = \vec{0}$$

for all its trajectories.[16]

Note that the zero state is stable if (and only if)

$$\lim_{t \to \infty} A^t = 0$$

(meaning that all entries of A^t approach zero). See Exercise 36.

Consider the examples shown in Figure 1.

---

[16]In this text, *stable* will always mean "asymptotically stable." Several other notions of stability are used in applied mathematics.

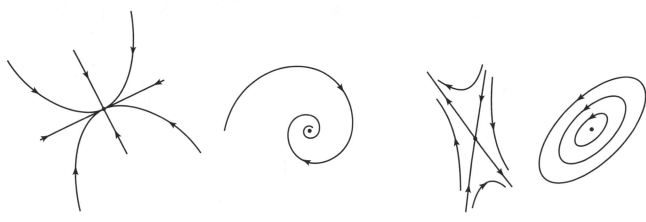

Figure I(a) Asymptotically stable.   **Figure I(b)** Not asymptotically stable.

Generalizing Example 1, we have the following result:

Theorem 7.6.2   Stability and eigenvalues

Consider a dynamical system $\vec{x}(t+1) = A\vec{x}(t)$. The zero state is asymptotically stable if (and only if) the modulus of all the complex eigenvalues of A is less than 1.

∎

Example 1 illustrates this fact only when A is diagonalizable (i.e., when there is a complex eigenbasis for A); recall that this is the case for most matrices A. In Exercises 45 through 50 of Section 8.1, we will discuss the nondiagonalizable case.

For an illustration of Theorem 7.6.2, see Figure 11 of Section 7.1, where we sketched the phase portraits of 2×2 matrices with two distinct positive eigenvalues.

We will now turn our attention to the phase portraits of 2×2 matrices with complex eigenvalues $p \pm iq$ (where $q \neq 0$).

EXAMPLE 2   Consider the dynamical system

$$\vec{x}(t+1) = \begin{bmatrix} p & -q \\ q & p \end{bmatrix} \vec{x}(t),$$

where p and q are real, and q is nonzero. Examine the stability of this system. Sketch phase portraits. Discuss your results in terms of Theorem 7.6.2.

Solution

As in Theorem 2.2.4, we can write

$$\begin{bmatrix} p & -q \\ q & p \end{bmatrix} = r \begin{bmatrix} \cos\theta & -\sin\theta \\ \sin\theta & \cos\theta \end{bmatrix},$$

representing the transformation as a rotation through an angle θ combined with a scaling by $r = \sqrt{p^2 + q^2}$. Then

$$\vec{x}(t) = \begin{bmatrix} p & -q \\ q & p \end{bmatrix}^t \vec{x}_0 = r^t \begin{bmatrix} \cos(\theta t) & -\sin(\theta t) \\ \sin(\theta t) & \cos(\theta t) \end{bmatrix} \vec{x}_0,$$

representing a rotation through an angle θt combined with a scaling by r^t.

Figure 2 illustrates that the zero state is stable if $r = \sqrt{p^2 + q^2} < 1$.

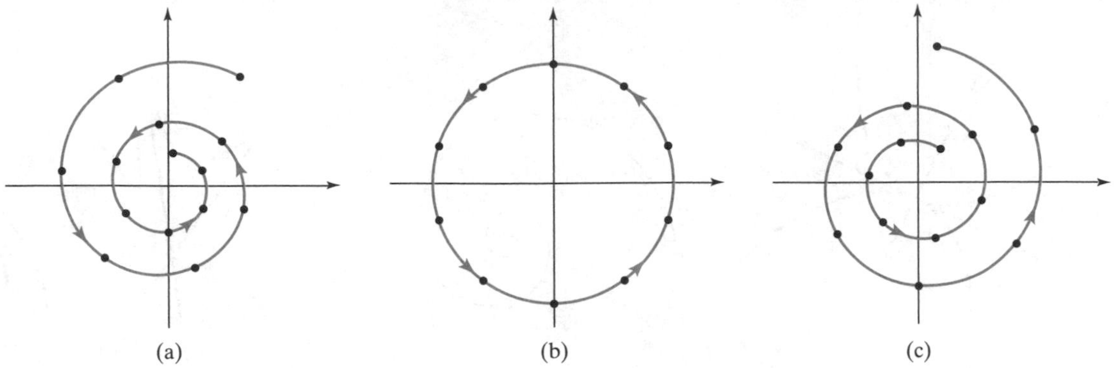

(a)   (b)   (c)

Figure 2 (a) $r < 1$: trajectories spiral inward. (b) $r = 1$: trajectories are circles.
(c) $r > 1$: trajectories spiral outward.

Alternatively, we can use Theorem 7.6.2 to examine the stability of the system. From Example 6 of Section 7.5, we know that the eigenvalues of $\begin{bmatrix} p & -q \\ q & p \end{bmatrix}$ are $\lambda_{1,2} = p \pm iq$, with $|\lambda_1| = |\lambda_2| = \sqrt{p^2 + q^2}$. By Theorem 7.6.2, the zero state is stable if $\sqrt{p^2 + q^2} < 1$. ∎

Let us generalize Example 2. If A is any 2×2 matrix with eigenvalues $\lambda_{1,2} = p \pm iq$, what does the phase portrait of the dynamical system $\vec{x}(t+1) = A\vec{x}(t)$ look like? Let $\vec{v} + i\vec{w}$ be an eigenvector of A with eigenvalue $p + iq$. From Theorem 7.5.3, we know that A is similar to the rotation–scaling matrix $\begin{bmatrix} p & -q \\ q & p \end{bmatrix}$, with

$$S^{-1}AS = \begin{bmatrix} p & -q \\ q & p \end{bmatrix} \quad \text{or} \quad A = S\begin{bmatrix} p & -q \\ q & p \end{bmatrix}S^{-1}, \quad \text{where} \quad S = \begin{bmatrix} \vec{w} & \vec{v} \end{bmatrix}.$$

Using the terminology introduced in Example 2, we find that

$$\vec{x}(t) = A^t\vec{x}_0 = S\begin{bmatrix} p & -q \\ q & p \end{bmatrix}^t S^{-1}\vec{x}_0 = r^t S\begin{bmatrix} \cos(\theta t) & -\sin(\theta t) \\ \sin(\theta t) & \cos(\theta t) \end{bmatrix} S^{-1}\vec{x}_0.$$

Theorem 7.6.3   **Dynamical systems with complex eigenvalues**

Consider the dynamical system $\vec{x}(t+1) = A\vec{x}(t)$, where A is a real 2×2 matrix with eigenvalues

$$\lambda_{1,2} = p \pm iq = r(\cos(\theta) \pm i\sin(\theta)), \quad \text{where} \quad q \neq 0.$$

Let $\vec{v} + i\vec{w}$ be an eigenvector of A with eigenvalue $p + iq$.
Then

$$\vec{x}(t) = r^t S\begin{bmatrix} \cos(\theta t) & -\sin(\theta t) \\ \sin(\theta t) & \cos(\theta t) \end{bmatrix} S^{-1}\vec{x}_0, \quad \text{where} \quad S = \begin{bmatrix} \vec{w} & \vec{v} \end{bmatrix}.$$

Note that $S^{-1}\vec{x}_0$ is the coordinate vector of $\vec{x}_0$ with respect to basis $\vec{w}, \vec{v}$. ∎

EXAMPLE 3   Consider the dynamical system

$$\vec{x}(t+1) = \begin{bmatrix} 3 & -5 \\ 1 & -1 \end{bmatrix} \vec{x}(t) \quad \text{with initial state} \quad \vec{x}_0 = \begin{bmatrix} 0 \\ 1 \end{bmatrix}.$$

Use Theorem 7.6.3 to find a closed formula for $\vec{x}(t)$, and sketch the trajectory.

Solution

In Example 8 of Section 7.5, we found the eigenvalues

$$\lambda_{1,2} = 1 \pm i.$$

The polar coordinates of eigenvalue $1 + i$ are $r = \sqrt{2}$ and $\theta = \frac{\pi}{4}$. Furthermore, we found that

$$S = \begin{bmatrix} \vec{w} & \vec{v} \end{bmatrix} = \begin{bmatrix} 0 & -5 \\ 1 & -2 \end{bmatrix}.$$

Since

$$S^{-1}\vec{x}_0 = \begin{bmatrix} 1 \\ 0 \end{bmatrix},$$

Theorem 7.6.3 gives

$$\vec{x}(t) = (\sqrt{2})^t \begin{bmatrix} 0 & -5 \\ 1 & -2 \end{bmatrix} \begin{bmatrix} \cos\left(\frac{\pi}{4}t\right) & -\sin\left(\frac{\pi}{4}t\right) \\ \sin\left(\frac{\pi}{4}t\right) & \cos\left(\frac{\pi}{4}t\right) \end{bmatrix} \begin{bmatrix} 1 \\ 0 \end{bmatrix}$$

$$= (\sqrt{2})^t \begin{bmatrix} -5\sin\left(\frac{\pi}{4}t\right) \\ \cos\left(\frac{\pi}{4}t\right) - 2\sin\left(\frac{\pi}{4}t\right) \end{bmatrix}.$$

We leave it to the reader to work out the details of this computation.

Next, let's think about the trajectory. We will develop the trajectory step by step:

- The points

$$\begin{bmatrix} \cos\left(\frac{\pi}{4}t\right) & -\sin\left(\frac{\pi}{4}t\right) \\ \sin\left(\frac{\pi}{4}t\right) & \cos\left(\frac{\pi}{4}t\right) \end{bmatrix} \begin{bmatrix} 1 \\ 0 \end{bmatrix} \quad \text{(for } t = 0, 1, 2, \ldots)$$

  are located on the unit circle, as shown in Figure 3a. Note that at $t = 8$ the system returns to its initial position, $\begin{bmatrix} 1 \\ 0 \end{bmatrix}$; the period of this system is 8.

- In Exercise 2.2.50, we saw that an invertible linear transformation maps the unit circle into an ellipse. Thus, the points

$$\begin{bmatrix} 0 & -5 \\ 1 & -2 \end{bmatrix} \begin{bmatrix} \cos\left(\frac{\pi}{4}t\right) & -\sin\left(\frac{\pi}{4}t\right) \\ \sin\left(\frac{\pi}{4}t\right) & \cos\left(\frac{\pi}{4}t\right) \end{bmatrix} \begin{bmatrix} 1 \\ 0 \end{bmatrix}$$

  are located on an ellipse, as shown in Figure 3b. The two column vectors of

$$S = \begin{bmatrix} 0 & -5 \\ 1 & -2 \end{bmatrix} = \begin{bmatrix} \vec{w} & \vec{v} \end{bmatrix}$$

  are shown in that figure as well. Again, the period of this system is 8.

- The exponential growth factor $(\sqrt{2})^t$ will produce longer and longer vectors

$$\vec{x}(t) = (\sqrt{2})^t \begin{bmatrix} 0 & -5 \\ 1 & -2 \end{bmatrix} \begin{bmatrix} \cos\left(\frac{\pi}{4}t\right) & -\sin\left(\frac{\pi}{4}t\right) \\ \sin\left(\frac{\pi}{4}t\right) & \cos\left(\frac{\pi}{4}t\right) \end{bmatrix} \begin{bmatrix} 1 \\ 0 \end{bmatrix}.$$

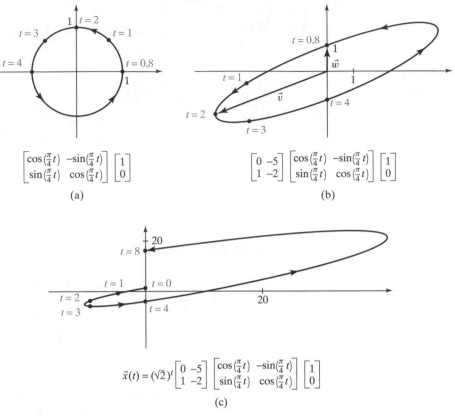

Figure 3

Thus, the trajectory *spirals outward*, as shown in Figure 3c. (We are using different scales in Figures 3a, b, and c.) Note that $\vec{x}(8) = (\sqrt{2})^8\vec{x}(0) = 16\vec{x}(0)$. ∎

We can generalize our findings in Example 3.

Theorem 7.6.4   **Phase portrait of a system with complex eigenvalues**

Consider a dynamical system

$$\vec{x}(t+1) = A\vec{x}(t),$$

where A is a real 2×2 matrix with eigenvalues $\lambda_{1,2} = p \pm iq$ (where $q \neq 0$). Let

$$r = |\lambda_1| = |\lambda_2| = \sqrt{p^2 + q^2}.$$

If $r = 1$, then the points $\vec{x}(t)$ are located on an ellipse; if r exceeds 1, then the trajectory spirals outward; and if r is less than 1, then the trajectory spirals inward, approaching the origin. ∎

Theorem 7.6.4 provides another illustration of Theorem 7.6.2: The zero state is stable if (and only if) $r = |\lambda_1| = |\lambda_2| < 1$.

If you have to sketch a trajectory of a system with complex eigenvalues without the aid of technology, it helps to compute and plot the first few points $\vec{x}(0)$, $\vec{x}(1), \vec{x}(2), \ldots$, until you see a trend.

EXERCISES 7.6

GOAL *Use eigenvalues to determine the stability of a dynamical system. Analyze the dynamical system $\vec{x}(t+1) = A\vec{x}(t)$, where A is a real 2×2 matrix with eigenvalues $p \pm iq$.*

For the matrices A in Exercises 1 through 10, determine whether the zero state is a stable equilibrium of the dynamical system $\vec{x}(t+1) = A\vec{x}(t)$.

1. $A = \begin{bmatrix} 0.9 & 0 \\ 0 & 0.8 \end{bmatrix}$

2. $A = \begin{bmatrix} -1.1 & 0 \\ 0 & 0.9 \end{bmatrix}$

3. $A = \begin{bmatrix} 0.8 & 0.7 \\ -0.7 & 0.8 \end{bmatrix}$

4. $A = \begin{bmatrix} -0.9 & -0.4 \\ 0.4 & -0.9 \end{bmatrix}$

5. $A = \begin{bmatrix} 0.5 & 0.6 \\ -0.3 & 1.4 \end{bmatrix}$

6. $A = \begin{bmatrix} -1 & 3 \\ -1.2 & 2.6 \end{bmatrix}$

7. $A = \begin{bmatrix} 2.4 & -2.5 \\ 1 & -0.6 \end{bmatrix}$

8. $A = \begin{bmatrix} 1 & -0.2 \\ 0.1 & 0.7 \end{bmatrix}$

9. $A = \begin{bmatrix} 0.8 & 0 & -0.6 \\ 0 & 0.7 & 0 \\ 0.6 & 0 & 0.8 \end{bmatrix}$

10. $A = \begin{bmatrix} 0.3 & 0.3 & 0.3 \\ 0.3 & 0.3 & 0.3 \\ 0.3 & 0.3 & 0.3 \end{bmatrix}$

Consider the matrices A in Exercises 11 through 16. For which real numbers k is the zero state a stable equilibrium of the dynamical system $\vec{x}(t+1) = A\vec{x}(t)$?

11. $A = \begin{bmatrix} k & 0 \\ 0 & 0.9 \end{bmatrix}$

12. $A = \begin{bmatrix} 0.6 & k \\ -k & 0.6 \end{bmatrix}$

13. $A = \begin{bmatrix} 0.7 & k \\ 0 & -0.9 \end{bmatrix}$

14. $A = \begin{bmatrix} k & k \\ k & k \end{bmatrix}$

15. $A = \begin{bmatrix} 1 & k \\ 0.01 & 1 \end{bmatrix}$

16. $A = \begin{bmatrix} 0.1 & k \\ 0.3 & 0.3 \end{bmatrix}$

For the matrices A in Exercises 17 through 24, find real closed formulas for the trajectory $\vec{x}(t+1) = A\vec{x}(t)$, where $\vec{x}(0) = \begin{bmatrix} 0 \\ 1 \end{bmatrix}$. Draw a rough sketch.

17. $A = \begin{bmatrix} 0.6 & -0.8 \\ 0.8 & 0.6 \end{bmatrix}$

18. $A = \begin{bmatrix} -0.8 & 0.6 \\ -0.8 & -0.8 \end{bmatrix}$

19. $A = \begin{bmatrix} 2 & -3 \\ 3 & 2 \end{bmatrix}$

20. $A = \begin{bmatrix} 4 & 3 \\ -3 & 4 \end{bmatrix}$

21. $A = \begin{bmatrix} 1 & 5 \\ -2 & 7 \end{bmatrix}$

22. $A = \begin{bmatrix} 7 & -15 \\ 6 & -11 \end{bmatrix}$

23. $A = \begin{bmatrix} -0.5 & 1.5 \\ -0.6 & 1.3 \end{bmatrix}$

24. $A = \begin{bmatrix} 1 & -3 \\ 1.2 & -2.6 \end{bmatrix}$

Consider an invertible $n \times n$ matrix A such that the zero state is a stable equilibrium of the dynamical system $\vec{x}(t+1) = A\vec{x}(t)$. What can you say about the stability of the systems listed in Exercises 25 through 30?

25. $\vec{x}(t+1) = A^{-1}\vec{x}(t)$

26. $\vec{x}(t+1) = A^T\vec{x}(t)$

27. $\vec{x}(t+1) = -A\vec{x}(t)$

28. $\vec{x}(t+1) = (A - 2I_n)\vec{x}(t)$

29. $\vec{x}(t+1) = (A + I_n)\vec{x}(t)$

30. $\vec{x}(t+1) = A^2\vec{x}(t)$

31. Let A be a real 2×2 matrix. Show that the zero state is a stable equilibrium of the dynamical system $\vec{x}(t+1) = A\vec{x}(t)$ if (and only if)

$$|\text{tr } A| - 1 < \det A < 1.$$

32. Let's revisit the introductory example of Section 7.5: The glucose regulatory system of a certain patient can be modeled by the equations

$$g(t+1) = 0.9g(t) - 0.4h(t)$$
$$h(t+1) = 0.1g(t) + 0.9h(t).$$

Find closed formulas for $g(t)$ and $h(t)$, and draw the trajectory. Does your trajectory look like the one on page 344?

33. Consider a real 2×2 matrix A with eigenvalues $p \pm iq$ and corresponding eigenvectors $\vec{v} \pm i\vec{w}$. Show that if a real vector $\vec{x}_0$ is written as $\vec{x}_0 = c_1(\vec{v}+i\vec{w})+c_2(\vec{v}-i\vec{w})$, then $c_2 = \bar{c}_1$.

34. Consider a dynamical system $\vec{x}(t+1) = A\vec{x}(t)$, where A is a real $n \times n$ matrix.
   a. If $|\det A| \geq 1$, what can you say about the stability of the zero state?
   b. If $|\det A| < 1$, what can you say about the stability of the zero state?

35. a. Consider a real $n \times n$ matrix with n distinct real eigenvalues $\lambda_1, \ldots, \lambda_n$, where $|\lambda_i| \leq 1$ for all $i = 1, \ldots, n$. Let $\vec{x}(t)$ be a trajectory of the dynamical system $\vec{x}(t+1) = A\vec{x}(t)$. Show that this trajectory is *bounded;* that is, there is a positive number M such that $\|\vec{x}(t)\| \leq M$ for all positive integers t.
   b. Are all trajectories of the dynamical system

$$\vec{x}(t+1) = \begin{bmatrix} 1 & 1 \\ 0 & 1 \end{bmatrix} \vec{x}(t)$$

   bounded? Explain.

36. Show that the zero state is a stable equilibrium of the dynamical system $\vec{x}(t+1) = A\vec{x}(t)$ if (and only if)

$$\lim_{t \to \infty} A^t = 0$$

(meaning that all entries of A^t approach zero).

37. Consider the national income of a country, which consists of consumption, investment, and government expenditures. Here we assume the government expenditure to be constant, at G_0, while the national income $Y(t)$, consumption $C(t)$, and investment $I(t)$ change over time. According to a simple model, we have

$$
\begin{vmatrix}
Y(t) & = C(t) + I(t) + G_0 \\
C(t+1) = \gamma Y(t) \\
I(t+1) = \alpha\big(C(t+1) - C(t)\big)
\end{vmatrix}
\quad
\begin{matrix}
(0 < \gamma < 1), \\
(\alpha > 0)
\end{matrix}
$$

where γ is the marginal propensity to consume and α is the acceleration coefficient. (See Paul E. Samuelson, "Interactions between the Multiplier Analysis and the Principle of Acceleration," *Review of Economic Statistics,* May 1939, pp. 75–78.)

a. Find the equilibrium solution of these equations, when $Y(t+1) = Y(t)$, $C(t+1) = C(t)$, and $I(t+1) = I(t)$.

b. Let $y(t), c(t)$, and $i(t)$ be the deviations of $Y(t), C(t)$, and $I(t)$, respectively, from the equilibrium state you found in part (a). These quantities are related by the equations

$$
\begin{vmatrix}
y(t) & = c(t) + i(t) \\
c(t+1) = \gamma y(t) \\
i(t+1) = \alpha\big(c(t+1) - c(t)\big)
\end{vmatrix}.
$$

(Verify this!) By substituting $y(t)$ into the second equation, set up equations of the form

$$
\begin{vmatrix}
c(t+1) = pc(t) + qi(t) \\
i(t+1) = rc(t) + si(t)
\end{vmatrix}.
$$

c. When $\alpha = 5$ and $\gamma = 0.2$, determine the stability of the zero state of this system.

d. When $\alpha = 1$ (and γ is arbitrary, $0 < \gamma < 1$), determine the stability of the zero state.

e. For each of the four sectors in the α–γ-plane, determine the stability of the zero state.

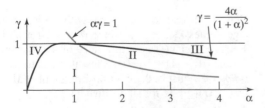

Discuss the various cases, in practical terms.

38. Consider an affine transformation

$$
T(\vec{x}) = A\vec{x} + \vec{b},
$$

where A is an $n \times n$ matrix and $\vec{b}$ is a vector in $\mathbb{R}^n$. (Compare this with Exercise 7.3.45.) Suppose that 1 is not an eigenvalue of A.

a. Find the vector $\vec{v}$ in $\mathbb{R}^n$ such that $T(\vec{v}) = \vec{v}$; this vector is called the *equilibrium state* of the dynamical system $\vec{x}(t+1) = T\big(\vec{x}(t)\big)$.

b. When is the equilibrium $\vec{v}$ in part (a) stable (meaning that $\lim_{t\to\infty} \vec{x}(t) = \vec{v}$ for all trajectories)?

39. Consider the dynamical system

$$
\begin{aligned}
x_1(t+1) &= 0.1x_1(t) + 0.2x_2(t) + 1, \\
x_2(t+1) &= 0.4x_1(t) + 0.3x_2(t) + 2.
\end{aligned}
$$

(See Exercise 7.3.45.) Find the equilibrium state of this system and determine its stability. (See Exercise 38.) Sketch a phase portrait.

40. Consider the matrix

$$
A =
\begin{bmatrix}
p & -q & -r & -s \\
q & p & s & -r \\
r & -s & p & q \\
s & r & -q & p
\end{bmatrix},
$$

where p, q, r, s are arbitrary real numbers. (Compare this with Exercise 5.3.64.)

a. Compute $A^T A$.

b. For which values of p, q, r, s is A invertible? Find the inverse if it exists.

c. Find the determinant of A.

d. Find the complex eigenvalues of A.

e. If $\vec{x}$ is a vector in $\mathbb{R}^4$, what is the relationship between $\|\vec{x}\|$ and $\|A\vec{x}\|$?

f. Consider the numbers

$$
59 = 3^2 + 3^2 + 4^2 + 5^2
$$

and

$$
37 = 1^2 + 2^2 + 4^2 + 4^2.
$$

Express the number

$$
2183
$$

as the sum of the squares of four integers:

$$
2183 = a^2 + b^2 + c^2 + d^2.
$$

[*Hint:* Part (e) is useful. Note that $2183 = 59 \cdot 37$.]

g. The French mathematician Joseph-Louis Lagrange (1736–1813) showed that any prime number can be expressed as the sum of the squares of four integers. Using this fact and your work in part (f) as a guide, show that any positive integer can be expressed in this way.

41. Find a 2×2 matrix A without real eigenvalues and a vector $\vec{x}_0$ in $\mathbb{R}^2$ such that for all positive integers t the point $A^t \vec{x}_0$ is located on the ellipse in the accompanying sketch.

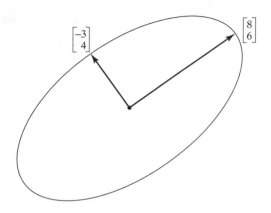

42. We quote from a text on computer graphics (M. Beeler et al., "HAKMEM," MIT Artificial Intelligence Report AIM-239, 1972):

> Here is an elegant way to draw almost circles on a point-plotting display.

```
CIRCLE ALGORITHM:

NEW X = OLD X - K*OLD Y;
NEW Y = OLD Y + K*NEW X.
```

This makes a very round ellipse centered at the origin with its size determined by the initial point. The circle algorithm was invented by mistake when I tried to save a register in a display hack!

(In the preceding formula, k is a small number.) Here, a dynamical system is defined in "computer lingo." In our terminology, the formulas are

$$x(t + 1) = x(t) - ky(t),$$
$$y(t + 1) = y(t) + kx(t + 1).$$

a. Find the matrix of this transformation. (Note the entry $x(t + 1)$ in the second formula.)

b. Explain why the trajectories are ellipses, as claimed.

Chapter Seven Exercises

TRUE OR FALSE?

1. The algebraic multiplicity of an eigenvalue cannot exceed its geometric multiplicity.

2. If an $n \times n$ matrix A is diagonalizable (over $\mathbb{R}$), then there must be a basis of $\mathbb{R}^n$ consisting of eigenvectors of A.

3. If the standard vectors $\vec{e}_1, \vec{e}_2, \dots, \vec{e}_n$ are eigenvectors of an $n \times n$ matrix A, then A must be diagonal.

4. If $\vec{v}$ is an eigenvector of A, then $\vec{v}$ must be an eigenvector of A^3 as well.

5. There exists a diagonalizable 5×5 matrix with only two distinct eigenvalues (over $\mathbb{C}$).

6. There exists a real 5×5 matrix without any real eigenvalues.

7. If 0 is an eigenvalue of a matrix A, then $\det A = 0$.

8. The eigenvalues of a 2×2 matrix A are the solutions of the equation $\lambda^2 - (\text{tr}A)\lambda + (\det A) = 0$.

9. The eigenvalues of any triangular matrix are its diagonal entries.

10. The trace of any square matrix is the sum of its diagonal entries.

11. Any rotation-scaling matrix in $\mathbb{R}^{2 \times 2}$ is diagonalizable over $\mathbb{C}$.

12. If A is a noninvertible $n \times n$ matrix, then the geometric multiplicity of eigenvalue 0 is $n - \text{rank}(A)$.

13. If matrix A is diagonalizable, then its transpose A^T must be diagonalizable as well.

14. If A and B are two 3×3 matrices such that $\text{tr } A = \text{tr } B$ and $\det A = \det B$, then A and B must have the same eigenvalues.

15. If 1 is the only eigenvalue of an $n \times n$ matrix A, then A must be I_n.

16. If A and B are $n \times n$ matrices, if α is an eigenvalue of A, and if β is an eigenvalue of B, then $\alpha\beta$ must be an eigenvalue of AB.

17. If 3 is an eigenvalue of an $n \times n$ matrix A, then 9 must be an eigenvalue of A^2.

18. The matrix of any orthogonal projection onto a subspace V of $\mathbb{R}^n$ is diagonalizable.

19. If matrices A and B have the same eigenvalues (over $\mathbb{C}$), with the same algebraic multiplicities, then matrices A and B must have the same trace.

20. If a real matrix A has only the eigenvalues 1 and -1, then A must be orthogonal.

21. If an invertible matrix A is diagonalizable, then A^{-1} must be diagonalizable as well.

22. If $\det(A) = \det(A^T)$, then matrix A must be symmetric.

23. If matrix $A = \begin{bmatrix} 7 & a & b \\ 0 & 7 & c \\ 0 & 0 & 7 \end{bmatrix}$ is diagonalizable, then a, b, and c must all be zero.

24. If two $n \times n$ matrices A and B are diagonalizable, then $A + B$ must be diagonalizable as well.

25. All diagonalizable matrices are invertible.

26. If vector $\vec{v}$ is an eigenvector of both A and B, then $\vec{v}$ must be an eigenvector of $A + B$.

27. If matrix A^2 is diagonalizable, then matrix A must be diagonalizable as well.

28. The determinant of a matrix is the product of its eigenvalues (over $\mathbb{C}$), counted with their algebraic multiplicities.

29. All lower triangular matrices are diagonalizable (over $\mathbb{C}$).

30. If two $n \times n$ matrices A and B are diagonalizable, then AB must be diagonalizable as well.

31. If $\vec{u}, \vec{v}, \vec{w}$ are eigenvectors of a 4×4 matrix A, with associated eigenvalues 3, 7, and 11, respectively, then vectors $\vec{u}, \vec{v}, \vec{w}$ must be linearly independent.

32. If a 4×4 matrix A is diagonalizable, then the matrix $A + 4I_4$ must be diagonalizable as well.

33. If an $n \times n$ matrix A is diagonalizable, then A must have n distinct eigenvalues.

34. If two 3×3 matrices A and B both have the eigenvalues 1, 2, and 3, then A must be similar to B.

35. If $\vec{v}$ is an eigenvector of A, then $\vec{v}$ must be an eigenvector of A^T as well.

36. All invertible matrices are diagonalizable.

37. If $\vec{v}$ and $\vec{w}$ are linearly independent eigenvectors of matrix A, then $\vec{v} + \vec{w}$ must be an eigenvector of A as well.

38. If a 2×2 matrix R represents a reflection about a line L, then R must be diagonalizable.

39. If A is a 2×2 matrix such that $\operatorname{tr} A = 1$ and $\det A = -6$, then A must be diagonalizable.

40. If a matrix is diagonalizable, then the algebraic multiplicity of each of its eigenvalues λ must equal the geometric multiplicity of λ.

41. All orthogonal matrices are diagonalizable (over $\mathbb{R}$).

42. If A is an $n \times n$ matrix and λ is an eigenvalue of the block matrix $M = \begin{bmatrix} A & A \\ 0 & A \end{bmatrix}$, then λ must be an eigenvalue of matrix A.

43. If two matrices A and B have the same characteristic polynomials, then they must be similar.

44. If A is a diagonalizable 4×4 matrix with $A^4 = 0$, then A must be the zero matrix.

45. If an $n \times n$ matrix A is diagonalizable (over $\mathbb{R}$), then every vector $\vec{v}$ in $\mathbb{R}^n$ can be expressed as a sum of eigenvectors of A.

46. If vector $\vec{v}$ is an eigenvector of both A and B, then $\vec{v}$ is an eigenvector of AB.

47. Similar matrices have the same characteristic polynomials.

48. If a matrix A has k distinct eigenvalues, then $\operatorname{rank}(A) \geq k$.

49. If the rank of a square matrix A is 1, then all the nonzero vectors in the image of A are eigenvectors of A.

50. If the rank of an $n \times n$ matrix A is 1, then A must be diagonalizable.

51. If A is a 4×4 matrix with $A^4 = 0$, then 0 is the only eigenvalue of A.

52. If two $n \times n$ matrices A and B are both diagonalizable, then they must commute.

53. If $\vec{v}$ is an eigenvector of A, then $\vec{v}$ must be in the kernel of A or in the image of A.

54. All symmetric 2×2 matrices are diagonalizable (over $\mathbb{R}$).

55. If A is a 2×2 matrix with eigenvalues 3 and 4 and if $\vec{u}$ is a unit eigenvector of A, then the length of vector $A\vec{u}$ cannot exceed 4.

56. If $\vec{u}$ is a nonzero vector in $\mathbb{R}^n$, then $\vec{u}$ must be an eigenvector of matrix $\vec{u}\vec{u}^T$.

57. If $\vec{v}_1, \vec{v}_2, \ldots, \vec{v}_n$ is an eigenbasis for both A and B, then matrices A and B must commute.

58. If $\vec{v}$ is an eigenvector of a 2×2 matrix $A = \begin{bmatrix} a & b \\ c & d \end{bmatrix}$, then $\vec{v}$ must be an eigenvector of its classical adjoint $\operatorname{adj}(A) = \begin{bmatrix} d & -b \\ -c & a \end{bmatrix}$ as well.

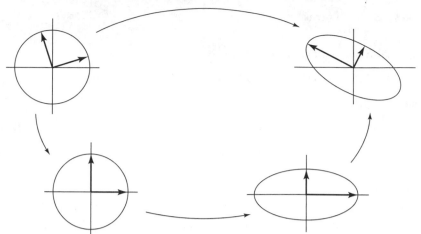

Symmetric Matrices and Quadratic Forms

8.1 Symmetric Matrices

In this chapter we will work with real numbers throughout, except for a brief digression into $\mathbb{C}$ in the discussion of Theorem 8.1.3.

Our work in the last chapter dealt with the following central question:

When is a given square matrix A *diagonalizable?* That is, when is there an *eigenbasis* for A?

In geometry, we prefer to work with *orthonormal* bases, which raises the following question:

For which matrices is there an *orthonormal* eigenbasis? Or, equivalently, for which matrices A is there an *orthogonal* matrix S such that $S^{-1}AS = S^T AS$ is diagonal?

(Recall that $S^{-1} = S^T$ for orthogonal matrices, by Theorem 5.3.7.) We say that A is *orthogonally diagonalizable* if there exists an orthogonal matrix S such that $S^{-1}AS = S^T AS$ is diagonal. Then, the question is

Which matrices are orthogonally diagonalizable?

Simple examples of orthogonally diagonalizable matrices are diagonal matrices (we can let $S = I_n$) and the matrices of orthogonal projections and reflections.

EXAMPLE I   If A is orthogonally diagonalizable, what is the relationship between A^T and A?

Solution

We have

$$S^{-1}AS = D \quad \text{or} \quad A = SDS^{-1} = SDS^T,$$

for an orthogonal S and a diagonal D. Then

$$A^T = (SDS^T)^T = SD^T S^T = SDS^T = A.$$

We find that A is symmetric:

$$A^T = A. \qquad \blacksquare$$

Surprisingly, the converse is true as well:

Theorem 8.1.1   **Spectral theorem**

A matrix A is *orthogonally diagonalizable* (i.e., there exists an orthogonal S such that $S^{-1}AS = S^T AS$ is diagonal) if and only if A is *symmetric* (i.e., $A^T = A$). $\blacksquare$

We will prove this theorem later in this section, based on two preliminary results, Theorems 8.1.2 and 8.1.3. First, we will illustrate the spectral theorem with an example.

EXAMPLE 2   For the symmetric matrix $A = \begin{bmatrix} 4 & 2 \\ 2 & 7 \end{bmatrix}$, find an orthogonal S such that $S^{-1}AS$ is diagonal.

Solution

We will first find an eigenbasis. The eigenvalues of A are 3 and 8, with corresponding eigenvectors $\begin{bmatrix} 2 \\ -1 \end{bmatrix}$ and $\begin{bmatrix} 1 \\ 2 \end{bmatrix}$, respectively. (See Figure 1.)

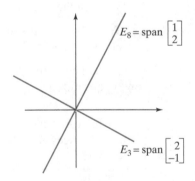

Figure 1

Note that the two eigenspaces, E_3 and E_8, are perpendicular. (This is no coincidence, as we will see in Theorem 8.1.2.) Therefore, we can find an orthonormal eigenbasis simply by dividing the given eigenvectors by their lengths:

$$\vec{v}_1 = \frac{1}{\sqrt{5}} \begin{bmatrix} 2 \\ -1 \end{bmatrix}, \quad \vec{v}_2 = \frac{1}{\sqrt{5}} \begin{bmatrix} 1 \\ 2 \end{bmatrix}.$$

If we define the orthogonal matrix

$$S = \begin{bmatrix} | & | \\ \vec{v}_1 & \vec{v}_2 \\ | & | \end{bmatrix} = \frac{1}{\sqrt{5}} \begin{bmatrix} 2 & 1 \\ -1 & 2 \end{bmatrix},$$

then $S^{-1}AS$ will be diagonal, namely, $S^{-1}AS = \begin{bmatrix} 3 & 0 \\ 0 & 8 \end{bmatrix}$. $\blacksquare$

The key observation we made in Example 2 generalizes as follows:

Theorem 8.1.2   Consider a symmetric matrix A. If $\vec{v}_1$ and $\vec{v}_2$ are eigenvectors of A with *distinct* eigenvalues λ_1 and λ_2, then $\vec{v}_1 \cdot \vec{v}_2 = 0$; that is, $\vec{v}_2$ is orthogonal to $\vec{v}_1$.

Proof   We compute the product

$$\vec{v}_1^T A \vec{v}_2$$

in two different ways:

$$\vec{v}_1^T A \vec{v}_2 = \vec{v}_1^T (\lambda_2 \vec{v}_2) = \lambda_2 (\vec{v}_1 \cdot \vec{v}_2)$$
$$\vec{v}_1^T A \vec{v}_2 = \vec{v}_1^T A^T \vec{v}_2 = (A \vec{v}_1)^T \vec{v}_2 = (\lambda_1 \vec{v}_1)^T \vec{v}_2 = \lambda_1 (\vec{v}_1 \cdot \vec{v}_2)$$

Comparing the results, we find

$$\lambda_1 (\vec{v}_1 \cdot \vec{v}_2) = \lambda_2 (\vec{v}_1 \cdot \vec{v}_2),$$

or

$$(\lambda_1 - \lambda_2)(\vec{v}_1 \cdot \vec{v}_2) = 0.$$

Since the first factor in this product, $\lambda_1 - \lambda_2$, is nonzero, the second factor, $\vec{v}_1 \cdot \vec{v}_2$, must be zero, as claimed.   ∎

Theorem 8.1.2 tells us that the eigenspaces of a symmetric matrix are perpendicular to one another. Here is another illustration of this property:

EXAMPLE 3   For the symmetric matrix

$$A = \begin{bmatrix} 1 & 1 & 1 \\ 1 & 1 & 1 \\ 1 & 1 & 1 \end{bmatrix},$$

find an orthogonal S such that $S^{-1} A S$ is diagonal.

Solution

The eigenvalues are 0 and 3, with

$$E_0 = \text{span} \left(\begin{bmatrix} -1 \\ 1 \\ 0 \end{bmatrix}, \begin{bmatrix} -1 \\ 0 \\ 1 \end{bmatrix} \right) \quad \text{and} \quad E_3 = \text{span} \begin{bmatrix} 1 \\ 1 \\ 1 \end{bmatrix}.$$

Note that the two eigenspaces are indeed perpendicular to one another, in accordance with Theorem 8.1.2. (See Figure 2.)

We can construct an orthonormal eigenbasis for A by picking an orthonormal basis of each eigenspace (using the Gram–Schmidt process in the case of E_0). See Figure 3.

In Figure 3, the vectors $\vec{v}_1$, $\vec{v}_2$ form an orthonormal basis of E_0, and $\vec{v}_3$ is a unit vector in E_3. Then $\vec{v}_1$, $\vec{v}_2$, $\vec{v}_3$ is an orthonormal eigenbasis for A. We can let $S = \begin{bmatrix} \vec{v}_1 & \vec{v}_2 & \vec{v}_3 \end{bmatrix}$ to diagonalize A orthogonally.

If we apply the Gram–Schmidt[1] process to the vectors

$$\begin{bmatrix} -1 \\ 1 \\ 0 \end{bmatrix}, \begin{bmatrix} -1 \\ 0 \\ 1 \end{bmatrix}$$

---

[1] Alternatively, we could find a unit vector $\vec{v}_1$ in E_0 and a unit vector $\vec{v}_3$ in E_3, and then let $\vec{v}_2 = \vec{v}_3 \times \vec{v}_1$.

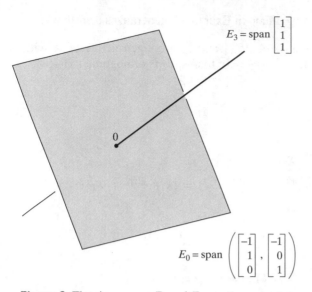

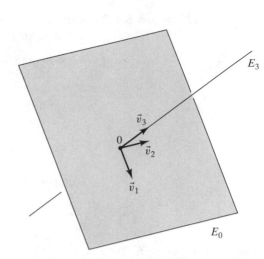

Figure 2 The eigenspaces E_0 and E_3 are orthogonal complements.

Figure 3

spanning E_0, we find

$$\vec{v}_1 = \frac{1}{\sqrt{2}} \begin{bmatrix} -1 \\ 1 \\ 0 \end{bmatrix} \quad \text{and} \quad \vec{v}_2 = \frac{1}{\sqrt{6}} \begin{bmatrix} -1 \\ -1 \\ 2 \end{bmatrix}.$$

The computations are left as an exercise. For E_3, we get

$$\vec{v}_3 = \frac{1}{\sqrt{3}} \begin{bmatrix} 1 \\ 1 \\ 1 \end{bmatrix}.$$

Therefore, the orthogonal matrix

$$S = \begin{bmatrix} | & | & | \\ \vec{v}_1 & \vec{v}_2 & \vec{v}_3 \\ | & | & | \end{bmatrix} = \begin{bmatrix} -1/\sqrt{2} & -1/\sqrt{6} & 1/\sqrt{3} \\ 1/\sqrt{2} & -1/\sqrt{6} & 1/\sqrt{3} \\ 0 & 2/\sqrt{6} & 1/\sqrt{3} \end{bmatrix}$$

diagonalizes the matrix A:

$$S^{-1}AS = \begin{bmatrix} 0 & 0 & 0 \\ 0 & 0 & 0 \\ 0 & 0 & 3 \end{bmatrix}. \qquad \blacksquare$$

By Theorem 8.1.2, if a symmetric matrix is diagonalizable, then it is orthogonally diagonalizable. We still have to show that symmetric matrices are diagonalizable in the first place (over $\mathbb{R}$). The key point is the following observation:

Theorem 8.1.3   A symmetric $n \times n$ matrix A has n real eigenvalues if they are counted with their algebraic multiplicities.

Proof   (This proof is for those who have studied Section 7.5.) By Theorem 7.5.4, we need to show that all the complex eigenvalues of matrix A are in fact real. Consider two complex conjugate eigenvalues $p \pm iq$ of A with corresponding eigenvectors

$\vec{v} \pm i\vec{w}$. (Compare this with Exercise 7.5.42b.) We wish to show that these eigenvalues are real; that is, $q = 0$. Note first that

$$(\vec{v} + i\vec{w})^T (\vec{v} - i\vec{w}) = \|\vec{v}\|^2 + \|\vec{w}\|^2.$$

(Verify this.) Now we compute the product

$$(\vec{v} + i\vec{w})^T A(\vec{v} - i\vec{w})$$

in two different ways:

$$\begin{aligned}
(\vec{v} + i\vec{w})^T A(\vec{v} - i\vec{w}) &= (\vec{v} + i\vec{w})^T (p - iq)(\vec{v} - i\vec{w}) \\
&= (p - iq)(\|\vec{v}\|^2 + \|\vec{w}\|^2) \\
(\vec{v} + i\vec{w})^T A(\vec{v} - i\vec{w}) &= \left(A(\vec{v} + i\vec{w})\right)^T (\vec{v} - i\vec{w}) = (p + iq)(\vec{v} + i\vec{w})^T (\vec{v} - i\vec{w}) \\
&= (p + iq)(\|\vec{v}\|^2 + \|\vec{w}\|^2).
\end{aligned}$$

Comparing the results, we find that $p + iq = p - iq$, so that $q = 0$, as claimed. ∎

The foregoing proof is not very enlightening. A more transparent proof would follow if we were to define the dot product for complex vectors, but to do so would lead us too far afield.

We are now ready to prove Theorem 8.1.1: Symmetric matrices are orthogonally diagonalizable.

Even though this is not logically necessary, let us first examine the case of a symmetric $n \times n$ matrix A with n distinct real eigenvalues. For each eigenvalue, we can choose an eigenvector of length 1. By Theorem 8.1.2, these eigenvectors will form an orthonormal eigenbasis, that is, the matrix A will be orthogonally diagonalizable, as claimed.

Proof
(of Theorem 8.1.1): This proof is somewhat technical; it may be skipped in a first reading of this text without harm.

We prove by induction on n that a symmetric $n \times n$ matrix A is orthogonally diagonalizable.[2]

For a 1×1 matrix A, we can let $S = [1]$.

Now assume that the claim is true for $n - 1$; we show that it holds for n. Pick a real eigenvalue λ of A (this is possible by Theorem 8.1.3), and choose an eigenvector $\vec{v}_1$ of length 1 for λ. We can find an orthonormal basis $\vec{v}_1, \vec{v}_2, \ldots, \vec{v}_n$ of $\mathbb{R}^n$. (Think about how you could construct such a basis.) Form the orthogonal matrix

$$P = \begin{bmatrix} | & | & & | \\ \vec{v}_1 & \vec{v}_2 & \cdots & \vec{v}_n \\ | & | & & | \end{bmatrix},$$

and compute

$$P^{-1}AP.$$

The first column of $P^{-1}AP$ is $\lambda\vec{e}_1$. (Why?) Also note that $P^{-1}AP = P^TAP$ is symmetric: $(P^TAP)^T = P^TA^TP = P^TAP$, because A is symmetric. Combining

---

[2]The *principle of mathematical induction,* applied to square matrices, can be stated as follows:
To show that a claim holds for all square matrices, show that it holds for
   a. 1×1 matrices, and
   b. $n \times n$ matrices, assuming that it holds for $(n-1) \times (n-1)$ matrices (for all $n \geq 2$).

these two statements, we conclude that $P^{-1}AP$ has the block form

$$P^{-1}AP = \begin{bmatrix} \lambda & 0 \\ 0 & B \end{bmatrix},$$ (I)

where B is a symmetric $(n-1) \times (n-1)$ matrix. By induction hypothesis, we assume that B is orthogonally diagonalizable; that is, there exists an orthogonal $(n-1) \times (n-1)$ matrix Q such that

$$Q^{-1}BQ = D$$

is a diagonal $(n-1) \times (n-1)$ matrix. Now introduce the orthogonal $n \times n$ matrix

$$R = \begin{bmatrix} 1 & 0 \\ 0 & Q \end{bmatrix}.$$

Then

$$R^{-1} \begin{bmatrix} \lambda & 0 \\ 0 & B \end{bmatrix} R = \begin{bmatrix} 1 & 0 \\ 0 & Q^{-1} \end{bmatrix} \begin{bmatrix} \lambda & 0 \\ 0 & B \end{bmatrix} \begin{bmatrix} 1 & 0 \\ 0 & Q \end{bmatrix} = \begin{bmatrix} \lambda & 0 \\ 0 & D \end{bmatrix}$$ (II)

is diagonal.

Combining equations (I) and (II), we find that

$$R^{-1}P^{-1}APR = \begin{bmatrix} \lambda & 0 \\ 0 & D \end{bmatrix}$$ (III)

is diagonal. Consider the orthogonal matrix $S = PR$. (Recall Theorem 5.3.4a: The product of orthogonal matrices is orthogonal.) Note that $S^{-1} = (PR)^{-1} = R^{-1}P^{-1}$. Therefore, equation (III) can be written

$$S^{-1}AS = \begin{bmatrix} \lambda & 0 \\ 0 & D \end{bmatrix},$$

proving our claim.   ∎

The method outlined in the proof of Theorem 8.1.1 is not a sensible way to find the matrix S in a numerical example. Rather, we can proceed as in Example 3:

Theorem 8.1.4   Orthogonal diagonalization of a symmetric matrix A

   a. Find the eigenvalues of A, and find a basis of each eigenspace.
   b. Using the Gram–Schmidt process, find an *orthonormal* basis of each eigenspace.
   c. Form an orthonormal eigenbasis $\vec{v}_1, \vec{v}_2, \ldots, \vec{v}_n$ for A by concatenating the orthonormal bases you found in part (b), and let

$$S = \begin{bmatrix} | & | & & | \\ \vec{v}_1 & \vec{v}_2 & \cdots & \vec{v}_n \\ | & | & & | \end{bmatrix}.$$

S is orthogonal (by Theorem 8.1.2), and $S^{-1}AS$ will be diagonal.   ∎

We conclude this section with an example of a geometric nature:

EXAMPLE 4   Consider an invertible symmetric 2×2 matrix A. Show that the linear transformation $T(\vec{x}) = A\vec{x}$ maps the unit circle into an ellipse, and find the lengths of the semimajor and the semiminor axes of this ellipse in terms of the eigenvalues of A. Compare this with Exercise 2.2.50.

Solution

The spectral theorem tells us that there exists an orthonormal eigenbasis $\vec{v}_1$, $\vec{v}_2$ for T, with associated real eigenvalues λ_1 and λ_2. These eigenvalues will be nonzero, since A is invertible. Arrange things so that $|\lambda_1| \geq |\lambda_2|$. The unit circle in $\mathbb{R}^2$ consists of all vectors of the form

$$\vec{v} = \cos(t)\vec{v}_1 + \sin(t)\vec{v}_2.$$

The image of the unit circle consists of the vectors

$$T(\vec{v}) = \cos(t)T(\vec{v}_1) + \sin(t)T(\vec{v}_2)$$
$$= \cos(t)\lambda_1\vec{v}_1 + \sin(t)\lambda_2\vec{v}_2,$$

an ellipse whose semimajor axis $\lambda_1\vec{v}_1$ has the length $\|\lambda_1\vec{v}_1\| = |\lambda_1|$, while the length of the semiminor axis is $\|\lambda_2\vec{v}_2\| = |\lambda_2|$. (See Figure 4.)

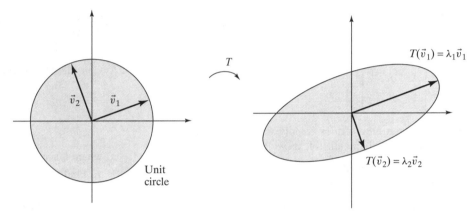

Figure 4

In the example illustrated in Figure 4, the eigenvalue λ_1 is positive, and λ_2 is negative. ∎

EXERCISES 8.1

GOAL *Find orthonormal eigenbases for symmetric matrices. Apply the spectral theorem.*

For each of the matrices in Exercises 1 through 6, find an orthonormal eigenbasis. Do not use technology.

1. $\begin{bmatrix} 1 & 0 \\ 0 & 2 \end{bmatrix}$ **2.** $\begin{bmatrix} 1 & 1 \\ 1 & 1 \end{bmatrix}$

3. $\begin{bmatrix} 6 & 2 \\ 2 & 3 \end{bmatrix}$ **4.** $\begin{bmatrix} 0 & 0 & 1 \\ 0 & 0 & 1 \\ 1 & 1 & 1 \end{bmatrix}$

5. $\begin{bmatrix} 0 & 1 & 1 \\ 1 & 0 & 1 \\ 1 & 1 & 0 \end{bmatrix}$ **6.** $\begin{bmatrix} 0 & 2 & 2 \\ 2 & 1 & 0 \\ 2 & 0 & -1 \end{bmatrix}$

For each of the matrices A in Exercises 7 through 11, find an orthogonal matrix S and a diagonal matrix D such that $S^{-1}AS = D$. Do not use technology.

7. $A = \begin{bmatrix} 3 & 2 \\ 2 & 3 \end{bmatrix}$ **8.** $A = \begin{bmatrix} 3 & 3 \\ 3 & -5 \end{bmatrix}$

9. $A = \begin{bmatrix} 0 & 0 & 3 \\ 0 & 2 & 0 \\ 3 & 0 & 0 \end{bmatrix}$ **10.** $A = \begin{bmatrix} 1 & -2 & 2 \\ -2 & 4 & -4 \\ 2 & -4 & 4 \end{bmatrix}$

11. $A = \begin{bmatrix} 1 & 0 & 1 \\ 0 & 1 & 0 \\ 1 & 0 & 1 \end{bmatrix}$

12. Let L from $\mathbb{R}^3$ to $\mathbb{R}^3$ be the reflection about the line spanned by

$$\vec{v} = \begin{bmatrix} 1 \\ 0 \\ 2 \end{bmatrix}.$$

a. Find an orthonormal eigenbasis $\mathfrak{B}$ for L.

b. Find the matrix B of L with respect to $\mathfrak{B}$.

c. Find the matrix A of L with respect to the standard basis of $\mathbb{R}^3$.

13. Consider a symmetric 3×3 matrix A with $A^2 = I_3$. Is the linear transformation $T(\vec{x}) = A\vec{x}$ necessarily the reflection about a subspace of $\mathbb{R}^3$?

14. In Example 3 of this section, we diagonalized the matrix

$$A = \begin{bmatrix} 1 & 1 & 1 \\ 1 & 1 & 1 \\ 1 & 1 & 1 \end{bmatrix}$$

by means of an orthogonal matrix S. Use this result to diagonalize the following matrices orthogonally (find S and D in each case):

a. $\begin{bmatrix} 2 & 2 & 2 \\ 2 & 2 & 2 \\ 2 & 2 & 2 \end{bmatrix}$ **b.** $\begin{bmatrix} -2 & 1 & 1 \\ 1 & -2 & 1 \\ 1 & 1 & -2 \end{bmatrix}$

c. $\begin{bmatrix} 0 & \frac{1}{2} & \frac{1}{2} \\ \frac{1}{2} & 0 & \frac{1}{2} \\ \frac{1}{2} & \frac{1}{2} & 0 \end{bmatrix}$

15. If A is invertible and orthogonally diagonalizable, is A^{-1} orthogonally diagonalizable as well?

16. a. Find the eigenvalues of the matrix

$$A = \begin{bmatrix} 1 & 1 & 1 & 1 & 1 \\ 1 & 1 & 1 & 1 & 1 \\ 1 & 1 & 1 & 1 & 1 \\ 1 & 1 & 1 & 1 & 1 \\ 1 & 1 & 1 & 1 & 1 \end{bmatrix}$$

with their multiplicities. Note that the algebraic multiplicity agrees with the geometric multiplicity. (Why?) (*Hint:* What is the kernel of A?)

b. Find the eigenvalues of the matrix

$$B = \begin{bmatrix} 3 & 1 & 1 & 1 & 1 \\ 1 & 3 & 1 & 1 & 1 \\ 1 & 1 & 3 & 1 & 1 \\ 1 & 1 & 1 & 3 & 1 \\ 1 & 1 & 1 & 1 & 3 \end{bmatrix}$$

with their multiplicities. Do not use technology.

c. Use your result in part (b) to find $\det(B)$.

17. Use the approach of Exercise 16 to find the determinant of the $n \times n$ matrix B that has p's on the diagonal and q's elsewhere:

$$B = \begin{bmatrix} p & q & \cdots & q \\ q & p & \cdots & q \\ \vdots & \vdots & \ddots & \vdots \\ q & q & \cdots & p \end{bmatrix}.$$

18. Consider unit vectors $\vec{v}_1, \ldots, \vec{v}_n$ in $\mathbb{R}^n$ such that the angle between $\vec{v}_i$ and $\vec{v}_j$ is $60°$ for all $i \neq j$. Find the

n-volume of the n-parallelepiped spanned by $\vec{v}_1, \ldots, \vec{v}_n$. {*Hint:* Let $A = \begin{bmatrix} \vec{v}_1 & \cdots & \vec{v}_n \end{bmatrix}$, and think about the matrix $A^T A$ and its determinant. Exercise 17 is useful.}

19. Consider a linear transformation L from $\mathbb{R}^m$ to $\mathbb{R}^n$. Show that there exists an orthonormal basis $\vec{v}_1, \vec{v}_2, \ldots, \vec{v}_m$ of $\mathbb{R}^m$ such that the vectors $L(\vec{v}_1), L(\vec{v}_2), \ldots, L(\vec{v}_m)$ are orthogonal. Note that some of the vectors $L(\vec{v}_i)$ may be zero. (*Hint:* Consider an orthonormal eigenbasis $\vec{v}_1, \vec{v}_2, \ldots, \vec{v}_m$ for the symmetric matrix $A^T A$.)

20. Consider a linear transformation T from $\mathbb{R}^m$ to $\mathbb{R}^n$, where $m \leq n$. Show that there exist an orthonormal basis $\vec{v}_1, \ldots, \vec{v}_m$ of $\mathbb{R}^m$ and an orthonormal basis $\vec{w}_1, \ldots, \vec{w}_n$ of $\mathbb{R}^n$ such that $T(\vec{v}_i)$ is a scalar multiple of $\vec{w}_i$, for $i = 1, \ldots, m$. (*Hint:* Exercise 19 is helpful.)

21. Consider a symmetric 3×3 matrix A with eigenvalues 1, 2, and 3. How many different orthogonal matrices S are there such that $S^{-1}AS$ is diagonal?

22. Consider the matrix

$$A = \begin{bmatrix} 0 & 2 & 0 & 0 \\ k & 0 & 2 & 0 \\ 0 & k & 0 & 2 \\ 0 & 0 & k & 0 \end{bmatrix},$$

where k is a constant.

a. Find a value of k such that the matrix A is diagonalizable.

b. Find a value of k such that A fails to be diagonalizable.

23. If an $n \times n$ matrix A is both symmetric and orthogonal, what can you say about the eigenvalues of A? What about the eigenspaces? Interpret the linear transformation $T(\vec{x}) = A\vec{x}$ geometrically in the cases $n = 2$ and $n = 3$.

24. Consider the matrix

$$A = \begin{bmatrix} 0 & 0 & 0 & 1 \\ 0 & 0 & 1 & 0 \\ 0 & 1 & 0 & 0 \\ 1 & 0 & 0 & 0 \end{bmatrix}.$$

Find an orthonormal eigenbasis for A.

25. Consider the matrix

$$\begin{bmatrix} 0 & 0 & 0 & 0 & 1 \\ 0 & 0 & 0 & 1 & 0 \\ 0 & 0 & 1 & 0 & 0 \\ 0 & 1 & 0 & 0 & 0 \\ 1 & 0 & 0 & 0 & 0 \end{bmatrix}.$$

Find an orthogonal 5×5 matrix S such that $S^{-1}AS$ is diagonal.

26. Let J_n be the $n \times n$ matrix with all ones on the "other diagonal" and zeros elsewhere. (In Exercises 24 and 25, we studied J_4 and J_5, respectively.) Find the eigenvalues of J_n, with their multiplicities.

27. Diagonalize the $n \times n$ matrix

$$\begin{bmatrix} 1 & 0 & 0 & \cdots & 0 & 1 \\ 0 & 1 & 0 & \cdots & 1 & 0 \\ 0 & 0 & 1 & \cdots & 0 & 0 \\ \vdots & \vdots & \vdots & \ddots & \vdots & \vdots \\ 0 & 1 & 0 & \cdots & 1 & 0 \\ 1 & 0 & 0 & \cdots & 0 & 1 \end{bmatrix}.$$

(All ones along both diagonals, and zeros elsewhere.)

28. Diagonalize the 13×13 matrix

$$\begin{bmatrix} 0 & 0 & 0 & \cdots & 0 & 1 \\ 0 & 0 & 0 & \cdots & 0 & 1 \\ 0 & 0 & 0 & \cdots & 0 & 1 \\ \vdots & \vdots & \vdots & \ddots & \vdots & \vdots \\ 0 & 0 & 0 & \cdots & 0 & 1 \\ 1 & 1 & 1 & \cdots & 1 & 1 \end{bmatrix}.$$

(All ones in the last row and the last column, and zeros elsewhere.)

29. Consider a symmetric matrix A. If the vector $\vec{v}$ is in the image of A and $\vec{w}$ is in the kernel of A, is $\vec{v}$ necessarily orthogonal to $\vec{w}$? Justify your answer.

30. Consider an orthogonal matrix R whose first column is $\vec{v}$. Form the symmetric matrix $A = \vec{v}\vec{v}^T$. Find an orthogonal matrix S and a diagonal matrix D such that $S^{-1}AS = D$. Describe S in terms of R.

31. *True or false?* If A is a symmetric matrix, then $\operatorname{rank}(A) = \operatorname{rank}(A^2)$.

32. Consider the $n \times n$ matrix with all ones on the main diagonal and all q's elsewhere. For which values of q is this matrix invertible? (*Hint:* Exercise 17 is helpful.)

33. For which angle(s) θ can you find three distinct unit vectors in $\mathbb{R}^2$ such that the angle between any two of them is θ? Draw a sketch.

34. For which angle(s) θ can you find four distinct unit vectors in $\mathbb{R}^3$ such that the angle between any two of them is θ? Draw a sketch.

35. Consider $n + 1$ distinct unit vectors in $\mathbb{R}^n$ such that the angle between any two of them is θ. Find θ.

36. Consider a symmetric $n \times n$ matrix A with $A^2 = A$. Is the linear transformation $T(\vec{x}) = A\vec{x}$ necessarily the orthogonal projection onto a subspace of $\mathbb{R}^n$?

37. If A is any symmetric 2×2 matrix with eigenvalues -2 and 3, and $\vec{u}$ is a unit vector in $\mathbb{R}^2$, what are the possible values of $\|A\vec{u}\|$? Explain your answer geometrically, using Example 4 as a guide.

38. If A is any symmetric 2×2 matrix with eigenvalues -2 and 3, and $\vec{u}$ is a unit vector in $\mathbb{R}^2$, what are the possible values of the dot product $\vec{u} \cdot A\vec{u}$? Illustrate your answer, in terms of the unit circle and its image under A.

39. If A is any symmetric 3×3 matrix with eigenvalues -2, 3, and 4, and $\vec{u}$ is a unit vector in $\mathbb{R}^3$, what are the possible values of the dot product $\vec{u} \cdot A\vec{u}$?

40. If A is any symmetric 3×3 matrix with eigenvalues -2, 3, and 4, and $\vec{u}$ is a unit vector in $\mathbb{R}^3$, what are the possible values of $\|A\vec{u}\|$? Explain your answer geometrically, in terms of the unit sphere and its image under A.

41. Show that for every symmetric $n \times n$ matrix A there exists a symmetric $n \times n$ matrix B such that $B^3 = A$.

42. Find a symmetric 2×2 matrix B such that
$$B^3 = \frac{1}{5}\begin{bmatrix} 12 & 14 \\ 14 & 33 \end{bmatrix}.$$

43. For $A = \begin{bmatrix} 2 & 11 & 11 \\ 11 & 2 & 11 \\ 11 & 11 & 2 \end{bmatrix}$ find a nonzero vector $\vec{v}$ in $\mathbb{R}^3$ such that $A\vec{v}$ is orthogonal to $\vec{v}$.

44. Consider an invertible symmetric $n \times n$ matrix A. When does there exist a nonzero vector $\vec{v}$ in $\mathbb{R}^n$ such that $A\vec{v}$ is orthogonal to $\vec{v}$? Give your answer in terms of the signs of the eigenvalues of A.

45. We say that an $n \times n$ matrix A is *triangulizable* if A is similar to an upper triangular $n \times n$ matrix B.
   a. Give an example of a matrix with real entries that fails to be triangulizable over $\mathbb{R}$.
   b. Show that any $n \times n$ matrix with complex entries is triangulizable over $\mathbb{C}$. (*Hint:* Give a proof by induction analogous to the proof of Theorem 8.1.1.)

46. a. Consider a complex upper triangular $n \times n$ matrix U with zeros on the diagonal. Show that U is *nilpotent* (i.e., that $U^n = 0$). Compare with Exercises 3.3.76 and 3.3.77.
   b. Consider a complex $n \times n$ matrix A that has zero as its only eigenvalue (with algebraic multiplicity n). Use Exercise 45 to show that A is nilpotent.

47. Let us first introduce two notations.
For a complex $n \times n$ matrix A, let $|A|$ be the matrix whose ijth entry is $|a_{ij}|$.
For two real $n \times n$ matrices A and B, we write $A \leq B$ if $a_{ij} \leq b_{ij}$ for all entries. Show that
   a. $|AB| \leq |A||B|$, for all complex $n \times n$ matrices A and B, and
   b. $|A^t| \leq |A|^t$, for all complex $n \times n$ matrices A and all positive integers t.

48. Let $U \geq 0$ be a real upper triangular $n \times n$ matrix with zeros on the diagonal. Show that

$$(I_n + U)^t \leq t^n(I_n + U + U^2 + \cdots + U^{n-1})$$

for all positive integers t. See Exercises 46 and 47.

49. Let R be a complex upper triangular $n \times n$ matrix with $|r_{ii}| < 1$ for $i = 1, \ldots, n$. Show that

$$\lim_{t \to \infty} R^t = 0,$$

meaning that the modulus of all entries of R^t approaches zero. *Hint*: We can write $|R| \leq \lambda(I_n + U)$, for some positive real number $\lambda < 1$ and an upper triangular matrix $U \geq 0$ with zeros on the diagonal. Exercises 47 and 48 are helpful.

50. a. Let A be a complex $n \times n$ matrix such that $|\lambda| < 1$ for all eigenvalues λ of A. Show that

$$\lim_{t \to \infty} A^t = 0,$$

meaning that the modulus of all entries of A^t approaches zero.

b. Prove Theorem 7.6.2.

8.2 Quadratic Forms

In this section, we will present an important application of the spectral theorem (Theorem 8.1.1).

In a multivariable calculus text, we find the following problem:

EXAMPLE I   Consider the function

$$q(x_1, x_2) = 8x_1^2 - 4x_1x_2 + 5x_2^2$$

from $\mathbb{R}^2$ to $\mathbb{R}$.

Determine whether $q(0, 0) = 0$ is the global maximum, the global minimum, or neither.

Recall that $q(0, 0)$ is called the global (or absolute) minimum if $q(0, 0) \leq q(x_1, x_2)$ for all real numbers x_1, x_2; the global maximum is defined analogously.

Solution

There are a number of ways to do this problem, some of which you may have seen in a previous course. Here we present an approach based on matrix techniques. We will first develop some theory and then do the example.

Note that we can write

$$q \begin{bmatrix} x_1 \\ x_2 \end{bmatrix} = 8x_1^2 - 4x_1x_2 + 5x_2^2$$

$$= \begin{bmatrix} x_1 \\ x_2 \end{bmatrix} \cdot \begin{bmatrix} 8x_1 - 2x_2 \\ -2x_1 + 5x_2 \end{bmatrix} \qquad \text{We "split" the term } -4x_1x_2 \text{ equally between the two components.}$$

More succinctly, we can write

$$q(\vec{x}) = \vec{x} \cdot A\vec{x}, \quad \text{where} \quad A = \begin{bmatrix} 8 & -2 \\ -2 & 5 \end{bmatrix},$$

or

$$q(\vec{x}) = \vec{x}^T A\vec{x}.$$

The matrix A is symmetric by construction. By the spectral theorem (Theorem 8.1.1), there exists an orthonormal eigenbasis $\vec{v}_1, \vec{v}_2$ for A. We find

$$\vec{v}_1 = \frac{1}{\sqrt{5}} \begin{bmatrix} 2 \\ -1 \end{bmatrix}, \quad \vec{v}_2 = \frac{1}{\sqrt{5}} \begin{bmatrix} 1 \\ 2 \end{bmatrix},$$

with associated eigenvalues $\lambda_1 = 9$ and $\lambda_2 = 4$. (Verify this.)

If we write $\vec{x} = c_1\vec{v}_1 + c_2\vec{v}_2$, we can express the value of the function as follows:

$$q(\vec{x}) = \vec{x} \cdot A\vec{x} = (c_1\vec{v}_1 + c_2\vec{v}_2) \cdot (c_1\lambda_1\vec{v}_1 + c_2\lambda_2\vec{v}_2) = \lambda_1 c_1^2 + \lambda_2 c_2^2 = 9c_1^2 + 4c_2^2.$$

(Recall that $\vec{v}_1 \cdot \vec{v}_1 = 1$, $\vec{v}_1 \cdot \vec{v}_2 = 0$, and $\vec{v}_2 \cdot \vec{v}_2 = 1$, since $\vec{v}_1, \vec{v}_2$ is an orthonormal basis of $\mathbb{R}^2$.)

The formula $q(\vec{x}) = 9c_1^2 + 4c_2^2$ shows that $q(\vec{x}) > 0$ for all nonzero $\vec{x}$, because at least one of the terms $9c_1^2$ and $4c_2^2$ is positive.

Thus $q(0, 0) = 0$ is the global minimum of the function.

The preceding work shows that the c_1–c_2 coordinate system defined by an orthonormal eigenbasis for A is "well adjusted" to the function q. The formula

$$9c_1^2 + 4c_2^2$$

is easier to work with than the original formula

$$8x_1^2 - 4x_1 x_2 + 5x_2^2,$$

because no term involves $c_1 c_2$:

$$q(x_1, x_2) = 8x_1^2 - 4x_1 x_2 + 5x_2^2$$
$$= 9c_1^2 + 4c_2^2$$

The two coordinate systems are shown in Figure 1. ∎

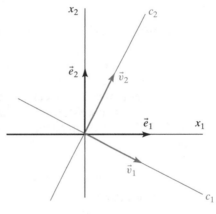

Figure 1

Let us present these ideas in greater generality:

Definition 8.2.1 Quadratic forms

A function $q(x_1, x_2, \ldots, x_n)$ from $\mathbb{R}^n$ to $\mathbb{R}$ is called a *quadratic form* if it is a linear combination of functions of the form $x_i x_j$ (where i and j may be equal). A quadratic form can be written as

$$q(\vec{x}) = \vec{x} \cdot A\vec{x} = \vec{x}^T A \vec{x},$$

for a unique symmetric $n \times n$ matrix A, called the matrix of q.

The uniqueness of matrix A will be shown in Exercise 52.

The set Q_n of quadratic forms $q(x_1, x_2, \ldots, x_n)$ is a *subspace* of the linear space of all functions from $\mathbb{R}^n$ to $\mathbb{R}$. In Exercise 42 you will be asked to think about the dimension of this space.

EXAMPLE 2 Consider the quadratic form

$$q(x_1, x_2, x_3) = 9x_1^2 + 7x_2^2 + 3x_3^2 - 2x_1 x_2 + 4x_1 x_3 - 6x_2 x_3.$$

Find the matrix of q.

Solution

As in Example 1, we let

$$a_{ii} = \text{(coefficient of } x_i^2),$$
$$a_{ij} = a_{ji} = \tfrac{1}{2}\text{ (coefficient of } x_i x_j), \quad \text{if } i \neq j.$$

Therefore,

$$A = \begin{bmatrix} 9 & -1 & 2 \\ -1 & 7 & -3 \\ 2 & -3 & 3 \end{bmatrix}.$$ ■

The observation we made in Example 1 can now be generalized as follows:

Theorem 8.2.2   Diagonalizing a quadratic form

Consider a quadratic form $q(\vec{x}) = \vec{x} \cdot A\vec{x}$, where A is a symmetric $n \times n$ matrix. Let $\mathfrak{B}$ be an orthonormal eigenbasis for A, with associated eigenvalues $\lambda_1, \ldots, \lambda_n$. Then

$$q(\vec{x}) = \lambda_1 c_1^2 + \lambda_2 c_2^2 + \cdots + \lambda_n c_n^2,$$

where the c_i are the coordinates of $\vec{x}$ with respect to $\mathfrak{B}$.[3] ■

Again, note that we have been able to get rid of the mixed terms: no summand involves $c_i c_j$ (with $i \neq j$) in the preceding formula. To justify the formula stated in Theorem 8.2.2, we can proceed as in Example 1. We leave the details as an exercise.

When we study a quadratic form q, we are often interested in finding out whether $q(\vec{x}) > 0$ for all nonzero $\vec{x}$ (as in Example 1). In this context, it is useful to introduce the following terminology:

Definition 8.2.3   Definiteness of a quadratic form

Consider a quadratic form $q(\vec{x}) = \vec{x} \cdot A\vec{x}$, where A is a symmetric $n \times n$ matrix.

We say that A is *positive definite* if $q(\vec{x})$ is positive for all nonzero $\vec{x}$ in $\mathbb{R}^n$, and we call A *positive semidefinite* if $q(\vec{x}) \geq 0$, for all $\vec{x}$ in $\mathbb{R}^n$.

Negative definite and negative semidefinite symmetric matrices are defined analogously.

Finally, we call A *indefinite* if q takes positive as well as negative values.

EXAMPLE 3   Consider an $n \times m$ matrix A. Show that the function $q(\vec{x}) = \|A\vec{x}\|^2$ is a quadratic form, find its matrix, and determine its definiteness.

Solution

We can write $q(\vec{x}) = (A\vec{x}) \cdot (A\vec{x}) = (A\vec{x})^T (A\vec{x}) = \vec{x}^T A^T A\vec{x} = \vec{x} \cdot (A^T A\vec{x})$. This shows that q is a quadratic form, with matrix $A^T A$. This quadratic form is positive semidefinite, because $q(\vec{x}) = \|A\vec{x}\|^2 \geq 0$ for all vectors $\vec{x}$ in $\mathbb{R}^m$. Note that $q(\vec{x}) = 0$ if and only if $\vec{x}$ is in the kernel of A. Therefore, the quadratic form is positive definite if and only if $\ker(A) = \{\vec{0}\}$. ■

---

[3] The basic properties of quadratic forms were first derived by the Dutchman Johan de Witt (1625–1672) in his *Elementa curvarum linearum*. De Witt was one of the leading statesmen of his time, guiding his country through two wars against England. He consolidated his nation's commercial and naval power. De Witt met an unfortunate end when he was literally torn to pieces by an angry mob. (He should have stayed with math!)

By Theorem 8.2.2, the definiteness of a symmetric matrix A is easy to determine from its eigenvalues:

Theorem 8.2.4 **Eigenvalues and definiteness**

A symmetric matrix A is *positive definite* if (and only if) all of its eigenvalues are positive. The matrix A is *positive semidefinite* if (and only if) all of its eigenvalues are positive or zero. ∎

These facts follow immediately from the formula

$$q(\vec{x}) = \lambda_1 c_1^2 + \cdots + \lambda_n c_n^2. \quad \text{(See Theorem 8.2.2.)}$$

The determinant of a positive definite matrix is positive, since the determinant is the product of the eigenvalues. The converse is not true, however: Consider a symmetric 3×3 matrix A with one positive and two negative eigenvalues. Then $\det A$ is positive, but $q(\vec{x}) = \vec{x} \cdot A\vec{x}$ is indefinite. In practice, the following criterion for positive definiteness is often used (a proof is outlined in Exercise 34):

Theorem 8.2.5 **Principal submatrices and definiteness**

Consider a symmetric $n \times n$ matrix A. For $m = 1, \ldots, n$, let $A^{(m)}$ be the $m \times m$ matrix obtained by omitting all rows and columns of A past the mth. These matrices $A^{(m)}$ are called the *principal submatrices* of A.

The matrix A is positive definite if (and only if) $\det(A^{(m)}) > 0$, for all $m = 1, \ldots, n$. ∎

Consider the matrix

$$A = \begin{bmatrix} 9 & -1 & 2 \\ -1 & 7 & -3 \\ 2 & -3 & 3 \end{bmatrix}$$

from Example 2:

$$\det(A^{(1)}) = \det \begin{bmatrix} 9 \end{bmatrix} = 9 > 0$$

$$\det(A^{(2)}) = \det \begin{bmatrix} 9 & -1 \\ -1 & 7 \end{bmatrix} = 62 > 0$$

$$\det(A^{(3)}) = \det(A) = 89 > 0$$

We can conclude that A is positive definite.

Alternatively, we could find the eigenvalues of A and use Theorem 8.2.4. Using technology, we find that $\lambda_1 \approx 10.7$, $\lambda_2 \approx 7.1$, and $\lambda_3 \approx 1.2$, confirming our result.

Principal Axes

When we study a function $f(x_1, x_2, \ldots, x_n)$ from $\mathbb{R}^n$ to $\mathbb{R}$, we are often interested in the solutions of the equations

$$f(x_1, x_2, \ldots, x_n) = k,$$

for a fixed k in $\mathbb{R}$, called the *level sets* of f (*level curves* for $n = 2$, *level surfaces* for $n = 3$).

Here we will think about the level curves of a quadratic form $q(x_1, x_2)$ of two variables. For simplicity, we focus on the level curve $q(x_1, x_2) = 1$.

Let us first consider the case when there is no mixed term in the formula. We trust that you had at least a brief encounter with those level curves in a previous course. Let us discuss the two major cases:

Case 1: $q(x_1, x_2) = ax_1^2 + bx_2^2 = 1$, where $b > a > 0$. This curve is an *ellipse*, as shown in Figure 2. The lengths of the semimajor and the semiminor axes are $1/\sqrt{a}$ and $1/\sqrt{b}$, respectively. This ellipse can be parameterized by

$$\begin{bmatrix} x_1 \\ x_2 \end{bmatrix} = \cos t \begin{bmatrix} 1/\sqrt{a} \\ 0 \end{bmatrix} + \sin t \begin{bmatrix} 0 \\ 1/\sqrt{b} \end{bmatrix}.$$

Case 2: $q(x_1, x_2) = ax_1^2 + bx_2^2 = 1$, where a is positive and b negative. This is a hyperbola, with x_1-intercepts $\begin{bmatrix} \pm 1/\sqrt{a} \\ 0 \end{bmatrix}$, as shown in Figure 3. What are the slopes of the asymptotes, in terms of a and b?

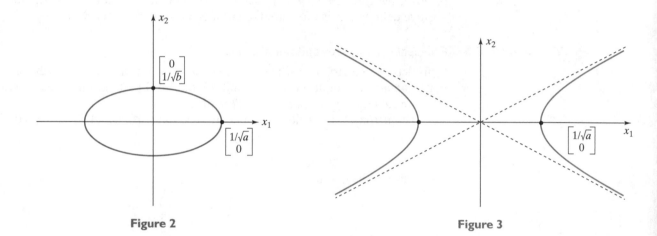

Figure 2　　　　　　　　　　　　　　Figure 3

Now consider the level curve

$$q(\vec{x}) = \vec{x} \cdot A\vec{x} = 1,$$

where A is an invertible symmetric 2×2 matrix. By Theorem 8.2.2, we can write this equation as

$$\lambda_1 c_1^2 + \lambda_2 c_2^2 = 1,$$

where c_1, c_2 are the coordinates of $\vec{x}$ with respect to an orthonormal eigenbasis for A, and λ_1, λ_2 are the associated eigenvalues.

This curve is an ellipse if both eigenvalues are positive and a hyperbola if one eigenvalue is positive and one negative. (What happens when both eigenvalues are negative?)

EXAMPLE 4　Sketch the curve

$$8x_1^2 - 4x_1x_2 + 5x_2^2 = 1.$$

(See Example 1.)

Solution

In Example 1, we found that we can write this equation as

$$9c_1^2 + 4c_2^2 = 1,$$

where c_1, c_2 are the coordinates of $\vec{x}$ with respect to the orthonormal eigenbasis

$$\vec{v}_1 = \frac{1}{\sqrt{5}}\begin{bmatrix} 2 \\ -1 \end{bmatrix}, \quad \vec{v}_2 = \frac{1}{\sqrt{5}}\begin{bmatrix} 1 \\ 2 \end{bmatrix},$$

for $A = \begin{bmatrix} 8 & -2 \\ -2 & 5 \end{bmatrix}$. We sketch this ellipse in Figure 4.

The c_1- and the c_2-axes are called the *principal axes* of the quadratic form $q(x_1, x_2) = 8x_1^2 - 4x_1x_2 + 5x_2^2$. Note that these are the eigenspaces of the matrix

$$A = \begin{bmatrix} 8 & -2 \\ -2 & 5 \end{bmatrix}$$

of the quadratic form. ∎

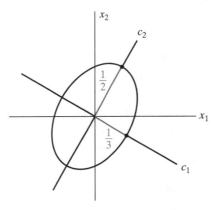

Figure 4

Definition 8.2.6 Principal axes

Consider a quadratic form $q(\vec{x}) = \vec{x} \cdot A\vec{x}$, where A is a symmetric $n \times n$ matrix with n distinct eigenvalues. Then the eigenspaces of A are called the *principal axes* of q. (Note that these eigenspaces will be one dimensional.)

Let's return to the case of a quadratic form of two variables. We can summarize our findings as follows:

Theorem 8.2.7 Ellipses and hyperbolas

Consider the curve C in $\mathbb{R}^2$ defined by

$$q(x_1, x_2) = ax_1^2 + bx_1x_2 + cx_2^2 = 1.$$

Let λ_1 and λ_2 be the eigenvalues of the matrix $\begin{bmatrix} a & b/2 \\ b/2 & c \end{bmatrix}$ of q.

If both λ_1 and λ_2 are positive, then C is an *ellipse*. If one eigenvalue is positive and the other is negative, then C is a hyperbola. ∎

EXERCISES 8.2

GOAL *Apply the concept of a quadratic form. Use an orthonormal eigenbasis for A to analyze the quadratic form* $q(\vec{x}) = \vec{x} \cdot A\vec{x}$.

For each of the quadratic forms q listed in Exercises 1 through 3, find the matrix of q.

1. $q(x_1, x_2) = 6x_1^2 - 7x_1x_2 + 8x_2^2$

2. $q(x_1, x_2) = x_1x_2$

3. $q(x_1, x_2, x_3) = 3x_1^2 + 4x_2^2 + 5x_3^2 + 6x_1x_3 + 7x_2x_3$

Determine the definiteness of the quadratic forms in Exercises 4 through 7.

4. $q(x_1, x_2) = 6x_1^2 + 4x_1x_2 + 3x_2^2$

5. $q(x_1, x_2) = x_1^2 + 4x_1x_2 + x_2^2$

6. $q(x_1, x_2) = 2x_1^2 + 6x_1x_2 + 4x_2^2$

7. $q(x_1, x_2, x_3) = 3x_2^2 + 4x_1x_3$

8. If A is a symmetric matrix, what can you say about the definiteness of A^2? When is A^2 positive definite?

9. Recall that a real square matrix A is called *skew symmetric* if $A^T = -A$.
  a. If A is skew-symmetric, is A^2 skew symmetric as well? Or is A^2 symmetric?
  b. If A is skew symmetric, what can you say about the definiteness of A^2? What about the eigenvalues of A^2?
  c. What can you say about the complex eigenvalues of a skew-symmetric matrix? Which skew-symmetric matrices are diagonalizable over $\mathbb{R}$?

10. Consider a quadratic form $q(\vec{x}) = \vec{x} \cdot A\vec{x}$ on $\mathbb{R}^n$ and a fixed vector $\vec{v}$ in $\mathbb{R}^n$. Is the transformation

$$L(\vec{x}) = q(\vec{x} + \vec{v}) - q(\vec{x}) - q(\vec{v})$$

linear? If so, what is its matrix?

11. If A is an invertible symmetric matrix, what is the relationship between the definiteness of A and A^{-1}?

12. Show that a quadratic form $q(\vec{x}) = \vec{x} \cdot A\vec{x}$ of two variables is indefinite if (and only if) $\det A < 0$. Here, A is a symmetric 2×2 matrix.

13. Show that the diagonal elements of a positive definite matrix A are positive.

14. Consider a 2×2 matrix $A = \begin{bmatrix} a & b \\ b & c \end{bmatrix}$, where a and $\det A$ are both positive. Without using Theorem 8.2.5, show that A is positive definite. [*Hint*: Show first that c is positive, and thus $\text{tr}(A)$ is positive. Then think about the signs of the eigenvalues.]

Sketch the curves defined in Exercises 15 through 20. In each case, draw and label the principal axes, label the intercepts of the curve with the principal axes, and give the formula of the curve in the coordinate system defined by the principal axes.

15. $6x_1^2 + 4x_1x_2 + 3x_2^2 = 1$ **16.** $x_1x_2 = 1$

17. $3x_1^2 + 4x_1x_2 = 1$ **18.** $9x_1^2 - 4x_1x_2 + 6x_2^2 = 1$

19. $x_1^2 + 4x_1x_2 + 4x_2^2 = 1$ **20.** $-3x_1^2 + 6x_1x_2 + 5x_2^2 = 1$

21. a. Sketch the following three surfaces:

$$x_1^2 + 4x_2^2 + 9x_3^2 = 1,$$
$$x_1^2 + 4x_2^2 - 9x_3^2 = 1,$$
$$-x_1^2 - 4x_2^2 + 9x_3^2 = 1.$$

Which of these are bounded? Which are connected? Label the points closest to and farthest from the origin (if there are any).

  b. Consider the surface

$$x_1^2 + 2x_2^2 + 3x_3^2 + x_1x_2 + 2x_1x_3 + 3x_2x_3 = 1.$$

Which of the three surfaces in part (a) does this surface qualitatively resemble most? Which points on this surface are closest to the origin? Give a rough approximation; you may use technology.

22. On the surface

$$-x_1^2 + x_2^2 - x_3^2 + 10x_1x_3 = 1,$$

find the two points closest to the origin.

23. Consider an $n \times n$ matrix M that is not symmetric, and define the function $g(\vec{x}) = \vec{x} \cdot M\vec{x}$ from $\mathbb{R}^n$ to $\mathbb{R}$. Is g necessarily a quadratic form? If so, find the matrix of g.

24. Consider a quadratic form

$$q(\vec{x}) = \vec{x} \cdot A\vec{x},$$

where A is a symmetric $n \times n$ matrix. Find $q(\vec{e}_1)$. Give your answer in terms of the entries of the matrix A.

25. Consider a quadratic form

$$q(\vec{x}) = \vec{x} \cdot A\vec{x},$$

where A is a symmetric $n \times n$ matrix. Let $\vec{v}$ be a unit eigenvector of A, with associated eigenvalue λ. Find $q(\vec{v})$.

26. Consider a quadratic form

$$q(\vec{x}) = \vec{x} \cdot A\vec{x},$$

where A is a symmetric $n \times n$ matrix. *True or false*? If there exists a nonzero vector $\vec{v}$ in $\mathbb{R}^n$ such that $q(\vec{v}) = 0$, then A fails to be invertible.

27. Consider a quadratic form $q(\vec{x}) = \vec{x} \cdot A\vec{x}$, where A is a symmetric $n \times n$ matrix with eigenvalues $\lambda_1 \geq \lambda_2 \geq \cdots \geq \lambda_n$. Let S^{n-1} be the set of all unit vectors in $\mathbb{R}^n$. Describe the image of S^{n-1} under q, in terms of the eigenvalues of A.

28. Show that any positive definite $n \times n$ matrix A can be written as $A = BB^T$, where B is an $n \times n$ matrix with orthogonal columns. (*Hint*: There exists an orthogonal matrix S such that $S^{-1}AS = S^T AS = D$ is a diagonal matrix with positive diagonal entries. Then $A = SDS^T$. Now write D as the square of a diagonal matrix.)

29. For the matrix $A = \begin{bmatrix} 8 & -2 \\ -2 & 5 \end{bmatrix}$ write $A = BB^T$ as discussed in Exercise 28. See Example 1.

30. Show that any positive definite matrix A can be written as $A = B^2$, where B is a positive definite matrix.

31. For the matrix $A = \begin{bmatrix} 8 & -2 \\ -2 & 5 \end{bmatrix}$ write $A = B^2$ as discussed in Exercise 30. See Example 1.

32. *Cholesky factorization for 2×2 matrices.* Show that any positive definite 2×2 matrix A can be written uniquely as $A = LL^T$, where L is a lower triangular 2×2 matrix with positive entries on the diagonal. {*Hint*: Solve the equation
$$\begin{bmatrix} a & b \\ b & c \end{bmatrix} = \begin{bmatrix} x & 0 \\ y & z \end{bmatrix} \begin{bmatrix} x & y \\ 0 & z \end{bmatrix}.\}$$

33. Find the Cholesky factorization (discussed in Exercise 32) for
$$A = \begin{bmatrix} 8 & -2 \\ -2 & 5 \end{bmatrix}.$$

34. A Cholesky factorization of a symmetric matrix A is a factorization of the form $A = LL^T$, where L is lower triangular with positive diagonal entries.

    Show that for a symmetric $n \times n$ matrix A the following are equivalent:

    (i) A is positive definite.

    (ii) All principal submatrices $A^{(m)}$ of A are positive definite. (See Theorem 8.2.5.)

    (iii) $\det(A^{(m)}) > 0$ for $m = 1, \ldots, n$.

    (iv) A has a Cholesky factorization $A = LL^T$.

    {*Hint*: Show that (i) implies (ii), (ii) implies (iii), (iii) implies (iv), and (iv) implies (i). The hardest step is the implication from (iii) to (iv): Arguing by induction on n, you may assume that $A^{(n-1)}$ has a Cholesky factorization $A^{(n-1)} = BB^T$. Now show that there exist a vector $\vec{x}$ in $\mathbb{R}^{n-1}$ and a scalar t such that
$$A = \begin{bmatrix} A^{(n-1)} & \vec{v} \\ \vec{v}^T & k \end{bmatrix} = \begin{bmatrix} B & 0 \\ \vec{x}^T & 1 \end{bmatrix} \begin{bmatrix} B^T & \vec{x} \\ 0 & t \end{bmatrix}.$$

Explain why the scalar t is positive. Therefore, we have the Cholesky factorization
$$A = \begin{bmatrix} B & 0 \\ \vec{x}^T & \sqrt{t} \end{bmatrix} \begin{bmatrix} B^T & \vec{x} \\ 0 & \sqrt{t} \end{bmatrix}.$$

This reasoning also shows that the Cholesky factorization of A is unique. Alternatively, you can use the LDL^T factorization of A to show that (iii) implies (iv). (See Exercise 5.3.63.)

    To show that (i) implies (ii), consider a nonzero vector $\vec{x}$ in $\mathbb{R}^m$, and define
$$\vec{y} = \begin{bmatrix} \vec{x} \\ 0 \\ \vdots \\ 0 \end{bmatrix}$$
in $\mathbb{R}^n$ (fill in $n - m$ zeros). Then
$$\vec{x}^T A^{(m)} \vec{x} = \vec{y}^T A\vec{y} > 0.\}$$

35. Find the Cholesky factorization of the matrix
$$A = \begin{bmatrix} 4 & -4 & 8 \\ -4 & 13 & 1 \\ 8 & 1 & 26 \end{bmatrix}.$$

36. Consider an invertible $n \times n$ matrix A. What is the relationship between the matrix R in the QR factorization of A and the matrix L in the Cholesky factorization of $A^T A$?

37. Consider the quadratic form
$$q(x_1, x_2) = ax_1^2 + bx_1x_2 + cx_2^2.$$

We define
$$q_{11} = \frac{\partial^2 q}{\partial x_1^2}, \quad q_{12} = q_{21} = \frac{\partial^2 q}{\partial x_1 \partial x_2}, \quad q_{22} = \frac{\partial^2 q}{\partial x_2^2}.$$

The discriminant D of q is defined as
$$D = \det \begin{bmatrix} q_{11} & q_{12} \\ q_{21} & q_{22} \end{bmatrix} = q_{11}q_{22} - (q_{12})^2.$$

The *second derivative test* tells us that if D and q_{11} are both positive, then $q(x_1, x_2)$ has a minimum at $(0, 0)$. Justify this fact, using the theory developed in this section.

38. For which values of the constants p and q is the $n \times n$ matrix
$$B = \begin{bmatrix} p & q & \cdots & q \\ q & p & \cdots & q \\ \vdots & \vdots & \ddots & \vdots \\ q & q & \cdots & p \end{bmatrix}$$
positive definite? (B has p's on the diagonal and q's elsewhere.) (*Hint*: Exercise 8.1.17 is helpful.)

39. For which angles θ can you find a basis of $\mathbb{R}^n$ such that the angle between any two vectors in this basis is θ?

40. Show that for every symmetric $n \times n$ matrix A there exists a constant k such that matrix $A + kI_n$ is positive definite.

41. Find the dimension of the space Q_2 of all quadratic forms in two variables.

42. Find the dimension of the space Q_n of all quadratic forms in n variables.

43. Consider the transformation $T\left(q(x_1, x_2)\right) = q(x_1, 0)$ from Q_2 to P_2. Is T a linear transformation? If so, find the image, rank, kernel, and nullity of T.

44. Consider the transformation $T\left(q(x_1, x_2)\right) = q(x_1, 1)$ from Q_2 to P_2. Is T a linear transformation? Is it an isomorphism?

45. Consider the transformation $T\left(q(x_1, x_2, x_3)\right)$ $= q(x_1, 1, 1)$ from Q_3 to P_2. Is T a linear transformation? If so, find the image, rank, kernel, and nullity of T.

46. Consider the linear transformation $T\left(q(x_1, x_2, x_3)\right)$ $= q(x_1, x_2, x_1)$ from Q_3 to Q_2. Find the image, kernel, rank, and nullity of this transformation.

47. Consider the function $T(A)(\vec{x}) = \vec{x}^T A \vec{x}$ from $\mathbb{R}^{n \times n}$ to Q_n. Show that T is a linear transformation. Find the image, kernel, rank, and nullity of T.

48. Consider the linear transformation $T\left(q(x_1, x_2)\right)$ $= q(x_2, x_1)$ from Q_2 to Q_2. Find all the eigenvalues and eigenfunctions of T. Is transformation T diagonalizable?

49. Consider the linear transformation $T\left(q(x_1, x_2)\right)$ $= q(x_1, 2x_2)$ from Q_2 to Q_2. Find all the eigenvalues and eigenfunctions of T. Is transformation T diagonalizable?

50. Consider the linear transformation

$$T\left(q(x_1, x_2)\right) = x_1 \frac{\partial q}{\partial x_2} + x_2 \frac{\partial q}{\partial x_1}$$

from Q_2 to Q_2. Find all the eigenvalues and eigenfunctions of T. Is transformation T diagonalizable?

51. What are the signs of the determinants of the principal submatrices of a negative definite matrix? (See Theorem 8.2.5.)

52. Consider a quadratic form q. If A is a symmetric matrix such that $q(\vec{x}) = \vec{x}^T A \vec{x}$ for all $\vec{x}$ in $\mathbb{R}^n$, show that $a_{ii} = q(\vec{e}_i)$ and $a_{ij} = \frac{1}{2}\left(q(\vec{e}_i + \vec{e}_j) - q(\vec{e}_i) - q(\vec{e}_j)\right)$ for $i \neq j$.

53. Consider a quadratic form $q(x_1, \dots, x_n)$ with symmetric matrix A. For two integers i and j with $1 \leq i < j \leq n$, we define the function

$$p(x, y) = q\left(0, \dots, 0, \underbrace{x}_{i\text{th}}, 0, \dots, 0, \underbrace{y}_{j\text{th}}, 0, \dots, 0\right).$$

a. Show that p is a quadratic form, with matrix $\begin{bmatrix} a_{ii} & a_{ij} \\ a_{ji} & a_{jj} \end{bmatrix}$.

b. If q is positive definite, show that p is positive definite as well.

c. If q is positive semidefinite, show that p is positive semidefinite as well.

d. Give an example where q is indefinite, but p is positive definite.

54. If A is a positive semidefinite matrix with $a_{11} = 0$, what can you say about the other entries in the first row and in the first column of A? (*Hint:* Exercise 53 is helpful.)

55. If A is a positive definite $n \times n$ matrix, show that the largest entry of A must be on the diagonal. (*Hint:* Use Exercise 53 to show that $a_{ij} < a_{ii}$ or $a_{ij} < a_{jj}$ for all $1 \leq i < j \leq n$).

56. If A is a real symmetric matrix, show that there exists an eigenvalue λ of A with $\lambda \geq a_{11}$. (*Hint:* Exercise 27 is helpful.)

In Exercises 57 through 61, consider a quadratic form q on $\mathbb{R}^3$ with symmetric matrix A, with the given properties. In each case, describe the level surface $q(\vec{x}) = 1$ geometrically.

57. q is positive definite.

58. q is positive semidefinite and rank $A = 2$.

59. q is positive semidefinite and rank $A = 1$.

60. q is indefinite and $\det A > 0$.

61. q is indefinite and $\det A < 0$.

62. Consider an indefinite quadratic form q on $\mathbb{R}^3$ with symmetric matrix A. If $\det A < 0$, describe the level surface $q(\vec{x}) = 0$.

63. Consider a positive definite quadratic form q on $\mathbb{R}^n$ with symmetric matrix A. We know that there exists an orthonormal eigenbasis $\vec{v}_1, \dots, \vec{v}_n$ for A, with associated positive eigenvalues $\lambda_1, \dots, \lambda_n$. Now consider the orthogonal eigenbasis $\vec{w}_1, \dots, \vec{w}_n$, where $\vec{w}_i = \frac{1}{\sqrt{\lambda_i}} \vec{v}_i$. Show that $q\left(c_1 \vec{w}_1 + \dots + c_n \vec{w}_n\right) = c_1^2 + \dots + c_n^2$.

64. For the quadratic form $q(x_1, x_2) = 8x_1^2 - 4x_1 x_2 + 5x_2^2$, find an orthogonal basis $\vec{w}_1, \vec{w}_2$ of $\mathbb{R}^2$ such that $q(c_1 \vec{w}_1 + c_2 \vec{w}_2) = c_1^2 + c_2^2$. Use your answer to sketch the level curve $q(\vec{x}) = 1$. Compare with Example 4 and Figure 4 in this section. Exercise 63 is helpful.

65. Show that for every indefinite quadratic form q on $\mathbb{R}^2$ there exists an orthogonal basis $\vec{w}_1, \vec{w}_2$ of $\mathbb{R}^2$ such that $q(c_1 \vec{w}_1 + c_2 \vec{w}_2) = c_1^2 - c_2^2$. *Hint:* Modify the approach outlined in Exercise 63.

66. For the quadratic form $q(x_1, x_2) = 3x_1^2 - 10x_1 x_2 + 3x_2^2$, find an orthogonal basis $\vec{w}_1, \vec{w}_2$ of $\mathbb{R}^2$ such that

$q(c_1\vec{w}_1 + c_2\vec{w}_2) = c_1^2 - c_2^2$. Use your answer to sketch the level curve $q(\vec{x}) = 1$. Exercise 65 is helpful.

67. Consider a quadratic form q on $\mathbb{R}^n$ with symmetric matrix A, with rank $A = r$. Suppose that A has p positive eigenvalues, if eigenvalues are counted with their multiplicities. Show that there exists an orthogonal basis $\vec{w}_1, \ldots, \vec{w}_n$ of $\mathbb{R}^n$ such that $q(c_1\vec{w}_1 + \cdots + c_n\vec{w}_n) = c_1^2 + \cdots + c_p^2 - c_{p+1}^2 - \cdots - c_r^2$. *Hint*: Modify the approach outlined in Exercises 63 and 65.

68. If q is a quadratic form on $\mathbb{R}^n$ with symmetric matrix A, and if $L(\vec{x}) = R\vec{x}$ is a linear transformation from $\mathbb{R}^m$ to $\mathbb{R}^n$, show that the composite function $p(\vec{x}) = q\big(L(\vec{x})\big)$

is a quadratic form on $\mathbb{R}^m$. Express the symmetric matrix of p in terms of R and A.

69. If A is a positive definite $n \times n$ matrix, and R is any real $n \times m$ matrix, what can you say about the definiteness of the matrix $R^T A R$? For which matrices R is $R^T A R$ positive definite?

70. If A is an indefinite $n \times n$ matrix, and R is a real $n \times m$ matrix of rank n, what can you say about the definiteness of the matrix $R^T A R$?

71. If A is an indefinite $n \times n$ matrix, and R is any real $n \times m$ matrix, what can you say about the definiteness of the matrix $R^T A R$?

8.3 Singular Values

In Exercise 47 of Section 2.2, we stated the following remarkable fact.

EXAMPLE I   Show that if $L(\vec{x}) = A\vec{x}$ is a linear transformation from $\mathbb{R}^2$ to $\mathbb{R}^2$, then there exist two orthogonal unit vectors $\vec{v}_1$ and $\vec{v}_2$ in $\mathbb{R}^2$ such that vectors $L(\vec{v}_1)$ and $L(\vec{v}_2)$ are orthogonal as well (although not necessarily unit vectors). See Figure 1. (*Hint*: Consider an orthonormal eigenbasis $\vec{v}_1, \vec{v}_2$ of the symmetric matrix $A^T A$.)

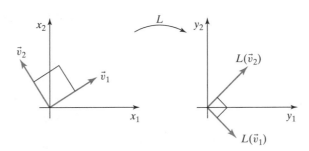

Figure I

Solution

This statement is clear for some classes of transformations. For example,

- If L is an orthogonal transformation, then any two orthogonal unit vectors $\vec{v}_1$ and $\vec{v}_2$ will do, by Theorem 5.3.2.
- If $L(\vec{x}) = A\vec{x}$, where A is symmetric, then we can choose two orthogonal unit eigenvectors, by the spectral theorem, Theorem 8.1.1. See also Example 4 of Section 8.1.

However, for an arbitrary linear transformation L, the statement isn't that obvious; think about the case of a shear, for example.

In Exercise 47 of Section 2.2, we suggested a proof based on the intermediate value theorem for continuous functions. Here we will present a proof in the spirit of linear algebra that generalizes more easily to higher-dimensional spaces.

Following the hint, we first note that matrix $A^T A$ is symmetric, since $(A^T A)^T = A^T (A^T)^T = A^T A$. The spectral theorem (Theorem 8.1.1) tells us that there exists an orthonormal eigenbasis $\vec{v}_1, \vec{v}_2$ for $A^T A$, with associated eigenvalues λ_1, λ_2. We can verify that vectors $L(\vec{v}_1) = A\vec{v}_1$ and $L(\vec{v}_2) = A\vec{v}_2$ are orthogonal, as claimed:

$$(A\vec{v}_1) \cdot (A\vec{v}_2) = (A\vec{v}_1)^T A\vec{v}_2 = \vec{v}_1^T A^T A\vec{v}_2 = \vec{v}_1^T (\lambda_2 \vec{v}_2) = \lambda_2 (\vec{v}_1 \cdot \vec{v}_2) = 0.$$

It is worth mentioning that $\vec{v}_1$ and $\vec{v}_2$ need not be eigenvectors of matrix A.   ■

EXAMPLE 2   Consider the linear transformation $L(\vec{x}) = A\vec{x}$, where $A = \begin{bmatrix} 6 & 2 \\ -7 & 6 \end{bmatrix}$.

a. Find an orthonormal basis $\vec{v}_1$, $\vec{v}_2$ of $\mathbb{R}^2$ such that vectors $L(\vec{v}_1)$ and $L(\vec{v}_2)$ are orthogonal.

b. Show that the image of the unit circle under transformation L is an ellipse. Find the lengths of the two semiaxes of this ellipse, in terms of the eigenvalues of matrix $A^T A$.

Solution

a. Using the ideas of Example 1, we will find an orthonormal eigenbasis for matrix $A^T A$:

$$A^T A = \begin{bmatrix} 6 & -7 \\ 2 & 6 \end{bmatrix} \begin{bmatrix} 6 & 2 \\ -7 & 6 \end{bmatrix} = \begin{bmatrix} 85 & -30 \\ -30 & 40 \end{bmatrix}.$$

The characteristic polynomial of $A^T A$ is

$$\lambda^2 - 125\lambda + 2500 = (\lambda - 100)(\lambda - 25),$$

so that the eigenvalues of $A^T A$ are $\lambda_1 = 100$ and $\lambda_2 = 25$. Now we can find the eigenspaces of $A^T A$.

$$E_{100} = \ker \begin{bmatrix} -15 & -30 \\ -30 & -60 \end{bmatrix} = \text{span} \begin{bmatrix} 2 \\ -1 \end{bmatrix}$$

and

$$E_{25} = \ker \begin{bmatrix} 60 & -30 \\ -30 & 15 \end{bmatrix} = \text{span} \begin{bmatrix} 1 \\ 2 \end{bmatrix}.$$

To find an ortho*normal* basis, we need to multiply these vectors by the reciprocals of their lengths:

$$\vec{v}_1 = \frac{1}{\sqrt{5}} \begin{bmatrix} 2 \\ -1 \end{bmatrix}, \quad \vec{v}_2 = \frac{1}{\sqrt{5}} \begin{bmatrix} 1 \\ 2 \end{bmatrix}.$$

b. The unit circle consists of the vectors of the form $\vec{x} = \cos(t)\vec{v}_1 + \sin(t)\vec{v}_2$, and the image of the unit circle consists of the vectors $L(\vec{x}) = \cos(t)L(\vec{v}_1) + \sin(t)L(\vec{v}_2)$. This image is the ellipse whose semimajor and semiminor axes are $L(\vec{v}_1)$ and $L(\vec{v}_2)$. What are the lengths of these axes?

$$\|L(\vec{v}_1)\|^2 = (A\vec{v}_1) \cdot (A\vec{v}_1) = \vec{v}_1^T A^T A\vec{v}_1 = \vec{v}_1^T (\lambda_1 \vec{v}_1) = \lambda_1 (\vec{v}_1 \cdot \vec{v}_1) = \lambda_1$$

Likewise,

$$\|L(\vec{v}_2)\|^2 = \lambda_2.$$

Thus,

$$\|L(\vec{v}_1)\| = \sqrt{\lambda_1} = \sqrt{100} = 10 \quad \text{and} \quad \|L(\vec{v}_2)\| = \sqrt{\lambda_2} = \sqrt{25} = 5.$$

See Figure 2. We can compute the lengths of vectors $L(\vec{v}_1)$ and $L(\vec{v}_2)$ directly, of course, but the way we did it before is more informative. For example,

$$L(\vec{v}_1) = A\vec{v}_1 = \frac{1}{\sqrt{5}} \begin{bmatrix} 6 & 2 \\ -7 & 6 \end{bmatrix} \begin{bmatrix} 2 \\ -1 \end{bmatrix} = \frac{1}{\sqrt{5}} \begin{bmatrix} 10 \\ -20 \end{bmatrix},$$

so that

$$\|L(\vec{v}_1)\| = \left\| \frac{1}{\sqrt{5}} \begin{bmatrix} 10 \\ -20 \end{bmatrix} \right\| = 10.$$ ■

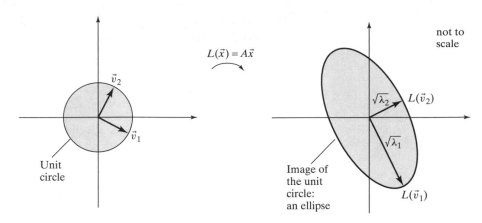

Figure 2

Part (b) of Example 2 shows that the square roots of the eigenvalues of matrix $A^T A$ play an important role in the geometrical interpretation of the transformation $L(\vec{x}) = A\vec{x}$. In Example 8.2.3 we have seen that the symmetric matrix $A^T A$ is positive semidefinite for any $n \times m$ matrix A, meaning that the eigenvalues of $A^T A$ are positive or zero.

Definition 8.3.1   Singular values

The *singular values* of an $n \times m$ matrix A are the square roots of the eigenvalues of the symmetric $m \times m$ matrix $A^T A$, listed with their algebraic multiplicities. It is customary to denote the singular values by $\sigma_1, \sigma_2, \ldots, \sigma_m$ and to list them in decreasing order:

$$\sigma_1 \geq \sigma_2 \geq \cdots \geq \sigma_m \geq 0.$$

The singular values of the matrix $A = \begin{bmatrix} 6 & 2 \\ -7 & 6 \end{bmatrix}$ considered in Example 2 are $\sigma_1 = \sqrt{\lambda_1} = 10$ and $\sigma_2 = \sqrt{\lambda_2} = 5$, since the eigenvalues of $A^T A$ are $\lambda_1 = 100$ and $\lambda_2 = 25$.

We can now generalize our work in Example 2.

Theorem 8.3.2   The image of the unit circle

Let $L(\vec{x}) = A\vec{x}$ be an invertible linear transformation from $\mathbb{R}^2$ to $\mathbb{R}^2$. The image of the unit circle under L is an ellipse E. The lengths of the semimajor and the semiminor axes of E are the singular values σ_1 and σ_2 of A, respectively.   ∎

Take another look at Figure 2.

We can generalize our findings in Examples 1 and 2 to matrices of arbitrary size.

Theorem 8.3.3   Let $L(\vec{x}) = A\vec{x}$ be a linear transformation from $\mathbb{R}^m$ to $\mathbb{R}^n$. Then there exists an orthonormal basis $\vec{v}_1, \vec{v}_2, \ldots, \vec{v}_m$ of $\mathbb{R}^m$ such that

a. Vectors $L(\vec{v}_1), L(\vec{v}_2), \ldots, L(\vec{v}_m)$ are orthogonal, and

b. The lengths of vectors $L(\vec{v}_1), L(\vec{v}_2), \ldots, L(\vec{v}_m)$ are the singular values $\sigma_1, \sigma_2, \ldots, \sigma_m$ of matrix A.

To construct $\vec{v}_1, \vec{v}_2, \ldots, \vec{v}_m$, find an orthonormal eigenbasis for matrix $A^T A$. Make sure that the corresponding eigenvalues $\lambda_1, \lambda_2, \ldots, \lambda_m$ appear in descending order:

$$\lambda_1 \geq \lambda_2 \geq \cdots \geq \lambda_m \geq 0,$$

so that $\lambda_i = \sigma_i^2$ for $i = 1, \ldots, m$. ∎

The proof is analogous to the special case $n = m = 2$ considered in Examples 1 and 2:

a. $L(\vec{v}_i) \cdot L(\vec{v}_j) = (A\vec{v}_i) \cdot (A\vec{v}_j) = (A\vec{v}_i)^T A\vec{v}_j = \vec{v}_i^T A^T A\vec{v}_j = \vec{v}_i^T (\lambda_j \vec{v}_j)$
$\quad = \lambda_j (\vec{v}_i \cdot \vec{v}_j) = 0$ when $i \neq j$, and

b. $\|L(\vec{v}_i)\|^2 = (A\vec{v}_i) \cdot (A\vec{v}_i) = \vec{v}_i^T A^T A\vec{v}_i = \vec{v}_i^T (\lambda_i \vec{v}_i) = \lambda_i (\vec{v}_i \cdot \vec{v}_i) = \lambda_i = \sigma_i^2$,
$\quad$ so that $\|L(\vec{v}_i)\| = \sigma_i$.

EXAMPLE 3   Consider the linear transformation

$$L(\vec{x}) = A\vec{x}, \quad \text{where} \quad A = \begin{bmatrix} 0 & 1 & 1 \\ 1 & 1 & 0 \end{bmatrix}.$$

a. Find the singular values of A.

b. Find orthonormal vectors $\vec{v}_1, \vec{v}_2, \vec{v}_3$ in $\mathbb{R}^3$ such that $L(\vec{v}_1), L(\vec{v}_2)$, and $L(\vec{v}_3)$ are orthogonal.

c. Sketch and describe the image of the unit sphere under the transformation L.

Solution

a. $A^T A = \begin{bmatrix} 0 & 1 \\ 1 & 1 \\ 1 & 0 \end{bmatrix} \begin{bmatrix} 0 & 1 & 1 \\ 1 & 1 & 0 \end{bmatrix} = \begin{bmatrix} 1 & 1 & 0 \\ 1 & 2 & 1 \\ 0 & 1 & 1 \end{bmatrix}$

The eigenvalues of $A^T A$ are $\lambda_1 = 3, \lambda_2 = 1, \lambda_3 = 0$. The singular values of A are

$$\sigma_1 = \sqrt{\lambda_1} = \sqrt{3}, \quad \sigma_2 = \sqrt{\lambda_2} = 1, \quad \sigma_3 = \sqrt{\lambda_3} = 0.$$

b. Find an orthonormal eigenbasis $\vec{v}_1, \vec{v}_2, \vec{v}_3$ for $A^T A$ (we omit the details):

$$E_3 = \text{span} \begin{bmatrix} 1 \\ 2 \\ 1 \end{bmatrix}, \quad E_1 = \text{span} \begin{bmatrix} 1 \\ 0 \\ -1 \end{bmatrix}, \quad E_0 = \ker(A^T A) = \text{span} \begin{bmatrix} 1 \\ -1 \\ 1 \end{bmatrix}$$

$$\vec{v}_1 = \frac{1}{\sqrt{6}} \begin{bmatrix} 1 \\ 2 \\ 1 \end{bmatrix}, \quad \vec{v}_2 = \frac{1}{\sqrt{2}} \begin{bmatrix} 1 \\ 0 \\ -1 \end{bmatrix}, \quad \vec{v}_3 = \frac{1}{\sqrt{3}} \begin{bmatrix} 1 \\ -1 \\ 1 \end{bmatrix}.$$

We compute $A\vec{v}_1, A\vec{v}_2, A\vec{v}_3$ and check orthogonality:

$$A\vec{v}_1 = \frac{1}{\sqrt{6}} \begin{bmatrix} 3 \\ 3 \end{bmatrix}, \quad A\vec{v}_2 = \frac{1}{\sqrt{2}} \begin{bmatrix} -1 \\ 1 \end{bmatrix}, \quad A\vec{v}_3 = \begin{bmatrix} 0 \\ 0 \end{bmatrix}.$$

We can also check that the length of $A\vec{v}_i$ is σ_i:

$$\|A\vec{v}_1\| = \sqrt{3} = \sigma_1, \quad \|A\vec{v}_2\| = 1 = \sigma_2, \quad \|A\vec{v}_3\| = 0 = \sigma_3.$$

c. The unit sphere in $\mathbb{R}^3$ consists of all vectors of the form

$$\vec{x} = c_1 \vec{v}_1 + c_2 \vec{v}_2 + c_3 \vec{v}_3, \quad \text{where} \quad c_1^2 + c_2^2 + c_3^2 = 1.$$

The image of the unit sphere consists of the vectors

$$L(\vec{x}) = c_1 L(\vec{v}_1) + c_2 L(\vec{v}_2),$$

where $c_1^2 + c_2^2 \leq 1$. [Recall that $L(\vec{v}_3) = \vec{0}$.]
The image is the full ellipse shaded in Figure 3.

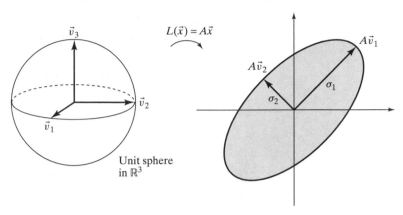

Figure 3

Example 3 shows that some of the singular values of a matrix may be zero. Suppose the singular values $\sigma_1, \ldots, \sigma_s$ of an $n \times m$ matrix A are nonzero, while $\sigma_{s+1}, \ldots, \sigma_m$ are zero. Choose vectors $\vec{v}_1, \ldots, \vec{v}_s, \vec{v}_{s+1}, \ldots, \vec{v}_m$ for A as introduced in Theorem 8.3.3. Note that $\|A\vec{v}_i\| = \sigma_i = 0$ and therefore $A\vec{v}_i = \vec{0}$ for $i = s + 1, \ldots, m$. We claim that the vectors $A\vec{v}_1, \ldots, A\vec{v}_s$ form a basis of the image of A. Indeed, these vectors are linearly independent (because they are orthogonal and nonzero), and they span the image, since any vector in the image of A can be written as

$$A\vec{x} = A(c_1\vec{v}_1 + \cdots + c_s\vec{v}_s + \cdots + c_m\vec{v}_m)$$
$$= c_1 A\vec{v}_1 + \cdots + c_s A\vec{v}_s.$$

This shows that $s = \dim(\text{im}A) = \text{rank}(A)$.

Theorem 8.3.4 **Singular values and rank**

If A is an $n \times m$ matrix of rank r, then the singular values $\sigma_1, \ldots, \sigma_r$ are nonzero, while $\sigma_{r+1}, \ldots, \sigma_m$ are zero. ∎

The Singular Value Decomposition

Just as we expressed the Gram–Schmidt process in terms of a matrix decomposition (the QR-factorization), we will now express Theorem 8.3.3 in terms of a matrix decomposition.

Consider a linear transformation $L(\vec{x}) = A\vec{x}$ from $\mathbb{R}^m$ to $\mathbb{R}^n$, and choose an orthonormal basis $\vec{v}_1, \ldots, \vec{v}_m$ of $\mathbb{R}^m$ as in Theorem 8.3.3. Let $r = \text{rank}(A)$. We know that the vectors $A\vec{v}_1, \ldots, A\vec{v}_r$ are orthogonal and nonzero, with $\|A\vec{v}_i\| = \sigma_i$. We introduce the unit vectors

$$\vec{u}_1 = \frac{1}{\sigma_1} A\vec{v}_1, \ldots, \vec{u}_r = \frac{1}{\sigma_r} A\vec{v}_r.$$

We can expand the sequence $\vec{u}_1, \ldots, \vec{u}_r$ to an orthonormal basis $\vec{u}_1, \ldots, \vec{u}_n$ of $\mathbb{R}^n$. Then we can write

$$A\vec{v}_i = \sigma_i\vec{u}_i \quad \text{for } i = 1, \ldots, r$$

and

$$A\vec{v}_i = \vec{0} \quad \text{for } i = r+1, \ldots, m.$$

We can express these equations in matrix form as follows:

$$A \underbrace{\begin{bmatrix} | & & | & | & & | \\ \vec{v}_1 & \cdots & \vec{v}_r & \vec{v}_{r+1} & \cdots & \vec{v}_m \\ | & & | & | & & | \end{bmatrix}}_{V} = \begin{bmatrix} | & & | & | & & | \\ \sigma_1\vec{u}_1 & \cdots & \sigma_r\vec{u}_r & \vec{0} & \cdots & \vec{0} \\ | & & | & | & & | \end{bmatrix}$$

$$= \begin{bmatrix} | & & | & | & & | \\ \vec{u}_1 & \cdots & \vec{u}_r & \vec{0} & \cdots & \vec{0} \\ | & & | & | & & | \end{bmatrix} \begin{bmatrix} \sigma_1 & & & \\ & \ddots & & 0 \\ & & \sigma_r & \\ 0 & & & 0 \end{bmatrix}$$

$$= \underbrace{\begin{bmatrix} | & & | & | & & | \\ \vec{u}_1 & \cdots & \vec{u}_r & \vec{u}_{r+1} & \cdots & \vec{u}_n \\ | & & | & | & & | \end{bmatrix}}_{U} \underbrace{\begin{bmatrix} \sigma_1 & & & \\ & \ddots & & 0 \\ & & \sigma_r & \\ 0 & & & 0 \end{bmatrix}}_{\Sigma},$$

or, more succinctly,

$$AV = U\Sigma.$$

Note that V is an orthogonal $m \times m$ matrix, U is an orthogonal $n \times n$ matrix, and Σ is an $n \times m$ matrix whose first r diagonal entries are $\sigma_1, \ldots, \sigma_r$, and all other entries are zero.

Multiplying the equation $AV = U\Sigma$ with V^T from the right, we find that $A = U\Sigma V^T$.

Theorem 8.3.5   Singular value decomposition (SVD)

Any $n \times m$ matrix A can be written as

$$A = U\Sigma V^T,$$

where U is an orthogonal $n \times n$ matrix; V is an orthogonal $m \times m$ matrix; and Σ is an $n \times m$ matrix whose first r diagonal entries are the nonzero singular values $\sigma_1, \ldots, \sigma_r$ of A, and all other entries are zero [where $r = \text{rank}(A)$].

Alternatively, this singular value decomposition can be written as

$$A = \sigma_1\vec{u}_1\vec{v}_1^T + \cdots + \sigma_r\vec{u}_r\vec{v}_r^T,$$

where the $\vec{u}_i$ and the $\vec{v}_i$ are the columns of U and V, respectively. (See Exercise 29.)

◼

A singular value decomposition of a 2×2 matrix A is presented in Figure 4.

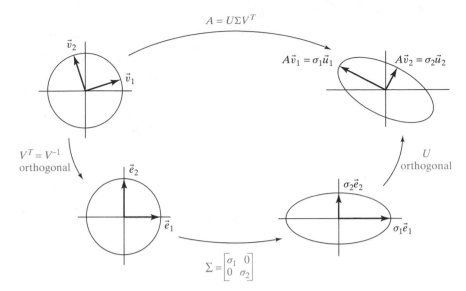

Figure 4

Here are two numerical examples.

EXAMPLE 4 Find an SVD for $A = \begin{bmatrix} 6 & 2 \\ -7 & 6 \end{bmatrix}$. (Compare with Example 2.)

Solution

In Example 2, we found $\vec{v}_1 = \dfrac{1}{\sqrt{5}} \begin{bmatrix} 2 \\ -1 \end{bmatrix}$ and $\vec{v}_2 = \dfrac{1}{\sqrt{5}} \begin{bmatrix} 1 \\ 2 \end{bmatrix}$, so that

$$V = \frac{1}{\sqrt{5}} \begin{bmatrix} 2 & 1 \\ -1 & 2 \end{bmatrix}.$$

The columns $\vec{u}_1$ and $\vec{u}_2$ of U are defined as

$$\vec{u}_1 = \frac{1}{\sigma_1} A \vec{v}_1 = \frac{1}{\sqrt{5}} \begin{bmatrix} 1 \\ -2 \end{bmatrix},$$

$$\vec{u}_2 = \frac{1}{\sigma_2} A \vec{v}_2 = \frac{1}{\sqrt{5}} \begin{bmatrix} 2 \\ 1 \end{bmatrix},$$

and therefore

$$U = \frac{1}{\sqrt{5}} \begin{bmatrix} 1 & 2 \\ -2 & 1 \end{bmatrix}.$$

Finally,

$$\Sigma = \begin{bmatrix} \sigma_1 & 0 \\ 0 & \sigma_2 \end{bmatrix} = \begin{bmatrix} 10 & 0 \\ 0 & 5 \end{bmatrix}.$$

You can check that

$$A = U \Sigma V^T.$$

EXAMPLE 5 Find an SVD for $A = \begin{bmatrix} 0 & 1 & 1 \\ 1 & 1 & 0 \end{bmatrix}$. (Compare with Example 3.)

Solution

Using our work in Example 3, we find that

$$
V = \begin{bmatrix} 1/\sqrt{6} & 1/\sqrt{2} & 1/\sqrt{3} \\ 2/\sqrt{6} & 0 & -1/\sqrt{3} \\ 1/\sqrt{6} & -1/\sqrt{2} & 1/\sqrt{3} \end{bmatrix},
$$

$$
U = \begin{bmatrix} 1/\sqrt{2} & -1/\sqrt{2} \\ 1/\sqrt{2} & 1/\sqrt{2} \end{bmatrix},
$$

and

$$
\Sigma = \begin{bmatrix} \sqrt{3} & 0 & 0 \\ 0 & 1 & 0 \end{bmatrix}.
$$

Check that $A = U\Sigma V^T$. ∎

Consider a singular value decomposition

$$
A = U\Sigma V^T,
$$

where

$$
V = \begin{bmatrix} | & & | \\ \vec{v}_1 & \cdots & \vec{v}_m \\ | & & | \end{bmatrix} \quad \text{and} \quad U = \begin{bmatrix} | & & | \\ \vec{u}_1 & \cdots & \vec{u}_n \\ | & & | \end{bmatrix}.
$$

We know that

$$
A\vec{v}_i = \sigma_i \vec{u}_i \quad \text{for } i = 1, \dots, r
$$

and

$$
A\vec{v}_i = \vec{0} \quad \text{for } i = r + 1, \dots, m.
$$

These equations tell us that

$$
\ker A = \text{span}(\vec{v}_{r+1}, \dots, \vec{v}_m)
$$

and

$$
\text{im } A = \text{span}(\vec{u}_1, \dots, \vec{u}_r).
$$

(Fill in the details.) We see that an SVD provides us with orthonormal bases for the kernel and image of A.

Likewise, we have

$$
A^T = V\Sigma^T U^T \quad \text{or} \quad A^T U = V\Sigma^T.
$$

Reading the last equation column by column, we find that

$$
A^T \vec{u}_i = \sigma_i \vec{v}_i, \quad \text{for } i = 1, \dots, r,
$$

and

$$
A^T \vec{u}_i = \vec{0}, \quad \text{for } i = r + 1, \dots, n.
$$

(Observe that the roles of vectors $\vec{u}_i$ and $\vec{v}_i$ are reversed.)

As before, we have

$$
\text{im}(A^T) = \text{span}(\vec{v}_1, \dots, \vec{v}_r)
$$

and

$$
\ker(A^T) = \text{span}(\vec{u}_{r+1}, \dots, \vec{u}_n).
$$

In Figure 5, we make an attempt to visualize these observations. We represent each of the kernels and images simply as a line.

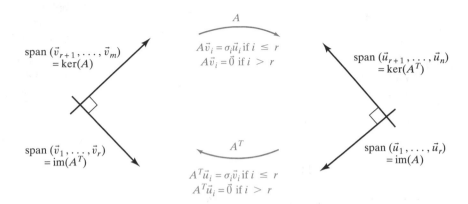

Figure 5

Note that $\text{im}(A)$ and $\text{ker}(A^T)$ are orthogonal complements, as observed in Theorem 5.4.1.

We conclude this section with a brief discussion of one of the many applications of the SVD—an application to data compression. We follow the exposition of Gilbert Strang (*Linear Algebra and Its Applications,* 4th ed., Brooks Cole, 2005).

Suppose a satellite transmits a picture containing 1000×1000 pixels. If the color of each pixel is digitized, this information can be represented in a 1000×1000 matrix A. How can we transmit the essential information contained in this picture without sending all 1,000,000 numbers?

Suppose we know an SVD

$$A = \sigma_1 \vec{u}_1 \vec{v}_1^T + \cdots + \sigma_r \vec{u}_r \vec{v}_r^T.$$

Even if the rank r of the matrix A is large, most of the singular values will typically be very small (relative to σ_1). If we neglect those, we get a good approximation $A \approx \sigma_1 \vec{u}_1 \vec{v}_1^T + \cdots + \sigma_s \vec{u}_s \vec{v}_s^T$, where s is much smaller than r. For example, if we choose $s = 10$, we need to transmit only the 20 vectors $\sigma_1 \vec{u}_1, \ldots, \sigma_{10} \vec{u}_{10}$ and $\vec{v}_1, \ldots, \vec{v}_{10}$ in $\mathbb{R}^{1000}$, that is, 20,000 numbers.

EXERCISES 8.3

GOAL *Find the singular values and a singular value decomposition of a matrix. Interpret the singular values of a 2 × 2 matrix in terms of the image of the unit circle.*

1. Find the singular values of $A = \begin{bmatrix} 1 & 0 \\ 0 & -2 \end{bmatrix}$.

2. Let A be an orthogonal 2×2 matrix. Use the image of the unit circle to find the singular values of A.

3. Let A be an orthogonal $n \times n$ matrix. Find the singular values of A algebraically.

4. Find the singular values of $A = \begin{bmatrix} 1 & 1 \\ 0 & 1 \end{bmatrix}$.

5. Find the singular values of $A = \begin{bmatrix} p & -q \\ q & p \end{bmatrix}$. Explain your answer geometrically.

6. Find the singular values of $A = \begin{bmatrix} 1 & 2 \\ 2 & 4 \end{bmatrix}$. Find a unit vector $\vec{v}_1$ such that $\|A\vec{v}_1\| = \sigma_1$. Sketch the image of the unit circle.

Find singular value decompositions for the matrices listed in Exercises 7 through 14. Work with paper and pencil. In each case, draw a sketch analogous to Figure 4 in the text, showing the effect of the transformation on the unit circle, in three steps.

7. $\begin{bmatrix} 1 & 0 \\ 0 & -2 \end{bmatrix}$ **8.** $\begin{bmatrix} p & -q \\ q & p \end{bmatrix}$ **9.** $\begin{bmatrix} 1 & 2 \\ 2 & 4 \end{bmatrix}$

10. $\begin{bmatrix} 6 & -7 \\ 2 & 6 \end{bmatrix}$ (See Example 4.)

11. $\begin{bmatrix} 1 & 0 \\ 0 & 2 \\ 0 & 0 \end{bmatrix}$

12. $\begin{bmatrix} 0 & 1 \\ 1 & 1 \\ 1 & 0 \end{bmatrix}$ (See Example 5.)

13. $\begin{bmatrix} 6 & 3 \\ -1 & 2 \end{bmatrix}$ **14.** $\begin{bmatrix} 2 & 3 \\ 0 & 2 \end{bmatrix}$

15. If A is an invertible 2×2 matrix, what is the relationship between the singular values of A and A^{-1}? Justify your answer in terms of the image of the unit circle.

16. If A is an invertible $n \times n$ matrix, what is the relationship between the singular values of A and A^{-1}?

17. Consider an $n \times m$ matrix A with rank$(A) = m$, and a singular value decomposition $A = U\Sigma V^T$. Show that the least-squares solution of a linear system $A\vec{x} = \vec{b}$ can be written as

$$\vec{x}^* = \frac{\vec{b} \cdot \vec{u}_1}{\sigma_1}\vec{v}_1 + \cdots + \frac{\vec{b} \cdot \vec{u}_m}{\sigma_m}\vec{v}_m.$$

18. Consider the 4×2 matrix

$$A = \frac{1}{10}\begin{bmatrix} 1 & 1 & 1 & 1 \\ 1 & 1 & -1 & -1 \\ 1 & -1 & 1 & -1 \\ 1 & -1 & -1 & 1 \end{bmatrix}\begin{bmatrix} 2 & 0 \\ 0 & 1 \\ 0 & 0 \\ 0 & 0 \end{bmatrix}\begin{bmatrix} 3 & -4 \\ 4 & 3 \end{bmatrix}.$$

Use the result of Exercise 17 to find the least-squares solution of the linear system

$$A\vec{x} = \vec{b}, \quad \text{where} \quad \vec{b} = \begin{bmatrix} 1 \\ 2 \\ 3 \\ 4 \end{bmatrix}.$$

Work with paper and pencil.

19. Consider an $n \times m$ matrix A of rank r, and a singular value decomposition $A = U\Sigma V^T$. Explain how you can express the least-squares solutions of a system $A\vec{x} = \vec{b}$ as linear combinations of the columns $\vec{v}_1, \ldots, \vec{v}_m$ of V.

20. a. Explain how any square matrix A can be written as

$$A = QS,$$

where Q is orthogonal and S is symmetric positive semidefinite. This is called the *polar decomposition* of A. (*Hint*: Write $A = U\Sigma V^T = UV^T V\Sigma V^T$.)

b. Is it possible to write $A = S_1 Q_1$, where Q_1 is orthogonal and S_1 is symmetric positive semidefinite?

21. Find a polar decomposition $A = QS$ as discussed in Exercise 20 for $A = \begin{bmatrix} 6 & 2 \\ -7 & 6 \end{bmatrix}$. Draw a sketch showing $S(C)$ and $A(C) = Q(S(C))$, where C is the unit circle centered at the origin. Compare with Examples 2 and 4 and with Figure 4.

22. Consider the standard matrix A representing the linear transformation

$$T(\vec{x}) = \vec{v} \times \vec{x} \quad \text{from } \mathbb{R}^3 \text{ to } \mathbb{R}^3,$$

where $\vec{v}$ is a given nonzero vector in $\mathbb{R}^3$.

a. Use the geometrical interpretation of the cross product to find an orthogonal projection T_1 onto a plane, a scaling T_2, and a rotation T_3 about a line such that $T(\vec{x}) = T_3(T_2(T_1(\vec{x})))$, for all $\vec{x}$ in $\mathbb{R}^3$. Describe the transformations T_1, T_2 and T_3 as precisely as you can: For T_1 give the plane onto which we project, for T_2 find the scaling factor, and for T_3 give the line about which we rotate and the angle of rotation. All of these answers, except for the angle of rotation, will be in terms of the given vector $\vec{v}$. Now let A_1, A_2, and A_3 be the standard matrices of these transformations T_1, T_2 and T_3, respectively. (You are not asked to find these matrices.) Explain how you can use the factorization $A = A_3 A_2 A_1$ to write a polar decomposition $A = QS$ of A. Express the matrices Q and S in terms of A_1, A_2, and A_3. See Exercise 20.

b. Find the $A_3 A_2 A_1$ and QS factorizations discussed in part (a) in the case

$$\vec{v} = \begin{bmatrix} 0 \\ 2 \\ 0 \end{bmatrix}.$$

23. Consider an SVD

$$A = U\Sigma V^T$$

of an $n \times m$ matrix A. Show that the columns of U form an orthonormal eigenbasis for AA^T. What are the associated eigenvalues? What does your answer tell you about the relationship between the eigenvalues of $A^T A$ and AA^T? (Compare this with Exercise 7.4.57.)

24. If A is a symmetric $n \times n$ matrix, what is the relationship between the eigenvalues of A and the singular values of A?

25. Let A be a 2×2 matrix and $\vec{u}$ a unit vector in $\mathbb{R}^2$. Show that

$$\sigma_2 \leq \|A\vec{u}\| \leq \sigma_1,$$

where σ_1, σ_2 are the singular values of A. Illustrate this inequality with a sketch, and justify it algebraically.

26. Let A be an $n \times m$ matrix and $\vec{v}$ a vector in $\mathbb{R}^m$. Show that

$$\sigma_m \|\vec{v}\| \leq \|A\vec{v}\| \leq \sigma_1 \|\vec{v}\|,$$

where σ_1 and σ_m are the largest and the smallest singular values of A, respectively. (Compare this with Exercise 25.)

27. Let λ be a real eigenvalue of an $n \times n$ matrix A. Show that

$$\sigma_n \le |\lambda| \le \sigma_1,$$

where σ_1 and σ_n are the largest and the smallest singular values of A, respectively.

28. If A is an $n \times n$ matrix, what is the product of its singular values $\sigma_1, \ldots, \sigma_n$? State the product in terms of the determinant of A. For a 2×2 matrix A, explain this result in terms of the image of the unit circle.

29. Show that an SVD

$$A = U \Sigma V^T$$

can be written as

$$A = \sigma_1 \vec{u}_1 \vec{v}_1^T + \cdots + \sigma_r \vec{u}_r \vec{v}_r^T.$$

30. Find a decomposition

$$A = \sigma_1 \vec{u}_1 \vec{v}_1^T + \sigma_2 \vec{u}_2 \vec{v}_2^T$$

for $A = \begin{bmatrix} 6 & 2 \\ -7 & 6 \end{bmatrix}$. (See Exercise 29 and Example 2.)

31. Show that any matrix of rank r can be written as the sum of r matrices of rank 1.

32. Consider an $n \times m$ matrix A, an orthogonal $n \times n$ matrix S, and an orthogonal $m \times m$ matrix R. Compare the singular values of A with those of SAR.

33. If the singular values of an $n \times n$ matrix A are all 1, is A necessarily orthogonal?

34. For which square matrices A is there a singular value decomposition $A = U \Sigma V^T$ with $U = V$?

35. Consider a singular value decomposition $A = U \Sigma V^T$ of an $n \times m$ matrix A with rank$(A) = m$. Let $\vec{v}_1, \ldots, \vec{v}_m$ be the columns of V and $\vec{u}_1, \ldots, \vec{u}_n$ the columns of U. Without using the results of Chapter 5, compute $(A^T A)^{-1} A^T \vec{u}_i$. Explain the result in terms of least-squares approximations.

36. Consider a singular value decomposition $A = U \Sigma V^T$ of an $n \times m$ matrix A with rank$(A) = m$. Let $\vec{u}_1, \ldots, \vec{u}_n$ be the columns of U. Without using the results of Chapter 5, compute $A(A^T A)^{-1} A^T \vec{u}_i$. Explain your result in terms of Theorem 5.4.7.

Chapter Eight Exercises

TRUE OR FALSE?

(Work with real numbers throughout.)

1. If A is an orthogonal matrix, then there must exist a symmetric invertible matrix S such that $S^{-1} A S$ is diagonal.

2. The singular value of the 2×1 matrix $\begin{bmatrix} 3 \\ 4 \end{bmatrix}$ is 5.

3. The function $q(x_1, x_2) = 3x_1^2 + 4x_1 x_2 + 5x_2$ is a quadratic form.

4. The singular values of any matrix A are the eigenvalues of matrix $A^T A$.

5. If matrix A is positive definite, then all the eigenvalues of A must be positive.

6. The function $q(\vec{x}) = \vec{x}^T \begin{bmatrix} 1 & 2 \\ 2 & 4 \end{bmatrix} \vec{x}$ is a quadratic form.

7. The singular values of any diagonal matrix D are the absolute values of the diagonal entries of D.

8. The equation $2x^2 + 5xy + 3y^2 = 1$ defines an ellipse.

9. All symmetric matrices are diagonalizable.

10. If the matrix $\begin{bmatrix} a & b \\ b & c \end{bmatrix}$ is positive definite, then a must be positive.

11. If the singular values of a 2×2 matrix A are 3 and 4, then there must exist a unit vector $\vec{u}$ in $\mathbb{R}^2$ such that $\|A\vec{u}\| = 4$.

12. The determinant of a negative definite 4×4 matrix must be positive.

13. If A is a symmetric matrix such that $A\vec{v} = 3\vec{v}$ and $A\vec{w} = 4\vec{w}$, then the equation $\vec{v} \cdot \vec{w} = 0$ must hold.

14. Matrix $\begin{bmatrix} -2 & 1 & 1 \\ 1 & -2 & 1 \\ 1 & 1 & -2 \end{bmatrix}$ is negative definite.

15. All skew-symmetric matrices are diagonalizable (over $\mathbb{R}$).

16. If A is any matrix, then matrix AA^T is diagonalizable.

17. All positive definite matrices are invertible.

18. Matrix $\begin{bmatrix} 3 & 2 & 1 \\ 2 & 3 & 2 \\ 1 & 2 & 3 \end{bmatrix}$ is diagonalizable.

19. The singular values of any triangular matrix are the absolute values of its diagonal entries.

20. If A is any matrix, then matrix $A^T A$ is the transpose of AA^T.

21. If $\vec{v}$ and $\vec{w}$ are linearly independent eigenvectors of a symmetric matrix A, then $\vec{w}$ must be orthogonal to $\vec{v}$.

22. For any $n \times m$ matrix A there exists an orthogonal $m \times m$ matrix S such that the columns of matrix AS are orthogonal.

23. If A is a symmetric $n \times n$ matrix such that $A^n = 0$, then A must be the zero matrix.

24. If $q(\vec{x})$ is a positive definite quadratic form, then so is $kq(\vec{x})$, for any scalar k.

25. If A is an invertible symmetric matrix, then A^2 must be positive definite.

26. If the two columns $\vec{v}$ and $\vec{w}$ of a 2×2 matrix A are orthogonal, then the singular values of A must be $\|\vec{v}\|$ and $\|\vec{w}\|$.

27. If A and S are invertible $n \times n$ matrices, then matrices A and $S^T A S$ must be similar.

28. If A is negative definite, then all the diagonal entries of A must be negative.

29. If the positive definite matrix A is similar to the symmetric matrix B, then B must be positive definite as well.

30. If A is a symmetric matrix, then there must exist an orthogonal matrix S such that SAS^T is diagonal.

31. If A and B are 2×2 matrices, then the singular values of matrices AB and BA must be the same.

32. If A is any orthogonal matrix, then matrix $A + A^{-1}$ is diagonalizable (over $\mathbb{R}$).

33. The product of two quadratic forms in 3 variables must be a quadratic form as well.

34. The function $q(\vec{x}) = \vec{x}^T \begin{bmatrix} 1 & 2 \\ 3 & 4 \end{bmatrix} \vec{x}$ is a quadratic form.

35. If the determinants of all the principal submatrices of a symmetric 3×3 matrix A are negative, then A must be negative definite.

36. If A and B are positive definite $n \times n$ matrices, then matrix $A + B$ must be positive definite as well.

37. If A is a positive definite $n \times n$ matrix and $\vec{x}$ is a nonzero vector in $\mathbb{R}^n$, then the angle between $\vec{x}$ and $A\vec{x}$ must be acute.

38. If the 2×2 matrix A has the singular values 2 and 3 and the 2×2 matrix B has the singular values 4 and 5, then both singular values of matrix AB must be ≤ 15.

39. The equation $A^T A = AA^T$ holds for all square matrices A.

40. For every symmetric $n \times n$ matrix A there exists a constant k such that $A + kI_n$ is positive definite.

41. If matrix $\begin{bmatrix} a & b & c \\ b & d & e \\ c & e & f \end{bmatrix}$ is positive definite, then af must exceed c^2.

42. If A is positive definite, then all the entries of A must be positive or zero.

43. If A is indefinite, then 0 must be an eigenvalue of A.

44. If A is a 2×2 matrix with singular values 3 and 5, then there must exist a unit vector $\vec{u}$ in $\mathbb{R}^2$ such that $\|A\vec{u}\| = 4$.

45. If A is skew symmetric, then A^2 must be negative semidefinite.

46. The product of the n singular values of an $n \times n$ matrix A must be $|\det A|$.

47. If $A = \begin{bmatrix} 1 & 2 \\ 2 & 3 \end{bmatrix}$, then there exist exactly 4 orthogonal 2×2 matrices S such that $S^{-1}AS$ is diagonal.

48. The sum of two quadratic forms in 3 variables must be a quadratic form as well.

49. The eigenvalues of a symmetric matrix A must be equal to the singular values of A.

50. Similar matrices must have the same singular values.

51. If A is a symmetric 2×2 matrix with eigenvalues 1 and 2, then the angle between $\vec{x}$ and $A\vec{x}$ must be less than $\pi/6$, for all nonzero vectors $\vec{x}$ in $\mathbb{R}^2$.

52. If both singular values of a 2×2 matrix A are less than 5, then all the entries of A must be less than 5.

53. If A is a positive definite matrix, then the largest entry of A must be on the diagonal.

54. If A and B are real symmetric matrices such that $A^3 = B^3$, then A must be equal to B.

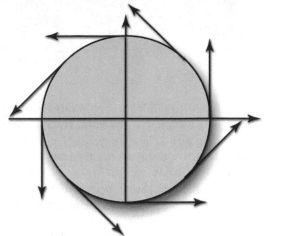

Linear Differential Equations

9.1 An Introduction to Continuous Dynamical Systems

There are two fundamentally different ways to model the evolution of a dynamical system over time: the *discrete* approach and the *continuous* approach. As a simple example, consider a dynamical system with only one component.

EXAMPLE 1 You want to open a savings account and you shop around for the best available interest rate. You learn that *DiscreetBank* pays 7%, compounded annually. Its competitor, the *Bank of Continuity*, offers 6% annual interest, compounded continuously. Everything else being equal, where should you open the account?

Solution

Let us examine what will happen to your investment at the two banks. At Discreet-Bank, the balance grows by 7% each year if no deposits or withdrawals are made.

$$
\boxed{\begin{array}{c}\text{new}\\\text{balance}\end{array}} = \boxed{\begin{array}{c}\text{old}\\\text{balance}\end{array}} + \boxed{\text{interest}}
$$

$$
\begin{array}{ccccc}
\downarrow & & \downarrow & & \downarrow \\
x(t+1) = & x(t) & + & 0.07x(t) \\
x(t+1) = & & 1.07x(t) &
\end{array}
$$

This equation describes a discrete linear dynamical system with one component. The balance after t years is

$$
x(t) = (1.07)^t x_0.
$$

The balance grows exponentially with time.

At the Bank of Continuity, by definition of continuous compounding, the balance $x(t)$ grows at an *instantaneous rate* of 6% of the current balance:

$$
\frac{dx}{dt} = 6\% \text{ of balance } x(t),
$$

or

$$\frac{dx}{dt} = 0.06x.$$

Here, we use a *differential equation* to model a continuous linear dynamical system with one component. We will solve the differential equation in two ways, by separating variables and by making an educated guess.

Let us try to guess the solution. We think about an easier problem first. Do we know a function $x(t)$ that is its own derivative: $dx/dt = x$? You may recall from calculus that $x(t) = e^t$ is such a function. [Some people *define* $x(t) = e^t$ in terms of this property.] More generally, the function $x(t) = Ce^t$ is its own derivative, for any constant C. How can we modify $x(t) = Ce^t$ to get a function whose derivative is 0.06 times itself? By the chain rule, $x(t) = Ce^{0.06t}$ will do:

$$\frac{dx}{dt} = \frac{d}{dt}(Ce^{0.06t}) = 0.06Ce^{0.06t} = 0.06x(t).$$

Note that $x(0) = Ce^0 = C$; that is, C is the initial value, x_0. We conclude that the balance after t years is

$$x(t) = e^{0.06t}x_0.$$

Again, the balance $x(t)$ grows exponentially.

Alternatively, we can solve the differential equation $dx/dt = 0.06x$ by separating variables. Write

$$\frac{dx}{x} = 0.06dt$$

and integrate both sides to get

$$\ln x = 0.06t + k,$$

for some constant k. Exponentiating gives

$$x = e^{\ln x} = e^{0.06t+k} = e^{0.06t}C,$$

where $C = e^k$.

Which bank offers the better deal? We have to compare the exponential functions $(1.07)^t$ and $e^{0.06t}$. Using a calculator (or a Taylor series), we compute

$$e^{0.06t} = (e^{0.06})^t \approx (1.0618)^t$$

to see that *DiscreetBank* offers the better deal. The extra interest from continuous compounding does not make up for the one-point difference in the nominal interest rate. ∎

We can generalize.

Theorem 9.1.1   Exponential growth and decay

Consider the linear differential equation

$$\frac{dx}{dt} = kx,$$

with initial value x_0 (k is an arbitrary constant). The solution is

$$x(t) = e^{kt}x_0.$$

The quantity x will grow or decay exponentially (depending on the sign of k). See Figure 1. ∎

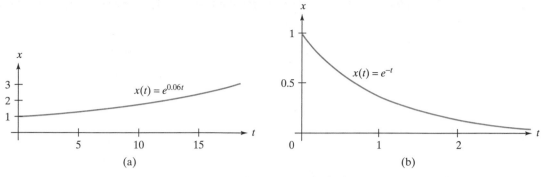

Figure 1 (a) $x(t) = e^{kt}$ with *positive* k. Exponential *growth*. (b) $x(t) = e^{kt}$ with *negative* k. Exponential *decay*.

Now consider a dynamical system with state vector $\vec{x}(t)$ and components $x_1(t)$, $\ldots$, $x_n(t)$. In Chapter 7, we use the *discrete* approach to model this dynamical system: we take a snapshot of the system at times $t = 1, 2, 3, \ldots$, and we describe the transformation the system undergoes between these snapshots. If $\vec{x}(t+1)$ depends linearly on $\vec{x}(t)$, we can write

$$\vec{x}(t + 1) = A\vec{x}(t),$$

or

$$\vec{x}(t) = A^t \vec{x}_0,$$

for some $n \times n$ matrix A.

In the *continuous* approach, we model the gradual change the system undergoes as time goes by. Mathematically speaking, we model the (instantaneous) *rates of change* of the components of the state vector $\vec{x}(t)$, or their derivatives

$$\frac{dx_1}{dt}, \quad \frac{dx_2}{dt}, \quad \ldots, \quad \frac{dx_n}{dt}.$$

If these rates depend linearly on $x_1, x_2, \ldots, x_n$, then we can write

$$\left|\begin{aligned}
\frac{dx_1}{dt} &= a_{11}x_1 + a_{12}x_2 + \cdots + a_{1n}x_n \\
\frac{dx_2}{dt} &= a_{21}x_1 + a_{22}x_2 + \cdots + a_{2n}x_n \\
&\vdots \qquad \vdots \qquad \vdots \qquad\qquad \vdots \\
\frac{dx_n}{dt} &= a_{n1}x_1 + a_{n2}x_2 + \cdots + a_{nn}x_n
\end{aligned}\right|,$$

or, in matrix form,

$$\frac{d\vec{x}}{dt} = A\vec{x},$$

where

$$A = \begin{bmatrix} a_{11} & a_{12} & \cdots & a_{1n} \\ a_{21} & a_{22} & \cdots & a_{2n} \\ \vdots & \vdots & \ddots & \vdots \\ a_{n1} & a_{n2} & \cdots & a_{nn} \end{bmatrix}.$$

The derivative of the parameterized curve $\vec{x}(t)$ is defined componentwise:

$$\frac{d\vec{x}}{dt} = \begin{bmatrix} \dfrac{dx_1}{dt} \\ \dfrac{dx_2}{dt} \\ \vdots \\ \dfrac{dx_n}{dt} \end{bmatrix}.$$

We summarize these observations:

Theorem 9.1.2   Linear dynamical systems: discrete versus continuous

A *linear dynamical system* can be modeled by

$$\vec{x}(t+1) = B\vec{x}(t) \quad \text{(discrete model)},$$

or

$$\frac{d\vec{x}}{dt} = A\vec{x} \quad \text{(continuous model)}.$$

A and B are $n \times n$ matrices, where n is the number of components of the system. ∎

We will first think about the equation

$$\frac{d\vec{x}}{dt} = A\vec{x}$$

from a graphical point of view when A is a 2×2 matrix. We are looking for the parameterized curve

$$\vec{x}(t) = \begin{bmatrix} x_1(t) \\ x_2(t) \end{bmatrix}$$

that represents the evolution of the system from a given initial value $\vec{x}_0$. Each point on the curve $\vec{x}(t)$ will represent the state of the system at a certain moment in time, as shown in Figure 2.

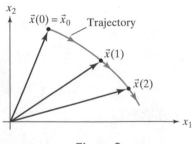

Figure 2

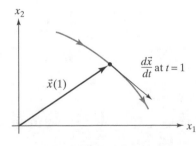

Figure 3

It is natural to think of the trajectory $\vec{x}(t)$ in Figure 2 as the path of a moving particle in the x_1-x_2-plane. As you may have seen in a previous course, the *velocity vector* $d\vec{x}/dt$ of this moving particle is tangent to the trajectory at each point.[1] See Figure 3.

---

[1]It is sensible to attach the velocity vector $d\vec{x}/dt$ at the endpoint of the state vector $\vec{x}(t)$, indicating the path the particle would take if it were to maintain its direction at time t.

In other words, to solve the system

$$\frac{d\vec{x}}{dt} = A\vec{x}$$

for a given initial state $\vec{x}_0$, we have to find the trajectory in the x_1-x_2-plane that starts at $\vec{x}_0$ and whose velocity vector at each point $\vec{x}$ is the vector $A\vec{x}$. The existence and uniqueness of such a trajectory seems intuitively obvious. Our intuition can be misleading in such matters, however, and it is comforting to know that we can establish the existence and uniqueness of the trajectory later. See Theorems 9.1.3 and 9.2.3 and Exercise 9.3.48.

We can represent $A\vec{x}$ graphically as a *vector field* in the x_1-x_2-plane: At the endpoint of each vector $\vec{x}$, we attach the vector $A\vec{x}$. To get a clearer picture, we often sketch merely a *direction field* for $A\vec{x}$, which means that we will not necessarily sketch the vectors $A\vec{x}$ to scale. (We care only about their direction.)

To find the trajectory $\vec{x}(t)$, we follow the vector field (or direction field); that is, we follow the arrows of the field, starting at the point representing the initial state $\vec{x}_0$. The trajectories are also called the *flow lines* of the vector field $A\vec{x}$.

To put it differently, imagine a traffic officer standing at each point of the plane, showing us in which direction to go and how fast to move (in other words, defining our velocity). As we follow these directions, we trace out a trajectory.

EXAMPLE 2   Consider the linear system $d\vec{x}/dt = A\vec{x}$, where $A = \begin{bmatrix} 1 & 2 \\ 4 & 3 \end{bmatrix}$. In Figure 4, we sketch a direction field for $A\vec{x}$. Draw rough trajectories for the three given initial values.

Solution

Sketch the flow lines for the three given points by following the arrows, as shown in Figure 5.

This picture does not tell the whole story about a trajectory $\vec{x}(t)$. We don't know the position $\vec{x}(t)$ of the moving particle at a specific time t. In other words, we know roughly which curve the particle follows, but we don't know how fast it moves along that curve. ∎

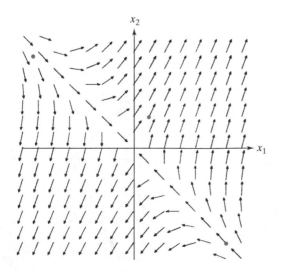

Figure 4

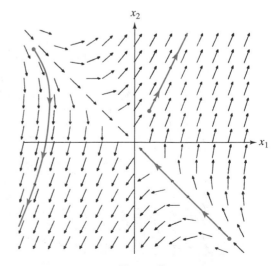

Figure 5

As we look at Figure 5, our eye's attention is drawn to two special lines, along which the vectors $A\vec{x}$ point either radially away from the origin or directly toward the origin. In either case, the vector $A\vec{x}$ is *parallel* to $\vec{x}$:

$$A\vec{x} = \lambda\vec{x},$$

for some scalar λ. This means that the nonzero vectors along these two special lines are just the eigenvectors of A, and the special lines themselves are the eigenspaces. See Figure 6.

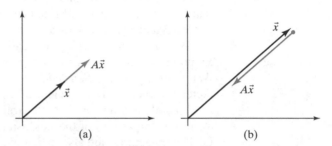

Figure 6 (a) $A\vec{x} = \lambda\vec{x}$, for a *positive* λ. (b) $A\vec{x} = \lambda\vec{x}$, for a *negative* λ.

In Examples 7.2.1 and 7.3.2 we have seen that the eigenvalues of $A = \begin{bmatrix} 1 & 2 \\ 4 & 3 \end{bmatrix}$ are 5 and -1, with corresponding eigenvectors $\begin{bmatrix} 1 \\ 2 \end{bmatrix}$ and $\begin{bmatrix} 1 \\ -1 \end{bmatrix}$. These results agree with our graphical work in Figures 4 and 5. See Figure 7.

As in the case of a discrete dynamical system, we can sketch a phase portrait for the system $d\vec{x}/dt = A\vec{x}$ that shows some representative trajectories. See Figure 8.

In summary, if the initial state vector $\vec{x}_0$ is an eigenvector, then the trajectory moves along the corresponding eigenspace, away from the origin if the eigenvalue is positive and toward the origin if the eigenvalue is negative. If the eigenvalue is zero, then $\vec{x}_0$ is an equilibrium solution: $\vec{x}(t) = \vec{x}_0$, for all times t.

How can we solve the system $d\vec{x}/dt = A\vec{x}$ analytically? We start with a simple case.

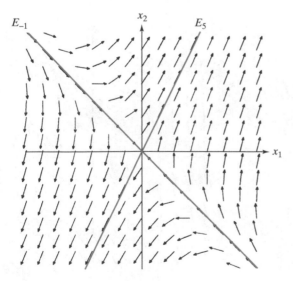

Figure 7

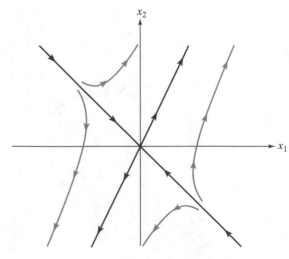

Figure 8

EXAMPLE 3   Find all solutions of the system

$$\frac{d\vec{x}}{dt} = \begin{bmatrix} 2 & 0 \\ 0 & 3 \end{bmatrix} \vec{x}.$$

Solution

The two differential equations

$$\frac{dx_1}{dt} = 2x_1$$

$$\frac{dx_2}{dt} = 3x_2$$

are unrelated, or *uncoupled*; we can solve them separately, using Theorem 9.1.1:

$$x_1(t) = e^{2t} x_1(0),$$
$$x_2(t) = e^{3t} x_2(0).$$

Thus,

$$\vec{x}(t) = \begin{bmatrix} e^{2t} x_1(0) \\ e^{3t} x_2(0) \end{bmatrix}.$$

Both components of $\vec{x}(t)$ grow exponentially, and the second one will grow faster than the first. In particular, if one of the components is initially 0, it remains 0 for all future times. In Figure 9, we sketch a rough phase portrait for this system. ■

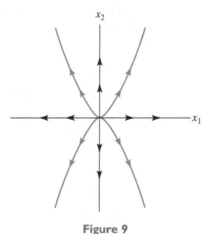

Figure 9

Now let's do a slightly harder example:

EXAMPLE 4   Find all solutions of the system

$$\frac{d\vec{x}}{dt} = A\vec{x}, \quad \text{where } A = \begin{bmatrix} 1 & 2 \\ -1 & 4 \end{bmatrix}$$

Solution

We have seen that the eigenvalues and eigenvectors of A tell us a lot about the behavior of the solutions of the system $d\vec{x}/dt = A\vec{x}$. The eigenvalues of A are

$\lambda_1 = 2$ and $\lambda_2 = 3$ with corresponding eigenvectors $\vec{v}_1 = \begin{bmatrix} 2 \\ 1 \end{bmatrix}$ and $\vec{v}_2 = \begin{bmatrix} 1 \\ 1 \end{bmatrix}$.

This means that $S^{-1}AS = B$, where $S = \begin{bmatrix} 2 & 1 \\ 1 & 1 \end{bmatrix}$ and $B = \begin{bmatrix} 2 & 0 \\ 0 & 3 \end{bmatrix}$, the matrix considered in Example 3.

We can write the system

$$\frac{d\vec{x}}{dt} = A\vec{x}$$

as

$$\frac{d\vec{x}}{dt} = SBS^{-1}\vec{x},$$

or

$$S^{-1}\frac{d\vec{x}}{dt} = BS^{-1}\vec{x},$$

or (see Exercise 51)

$$\frac{d}{dt}(S^{-1}\vec{x}) = B(S^{-1}\vec{x}).$$

Let us introduce the notation $\vec{c}(t) = S^{-1}\vec{x}(t)$; note that $\vec{c}(t)$ is the coordinate vector of $\vec{x}(t)$ with respect to the eigenbasis $\vec{v}_1, \vec{v}_2$. Then the system takes the form

$$\frac{d\vec{c}}{dt} = B\vec{c},$$

which is just the equation we solved in Example 3. We found that the solutions are of the form

$$\vec{c}(t) = \begin{bmatrix} e^{2t}c_1 \\ e^{3t}c_2 \end{bmatrix},$$

where c_1 and c_2 are arbitrary constants. Therefore, the solutions of the original system

$$\frac{d\vec{x}}{dt} = A\vec{x}$$

are

$$\vec{x}(t) = S\vec{c}(t) = \begin{bmatrix} 2 & 1 \\ 1 & 1 \end{bmatrix} \begin{bmatrix} e^{2t}c_1 \\ e^{3t}c_2 \end{bmatrix} = c_1 e^{2t} \begin{bmatrix} 2 \\ 1 \end{bmatrix} + c_2 e^{3t} \begin{bmatrix} 1 \\ 1 \end{bmatrix}.$$

We can write this formula in more general terms as

$$\vec{x}(t) = c_1 e^{\lambda_1 t}\vec{v}_1 + c_2 e^{\lambda_2 t}\vec{v}_2.$$

Note that c_1 and c_2 are the coordinates of $\vec{x}(0)$ with respect to the basis $\vec{v}_1, \vec{v}_2$, since

$$\vec{x}(0) = c_1\vec{v}_1 + c_2\vec{v}_2.$$

It is informative to consider a few special trajectories: If $c_1 = 1$ and $c_2 = 0$, the trajectory

$$\vec{x}(t) = e^{2t} \begin{bmatrix} 2 \\ 1 \end{bmatrix}$$

moves along the eigenspace E_2 spanned by $\begin{bmatrix} 2 \\ 1 \end{bmatrix}$, as expected. Likewise, if $c_1 = 0$ and $c_2 = 1$, we have the trajectory

$$\vec{x}(t) = e^{3t} \begin{bmatrix} 1 \\ 1 \end{bmatrix}$$

moving along the eigenspace E_3.

If $c_2 \neq 0$, then the entries of $c_2 e^{3t} \begin{bmatrix} 1 \\ 1 \end{bmatrix}$ will become much larger (in absolute value) than the entries of $c_1 e^{2t} \begin{bmatrix} 2 \\ 1 \end{bmatrix}$ as t goes to infinity. The *dominant term* $c_2 e^{3t} \begin{bmatrix} 1 \\ 1 \end{bmatrix}$, associated with the larger eigenvalue, determines the behavior of the system in the distant future. The state vector $\vec{x}(t)$ is almost parallel to E_3 for large t. For large negative t, on the other hand, the state vector is very small and almost parallel to E_2.

In Figure 10, we sketch a rough phase portrait for the system $d\vec{x}/dt = A\vec{x}$.

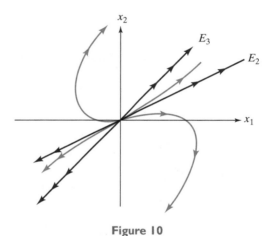

Figure 10

This is a linear distortion of the phase portrait we sketched in Figure 9. More precisely, the matrix $S = \begin{bmatrix} 2 & 1 \\ 1 & 1 \end{bmatrix}$ transforms the phase portraits in Figure 9 into the phase portrait sketched in Figure 10 (transforming $\vec{e}_1$ into the eigenvector $\begin{bmatrix} 2 \\ 1 \end{bmatrix}$ and $\vec{e}_2$ into $\begin{bmatrix} 1 \\ 1 \end{bmatrix}$).   ■

Our work in Examples 3 and 4 generalizes readily to any $n \times n$ matrix A that is diagonalizable over $\mathbb{R}$ (i.e., for which there is an eigenbasis in $\mathbb{R}^n$):

Theorem 9.1.3   Continuous dynamical systems

Consider the system $d\vec{x}/dt = A\vec{x}$. Suppose there is a real eigenbasis $\vec{v}_1, \ldots, \vec{v}_n$ for A, with associated eigenvalues $\lambda_1, \ldots, \lambda_n$. Then the general solution of the system is

$$\vec{x}(t) = c_1 e^{\lambda_1 t} \vec{v}_1 + \cdots + c_n e^{\lambda_n t} \vec{v}_n.$$

The scalars $c_1, c_2, \ldots, c_n$ are the coordinates of $\vec{x}_0$ with respect to the basis $\vec{v}_1, \vec{v}_2, \ldots, \vec{v}_n$.

We can write the preceding equation in matrix form as

$$\vec{x}(t) = \begin{bmatrix} | & | & & | \\ \vec{v}_1 & \vec{v}_2 & \cdots & \vec{v}_n \\ | & | & & | \end{bmatrix} \begin{bmatrix} e^{\lambda_1 t} & 0 & \cdots & 0 \\ 0 & e^{\lambda_2 t} & \cdots & 0 \\ \vdots & \vdots & \ddots & \vdots \\ 0 & 0 & \cdots & e^{\lambda_n t} \end{bmatrix} \begin{bmatrix} c_1 \\ c_2 \\ \vdots \\ c_n \end{bmatrix}$$

$$= S \begin{bmatrix} e^{\lambda_1 t} & 0 & \cdots & 0 \\ 0 & e^{\lambda_2 t} & \cdots & 0 \\ \vdots & \vdots & \ddots & \vdots \\ 0 & 0 & \cdots & e^{\lambda_n t} \end{bmatrix} S^{-1} \vec{x}_0,$$

where $S = \begin{bmatrix} | & | & & | \\ \vec{v}_1 & \vec{v}_2 & \cdots & \vec{v}_n \\ | & | & & | \end{bmatrix}$. ∎

We can think of the general solution as a linear combination of the solutions $e^{\lambda_i t} \vec{v}_i$ associated with the eigenvectors $\vec{v}_i$. Note the analogy between this solution and the general solution of the *discrete* dynamical system $\vec{x}(t + 1) = A\vec{x}(t)$,

$$\vec{x}(t) = c_1 \lambda_1^t \vec{v}_1 + \cdots + c_n \lambda_n^t \vec{v}_n.$$

See Theorem 7.1.3.

The terms λ_i^t are replaced by $e^{\lambda_i t}$. We have already observed this fact in a dynamical system with only one component. (See Example 1.)

We can state Theorem 9.1.3 in the language of linear spaces. The solutions of the system $d\vec{x}/dt = A\vec{x}$ form a subspace of the space $F(\mathbb{R}, \mathbb{R}^n)$ of all functions from $\mathbb{R}$ to $\mathbb{R}^n$. (See Exercises 22 and 23.) This space is n-dimensional, with basis $e^{\lambda_1 t} \vec{v}_1, e^{\lambda_2 t} \vec{v}_2, \ldots, e^{\lambda_n t} \vec{v}_n$.

EXAMPLE 5 Consider a system $d\vec{x}/dt = A\vec{x}$, where A is diagonalizable over $\mathbb{R}$. When is the zero state a stable equilibrium solution? Give your answer in terms of the eigenvalues of A.

Solution

Note that $\lim\limits_{t \to \infty} e^{\lambda t} = 0$ if (and only if) λ is negative. Therefore, we observe stability if (and only if) all eigenvalues of A are negative. ∎

Consider an invertible 2×2 matrix A with two distinct eigenvalues $\lambda_1 > \lambda_2$. Then the phase portrait of $d\vec{x}/dt = A\vec{x}$ looks qualitatively like one of the three sketches in Figure 11. We observe stability only in Figure 11c.

Consider a trajectory that does not run along one of the eigenspaces. In all three cases, the state vector $\vec{x}(t)$ is almost parallel to the dominant eigenspace E_{λ_1} for large t. For large negative t, on the other hand, the state vector is almost parallel to E_{λ_2}. Compare with Figure 7.1.11.

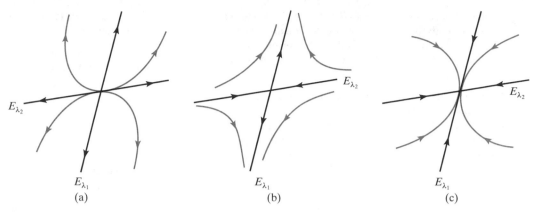

Figure 11 (a) $\lambda_1 > \lambda_2 > 0$. (b) $\lambda_1 > 0 > \lambda_2$. (c) $0 > \lambda_1 > \lambda_2$.

EXERCISES 9.1

GOAL *Use the concept of a continuous dynamical system. Solve the differential equation $dx/dt = kx$. Solve the system $d\vec{x}/dt = A\vec{x}$ when A is diagonalizable over $\mathbb{R}$, and sketch the phase portrait for 2×2 matrices A.*

Solve the initial value problems posed in Exercises 1 through 5. Graph the solution.

1. $\dfrac{dx}{dt} = 5x$ with $x(0) = 7$

2. $\dfrac{dx}{dt} = -0.71x$ with $x(0) = -e$

3. $\dfrac{dP}{dt} = 0.03P$ with $P(0) = 7$

4. $\dfrac{dy}{dt} = 0.8t$ with $y(0) = -0.8$

5. $\dfrac{dy}{dt} = 0.8y$ with $y(0) = -0.8$

Solve the nonlinear differential equations in Exercises 6 through 11 using the method of separation of variables: Write the differential equation $dx/dt = f(x)$ as $dx/f(x) = dt$ and integrate both sides.

6. $\dfrac{dx}{dt} = \dfrac{1}{x}, x(0) = 1$

7. $\dfrac{dx}{dt} = x^2, x(0) = 1$. Describe the behavior of your solution as t increases.

8. $\dfrac{dx}{dt} = \sqrt{x}, x(0) = 4$

9. $\dfrac{dx}{dt} = x^k$ (with $k \neq 1$), $x(0) = 1$

10. $\dfrac{dx}{dt} = \dfrac{1}{\cos(x)}, x(0) = 0$

11. $\dfrac{dx}{dt} = 1 + x^2, x(0) = 0$

12. Find a differential equation of the form $dx/dt = kx$ for which $x(t) = 3^t$ is a solution.

13. In 1778, a wealthy Pennsylvanian merchant named Jacob DeHaven lent $450,000 to the Continental Congress to support the troops at Valley Forge. The loan was never repaid. Mr. DeHaven's descendants have taken the U.S. government to court to collect what they believe they are owed. The going interest rate at the time was 6%. How much were the DeHavens owed in 1990

   a. if interest is compounded yearly?

   b. if interest is compounded continuously?

   (*Source:* Adapted from *The New York Times,* May 27, 1990.)

14. The carbon in living matter contains a minute proportion of the radioactive isotope C-14. This radiocarbon arises from cosmic-ray bombardment in the upper atmosphere and enters living systems by exchange processes. After the death of an organism, exchange stops, and the carbon decays. Therefore, carbon dating enables us to calculate the time at which an organism died. Let $x(t)$ be the proportion of the original C-14 still present t years after death. By definition, $x(0) = 1 = 100\%$. We are told that $x(t)$ satisfies the differential equation

$$\frac{dx}{dt} = -\frac{1}{8270}x.$$

   a. Find a formula for $x(t)$. Determine the half-life of C-14 (that is, the time it takes for half of the C-14 to decay).

   b. *The Iceman.* In 1991, the body of a man was found in melting snow in the Alps of Northern Italy. A well-known historian in Innsbruck, Austria, determined that the man had lived in the Bronze Age, which started about 2000 B.C. in that region. Examination of tissue samples performed independently at Zürich and Oxford revealed that 47% of the C-14 present in the body at the time of his death had decayed. When did this man die? Is the result of the carbon dating compatible with the estimate of the Austrian historian?

15. Justify the "Rule of 69": If a quantity grows at a constant instantaneous rate of $k\%$, then its doubling time is about $69/k$. *Example:* In 2008 the population of Madagascar was about 20 million, growing at an annual rate of about 3%, with a doubling time of about $69/3 = 23$ years.

Consider the system

$$\frac{d\vec{x}}{dt} = \begin{bmatrix} \lambda_1 & 0 \\ 0 & \lambda_2 \end{bmatrix} \vec{x}.$$

For the values of λ_1 and λ_2 given in Exercises 16 through 19, sketch the trajectories for all nine initial values shown in the following figure. For each of the points, trace out both the future and the past of the system.

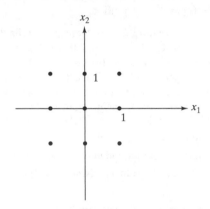

16. $\lambda_1 = 1, \lambda_2 = -1$ 17. $\lambda_1 = 1, \lambda_2 = 2$

18. $\lambda_1 = -1, \lambda_2 = -2$ 19. $\lambda_1 = 0, \lambda_2 = 1$

20. Consider the system $d\vec{x}/dt = A\vec{x}$ with $A = \begin{bmatrix} 0 & -1 \\ 1 & 0 \end{bmatrix}$.
    Sketch a direction field for $A\vec{x}$. Based on your sketch, describe the trajectories geometrically. From your sketch, can you guess a formula for the solution with $\vec{x}_0 = \begin{bmatrix} 1 \\ 0 \end{bmatrix}$?
    Verify your guess by substituting into the equations.

21. Consider the system $d\vec{x}/dt = A\vec{x}$ with $A = \begin{bmatrix} 0 & 1 \\ 0 & 0 \end{bmatrix}$.
    Sketch a direction field of $A\vec{x}$. Based on your sketch, describe the trajectories geometrically. Can you find the solutions analytically?

22. Consider a linear system $d\vec{x}/dt = A\vec{x}$ of arbitrary size. Suppose $\vec{x}_1(t)$ and $\vec{x}_2(t)$ are solutions of the system. Is the sum $\vec{x}(t) = \vec{x}_1(t) + \vec{x}_2(t)$ a solution as well? How do you know?

23. Consider a linear system $d\vec{x}/dt = A\vec{x}$ of arbitrary size. Suppose $\vec{x}_1(t)$ is a solution of the system and k is an arbitrary constant. Is $\vec{x}(t) = k\vec{x}_1(t)$ a solution as well? How do you know?

24. Let A be an $n \times n$ matrix and k a scalar. Consider the following two systems:

$$\frac{d\vec{x}}{dt} = A\vec{x}, \tag{I}$$

$$\frac{d\vec{c}}{dt} = (A + kI_n)\vec{c}. \tag{II}$$

Show that if $\vec{x}(t)$ is a solution of system (I), then $\vec{c}(t) = e^{kt}\vec{x}(t)$ is a solution of system (II).

25. Let A be an $n \times n$ matrix and k a scalar. Consider the following two systems:

$$\frac{d\vec{x}}{dt} = A\vec{x}, \tag{I}$$

$$\frac{d\vec{c}}{dt} = kA\vec{c}. \tag{II}$$

Show that if $\vec{x}(t)$ is a solution of system (I), then $\vec{c}(t) = \vec{x}(kt)$ is a solution of system (II). Compare the vector fields of the two systems.

In Exercises 26 through 31, solve the system with the given initial value.

26. $\dfrac{d\vec{x}}{dt} = \begin{bmatrix} 1 & 2 \\ 3 & 0 \end{bmatrix} \vec{x}$ with $\vec{x}(0) = \begin{bmatrix} 7 \\ 2 \end{bmatrix}$

27. $\dfrac{d\vec{x}}{dt} = \begin{bmatrix} -4 & 3 \\ 2 & -3 \end{bmatrix} \vec{x}$ with $\vec{x}(0) = \begin{bmatrix} 1 \\ 0 \end{bmatrix}$

28. $\dfrac{d\vec{x}}{dt} = \begin{bmatrix} 4 & 3 \\ 4 & 8 \end{bmatrix} \vec{x}$ with $\vec{x}(0) = \begin{bmatrix} 1 \\ 1 \end{bmatrix}$

29. $\dfrac{d\vec{x}}{dt} = \begin{bmatrix} 1 & 2 \\ 2 & 4 \end{bmatrix} \vec{x}$ with $\vec{x}(0) = \begin{bmatrix} 5 \\ 0 \end{bmatrix}$

30. $\dfrac{d\vec{x}}{dt} = \begin{bmatrix} 1 & 2 \\ 2 & 4 \end{bmatrix} \vec{x}$ with $\vec{x}(0) = \begin{bmatrix} 2 \\ -1 \end{bmatrix}$

31. $\dfrac{d\vec{x}}{dt} = \begin{bmatrix} 2 & 1 & 1 \\ 1 & 3 & 3 \\ 3 & 2 & 2 \end{bmatrix} \vec{x}$ with $\vec{x}(0) = \begin{bmatrix} 1 \\ -2 \\ 1 \end{bmatrix}$

Sketch rough phase portraits for the dynamical systems given in Exercises 32 through 39.

32. $\dfrac{d\vec{x}}{dt} = \begin{bmatrix} 1 & 2 \\ 3 & 0 \end{bmatrix} \vec{x}$ 33. $\dfrac{d\vec{x}}{dt} = \begin{bmatrix} -4 & 3 \\ 2 & -3 \end{bmatrix} \vec{x}$

34. $\dfrac{d\vec{x}}{dt} = \begin{bmatrix} 4 & 3 \\ 4 & 8 \end{bmatrix} \vec{x}$ 35. $\dfrac{d\vec{x}}{dt} = \begin{bmatrix} 1 & 2 \\ 2 & 4 \end{bmatrix} \vec{x}$

36. $\vec{x}(t + 1) = \begin{bmatrix} 0.9 & 0.2 \\ 0.2 & 1.2 \end{bmatrix} \vec{x}(t)$

37. $\vec{x}(t + 1) = \begin{bmatrix} 1 & 0.3 \\ -0.2 & 1.7 \end{bmatrix} \vec{x}(t)$

38. $\vec{x}(t+1) = \begin{bmatrix} 1.1 & 0.2 \\ -0.4 & 0.5 \end{bmatrix} \vec{x}(t)$

39. $\vec{x}(t+1) = \begin{bmatrix} 0.8 & -0.4 \\ 0.3 & 1.6 \end{bmatrix} \vec{x}(t)$

40. Find a 2×2 matrix A such that the system $d\vec{x}/dt = A\vec{x}$ has

$$\vec{x}(t) = \begin{bmatrix} 2e^{2t} + 3e^{3t} \\ 3e^{2t} + 4e^{3t} \end{bmatrix}$$

as one of its solutions.

41. Consider a noninvertible 2×2 matrix A with two distinct eigenvalues. (Note that one of the eigenvalues must be 0.) Choose two eigenvectors $\vec{v}_1$ and $\vec{v}_2$ with eigenvalues $\lambda_1 = 0$ and λ_2 as shown in the accompanying figure. Suppose λ_2 is *negative*. Sketch a phase portrait for the system $d\vec{x}/dt = A\vec{x}$, clearly indicating the shape and long-term behavior of the trajectories.

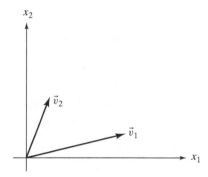

42. Consider the interaction of two species of animals in a habitat. We are told that the change of the populations $x(t)$ and $y(t)$ can be modeled by the equations

$$\begin{vmatrix} \dfrac{dx}{dt} = 1.4x - 1.2y \\ \dfrac{dy}{dt} = 0.8x - 1.4y \end{vmatrix},$$

where time t is measured in years.

a. What kind of interaction do we observe (symbiosis, competition, or predator-prey)?

b. Sketch a phase portrait for this system. From the nature of the problem, we are interested only in the first quadrant.

c. What will happen in the long term? Does the outcome depend on the initial populations? If so, how?

43. Answer the questions posed in Exercise 42 for the following system:

$$\begin{vmatrix} \dfrac{dx}{dt} = 5x - y \\ \dfrac{dy}{dt} = -2x + 4y \end{vmatrix}.$$

44. Answer the questions posed in Exercise 42 for the following system:

$$\begin{vmatrix} \dfrac{dx}{dt} = x + 4y \\ \dfrac{dy}{dt} = 2x - y \end{vmatrix}.$$

45. Two herds of vicious animals are fighting each other to the death. During the fight, the populations $x(t)$ and $y(t)$ of the two species can be modeled by the following system:[2]

$$\begin{vmatrix} \dfrac{dx}{dt} = -4y \\ \dfrac{dy}{dt} = -x \end{vmatrix}.$$

a. What is the significance of the constants -4 and -1 in these equations? Which species has the more vicious (or more efficient) fighters?

b. Sketch a phase portrait for this system.

c. Who wins the fight (in the sense that some individuals of that species are left while the other herd is eradicated)? How does your answer depend on the initial populations?

46. Repeat Exercise 45 for the system

$$\begin{vmatrix} \dfrac{dx}{dt} = -py \\ \dfrac{dy}{dt} = -qx \end{vmatrix},$$

where p and q are two positive constants.[3]

47. The interaction of two populations of animals is modeled by the differential equations

$$\begin{vmatrix} \dfrac{dx}{dt} = -x + ky \\ \dfrac{dy}{dt} = kx - 4y \end{vmatrix},$$

for some positive constant k.

a. What kind of interaction do we observe? What is the practical significance of the constant k?

b. Find the eigenvalues of the coefficient matrix of the system. What can you say about the signs of these eigenvalues? How does your answer depend on the value of the constant k?

c. For each case you discussed in part (b), sketch a rough phase portrait. What does each phase portrait tell you about the future of the two populations?

---

[2] This is the simplest in a series of combat models developed by F. W. Lanchester during World War I (F. W. Lanchester, *Aircraft in Warfare, the Dawn of the Fourth Arm,* Tiptree, Constable and Co., Ltd., 1916).

[3] The result is known as Lanchester's square law.

48. Repeat Exercise 47 for the system

$$\left| \begin{array}{l} \dfrac{dx}{dt} = -x + ky \\[2mm] \dfrac{dy}{dt} = \quad x - 4y \end{array} \right|,$$

where k is a positive constant.

49. Here is a continuous model of a person's glucose regulatory system. (Compare this with Exercise 7.1.52.) Let $g(t)$ and $h(t)$ be the excess glucose and insulin concentrations in a person's blood. We are told that

$$\left| \begin{array}{l} \dfrac{dg}{dt} = \ -g - 0.2h \\[2mm] \dfrac{dh}{dt} = \ 0.6g - 0.2h \end{array} \right|,$$

where time t is measured in hours. After a heavy holiday dinner, we measure $g(0) = 30$ and $h(0) = 0$. Find closed formulas for $g(t)$ and $h(t)$. Sketch the trajectory.

50. Consider a linear system $d\vec{x}/dt = A\vec{x}$, where A is a 2×2 matrix that is diagonalizable over $\mathbb{R}$. When is the zero state a stable equilibrium solution? Give your answer in terms of the determinant and the trace of A.

51. Let $\vec{x}(t)$ be a differentiable curve in $\mathbb{R}^n$ and S an $n \times n$ matrix. Show that

$$\frac{d}{dt}(S\vec{x}) = S\frac{d\vec{x}}{dt}.$$

52. Find all solutions of the system

$$\frac{d\vec{x}}{dt} = \begin{bmatrix} \lambda & 1 \\ 0 & \lambda \end{bmatrix} \vec{x},$$

where λ is an arbitrary constant. *Hint*: Exercises 21 and 24 are helpful. Sketch a phase portrait. For which choices of λ is the zero state a stable equilibrium solution?

53. Solve the initial value problem

$$\frac{d\vec{x}}{dt} = \begin{bmatrix} p & -q \\ q & p \end{bmatrix} \vec{x} \quad \text{with} \quad \vec{x}_0 = \begin{bmatrix} 1 \\ 0 \end{bmatrix}.$$

Sketch the trajectory for the cases when p is positive, negative, or 0. In which cases does the trajectory approach the origin? (*Hint*: Exercises 20, 24, and 25 are helpful.)

54. Consider a door that opens to only one side (as most doors do). A spring mechanism closes the door automatically. The state of the door at a given time t (measured in seconds) is determined by the *angular displacement* $\theta(t)$ (measured in radians) and the *angular velocity* $\omega(t) = d\theta/dt$. Note that θ is always positive or zero (since the door opens to only one side), but ω can be positive or negative (depending on whether the door is opening or closing).

When the door is moving freely (nobody is pushing or pulling), its movement is subject to the following differential equations:

$$\left| \begin{array}{l} \dfrac{d\theta}{dt} = \qquad\qquad \omega \\[2mm] \dfrac{d\omega}{dt} = -2\theta - 3\omega \end{array} \right.$$
 (the definition of ω)

(-2θ reflects the force of the spring, and -3ω models friction).

a. Sketch a phase portrait for this system.

b. Discuss the movement of the door represented by the qualitatively different trajectories. For which initial states does the door slam (i.e., reach $\theta = 0$ with velocity $\omega < 0$)?

55. Answer the questions posed in Exercise 54 for the system

$$\left| \begin{array}{l} \dfrac{d\theta}{dt} = \qquad\qquad \omega \\[2mm] \dfrac{d\omega}{dt} = -p\theta - q\omega \end{array} \right|,$$

where p and q are positive, and $q^2 > 4p$.

9.2 The Complex Case: Euler's Formula

Consider a linear system

$$\frac{d\vec{x}}{dt} = A\vec{x},$$

where the $n \times n$ matrix A is diagonalizable over $\mathbb{C}$: There exists a complex eigenbasis $\vec{v}_1, \ldots, \vec{v}_n$ for A, with associated complex eigenvalues $\lambda_1, \ldots, \lambda_n$. You may wonder whether the formula

$$\vec{x}(t) = c_1 e^{\lambda_1 t}\vec{v}_1 + \cdots + c_n e^{\lambda_n t}\vec{v}_n$$

(with complex c_i) produces the general complex solution of the system, just as in the real case (Theorem 9.1.3).

Before we can make sense out of the formula above, we have to think about the idea of a complex-valued function and in particular about the exponential function $e^{\lambda t}$ for complex λ.

Complex-Valued Functions

A complex-valued function $z = f(t)$ is a function from $\mathbb{R}$ to $\mathbb{C}$ (with domain $\mathbb{R}$ and target space $\mathbb{C}$): The input t is real, and the output z is complex. Here are two examples:

$$z = t + it^2,$$
$$z = \cos t + i \sin t.$$

For each t, the output z can be represented as a point in the complex plane. As we let t vary, we trace out a *trajectory* in the complex plane. In Figure 1, we sketch the trajectories of the two complex-valued functions just defined.

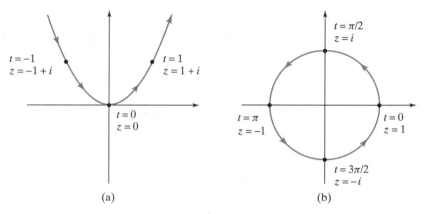

(a)   (b)

Figure I (a) The trajectory of $z = t + it^2$. (b) The trajectory of $z = \cos t + i \sin t$.

We can write a complex-valued function $z(t)$ in terms of its real and imaginary parts:

$$z(t) = x(t) + iy(t).$$

(Consider the two preceding examples.) If $x(t)$ and $y(t)$ are differentiable real-valued functions, then the *derivative* of the complex-valued function $z(t)$ is defined by

$$\frac{dz}{dt} = \frac{dx}{dt} + i\frac{dy}{dt}.$$

For example, if

$$z(t) = t + it^2,$$

then

$$\frac{dz}{dt} = 1 + 2it.$$

If

$$z(t) = \cos t + i \sin t,$$

then

$$\frac{dz}{dt} = -\sin t + i \cos t.$$

Please verify that the basic rules of differential calculus (the sum, product, and quotient rules) apply to complex-valued functions. The chain rule holds in the following form: If $z = f(t)$ is a differentiable complex-valued function and $t = g(s)$ is a differentiable function from $\mathbb{R}$ to $\mathbb{R}$, then

$$\frac{dz}{ds} = \frac{dz}{dt}\frac{dt}{ds}.$$

The derivative dz/dt of a complex-valued function $z(t)$, for a given t, can be visualized as a tangent vector to the trajectory at $z(t)$, as shown in Figure 2.

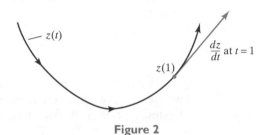

Figure 2

Next let's think about the complex-valued exponential function $z = e^{\lambda t}$, where λ is complex and t real. How should the function $z = e^{\lambda t}$ be defined? We can get some inspiration from the real case: The exponential function $x = e^{kt}$ (for real k) is the unique function such that $dx/dt = kx$ and $x(0) = 1$. (Compare this with Theorem 9.1.1.)

We can use this fundamental property of real exponential functions to define the complex exponential functions:

Definition 9.2.1   Complex exponential functions

If λ is a complex number, then $z = e^{\lambda t}$ is the unique complex-valued function such that

$$\frac{dz}{dt} = \lambda z \quad \text{and} \quad z(0) = 1.$$

(The existence of such a function, for any λ, will be established later; the proof of uniqueness is left as Exercise 38.)

It follows that the unique complex-valued function $z(t)$ with

$$\frac{dz}{dt} = \lambda z \quad \text{and} \quad z(0) = z_0$$

is

$$z(t) = e^{\lambda t}z_0,$$

for an arbitrary complex initial value z_0.

Let us first consider the simplest case, $z = e^{it}$, where $\lambda = i$. We are looking for a complex-valued function $z(t)$ such that $dz/dt = iz$ and $z(0) = 1$.

From a graphical point of view, we are looking for the trajectory $z(t)$ in the complex plane that starts at $z = 1$ and whose tangent vector $dz/dt = iz$ is perpendicular to z at each point. (See Example 1 of Section 7.5.) In other words, we are looking for the flow line of the vector field in Figure 3 starting at $z = 1$.

The *unit circle*, with parameterization $z(t) = \cos t + i \sin t$, satisfies

$$\frac{dz}{dt} = -\sin t + i \cos t = iz(t),$$

and $z(0) = 1$. See Figure 4.

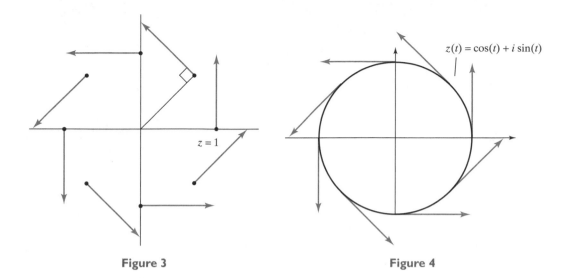

Figure 3   **Figure 4**

We have shown the following fundamental result:

Theorem 9.2.2   Euler's formula

$$e^{it} = \cos t + i \sin t$$

The case $t = \pi$ leads to the intriguing formula $e^{i\pi} = -1$; this has been called the most beautiful formula in all of mathematics.[4]
Euler's formula can be used to write the polar form of a complex number more succinctly:

$$z = r(\cos\theta + i \sin\theta) = re^{i\theta}$$

Now consider $z = e^{\lambda t}$, where λ is an arbitrary complex number, $\lambda = p + iq$. By manipulating exponentials as if they were real, we find that

$$e^{\lambda t} = e^{(p+iq)t} = e^{pt}e^{iqt} = e^{pt}\big(\cos(qt) + i \sin(qt)\big).$$

We can validate this result by checking that the complex-valued function

$$z(t) = e^{pt}\big(\cos(qt) + i \sin(qt)\big)$$

does indeed satisfy the definition of $e^{\lambda t}$, namely, $dz/dt = \lambda z$ and $z(0) = 1$:

$$\frac{dz}{dt} = pe^{pt}\big(\cos(qt) + i \sin(qt)\big) + e^{pt}\big(-q \sin(qt) + iq \cos(qt)\big)$$

$$= (p + iq)e^{pt}\big(\cos(qt) + i \sin(qt)\big) = \lambda z.$$

Figure 5 Euler's likeness and his celebrated formula are shown on a Swiss postage stamp.

EXAMPLE 1   Sketch the trajectory of the complex-valued function $z(t) = e^{(0.1+i)t}$ in the complex plane.

Solution

$$z(t) = e^{0.1t}e^{it} = e^{0.1t}(\cos t + i \sin t)$$

---

[4]Benjamin Peirce (1809–1880), a Harvard mathematician, after observing that $e^{i\pi} = -1$, used to turn to his students and say, "Gentlemen, that is surely true, it is absolutely paradoxical, we cannot understand it, and we don't know what it means, but we have proved it, and therefore we know it must be the truth." Do you not now think that we understand not only that the formula is true but also what it means?

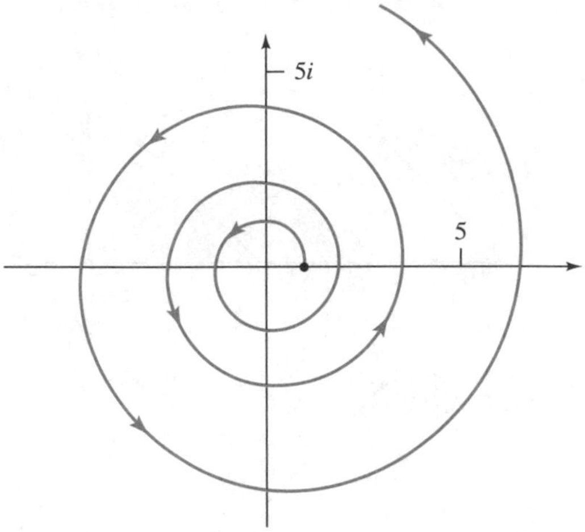

Figure 6

The trajectory spirals outward as shown in Figure 6, since the function $e^{0.1t}$ grows exponentially. ∎

EXAMPLE 2   For which complex numbers λ is $\lim\limits_{t \to \infty} e^{\lambda t} = 0$?

Solution

Recall that

$$e^{\lambda t} = e^{(p+iq)t} = e^{pt}\big(\cos(qt) + i\sin(qt)\big),$$

so that $|e^{\lambda t}| = e^{pt}$. This quantity approaches zero if (and only if) p is negative (i.e., if e^{pt} decays exponentially).

We summarize: $\lim\limits_{t \to \infty} e^{\lambda t} = 0$ if (and only if) the real part of λ is negative. ∎

We are now ready to tackle the problem posed at the beginning of this section: Consider a system $d\vec{x}/dt = A\vec{x}$, where the $n \times n$ matrix A has a complex eigenbasis $\vec{v}_1, \ldots, \vec{v}_n$, with eigenvalues $\lambda_1, \ldots, \lambda_n$. Find all complex solutions $\vec{x}(t)$ of this system. By a complex solution we mean a function from $\mathbb{R}$ to $\mathbb{C}^n$ (that is, t is real and $\vec{x}$ is in $\mathbb{C}^n$). In other words, the component functions $x_1(t), \ldots, x_n(t)$ of $\vec{x}(t)$ are complex-valued functions.

As you review our work in the last section, you will find that the approach we took to the *real* case applies to the *complex* case as well, without modifications:

Theorem 9.2.3   Continuous dynamical systems with complex eigenvalues

Consider a linear system

$$\frac{d\vec{x}}{dt} = A\vec{x}.$$

Suppose there exists a complex eigenbasis $\vec{v}_1, \ldots, \vec{v}_n$ for A, with associated complex eigenvalues $\lambda_1, \ldots, \lambda_n$. Then the general complex solution of the system is

$$\vec{x}(t) = c_1 e^{\lambda_1 t}\vec{v}_1 + \cdots + c_n e^{\lambda_n t}\vec{v}_n,$$

where the c_i are arbitrary complex numbers.

We can write this solution in matrix form, as in Theorem 9.1.3. ∎

We can check that the given curve $\vec{x}(t)$ satisfies the equation $d\vec{x}/dt = A\vec{x}$: We have

$$\frac{d\vec{x}}{dt} = c_1\lambda_1 e^{\lambda_1 t}\vec{v}_1 + \cdots + c_n\lambda_n e^{\lambda_n t}\vec{v}_n$$

(by Definition 9.2.1), and

$$A\vec{x} = c_1 e^{\lambda_1 t}\lambda_1\vec{v}_1 + \cdots + c_n e^{\lambda_n t}\lambda_n\vec{v}_n,$$

because the $\vec{v}_i$ are eigenvectors. The two answers match.

When is the zero state a stable equilibrium solution for the system $d\vec{x}/dt = A\vec{x}$? Considering Example 3 and the form of the solution given in Theorem 9.2.3, we can conclude that this is the case if (and only if) the real parts of all eigenvalues are negative (at least when A is diagonalizable over $\mathbb{C}$). The nondiagonalizable case is left as Exercise 9.3.48.

Theorem 9.2.4   **Stability of a continuous dynamical system**

For a system

$$\frac{d\vec{x}}{dt} = A\vec{x},$$

the zero state is an asymptotically stable equilibrium solution if (and only if) the real parts of all eigenvalues of A are negative. ∎

EXAMPLE 3   Consider the system $d\vec{x}/dt = A\vec{x}$, where A is a (real) 2×2 matrix. When is the zero state a stable equilibrium solution for this system? Give your answer in terms of the trace and the determinant of A.

Solution

We observe stability either if A has two negative eigenvalues or if A has two conjugate eigenvalues $p \pm iq$, where p is negative. In both cases, tr A is negative and det A is positive. Check that in all other cases tr $A \geq 0$ or det $A \leq 0$. ∎

Theorem 9.2.5   **Determinant, trace, and stability**

Consider the system

$$\frac{d\vec{x}}{dt} = A\vec{x},$$

where A is a real 2×2 matrix. Then the zero state is an asymptotically stable equilibrium solution if (and only if) tr $A < 0$ and det $A > 0$. ∎

As a special case of Theorem 9.2.3, let's consider the system

$$\frac{d\vec{x}}{dt} = A\vec{x},$$

where A is a real 2×2 matrix with eigenvalues $\lambda_{1,2} = p \pm iq$ (where $q \neq 0$) and corresponding eigenvectors $\vec{v}_{1,2} = \vec{v} \pm i\vec{w}$.

Theorems 9.1.3 and 9.2.3 tell us that

$$\vec{x}(t) = P \begin{bmatrix} e^{\lambda_1 t} & 0 \\ 0 & e^{\lambda_2 t} \end{bmatrix} P^{-1}\vec{x}_0 = P \begin{bmatrix} e^{(p+iq)t} & 0 \\ 0 & e^{(p-iq)t} \end{bmatrix} P^{-1}\vec{x}_0$$

$$= e^{pt} P \begin{bmatrix} \cos(qt) + i\sin(qt) & 0 \\ 0 & \cos(qt) - i\sin(qt) \end{bmatrix} P^{-1}\vec{x}_0,$$

where $P = \begin{bmatrix} \vec{v} + i\vec{w} & \vec{v} - i\vec{w} \end{bmatrix}$. Note that we have used Euler's formula (Theorem 9.2.2).

We can write this formula in terms of real quantities. By Example 6 of Section 7.5,

$$\begin{bmatrix} \cos(qt) + i\sin(qt) & 0 \\ 0 & \cos(qt) - i\sin(qt) \end{bmatrix} = R^{-1} \begin{bmatrix} \cos(qt) & -\sin(qt) \\ \sin(qt) & \cos(qt) \end{bmatrix} R,$$

where

$$R = \begin{bmatrix} i & -i \\ 1 & 1 \end{bmatrix}.$$

Thus,

$$\vec{x}(t) = e^{pt} P R^{-1} \begin{bmatrix} \cos(qt) & -\sin(qt) \\ \sin(qt) & \cos(qt) \end{bmatrix} R P^{-1} \vec{x}_0$$

$$= e^{pt} S \begin{bmatrix} \cos(qt) & -\sin(qt) \\ \sin(qt) & \cos(qt) \end{bmatrix} S^{-1} \vec{x}_0,$$

where

$$S = P R^{-1} = \frac{1}{2i} \begin{bmatrix} \vec{v} + i\vec{w} & \vec{v} - i\vec{w} \end{bmatrix} \begin{bmatrix} 1 & i \\ -1 & i \end{bmatrix} = \begin{bmatrix} \vec{w} & \vec{v} \end{bmatrix},$$

and

$$S^{-1} = (P R^{-1})^{-1} = R P^{-1}.$$

Recall that we have performed the same computations in Example 7 of Section 7.5.

Theorem 9.2.6 Continuous dynamical systems with eigenvalues $p \pm iq$

Consider the linear system

$$\frac{d\vec{x}}{dt} = A\vec{x},$$

where A is a real 2×2 matrix with complex eigenvalues $p \pm iq$ (and $q \neq 0$). Consider an eigenvector $\vec{v} + i\vec{w}$ with eigenvalue $p + iq$. Then

$$\vec{x}(t) = e^{pt} S \begin{bmatrix} \cos(qt) & -\sin(qt) \\ \sin(qt) & \cos(qt) \end{bmatrix} S^{-1} \vec{x}_0,$$

where $S = \begin{bmatrix} \vec{w} & \vec{v} \end{bmatrix}$. Recall that $S^{-1}\vec{x}_0$ is the coordinate vector of $\vec{x}_0$ with respect to basis $\vec{w}, \vec{v}$.

The trajectories are either ellipses (linearly distorted circles), if $p = 0$, or spirals, spiraling outward if p is positive and inward if p is negative. In the case of an ellipse, the trajectories have a period of $2\pi/q$. ∎

Note the analogy between Theorem 9.2.6 and the formula

$$\vec{x}(t) = r^t S \begin{bmatrix} \cos(\theta t) & -\sin(\theta t) \\ \sin(\theta t) & \cos(\theta t) \end{bmatrix} S^{-1} \vec{x}_0$$

in the case of the discrete system $\vec{x}(t+1) = A\vec{x}(t)$ (Theorem 7.6.3).

EXAMPLE 4 Solve the system

$$\frac{d\vec{x}}{dt} = \begin{bmatrix} 3 & -2 \\ 5 & -3 \end{bmatrix} \vec{x} \quad \text{with} \quad \vec{x}_0 = \begin{bmatrix} 0 \\ 1 \end{bmatrix}.$$

Solution

The eigenvalues are $\lambda_{1,2} = \pm i$, so that $p = 0$ and $q = 1$. This tells us that the trajectory is an ellipse. To determine the direction of the trajectory (clockwise or counterclockwise) and its rough shape, we can draw the direction field $A\vec{x}$ for a few simple points $\vec{x}$, say, $\vec{x} = \pm\vec{e}_1$ and $\vec{x} = \pm\vec{e}_2$, and sketch the flow line starting at $\begin{bmatrix} 0 \\ 1 \end{bmatrix}$. See Figure 7.

Now let us find a formula for the trajectory.

$$E_i = \ker\begin{bmatrix} 3-i & -2 \\ 5 & -3-i \end{bmatrix} = \operatorname{span}\begin{bmatrix} -2 \\ i-3 \end{bmatrix}$$

$$\begin{bmatrix} -2 \\ i-3 \end{bmatrix} = \underbrace{\begin{bmatrix} -2 \\ -3 \end{bmatrix}}_{\vec{v}} + i\underbrace{\begin{bmatrix} 0 \\ 1 \end{bmatrix}}_{\vec{w}}.$$

Therefore,

$$\vec{x}(t) = e^{pt} S\begin{bmatrix} \cos(qt) & -\sin(qt) \\ \sin(qt) & \cos(qt) \end{bmatrix} S^{-1}\vec{x}_0 = \begin{bmatrix} 0 & -2 \\ 1 & -3 \end{bmatrix}\begin{bmatrix} \cos t & -\sin t \\ \sin t & \cos t \end{bmatrix}\begin{bmatrix} 1 \\ 0 \end{bmatrix}$$

$$= \begin{bmatrix} -2\sin t \\ \cos t - 3\sin t \end{bmatrix} = \cos t\begin{bmatrix} 0 \\ 1 \end{bmatrix} + \sin t\begin{bmatrix} -2 \\ -3 \end{bmatrix}.$$

You can check that

$$\frac{d\vec{x}}{dt} = A\vec{x} \quad \text{and} \quad \vec{x}(0) = \begin{bmatrix} 0 \\ 1 \end{bmatrix}.$$

The trajectory is the ellipse shown in Figure 8. ∎

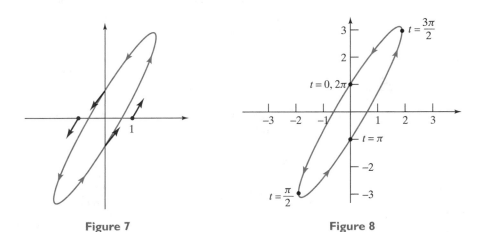

Figure 7   Figure 8

Consider a 2×2 matrix A. The various scenarios for the system $d\vec{x}/dt = A\vec{x}$ can be conveniently represented in the tr A-det A plane, where a 2×2 matrix A is represented by the point (tr A, det A). Recall that the characteristic polynomial is

$$\lambda^2 - (\operatorname{tr} A)\lambda + \det A$$

and the eigenvalues are

$$\lambda_{1,2} = \frac{1}{2}\left(\operatorname{tr} A \pm \sqrt{(\operatorname{tr}A)^2 - 4\det A}\right).$$

Therefore, the eigenvalues of A are real if (and only if) the point $(\operatorname{tr} A, \det A)$ is located below or on the parabola

$$\det A = \left(\frac{\operatorname{tr} A}{2}\right)^2$$

in the tr A–det A plane. See Figure 9.

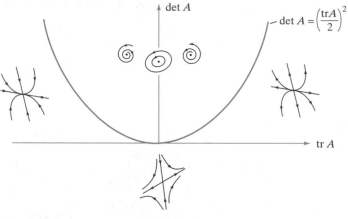

Figure 9

Note that there are five major cases, corresponding to the regions in Figure 9, and some exceptional cases, corresponding to the dividing lines.

What does the phase portrait look like when $\det A = 0$ and $\operatorname{tr} A \neq 0$?

In Figure 10 we take another look at the five major types of phase portraits. Both in the discrete and in the continuous case, we sketch the phase portraits produced by various eigenvalues. We include the case of an ellipse, since it is important in applications.

EXERCISES 9.2

GOAL *Use the definition of the complex-valued exponential function $z = e^{\lambda t}$. Solve the system*

$$\frac{d\vec{x}}{dt} = A\vec{x}$$

for a 2 × 2 matrix A with complex eigenvalues $p \pm iq$.

1. Find $e^{2\pi i}$.

2. Find $e^{(1/2)\pi i}$.

3. Write $z = -1 + i$ in polar form as $z = re^{i\theta}$.

4. Sketch the trajectory of the complex-valued function

$$z = e^{3it}.$$

What is the period?

5. Sketch the trajectory of the complex-valued function

$$z = e^{(-0.1-2i)t}.$$

6. Find all complex solutions of the system

$$\frac{d\vec{x}}{dt} = \begin{bmatrix} 3 & -2 \\ 5 & -3 \end{bmatrix} \vec{x}$$

in the form given in Theorem 9.2.3. What solution do you get if you let $c_1 = c_2 = 1$?

7. Determine the stability of the system

$$\frac{d\vec{x}}{dt} = \begin{bmatrix} -1 & 2 \\ 3 & -4 \end{bmatrix} \vec{x}.$$

8. Consider a system

$$\frac{d\vec{x}}{dt} = A\vec{x},$$

where A is a symmetric matrix. When is the zero state a stable equilibrium solution? Give your answer in terms of the definiteness of the matrix A.

Discrete	Continuous	Phase Portrait
$\lambda_1 > \lambda_2 > 1$	$\lambda_1 > \lambda_2 > 0$	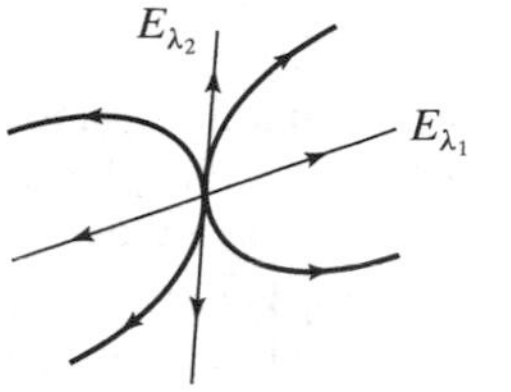
$\lambda_1 > 1 > \lambda_2 > 0$	$\lambda_1 > 0 > \lambda_2$	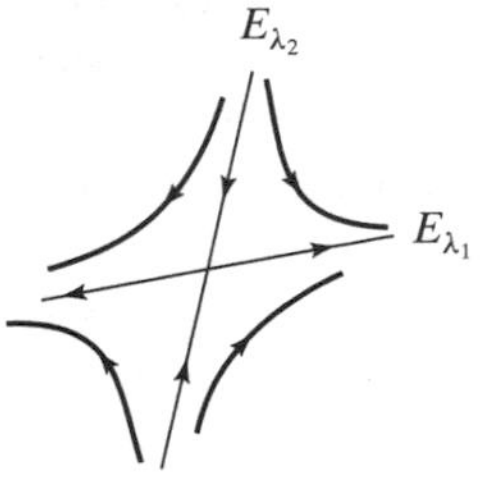
$1 > \lambda_1 > \lambda_2 > 0$	$0 > \lambda_1 > \lambda_2$	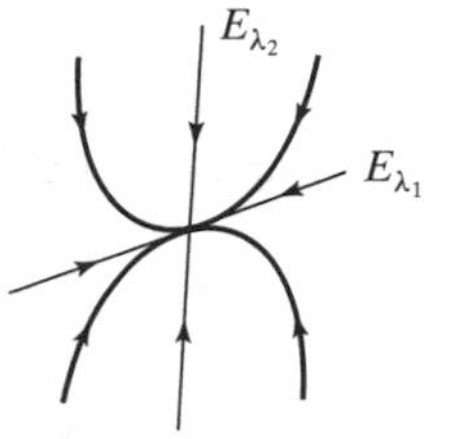
$\lambda_{1,2} = p \pm iq$ $p^2 + q^2 > 1$	$\lambda_{1,2} = p \pm iq$ $p > 0$	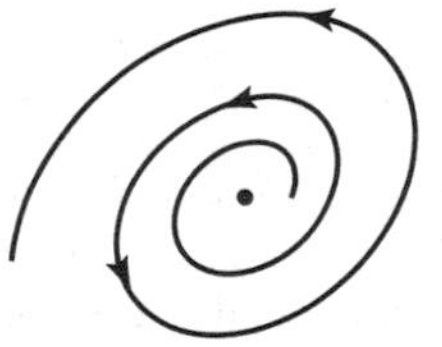
$\lambda_{1,2} = p \pm iq$ $p^2 + q^2 < 1$	$\lambda_{1,2} = p \pm iq$ $p < 0$	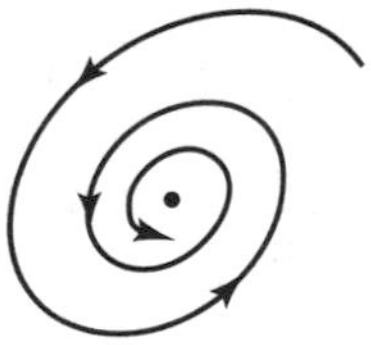
$\lambda_{1,2} = p \pm iq$ $p^2 + q^2 = 1$	$\lambda_{1,2} = \pm iq$	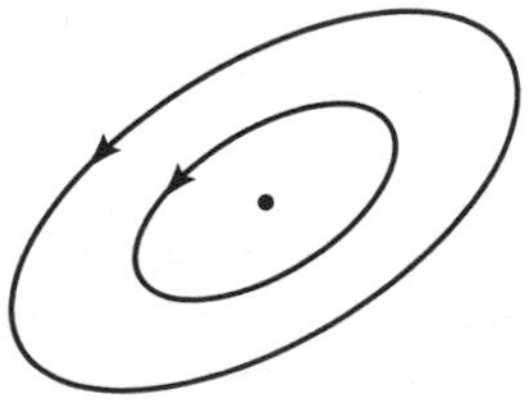

Figure 10 The major types of phase portraits.

9. Consider a system

$$\frac{d\vec{x}}{dt} = A\vec{x},$$

where A is a 2×2 matrix with tr $A < 0$. We are told that A has no real eigenvalues. What can you say about the stability of the system?

10. Consider a quadratic form $q(\vec{x}) = \vec{x} \cdot A\vec{x}$ of two variables, x_1 and x_2. Consider the system of differential equations

$$\left| \begin{array}{l} \dfrac{dx_1}{dt} = \dfrac{\partial q}{\partial x_1} \\[2mm] \dfrac{dx_2}{dt} = \dfrac{\partial q}{\partial x_2} \end{array} \right.,$$

or, more succinctly,

$$\frac{d\vec{x}}{dt} = \text{grad } q.$$

a. Show that the system $d\vec{x}/dt = \text{grad } q$ is linear by finding a matrix B (in terms of the symmetric matrix A) such that $\text{grad } q = B\vec{x}$.

b. When q is negative definite, draw a sketch showing possible level curves of q. On the same sketch, draw a few trajectories of the system $d\vec{x}/dt = \text{grad } q$. What does your sketch suggest about the stability of the system $d\vec{x}/dt = \text{grad } q$?

c. Do the same as in part (b) for an indefinite quadratic form.

d. Explain the relationship between the definiteness of the form q and the stability of the system $d\vec{x}/dt = \text{grad } q$.

11. Do parts (a) and (d) of Exercise 10 for a quadratic form of n variables.

12. Determine the stability of the system

$$\frac{d\vec{x}}{dt} = \begin{bmatrix} 0 & 1 & 0 \\ 0 & 0 & 1 \\ -1 & -1 & -2 \end{bmatrix} \vec{x}.$$

13. If the system $d\vec{x}/dt = A\vec{x}$ is stable, is $d\vec{x}/dt = A^{-1}\vec{x}$ stable as well? How can you tell?

14. *Negative Feedback Loops.* Suppose some quantities $x_1(t), x_2(t), \dots, x_n(t)$ can be modeled by differential equations of the form

$$\begin{vmatrix} \dfrac{dx_1}{dt} = -k_1 x_1 & & & - bx_n \\ \dfrac{dx_2}{dt} = & x_1 - k_2 x_2 & & \\ \vdots & & \ddots & \\ \dfrac{dx_n}{dt} = & & x_{n-1} - k_n x_n \end{vmatrix},$$

where b is positive and the k_i are positive. (The matrix of this system has negative numbers on the diagonal, 1's directly below the diagonal, and a negative number in the top right corner.) We say that the quantities $x_1, \dots, x_n$ describe a (linear) negative feedback loop.

a. Describe the significance of the entries in this system, in practical terms.

b. Is a negative feedback loop with two components ($n = 2$) necessarily stable?

c. Is a negative feedback loop with three components necessarily stable?

15. Consider a noninvertible 2×2 matrix A with a positive trace. What does the phase portrait of the system $d\vec{x}/dt = A\vec{x}$ look like?

16. Consider the system

$$\frac{d\vec{x}}{dt} = \begin{bmatrix} 0 & 1 \\ a & b \end{bmatrix} \vec{x},$$

where a and b are arbitrary constants. For which values of a and b is the zero state a stable equilibrium solution?

17. Consider the system

$$\frac{d\vec{x}}{dt} = \begin{bmatrix} -1 & k \\ k & -1 \end{bmatrix} \vec{x},$$

where k is an arbitrary constant. For which values of k is the zero state a stable equilibrium solution?

18. Consider a diagonalizable 3×3 matrix A such that the zero state is a stable equilibrium solution of the system $d\vec{x}/dt = A\vec{x}$. What can you say about the determinant and the trace of A?

19. *True or False?* If the trace and the determinant of a 3×3 matrix A are both negative, then the origin is a stable equilibrium solution of the system $d\vec{x}/dt = A\vec{x}$. Justify your answer.

20. Consider a 2×2 matrix A with eigenvalues $\pm \pi i$. Let $\vec{v} + i\vec{w}$ be an eigenvector of A with eigenvalue πi. Solve the initial value problem

$$\frac{d\vec{x}}{dt} = A\vec{x}, \quad \text{with} \quad \vec{x}_0 = \vec{w}.$$

Draw the solution in the accompanying figure. Mark the vectors $\vec{x}(0)$, $\vec{x}(\frac{1}{2})$, $\vec{x}(1)$, and $\vec{x}(2)$.

21. Ngozi opens a bank account with an initial balance of 1,000 Nigerian naira. Let $b(t)$ be the balance in the account at time t; we are told that $b(0) = 1,000$. The bank is paying interest at a continuous rate of 5% per year. Ngozi makes deposits into the account at a continuous rate of $s(t)$ (measured in naira per year). We are told that $s(0) = 1,000$ and that $s(t)$ is increasing at a continuous rate of 7% per year. (Ngozi can save more as her income goes up over time.)

a. Set up a linear system of the form

$$\begin{vmatrix} \dfrac{db}{dt} = ?b + ?s \\ \dfrac{ds}{dt} = ?b + ?s \end{vmatrix}$$

(Time is measured in years.)

b. Find $b(t)$ and $s(t)$.

For each of the linear systems in Exercises 22 through 26, find the matching phase portrait below.

22. $\vec{x}(t + 1) = \begin{bmatrix} 3 & 0 \\ -2.5 & 0.5 \end{bmatrix} \vec{x}(t)$

23. $\vec{x}(t + 1) = \begin{bmatrix} -1.5 & -1 \\ 2 & 0.5 \end{bmatrix} \vec{x}(t)$

24. $\dfrac{d\vec{x}}{dt} = \begin{bmatrix} 3 & 0 \\ -2.5 & 0.5 \end{bmatrix} \vec{x}$

25. $\dfrac{d\vec{x}}{dt} = \begin{bmatrix} -1.5 & -1 \\ 2 & 0.5 \end{bmatrix} \vec{x}$

26. $\dfrac{d\vec{x}}{dt} = \begin{bmatrix} -2 & 0 \\ 3 & 1 \end{bmatrix} \vec{x}$

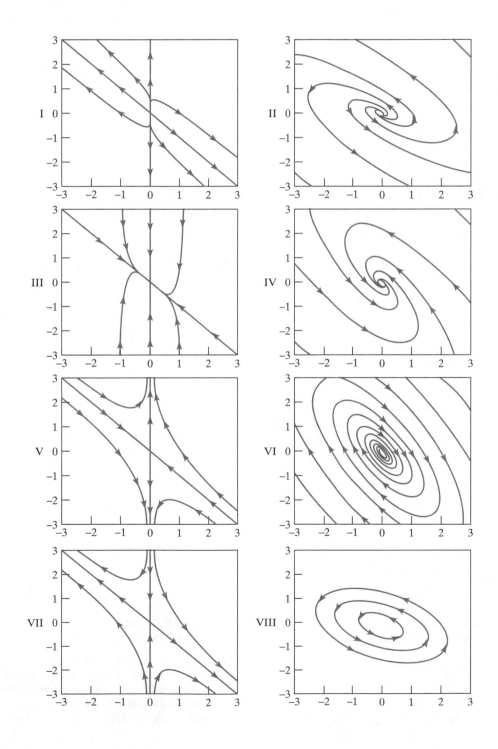

$y^2 - 6x + 20$

Find all real solutions of the systems in Exercises 27 through 30.

27. $\dfrac{d\vec{x}}{dt} = \begin{bmatrix} 0 & -3 \\ 3 & 0 \end{bmatrix} \vec{x}$ **28.** $\dfrac{d\vec{x}}{dt} = \begin{bmatrix} 0 & 4 \\ -9 & 0 \end{bmatrix} \vec{x}$

29. $\dfrac{d\vec{x}}{dt} = \begin{bmatrix} 2 & 4 \\ -4 & 2 \end{bmatrix} \vec{x}$ **30.** $\dfrac{d\vec{x}}{dt} = \begin{bmatrix} -11 & 15 \\ -6 & 7 \end{bmatrix} \vec{x}$

Solve the systems in Exercises 31 through 34. Give the solution in real form. Sketch the solution.

31. $\dfrac{d\vec{x}}{dt} = \begin{bmatrix} -1 & -2 \\ 2 & -1 \end{bmatrix} \vec{x}$ with $\vec{x}(0) = \begin{bmatrix} 1 \\ -1 \end{bmatrix}$

32. $\dfrac{d\vec{x}}{dt} = \begin{bmatrix} 0 & 1 \\ -4 & 0 \end{bmatrix} \vec{x}$ with $\vec{x}(0) = \begin{bmatrix} 1 \\ 0 \end{bmatrix}$

33. $\dfrac{d\vec{x}}{dt} = \begin{bmatrix} -1 & 1 \\ -2 & 1 \end{bmatrix} \vec{x}$ with $\vec{x}(0) = \begin{bmatrix} 0 \\ 1 \end{bmatrix}$

34. $\dfrac{d\vec{x}}{dt} = \begin{bmatrix} 7 & 10 \\ -4 & -5 \end{bmatrix} \vec{x}$ with $\vec{x}(0) = \begin{bmatrix} 1 \\ 0 \end{bmatrix}$

35. Prove the product rule for derivatives of complex-valued functions.

36. Consider the following mass-spring system:

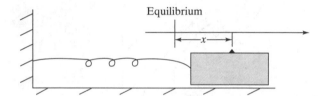

Equilibrium

Let $x(t)$ be the deviation of the block from the equilibrium position at time t. Consider the velocity $v(t) = dx/dt$ of the block. There are two forces acting on the mass: The spring force F_s, which is assumed to be proportional to the displacement x, and the force F_f of friction, which is assumed to be proportional to the velocity,

$$F_s = -px \quad \text{and} \quad F_f = -qv,$$

where $p > 0$ and $q \geq 0$. (q is 0 if the oscillation is frictionless.) Therefore, the total force acting on the mass is

$$F = F_s + F_f = -px - qv.$$

By Newton's second law of motion, we have

$$F = ma = m\frac{dv}{dt},$$

where a represents acceleration and m the mass of the block. Combining the last two equations, we find that

$$m\frac{dv}{dt} = -px - qv,$$

or

$$\frac{dv}{dt} = -\frac{p}{m}x - \frac{q}{m}v.$$

Let $b = p/m$ and $c = q/m$ for simplicity. Then the dynamics of this mass-spring system are described by the system

$$\left| \begin{aligned} \frac{dx}{dt} &= v \\ \frac{dv}{dt} &= -bx - cv \end{aligned} \right| \quad (b > 0, \ c \geq 0).$$

Sketch a phase portrait for this system in each of the following cases, and describe briefly the significance of your trajectories in terms of the movement of the block. Comment on the stability in each case.

a. $c = 0$ (frictionless). Find the period.

b. $c^2 < 4b$ (underdamped).

c. $c^2 > 4b$ (overdamped).

37. a. For a differentiable complex-valued function $z(t)$, find the derivative of

$$\frac{1}{z(t)}.$$

b. Prove the quotient rule for derivatives of complex-valued functions.

In both parts of this exercise, you may use the product rule. (See Exercise 35.)

38. Let $z_1(t)$ and $z_2(t)$ be two complex-valued solutions of the initial value problem

$$\frac{dz}{dt} = \lambda z, \quad \text{with} \quad z(0) = 1,$$

where λ is a complex number. Suppose that $z_2(t) \neq 0$ for all t.

a. Using the quotient rule (Exercise 37), show that the derivative of

$$\frac{z_1(t)}{z_2(t)}$$

is zero. Conclude that $z_1(t) = z_2(t)$ for all t.

b. Show that the initial value problem

$$\frac{dz}{dt} = \lambda z, \quad \text{with} \quad z(0) = 1,$$

has a unique complex-valued solution $z(t)$. *Hint*: One solution is given in the text.

39. Solve the system

$$\frac{d\vec{x}}{dt} = \begin{bmatrix} \lambda & 1 & 0 \\ 0 & \lambda & 1 \\ 0 & 0 & \lambda \end{bmatrix} \vec{x}.$$

Compare this with Exercise 9.1.24. When is the zero state a stable equilibrium solution?

40. An eccentric mathematician is able to gain autocratic power in a small Alpine country. In her first decree, she announces the introduction of a new currency, the Euler, which is measured in complex units. Banks are ordered to pay only imaginary interest on deposits.

a. If you invest 1,000 Euler at $5i\%$ interest, compounded annually, how much money do you have after 1 year, after 2 years, after t years? Describe the effect of compounding in this case. Sketch a trajectory showing the evolution of the balance in the complex plane.

b. Do part (a) in the case when the $5i\%$ interest is compounded continuously.

c. Suppose people's social standing is determined by the modulus of the balance of their bank account. Under these circumstances, would you choose an account with annual compounding or with continuous compounding of interest?

(*Source:* This problem is based on an idea of Professor D. Mumford, Brown University.)

9.3 Linear Differential Operators and Linear Differential Equations

In this final section, we will study an important class of linear transformations from C^∞ to C^∞. Here, C^∞ denotes the linear space of complex-valued smooth functions (i.e., functions from $\mathbb{R}$ to $\mathbb{C}$), which we consider as a linear space over $\mathbb{C}$.

Definition 9.3.1 Linear differential operators and linear differential equations

A transformation T from C^∞ to C^∞ of the form

$$T(f) = f^{(n)} + a_{n-1}f^{(n-1)} + \cdots + a_1 f' + a_0 f$$

is called an nth-order *linear differential operator*.[5,6] Here $f^{(k)}$ denotes the kth derivative of function f, and the coefficients a_k are complex scalars.

If T is an nth-order linear differential operator and g is a smooth function, then the equation

$$T(f) = g \quad \text{or} \quad f^{(n)} + a_{n-1}f^{(n-1)} + \cdots + a_1 f' + a_0 f = g$$

is called an nth-order *linear differential equation* (DE). The DE is called *homogeneous* if $g = 0$ and inhomogeneous otherwise.

Verify that a linear differential operator is indeed a linear transformation. Examples of linear differential operators are

$$D(f) = f',$$
$$T(f) = f'' - 5f' + 6f, \quad \text{and}$$
$$L(f) = f''' - 6f'' + 5f,$$

of first, second, and third order, respectively.

Examples of linear DEs are

$$f'' - f' - 6f = 0 \quad \text{(second order, homogeneous)}$$

and

$$f'(t) - 5f(t) = \sin t \quad \text{(first order, inhomogeneous).}$$

Note that solving a homogeneous DE $T(f) = 0$ amounts to finding the kernel of T.

We will first think about the relationship between the solutions of the DEs $T(f) = 0$ and $T(f) = g$.

---

[5]More precisely, this is a linear differential operator *with constant coefficients*. More advanced texts consider the case when the a_k are functions.

[6]The term *operator* is often used for a transformation whose domain and target space consist of functions.

More generally, consider a linear transformation T from V to W, where V and W are arbitrary linear spaces. What is the relationship between the kernel of T and the solutions f of the equation $T(f) = g$, provided that this equation has solutions at all? (Compare this with Exercise 1.3.48.)

Here is a simple example:

EXAMPLE I   Consider the linear transformation $T(\vec{x}) = \begin{bmatrix} 1 & 2 & 3 \\ 2 & 4 & 6 \end{bmatrix} \vec{x}$ from $\mathbb{R}^3$ to $\mathbb{R}^2$. Describe the relationship between the kernel of T and the solutions of the linear system $T(\vec{x}) = \begin{bmatrix} 6 \\ 12 \end{bmatrix}$, both algebraically and geometrically.

Solution

We find that the kernel of T consists of all vectors of the form

$$\begin{bmatrix} -2x_2 - 3x_3 \\ x_2 \\ x_3 \end{bmatrix} = x_2 \begin{bmatrix} -2 \\ 1 \\ 0 \end{bmatrix} + x_3 \begin{bmatrix} -3 \\ 0 \\ 1 \end{bmatrix},$$

with basis

$$\begin{bmatrix} -2 \\ 1 \\ 0 \end{bmatrix}, \begin{bmatrix} -3 \\ 0 \\ 1 \end{bmatrix}.$$

The solution set of the system $T(\vec{x}) = \begin{bmatrix} 6 \\ 12 \end{bmatrix}$ consists of all vectors of the form

$$\begin{bmatrix} 6 - 2x_2 - 3x_3 \\ x_2 \\ x_3 \end{bmatrix} = \underbrace{x_2 \begin{bmatrix} -2 \\ 1 \\ 0 \end{bmatrix} + x_3 \begin{bmatrix} -3 \\ 0 \\ 1 \end{bmatrix}}_{\substack{\text{A vector in the kernel of } T}} + \underbrace{\begin{bmatrix} 6 \\ 0 \\ 0 \end{bmatrix}}_{\substack{\text{A particular solution} \\ \text{of the system} \\ T(\vec{x}) = \begin{bmatrix} 6 \\ 12 \end{bmatrix}}}$$

The kernel of T and the solution set of $T(\vec{x}) = \begin{bmatrix} 6 \\ 12 \end{bmatrix}$ form two parallel planes in $\mathbb{R}^3$, as shown in Figure 1.   ∎

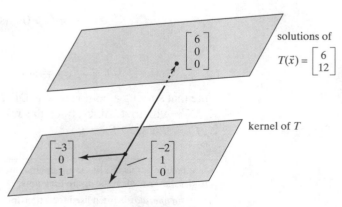

Figure I

These observations generalize as follows:

Theorem 9.3.2 Consider a linear transformation T from V to W, where V and W are arbitrary linear spaces. Suppose we have a basis $f_1, f_2, \ldots, f_n$ of the kernel of T. Consider an equation $T(f) = g$ with a particular solution f_p. Then the solutions f of the equation $T(f) = g$ are of the form

$$f = c_1 f_1 + c_2 f_2 + \cdots + c_n f_n + f_p,$$

where the c_i are arbitrary constants. ◼

Note that $T(f) = T(c_1 f_1 + \cdots + c_n f_n) + T(f_p) = 0 + g = g$, so that f is indeed a solution. Verify that all solutions are of this form.

What is the significance of Theorem 9.3.2 for linear differential equations? At the end of this section, we will demonstrate the following fundamental result:

Theorem 9.3.3 The kernel of an nth-order linear differential operator is n-dimensional. ◼

Theorem 9.3.2 now provides us with the following strategy for solving linear differential equations:

Theorem 9.3.4 Strategy for solving linear differential equations

To solve an nth-order linear DE

$$T(f) = g,$$

we have to find

  a. a basis $f_1, \ldots, f_n$ of $\ker(T)$, and
  b. a particular solution f_p of the DE $T(f) = g$.

Then the solutions f are of the form

$$f = c_1 f_1 + \cdots + c_n f_n + f_p,$$

where the c_i are arbitrary constants. ◼

EXAMPLE 2 Find all solutions of the DE

$$f''(t) + f(t) = e^t.$$

We are told that $f_p(t) = \frac{1}{2} e^t$ is a particular solution (verify this).

Solution

Consider the linear differential operator $T(f) = f'' + f$. A basis of the kernel of T is $f_1(t) = \cos t$ and $f_2(t) = \sin t$ (compare with Example 1 of Section 4.1).

Therefore, the solutions f of the DE $f'' + f = e^t$ are of the form

$$f(t) = c_1 \cos t + c_2 \sin t + \frac{1}{2} e^t,$$

where c_1 and c_2 are arbitrary constants. ◼

We now present an approach that allows us to find solutions to homogeneous linear DEs more systematically.

The Eigenfunction Approach to Solving Linear DEs

Definition 9.3.5   Eigenfunctions

Consider a linear differential operator T from C^∞ to C^∞. A smooth function f is called an *eigenfunction* of T if $T(f) = \lambda f$ for some complex scalar λ; this scalar λ is called the eigenvalue associated with the eigenfunction f.

EXAMPLE 3   Find all eigenfunctions and eigenvalues of the operator $D(f) = f'$.

Solution

We have to solve the differential equation

$$D(f) = \lambda f \quad \text{or} \quad f' = \lambda f.$$

For a given λ, the solutions are all exponential functions of the form $f(t) = Ce^{\lambda t}$. This means that all complex numbers are eigenvalues of D, and the eigenspace associated with the eigenvalue λ is one dimensional, spanned by $e^{\lambda t}$. (Compare this with Definition 9.2.1.)   ∎

It follows that the exponential functions are eigenfunctions for all linear differential operators: If

$$T(f) = f^{(n)} + a_{n-1} f^{(n-1)} + \cdots + a_1 f' + a_0 f,$$

then

$$T(e^{\lambda t}) = (\lambda^n + a_{n-1}\lambda^{n-1} + \cdots + a_1\lambda + a_0)e^{\lambda t}.$$

This observation motivates the following definition:

Definition 9.3.6   Characteristic polynomial

Consider the linear differential operator

$$T(f) = f^{(n)} + a_{n-1} f^{(n-1)} + \cdots + a_1 f' + a_0 f.$$

The *characteristic polynomial* of T is defined as

$$p_T(\lambda) = \lambda^n + a_{n-1}\lambda^{n-1} + \cdots + a_1\lambda + a_0.$$

Theorem 9.3.7   If T is a linear differential operator, then $e^{\lambda t}$ is an eigenfunction of T, with associated eigenvalue $p_T(\lambda)$, for all λ:

$$T(e^{\lambda t}) = p_T(\lambda)e^{\lambda t}.$$

In particular, if $p_T(\lambda) = 0$, then $e^{\lambda t}$ is in the kernel of T.   ∎

EXAMPLE 4   Find all exponential functions $e^{\lambda t}$ in the kernel of the linear differential operator

$$T(f) = f'' + f' - 6f.$$

Solution

The characteristic polynomial is $p_T(\lambda) = \lambda^2 + \lambda - 6 = (\lambda + 3)(\lambda - 2)$, with roots 2 and -3. Therefore, the functions e^{2t} and e^{-3t} are in the kernel of T. We can check this:

$$T(e^{2t}) = 4e^{2t} + 2e^{2t} - 6e^{2t} = 0,$$
$$T(e^{-3t}) = 9e^{-3t} - 3e^{-3t} - 6e^{-3t} = 0.$$ ∎

Since most polynomials of degree n have n distinct complex roots, we can find n distinct exponential functions $e^{\lambda_1 t}, \ldots, e^{\lambda_n t}$ in the kernel of most nth-order linear differential operators. Note that these functions are linearly independent. (They are eigenfunctions of D with distinct eigenvalues; the proof of Theorem 7.3.4 applies.)

Now we can use Theorem 9.3.3.

Theorem 9.3.8   **The kernel of a linear differential operator**

Consider an nth-order linear differential operator T whose characteristic polynomial $p_T(\lambda)$ has n distinct roots $\lambda_1, \ldots, \lambda_n$. Then the exponential functions

$$e^{\lambda_1 t}, e^{\lambda_2 t}, \ldots, e^{\lambda_n t}$$

form a basis of the kernel of T; that is, they form a basis of the solution space of the homogeneous DE

$$T(f) = 0.$$ ∎

See Exercise 38 for the case of an nth-order linear differential operator whose characteristic polynomial has fewer than n distinct roots.

EXAMPLE 5   Find all solutions f of the differential equation

$$f'' + 2f' - 3f = 0.$$

Solution

The characteristic polynomial of the operator $T(f) = f'' + 2f' - 3f$ is $p_T(\lambda) = \lambda^2 + 2\lambda - 3 = (\lambda + 3)(\lambda - 1)$, with roots 1 and -3. The exponential functions e^t and e^{-3t} form a basis of the solution space; that is, the solutions are of the form

$$f(t) = c_1 e^t + c_2 e^{-3t}.$$ ∎

EXAMPLE 6   Find all solutions f of the differential equation

$$f'' - 6f' + 13f = 0.$$

Solution

The characteristic polynomial is $p_T(\lambda) = \lambda^2 - 6\lambda + 13$, with complex roots $3 \pm 2i$. The exponential functions

$$e^{(3+2i)t} = e^{3t}\big(\cos(2t) + i\sin(2t)\big)$$

and

$$e^{(3-2i)t} = e^{3t}\big(\cos(2t) - i\sin(2t)\big)$$

form a basis of the solution space. We may wish to find a basis of the solution space consisting of real-valued functions. The following observation is helpful: If $f(t) = g(t) + ih(t)$ is a solution of the DE $T(f) = 0$, then $T(f) = T(g) + iT(h) = 0$, so that g and h are solutions as well. We can apply this remark to the real and the imaginary parts of the solution $e^{(3+2i)t}$: The functions

$$e^{3t}\cos(2t) \quad \text{and} \quad e^{3t}\sin(2t)$$

are a basis of the solution space (they are clearly linearly independent), and the general solution is

$$f(t) = c_1 e^{3t}\cos(2t) + c_2 e^{3t}\sin(2t) = e^{3t}\big(c_1\cos(2t) + c_2\sin(2t)\big).$$ ∎

Theorem 9.3.9   Consider a differential equation

$$T(f) = f'' + af' + bf = 0,$$

where the coefficients a and b are real. Suppose the zeros of $p_T(\lambda)$ are $p \pm iq$, with $q \neq 0$. Then the solutions of the given DE are

$$f(t) = e^{pt}\big(c_1 \cos(qt) + c_2 \sin(qt)\big),$$

where c_1 and c_2 are arbitrary constants.

The special case when $a = 0$ and $b > 0$ is important in many applications. Then $p = 0$ and $q = \sqrt{b}$, so that the solutions of the DE

$$f'' + bf = 0$$

are

$$f(t) = c_1 \cos(\sqrt{b}t) + c_2 \sin(\sqrt{b}t). \qquad \blacksquare$$

Note that the function

$$f(t) = e^{pt}\big(c_1 \cos(qt) + c_2 \sin(qt)\big)$$

is the product of an exponential and a sinusoidal function. The case when p is negative comes up frequently in physics, when we model a *damped oscillator*. See Figure 2.

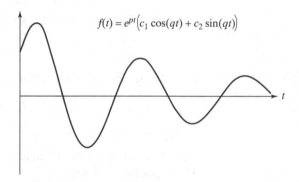

$$f(t) = e^{pt}\big(c_1 \cos(qt) + c_2 \sin(qt)\big)$$

Figure 2

What about nonhomogeneous differential equations? Let us discuss an example that is particularly important in applications.

EXAMPLE 7   Consider the differential equation

$$f''(t) + f'(t) - 6f(t) = 8\cos(2t).$$

a. Let V be the linear space consisting of all functions of the form $c_1 \cos(2t) + c_2 \sin(2t)$. Show that the linear differential operator $T(f) = f'' + f' - 6f$ defines an isomorphism from V to V.

b. Part (a) implies that the DE $T(f) = 8\cos(2t)$ has a unique particular solution $f_p(t)$ in V. Find this solution.

c. Find all solutions of the DE $T(f) = 8\cos(2t)$.

Solution

a. Consider the matrix A of T with respect to the basis $\cos(2t)$, $\sin(2t)$. A straightforward computation shows that

$$A = \begin{bmatrix} -10 & 2 \\ -2 & -10 \end{bmatrix},$$

representing a rotation combined with a scaling. Since A is invertible, T defines an isomorphism from V to V.

b. If we work in coordinates with respect to the basis $\cos(2t)$, $\sin(2t)$, the DE $T(f) = 8\cos(2t)$ takes the form $A\vec{x} = \begin{bmatrix} 8 \\ 0 \end{bmatrix}$, with the solution

$$\vec{x} = A^{-1} \begin{bmatrix} 8 \\ 0 \end{bmatrix} = \frac{1}{104} \begin{bmatrix} -10 & -2 \\ 2 & -10 \end{bmatrix} \begin{bmatrix} 8 \\ 0 \end{bmatrix} = \begin{bmatrix} -10/13 \\ 2/13 \end{bmatrix}.$$

The particular solution in V is

$$f_p(t) = -\frac{10}{13}\cos(2t) + \frac{2}{13}\sin(2t).$$

A more straightforward way to find $f_p(t)$ is to set $f_p(t) = P\cos(2t) + Q\sin(2t)$ and substitute this trial solution into the DE to determine P and Q. This approach is referred to as the *method of undetermined coefficients*.

c. In Example 4, we have seen that the functions $f_1(t) = e^{2t}$ and $f_2(t) = e^{-3t}$ form a basis of the kernel of T. By Theorem 9.3.4, the solutions of the DE are of the form

$$f(t) = c_1 f_1(t) + c_2 f_2(t) + f_p(t)$$
$$= c_1 e^{2t} + c_2 e^{-3t} - \frac{10}{13}\cos(2t) + \frac{2}{13}\sin(2t). \qquad \blacksquare$$

Let us summarize the methods developed in Example 7:

Theorem 9.3.10   Consider the linear differential equation

$$f''(t) + af'(t) + bf(t) = C\cos(\omega t),$$

where a, b, C, and ω are real numbers. Suppose that $a \neq 0$ or $b \neq \omega^2$. This DE has a particular solution of the form

$$f_p(t) = P\cos(\omega t) + Q\sin(\omega t).$$

Now use Theorems 9.3.4 and 9.3.8 to find all solutions f of the DE. $\qquad \blacksquare$

What goes wrong when $a = 0$ and $b = \omega^2$?

The Operator Approach to Solving Linear DEs

We will now present an alternative, deeper approach to DEs, which allows us to solve any linear DE (at least if we can find the zeros of the characteristic polynomial). This approach will lead us to a better understanding of the kernel and image of a linear differential operator; in particular, it will enable us to prove Theorem 9.3.3.

Let us first introduce a more succinct notation for linear differential operators. Recall the notation $Df = f'$ for the derivative operator. We let

$$D^m = \underbrace{D \circ D \circ \cdots \circ D}_{m \text{ times}};$$

that is,

$$D^m f = f^{(m)}.$$

Then the operator

$$T(f) = f^{(n)} + a_{n-1} f^{(n-1)} + \cdots + a_1 f' + a_0 f$$

can be written more succinctly as

$$T = D^n + a_{n-1} D^{n-1} + \cdots + a_1 D + a_0,$$

the characteristic polynomial $p_T(\lambda)$ "evaluated at D."

For example, the operator

$$T(f) = f'' + f' - 6f$$

can be written as

$$T = D^2 + D - 6.$$

Treating T formally as a polynomial in D, we can write

$$T = (D + 3) \circ (D - 2).$$

We can verify that this formula gives us a decomposition of the operator T:

$$((D+3) \circ (D-2)) f = (D+3)(f' - 2f) = f'' - 2f' + 3f' - 6f = (D^2 + D - 6) f.$$

This works because D is linear: We have $D(f' - 2f) = f'' - 2f'$.

The fundamental theorem of algebra (Theorem 7.5.2) now tells us the following:

Theorem 9.3.11   An nth-order linear differential operator T can be expressed as the composite of n first-order linear differential operators:

$$T = D^n + a_{n-1} D^{n-1} + \cdots + a_1 D + a_0$$
$$= (D - \lambda_1)(D - \lambda_2) \ldots (D - \lambda_n),$$

where the λ_i are complex numbers.   ∎

We can therefore hope to understand all linear differential operators by studying first-order operators.

EXAMPLE 8   Find the kernel of the operator $T = D - a$, where a is a complex number. Do not use Theorem 9.3.3.

Solution

We have to solve the homogeneous differential equation $T(f) = 0$ or $f'(t) - af(t) = 0$ or $f'(t) = af(t)$. By definition of an exponential function, the solutions are the functions of the form $f(t) = Ce^{at}$, where C is an arbitrary constant. (See Definition 9.2.1.)   ∎

Theorem 9.3.12   The kernel of the operator

$$T = D - a$$

is one dimensional, spanned by

$$f(t) = e^{at}.$$ ∎

Next we think about the nonhomogeneous equation

$$(D - a)f = g,$$

or

$$f'(t) - af(t) = g(t),$$

where $g(t)$ is a smooth function. It will turn out to be useful to multiply both sides of this equation with the function e^{-at}:

$$e^{-at}f'(t) - ae^{-at}f(t) = e^{-at}g(t).$$

We recognize the left-hand side of this equation as the derivative of the function $e^{-at}f(t)$, so that we can write

$$\left(e^{-at}f(t)\right)' = e^{-at}g(t).$$

Integrating, we get

$$e^{-at}f(t) = \int e^{-at}g(t)\,dt$$

and

$$f(t) = e^{at}\int e^{-at}g(t)\,dt,$$

where $\int e^{-at}g(t)\,dt$ denotes the indefinite integral, that is, the family of all antiderivatives of the functions $e^{-at}g(t)$, involving a parameter C.

Theorem 9.3.13 First-order linear differential equations

Consider the differential equation

$$f'(t) - af(t) = g(t),$$

where $g(t)$ is a smooth function and a a constant.
    Then

$$f(t) = e^{at}\int e^{-at}g(t)\,dt.$$ ∎

Theorem 9.3.13 shows that the differential equation $(D - a)f = g$ has solutions f, for any smooth function g; this means that

$$\text{im}(D - a) = C^{\infty}.$$

EXAMPLE 9 Find the solutions f of the DE

$$f' - af = ce^{at},$$

where c is an arbitrary constant.

Solution
Using Theorem 9.3.13 we find that

$$f(t) = e^{at}\int e^{-at}ce^{at}\,dt = e^{at}\int c\,dt = e^{at}(ct + C),$$

where C is another arbitrary constant. ∎

Now consider an nth-order DE $T(f) = g$, where

$$T = D^n + a_{n-1}D^{n-1} + \cdots + a_1 D + a_0,$$
$$= (D - \lambda_1)(D - \lambda_2) \cdots (D - \lambda_{n-1})(D - \lambda_n).$$

We can break this DE down into n first-order DEs:

$$f \xrightarrow{D - \lambda_n} f_{n-1} \xrightarrow{D - \lambda_{n-1}} f_{n-2} \quad \cdots \quad f_2 \xrightarrow{D - \lambda_2} f_1 \xrightarrow{D - \lambda_1} g$$

We can successively solve the first-order DEs:

$$(D - \lambda_1)f_1 = g,$$
$$(D - \lambda_2)f_2 = f_1,$$
$$\vdots$$
$$(D - \lambda_{n-1})f_{n-1} = f_{n-2},$$
$$(D - \lambda_n)f = f_{n-1}.$$

In particular, the DE $T(f) = g$ does have solutions f.

Theorem 9.3.14   The image of a linear differential operator

The image of all linear differential operators (from C^∞ to C^∞) is C^∞; that is, any linear DE $T(f) = g$ has solutions f.   ■

EXAMPLE 10   Find all solutions of the DE

$$T(f) = f'' - 2f' + f = 0.$$

Note that $p_T(\lambda) = \lambda^2 - 2\lambda + 1 = (\lambda - 1)^2$ has only one root, 1, so that we cannot use Theorem 9.3.8.

Solution

We break the DE down into two first-order DEs, as discussed earlier:

$$f \xrightarrow{D-1} f_1 \xrightarrow{D-1} 0$$

The DE $(D-1)f_1 = 0$ has the general solution $f_1(t) = c_1 e^t$, where c_1 is an arbitrary constant.

Then the DE $(D-1)f = f_1 = c_1 e^t$ has the general solution $f(t) = e^t(c_1 t + c_2)$, where c_2 is another arbitrary constant. (See Example 9.)

The functions e^t and te^t form a basis of the solution space (i.e., of the kernel of T). Note that the kernel is two dimensional, since we pick up an arbitrary constant each time we solve a first-order DE.   ■

Now we can explain why the kernel of an nth-order linear differential operator T is n-dimensional. Roughly speaking, this is true because the general solution of the DE $T(f) = 0$ contains n arbitrary constants. (We pick up one each time we solve a first-order linear DE.)

Here is a formal proof of Theorem 9.3.3. We will argue by induction on n. Theorem 9.3.12 takes care of the case $n = 1$. By Theorem 9.3.11, we can write an nth-order linear differential operator T as $T = (D - \lambda) \circ L$, where L is of order $n - 1$. Arguing by induction, we assume that the kernel of L is $(n-1)$-dimensional. Since $\dim(\ker(D - \lambda)) = 1$ and $\operatorname{im}L = C^\infty$, by Theorem 9.3.14, we can conclude that $\dim(\ker T) = \dim(\ker(D - \lambda)) + \dim(\ker L) = n$, by Exercise 4.2.84.

Let's summarize the main techniques we discussed in this section.

SUMMARY 9.3.15 | **Strategy for linear differential equations**

Suppose you have to solve an nth-order linear differential equation $T(f) = g$.

Step 1 Find n linearly independent solutions of the DE $T(f) = 0$.
- Write the characteristic polynomial $p_T(\lambda)$ of T [replacing $f^{(k)}$ by λ^k].
- Find the solutions $\lambda_1, \lambda_2, \ldots, \lambda_n$ of the equation $p_T(\lambda) = 0$.
- If λ is a solution of the equation $p_T(\lambda) = 0$, then $e^{\lambda t}$ is a solution of $T(f) = 0$.
- If λ is a solution of $p_T(\lambda) = 0$ with multiplicity m, then $e^{\lambda t}, te^{\lambda t}, t^2 e^{\lambda t}, \ldots,$ $t^{m-1} e^{\lambda t}$ are solutions of the DE $T(f) = 0$. (See Exercise 38.)
- If $p \pm iq$ are complex solutions of $p_T(\lambda) = 0$, then $e^{pt} \cos(qt)$ and $e^{pt} \sin(qt)$ are real solutions of the DE $T(f) = 0$.

Step 2 If the DE is inhomogeneous (i.e., if $g \neq 0$), find one particular solution f_p of the DE $T(f) = g$.
- If g is of the form $g(t) = A\cos(\omega t) + B\sin(\omega t)$, look for a particular solution of the same form, $f_p(t) = P\cos(\omega t) + Q\sin(\omega t)$.
- If g is constant, look for a constant particular solution $f_p(t) = c$.[7]
- If the DE is of first order, of the form $f'(t) - af(t) = g(t)$, use the formula $f(t) = e^{at} \int e^{-at} g(t)\, dt$.
- If none of the preceding techniques applies, write $T = (D - \lambda_1)$ $(D - \lambda_2) \cdots (D - \lambda_n)$, and solve the corresponding first-order DEs.

Step 3 The solutions of the DE $T(f) = g$ are of the form

$$f(t) = c_1 f_1(t) + c_2 f_2(t) + \cdots + c_n f_n(t) + f_p(t),$$

where $f_1, f_2, \ldots, f_n$ are the solutions from Step 1 and f_p is the solution from Step 2.

Take another look at Examples 2, 5, 6, 7, 9, and 10.

EXAMPLE 11 Find all solutions of the DE

$$f'''(t) + f''(t) - f'(t) - f(t) = 10.$$

Solution

We will follow the approach just outlined.

Step 1
- $p_T(\lambda) = \lambda^3 + \lambda^2 - \lambda - 1$
- We recognize $\lambda = 1$ as a root, and we can use long division to factor:

$$p_T(\lambda) = \lambda^3 + \lambda^2 - \lambda - 1 = (\lambda - 1)(\lambda^2 + 2\lambda + 1) = (\lambda - 1)(\lambda + 1)^2.$$

- Since $\lambda = 1$ is a solution of the equation $p_T(\lambda) = 0$, we let $f_1(t) = e^t$.
- Since $\lambda = -1$ is a solution of $p_T(\lambda) = 0$ with multiplicity 2, we let $f_2(t)$ $= e^{-t}$ and $f_3(t) = te^{-t}$.

---

[7]More generally, it is often helpful to look for a particular solution of the same form as $g(t)$, for example, a polynomial of a certain degree, or an exponential function Ce^{kt}. This technique is explored more fully in a course on differential equations.

Step 2 Since $g(t) = 10$ is a constant, we look for a constant solution, $f_p(t) = c$. Plugging into the DE, we find that $c = -10$. Thus we can let $f_p(t) = -10$.

Step 3 The solutions can be written in the form

$$f(t) = c_1 e^t + c_2 e^{-t} + c_3 t e^{-t} - 10,$$

where $c_1, c_2,$ and c_3 are arbitrary constants.   ∎

EXERCISES 9.3

GOAL   *Solve linear differential equations.*

Find all real solutions of the differential equations in Exercises 1 through 22.

1. $f'(t) - 5f(t) = 0$

2. $\dfrac{dx}{dt} + 3x = 7$

3. $f'(t) + 2f(t) = e^{3t}$

4. $\dfrac{dx}{dt} - 2x = \cos(3t)$

5. $f'(t) - f(t) = t$

6. $f'(t) - 2f(t) = e^{2t}$

7. $f''(t) + f'(t) - 12f(t) = 0$

8. $\dfrac{d^2x}{dt^2} + 3\dfrac{dx}{dt} - 10x = 0$

9. $f''(t) - 9f(t) = 0$

10. $f''(t) + f(t) = 0$

11. $\dfrac{d^2x}{dt^2} - 2\dfrac{dx}{dt} + 2x = 0$

12. $f''(t) - 4f'(t) + 13f(t) = 0$

13. $f''(t) + 2f'(t) + f(t) = 0$

14. $f''(t) + 3f'(t) = 0$

15. $f''(t) = 0$

16. $f''(t) + 4f'(t) + 13f(t) = \cos(t)$

17. $f''(t) + 2f'(t) + f(t) = \sin(t)$

18. $f''(t) + 3f'(t) + 2f(t) = \cos(t)$

19. $\dfrac{d^2x}{dt^2} + 2x = \cos(t)$

20. $f'''(t) - 3f''(t) + 2f'(t) = 0$

21. $f'''(t) + 2f''(t) - f'(t) - 2f(t) = 0$

22. $f'''(t) - f''(t) - 4f'(t) + 4f(t) = 0$

Solve the initial value problems in Exercises 23 through 29.

23. $f'(t) - 5f(t) = 0, \ f(0) = 3$

24. $\dfrac{dx}{dt} + 3x = 7, \ x(0) = 0$

25. $f'(t) + 2f(t) = 0, \ f(1) = 1$

26. $f''(t) - 9f(t) = 0, \ f(0) = 0, \ f'(0) = 1$

27. $f''(t) + 9f(t) = 0, \ f(0) = 0, \ f\left(\frac{\pi}{2}\right) = 1$

28. $f''(t) + f'(t) - 12f(t) = 0, \ f(0) = f'(0) = 0$

29. $f''(t) + 4f(t) = \sin(t), \ f(0) = f'(0) = 0$

30. The temperature of a hot cup of coffee can be modeled by the DE

$$T'(t) = -k\big(T(t) - A\big).$$

   a. What is the significance of the constants k and A?

   b. Solve the DE for $T(t)$, in terms of k, A, and the initial temperature T_0.

31. The speed $v(t)$ of a falling object can sometimes be modeled by

$$m\frac{dv}{dt} = mg - kv,$$

   or

$$\frac{dv}{dt} + \frac{k}{m}v = g,$$

   where m is the mass of the body, g the gravitational acceleration, and k a constant related to the air resistance. Solve this DE when $v(0) = 0$. Describe the long term behavior of $v(t)$. Sketch a graph.

32. Consider the balance $B(t)$ of a bank account, with initial balance $B(0) = B_0$. We are withdrawing money at a continuous rate r (in Euro/year). The interest rate is k (%/year), compounded continuously. Set up a differential equation for $B(t)$, and solve it in terms of B_0, r, and k. What will happen in the long run? Describe all possible scenarios. Sketch a graph for $B(t)$ in each case.

33. Consider a pendulum of length L. Let $x(t)$ be the angle the pendulum makes with the vertical (measured in radians). For small angles, the motion is well approximated by the DE

$$\frac{d^2x}{dt^2} = -\frac{g}{L}x,$$

   where g is the acceleration due to gravity ($g \approx 9.81$ m/sec^2). How long does the pendulum have to be so that it swings from one extreme position to the other in exactly one second?

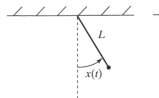

| Note: $x(t)$ is negative when the pendulum is on the left. | The two extreme positions of the pendulum. |

Historical note: The result of this exercise was considered as a possible *definition* of the meter. The French committee reforming the measures in the 1790s finally adopted another definition: A meter was set to be the 10,000,000th part of the distance from the North Pole to the Equator, measured along the meridian through Paris. (In 1983 a new definition of the meter was adopted, based on the speed of light.)

34. Consider a wooden block in the shape of a cube whose edges are 10 cm long. The density of the wood is 0.8 g/cm^3. The block is submersed in water; a guiding mechanism guarantees that the top and the bottom surfaces of the block are parallel to the surface of the water at all times. Let $x(t)$ be the depth of the block in the water at time t. Assume that x is between 0 and 10 at all times.

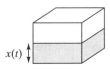

a. Two forces are acting on the block: its weight and the buoyancy (the weight of the displaced water). Recall that the density of water is 1 g/cm^3. Find formulas for these two forces.

b. Set up a differential equation for $x(t)$. Find the solution, assuming that the block is initially completely submersed ($x(0) = 10$) and at rest.

c. How does the period of the oscillation change if you change the dimensions of the block? (Consider a larger or smaller cube.) What if the wood has a different density or if the initial state is different? What if you conduct the experiment on the moon?

35. The displacement $x(t)$ of a certain oscillator can be modeled by the DE

$$\frac{d^2x}{dt^2} + 3\frac{dx}{dt} + 2x = 0.$$

a. Find all solutions of this DE.

b. Find the solution with initial values $x(0) = 1$, $x'(0) = 0$. Graph the solution.

c. Find the solution with initial values $x(0) = 1$, $x'(0) = -3$. Graph the solution.

d. Describe the qualitative difference of the solutions in parts (b) and parts (c), in terms of the motion of

the oscillator. How many times will the oscillator go through the equilibrium state $x = 0$ in each case?

36. The displacement $x(t)$ of a certain oscillator can be modeled by the DE

$$\frac{d^2x}{dt^2} + 2\frac{dx}{dt} + 101x = 0.$$

Find all solutions of this DE, and graph a typical solution. How many times will the oscillator go through the equilibrium state $x = 0$?

37. The displacement $x(t)$ of a certain oscillator can be modeled by the DE

$$\frac{d^2x}{dt^2} + 6\frac{dx}{dt} + 9x = 0.$$

Find the solution $x(t)$ for the initial values $x(0) = 0$, $x'(0) = 1$. Sketch the graph of the solution. How many times will the oscillator go through the equilibrium state $x = 0$ in this case?

38. a. If $p(t)$ is a polynomial and λ a scalar, show that

$$(D - \lambda)\left(p(t)e^{\lambda t}\right) = p'(t)e^{\lambda t}.$$

b. If $p(t)$ is a polynomial of degree less than m, what is

$$(D - \lambda)^m \left(p(t)e^{\lambda t}\right)?$$

c. Find a basis of the kernel of the linear differential operator $(D - \lambda)^m$.

d. If $\lambda_1, \ldots, \lambda_r$ are distinct scalars and $m_1, \ldots, m_r$ are positive integers, find a basis of the kernel of the linear differential operator

$$(D - \lambda_1)^{m_1} \ldots (D - \lambda_r)^{m_r}.$$

39. Find all solutions of the linear DE

$$f'''(t) + 3f''(t) + 3f'(t) + f(t) = 0.$$

(*Hint*: Use Exercise 38.)

40. Find all solutions of the linear DE

$$\frac{d^3x}{dt^3} + \frac{d^2x}{dt^2} - \frac{dx}{dt} - x = 0.$$

(*Hint*: Use Exercise 38.)

41. If T is an nth-order linear differential operator and λ is an arbitrary scalar, is λ necessarily an eigenvalue of T? If so, what is the dimension of the eigenspace associated with λ?

42. Let C^∞ be the space of all *real-valued* smooth functions.

a. Consider the linear differential operator $T = D^2$ from C^∞ to C^∞. Find all (real) eigenvalues of T. For each eigenvalue, find a basis of the associated eigenspace.

b. Let V be the subspace of C^∞ consisting of all *periodic* functions $f(t)$ with period one [i.e.,

$f(t+1) = f(t)$, for all t]. Consider the linear differential operator $L = D^2$ from V to V. Find all (real) eigenvalues and eigenfunctions of L.

43. The displacement of a certain forced oscillator can be modeled by the DE

$$\frac{d^2x}{dt^2} + 5\frac{dx}{dt} + 6x = \cos(t).$$

a. Find all solutions of this DE.

b. Describe the long-term behavior of this oscillator.

44. The displacement of a certain forced oscillator can be modeled by the DE

$$\frac{d^2x}{dt^2} + 4\frac{dx}{dt} + 5x = \cos(3t).$$

a. Find all solutions of this DE.

b. Describe the long-term behavior of this oscillator.

45. Use Theorem 9.3.13 to solve the initial value problem

$$\frac{d\vec{x}}{dt} = \begin{bmatrix} 1 & 2 \\ 0 & 1 \end{bmatrix} \vec{x}, \quad \text{with} \quad \vec{x}(0) = \begin{bmatrix} 1 \\ -1 \end{bmatrix}.$$

[*Hint*: Find first $x_2(t)$ and then $x_1(t)$.]

46. Use Theorem 9.3.13 to solve the initial value problem

$$\frac{d\vec{x}}{dt} = \begin{bmatrix} 2 & 3 & 1 \\ 0 & 1 & 2 \\ 0 & 0 & 1 \end{bmatrix}, \quad \text{with} \quad \vec{x}(0) = \begin{bmatrix} 2 \\ 1 \\ -1 \end{bmatrix}.$$

[*Hint*: Find first $x_3(t)$, then $x_2(t)$, and finally $x_1(t)$.]

47. Consider the initial value problem

$$\frac{d\vec{x}}{dt} = A\vec{x}, \quad \text{with} \quad \vec{x}(0) = \vec{x}_0,$$

where A is an upper triangular $n \times n$ matrix with m distinct diagonal entries $\lambda_1, \ldots, \lambda_m$. See the examples in Exercises 45 and 46.

a. Show that this problem has a unique solution $\vec{x}(t)$, whose components $x_i(t)$ are of the form

$$x_i(t) = p_1(t)e^{\lambda_1 t} + \cdots + p_m(t)e^{\lambda_m t},$$

for some polynomials $p_j(t)$. *Hint*: Find first $x_n(t)$, then $x_{n-1}(t)$, and so on.

b. Show that the zero state is a stable equilibrium solution of this system if (and only if) the real part of all the λ_i is negative.

48. Consider an $n \times n$ matrix A with m distinct eigenvalues $\lambda_1, \ldots, \lambda_m$.

a. Show that the initial value problem

$$\frac{d\vec{x}}{dt} = A\vec{x}, \quad \text{with} \quad \vec{x}(0) = \vec{x}_0,$$

has a unique solution $\vec{x}(t)$.

b. Show that the zero state is a stable equilibrium solution of the system

$$\frac{d\vec{x}}{dt} = A\vec{x}$$

if and only if the real part of all the λ_i is negative. (*Hint*: Exercise 47 and Exercise 8.1.45 are helpful.)

Vectors

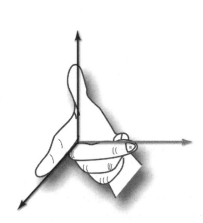

$$\mathbf{H}$$ ere we will provide a concise summary of basic facts on vectors. In Section 1.2, vectors are defined as matrices with only one column: $\vec{v} = \begin{bmatrix} v_1 \\ v_2 \\ \vdots \\ v_n \end{bmatrix}$. The scalars v_i are called the *components* of the vector.[1] The set of all vectors with n components is denoted by $\mathbb{R}^n$.

You may be accustomed to a different notation for vectors. Writing the components in a column is the most convenient notation for linear algebra.

Vector Algebra

Definition A.1   Vector addition and scalar multiplication

a. The sum of two vectors $\vec{v}$ and $\vec{w}$ in $\mathbb{R}^n$ is defined "componentwise":

$$\vec{v} + \vec{w} = \begin{bmatrix} v_1 \\ v_2 \\ \vdots \\ v_n \end{bmatrix} + \begin{bmatrix} w_1 \\ w_2 \\ \vdots \\ w_n \end{bmatrix} = \begin{bmatrix} v_1 + w_1 \\ v_2 + w_2 \\ \vdots \\ v_n + w_n \end{bmatrix}$$

b. *Scalar multiplication* The product of a scalar k and a vector $\vec{v}$ is defined componentwise as well:

$$k\vec{v} = k \begin{bmatrix} v_1 \\ v_2 \\ \vdots \\ v_n \end{bmatrix} = \begin{bmatrix} kv_1 \\ kv_2 \\ \vdots \\ kv_n \end{bmatrix}$$

[1] In vector and matrix algebra, the term "scalar" is synonymous with (real) number.

EXAMPLE 1 $\begin{bmatrix} 1 \\ 2 \\ 3 \\ 4 \end{bmatrix} + \begin{bmatrix} 4 \\ 2 \\ 0 \\ -1 \end{bmatrix} = \begin{bmatrix} 5 \\ 4 \\ 3 \\ 3 \end{bmatrix}$ ■

EXAMPLE 2 $3 \begin{bmatrix} 1 \\ 2 \\ 0 \\ -1 \end{bmatrix} = \begin{bmatrix} 3 \\ 6 \\ 0 \\ -3 \end{bmatrix}$ ■

The *negative* or *opposite* of a vector $\vec{v}$ in $\mathbb{R}^n$ is defined as

$$-\vec{v} = (-1)\vec{v}.$$

The *difference* $\vec{v} - \vec{w}$ of two vectors $\vec{v}$ and $\vec{w}$ in $\mathbb{R}^n$ is defined componentwise. Alternatively, we can express the difference of two vectors as

$$\vec{v} - \vec{w} = \vec{v} + (-\vec{w}).$$

The vector in $\mathbb{R}^n$ that consists of n zeros is called the *zero vector* in $\mathbb{R}^n$:

$$\vec{0} = \begin{bmatrix} 0 \\ 0 \\ \vdots \\ 0 \end{bmatrix}$$

Theorem A.2   **Rules of vector algebra**

The following formulas hold for all vectors $\vec{u}, \vec{v}, \vec{w}$ in $\mathbb{R}^n$ and for all scalars c and k:

1. $(\vec{u} + \vec{v}) + \vec{w} = \vec{u} + (\vec{v} + \vec{w})$: Addition is *associative*.
2. $\vec{v} + \vec{w} = \vec{w} + \vec{v}$: Addition is *commutative*.
3. $\vec{v} + \vec{0} = \vec{v}$
4. For each $\vec{v}$ in $\mathbb{R}^n$, there exists a unique $\vec{x}$ in $\mathbb{R}^n$ such that $\vec{v} + \vec{x} = \vec{0}$, namely, $\vec{x} = -\vec{v}$.
5. $k(\vec{v} + \vec{w}) = k\vec{v} + k\vec{w}$
6. $(c + k)\vec{v} = c\vec{v} + k\vec{v}$
7. $c(k\vec{v}) = (ck)\vec{v}$
8. $1\vec{v} = \vec{v}$ ■

These rules follow from the corresponding rules for scalars (commutativity, associativity, distributivity); for example:

$$\vec{v} + \vec{w} = \begin{bmatrix} v_1 \\ v_2 \\ \vdots \\ v_n \end{bmatrix} + \begin{bmatrix} w_1 \\ w_2 \\ \vdots \\ w_n \end{bmatrix} = \begin{bmatrix} v_1 + w_1 \\ v_2 + w_2 \\ \vdots \\ v_n + w_n \end{bmatrix} = \begin{bmatrix} w_1 + v_1 \\ w_2 + v_2 \\ \vdots \\ w_n + v_n \end{bmatrix}$$

$$= \begin{bmatrix} w_1 \\ w_2 \\ \vdots \\ w_n \end{bmatrix} + \begin{bmatrix} v_1 \\ v_2 \\ \vdots \\ v_n \end{bmatrix} = \vec{w} + \vec{v}$$

Geometrical Representation of Vectors

The *standard representation* of a vector

$$\vec{x} = \begin{bmatrix} x_1 \\ x_2 \end{bmatrix}$$

in the Cartesian coordinate plane is as an *arrow* (a directed line segment) connecting the origin to the point (x_1, x_2), as shown in Figure 1.

Occasionally, it is helpful to translate (or shift) the vector in the plane (preserving its direction and length), so that it will connect some point (a_1, a_2) to the point $(a_1 + x_1, a_2 + x_2)$. See Figure 2.

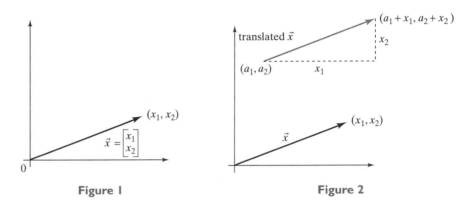

Figure 1   **Figure 2**

In this text, we consider the *standard representation* of vectors, unless we explicitly state that the vector has been translated.

A vector in $\mathbb{R}^2$ (in standard representation) is uniquely determined by its endpoint. Conversely, with each point in the plane we can associate its *position vector*, which connects the origin to the given point. See Figure 3.

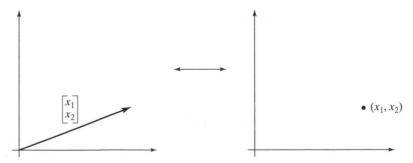

Figure 3 The components of a vector in standard representation are the coordinates of its endpoint.

We need not clearly distinguish between a vector and its endpoint; we can identify them as long as we consistently use the standard representation of vectors.

For example, we will talk about "the vectors on a line L" when we really mean the vectors whose endpoints are on the line L (in standard representation). Likewise, we can talk about "the vectors in a region R" in the plane. See Figure 4.

Adding vectors in $\mathbb{R}^2$ can be represented by means of a parallelogram, as shown in Figure 5.

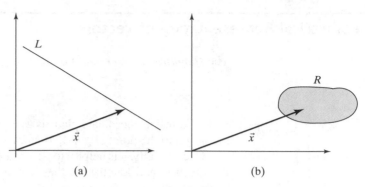

Figure 4 (a) $\vec{x}$ is a vector on the line L. (b) $\vec{x}$ is a vector in the region R.

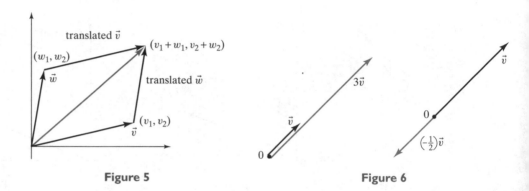

Figure 5   **Figure 6**

If k is a positive scalar, then $k\vec{v}$ is obtained by stretching the vector $\vec{v}$ by a factor of k, leaving its direction unchanged. If k is negative, then the direction is reversed. See Figure 6.

Definition A.3   We say that two vectors $\vec{v}$ and $\vec{w}$ in $\mathbb{R}^n$ are *parallel* if one of them is a scalar multiple of the other.

EXAMPLE 3   The vectors

$$\begin{bmatrix} 1 \\ 3 \\ 2 \\ -2 \end{bmatrix} \quad \text{and} \quad \begin{bmatrix} 3 \\ 9 \\ 6 \\ -6 \end{bmatrix}$$

are parallel, since

$$\begin{bmatrix} 3 \\ 9 \\ 6 \\ -6 \end{bmatrix} = 3 \begin{bmatrix} 1 \\ 3 \\ 2 \\ -2 \end{bmatrix}.$$

∎

EXAMPLE 4   The vectors

$$\begin{bmatrix} 1 \\ 2 \\ 3 \\ 4 \end{bmatrix} \quad \text{and} \quad \begin{bmatrix} 0 \\ 0 \\ 0 \\ 0 \end{bmatrix}$$

are parallel, since

$$\begin{bmatrix} 0 \\ 0 \\ 0 \\ 0 \end{bmatrix} = 0 \begin{bmatrix} 1 \\ 2 \\ 3 \\ 4 \end{bmatrix}.$$

∎

Let us briefly review Cartesian coordinates in *space*: If we choose an origin 0 and three mutually perpendicular coordinate axes through 0, we can describe any point in space by a triple of numbers, (x_1, x_2, x_3). See Figure 7.

The standard representation of the vector

$$\vec{x} = \begin{bmatrix} x_1 \\ x_2 \\ x_3 \end{bmatrix}$$

is the arrow connecting the origin to the point (x_1, x_2, x_3), as shown in Figure 8.

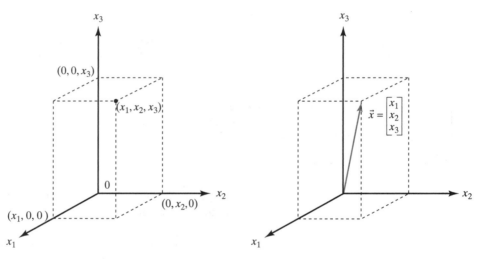

Figure 7 **Figure 8**

Dot Product, Length, Orthogonality

Definition A.4 Consider two vectors $\vec{v}$ and $\vec{w}$ with components $v_1, v_2, \ldots, v_n$ and $w_1, w_2, \ldots, w_n$, respectively. Here $\vec{v}$ and $\vec{w}$ may be column or row vectors, and they need not be of the same type (these conventions are convenient in linear algebra). The *dot product* of $\vec{v}$ and $\vec{w}$ is defined as

$$\vec{v} \cdot \vec{w} = v_1 w_1 + v_2 w_2 + \cdots + v_n w_n.$$

We can interpret the dot product geometrically: If $\vec{v}$ and $\vec{w}$ are two nonzero vectors in $\mathbb{R}^n$, then

$$\vec{v} \cdot \vec{w} = \|\vec{v}\| \cos\theta \|\vec{w}\|,$$

where θ is the angle enclosed by vectors $\vec{v}$ and $\vec{w}$. (See Definition 5.1.12.)

Note that the dot product of two vectors is a *scalar*.

EXAMPLE 5
$$\begin{bmatrix} 1 \\ 2 \\ 1 \end{bmatrix} \cdot \begin{bmatrix} 3 \\ -1 \\ -1 \end{bmatrix} = 1 \cdot 3 + 2 \cdot (-1) + 1 \cdot (-1) = 0 \qquad \blacksquare$$

EXAMPLE 6
$$\begin{bmatrix} 1 & 2 & 3 & 4 \end{bmatrix} \cdot \begin{bmatrix} 3 \\ 1 \\ 0 \\ -1 \end{bmatrix} = 3 + 2 + 0 - 4 = 1 \qquad \blacksquare$$

Theorem A.5   Rules for dot products

The following equations hold for all column or row vectors $\vec{u}, \vec{v}, \vec{w}$ with n components, and for all scalars k:

1. $\vec{v} \cdot \vec{w} = \vec{w} \cdot \vec{v}$
2. $(\vec{u} + \vec{v}) \cdot \vec{w} = \vec{u} \cdot \vec{w} + \vec{v} \cdot \vec{w}$
3. $(k\vec{v}) \cdot \vec{w} = k(\vec{v} \cdot \vec{w})$
4. $\vec{v} \cdot \vec{v} > 0$ for all nonzero $\vec{v}$ $\blacksquare$

The verification of these rules is straightforward. Let us justify rule (d): since $\vec{v}$ is nonzero, at least one of the components v_i is nonzero, so that v_i^2 is positive. Then

$$\vec{v} \cdot \vec{v} = v_1^2 + v_2^2 + \cdots + v_i^2 + \cdots + v_n^2$$

is positive as well.

Let us think about the *length* of a vector. The length of a vector

$$\vec{x} = \begin{bmatrix} x_1 \\ x_2 \end{bmatrix}$$

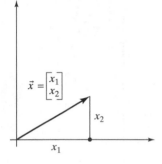

$$\vec{x} = \begin{bmatrix} x_1 \\ x_2 \end{bmatrix}$$

Figure 9

in $\mathbb{R}^2$ is $\sqrt{x_1^2 + x_2^2}$ by the Pythagorean theorem. See Figure 9.

This length is often denoted by $\|\vec{x}\|$. Note that we have

$$\vec{x} \cdot \vec{x} = \begin{bmatrix} x_1 \\ x_2 \end{bmatrix} \cdot \begin{bmatrix} x_1 \\ x_2 \end{bmatrix} = x_1^2 + x_2^2 = \|\vec{x}\|^2;$$

therefore,

$$\|\vec{x}\| = \sqrt{\vec{x} \cdot \vec{x}}.$$

Verify that this formula holds for vectors $\vec{x}$ in $\mathbb{R}^3$ as well.

We can use this formula to *define* the length of a vector in $\mathbb{R}^n$:

Definition A.6   The *length* (or *norm*) $\|\vec{x}\|$ of a vector $\vec{x}$ in $\mathbb{R}^n$ is

$$\|\vec{x}\| = \sqrt{\vec{x} \cdot \vec{x}} = \sqrt{x_1^2 + x_2^2 + \cdots + x_n^2}.$$

EXAMPLE 7   Find $\|\vec{x}\|$ for

$$\vec{x} = \begin{bmatrix} 7 \\ 1 \\ 7 \\ -1 \end{bmatrix}.$$

Solution

$$\|\vec{x}\| = \sqrt{\vec{x} \cdot \vec{x}} = \sqrt{49 + 1 + 49 + 1} = 10 \qquad \blacksquare$$

Definition A.7   A vector $\vec{u}$ in $\mathbb{R}^n$ is called a *unit vector* if $\|\vec{u}\| = 1$; that is, the length of the vector $\vec{u}$ is 1.

Consider two perpendicular vectors $\vec{x}$ and $\vec{y}$ in $\mathbb{R}^2$, as shown in Figure 10.

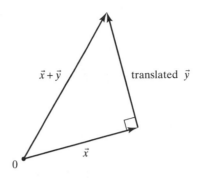

Figure 10

By the theorem of Pythagoras,

$$\|\vec{x} + \vec{y}\|^2 = \|\vec{x}\|^2 + \|\vec{y}\|^2,$$

or

$$(\vec{x} + \vec{y}) \cdot (\vec{x} + \vec{y}) = \vec{x} \cdot \vec{x} + \vec{y} \cdot \vec{y}.$$

By Theorem A.5,

$$\vec{x} \cdot \vec{x} + 2(\vec{x} \cdot \vec{y}) + \vec{y} \cdot \vec{y} = \vec{x} \cdot \vec{x} + \vec{y} \cdot \vec{y},$$

or

$$\vec{x} \cdot \vec{y} = 0.$$

You can read these equations backward to show that $\vec{x} \cdot \vec{y} = 0$ if and only if $\vec{x}$ and $\vec{y}$ are perpendicular. This reasoning applies to vectors in $\mathbb{R}^3$ as well.

We can use this characterization to *define* perpendicular vectors in $\mathbb{R}^n$:

Definition A.8   Two vectors $\vec{v}$ and $\vec{w}$ in $\mathbb{R}^n$ are called *perpendicular* (or *orthogonal*) if $\vec{v} \cdot \vec{w} = 0$.

Cross Product

Here we present the cross product for vectors in $\mathbb{R}^3$ only; for a generalization to $\mathbb{R}^n$, see Exercises 6.2.44 and 6.3.17.

In Chapter 6, we discuss the cross product in the context of linear algebra.

Definition A.9   Cross product in $\mathbb{R}^3$

The cross product $\vec{v} \times \vec{w}$ of two vectors $\vec{v}$ and $\vec{w}$ in $\mathbb{R}^3$ is the vector in $\mathbb{R}^3$ with the following three properties:

- $\vec{v} \times \vec{w}$ is orthogonal to both $\vec{v}$ and $\vec{w}$.
- $\|\vec{v} \times \vec{w}\| = \|\vec{v}\| \sin\theta \|\vec{w}\|$, where θ is the angle between $\vec{v}$ and $\vec{w}$, with $0 \le \theta \le \pi$. This means that the magnitude of the vector $\vec{v} \times \vec{w}$ is the area of the parallelogram spanned by $\vec{v}$ and $\vec{w}$, as illustrated in Figure 11a.

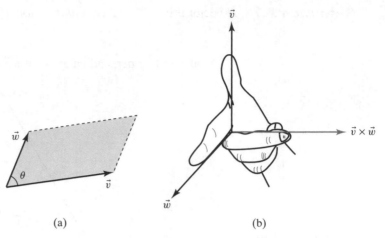

Figure 11 (a) $\|\vec{v} \times \vec{w}\|$ is the shaded area. (b) A right-handed system.

- The direction of $\vec{v} \times \vec{w}$ follows the *right-hand rule,* as illustrated in Figure 11b.

Theorem A.10   **Properties of the cross product**

The following equations hold for all vectors $\vec{u}, \vec{v}, \vec{w}$ in $\mathbb{R}^3$ and for all scalars k.

a. $\vec{w} \times \vec{v} = -(\vec{v} \times \vec{w})$: The cross product is *anticommutative.*

b. $(k\vec{v}) \times \vec{w} = k(\vec{v} \times \vec{w}) = \vec{v} \times (k\vec{w})$

c. $\vec{v} \times (\vec{u} + \vec{w}) = \vec{v} \times \vec{u} + \vec{v} \times \vec{w}$

d. $\vec{v} \times \vec{w} = \vec{0}$ if (and only if) $\vec{v}$ is parallel to $\vec{w}$

e. $\vec{v} \times \vec{v} = \vec{0}$

f. $\vec{e}_1 \times \vec{e}_2 = \vec{e}_3, \vec{e}_2 \times \vec{e}_3 = \vec{e}_1, \vec{e}_3 \times \vec{e}_1 = \vec{e}_2$
   (and $\vec{e}_2 \times \vec{e}_1 = -\vec{e}_3, \vec{e}_3 \times \vec{e}_2 = -\vec{e}_1, \vec{e}_1 \times \vec{e}_3 = -\vec{e}_2$) ∎

Note that the cross product fails to be associative: $\vec{u} \times (\vec{v} \times \vec{w}) \neq (\vec{u} \times \vec{v}) \times \vec{w}$, in general. For example, $(\vec{e}_1 \times \vec{e}_1) \times \vec{e}_2 = \vec{0}$ but $\vec{e}_1 \times (\vec{e}_1 \times \vec{e}_2) = -\vec{e}_2$.

Properties (b) and (c) stated in Theorem A.10 imply that the function $T(\vec{x}) = \vec{v} \times \vec{x}$ is a linear transformation from $\mathbb{R}^3$ to $\mathbb{R}^3$, for any fixed vector $\vec{v}$ in $\mathbb{R}^3$.

The following diagram can serve as a memory aid for property (f):

$$\begin{array}{ccc} & \vec{e}_3 & \\ \swarrow & & \nwarrow \\ \vec{e}_1 & \longrightarrow & \vec{e}_2 \end{array}$$

We can use the properties stated in Theorem A.10 to express the cross products in components.

$$\begin{bmatrix} v_1 \\ v_2 \\ v_3 \end{bmatrix} \times \begin{bmatrix} w_1 \\ w_2 \\ w_3 \end{bmatrix} = (v_1\vec{e}_1 + v_2\vec{e}_2 + v_3\vec{e}_3) \times (w_1\vec{e}_1 + w_2\vec{e}_2 + w_3\vec{e}_3)$$

$$= (v_2w_3 - v_3w_2)\vec{e}_1 + (v_3w_1 - v_1w_3)\vec{e}_2 + (v_1w_2 - v_2w_1)\vec{e}_3$$

$$= \begin{bmatrix} v_2w_3 - v_3w_2 \\ v_3w_1 - v_1w_3 \\ v_1w_2 - v_2w_1 \end{bmatrix}$$

Theorem A.11   The cross product in components

$$\begin{bmatrix} v_1 \\ v_2 \\ v_3 \end{bmatrix} \times \begin{bmatrix} w_1 \\ w_2 \\ w_3 \end{bmatrix} = \begin{bmatrix} v_2 w_3 - v_3 w_2 \\ v_3 w_1 - v_1 w_3 \\ v_1 w_2 - v_2 w_1 \end{bmatrix}$$

EXAMPLE 8

$$\begin{bmatrix} 2 \\ 3 \\ 4 \end{bmatrix} \times \begin{bmatrix} 5 \\ 6 \\ 7 \end{bmatrix} = \begin{bmatrix} 3 \cdot 7 - 4 \cdot 6 \\ 4 \cdot 5 - 2 \cdot 7 \\ 2 \cdot 6 - 3 \cdot 5 \end{bmatrix} = \begin{bmatrix} -3 \\ 6 \\ -3 \end{bmatrix}$$

CHAPTER I

1.1 Answers to more theoretical questions are omitted.

1. $(x, y) = (-1, 1)$

3. No solutions

5. $(x, y) = (0, 0)$

7. No solutions

9. $(x, y, z) = (t, \frac{1}{2} - 2t, t)$, where t is arbitrary

11. $(x, y) = (4, 1)$ 13. No solutions

15. $(x, y, z) = (0, 0, 0)$

17. $(x, y) = (-5a + 2b, 3a - b)$

19. a. Products are competing.
    b. $P_1 = 26, P_2 = 46$

21. $a = 400, b = 300$

23. a. $(x, y) = (t, 2t)$;
    b. $(x, y) = (t, -3t)$;
    c. $(x, y) = (0, 0)$

25. a. If $k = 7$
    b. If $k = 7$, there are infinitely many solutions.
    c. If $k = 7$, the solutions are $(x, y, z) = (1 - t, 2t - 3, t)$.

27. 7 children (3 boys and 4 girls)

29. $f(t) = 1 - 5t + 3t^2$ 31. $f(t) = 2t^2 - 3t + 4$

33. $f(t) = at^2 + (1 - 4a)t + 3a$, for arbitrary a

35. $f(t) = 2e^{3t} - e^{2t}$

37. $-20 - 2x - 4y + x^2 + y^2 = 0$, the circle centered at $(1, 2)$ with radius 5

39. If $a - 2b + c = 0$

41. a. The intercepts of the line $x + y = 1$ are $(1, 0)$ and $(0, 1)$. The intercepts of the line $x + \frac{t}{2}y = t$ are $(t, 0)$ and $(0, 2)$. The lines intersect if $t \neq 2$.
    b. $x = -\dfrac{t}{t - 2}, y = \dfrac{2t - 2}{t - 2}$

43. There are many correct answers. Example:
$$\begin{vmatrix} x & -5z = -4 \\ y & -3z = -2 \end{vmatrix}$$

47. Twenty \$1 bills, eight \$5 bills, and four \$10 bills.

1.2 Answers to more theoretical questions are omitted.

1. $\begin{bmatrix} x \\ y \\ z \end{bmatrix} = \begin{bmatrix} 10t + 13 \\ -8t - 8 \\ t \end{bmatrix}$

3. $\begin{bmatrix} x \\ y \\ z \end{bmatrix} = \begin{bmatrix} 4 - 2s - 3t \\ s \\ t \end{bmatrix}$

5. $\begin{bmatrix} x_1 \\ x_2 \\ x_3 \\ x_4 \end{bmatrix} = \begin{bmatrix} -t \\ t \\ -t \\ t \end{bmatrix}$

7. $\begin{bmatrix} x_1 \\ x_2 \\ x_3 \\ x_4 \\ x_5 \end{bmatrix} = \begin{bmatrix} -2t \\ t \\ 0 \\ 0 \\ 0 \end{bmatrix}$

9. $\begin{bmatrix} x_1 \\ x_2 \\ x_3 \\ x_4 \\ x_5 \\ x_6 \end{bmatrix} = \begin{bmatrix} t - s - 2r \\ r \\ -t + s + 1 \\ t - 2s + 2 \\ s \\ t \end{bmatrix}$

11. $\begin{bmatrix} x_1 \\ x_2 \\ x_3 \\ x_4 \end{bmatrix} = \begin{bmatrix} -2t \\ 3t + 4 \\ t \\ -2 \end{bmatrix}$

13. No solutions

15. $\begin{bmatrix} x \\ y \\ z \end{bmatrix} = \begin{bmatrix} 4 \\ 2 \\ 1 \end{bmatrix}$

17. $\begin{bmatrix} x_1 \\ x_2 \\ x_3 \\ x_4 \\ x_5 \end{bmatrix} = \begin{bmatrix} -8221/4340 \\ 8591/8680 \\ 4695/434 \\ -459/434 \\ 699/434 \end{bmatrix}$

19. $\begin{bmatrix} 0 \\ 0 \\ 0 \\ 0 \end{bmatrix}$ and $\begin{bmatrix} 1 \\ 0 \\ 0 \\ 0 \end{bmatrix}$

21. 4 types

25. Yes; perform the operations backwards.

27. No; you cannot make the last column zero by elementary row operations.

29. $a = 2, b = c = d = 1$

31. $f(t) = 1 - 5t + 4t^2 + 3t^3 - 2t^4$

33. $f(t) = -5 + 13t - 10t^2 + 3t^3$

35. $\begin{bmatrix} -t \\ 6t \\ -9t \\ 4t \end{bmatrix}$, where t is arbitrary.

37. $\begin{bmatrix} x_1 \\ x_2 \\ x_3 \end{bmatrix} = \begin{bmatrix} 500 \\ 300 \\ 400 \end{bmatrix}$

39. a. Neither the manufacturing nor the energy sector makes demands on agriculture.
    b. $x_1 \approx 18.67, x_2 \approx 22.60, x_3 \approx 3.63$

41. $m_1 = \frac{2}{3}m_2$

43. $a \approx 12.17$, $b \approx -1.15$, $c \approx 0.18$. The longest day is about 13.3 hours.

45. a. If k is neither 1 nor 2.
    b. If $k = 1$.   c. If $k = 2$.

47. a. $x_1 = 3x_3 - 2x_4$, $x_2 = 2x_3 - x_4$, for arbitrary x_3 and x_4.
    b. Yes, $x_1 = 1$, $x_2 = 5$, $x_3 = 9$, $x_4 = 13$.

49. $C = 25$

51. $xy = 0$, the union of the two coordinate axes

53. $a(xy - y) + b(y^2 - y) = 0$, where $a \neq 0$ or $b \neq 0$

55. $a(x^2 - x) + b(y^2 - y) = 0$, where $a \neq 0$ or $b \neq 0$

57. $25 - 10x - 10y + x^2 + y^2 = 0$, the circle of radius 5 centered at $(5, 5)$

59. No solutions

61. Statistics: \$86; Set Theory: \$92; Psychology: \$55.

63. Beginning: 120 liberal, 140 conservative.
    End: 140 liberal, 120 conservative.

65. Cow: 34/21 liang; sheep: 20/21 liang.

67. Swallow: 24/19 liang; sparrow: 32/19 liang.

69. A: 265; B: 191; C: 148; D: 129; E: 76.

71. Gaussian elimination shows that
    Pigeons $= -250 + \frac{5}{9}$(Swans) $+ 20$(Peacocks)   and
    Sarasas $= 350 - \frac{14}{9}$(Swans) $- 21$(Peacocks)

    One solution (the one given by Mahavira) is: 15 pigeons, 28 sarasabirds, 45 swans, and 12 peacocks (spending 9, 20, 35, and 36 panas, respectively).

73. 53 sheep, 42 goats, and 5 hogs.

75.

	Full	Half	Empty
1st Son	p	$10 - 2p$	p
2nd Son	q	$10 - 2q$	q
3rd Son	$10 - p - q$	$2p + 2q - 10$	$10 - p - q$

Here, p and q are integers between 0 and 5 such that $p + q \geq 5$.

1.3 Answers to more theoretical questions are omitted.

1. a. No solutions   b. One solution
   c. Infinitely many solutions

3. Rank is 1

5. a. $x \begin{bmatrix} 1 \\ 3 \end{bmatrix} + y \begin{bmatrix} 2 \\ 1 \end{bmatrix} = \begin{bmatrix} 7 \\ 11 \end{bmatrix}$
   b. $x = 3$, $y = 2$

7. One solution

9. $\begin{bmatrix} 1 & 2 & 3 \\ 4 & 5 & 6 \\ 7 & 8 & 9 \end{bmatrix} \begin{bmatrix} x \\ y \\ z \end{bmatrix} = \begin{bmatrix} 1 \\ 4 \\ 9 \end{bmatrix}$

11. Undefined

13. $\begin{bmatrix} 29 \\ 65 \end{bmatrix}$ 15. 70

17. Undefined   19. $\begin{bmatrix} 0 \\ 0 \\ 0 \end{bmatrix}$

21. $\begin{bmatrix} 158 \\ 70 \\ 81 \\ 123 \end{bmatrix}$ 23. $\begin{bmatrix} 1 & 0 & 0 \\ 0 & 1 & 0 \\ 0 & 0 & 1 \\ 0 & 0 & 0 \end{bmatrix}$

25. The system $A\vec{x} = \vec{c}$ has infinitely many solutions or none.

27. $\begin{bmatrix} 1 & 0 & 0 & 0 \\ 0 & 1 & 0 & 0 \\ 0 & 0 & 1 & 0 \\ 0 & 0 & 0 & 1 \end{bmatrix}$ 29. $\begin{bmatrix} \frac{2}{5} & 0 & 0 \\ 0 & 0 & 0 \\ 0 & 0 & -\frac{1}{9} \end{bmatrix}$

31. $\begin{bmatrix} 1 & 2 & 1 \\ 0 & 3 & 1 \\ 0 & 0 & -\frac{1}{9} \end{bmatrix}$, for example

33. $A\vec{x} = \vec{x}$

35. $A\vec{e}_i$ is the ith column of A.

37. $\vec{x} = \begin{bmatrix} 2 - 2t \\ t \\ 1 \end{bmatrix}$, where t is arbitrary

39. $\begin{bmatrix} 1 & 0 & 0 & * \\ 0 & 1 & 0 & * \\ 0 & 0 & 1 & * \end{bmatrix}$

41. One solution   43. No solutions

47. a. $\vec{x} = 0$ is a solution.
    b. By part (a) and Theorem 1.3.3
    c. $A(\vec{x}_1 + \vec{x}_2) = A\vec{x}_1 + A\vec{x}_2 = \vec{0} + \vec{0} = \vec{0}$
    d. $A(k\vec{x}) = k(A\vec{x}) = k\vec{0} = \vec{0}$

49. a. Infinitely many solutions or none
    b. One solution or none
    c. No solutions
    d. Infinitely many solutions

51. If $m = r$ and $s = p$

53. Yes

55. Yes; $\begin{bmatrix} 7 \\ 8 \\ 9 \end{bmatrix} = -1 \begin{bmatrix} 1 \\ 2 \\ 3 \end{bmatrix} + 2 \begin{bmatrix} 4 \\ 5 \\ 6 \end{bmatrix}$

57. $\begin{bmatrix} 7 \\ 11 \end{bmatrix} = \begin{bmatrix} 3 \\ 9 \end{bmatrix} + \begin{bmatrix} 4 \\ 2 \end{bmatrix}$

59. $c = 9$, $d = 11$

61. For $c = 2$ and for $c = 3$

63. Line through the end point of $\vec{v}$ in the direction of $\vec{w}$

65. Parallelogram with its vertices at the origin and at the endpoints of vectors $\vec{v}$, $\vec{w}$, and $\vec{v} + \vec{w}$

67. Triangle with its vertices at the origin and at the endpoints of vectors $\vec{v}$ and $\vec{w}$

69. $\left(\frac{b+c-a}{2}, \frac{c-b+a}{2}, \frac{b-c+a}{2} \right)$

CHAPTER 2

2.1 Answers to more theoretical questions are omitted.

1. Not linear   3. Not linear

5. $A = \begin{bmatrix} 7 & 6 & -13 \\ 11 & 9 & 17 \end{bmatrix}$

7. T is linear; $A = \begin{bmatrix} \vec{v}_1 & \vec{v}_2 & \cdots & \vec{v}_m \end{bmatrix}$

9. Not invertible

11. The inverse is $\begin{bmatrix} 3 & -2/3 \\ -1 & 1/3 \end{bmatrix}$.

15. A is invertible if $a \neq 0$ or $b \neq 0$. In this case,
$A^{-1} = \frac{1}{a^2 + b^2} \begin{bmatrix} a & b \\ -b & a \end{bmatrix}$

17. Reflection about the origin; this transformation is its own inverse.

19. Orthogonal projection onto the $\vec{e}_1$ axis; not invertible.

21. Rotation through an angle of $90°$ in the clockwise direction; invertible.

23. Clockwise rotation through an angle of $90°$, followed by a scaling by a factor of 2; invertible.

25. Scaling by a factor of 2

27. Reflection about the $\vec{e}_1$ axis

29. Reflection about the origin

31. Reflection about the $\vec{e}_2$ axis, represented by the matrix $\begin{bmatrix} -1 & 0 \\ 0 & 1 \end{bmatrix}$

33. $\frac{\sqrt{2}}{2} \begin{bmatrix} 1 & -1 \\ 1 & 1 \end{bmatrix}$ 35. $\begin{bmatrix} 1 & 2 \\ 2 & 1 \end{bmatrix}$

37. $T(\vec{x}) = T(\vec{v}) + k\left(T(\vec{w}) - T(\vec{v})\right)$ is on the line segment.

41. $y = c_1 x_1 + c_2 x_2$; the graph is a plane through the origin in $\mathbb{R}^3$.

43. a. T is linear, represented by the matrix $\begin{bmatrix} 2 & 3 & 4 \end{bmatrix}$.

45. Yes; use Theorem 2.1.3.

47. Write $\vec{w} = c_1 \vec{v}_1 + c_2 \vec{v}_2$; then $T(\vec{w}) = c_1 T(\vec{v}_1) + c_2 T(\vec{v}_2)$

49. a. 37 SFr2 coins, and 14 SFr5 coins.

    b. $A = \begin{bmatrix} 2 & 5 \\ 1 & 1 \end{bmatrix}$

c. Yes. $A^{-1} = \begin{bmatrix} -\frac{1}{3} & \frac{5}{3} \\ \frac{1}{3} & -\frac{2}{3} \end{bmatrix}$

51. a. $\begin{bmatrix} \frac{5}{9} & -\frac{160}{9} \\ 0 & 1 \end{bmatrix}$

    b. Yes. $\begin{bmatrix} \frac{9}{5} & 32 \\ 0 & 1 \end{bmatrix}$, $F = \frac{9}{5}C + 32$

53. $A = \begin{bmatrix} 1 & 5/8 & 1/170 & 5/4 \\ 8/5 & 1 & 4/425 & 2 \\ 170 & 425/4 & 1 & 425/2 \\ 4/5 & 1/2 & 2/425 & 1 \end{bmatrix}$

2.2 Answers to more theoretical questions are omitted.

1. The image consists of the vectors $\begin{bmatrix} 3 \\ 1 \end{bmatrix}$ and $\begin{bmatrix} 2 \\ 4 \end{bmatrix}$.

3. The parallelogram in $\mathbb{R}^3$ defined by the vectors $T(\vec{e}_1)$ and $T(\vec{e}_2)$

5. About 2.5 radians   7. $\frac{1}{9} \begin{bmatrix} 11 \\ 1 \\ 11 \end{bmatrix}$

9. A vertical shear

11. $\frac{1}{25} \begin{bmatrix} 7 & 24 \\ 24 & -7 \end{bmatrix}$ 13. $\begin{bmatrix} 2u_1^2 - 1 & 2u_1 u_2 \\ 2u_1 u_2 & 2u_2^2 - 1 \end{bmatrix}$

15. $a_{ii} = 2u_i^2 - 1$, and $a_{ij} = 2u_i u_j$ when $i \neq j$

17. If $b \neq 0$, then let $\vec{v} = \begin{bmatrix} b \\ 1 - a \end{bmatrix}$ and $\vec{w} = \begin{bmatrix} -b \\ 1 + a \end{bmatrix}$, for example.

19. $\begin{bmatrix} 1 & 0 & 0 \\ 0 & 1 & 0 \\ 0 & 0 & 0 \end{bmatrix}$ 21. $\begin{bmatrix} 0 & -1 & 0 \\ 1 & 0 & 0 \\ 0 & 0 & 1 \end{bmatrix}$

23. $\begin{bmatrix} 1 & 0 & 0 \\ 0 & 0 & 1 \\ 0 & 1 & 0 \end{bmatrix}$

25. $\begin{bmatrix} 1 & -k \\ 0 & 1 \end{bmatrix}$; you shear back.

27. A is a reflection; B is a scaling; C is a projection; D is a rotation; E is a shear.

31. $A = \begin{bmatrix} -2 & 0 & 0 \\ 1 & 0 & 0 \\ 0 & 0 & 0 \end{bmatrix}$

33. Use a parallelogram.   35. Yes

37. a. 1   b. 0
    c. $-2 \leq \text{tr}(A) \leq 2$ d. 2

39. a. Projection with scaling
    b. Shear with scaling
    c. Reflection with scaling.

41. $\text{ref}_Q \vec{x} = -\text{ref}_P \vec{x}$

43. $\begin{bmatrix} \cos\theta & \sin\theta \\ -\sin\theta & \cos\theta \end{bmatrix}$, a clockwise rotation through the angle θ.

45. $A^{-1} = A$. The inverse of a reflection is the reflection itself.

47. Write $T(\vec{x}) = \begin{bmatrix} a & b \\ c & d \end{bmatrix} \vec{x}$. Express $f(t)$ in terms of a, b, c, d.

49. The image is an ellipse with semimajor axes $\pm 5\vec{e}_1$ and semiminor axes $\pm 2\vec{e}_2$.

51. The curve C is the image of the unit circle under the transformation with matrix $\begin{bmatrix} \vec{w}_1 & \vec{w}_2 \end{bmatrix}$.

2.3 Answers to more theoretical questions are omitted.

1. $\begin{bmatrix} 4 & 6 \\ 3 & 4 \end{bmatrix}$　　　　3. Undefined

5. $\begin{bmatrix} a & b \\ c & d \\ 0 & 0 \end{bmatrix}$　　　　7. $\begin{bmatrix} -1 & 1 & 0 \\ 5 & 3 & 4 \\ -6 & -2 & -4 \end{bmatrix}$

9. $\begin{bmatrix} 0 & 0 \\ 0 & 0 \end{bmatrix}$　　11. $[10]$　　13. $[h]$

15. $\left[\begin{array}{c|c} 1 & 0 \\ 2 & 0 \\ \hline 19 & 16 \end{array}\right]$

17. All matrices of the form $\begin{bmatrix} a & 0 \\ 0 & b \end{bmatrix}$.

19. All matrices of the form $\begin{bmatrix} a & b \\ -b & a \end{bmatrix}$.

21. All matrices of the form $\begin{bmatrix} a & b \\ b & a-b \end{bmatrix}$.

23. All matrices of the form $\begin{bmatrix} a & b \\ \frac{2}{3}b & \frac{5}{3}b+a \end{bmatrix}$.

25. All matrices of the form $\begin{bmatrix} a & 0 & b \\ 0 & c & 0 \\ d & 0 & e \end{bmatrix}$.

29. a. The matrices $D_\alpha D_\beta$ and $D_\beta D_\alpha$ both represent the counterclockwise rotation through the angle $\alpha + \beta$.

　　b. $D_\alpha D_\beta = D_\beta D_\alpha$
　　　$= \begin{bmatrix} \cos(\alpha+\beta) & -\sin(\alpha+\beta) \\ \sin(\alpha+\beta) & \cos(\alpha+\beta) \end{bmatrix}$

31. (ith row of AB) = (ith row of A)B

33. $A^n = I_2$ for even n and $A^n = -I_2$ for odd n. A represents the rotation through π.

35. $A^n = I_2$ for even n and $A^n = A$ for odd n. A represents the reflection about the line spanned by $\begin{bmatrix} 1 \\ 1 \end{bmatrix}$.

37. $A^n = \begin{bmatrix} 1 & 0 \\ -n & 1 \end{bmatrix}$. A represents a vertical shear.

39. A^n represents the rotation through $n\pi/4$ in the clockwise direction, so that $A^8 = I_2$. Now $A^{1001} = (A^8)^{125} A = A$.

41. $A^n = I_2$ for even n and $A^n = A$ for odd n. A represents the reflection about a line.

43. $A = -I_2$, for example, or any matrix representing a reflection about a line.

45. The matrix representing the rotation through $2\pi/3$, for example, $A = \begin{bmatrix} \cos(2\pi/3) & -\sin(2\pi/3) \\ \sin(2\pi/3) & \cos(2\pi/3) \end{bmatrix} = \frac{1}{2}\begin{bmatrix} -1 & -\sqrt{3} \\ \sqrt{3} & -1 \end{bmatrix}$

47. Any projection or reflection matrix with nonzero entries will do, for example, $A = \begin{bmatrix} 0.6 & 0.8 \\ 0.8 & -0.6 \end{bmatrix}$

49. $AF = \begin{bmatrix} 1 & 0 \\ 0 & -1 \end{bmatrix}$, $FA = \begin{bmatrix} -1 & 0 \\ 0 & 1 \end{bmatrix}$. We compose a rotation with a reflection to obtain a reflection.

51. $FJ = JF = \begin{bmatrix} -1 & -1 \\ 1 & -1 \end{bmatrix}$. Note that F represents the rotation through $\pi/2$ while J represents the rotation through $\pi/4$ combined with a scaling by $\sqrt{2}$. The products FJ and JF both represent the rotation through $3\pi/4$ combined with a scaling by $\sqrt{2}$.

53. $CD = \begin{bmatrix} 0 & -1 \\ 1 & 0 \end{bmatrix}$ and $DC = \begin{bmatrix} 0 & 1 \\ -1 & 0 \end{bmatrix}$. We compose two reflections to obtain a rotation.

55. $X = \begin{bmatrix} -2s & -2t \\ s & t \end{bmatrix}$, where s and t are arbitrary constants

57. $X = \begin{bmatrix} -5 & 2 \\ 3 & -1 \end{bmatrix}$

59. No such matrix X exists.

61. $X = \begin{bmatrix} 1+s & -2+t \\ -2s & 1-2t \\ s & t \end{bmatrix}$, where s and t are arbitrary constants

63. No such matrix X exists.

65. $X = \begin{bmatrix} 0 & t \\ 0 & 0 \end{bmatrix}$, where t is an arbitrary constant

67. $(A - I_2)^2 = 0$, since $2A\vec{x}$ is a diagonal of the parallelogram spanned by $\vec{x}$ and $A^2\vec{x}$, so that $2A\vec{x} = \vec{x} + A^2\vec{x}$ or $(A^2 - 2A + I_2)\vec{x} = \vec{0}$, for all $\vec{x}$

69. There is one and only one such X.

2.4 Answers to more theoretical questions are omitted.

1. $\begin{bmatrix} 8 & -3 \\ -5 & 2 \end{bmatrix}$　　　　3. $\begin{bmatrix} -\frac{1}{2} & 1 \\ \frac{1}{2} & 0 \end{bmatrix}$

5. Not invertible   7. Not invertible

9. Not invertible   11. $\begin{bmatrix} 1 & 0 & -1 \\ 0 & 1 & 0 \\ 0 & 0 & 1 \end{bmatrix}$

13. $\begin{bmatrix} 1 & 0 & 0 & 0 \\ -2 & 1 & 0 & 0 \\ 1 & -2 & 1 & 0 \\ 0 & 1 & -2 & 1 \end{bmatrix}$

15. $\begin{bmatrix} -6 & 9 & -5 & 1 \\ 9 & -1 & -5 & 2 \\ -5 & -5 & 9 & -3 \\ 1 & 2 & -3 & 1 \end{bmatrix}$

17. Not invertible

19. $x_1 = 3y_1 - 2.5y_2 + 0.5y_3$
    $x_2 = -3y_1 + 4y_2 - y_3$
    $x_3 = y_1 - 1.5y_2 + 0.5y_3$

21. Not invertible   23. Invertible

25. Invertible   27. Not invertible

29. For all k except $k = 1$ and $k = 2$

31. It's never invertible.

33. If $a^2 + b^2 = 1$

35. a. Invertible if a, d, f are all nonzero
    b. Invertible if all diagonal entries are nonzero
    c. Yes; use Theorem 2.4.5.
    d. Invertible if all diagonal entries are nonzero

37. $(cA)^{-1} = \dfrac{1}{c} A^{-1}$

39. M is invertible; if $m_{ij} = k$ (where $i \neq j$), then the ijth entry of M^{-1} is $-k$; all other entries are the same.

41. The transformations in parts (a), (c), (d) are invertible, while the projection in part (b) is not.

43. Yes; $\vec{x} = B^{-1}(A^{-1}\vec{y})$

45. a. $3^3 = 27$ b. n^3
    c. $\dfrac{12^3}{3^3} = 64$ (seconds)

47. $f(x) = x^2$ is not invertible, but the equation $f(x) = 0$ has the unique solution $x = 0$.

51. a. Since rank$(A) < n$, all the entries in the last row of E will be zero. We can let $\vec{c} = \vec{e}_n$; the system $E\vec{x} = \vec{e}_n$ will be inconsistent. Reversing the row reduction from A to E, we can transform the system $E\vec{x} = \vec{e}_n$ into an inconsistent system $A\vec{x} = \vec{b}$.
    b. We have rank$(A) \leq m < n$. Now use part a.

53. a. $\lambda_1 = 2, \lambda_2 = 6$.
    b. We chose $\lambda = 2$. $A - \lambda I_2 = \begin{bmatrix} 1 & 1 \\ 3 & 3 \end{bmatrix}, \vec{x} = \begin{bmatrix} -1 \\ 1 \end{bmatrix}$
    c. Check that $\begin{bmatrix} 3 & 1 \\ 3 & 5 \end{bmatrix} \vec{x} = 2\vec{x}$.

55. $\det A = \left\| \begin{bmatrix} 2 \\ 0 \end{bmatrix} \right\| \sin(\pi/2) \left\| \begin{bmatrix} 0 \\ 2 \end{bmatrix} \right\| = 4$, and $A^{-1} = \begin{bmatrix} 1/2 & 0 \\ 0 & 1/2 \end{bmatrix}$, a scaling by $^1/_2$

57. $\det A = \left\| \begin{bmatrix} \cos\alpha \\ \sin\alpha \end{bmatrix} \right\| \sin(-\pi/2) \left\| \begin{bmatrix} \sin\alpha \\ -\cos\alpha \end{bmatrix} \right\| = -1$, and $A^{-1} = A$, a reflection

59. $\det A = \left\| \begin{bmatrix} 0.6 \\ 0.8 \end{bmatrix} \right\| \sin(\pi/2) \left\| \begin{bmatrix} -0.8 \\ 0.6 \end{bmatrix} \right\| = 1$, and $A^{-1} = \begin{bmatrix} 0.6 & 0.8 \\ -0.8 & 0.6 \end{bmatrix}$, a rotation

61. $\det A = \left\| \begin{bmatrix} 1 \\ -1 \end{bmatrix} \right\| \sin(\pi/2) \left\| \begin{bmatrix} 1 \\ 1 \end{bmatrix} \right\| = 2$, and $A^{-1} = \dfrac{1}{2} \begin{bmatrix} 1 & -1 \\ 1 & 1 \end{bmatrix}$, a rotation through $\pi/4$ combined with a scaling by $\sqrt{2}/2$

63. $\det A = \left\| \begin{bmatrix} -3 \\ 4 \end{bmatrix} \right\| \sin(-\pi/2) \left\| \begin{bmatrix} 4 \\ 3 \end{bmatrix} \right\| = -25$, and $A^{-1} = \dfrac{1}{25} \begin{bmatrix} -3 & 4 \\ 4 & 3 \end{bmatrix}$, a reflection combined with a scaling by $1/5$

65. $\det A = \left\| \begin{bmatrix} 1 \\ 1 \end{bmatrix} \right\| \sin(\pi/4) \left\| \begin{bmatrix} 0 \\ 1 \end{bmatrix} \right\| = 1$, and $A^{-1} = \begin{bmatrix} 1 & 0 \\ -1 & 1 \end{bmatrix}$, a vertical shear

67. False   69. False   71. True

73. True   75. True

77. $A = BS^{-1}$ 79. $A = \dfrac{1}{5} \begin{bmatrix} 9 & 3 \\ -2 & 16 \end{bmatrix}$

81. $\begin{cases} \text{matrix of } T: & \begin{bmatrix} 0 & 0 & 1 \\ -1 & 0 & 0 \\ 0 & -1 & 0 \end{bmatrix} \\ \text{matrix of } L: & \begin{bmatrix} 0 & 1 & 0 \\ 1 & 0 & 0 \\ 0 & 0 & 1 \end{bmatrix} \end{cases}$

83. Yes; yes; each elementary row operation can be "undone" by an elementary row operation.

85. a. Use Exercise 84; let $S = E_1 E_2 \cdots E_p$
    b. $S = \begin{bmatrix} 1 & 0 \\ -4 & 1 \end{bmatrix} \begin{bmatrix} \frac{1}{2} & 0 \\ 0 & 1 \end{bmatrix} = \begin{bmatrix} \frac{1}{2} & 0 \\ -2 & 1 \end{bmatrix}$

87. $\begin{bmatrix} k & 0 \\ 0 & 1 \end{bmatrix}, \begin{bmatrix} 1 & 0 \\ 0 & k \end{bmatrix}, \begin{bmatrix} 1 & c \\ 0 & 1 \end{bmatrix}, \begin{bmatrix} 1 & 0 \\ c & 1 \end{bmatrix}, \begin{bmatrix} 0 & 1 \\ 1 & 0 \end{bmatrix}$,
    where k is nonzero and c is arbitrary. Cases 3 and 4 represent shears, and case 5 is a reflection.

89. Yes; use Theorem 2.3.4.

91. a. $\vec{y} = \begin{bmatrix} -3 \\ 5 \\ 2 \\ 0 \end{bmatrix}$ b. $\vec{x} = \begin{bmatrix} 1 \\ -1 \\ 2 \\ 0 \end{bmatrix}$

93. a. Write $A = \begin{bmatrix} A^{(m)} & A_2 \\ A_3 & A_4 \end{bmatrix}$, $L = \begin{bmatrix} L^{(m)} & 0 \\ L_3 & L_4 \end{bmatrix}$,

$U = \begin{bmatrix} U^{(m)} & U_2 \\ 0 & U_4 \end{bmatrix}$. Use Theorem 2.3.9.

c. Solve the equation

$$A = \begin{bmatrix} A^{(n-1)} & \vec{v} \\ \vec{w} & k \end{bmatrix} = \begin{bmatrix} L' & 0 \\ \vec{x} & t \end{bmatrix} \begin{bmatrix} U' & \vec{y} \\ 0 & 1 \end{bmatrix}$$

for $\vec{x}$, $\vec{y}$, and t.

95. A is invertible if both A_{11} and A_{22} are invertible. In this case,

$$A^{-1} = \begin{bmatrix} A_{11}^{-1} & 0 \\ 0 & A_{22}^{-1} \end{bmatrix}$$

97. $\text{rank}(A) = \text{rank}(A_{11}) + \text{rank}(A_{23})$

99. Only $A = I_n$

101. (ijth entry of AB) $= \sum_{k=1}^{n} a_{ik}b_{kj} \le s \sum_{k=1}^{n} b_{kj} \le sr$

107. $g(f(x)) = x$, for all x.

$f(g(x)) = \begin{cases} x & \text{if } x \text{ is even} \\ x+1 & \text{if } x \text{ is odd} \end{cases}$

The functions f and g are not invertible.

CHAPTER 3

3.1 Answers to more theoretical questions are omitted.

1. $\ker(A) = \{\vec{0}\}$

3. $\vec{e}_1, \vec{e}_2$

5. $\begin{bmatrix} 1 \\ -2 \\ 1 \end{bmatrix}$

7. $\ker(A) = \{\vec{0}\}$

9. $\ker(A) = \{\vec{0}\}$

11. $\begin{bmatrix} -2 \\ 3 \\ 1 \\ 0 \end{bmatrix}$

13. $\begin{bmatrix} -2 \\ 1 \\ 0 \\ 0 \\ 0 \\ 0 \end{bmatrix}, \begin{bmatrix} -3 \\ 0 \\ -2 \\ -1 \\ 1 \\ 0 \end{bmatrix}, \begin{bmatrix} 0 \\ 0 \\ 0 \\ 0 \\ 0 \\ 1 \end{bmatrix}$

15. $\begin{bmatrix} 1 \\ 1 \end{bmatrix}, \begin{bmatrix} 1 \\ 2 \end{bmatrix}$

17. All of $\mathbb{R}^2$

19. The line spanned by $\begin{bmatrix} 1 \\ -2 \end{bmatrix}$

21. All of $\mathbb{R}^3$

23. kernel is $\{\vec{0}\}$, image is all of $\mathbb{R}^2$

25. Same as Exercise 23

27. $f(x) = x^3 - x$

29. $f\begin{bmatrix} \phi \\ \theta \end{bmatrix} = \begin{bmatrix} \sin\phi\cos\theta \\ \sin\phi\sin\theta \\ \cos\phi \end{bmatrix}$

(compare with spherical coordinates)

31. $A = \begin{bmatrix} -2 & -3 \\ 0 & 1 \\ 1 & 0 \end{bmatrix}$

33. $T\begin{bmatrix} x \\ y \\ z \end{bmatrix} = x + 2y + 3z$

35. $\ker(T)$ is the plane with normal vector $\vec{v}$; $\text{im}(T) = \mathbb{R}$.

37. $\text{im}(A) = \text{span}(\vec{e}_1, \vec{e}_2)$; $\ker(A) = \text{span}(\vec{e}_1)$; $\text{im}(A^2) = \text{span}(\vec{e}_1)$; $\ker(A^2) = \text{span}(\vec{e}_1, \vec{e}_2)$; $A^3 = 0$ so $\ker(A^3) = \mathbb{R}^3$ and $\text{im}(A^3) = \{\vec{0}\}$

39. a. $\ker(B)$ is contained in $\ker(AB)$, but they need not be equal.

b. $\text{im}(AB)$ is contained in $\text{im}(A)$, but they need not be equal.

41. a. $\text{im}(A)$ is the line spanned by $\begin{bmatrix} 3 \\ 4 \end{bmatrix}$, and $\ker(A)$ is the perpendicular line, spanned by $\begin{bmatrix} -4 \\ 3 \end{bmatrix}$.

b. $A^2 = A$; if $\vec{v}$ is in $\text{im}(A)$, then $A\vec{v} = \vec{v}$.

c. Orthogonal projection onto the line spanned by $\begin{bmatrix} 3 \\ 4 \end{bmatrix}$.

43. Suppose A is an $n \times m$ matrix of rank r. Let B be the matrix you get when you omit the first r rows and the first m columns of rref $\begin{bmatrix} A & \vdots & I_n \end{bmatrix}$. (What can you do when $r = n$?)

45. There are $m - r$ nonleading variables, which can be chosen freely. The general vector in the kernel can be written as a linear combination of $m - r$ vectors, with the nonleading variables as coefficients.

47. $\text{im}(T) = L_2$ and $\ker(T) = L_1$

51. $\ker(AB) = \{\vec{0}\}$

53. a. $\begin{bmatrix} 1 \\ 1 \\ 1 \\ 1 \\ 0 \\ 0 \\ 0 \end{bmatrix}, \begin{bmatrix} 0 \\ 1 \\ 1 \\ 0 \\ 1 \\ 0 \\ 0 \end{bmatrix}, \begin{bmatrix} 1 \\ 0 \\ 1 \\ 0 \\ 0 \\ 1 \\ 0 \end{bmatrix}, \begin{bmatrix} 1 \\ 1 \\ 0 \\ 0 \\ 0 \\ 0 \\ 1 \end{bmatrix}$

b. $\ker(H) = \text{span}(\vec{v}_1, \vec{v}_2, \vec{v}_3, \vec{v}_4)$, by part (a), and $\text{im}(M) = \text{span}(\vec{v}_1, \vec{v}_2, \vec{v}_3, \vec{v}_4)$, by Theorem 3.1.3. Thus $\ker(H) = \text{im}(M)$. $H(M\vec{x}) = \vec{0}$, since $M\vec{x}$ is in $\text{im}(M) = \ker(H)$.

3.2 Answers to more theoretical questions are omitted.

1. Not a subspace.   3. W is a subspace.

7. Yes   9. Dependent

11. Independent   13. Dependent

15. Dependent   17. Independent

19. $\begin{bmatrix} 2 \\ 0 \\ 0 \\ 0 \end{bmatrix}$ and $\begin{bmatrix} 3 \\ 4 \\ 5 \\ 0 \end{bmatrix}$ are redundant.

21. $\vec{v}_2 = \vec{v}_1$, or, $\vec{v}_1 - \vec{v}_2 = \vec{0}$, so that $\begin{bmatrix} 1 \\ -1 \end{bmatrix}$ is in the kernel.

23. $\vec{v}_1 = \vec{0}$, so that $\begin{bmatrix} 1 \\ 0 \end{bmatrix}$ is in the kernel.

25. $\vec{v}_3 = \vec{v}_1$, or, $\vec{v}_1 - \vec{v}_3 = \vec{0}$, so that $\begin{bmatrix} 1 \\ 0 \\ -1 \end{bmatrix}$ is in the kernel.

27. $\begin{bmatrix} 1 \\ 1 \\ 1 \end{bmatrix}, \begin{bmatrix} 1 \\ 2 \\ 3 \end{bmatrix}$ 29. $\begin{bmatrix} 1 \\ 4 \end{bmatrix}, \begin{bmatrix} 2 \\ 5 \end{bmatrix}$

31. $\begin{bmatrix} 1 \\ 2 \\ 3 \\ 5 \end{bmatrix}, \begin{bmatrix} 5 \\ 6 \\ 7 \\ 8 \end{bmatrix}$ 33. $\begin{bmatrix} 1 \\ 0 \\ 0 \\ 0 \end{bmatrix}, \begin{bmatrix} 0 \\ 1 \\ 0 \\ 0 \end{bmatrix}, \begin{bmatrix} 0 \\ 0 \\ 1 \\ 0 \end{bmatrix}$

35. Suppose there is a non-trivial relation $c_1 \vec{v}_1 + \cdots + c_i \vec{v}_i + \cdots + c_m \vec{v}_m = \vec{0}$, with $c_i \neq 0$. We can solve this equation for $\vec{v}_i$ and thus express $\vec{v}_i$ as a linear combination of the other vectors in the list.
    Conversely, if $\vec{v}_i$ is a linear combination of the other vectors, $\vec{v}_i = \cdots$, then we can substract $\vec{v}_i$ from both sides of the equation to obtain a non-trivial relation (the coefficient of $\vec{v}_i$ will be -1).

37. The vectors $T(\vec{v}_1), \ldots, T(\vec{v}_m)$ are not necessarily independent.

39. The vectors $\vec{v}_1, \ldots, \vec{v}_m, \vec{v}$ are linearly independent.

41. The columns of B are linearly independent, while the columns of A are dependent.

43. The vectors are linearly independent.

45. Yes   47. $\begin{bmatrix} 1 & 0 & 0 \\ 0 & 1 & 0 \\ 0 & 0 & 1 \\ 0 & 0 & 0 \end{bmatrix}$

49. $L = \text{im} \begin{bmatrix} 1 \\ 1 \\ 1 \end{bmatrix} = \ker \begin{bmatrix} 1 & 0 & -1 \\ 0 & 1 & -1 \end{bmatrix}$

51. a. Consider a relation $c_1 \vec{v}_1 + \cdots + c_p \vec{v}_p + d_1 \vec{w}_1 + \cdots + d_q \vec{w}_q = \vec{0}$. Then $c_1 \vec{v}_1 + \cdots + c_p \vec{v}_p = -d_1 \vec{w}_1 - \cdots - d_q \vec{w}_q$ is $\vec{0}$, because this vector is both in V and in W. The claim follows.

b. From part (a) we know that the vectors $\vec{v}_1, \ldots, \vec{v}_p, \vec{w}_1, \ldots, \vec{w}_q$ are linearly independent. Consider a vector $\vec{x}$ in $V + W$. By definition of $V + W$ we can write $\vec{x} = \vec{v} + \vec{w}$ for a $\vec{v}$ in V and a $\vec{w}$ in W. The $\vec{v}$ is a linear combination of the $\vec{v}_i$, and $\vec{w}$ is a linear combination of the $\vec{w}_j$. This shows that the vectors $\vec{v}_1, \ldots, \vec{v}_p, \vec{w}_1, \ldots, \vec{w}_q$ span $V + W$.

55. $\begin{bmatrix} -2 \\ 1 \\ 0 \\ 0 \\ 0 \end{bmatrix}, \begin{bmatrix} -3 \\ 0 \\ 1 \\ 0 \\ 0 \end{bmatrix}, \begin{bmatrix} -4 \\ 0 \\ 0 \\ 1 \\ 0 \end{bmatrix}, \begin{bmatrix} -5 \\ 0 \\ 0 \\ 0 \\ 1 \end{bmatrix}$

57. For $j = 1, 3, 6$, and 7, corresponding to the columns that do not contain leading 1's.

3.3 Answers to more theoretical questions are omitted.

1. $\vec{v}_2 = 3\vec{v}_1$; basis of image: $\begin{bmatrix} 1 \\ 2 \end{bmatrix}$;
    basis of kernel: $\begin{bmatrix} -3 \\ 1 \end{bmatrix}$.

3. No redundant vectors; basis of image: $\begin{bmatrix} 1 \\ 3 \end{bmatrix}, \begin{bmatrix} 2 \\ 4 \end{bmatrix}$;
    basis of kernel: $\varnothing$

5. $\vec{v}_3 = 3\vec{v}_1$; basis of image: $\begin{bmatrix} 1 \\ 2 \end{bmatrix}, \begin{bmatrix} -2 \\ 4 \end{bmatrix}$;
    basis of kernel: $\begin{bmatrix} -3 \\ 0 \\ 1 \end{bmatrix}$.

7. $\vec{v}_2 = 2\vec{v}_1$; basis of image: $\begin{bmatrix} 1 \\ 1 \end{bmatrix}, \begin{bmatrix} 3 \\ 4 \end{bmatrix}$;
    basis of kernel: $\begin{bmatrix} -2 \\ 1 \\ 0 \end{bmatrix}$.

9. $\vec{v}_2 = 2\vec{v}_1$; basis of image: $\begin{bmatrix} 1 \\ 1 \\ 1 \end{bmatrix}, \begin{bmatrix} 1 \\ 2 \\ 3 \end{bmatrix}$;
    basis of kernel: $\begin{bmatrix} -2 \\ 1 \\ 0 \end{bmatrix}$.

11. $\vec{v}_3 = \vec{v}_1$; basis of image: $\begin{bmatrix} 1 \\ 0 \\ 0 \end{bmatrix}, \begin{bmatrix} 0 \\ 1 \\ 1 \end{bmatrix}$;
    basis of kernel: $\begin{bmatrix} -1 \\ 0 \\ 1 \end{bmatrix}$.

13. $\vec{v}_2 = 2\vec{v}_1, \vec{v}_3 = 3\vec{v}_1$; basis of image: $\begin{bmatrix} 1 \end{bmatrix}$;
    basis of kernel: $\begin{bmatrix} -2 \\ 1 \\ 0 \end{bmatrix}, \begin{bmatrix} -3 \\ 0 \\ 1 \end{bmatrix}$.

15. $\vec{v}_3 = 2\vec{v}_1 + 2\vec{v}_2, \vec{v}_4 = \vec{0}$;

basis of image: $\begin{bmatrix} 1 \\ 0 \\ 1 \\ 0 \end{bmatrix}, \begin{bmatrix} 0 \\ 1 \\ 0 \\ 1 \end{bmatrix}$;

basis of kernel: $\begin{bmatrix} -2 \\ -2 \\ 1 \\ 0 \end{bmatrix}, \begin{bmatrix} 0 \\ 0 \\ 0 \\ 1 \end{bmatrix}$.

17. $\vec{v}_1 = \vec{0}, \vec{v}_3 = 2\vec{v}_2, \vec{v}_5 = 3\vec{v}_2 + 4\vec{v}_4$;

basis of image: $\begin{bmatrix} 1 \\ 0 \end{bmatrix}, \begin{bmatrix} 0 \\ 1 \end{bmatrix}$;

basis of kernel: $\begin{bmatrix} 1 \\ 0 \\ 0 \\ 0 \\ 0 \end{bmatrix}, \begin{bmatrix} 0 \\ -2 \\ 1 \\ 0 \\ 0 \end{bmatrix}, \begin{bmatrix} 0 \\ -3 \\ 0 \\ -4 \\ 1 \end{bmatrix}$.

19. $\vec{v}_3 = 5\vec{v}_1 + 4\vec{v}_2, \vec{v}_4 = 3\vec{v}_1 + 2\vec{v}_2$;

basis of image: $\begin{bmatrix} 1 \\ 0 \\ 0 \\ 0 \end{bmatrix}, \begin{bmatrix} 0 \\ 1 \\ 0 \\ 0 \end{bmatrix}, \begin{bmatrix} 0 \\ 0 \\ 1 \\ 0 \end{bmatrix}$;

basis of kernel: $\begin{bmatrix} -5 \\ -4 \\ 1 \\ 0 \\ 0 \end{bmatrix}, \begin{bmatrix} -3 \\ -2 \\ 0 \\ 1 \\ 0 \end{bmatrix}$.

21. $\text{rref}(A) = \begin{bmatrix} 1 & 0 & -3 \\ 0 & 1 & 4 \\ 0 & 0 & 0 \end{bmatrix}$;

basis of image: $\begin{bmatrix} 1 \\ 4 \\ 7 \end{bmatrix}, \begin{bmatrix} 3 \\ 5 \\ 6 \end{bmatrix}$;

basis of kernel: $\begin{bmatrix} 3 \\ -4 \\ 1 \end{bmatrix}$.

23. $\text{rref}(A) = \begin{bmatrix} 1 & 0 & 2 & 4 \\ 0 & 1 & -3 & -1 \\ 0 & 0 & 0 & 0 \\ 0 & 0 & 0 & 0 \end{bmatrix}$;

basis of image: $\begin{bmatrix} 1 \\ 0 \\ 3 \\ 0 \end{bmatrix}, \begin{bmatrix} 0 \\ 1 \\ 4 \\ -1 \end{bmatrix}$;

basis of kernel: $\begin{bmatrix} -2 \\ 3 \\ 1 \\ 0 \end{bmatrix}, \begin{bmatrix} -4 \\ 1 \\ 0 \\ 1 \end{bmatrix}$.

25. $\text{rref}(A) = \begin{bmatrix} 1 & 2 & 0 & 5 & 0 \\ 0 & 0 & 1 & -1 & 0 \\ 0 & 0 & 0 & 0 & 1 \\ 0 & 0 & 0 & 0 & 0 \\ 0 & 0 & 0 & 0 & 0 \end{bmatrix}$;

basis of image: $\begin{bmatrix} 1 \\ 3 \\ 1 \\ 2 \end{bmatrix}, \begin{bmatrix} 3 \\ 9 \\ 4 \\ 9 \end{bmatrix}, \begin{bmatrix} 1 \\ 3 \\ 2 \\ 2 \end{bmatrix}$;

basis of kernel: $\begin{bmatrix} -2 \\ 1 \\ 0 \\ 0 \\ 0 \end{bmatrix}, \begin{bmatrix} -5 \\ 0 \\ 1 \\ 1 \\ 0 \end{bmatrix}$.

27. They do.

29. $\begin{bmatrix} -3 \\ 2 \\ 0 \end{bmatrix}, \begin{bmatrix} -1 \\ 0 \\ 2 \end{bmatrix}$

31. $A = \begin{bmatrix} 1 & -2 & -4 \\ 1 & 0 & 0 \\ 0 & 1 & 0 \\ 0 & 0 & 1 \end{bmatrix}$

33. The dimension of a hyperplane in $\mathbb{R}^n$ is $n - 1$.

35. The dimension is $n - 1$.

37. $A = \begin{bmatrix} 1 & 0 & 0 & 0 & 0 \\ 0 & 1 & 0 & 0 & 0 \\ 0 & 0 & 0 & 0 & 0 \\ 0 & 0 & 0 & 0 & 0 \end{bmatrix}$

39. $\text{ker}(C)$ is at least 1-dimensional, and $\text{ker}(C)$ is contained in $\text{ker}(A)$.

41. To fit a conic through a given point $P_j(x_j, y_j)$, we need to solve the equation $c_1 + c_2 x_j + c_3 y_j + c_4 x_j^2 + c_5 x_j y_j + c_6 y_j^2 = 0$, a homogeneous linear equation in the six unknowns $c_1, \ldots, c_6$. Thus, fitting a conic to four given points $P_1(x_1, y_1), \ldots, P_4(x_4, y_4)$ amounts to solving a system of four homogeneous linear equations with six unknowns. This in turn amounts to finding the kernel of a 4×6 matrix A. This kernel is at least two-dimensional. Since every one-dimensional subspace of $\text{ker } A$ defines a unique conic (see Exercise 40), there will be infinitely many such conics.

43. Building on our work in Exercise 41, we observe that fitting a conic through 6 points amounts to finding the kernel of a 6×6 matrix A. There will be no such conic if $\text{ker } A = \{\vec{0}\}$, one conic if $\dim \text{ker } A = 1$, and infinitely many conics if $\dim \text{ker } A > 1$. To give an example for each case, recall Exercise 1.2.51, where we showed that the unique conic $xy = 0$ runs through the points $(0, 0), (1, 0), (2, 0), (0, 1), (0, 2)$. Thus, there is no conic through the points $(0, 0), (1, 0), (2, 0), (0, 1), (0, 2), (1, 1)$, whereas the only conic

through $(0, 0)$, $(1, 0)$, $(2, 0)$, $(0, 1)$, $(0, 2)$, $(0, 3)$ is $xy = 0$. There are infinitely many conics through $(0, 0)$, $(1, 0)$, $(2, 0)$, $(3, 0)$, $(4, 0)$, $(5, 0)$.

45. $x^2y - xy^2 = xy(x - y) = 0$

47. No such cubic exists, since the unique cubic through the first 9 points does not pass through $(2, 1)$. See Exercise 45.

49. $a(xy^2 - xy) + b(x^2y - xy) = axy(y - 1)$
 $+ bxy(x - 1) = 0$, where $a \neq 0$ or $b \neq 0$

51. $2y - 3y^2 + y^3 = y(y - 1)(y - 2) = 0$

53. $2x - 2y - 3x^2 + 3y^2 + x^3 - y^3 = 0$, or,
 $x(x - 1)(x - 2) = y(y - 1)(y - 2)$

55. See Exercises 41 and 54. Since the kernel of the 8×10 matrix A is at least two-dimensional, and, because every one-dimensional subspace of ker A defines a unique cubic (compare with Exercise 40), there will be infinitely many such cubics.

57. There may be no such cubic [as in Exercise 47], exactly one [take the 9 points in Exercise 45 and add $(-1, -1)$], or infinitely many [as in Exercise 49].

61. A basis of V is also a basis of W, by Theorem 3.3.4c.

63. $\dim(V + W) = \dim(V) + \dim(W)$, by Exercise 3.2.51.

65. The first p columns of rref(A) contain leading 1's because the $\vec{v}_i$ are linearly independent. Now apply Theorem 3.3.5.

69. $\begin{bmatrix} 0 & 1 & 0 & 2 & 0 \end{bmatrix}$, $\begin{bmatrix} 0 & 0 & 1 & 3 & 0 \end{bmatrix}$,
 $\begin{bmatrix} 0 & 0 & 0 & 0 & 1 \end{bmatrix}$

71. a. A and E have the same row space, since elementary row operations leave the row space unchanged.
 b. rank$(A) = \dim(\text{rowspace}(A))$, by part (a) and Exercise 70.

75. Suppose rank$(A) = n$. The submatrix of A consisting of the n pivot columns of A is invertible, since the pivot columns are linearly independent.
 Conversely, if A has an invertible $n \times n$ submatrix, then the columns of that submatrix span $\mathbb{R}^n$, so im$(A) = \mathbb{R}^n$ and rank$(A) = n$.

77. Let m be the smallest number such that $A^m = 0$. By Exercise 76 there are m linearly independent vectors in $\mathbb{R}^n$; therefore, $m \leq n$, and $A^n = A^m A^{n-m} = 0$.

81. a. 3, 4, or 5   b. 0, 1, or 2

83. a. rank$(AB) \leq$ rank(A) b. rank$(AB) \leq$ rank(B)

3.4 Answers to more theoretical questions are omitted.

1. $[\vec{x}]_{\mathcal{B}} = \begin{bmatrix} 2 \\ 3 \end{bmatrix}$ 3. $[\vec{x}]_{\mathcal{B}} = \begin{bmatrix} 0 \\ 1 \end{bmatrix}$

5. $[\vec{x}]_{\mathcal{B}} = \begin{bmatrix} -4 \\ 3 \end{bmatrix}$ 7. $[\vec{x}]_{\mathcal{B}} = \begin{bmatrix} 3 \\ 4 \end{bmatrix}$

9. $\vec{x}$ isn't in V.

11. $[\vec{x}]_{\mathcal{B}} = \begin{bmatrix} 1/2 \\ 1/2 \end{bmatrix}$

13. $[\vec{x}]_{\mathcal{B}} = \begin{bmatrix} 1 \\ -1 \\ 0 \end{bmatrix}$

15. $[\vec{x}]_{\mathcal{B}} = \begin{bmatrix} 8 \\ -12 \\ 5 \end{bmatrix}$

17. $[\vec{x}]_{\mathcal{B}} = \begin{bmatrix} 1 \\ 1 \\ -1 \end{bmatrix}$

19. $B = \begin{bmatrix} 1 & 0 \\ 0 & -1 \end{bmatrix}$

21. $B = \begin{bmatrix} 7 & 0 \\ 0 & 0 \end{bmatrix}$

23. $B = \begin{bmatrix} 2 & 0 \\ 0 & -1 \end{bmatrix}$

25. $B = \begin{bmatrix} -1 & -1 \\ 4 & 6 \end{bmatrix}$

27. $B = \begin{bmatrix} 9 & 0 & 0 \\ 0 & 0 & 0 \\ 0 & 0 & 0 \end{bmatrix}$

29. $B = \begin{bmatrix} 0 & 0 & 0 \\ 0 & 1 & 0 \\ 0 & 0 & 2 \end{bmatrix}$

31. $B = \begin{bmatrix} 0 & 0 & 1 \\ 0 & 0 & 0 \\ -1 & 0 & 0 \end{bmatrix}$

33. $B = \begin{bmatrix} 0 & 0 & 0 \\ 0 & 1 & 0 \\ 0 & 0 & 0 \end{bmatrix}$

35. $B = \begin{bmatrix} 1 & 0 & 0 \\ -2 & 1 & 0 \\ 0 & 0 & 1 \end{bmatrix}$

37. $\begin{bmatrix} 1 \\ 2 \end{bmatrix}, \begin{bmatrix} -2 \\ 1 \end{bmatrix}$, for example

39. $\begin{bmatrix} 1 \\ 2 \\ 3 \end{bmatrix}, \begin{bmatrix} -2 \\ 1 \\ 0 \end{bmatrix}, \begin{bmatrix} -3 \\ 0 \\ 1 \end{bmatrix}$, for example

41. $\begin{bmatrix} 3 \\ 1 \\ 2 \end{bmatrix}, \begin{bmatrix} -1 \\ 3 \\ 0 \end{bmatrix}, \begin{bmatrix} 0 \\ -2 \\ 1 \end{bmatrix}$, for example

43. $\vec{x} = \begin{bmatrix} 4 \\ -3 \\ 2 \end{bmatrix}$

45. If $\vec{v}$ is any vector in the plane that is not parallel to $\vec{x}$, then $\vec{v}, \frac{1}{3}(\vec{x} - 2\vec{v})$ is a basis with the desired property. For example, $\vec{v} = \begin{bmatrix} 3 \\ 2 \\ 0 \end{bmatrix}$ gives the basis $\begin{bmatrix} 3 \\ 2 \\ 0 \end{bmatrix}, \frac{1}{3}\begin{bmatrix} -4 \\ -4 \\ -1 \end{bmatrix}$.

47. $A = \begin{bmatrix} d & c \\ b & a \end{bmatrix}$

49. $\begin{bmatrix} -1 \\ -1 \end{bmatrix}$

53. $\vec{x} = \begin{bmatrix} 40 \\ 58 \end{bmatrix}$

55. $\frac{1}{2}\begin{bmatrix} -1 & 2 \\ 1 & 0 \end{bmatrix}$

57. Consider a basis with two vectors parallel to the plane, and one vector perpendicular.

59. Yes

61. $\begin{bmatrix} -9 \\ 6 \end{bmatrix}$, $\begin{bmatrix} 0 \\ 1 \end{bmatrix}$, for example

63. Yes

67. $B = \begin{bmatrix} 0 & bc - ad \\ 1 & a + d \end{bmatrix}$

69. If $S = \begin{bmatrix} 1 & 2 \\ 2 & 1 \end{bmatrix}$, then $S^{-1}AS = \begin{bmatrix} 3 & 0 \\ 0 & -1 \end{bmatrix}$.

71. a. If $B\vec{x} = S^{-1}AS\vec{x} = \vec{0}$, then $A(S\vec{x}) = \vec{0}$.

    b. Since we have the p linearly independent vectors $S\vec{v}_1, S\vec{v}_2, \ldots, S\vec{v}_p$ in ker(A), we know that dim(ker B) $= p \le$ dim(ker A), by Theorem 3.3.4a. Reversing the roles of A and B, we find that dim(ker A) $\le$ dim(ker B). Thus, nullity(A) = dim(ker A) = dim(ker B) = nullity(B).

73. $\begin{bmatrix} 0.36 & 0.48 & 0.8 \\ 0.48 & 0.64 & -0.6 \\ -0.8 & 0.6 & 0 \end{bmatrix}$

75. $\begin{bmatrix} 0 & -1 \\ 1 & 0 \end{bmatrix}$

77. $b_{ij} = a_{n+1-i,n+1-j}$

CHAPTER 4

4.1 Answers to more theoretical questions are omitted.

1. Not a subspace

3. Subspace with basis $1 - t, 2 - t^2$

5. Subspace with basis t 7. Subspace

9. Not a subspace   11. Not a subspace

13. Not a subspace   15. Subspace

17. Matrices with one entry equal to 1 and all other entries equal to 0. The dimension is mn.

19. A basis is $\begin{bmatrix} 1 \\ 0 \end{bmatrix}$, $\begin{bmatrix} i \\ 0 \end{bmatrix}$, $\begin{bmatrix} 0 \\ 1 \end{bmatrix}$, $\begin{bmatrix} 0 \\ i \end{bmatrix}$, so that the dimension is 4.

21. A basis is $\begin{bmatrix} 1 & 0 \\ 0 & 0 \end{bmatrix}$, $\begin{bmatrix} 0 & 0 \\ 0 & 1 \end{bmatrix}$, so that the dimension is 2.

23. A basis is $\begin{bmatrix} 1 & 0 \\ 0 & 0 \end{bmatrix}$, $\begin{bmatrix} 0 & 0 \\ 1 & 0 \end{bmatrix}$, $\begin{bmatrix} 0 & 0 \\ 0 & 1 \end{bmatrix}$, so that the dimension is 3.

25. A basis is $1 - t, 1 - t^2$, so that the dimension is 2.

27. A basis is $\begin{bmatrix} 1 & 0 \\ 0 & 0 \end{bmatrix}$, $\begin{bmatrix} 0 & 0 \\ 0 & 1 \end{bmatrix}$, so that the dimension is 2.

29. A basis is $\begin{bmatrix} -1 & 1 \\ 0 & 0 \end{bmatrix}$, $\begin{bmatrix} 0 & 0 \\ -1 & 1 \end{bmatrix}$, so that the dimension is 2.

31. A basis is $\begin{bmatrix} 1 & 0 \\ 1 & 0 \end{bmatrix}$, $\begin{bmatrix} 0 & -1 \\ 0 & 1 \end{bmatrix}$, so that the dimension is 2.

33. Only the zero matrix has this property, so that the basis is $\varnothing$, and the dimension is 0.

35. A basis is $\begin{bmatrix} 1 & 0 & 0 \\ 0 & 0 & 0 \\ 0 & 0 & 0 \end{bmatrix}$, $\begin{bmatrix} 0 & 0 & 0 \\ 0 & 1 & 0 \\ 0 & 0 & 0 \end{bmatrix}$, $\begin{bmatrix} 0 & 0 & 0 \\ 0 & 0 & 0 \\ 0 & 0 & 1 \end{bmatrix}$, and the dimension is 3.

37. 3, 5, or 9

39. $\displaystyle\sum_{k=1}^{n} k = \frac{n(n + 1)}{2}$

41. 0, 3, 6, or 9

43. 2   45. dim$(V) = 3$

47. Yes and yes   49. Yes

51. $f(x) = ae^{3x} + be^{4x}$

4.2 Answers to more theoretical questions are omitted.

1. Nonlinear

3. Linear, not an isomorphism

5. Nonlinear   7. Isomorphism

9. Isomorphism   11. Isomorphism

13. Linear, not an isomorphism

15. Isomorphism

17. Linear, not an isomorphism

19. Isomorphism   21. Isomorphism

23. Linear, not an isomorphism

25. Linear, not an isomorphism

27. Isomorphism

29. Linear, not an isomorphism

31. Linear, not an isomorphism

33. Linear, not an isomorphism

35. Linear, not an isomorphism

37. Linear, not an isomorphism

39. Linear, not an isomorphism

41. Nonlinear   43. Isomorphism

45. Linear, not an isomorphism

47. Linear, not an isomorphism

49. Linear, not an isomorphism

51. ker(T) consists of all matrices of the form $\begin{bmatrix} a & b \\ 0 & a \end{bmatrix}$, so that the nullity is 2.

53. The image consists of all linear functions, of the form $mt + b$, so that the rank is 2. The kernel consists of the constant functions, so that the nullity is 1.

55. The image consists of all infinite sequences, and the kernel consists of all sequences of the form $(0, x_1, 0, x_3, 0, x_5, \ldots)$.

57. The kernel consists of all functions of the form $ae^{2t} + be^{3t}$, so that the nullity is 2.

59. The kernel has the basis $t - 7$, $(t - 7)^2$, so that the nullity is 2. The image is all of $\mathbb{R}$, so that the rank is 1.

61. The kernel consists of the zero function alone, and the image consists of all polynomials $g(t)$ whose constant term is zero [that is, $g(0) = 0$].

63. Impossible, since $\dim(P_3) \neq \dim(\mathbb{R}^3)$.

65. b. $\ker(T)$ consists of the zero matrix alone.
    d. This dimension is mn.

67. For all k except $k = 2$ and $k = 4$.

69. No; if $B = S^{-1}AS$, then $T(S) = 0$.

71. Yes, there is exactly one such polynomial.

73. Yes   77. Yes, and yes.

79. Surprisingly, yes.

83. The transformation T induces a transformation $\tilde{T}$ from $\ker(L \circ T)$ to $\ker L$, with $\ker \tilde{T} = \ker T$. Applying the rank-nullity theorem as stated in Exercise 82 to $\tilde{T}$, we find that $\dim \ker(L \circ T) = \dim \ker \tilde{T} + \dim \operatorname{im} \tilde{T} \leq \dim \ker T + \dim \ker L$, since $\operatorname{im} \tilde{T}$ is a subspace of the kernel of L.

4.3 Answers to more theoretical questions are omitted.

1. Yes   3. Yes

5. $\begin{bmatrix} 1 & 0 & 0 \\ 0 & 1 & 2 \\ 0 & 0 & 3 \end{bmatrix}$ 7. $\begin{bmatrix} 0 & 0 & 0 \\ 0 & 0 & 4 \\ 0 & 0 & 0 \end{bmatrix}$

9. $\begin{bmatrix} 1 & 0 & 0 \\ 0 & 2 & 0 \\ 0 & 0 & 1 \end{bmatrix}$ 11. $\begin{bmatrix} 1 & 0 & 0 \\ 0 & 1 & 0 \\ 0 & 0 & 3 \end{bmatrix}$

13. $\begin{bmatrix} 1 & 0 & 1 & 0 \\ 0 & 1 & 0 & 1 \\ 2 & 0 & 2 & 0 \\ 0 & 2 & 0 & 2 \end{bmatrix}$ 15. $\begin{bmatrix} 1 & 0 \\ 0 & -1 \end{bmatrix}$

17. $\begin{bmatrix} 0 & -1 \\ 1 & 0 \end{bmatrix}$ 19. $\begin{bmatrix} p & -q \\ q & p \end{bmatrix}$

21. $\begin{bmatrix} -3 & 1 & 0 \\ 0 & -3 & 2 \\ 0 & 0 & -3 \end{bmatrix}$ 23. $\begin{bmatrix} 1 & 3 & 9 \\ 0 & 0 & 0 \\ 0 & 0 & 0 \end{bmatrix}$

25. $\begin{bmatrix} 1 & 0 & 0 \\ 0 & -1 & 0 \\ 0 & 0 & 1 \end{bmatrix}$ 27. $\begin{bmatrix} 1 & -1 & 1 \\ 0 & 2 & -4 \\ 0 & 0 & 4 \end{bmatrix}$

29. $\begin{bmatrix} 2 & 2 & 8/3 \\ 0 & 0 & 0 \\ 0 & 0 & 0 \end{bmatrix}$ 31. $\begin{bmatrix} 0 & 1 & 0 \\ 0 & 0 & 2 \\ 0 & 0 & 0 \end{bmatrix}$

33. $\begin{bmatrix} 1 & 0 & 0 \\ 0 & 1 & 0 \\ 0 & 0 & 0 \end{bmatrix}$ 35. $\begin{bmatrix} 0 & 0 & 1 & 0 \\ -1 & 0 & 0 & 1 \\ 0 & 0 & 0 & 0 \\ 0 & 0 & -1 & 0 \end{bmatrix}$

37. $\begin{bmatrix} -2 & 0 & 0 & 0 \\ 0 & 2 & 0 & 0 \\ 0 & 0 & 0 & 0 \\ 0 & 0 & 0 & 0 \end{bmatrix}$ 39. $\begin{bmatrix} -1 & 0 & 1 & 0 \\ 0 & 1 & 0 & 1 \\ 1 & 0 & -1 & 0 \\ 0 & 1 & 0 & 1 \end{bmatrix}$

41. a. $S = \begin{bmatrix} 1 & 0 & 0 \\ 0 & 1 & 1 \\ 0 & 0 & 1 \end{bmatrix}$

   c. $S^{-1} = \begin{bmatrix} 1 & 0 & 0 \\ 0 & 1 & -1 \\ 0 & 0 & 1 \end{bmatrix}$

43. a. $S = \begin{bmatrix} 1 & 0 & 0 \\ -1 & 1 & 1 \\ 0 & 1 & 0 \end{bmatrix}$

   c. $S^{-1} = \begin{bmatrix} 1 & 0 & 0 \\ 0 & 0 & 1 \\ 1 & 1 & -1 \end{bmatrix}$

45. a. $S = \begin{bmatrix} 1 & 1 \\ 1 & -1 \end{bmatrix}$ c. $S^{-1} = \dfrac{1}{2} \begin{bmatrix} 1 & 1 \\ 1 & -1 \end{bmatrix}$

47. a. $S = \begin{bmatrix} 1 & -1 & 1 \\ 0 & 1 & -2 \\ 0 & 0 & 1 \end{bmatrix}$

   c. $S^{-1} = \begin{bmatrix} 1 & 1 & 1 \\ 0 & 1 & 2 \\ 0 & 0 & 1 \end{bmatrix}$

49. $\begin{bmatrix} 2 & 2 \\ -2 & 2 \end{bmatrix}$ 51. $\begin{bmatrix} 0 & -1 \\ 1 & 0 \end{bmatrix}$

53. $\begin{bmatrix} \cos(\theta) & -\sin(\theta) \\ \sin(\theta) & \cos(\theta) \end{bmatrix}$ 55. $\begin{bmatrix} 2/9 & -14/9 \\ -1/9 & 7/9 \end{bmatrix}$

57. $\begin{bmatrix} -1 & 3 \\ -1 & 0 \end{bmatrix}$

59. $T(f(t)) = t \cdot f(t)$ from P to P, for example.

61. a. $S = \begin{bmatrix} -3 & 4 \\ 4 & 3 \end{bmatrix}$ b. $S^{-1} = \dfrac{1}{25} \begin{bmatrix} -3 & 4 \\ 4 & 3 \end{bmatrix}$

63. a. $\vec{b}_1 = \begin{bmatrix} 2 \\ 0 \\ 1 \end{bmatrix}, \vec{b}_2 = \begin{bmatrix} 0 \\ 2 \\ 3 \end{bmatrix}$, for example.

   b. $S = \begin{bmatrix} -1 & -1 \\ 1 & 3 \end{bmatrix}$ c. $S^{-1} = \dfrac{1}{2} \begin{bmatrix} -3 & -1 \\ 1 & 1 \end{bmatrix}$

65. a. $P^2 = \begin{bmatrix} a^2 + bc & b(a+d) \\ c(a+d) & bc + d^2 \end{bmatrix}$ and

   $[P^2]_{\mathfrak{B}} = \begin{bmatrix} bc - ad \\ a + d \end{bmatrix}$

   b. $B = \begin{bmatrix} 0 & bc - ad \\ 1 & a + d \end{bmatrix}$. T is an isomorphism if B is invertible.

   c. In this case, $\operatorname{im}(T)$ is spanned by P and $\ker(T)$ is spanned by $(a+d)I_2 - P = \begin{bmatrix} d & -b \\ -c & a \end{bmatrix}$.

67. a. $\begin{bmatrix} 0 & 0 & 0 & 2 \\ 0 & 0 & -2 & 0 \\ 0 & 0 & 0 & 0 \\ 0 & 0 & 0 & 0 \end{bmatrix}$ b. $f(t) = \frac{1}{2}t\sin(t)$

CHAPTER 5

5.1 Answers to more theoretical questions are omitted.

1. $\sqrt{170}$ 3. $\sqrt{54}$

5. $\arccos\left(\dfrac{20}{\sqrt{406}}\right) \approx 0.12$ (radians)

7. obtuse 9. acute

11. $\arccos\left(\dfrac{1}{\sqrt{n}}\right) \rightarrow \dfrac{\pi}{2}$ (as $n \rightarrow \infty$)

13. $2\arccos(0.8) \approx 74°$

15. $\begin{bmatrix} -2 \\ 1 \\ 0 \\ 0 \end{bmatrix}, \begin{bmatrix} -3 \\ 0 \\ 1 \\ 0 \end{bmatrix}, \begin{bmatrix} -4 \\ 0 \\ 0 \\ 1 \end{bmatrix}$

17. $\begin{bmatrix} 1 \\ -2 \\ 1 \\ 0 \end{bmatrix}, \begin{bmatrix} 2 \\ -3 \\ 0 \\ 1 \end{bmatrix}$

19. a. Orthogonal projection onto $L^\perp$
     b. Reflection about $L^\perp$
     c. Reflection about L

21. For example: $b = d = e = g = 0$, $a = \dfrac{1}{2}$,
    $c = \dfrac{\sqrt{3}}{2}, f = -\dfrac{\sqrt{3}}{2}$

25. a. $\|k\vec{v}\| = \sqrt{(k\vec{v})\cdot(k\vec{v})} = \sqrt{k^2(\vec{v}\cdot\vec{v})} = \sqrt{k^2}\sqrt{\vec{v}\cdot\vec{v}} = |k|\,\|\vec{v}\|$

     b. By part (a), $\left\|\dfrac{1}{\|\vec{v}\|}\vec{v}\right\| = \dfrac{1}{\|\vec{v}\|}\|\vec{v}\| = 1$.

27. $\begin{bmatrix} 8 \\ 0 \\ 2 \\ -2 \end{bmatrix}$

29. By Pythagoras, $\|\vec{x}\| = \sqrt{49 + 9 + 4 + 1 + 1} = 8$.

31. $p \le \|\vec{x}\|^2$. Equality holds if (and only if) $\vec{x}$ is a linear combination of the vectors $\vec{u}_i$.

33. The vector whose n components are all $1/n$

35. $-\dfrac{1}{\sqrt{14}}\begin{bmatrix} 1 \\ 2 \\ 3 \end{bmatrix}$

37. $R(\vec{x}) = 2(\vec{u}_1 \cdot \vec{x})\vec{u}_1 + 2(\vec{u}_2 \cdot \vec{x})\vec{u}_2 - \vec{x}$

39. No; if $\vec{u}$ is a unit vector in L, then $\vec{x} \cdot \text{proj}_L\vec{x} = \vec{x} \cdot (\vec{u} \cdot \vec{x})\vec{u} = (\vec{u} \cdot \vec{x})^2 \ge 0$.

41. $\arccos(20/21) \approx 0.31$ radians

43. $\frac{5}{9}\vec{v}_2$

45. $\frac{25}{41}\vec{v}_2 - \frac{1}{41}\vec{v}_3$

5.2 Answers to more theoretical questions are omitted.

1. $\begin{bmatrix} 2/3 \\ 1/3 \\ -2/3 \end{bmatrix}$

3. $\begin{bmatrix} 4/5 \\ 0 \\ 3/5 \end{bmatrix}, \begin{bmatrix} 3/5 \\ 0 \\ -4/5 \end{bmatrix}$

5. $\begin{bmatrix} 2/3 \\ 2/3 \\ 1/3 \end{bmatrix}, \dfrac{1}{\sqrt{18}}\begin{bmatrix} -1 \\ -1 \\ 4 \end{bmatrix}$

7. $\begin{bmatrix} 2/3 \\ 2/3 \\ 1/3 \end{bmatrix}, \begin{bmatrix} -2/3 \\ 1/3 \\ 2/3 \end{bmatrix}, \begin{bmatrix} 1/3 \\ -2/3 \\ 2/3 \end{bmatrix}$

9. $\begin{bmatrix} 1/2 \\ 1/2 \\ 1/2 \\ 1/2 \end{bmatrix}, \begin{bmatrix} -1/10 \\ 7/10 \\ -7/10 \\ 1/10 \end{bmatrix}$

11. $\begin{bmatrix} 4/5 \\ 0 \\ 0 \\ 3/5 \end{bmatrix}, \begin{bmatrix} -3/15 \\ 2/15 \\ 14/15 \\ 4/15 \end{bmatrix}$

13. $\begin{bmatrix} 1/2 \\ 1/2 \\ 1/2 \\ 1/2 \end{bmatrix}, \begin{bmatrix} 1/2 \\ -1/2 \\ -1/2 \\ 1/2 \end{bmatrix}, \begin{bmatrix} 1/2 \\ 1/2 \\ -1/2 \\ -1/2 \end{bmatrix}$

15. $\begin{bmatrix} 2/3 \\ 1/3 \\ -2/3 \end{bmatrix} [3]$

17. $\begin{bmatrix} 4/5 & 3/5 \\ 0 & 0 \\ 3/5 & -4/5 \end{bmatrix}\begin{bmatrix} 5 & 5 \\ 0 & 35 \end{bmatrix}$

19. $\begin{bmatrix} 2/3 & -1/\sqrt{18} \\ 2/3 & -1/\sqrt{18} \\ 1/3 & 4/\sqrt{18} \end{bmatrix}\begin{bmatrix} 3 & 3 \\ 0 & \sqrt{18} \end{bmatrix}$

21. $\dfrac{1}{3}\begin{bmatrix} 2 & -2 & 1 \\ 2 & 1 & -2 \\ 1 & 2 & 2 \end{bmatrix}\begin{bmatrix} 3 & 0 & 12 \\ 0 & 3 & -12 \\ 0 & 0 & 6 \end{bmatrix}$

23. $\begin{bmatrix} 1/2 & -1/10 \\ 1/2 & 7/10 \\ 1/2 & -7/10 \\ 1/2 & 1/10 \end{bmatrix}\begin{bmatrix} 2 & 4 \\ 0 & 10 \end{bmatrix}$

25. $\begin{bmatrix} 4/5 & -3/15 \\ 0 & 2/15 \\ 0 & 14/15 \\ 3/5 & 4/15 \end{bmatrix}\begin{bmatrix} 5 & 10 \\ 0 & 15 \end{bmatrix}$

27. $\dfrac{1}{2}\begin{bmatrix} 1 & 1 & 1 \\ 1 & -1 & 1 \\ 1 & -1 & -1 \\ 1 & 1 & -1 \end{bmatrix}\begin{bmatrix} 2 & 1 & 1 \\ 0 & 1 & -2 \\ 0 & 0 & 1 \end{bmatrix}$

29. $\begin{bmatrix} -3/5 \\ 4/5 \end{bmatrix}, \begin{bmatrix} 4/5 \\ 3/5 \end{bmatrix}$ 31. $\vec{e}_1, \vec{e}_2, \vec{e}_3$

33. $\dfrac{1}{\sqrt{2}}\begin{bmatrix} 1 \\ 0 \\ 0 \\ -1 \end{bmatrix}, \dfrac{1}{\sqrt{2}}\begin{bmatrix} 0 \\ 1 \\ -1 \\ 0 \end{bmatrix}$

35. $\begin{bmatrix} 1/3 \\ 2/3 \\ 2/3 \end{bmatrix}, \begin{bmatrix} 2/3 \\ 1/3 \\ -2/3 \end{bmatrix}$ 37. $\dfrac{1}{2}\begin{bmatrix} 1 & 1 \\ 1 & -1 \\ 1 & -1 \\ 1 & 1 \end{bmatrix}\begin{bmatrix} 3 & 4 \\ 0 & 5 \end{bmatrix}$

39. $\dfrac{1}{\sqrt{14}}\begin{bmatrix} 1 \\ 2 \\ 3 \end{bmatrix}, \dfrac{1}{\sqrt{3}}\begin{bmatrix} 1 \\ 1 \\ -1 \end{bmatrix}, \dfrac{1}{\sqrt{42}}\begin{bmatrix} 5 \\ -4 \\ 1 \end{bmatrix}$

41. Q is diagonal with $q_{ii} = 1$ if $a_{ii} > 0$ and $q_{ii} = -1$ if $a_{ii} < 0$. You can get R from A by multiplying the ith row of A with -1 whenever a_{ii} is negative.

43. Write the QR factorization of A in partitioned form as $A = \begin{bmatrix} A_1 & A_2 \end{bmatrix} = \begin{bmatrix} Q_1 & Q_2 \end{bmatrix}\begin{bmatrix} R_1 & R_2 \\ 0 & R_4 \end{bmatrix}$. Then $A_1 = Q_1 R_1$ is the QR factorization of A_1.

45. Yes

5.3 Answers to more theoretical questions are omitted.

1. Not orthogonal   3. Orthogonal

5. Not orthogonal   7. Orthogonal

9. Orthogonal   11. Orthogonal

13. Symmetric

15. Not necessarily symmetric

17. Symmetric   19. Symmetric

21. Symmetric

23. Not necessarily symmetric

25. Symmetric

27. $(A\vec{v}) \cdot \vec{w} = (A\vec{v})^T \vec{w} = \vec{v}^T A^T \vec{w} = \vec{v} \cdot (A^T \vec{w})$.

29. $\angle(L(\vec{v}), L(\vec{w})) = \arccos \dfrac{L(\vec{v}) \cdot L(\vec{w})}{\|L(\vec{v})\|\|L(\vec{w})\|} = \arccos \dfrac{\vec{v} \cdot \vec{w}}{\|\vec{v}\|\|\vec{w}\|} = \angle(\vec{v}, \vec{w})$. [The equation $L(\vec{v}) \cdot L(\vec{w}) = \vec{v} \cdot \vec{w}$ is shown in Exercise 28.]

31. Yes, since $AA^T = I_n$.

33. The first column is a unit vector; we can write it as $\vec{v}_1 = \begin{bmatrix} \cos(\theta) \\ \sin(\theta) \end{bmatrix}$ for some θ. The second column is a unit vector orthogonal to $\vec{v}_1$; there are two choices: $\begin{bmatrix} -\sin(\theta) \\ \cos(\theta) \end{bmatrix}$ and $\begin{bmatrix} \sin(\theta) \\ -\cos(\theta) \end{bmatrix}$. Solution:

$\begin{bmatrix} \cos(\theta) & -\sin(\theta) \\ \sin(\theta) & \cos(\theta) \end{bmatrix}$ and $\begin{bmatrix} \cos(\theta) & \sin(\theta) \\ \sin(\theta) & -\cos(\theta) \end{bmatrix}$, for arbitrary θ.

35. For example, $T(\vec{x}) = \dfrac{1}{3}\begin{bmatrix} 1 & -2 & 2 \\ -2 & 1 & 2 \\ 2 & 2 & 1 \end{bmatrix}\vec{x}$.

37. No, by Theorem 5.3.2.

39. (ijth entry of A) $= u_i u_j$

41. All entries of A are $\dfrac{1}{n}$.

43. A represents the reflection about the line spanned by $\vec{u}$ (compare with Example 2). and B represents the reflection about the plane with normal vector $\vec{u}$.

45. $\dim(\ker(A)) = m - \operatorname{rank}(A)$ (by Theorem 3.3.7) and $\dim(\ker(A^T)) = n - \operatorname{rank}(A^T) = n - \operatorname{rank}(A)$ (by Theorem 5.3.9c). Therefore, the dimensions of the two kernels are equal if (and only if) $m = n$, that is, if A is a square matrix.

47. $A^T A = (QR)^T QR = R^T Q^T QR = R^T R$

49. By Exercise 5.2.45, we can write $A^T = QL$ where Q is orthogonal and L is lower triangular. Then $A = (QL)^T = L^T Q^T$ does the job.

51. a. $I_m = Q_1^T Q_1 = S^T Q_2^T Q_2 S = S^T S$, so that S is orthogonal.

    b. $R_2 R_1^{-1}$ is both orthogonal [by part (a)] and upper triangular, with positive diagonal entries. By Exercise 50a, we have $R_2 R_1^{-1} = I_m$ so that $R_2 = R_1$ and $Q_1 = Q_2$, as claimed.

53. $\begin{bmatrix} 0 & 1 & 0 \\ -1 & 0 & 0 \\ 0 & 0 & 0 \end{bmatrix}, \begin{bmatrix} 0 & 0 & 1 \\ 0 & 0 & 0 \\ -1 & 0 & 0 \end{bmatrix}, \begin{bmatrix} 0 & 0 & 0 \\ 0 & 0 & 1 \\ 0 & -1 & 0 \end{bmatrix}$; dimension 3

55. $\dfrac{n(n+1)}{2}$ 57. Yes, and yes

59. The kernel consists of the symmetric $n \times n$ matrices, and the image consists of the skew-symmetric matrices.

61. $\begin{bmatrix} 0 & 0 & 0 & 0 \\ 0 & 0 & 0 & 0 \\ 0 & 0 & 0 & 0 \\ 0 & 0 & 0 & 2 \end{bmatrix}$

63. If $A = LDU$, then $A^T = U^T DL^T$ is the LDU factorization of A^T. Since $A = A^T$ the two factorizations are identical, so that $U = L^T$, as claimed.

65. $\begin{bmatrix} 0.6 & -0.8 \\ 0.8 & 0.6 \end{bmatrix}, \begin{bmatrix} 0.8 & -0.6 \\ 0.6 & 0.8 \end{bmatrix}, \begin{bmatrix} 0.6 & 0.8 \\ 0.8 & -0.6 \end{bmatrix}$, and $\begin{bmatrix} 0.8 & 0.6 \\ 0.6 & -0.8 \end{bmatrix}$

5.4 Answers to more theoretical questions are omitted.

1. $\operatorname{im}(A) = \operatorname{span}\begin{bmatrix} 2 \\ 3 \end{bmatrix}$ and $\ker(A^T) = \operatorname{span}\begin{bmatrix} -3 \\ 2 \end{bmatrix}$.

3. The vectors form a basis of $\mathbb{R}^n$.

5. $V^\perp = \big(\ker(A)\big)^\perp = \operatorname{im}(A^T)$, where
$$A = \begin{bmatrix} 1 & 1 & 1 & 1 \\ 1 & 2 & 5 & 4 \end{bmatrix}.$$

Basis of $V^\perp$: $\begin{bmatrix} 1 \\ 1 \\ 1 \\ 1 \end{bmatrix}, \begin{bmatrix} 1 \\ 2 \\ 5 \\ 4 \end{bmatrix}$.

7. $\operatorname{im}(A) = \big(\ker(A)\big)^\perp$

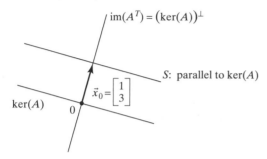

$\operatorname{im}(A^T) = \big(\ker(A)\big)^\perp$

S: parallel to $\ker(A)$

$\vec{x}_0 = \begin{bmatrix} 1 \\ 3 \end{bmatrix}$

$\ker(A)$

9. $\|\vec{x}_0\| < \|\vec{x}\|$ for all other vectors $\vec{x}$ in S.

11. b. $L\big(L^+(\vec{y})\big) = \vec{y}$

c. $L^+\big(L(\vec{x})\big) = \operatorname{proj}_V \vec{x}$, where $V = \big(\ker(A)\big)^\perp = \operatorname{im}(A^T)$

d. $\operatorname{im}(L^+) = \operatorname{im}(A^T)$ and $\ker(L^+) = \{\vec{0}\}$

e. $L^+(\vec{y}) = \begin{bmatrix} 1 & 0 \\ 0 & 1 \\ 0 & 0 \end{bmatrix} \vec{y}$

13. b. $L^+\big(L(\vec{x})\big) = \operatorname{proj}_V \vec{x}$, where $V = \big(\ker(A)\big)^\perp = \operatorname{im}(A^T)$

c. $L\big(L^+(\vec{y})\big) = \operatorname{proj}_W \vec{y}$, where $W = \operatorname{im}(A) = \big(\ker(A^T)\big)^\perp$

d. $\operatorname{im}(L^+) = \operatorname{im}(A^T)$ and $\ker(L^+) = \ker(A^T)$

e. $L^+(\vec{y}) = \begin{bmatrix} \frac{1}{2} & 0 \\ 0 & 0 \\ 0 & 0 \end{bmatrix} \vec{y}$

15. Let $B = (A^T A)^{-1} A^T$.

17. Yes; note that $\ker(A) = \ker(A^T A)$.

19. $\begin{bmatrix} 1 \\ 1 \end{bmatrix}$

21. $\vec{x}^* = \begin{bmatrix} -1 \\ 2 \end{bmatrix}$, $\|\vec{b} - A\vec{x}^*\| = 42$

23. $\begin{bmatrix} 0 \\ 0 \end{bmatrix}$

25. $\begin{bmatrix} 1 - 3t \\ t \end{bmatrix}$, for arbitrary t

27. $\begin{bmatrix} 7 \\ 11 \end{bmatrix}$

29. $x_1^* = x_2^* \approx \frac{1}{2}$

31. $3 + 1.5t$

33. approximately $1.5 + 0.1 \sin(t) - 1.41 \cos(t)$

37. a. Try to solve the system $\begin{vmatrix} c_0 + 35c_1 = \log(35) \\ c_0 + 46c_1 = \log(46) \\ c_0 + 59c_1 = \log(77) \\ c_0 + 69c_1 = \log(133) \end{vmatrix}$.

Least-squares solution $\begin{bmatrix} c_0^* \\ c_1^* \end{bmatrix} \approx \begin{bmatrix} 0.915 \\ 0.017 \end{bmatrix}$. Use approximation $\log(d) = 0.915 + 0.017t$.

b. Exponentiate the equation in part (a): $d = 10^{\log d} = 10^{0.915 + 0.017t} \approx 8.221 \cdot 10^{0.017t} \approx 8.221 \cdot 1.04^t$

c. Predicts 259 displays for the A320; there are much fewer since the A320 is highly computerized.

39. a. Try to solve the system
$$\begin{vmatrix} c_0 + \log(600{,}000)c_1 = \log(250) \\ c_0 + \log(200{,}000)c_1 = \log(60) \\ c_0 + \log(60{,}000)c_1 = \log(25) \\ c_0 + \log(10{,}000)c_1 = \log(12) \\ c_0 + \log(2{,}500)c_1 = \log(5) \end{vmatrix}.$$

5.5 Answers to more theoretical questions are omitted.

3. a. If S is invertible.

b. If S is orthogonal.

5. Yes.

7. For positive k.

9. True.

11. The angle is δ.

13. The two norms are equal, by Theorem 5.5.6.

15. If $b = c$ and $b^2 < d$.

17. If $\ker(T) = \{0\}$.

19. The matrices $A = \begin{bmatrix} a & b \\ c & d \end{bmatrix}$ such that $b = c, a > 0$, and $b^2 < ad$.

21. Yes, $\langle \vec{v}, \vec{w} \rangle = 2(\vec{v} \cdot \vec{w})$.

23. $1, 2t - 1$.

25. $\sqrt{1 + \dfrac{1}{4} + \dfrac{1}{9} + \cdots} = \dfrac{\pi}{\sqrt{6}}$.

27. $a_0 = \dfrac{1}{\sqrt{2}}$. $c_k = 0$ for all k.

$b_k = \begin{cases} \dfrac{2}{k\pi} & \text{if } k \text{ is odd} \\ 0 & \text{if } k \text{ is even.} \end{cases}$

29. $\displaystyle\sum_{k \text{ odd}} \dfrac{1}{k^2} = \dfrac{\pi^2}{8}$

33. b. $\|f\|^2 = \langle f, f \rangle = \int_a^b w(t)\,dt = 1$, so that $\|f\| = 1$

35. a. $\|t\|_{32} = \sqrt{\dfrac{1}{2} \int_{-1}^1 t^2\,dt} = \sqrt{\dfrac{1}{3}}$ and $\|t\|_{34} =$
$\sqrt{\dfrac{2}{\pi} \int_{-1}^1 \sqrt{1 - t^2}\, t^2\,dt} = \sqrt{\dfrac{2}{\pi} \cdot \dfrac{\pi}{8}} = \dfrac{1}{2}$

b. For $f(t) = \sqrt{1-t^2}$ we have $\|f\|_{32} = \sqrt{\dfrac{2}{3}}$ and

$$\|f\|_{34} = \sqrt{\dfrac{3}{4}}$$

CHAPTER 6

6.1 Answers to more theoretical questions are omitted.

1. 0   3. -2 5. 110

7. 0   9. -36 11. $k \neq 3/2$

13. $k \neq 0$ 15. $k \neq 1/2$

17. If k is neither 1 nor -1.

19. If k is neither 0 nor 1.

21. If k is neither 1 nor -2.

23. If λ is 1 or 4.   25. If λ is 2 or 8.

27. If λ is 2, 3, or 4.   29. If λ is 3 or 8.

31. 24   33. 99   35. 18

37. 55   39. 120   41. 24

43. $\det(-A) = (-1)^n \det(A)$

45. They are the same.

49. The kernel is the plane $\mathrm{span}(\vec{v}, \vec{w})$ and the image is $\mathbb{R}$.

51. Let a_{ii} be the first diagonal entry that does not belong to the pattern. The pattern must contain a number in the ith row to the right of a_{ii} as well as a number in the ith column below a_{ii}.

53. Only one pattern has a nonzero product, and that product is 1. Since there are n^2 inversions in that pattern, we have $\det A = (-1)^{n^2} = (-1)^n$.

55. Yes, since the determinants of all principal submatrices are nonzero. See Exercise 2.4.93.

57. Only one pattern has a nonzero product, and that product is 1. Thus $\det A = 1$ or $\det A = -1$.

59. a. Yes   b. No   c. No

61. Fails to be alternating, since $F \begin{bmatrix} 0 & 1 & 0 \\ 0 & 0 & 1 \\ 1 & 0 & 0 \end{bmatrix} = 1$ but

$$F \begin{bmatrix} 0 & 0 & 1 \\ 0 & 1 & 0 \\ 1 & 0 & 0 \end{bmatrix} = 0.$$

6.2 Answers to more theoretical questions are omitted.

1. 6   3. -24 5. -24

7. 1   9. 24   11. -72

13. 8   15. 8   17. 8

19. -1 21. 1

23. $(-1)^{n(n-1)/2}$. This is 1 if either n or $(n-1)$ is divisible by 4, and -1 otherwise.

25. 16   27. $a^2 + b^2$

29. $\det(P_1) = 1$ and $\det(P_n) = \det(P_{n-1})$, by expansion down the first column, so $\det(P_n) = 1$ for all n.

31. a. $\det \begin{bmatrix} 1 & 1 \\ a_0 & a_1 \end{bmatrix} = a_1 - a_0$

b. Use Laplace Expansion down the last column to see that $f(t)$ is a polynomial of degree $\leq n$. The coefficient k of t^n is $\prod\limits_{n-1 \geq i > j} (a_i - a_j)$. Now $\det(A) = f(a_n) = k(a_n - a_0)(a_n - a_1)\ldots(a_n - a_{n-1}) = \prod\limits_{n \geq i > j} (a_i - a_j)$, as claimed.

33. $\prod\limits_{i=1}^{n} a_i \cdot \prod\limits_{i > j} (a_i - a_j)$ (use linearity in the columns and Exercise 31)

35. $\begin{bmatrix} x_1 \\ x_2 \end{bmatrix} = \begin{bmatrix} a_1 \\ a_2 \end{bmatrix}$ and $\begin{bmatrix} x_1 \\ x_2 \end{bmatrix} = \begin{bmatrix} b_1 \\ b_2 \end{bmatrix}$ are solutions. The equation is of the form $px_1 + qx_2 + b = 0$, that is, it defines a line.

37. ± 1

39. $\det(A^T A) = \left(\det(A)\right)^2 > 0$

41. $\det(A) = \det(A^T) = \det(-A) = (-1)^n \det(A) = -\det(A)$, so $\det(A) = 0$

43. $A^T A = \begin{bmatrix} \|\vec{v}\|^2 & \vec{v} \cdot \vec{w} \\ \vec{v} \cdot \vec{w} & \|\vec{w}\|^2 \end{bmatrix}$,

so $\det(A^T A) = \|\vec{v}\|^2 \|\vec{w}\|^2 - (\vec{v} \cdot \vec{w})^2 \geq 0$, by the Cauchy-Schwarz inequality. Equality holds only if $\vec{v}$ and $\vec{w}$ are parallel.

45. Expand down the first column: $f(x) = -x \det(A_{41}) + \text{constant}$, so $f'(x) = -\det(A_{41}) = -24$.

47. T is linear in the rows and columns.

49. $A = \begin{bmatrix} 1 & 1 & 1 \\ 1 & 2 & 2 \\ 1 & 1 & 14 \end{bmatrix}$, for example. Start with a triangular matrix with determinant 13, such as $\begin{bmatrix} 1 & 1 & 1 \\ 0 & 1 & 1 \\ 0 & 0 & 13 \end{bmatrix}$, and add the first row to the second and to the third to make all entries nonzero.

51. $\det(A) = (-1)^n$

53. a. Note that $\det(A)\det(A^{-1}) = 1$, and both factors are integers.

b. Use the formula for the inverse of a 2×2 matrix (Theorem 2.4.9b).

59. No

61. Take the determinant of both sides of

$$\begin{bmatrix} I_n & 0 \\ -C & A \end{bmatrix} \begin{bmatrix} A & B \\ C & D \end{bmatrix} = \begin{bmatrix} A & B \\ 0 & AD - CB \end{bmatrix},$$

and divide by $\det(A)$.

65. a. $d_n = d_{n-1} + d_{n-2}$, a Fibonacci sequence
    b. $d_1 = 1, d_2 = 2, d_3 = 3, d_4 = 5, \ldots, d_{10} = 89$
    c. Invertible for all positive integers n.

6.3 Answers to more theoretical questions are omitted.

1. 50   3. 13   7. 110

11. $|\det(A)| = 12$, the expansion factor of T on the parallelogram defined by $\vec{v}_1$ and $\vec{v}_2$.

13. $\sqrt{20}$

15. We need to show that if $\vec{v}_1, \ldots, \vec{v}_m$ are linearly dependent, then (a) $V(\vec{v}_1, \ldots, \vec{v}_m) = 0$ and (b) $\det(A^T A) = 0$.
    a. One of the $\vec{v}_i$ is redundant, so that $\vec{v}_i^{\perp} = \vec{0}$ and $V(\vec{v}_1, \ldots, \vec{v}_m) = 0$, by Definition 6.3.5.
    b. $\ker(A) \neq \{\vec{0}\}$ and $\ker(A) \subseteq \ker(A^T A)$, so that $\ker(A^T A) \neq \{\vec{0}\}$. Thus $A^T A$ fails to be invertible.

17. a. $V(\vec{v}_1, \vec{v}_2, \vec{v}_3, \vec{v}_1 \times \vec{v}_2 \times \vec{v}_3)$
       $= V(\vec{v}_1, \vec{v}_2, \vec{v}_3) \|\vec{v}_1 \times \vec{v}_2 \times \vec{v}_3\|$ because $\vec{v}_1 \times \vec{v}_2 \times \vec{v}_3$ is orthogonal to $\vec{v}_1, \vec{v}_2$, and $\vec{v}_3$.
    b. $V(\vec{v}_1, \vec{v}_2, \vec{v}_3, \vec{v}_1 \times \vec{v}_2 \times \vec{v}_3)$
       $= \left|\det\begin{bmatrix} \vec{v}_1 \times \vec{v}_2 \times \vec{v}_3 & \vec{v}_1 & \vec{v}_2 & \vec{v}_3 \end{bmatrix}\right|$
       $= \|\vec{v}_1 \times \vec{v}_2 \times \vec{v}_3\|^2$, by definition of the cross product.
    c. $V(\vec{v}_1, \vec{v}_2, \vec{v}_3) = \|\vec{v}_1 \times \vec{v}_2 \times \vec{v}_3\|$, by parts (a) and (b).

19. $\det\begin{bmatrix} \vec{v}_1 & \vec{v}_2 & \vec{v}_3 \end{bmatrix} = \vec{v}_1 \cdot (\vec{v}_2 \times \vec{v}_3)$ is positive if (and only if) $\vec{v}_1$ and $\vec{v}_2 \times \vec{v}_3$ enclose an acute angle.

21. a. reverses   b. preserves   c. reverses

23. $x_1 = \dfrac{\det\begin{bmatrix} 1 & -3 \\ 0 & 7 \end{bmatrix}}{\det\begin{bmatrix} 5 & -3 \\ -6 & 7 \end{bmatrix}} = \dfrac{7}{17};$

$x_2 = \dfrac{\det\begin{bmatrix} 5 & 1 \\ -6 & 0 \end{bmatrix}}{\det\begin{bmatrix} 5 & -3 \\ -6 & 7 \end{bmatrix}} = \dfrac{6}{17}$

25. $\operatorname{adj}(A) = \begin{bmatrix} 1 & 0 & -1 \\ 0 & -1 & 0 \\ -2 & 0 & 1 \end{bmatrix};$

$A^{-1} = \dfrac{1}{\det(A)} \operatorname{adj}(A) = -\operatorname{adj}(A)$

$= \begin{bmatrix} -1 & 0 & 1 \\ 0 & 1 & 0 \\ 2 & 0 & -1 \end{bmatrix}$

27. $x = \dfrac{a}{a^2 + b^2} > 0; \ y = \dfrac{-b}{a^2 + b^2} < 0; \ x$ decreases as b increases.

29. $dx_1 = -D^{-1} R_2 (1 - R_1)(1 - \alpha)^2 de_2,$
    $dy_1 = D^{-1}(1 - \alpha) R_2 \big(R_1 (1 - \alpha) + \alpha \big) de_2 > 0,$
    $dp = D^{-1} R_1 R_2 de_2 > 0.$

31. $\begin{bmatrix} -6 & 0 & 1 \\ -3 & 5 & -2 \\ 4 & -5 & 1 \end{bmatrix}$ 33. $\begin{bmatrix} 24 & 0 & 0 & 0 \\ 0 & 12 & 0 & 0 \\ 0 & 0 & 8 & 0 \\ 0 & 0 & 0 & 6 \end{bmatrix}$

35. $\det(\operatorname{adj} A) = (\det A)^{n-1}$

37. $\operatorname{adj}(A^{-1}) = (\operatorname{adj} A)^{-1} = (\det A)^{-1} A$

39. Yes. Use Exercises 38: If $AS = SB$, then $(\operatorname{adj} S)(\operatorname{adj} A) = (\operatorname{adj} B)(\operatorname{adj} S)$

43. $A(\operatorname{adj} A) = (\operatorname{adj} A) A = (\det A) I_n = 0$

45. $\begin{bmatrix} a & -b \\ b & a \end{bmatrix}$

CHAPTER 7

7.1 Answers to more theoretical questions are omitted.

1. Yes; the eigenvalue is λ^3

3. Yes; the eigenvalue is $\lambda + 2$

5. Yes

7. $\ker(A - \lambda I_n) \neq \{\vec{0}\}$ because $(A - \lambda I_n)\vec{v} = \vec{0}$. The matrix $A - \lambda I_n$ fails to be invertible.

9. $\begin{bmatrix} a & b \\ 0 & d \end{bmatrix}$ 11. $\begin{bmatrix} a & \frac{-2-2a}{3} \\ c & \frac{-3-2c}{3} \end{bmatrix}$

13. All vectors of the form $\begin{bmatrix} 3t \\ 5t \end{bmatrix}$, where $t \neq 0$ (solve the linear system $A\vec{x} = 4\vec{x}$).

15. The nonzero vectors in L are the eigenvectors with eigenvalue 1, and the nonzero vectors in $L^{\perp}$ have eigenvalue -1. Construct a basis of eigenvectors by picking one of each.

17. No eigenvectors and eigenvalues (compare with Example 3).

19. The nonzero vectors in L are the eigenvectors with eigenvalue 1, and the nonzero vectors in the plane $L^{\perp}$ have eigenvalue 0. Construct a basis of eigenvectors by picking one nonzero vector in L and two linearly independent vectors in $L^{\perp}$. (Compare with Example 2.)

21. All nonzero vectors in $\mathbb{R}^3$ are eigenvectors with eigenvalue 5. Any basis of $\mathbb{R}^3$ is a basis of eigenvectors.

23. a. $S^{-1}\vec{v}_i = \vec{e}_i$ because $S\vec{e}_i = \vec{v}_i$
    b. $S^{-1} A S = S^{-1} A \begin{bmatrix} \vec{v}_1 \ldots \vec{v}_n \end{bmatrix} = S^{-1} \begin{bmatrix} \lambda_1 \vec{v}_1 \ldots \lambda_n \vec{v}_n \end{bmatrix} = \begin{bmatrix} \lambda_1 \vec{e}_1 \ldots \lambda_n \vec{e}_n \end{bmatrix} = \begin{bmatrix} \lambda_1 & & 0 \\ & \ddots & \\ 0 & & \lambda_n \end{bmatrix}$, the diagonal matrix with diagonal entries $\lambda_1, \ldots, \lambda_n$.

25.

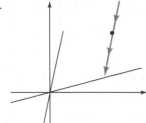

27.

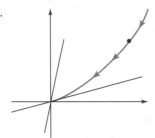

29.

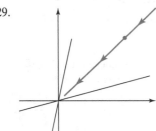

31.

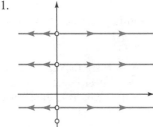

33. $\vec{x}(t) = 2^t \begin{bmatrix} 1 \\ 1 \end{bmatrix} + 6^t \begin{bmatrix} -1 \\ 1 \end{bmatrix}$. We need a matrix A with eigenvectors $\begin{bmatrix} 1 \\ 1 \end{bmatrix}$, $\begin{bmatrix} -1 \\ 1 \end{bmatrix}$, with associated eigenvalues 2 and 6, respectively. Let $A \begin{bmatrix} 1 & -1 \\ 1 & 1 \end{bmatrix} = \begin{bmatrix} 2 & -6 \\ 2 & 6 \end{bmatrix}$ and solve for A. We find $A = \begin{bmatrix} 4 & -2 \\ -2 & 4 \end{bmatrix}$.

35. If λ is an eigenvalue of $S^{-1}AS$, with corresponding eigenvector $\vec{v}$, then

$$S^{-1}AS\vec{v} = \lambda\vec{v},$$

so

$$AS\vec{v} = S\lambda\vec{v} = \lambda S\vec{v},$$

and λ is an eigenvalue of A ($S\vec{v}$ is an eigenvector). Likewise, if $\vec{w}$ is an eigenvector of A, then $S^{-1}\vec{w}$ is an eigenvector of $S^{-1}AS$ with the same eigenvalue.

37. a. A represents a reflection about a line followed by a scaling by a factor of $\sqrt{3^2 + 4^2} = 5$. The eigenvalues are therefore 5 and -5.

b. Solving the linear systems $A\vec{x} = 5\vec{x}$ and $A\vec{x} = -5\vec{x}$ we find the basis $\begin{bmatrix} 2 \\ 1 \end{bmatrix}$, $\begin{bmatrix} -1 \\ 2 \end{bmatrix}$.

39. V consists of all lower triangular 2×2 matrices, and $\dim(V) = 3$.

41. A basis is $\begin{bmatrix} 2 & -1 \\ 2 & -1 \end{bmatrix}$, $\begin{bmatrix} -1 & 1 \\ -2 & 2 \end{bmatrix}$, and $\dim(V) = 2$.

43. V consists of all diagonal matrices, so that $\dim(V) = n$.

45. $\dim(V) = 3$.

47. The subspaces spanned by an eigenvector.

49. $c(t) = 300(1.1)^t - 200(0.9)^t$
$r(t) = 900(1.1)^t - 100(0.9)^t$

51. a. $c(t) = 100(1.5)^t$, $r(t) = 200(1.5)^t$
b. $c(t) = 100(0.75)^t$, $r(t) = 100(0.75)^t$
c. $c(t) = 300(0.75)^t + 200(1.5)^t$,
$r(t) = 300(0.75)^t + 400(1.5)^t$

53. $\begin{bmatrix} a(t+1) \\ b(t+1) \\ c(t+1) \end{bmatrix} = \frac{1}{2} \underbrace{\begin{bmatrix} 0 & 1 & 1 \\ 1 & 0 & 1 \\ 1 & 1 & 0 \end{bmatrix}}_{A} \begin{bmatrix} a(t) \\ b(t) \\ c(t) \end{bmatrix}$. The three given vectors are eigenvectors of A, with eigenvalues $1, -\frac{1}{2}, -\frac{1}{2}$, respectively.

a. $a(t) = 3 + 3\left(-\frac{1}{2}\right)^t$, $b(t) = 3 - 2\left(-\frac{1}{2}\right)^t$,
$c(t) = 3 - \left(-\frac{1}{2}\right)^t$

b. Benjamin will have the most.

7.2 Answers to more theoretical questions are omitted.

1. 1, 3   3. 1, 3   5. None

7. 1, 1, 1   9. 1, 2, 2

11. -1 13. 1

15. Eigenvalues $\lambda_{1,2} = 1 \pm \sqrt{k}$. Two distinct real eigenvalues if k is positive; none, if k is negative.

17. A represents a reflection followed by a scaling, with a scaling factor of $\sqrt{a^2 + b^2}$. The eigenvalues are $\pm\sqrt{a^2 + b^2}$.

19. True [the discriminant $\text{tr}(A)^2 - 4\det(A)$ is positive]

21. Write $f_A(\lambda) = (\lambda_1 - \lambda) \cdots (\lambda_n - \lambda)$ to show that the coefficient of $(-\lambda)^{n-1}$ is $\lambda_1 + \cdots + \lambda_n$. But that coefficient is $\text{tr}(A)$, by Theorem 7.2.5.

23. A and B have the same characteristic polynomial and the same eigenvalues, with the same algebraic multiplicities.

25. $A \begin{bmatrix} b \\ c \end{bmatrix} = \begin{bmatrix} b \\ c \end{bmatrix}$ and $A \begin{bmatrix} 1 \\ -1 \end{bmatrix} = (a - b) \begin{bmatrix} 1 \\ -1 \end{bmatrix}$. Note that $|a - b| < 1$. Phase portrait when $a > b$:

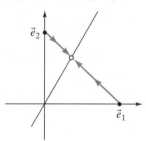

line spanned by $\begin{bmatrix} b \\ c \end{bmatrix}$

line spanned by $\begin{bmatrix} 1 \\ -1 \end{bmatrix}$

27. a. $\vec{x}(t) = \dfrac{1}{3}\begin{bmatrix} 1 \\ 2 \end{bmatrix} + \dfrac{2}{3}\left(\dfrac{1}{4}\right)^t\begin{bmatrix} 1 \\ -1 \end{bmatrix}$ for $\vec{x}_0 = \vec{e}_1$

$\vec{x}(t) = \dfrac{1}{3}\begin{bmatrix} 1 \\ 2 \end{bmatrix} - \dfrac{1}{3}\left(\dfrac{1}{4}\right)^t\begin{bmatrix} 1 \\ -1 \end{bmatrix}$ for $\vec{x}_0 = \vec{e}_2$

$\vec{e}_2$

$\vec{e}_1$

b. $A^t = \begin{bmatrix} A^t\vec{e}_1 & A^t\vec{e}_2 \end{bmatrix}$ approaches $\dfrac{1}{3}\begin{bmatrix} 1 & 1 \\ 2 & 2 \end{bmatrix}$, by part (a).

c. $A^t \to \dfrac{1}{b+c}\begin{bmatrix} b & b \\ c & c \end{bmatrix}$

29. $A\vec{e} = \vec{e}$, so that $\vec{e}$ is an eigenvector with associated eigenvalue 1.

31. A and A^T have the same eigenvalues, by Exercise 22. Since the row sums of A^T are 1, we can use the results of Exercises 29 and 30: 1 is an eigenvalue of A; if λ is an eigenvalue of A, then $|\lambda| \le 1$. $\vec{e}$ need not be an eigenvector of A; consider $A = \begin{bmatrix} 0.9 & 0.9 \\ 0.1 & 0.1 \end{bmatrix}$

33. a. $f_A(\lambda) = -\lambda^3 + c\lambda^2 + b\lambda + a$

b. $M = \begin{bmatrix} 0 & 1 & 0 \\ 0 & 0 & 1 \\ \pi & -5 & 17 \end{bmatrix}$

35. $A = \begin{bmatrix} 0 & -1 & 0 & 0 \\ 1 & 0 & 0 & 0 \\ 0 & 0 & 0 & -1 \\ 0 & 0 & 1 & 0 \end{bmatrix}$

37. We can write $f_A(\lambda) = (\lambda - \lambda_0)^2 g(\lambda)$. By the product rule, $f_A'(\lambda) = 2(\lambda - \lambda_0)g(\lambda) + (\lambda - \lambda_0)^2 g'(\lambda)$, so that $f_A'(\lambda_0) = 0$.

39. It's a straightforward computation.

41. $\text{tr}\left(S^{-1}(AS)\right) = \text{tr}\left((AS)S^{-1}\right) = \text{tr}(A)$

43. No, since $\text{tr}(AB - BA) = 0$ and $\text{tr}(I_n) = n$

45. For $k = 3$

47. If we write $M = \begin{bmatrix} \vec{v} & \vec{w} \end{bmatrix}$, then it is required that $A\vec{v} = 2\vec{v}$ and $A\vec{w} = 3\vec{w}$. Thus a nonzero M with the given property exists if 2 or 3 is an eigenvalue of A.

49. If 2, 3, or 4 is an eigenvalue of A.

7.3 Answers to more theoretical questions are omitted.

1. Eigenbasis: $\begin{bmatrix} 1 \\ 0 \end{bmatrix}, \begin{bmatrix} 4 \\ 1 \end{bmatrix}$, with eigenvalues 7, 9.

3. Eigenbasis: $\begin{bmatrix} 3 \\ -2 \end{bmatrix}, \begin{bmatrix} 1 \\ 1 \end{bmatrix}$, with eigenvalues 4, 9.

5. No real eigenvalues.

7. Eigenbasis: $\vec{e}_1, \vec{e}_2, \vec{e}_3$, with eigenvalues 1, 2, 3.

9. Eigenbasis: $\begin{bmatrix} 1 \\ 0 \\ 0 \end{bmatrix}, \begin{bmatrix} 0 \\ 1 \\ 0 \end{bmatrix}, \begin{bmatrix} -1 \\ 0 \\ 1 \end{bmatrix}$, with eigenvalues 1, 1, 0.

11. Eigenbasis: $\begin{bmatrix} 1 \\ 1 \\ 1 \end{bmatrix}, \begin{bmatrix} 1 \\ -1 \\ 0 \end{bmatrix}, \begin{bmatrix} 1 \\ 0 \\ -1 \end{bmatrix}$, with eigenvalues 3, 0, 0.

13. Eigenbasis: $\begin{bmatrix} 0 \\ 1 \\ 0 \end{bmatrix}, \begin{bmatrix} 1 \\ -3 \\ 1 \end{bmatrix}, \begin{bmatrix} 1 \\ -1 \\ 2 \end{bmatrix}$, with eigenvalues 0, 1, -1.

15. Eigenvectors: $\begin{bmatrix} 0 \\ 1 \\ 0 \end{bmatrix}, \begin{bmatrix} 1 \\ -1 \\ 2 \end{bmatrix}$, with eigenvalues 0, 1. No eigenbasis.

17. Eigenbasis: $\vec{e}_2, \vec{e}_4, \vec{e}_1, \vec{e}_3 - \vec{e}_2$, with eigenvalues 1, 1, 0, 0.

19. The geometric multiplicity of the eigenvalue 1 is 1 if $a \ne 0$ and $c \ne 0$; 3 if $a = b = c = 0$; and 2 otherwise. Therefore, there is an eigenbasis only if $A = I_3$.

21. We want $A\begin{bmatrix} 1 \\ 2 \end{bmatrix} = \begin{bmatrix} 1 \\ 2 \end{bmatrix}$ and $A\begin{bmatrix} 2 \\ 3 \end{bmatrix} = 2\begin{bmatrix} 2 \\ 3 \end{bmatrix} = \begin{bmatrix} 4 \\ 6 \end{bmatrix}$, that is, $A\begin{bmatrix} 1 & 2 \\ 2 & 3 \end{bmatrix} = \begin{bmatrix} 1 & 4 \\ 2 & 6 \end{bmatrix}$. The unique solution is $A = \begin{bmatrix} 5 & -2 \\ 6 & -2 \end{bmatrix}$.

23. The only eigenvalue of A is 1, with $E_1 = \text{span}(\vec{e}_1)$. There is no eigenbasis. A represents a horizontal shear.

25. The geometric multiplicity is always 1.

27. $f_A(\lambda) = \lambda^2 - 5\lambda + 6 = (\lambda - 2)(\lambda - 3)$, so that the eigenvalues are 2, 3.

29. Both multiplicities are $n - r$.

31. They are the same.

33. If $B = S^{-1}AS$, then $B - \lambda I_n = S^{-1}(A - \lambda I_n)S$.

35. No (consider the eigenvalues).

37. a. $A\vec{v} \cdot \vec{w} = (A\vec{v})^T \vec{w} = \vec{v}^T A^T \vec{w} = \vec{v}^T A\vec{w} = \vec{v} \cdot A\vec{w}$

 b. Suppose $A\vec{v} = \lambda\vec{v}$ and $A\vec{w} = \mu\vec{w}$. Then $A\vec{v}\cdot\vec{w} = \lambda(\vec{v} \cdot \vec{w})$ and $\vec{v} \cdot A\vec{w} = \mu(\vec{v} \cdot \vec{w})$. By part (a), $\lambda(\vec{v} \cdot \vec{w}) = \mu(\vec{v} \cdot \vec{w})$, so that $(\lambda - \mu)(\vec{v} \cdot \vec{w}) = 0$. Since $\lambda \neq \mu$ it follows that $\vec{v} \cdot \vec{w} = 0$, as claimed.

39. a. $E_1 = V$ and $E_0 = V^\perp$, so that the geometric multiplicity of 1 is m and that of 0 is $n - m$. The algebraic multiplicities are the same (see Exercise 31).

 b. $E_1 = V$ and $E_{-1} = V^\perp$, so that the multiplicity of 1 is m and that of -1 is $n - m$.

41. Eigenbasis for A: $\begin{bmatrix} 9 \\ 6 \\ 2 \end{bmatrix}$, $\begin{bmatrix} 2 \\ -2 \\ 1 \end{bmatrix}$, $\begin{bmatrix} 1 \\ -2 \\ 2 \end{bmatrix}$, with eigenvalues 1.2, -0.8, -0.4; $\vec{x}_0 = 50\vec{v}_1 + 50\vec{v}_2 + 50\vec{v}_3$.

$$j(t) = 450(1.2)^t + 100(-0.8)^t + 50(-0.4)^t$$
$$m(t) = 300(1.2)^t - 100(-0.8)^t - 100(-0.4)^t$$
$$a(t) = 100(1.2)^t + 50(-0.8)^t + 100(-0.4)^t$$

The populations approach the proportion 9:6:2.

43. a. $A = \dfrac{1}{2}\begin{bmatrix} 0 & 1 & 1 \\ 1 & 0 & 1 \\ 1 & 1 & 0 \end{bmatrix}$

 c. $\vec{x}(t) = \left(1 + \dfrac{c_0}{3}\right)\begin{bmatrix} 1 \\ 1 \\ 1 \end{bmatrix} + \left(-\dfrac{1}{2}\right)^t \begin{bmatrix} 0 \\ 1 \\ -1 \end{bmatrix} + \left(-\dfrac{1}{2}\right)^t \dfrac{c_0}{3}\begin{bmatrix} -1 \\ -1 \\ 2 \end{bmatrix}$.

 Carl wins if he chooses $c_0 < 1$.

45. a. $A = \begin{bmatrix} 0.1 & 0.2 \\ 0.4 & 0.3 \end{bmatrix}$, $\vec{b} = \begin{bmatrix} 1 \\ 2 \end{bmatrix}$

 b. $B = \begin{bmatrix} A & \vec{b} \\ 0 & 1 \end{bmatrix}$

 c. The eigenvalues of A are 0.5 and -0.1, those of B are 0.5, -0.1, 1. If $\vec{v}$ is an eigenvector of A, then $\begin{bmatrix} \vec{v} \\ 0 \end{bmatrix}$ is an eigenvector of B. Furthermore, $\begin{bmatrix} (I_2 - A)^{-1}\vec{b} \\ 1 \end{bmatrix} = \begin{bmatrix} 2 \\ 4 \\ 1 \end{bmatrix}$ is an eigenvector of B with eigenvalue 1.

 d. Will approach $(I_2 - A)^{-1}\vec{b} = \begin{bmatrix} 2 \\ 4 \end{bmatrix}$, for any initial value.

47. Let $\vec{x}(t) = \begin{bmatrix} r(t) \\ p(t) \\ w(t) \end{bmatrix}$. Then $\vec{x}(t + 1) = A\vec{x}(t)$ where

$A = \begin{bmatrix} \frac{1}{2} & \frac{1}{4} & 0 \\ \frac{1}{2} & \frac{1}{2} & \frac{1}{2} \\ 0 & \frac{1}{4} & \frac{1}{2} \end{bmatrix}$. Eigenbasis for A: $\begin{bmatrix} 1 \\ 2 \\ 1 \end{bmatrix}$, $\begin{bmatrix} 1 \\ 0 \\ -1 \end{bmatrix}$, $\begin{bmatrix} 1 \\ -2 \\ 1 \end{bmatrix}$, with eigenvalues 1, $\frac{1}{2}$, 0.

$\vec{x}_0 = \vec{e}_1 = \dfrac{1}{4}\begin{bmatrix} 1 \\ 2 \\ 1 \end{bmatrix} + \dfrac{1}{2}\begin{bmatrix} 1 \\ 0 \\ -1 \end{bmatrix} + \dfrac{1}{4}\begin{bmatrix} 1 \\ -2 \\ 1 \end{bmatrix}$,

so $\vec{x}(t) = \dfrac{1}{4}\begin{bmatrix} 1 \\ 2 \\ 1 \end{bmatrix} + \left(\dfrac{1}{2}\right)^{t+1}\begin{bmatrix} 1 \\ 0 \\ -1 \end{bmatrix}$ for $t > 0$.

The proportion in the long run is 1:2:1.

49. 1 [rref(A) is likely to be the matrix with all 1's directly above the main diagonal and 0's everywhere else]

51. $f_A(\lambda) = -\lambda^3 + c\lambda^2 + b\lambda + a$

53. a. $B = \begin{bmatrix} 0 & 0 & a & * & * \\ 1 & 0 & b & * & * \\ 0 & 1 & c & * & * \\ 0 & 0 & 0 & w & x \\ 0 & 0 & 0 & y & z \end{bmatrix} = \begin{bmatrix} B_1 & B_2 \\ 0 & B_3 \end{bmatrix}$

 b. Note that A is similar to B. Thus
 $f_A(\lambda) = f_B(\lambda) = f_{B_3}(\lambda)f_{B_1}(\lambda)$
 $= h(\lambda)(-\lambda^3 + c\lambda^2 + b\lambda + a)$, where $h(\lambda) = f_{B_3}(\lambda)$. (See Exercise 51.)

 c. $f_A(A)\vec{v} = h(A)(-A^3 + cA^2 + bA + aI_5)\vec{v} = h(A)\underbrace{(-A^3\vec{v} + cA^2\vec{v} + bA\vec{v} + a\vec{v})}_{\vec{0}} = \vec{0}$

7.4 Answers to more theoretical questions are omitted.

1. $S = I_2, D = A$

3. $S = \begin{bmatrix} 1 & 1 \\ -1 & 2 \end{bmatrix}, D = \begin{bmatrix} 0 & 0 \\ 0 & 3 \end{bmatrix}$, for example. If you found a different solution, check that $AS = SD$.

5. Not diagonalizable

7. $S = \begin{bmatrix} 4 & 1 \\ 1 & -1 \end{bmatrix}, D = \begin{bmatrix} 2 & 0 \\ 0 & -3 \end{bmatrix}$

9. Not diagonalizable   11. Not diagonalizable

13. $S = \begin{bmatrix} 1 & 1 & 1 \\ 0 & 1 & 2 \\ 0 & 0 & 1 \end{bmatrix}, D = \begin{bmatrix} 1 & 0 & 0 \\ 0 & 2 & 0 \\ 0 & 0 & 3 \end{bmatrix}$

15. $S = \begin{bmatrix} 2 & 0 & 1 \\ 1 & 0 & 1 \\ 0 & 1 & 0 \end{bmatrix}, D = \begin{bmatrix} 1 & 0 & 0 \\ 0 & 1 & 0 \\ 0 & 0 & -1 \end{bmatrix}$

17. $S = \begin{bmatrix} -1 & -1 & 1 \\ 1 & 0 & 1 \\ 0 & 1 & 1 \end{bmatrix}$, $D = \begin{bmatrix} 0 & 0 & 0 \\ 0 & 0 & 0 \\ 0 & 0 & 3 \end{bmatrix}$

19. Not diagonalizable; eigenvalue 1 has algebraic multiplicity 2, but geometric multiplicity 1.

21. Diagonalizable for all a

23. Diagonalizable for positive a

25. Diagonalizable for all a, b, and c.

27. Diagonalizable only if $a = b = c = 0$

29. Never diagonalizable

31. $A^t = \dfrac{1}{3} \begin{bmatrix} 5^t + 2(-1)^t & 5^t - (-1)^t \\ 2(5^t) - 2(-1)^t & 2(5^t) + (-1)^t \end{bmatrix}$

33. $A^t = 7^{t-1} A$

35. Yes, since $\begin{bmatrix} -1 & 6 \\ -2 & 6 \end{bmatrix}$ is diagonalizable, with eigenvalues 3 and 2.

37. Yes. Both matrices have the characteristic polynomial $\lambda^2 - 7\lambda + 7$, so that they both have the eigenvalues $\lambda_{1,2} = \dfrac{7 \pm \sqrt{21}}{2}$. Thus, A and B are both similar to $\begin{bmatrix} \lambda_1 & 0 \\ 0 & \lambda_2 \end{bmatrix}$, and therefore, A is similar to B.

39. All real numbers λ are eigenvalues, with corresponding eigenfunctions $Ce^{(\lambda+1)t}$.

41. The symmetric matrices are eigenmatrices with eigenvalue 2, and the skew-symmetric matrices have eigenvalue 0. Yes, L is diagonalizable, since the sum of the dimensions of the eigenspaces is 4.

43. 1 and i are "eigenvectors" with eigenvalues 1 and -1, respectively. Yes, T is diagonalizable; 1, i is an eigenbasis.

45. No eigensequences

47. Nonzero polynomials of the form $a + cx^2$ are eigenfunctions with eigenvalue 1, and bx (with $b \neq 0$) has eigenvalue -1. Yes, T is diagonalizable, with eigenbasis 1, x, x^2.

49. 1, $2x - 1$, and $(2x - 1)^2$ are eigenfunctions with eigenvalues 1, 3, and 9, respectively. These functions form an eigenbasis, so that T is indeed diagonalizable.

51. The only eigenfunctions are the nonzero constant functions, with eigenvalue 0.

55. $A = \begin{bmatrix} 0 & 1 \\ 0 & 0 \end{bmatrix}$, $B = \begin{bmatrix} 0 & 0 \\ 0 & 1 \end{bmatrix}$, for example.

59. Exercise 58 implies that A and B are both similar to the matrix $\begin{bmatrix} 0 & 1 & 0 \\ 0 & 0 & 0 \\ 0 & 0 & 0 \end{bmatrix}$, so that A is similar to B.

65. A basis of V is $\begin{bmatrix} 1 & 0 \\ 2 & 0 \end{bmatrix}$, $\begin{bmatrix} 0 & 1 \\ 0 & -1 \end{bmatrix}$, and $\dim(V) = 2$.

67. The dimension is $3^2 + 2^2 = 13$.

71. The eigenvalues are 1 and 2, and $(A - I_3)(A - 2I_3) = 0$. Thus A is diagonalizable.

73. If $\lambda_1, \ldots, \lambda_m$ are the distinct eigenvalues of A, then $f_A(\lambda) = (\lambda - \lambda_1) \cdots (\lambda - \lambda_m) h(\lambda)$ for some polynomial $h(\lambda)$, so that $f_A(A) = \underbrace{(A - \lambda_1 I_n) \cdots (A - \lambda_m I_n)}_{0} h(A) = 0$, by Exercise 70.

7.5 Answers to more theoretical questions are omitted.

1. $\sqrt{18} \left(\cos\left(-\dfrac{\pi}{4} \right) + i \sin\left(-\dfrac{\pi}{4} \right) \right)$

3. $\cos\left(\dfrac{2\pi k}{n} \right) + i \sin\left(\dfrac{2\pi k}{n} \right)$, for $k = 0, \ldots, n - 1$

5. If $z = r\big(\cos(\phi) + i \sin(\phi) \big)$, then
$$w = \sqrt[n]{r} \left(\cos\left(\dfrac{\phi + 2\pi k}{n} \right) + i \sin\left(\dfrac{\phi + 2\pi k}{n} \right) \right),$$
for $k = 0, \ldots, n - 1$.

7. Clockwise rotation through an angle of $\frac{\pi}{4}$ followed by a scaling by a factor of $\sqrt{2}$.

9. Spirals outward since $|z| > 1$.

11. $f(\lambda) = (\lambda - 1)(\lambda - 1 - 2i)(\lambda - 1 + 2i)$

13. $S = \begin{bmatrix} 2 & 0 \\ 0 & 1 \end{bmatrix}$, for example

15. $S = \begin{bmatrix} 0 & 1 \\ 1 & 2 \end{bmatrix}$, for example

17. $S = \begin{bmatrix} 2 & 0 \\ -1 & 2 \end{bmatrix}$, for example

19. a. $\text{tr}(A) = m$, $\det(A) = 0$
    b. $\text{tr}(B) = 2m - n$, $\det(B) = (-1)^{n-m}$ (compare with Exercise 7.3.39)

21. $2 \pm 3i$

23. $1, -\dfrac{1}{2} \pm \dfrac{\sqrt{3}}{2} i$

25. $\pm 1, \pm i$

27. $-1, -1, 3$

29. $\text{tr}(A) = \lambda_1 + \lambda_2 + \lambda_3 = 0$ and $\det(A) = \lambda_1 \lambda_2 \lambda_3 = bcd > 0$. Therefore, there are one positive and two negative eigenvalues; the positive one is largest in absolute value.

31. $\frac{1}{15} A$ is a regular transition matrix (compare with Exercise 30), so that $\lim_{t \to \infty} (\frac{1}{15} A)^t$ exists and has identical columns (see Exercise 30). Therefore, the columns of A^t are nearly identical for large t.

33. c. *Hint:* Let $\lambda_1, \lambda_2, \ldots, \lambda_5$ be the eigenvalues, with $\lambda_1 > |\lambda_j|$, for $j = 2, \ldots, 5$. Let $\vec{v}_1, \vec{v}_2, \ldots, \vec{v}_5$

be corresponding eigenvectors. Write $\vec{e}_i = c_1\vec{v}_1 + \cdots + c_5\vec{v}_5$. Then ith column of $A^t = A^t\vec{e}_i = c_1\lambda_1^t\vec{v}_1 + \cdots + c_5\lambda_5^t\vec{v}_5$ is nearly parallel to $\vec{v}_1$ for large t.

45. If a is nonzero.

47. If a is nonzero.

49. If a is neither 1 nor 2.

51. $\mathbb{Q}$ is a field.

53. The binary digits form a field.

55. H is not a field (multiplication is noncommutative).

7.6 Answers to more theoretical questions are omitted.

1. Stable   3. Not stable

5. Not stable   7. Not stable

9. Not stable   11. For $|k| < 1$

13. For all k 15. Never stable

17. $\vec{x}(t) = \begin{bmatrix} -\sin(\phi t) \\ \cos(\phi t) \end{bmatrix}$, where $\phi = \arctan\left(\frac{4}{3}\right)$; a circle.

19. $\vec{x}(t) = \sqrt{13}^t \begin{bmatrix} -\sin(\phi t) \\ \cos(\phi t) \end{bmatrix}$, where $\phi = \arctan\left(\frac{3}{2}\right)$; spirals outward.

21. $\vec{x}(t) = \sqrt{17}^t \begin{bmatrix} 5\sin(\phi t) \\ \cos(\phi t) + 3\sin(\phi t) \end{bmatrix}$, where $\phi = \arctan\left(\frac{1}{4}\right)$; spirals outward.

23. $\vec{x}(t) = \left(\frac{1}{2}\right)^t \begin{bmatrix} 5\sin(\phi t) \\ \cos(\phi t) + 3\sin(\phi t) \end{bmatrix}$, where $\phi = \arctan\left(\frac{3}{4}\right)$; spirals inward.

25. Not stable   27. Stable

29. May or may not be stable; consider $A = \pm\frac{1}{2}I_2$

33. The matrix represents a rotation followed by a scaling with a scaling factor of $\sqrt{0.99^2 + 0.01^2} < 1$. Trajectory spirals inward.

35. a. Choose an eigenbasis $\vec{v}_1, \ldots, \vec{v}_n$ and write
$$\vec{x}_0 = c_1\vec{v}_1 + \cdots + c_n\vec{v}_n.$$
Then
$$\vec{x}(t) = c_1\lambda_1^t\vec{v}_1 + \cdots + c_n\lambda_n^t\vec{v}_n$$
and
$$\|\vec{x}(t)\| \le |c_1|\|\vec{v}_1\| + \cdots + |c_n|\|\vec{v}_n\| = M$$
(use the triangle inequality $\|\vec{u} + \vec{w}\| \le \|\vec{u}\| + \|\vec{w}\|$, and observe that $|\lambda_i^t| \le 1$)

b. The trajectory $\vec{x}(t) = \begin{bmatrix} 1 & 1 \\ 0 & 1 \end{bmatrix}^t \begin{bmatrix} 0 \\ 1 \end{bmatrix} = \begin{bmatrix} t \\ 1 \end{bmatrix}$ is not bounded. This does not contradict part (a), since there is no eigenbasis for the matrix $\begin{bmatrix} 1 & 1 \\ 0 & 1 \end{bmatrix}$.

39. $\begin{bmatrix} 2 \\ 4 \end{bmatrix}$ is a stable equilibrium.

CHAPTER 8

8.1 Answers to more theoretical questions are omitted.

1. $\begin{bmatrix} 1 \\ 0 \end{bmatrix}, \begin{bmatrix} 0 \\ 1 \end{bmatrix}$

3. $\frac{1}{\sqrt{5}}\begin{bmatrix} 2 \\ 1 \end{bmatrix}, \frac{1}{\sqrt{5}}\begin{bmatrix} -1 \\ 2 \end{bmatrix}$

5. $\frac{1}{\sqrt{2}}\begin{bmatrix} -1 \\ 1 \\ 0 \end{bmatrix}, \frac{1}{\sqrt{6}}\begin{bmatrix} -1 \\ -1 \\ 2 \end{bmatrix}, \frac{1}{\sqrt{3}}\begin{bmatrix} 1 \\ 1 \\ 1 \end{bmatrix}$

7. $S = \frac{1}{\sqrt{2}}\begin{bmatrix} 1 & -1 \\ 1 & 1 \end{bmatrix}, D = \begin{bmatrix} 5 & 0 \\ 0 & 1 \end{bmatrix}$

9. $S = \begin{bmatrix} 1/\sqrt{2} & -1/\sqrt{2} & 0 \\ 0 & 0 & 1 \\ 1/\sqrt{2} & 1/\sqrt{2} & 0 \end{bmatrix}, D = \begin{bmatrix} 3 & 0 & 0 \\ 0 & -3 & 0 \\ 0 & 0 & 2 \end{bmatrix}$

11. Same S as in 9, $D = \begin{bmatrix} 2 & 0 & 0 \\ 0 & 0 & 0 \\ 0 & 0 & 1 \end{bmatrix}$

13. Yes (reflection about E_1)

15. Yes (can use the same orthonormal eigenbasis)

17. Let A be the $n \times n$ matrix whose entries are all 1. The eigenvalues of A are 0 (with multiplicity $n-1$) and n. Now $B = qA + (p-q)I_n$, so that the eigenvalues of B are $p - q$ (with multiplicity $n-1$) and $qn + p - q$. Therefore, $\det(B) = (p - q)^{n-1}(qn + p - q)$.

21. $48 = 6 \cdot 4 \cdot 2$ (note that A has 6 unit eigenvectors)

23. The only possible eigenvalues are 1 and -1 (because A is orthogonal), and the eigenspaces E_1 and E_{-1} are orthogonal complements (because A is symmetric). A represents the reflection about a subspace of $\mathbb{R}^n$.

25. $S = \frac{1}{\sqrt{2}}\begin{bmatrix} 1 & 1 & 0 & 0 & 0 \\ 0 & 0 & 1 & 1 & 0 \\ 0 & 0 & 0 & 0 & \sqrt{2} \\ 0 & 0 & 1 & -1 & 0 \\ 1 & -1 & 0 & 0 & 0 \end{bmatrix}$

27. If n is even, we have the eigenbasis $\vec{e}_1 - \vec{e}_n$, $\vec{e}_2 - \vec{e}_{n-1}, \ldots, \vec{e}_{n/2} - \vec{e}_{n/2+1}, \vec{e}_1 + \vec{e}_n, \vec{e}_2 + \vec{e}_{n-1}$, $\ldots, \vec{e}_{n/2} + \vec{e}_{n/2+1}$, with associated eigenvalues 0 ($n/2$ times) and 2 ($n/2$ times).

29. Yes   31. True   33. $\theta = \frac{2}{3}\pi = 120°$.

35. $\theta = \arccos(-\frac{1}{n})$. *Hint*: If $\vec{v}_0, \ldots, \vec{v}_n$ are such vectors, let $A = \begin{bmatrix} \vec{v}_0 & \cdots & \vec{v}_n \end{bmatrix}$. Then the noninvertible matrix $A^T A$ has 1's on the diagonal and $\cos(\theta)$ everywhere else. Now use Exercise 17.

37. In Example 4 we see that the image of the unit circle will be an ellipse with semimajor axis 3 and semiminor axis 2. Thus $2 \le \|A\vec{u}\| \le 3$.

39. Let $\vec{v}_1$, $\vec{v}_2$, $\vec{v}_3$ be an orthonormal eigenbasis with associated eigenvalues $-2, 3$, and 4, respectively. Consider a unit vector $\vec{u} = c_1\vec{v}_1 + c_2\vec{v}_2 + c_3\vec{v}_3$. Then $A\vec{u} = -2c_1\vec{v}_1 + 3c_2\vec{v}_2 + 4c_3\vec{v}_3$ and $\vec{u} \cdot A\vec{u} = -2c_1^2 + 3c_2^2 + 4c_3^2 \le 4c_1^2 + 4c_2^2 + 4c_3^2 = 4$. Likewise, $\vec{u} \cdot A\vec{u} = -2c_1^2 + 3c_2^2 + 4c_3^2 \ge -2c_1^2 - 2c_2^2 - 2c_3^2 = -2$. Thus $-2 \le \vec{u} \cdot A\vec{u} \le 4$.

41. There exist an orthogonal S and a diagonal D such that $A = SDS^{-1}$. Taking cube roots of the diagonal entries of D, we can write $D = D_0^3$ for some diagonal D_0. Now $A = SDS^{-1} = SD_0^3S^{-1} = (SD_0S^{-1})^3 = B^3$, where $B = SD_0S^{-1}$.

43. Consider the eigenvectors $\vec{v}_1 = \begin{bmatrix} 1 \\ 1 \\ 1 \end{bmatrix}$ and $\vec{v}_2 = \begin{bmatrix} 1 \\ -1 \\ 0 \end{bmatrix}$, with eigenvalues 24 and -9, respectively.

There must exist a nonzero solution of the form $\vec{v} = a\vec{v}_1 + b\vec{v}_2$. Now $\vec{v} \cdot A\vec{v} = (a\vec{v}_1 + b\vec{v}_2) \cdot (24a\vec{v}_1 - 9b\vec{v}_2) = 72a^2 - 18b^2 = 0$ when $b = \pm 2a$. Let $a = 1$ and $b = 2$ to find the solution $\vec{v} = \begin{bmatrix} 3 \\ -1 \\ 1 \end{bmatrix}$.

47. a. ijth entry of $|AB| = \left| \sum_{k=1}^{n} a_{ik}b_{kj} \right|$
    $\le \sum_{k=1}^{n} |a_{ik}||b_{kj}|$
    $= ij$th entry of $|A||B|$

    b. By induction on t, using part (a): $|A^t| = |A^{t-1}A| \le |A^{t-1}||A| \le |A|^{t-1}|A| = |A|^t$

49. Let λ be the maximum of all $|r_{ii}|$, for $i = 1, \dots, n$. Note that $\lambda < 1$. Then $|R| \le \lambda(I_n + U)$, where U is upper triangular with $u_{ii} = 0$ and $u_{ij} = |r_{ij}|/\lambda$ if $j > i$. Note that $U^n = 0$ (see Exercise 46a). Now $|R^t| \le |R|^t \le \lambda^t(I_n + U)^t \le \lambda^t t^n(I_n + U + \cdots + U^{n-1})$. From calculus we know that $\lim_{t\to\infty} \lambda^t t^n = 0$.

8.2 Answers to more theoretical questions are omitted.

1. $\begin{bmatrix} 6 & -3.5 \\ -3.5 & 8 \end{bmatrix}$ 3. $\begin{bmatrix} 3 & 0 & 3 \\ 0 & 4 & 3.5 \\ 3 & 3.5 & 5 \end{bmatrix}$

5. Indefinite 7. Indefinite

9. a. A^2 is symmetric
   b. $A^2 = -A^T A$ is negative semidefinite, so that its eigenvalues are ≤ 0.
   c. The eigenvalues of A are imaginary (that is, of the form bi, for a real b). The zero matrix is the only skew-symmetric matrix that is diagonalizable over $\mathbb{R}$.

11. The same (the eigenvalues of A and A^{-1} have the same signs).

13. $a_{ii} = q(\vec{e}_i) > 0$.

15. Ellipse; principal axes spanned by $\begin{bmatrix} 2 \\ 1 \end{bmatrix}$ and $\begin{bmatrix} -1 \\ 2 \end{bmatrix}$; equation $7c_1^2 + 2c_2^2 = 1$

17. Hyperbola; principal axes spanned by $\begin{bmatrix} 2 \\ 1 \end{bmatrix}$ and $\begin{bmatrix} -1 \\ 2 \end{bmatrix}$, equation $4c_1^2 - c_2^2 = 1$.

19. A pair of lines; principal axes spanned by $\begin{bmatrix} 2 \\ -1 \end{bmatrix}$ and $\begin{bmatrix} 1 \\ 2 \end{bmatrix}$; equation $5c_2^2 = 1$.
    Note that we can write $x_1^2 + 4x_1x_2 + 4x_2^2 = (x_1 + 2x_2)^2 = 1$, so that $x_1 + 2x_2 = \pm 1$

21. a. The first is an ellipsoid, the second a hyperboloid of one sheet, and the third a hyperboloid of two sheets (see any text in multivariable calculus). Only the ellipsoid is bounded, and the first two surfaces are connected.

    b. The matrix A of this quadratic form has positive eigenvalues $\lambda_1 \approx 0.56$, $\lambda_2 \approx 4.44$, and $\lambda_3 = 1$, with corresponding unit eigenvectors

    $$\vec{v}_1 \approx \begin{bmatrix} 0.86 \\ 0.19 \\ -0.47 \end{bmatrix}, \quad \vec{v}_2 \approx \begin{bmatrix} 0.31 \\ 0.54 \\ 0.78 \end{bmatrix},$$
    $$\vec{v}_3 \approx \begin{bmatrix} 0.41 \\ -0.82 \\ 0.41 \end{bmatrix}.$$

    Since all eigenvalues are positive, the surface is an ellipsoid. The points farthest from the origin are

    $$\pm \frac{1}{\sqrt{\lambda_1}}\vec{v}_1 \approx \pm \begin{bmatrix} 1.15 \\ 0.26 \\ -0.63 \end{bmatrix}$$

    and those closest are

    $$\pm \frac{1}{\sqrt{\lambda_2}}\vec{v}_2 \approx \pm \begin{bmatrix} 0.15 \\ 0.26 \\ 0.37 \end{bmatrix}.$$

23. Yes; $A = \frac{1}{2}(M + M^T)$

25. $q(\vec{v}) = \vec{v} \cdot \lambda\vec{v} = \lambda$.

27. The closed interval $[\lambda_n, \lambda_1]$

29. $B = \frac{1}{\sqrt{5}} \begin{bmatrix} 6 & 2 \\ -3 & 4 \end{bmatrix}$ 31. $B = \frac{1}{5} \begin{bmatrix} 14 & -2 \\ -2 & 11 \end{bmatrix}$

33. $L = \frac{1}{\sqrt{2}} \begin{bmatrix} 4 & 0 \\ -1 & 3 \end{bmatrix}$ 35. $L = \begin{bmatrix} 2 & 0 & 0 \\ -2 & 3 & 0 \\ 4 & 3 & 1 \end{bmatrix}$

39. For $0 < \theta < \arccos\left(-\dfrac{1}{n-1}\right)$

41. 3

43. $\mathrm{im}(T) = \mathrm{span}(x_1^2)$, $\mathrm{rank}(T) = 1$,
$\ker(T) = \mathrm{span}(x_1 x_2, x_2^2)$, $\mathrm{nullity}(T) = 2$

45. $\mathrm{im}(T) = P_2$, $\mathrm{rank}(T) = 3$,
$\ker(T) = \mathrm{span}(x_3^2 - x_2^2, x_1 x_3 - x_1 x_2, x_2 x_3 - x_2^2)$,
$\mathrm{nullity}(T) = 3$

47. The determinant of the mth principal submatrix is positive if m is even, and negative if m is odd.

55. Note that $\det \begin{bmatrix} a_{ii} & a_{ij} \\ a_{ji} & a_{jj} \end{bmatrix} = a_{ii} a_{jj} - a_{ij}^2 > 0$, so that $a_{ii} > a_{ij}$ or $a_{jj} > a_{ij}$.

57. $q(\vec{x}) = \lambda_1 c_1^2 + \lambda_2 c_2^2 + \lambda_3 c_3^2 = 1$, with positive λ_i, defines an ellipsoid.

59. $q(\vec{x}) = \lambda_1 c_1^2 = 1$, with positive λ_1, defines a pair of parallel planes, $c_1 = \pm \dfrac{1}{\sqrt{\lambda_1}}$.

61. $q(\vec{x}) = \lambda_1 c_1^2 + \lambda_2 c_2^2 + \lambda_3 c_3^2 = 1$, with $\lambda_1 > 0$, $\lambda_2 > 0$, $\lambda_3 < 0$ defines a hyperboloid of one sheet.

63. $q(c_1 \vec{w}_1 + \cdots + c_n \vec{w}_n) = (c_1 \vec{w}_1 + \cdots + c_n \vec{w}_n) \cdot (c_1 \lambda_1 \vec{w}_1 + \cdots + c_n \lambda_n \vec{w}_n) = c_1^2 \lambda_1 \|\vec{w}_1\|^2 + \cdots + c_n^2 \lambda_n \|\vec{w}_n\|^2 = c_1^2 + \cdots + c_n^2$ since $\|\vec{w}_i\|^2 = \dfrac{1}{\lambda_i}$, by construction.

65. Adapt the method outlined in Exercise 63. Consider an orthonormal eigenbasis $\vec{v}_1, \vec{v}_2$ for A with associated eigenvalues $\lambda_1 > 0$ and $\lambda_2 < 0$. Now let $\vec{w}_1 = \vec{v}_1/\sqrt{\lambda_1}$ and $\vec{w}_2 = \vec{v}_2/\sqrt{-\lambda_2}$, so that $\|\vec{w}_1\|^2 = 1/\lambda_1$ and $\|\vec{w}_2\|^2 = -1/\lambda_2$. Then $q(c_1 \vec{w}_1 + c_2 \vec{w}_2) = (c_1 \vec{w}_1 + c_2 \vec{w}_2) \cdot (\lambda_1 c_1 \vec{w}_1 + \lambda_2 c_2 \vec{w}_2) = \lambda_1 c_1^2 \|\vec{w}_1\|^2 + \lambda_2 c_2^2 \|\vec{w}_2\|^2 = c_1^2 - c_2^2$.

67. Adapt the method outlined in Exercises 63 and 65. Consider an orthonormal eigenbasis $\vec{v}_1, \ldots, \vec{v}_p, \ldots, \vec{v}_r, \ldots, \vec{v}_n$ for A such that the associated eigenvalues λ_j are positive for $j = 1, \ldots, p$, negative for $j = p + 1, \ldots, r$, and zero for $j = r + 1, \ldots, n$. Let $\vec{w}_j = \vec{v}_j/\sqrt{|\lambda_j|}$ for $j = 1, \ldots, r$ and $\vec{w}_j = \vec{v}_j$ for $j = r + 1, \ldots, n$.

69. Note that $\vec{x}^T R^T A R \vec{x} = (R\vec{x})^T A (R\vec{x}) \geq 0$ for all $\vec{x}$ in $\mathbb{R}^m$. Thus $R^T A R$ is positive semidefinite. $R^T A R$ is positive definite if (and only if) $\ker R = \{\vec{0}\}$.

71. Anything can happen: The matrix $R^T A R$ may be positive definite, positive semidefinite, negative definite, negative semidefinite, or indefinite.

8.3 Answers to more theoretical questions are omitted.

1. $\sigma_1 = 2, \sigma_2 = 1$

3. All singular values are 1 (since $A^T A = I_n$)

5. $\sigma_1 = \sigma_2 = \sqrt{p^2 + q^2}$

7. $\begin{bmatrix} 0 & 1 \\ -1 & 0 \end{bmatrix} \begin{bmatrix} 2 & 0 \\ 0 & 1 \end{bmatrix} \begin{bmatrix} 0 & 1 \\ 1 & 0 \end{bmatrix}$

9. $\dfrac{1}{\sqrt{5}} \begin{bmatrix} 1 & -2 \\ 2 & 1 \end{bmatrix} \begin{bmatrix} 5 & 0 \\ 0 & 0 \end{bmatrix} \dfrac{1}{\sqrt{5}} \begin{bmatrix} 1 & 2 \\ -2 & 1 \end{bmatrix}$

11. $\begin{bmatrix} 0 & 1 & 0 \\ 1 & 0 & 0 \\ 0 & 0 & 1 \end{bmatrix} \begin{bmatrix} 2 & 0 \\ 0 & 1 \\ 0 & 0 \end{bmatrix} \begin{bmatrix} 0 & 1 \\ 1 & 0 \end{bmatrix}$

13. $I_2 \begin{bmatrix} 3\sqrt{5} & 0 \\ 0 & \sqrt{5} \end{bmatrix} \dfrac{1}{\sqrt{5}} \begin{bmatrix} 2 & 1 \\ -1 & 2 \end{bmatrix}$

15. Singular values of A^{-1} are the reciprocals of those of A.

21. $\begin{bmatrix} 0.8 & 0.6 \\ -0.6 & 0.8 \end{bmatrix} \begin{bmatrix} 9 & -2 \\ -2 & 6 \end{bmatrix}$

23. $AA^T \vec{u}_i = \begin{cases} \sigma_i^2 \vec{u}_i & \text{for } i = 1, \ldots, r \\ \vec{0} & \text{for } i = r + 1, \ldots, n \end{cases}$
The nonzero eigenvalues of $A^T A$ and AA^T are the same.

25. Choose vectors $\vec{v}_1$ and $\vec{v}_2$ as in Theorem 8.3.3. Write
$$\vec{u} = c_1 \vec{v}_1 + c_2 \vec{v}_2.$$
Note that
$$\|\vec{u}\|^2 = c_1^2 + c_2^2 = 1.$$
Now
$$A\vec{u} = c_1 A\vec{v}_1 + c_2 A\vec{v}_2,$$
so that
$$\|A\vec{u}\|^2 = c_1^2 \|A\vec{v}_1\|^2 + c_2^2 \|A\vec{v}_2\|^2$$
$$= c_1^2 \sigma_1^2 + c_2^2 \sigma_2^2$$
$$\leq (c_1^2 + c_2^2)\sigma_1^2$$
$$= \sigma_1^2$$
We conclude that $\|A\vec{u}\| \leq \sigma_1$. The proof of $\sigma_2 \leq \|A\vec{u}\|$ is analogous.

27. Apply Exercise 26 to a unit eigenvector $\vec{v}$ with associated eigenvalue λ.

33. No; consider $A = \begin{bmatrix} 0 & 1 \\ 2 & 0 \end{bmatrix}$.

35. $(A^T A)^{-1} A^T \vec{u}_i = \begin{cases} \dfrac{1}{\sigma_i} \vec{v}_i & \text{for } i = 1, \ldots, m \\ \vec{0} & \text{for } i = m + 1, \ldots, n \end{cases}$

CHAPTER 9

9.1 Answers to more theoretical questions are omitted.

1. $x(t) = 7e^{5t}$ 3. $P(t) = 7e^{0.03t}$

5. $y(t) = -0.8 e^{0.8t}$

7. $x(t) = \dfrac{1}{1-t}$, has a vertical asymptote at $t = 1$.

9. $x(t) = \left((1 - k)t + 1\right)^{1/(1-k)}$

11. $x(t) = \tan(t)$

13. a. about 104 billion dollars
    b. about 150 billion dollars

15. The solution of the equation $e^{kT/100} = 2$ is
$$T = \dfrac{100 \ln(2)}{k} \approx \dfrac{69}{k}$$

17.

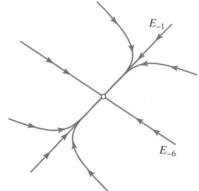

19.

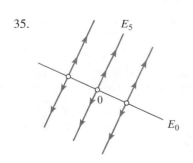

21. $\vec{x}(t) = \begin{bmatrix} 1 & t \\ 0 & 1 \end{bmatrix} \vec{x}_0$ 23. $\vec{x}(t)$ is a solution.

27. $\vec{x}(t) = 0.2e^{-6t} \begin{bmatrix} 3 \\ -2 \end{bmatrix} + 0.4e^{-t} \begin{bmatrix} 1 \\ 1 \end{bmatrix}$

29. $\vec{x}(t) = 2 \begin{bmatrix} 2 \\ -1 \end{bmatrix} + e^{5t} \begin{bmatrix} 1 \\ 2 \end{bmatrix}$

31. $\vec{x}(t) = e^t \begin{bmatrix} 1 \\ -2 \\ 1 \end{bmatrix}$

33.

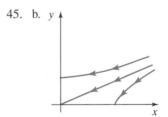

35.

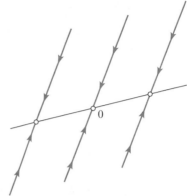

37. $E_{1.1} = \text{span} \begin{bmatrix} 3 \\ 1 \end{bmatrix}$ and $E_{1.6} = \text{span} \begin{bmatrix} 1 \\ 2 \end{bmatrix}$. Looks roughly like the phase portrait in Figure 10.

39. $E_1 = \text{span} \begin{bmatrix} 2 \\ -1 \end{bmatrix}$ and $E_{1.4} = \text{span} \begin{bmatrix} 2 \\ -3 \end{bmatrix}$. Looks roughly like the phase portrait in Exercise 35.

41.

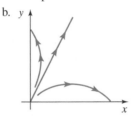

43. a. Competition

b. y

c. Species 1 "wins" if $\dfrac{y(0)}{x(0)} < 2$.

45. b. y

c. Species 1 "wins" if $\dfrac{y(0)}{x(0)} < \dfrac{1}{2}$

47. a. Symbiosis

b. The eigenvalues are $\frac{1}{2}(-5 \pm \sqrt{9 + 4k^2})$. There are two negative eigenvalues if $k < 2$; if $k > 2$ there is a negative and a positive eigenvalue.

49. $g(t) = 45e^{-0.8t} - 15e^{-0.4t}$ and $h(t) = -45e^{-0.8t} + 45e^{-0.4t}$

53. $\vec{x}(t) = e^{pt} \begin{bmatrix} \cos(qt) \\ \sin(qt) \end{bmatrix}$, a spiral if $p \neq 0$ and a circle if $p = 0$. Approaches the origin if p is negative.

55. Eigenvalues $\lambda_{1,2} = \frac{1}{2}(-q \pm \sqrt{q^2 - 4p})$; both eigenvalues are negative

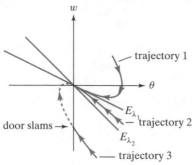

door slams if $\dfrac{\omega(0)}{\theta(0)} < \lambda_2$.

9.2 Answers to more theoretical questions are omitted.

1. 1

3. $\sqrt{2}e^{3\pi i/4}$

5. $e^{-0.1t}\big(\cos(2t) - i\sin(2t)\big)$; spirals inward, in the clockwise direction

7. Not stable

9. Stable

11. a. $B = 2A$

d. The zero state is a stable equilibrium of the system $\dfrac{d\vec{x}}{dt} = \operatorname{grad}(q)$ if (and only if) q is negative definite (then, the eigenvalues of A and B are all negative).

13. The eigenvalues of A^{-1} are the reciprocals of the eigenvalues of A; the real parts have the same sign.

15.

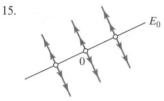

17. If $|k| < 1$

19. False; consider A with eigenvalues $1, 2, -4$.

21. a. $\begin{vmatrix} \dfrac{db}{dt} = 0.5b + & s \\ \dfrac{ds}{dt} = & 0.07s \end{vmatrix}$

b. $b(t) = 50{,}000e^{0.07t} - 49{,}000e^{0.5t}$
$s(t) = 1{,}000e^{0.07t}$

27. $\vec{x}(t) = \begin{bmatrix} \cos(3t) & -\sin(3t) \\ \sin(3t) & \cos(3t) \end{bmatrix}\begin{bmatrix} a \\ b \end{bmatrix}$, where a, b are arbitrary constants

29. Eigenvalue $2 + 4i$ with corresponding eigenvector $\begin{bmatrix} i \\ -1 \end{bmatrix}$. Use Theorem 9.2.6, with $p = 2$, $q = 4$,
$\vec{w} = \begin{bmatrix} 1 \\ 0 \end{bmatrix}$, $\vec{v} = \begin{bmatrix} 0 \\ -1 \end{bmatrix}$.
$\vec{x}(t) = e^{2t}\begin{bmatrix} 1 & 0 \\ 0 & -1 \end{bmatrix}\begin{bmatrix} \cos(4t) & -\sin(4t) \\ \sin(4t) & \cos(4t) \end{bmatrix}\begin{bmatrix} a \\ b \end{bmatrix}$

31. Eigenvalue $-1 + 2i$ with corresponding eigenvector $\begin{bmatrix} i \\ 1 \end{bmatrix}$. $\vec{x}(t) = e^{-t}\begin{bmatrix} \cos(2t) & -\sin(2t) \\ \sin(2t) & \cos(2t) \end{bmatrix}\begin{bmatrix} 1 \\ -1 \end{bmatrix} =$

$e^{-t}\begin{bmatrix} \cos(2t) + \sin(2t) \\ \sin(2t) - \cos(2t) \end{bmatrix}$. Spirals inward, in the counterclockwise direction.

33. Eigenvalue i with corresponding eigenvector $\begin{bmatrix} 1 \\ 1+i \end{bmatrix}$. $\vec{x}(t) = \begin{bmatrix} 0 & 1 \\ 1 & 1 \end{bmatrix}\begin{bmatrix} \cos(t) \\ \sin(t) \end{bmatrix} = \begin{bmatrix} \sin(t) \\ \sin(t) + \cos(t) \end{bmatrix}$. An ellipse with clockwise orientation.

39. The system $\dfrac{d\vec{c}}{dt} = \begin{bmatrix} 0 & 1 & 0 \\ 0 & 0 & 1 \\ 0 & 0 & 0 \end{bmatrix}\vec{c}$ has the solutions

$\vec{c}(t) = \begin{bmatrix} k_1 + k_2 t + k_3 t^2/2 \\ k_2 + k_3 t \\ k_3 \end{bmatrix}$,

where k_1, k_2, k_3 are arbitrary constants. The solutions of the given system are $\vec{x}(t) = e^{\lambda t}\vec{c}(t)$, by Exercise 9.1.24. The zero state is a stable equilibrium solution if (and only if) the real part of λ is negative.

9.3 Answers to more theoretical questions are omitted.

1. Ce^{5t}

3. $\frac{1}{5}e^{3t} + Ce^{-2t}$ (use Theorem 9.3.13)

5. $-1 - t + Ce^t$ 7. $c_1 e^{-4t} + c_2 e^{3t}$

9. $c_1 e^{3t} + c_2 e^{-3t}$

11. $e^t\big(c_1 \cos(t) + c_2 \sin(t)\big)$

13. $e^{-t}(c_1 + c_2 t)$ (compare with Example 10).

15. $c_1 + c_2 t$

17. $e^{-t}(c_1 + c_2 t) - \frac{1}{2}\cos(t)$

19. $\cos(t) + c_1 \cos(\sqrt{2}t) + c_2 \sin(\sqrt{2}t)$

21. $c_1 e^t + c_2 e^{-t} + c_3 e^{-2t}$ 23. $3e^{5t}$

25. e^{-2t+2} 27. $-\sin(3t)$

29. $\frac{1}{3}\sin(t) - \frac{1}{6}\sin(2t)$

31. $v(t) = \dfrac{mg}{k}(1 - e^{-kt/m})$

$\displaystyle\lim_{t\to\infty} v(t) = \dfrac{mg}{k} =$ terminal velocity.

35. a. $c_1 e^{-t} + c_2 e^{-2t}$

b. $2e^{-t} - e^{-2t}$

c. $-e^{-t} + 2e^{-2t}$

d. In part (c) the oscillator goes through the equilibrium state once; in part (b) it never reaches it.

37. $x(t) = te^{-3t}$ 39. $e^{-t}(c_1 + c_2 t + c_3 t^2)$

41. λ is an eigenvalue with $\dim(E_\lambda) = n$, because E_λ is the kernel of the nth-order linear differential operator $T(x) - \lambda x$.

43. $\frac{1}{10}\cos(t) + \frac{1}{10}\sin(t) + c_1 e^{-2t} + c_2 e^{-3t}$

45. $e^t\begin{bmatrix} 1 - 2t \\ -1 \end{bmatrix}$

SUBJECT INDEX

A

Affine system, 329
Algebraic multiplicity of an eigenvalue, 312
   and complex eigenvalues, 352
   and geometric multiplicity, 326
Algorithm, 18
Alternating property of determinant, 251, 265
Angle, 196
   and orthogonal transformations, 211
Approximations,
   Fourier 232, 239
   least-squares 17, 220
Argument (of a complex number), 346
   of a product, 348
Associative law for matrix multiplication, 75
Asymptotically stable equilibrium, 358
   (*see* stable equilibrium)
Augmented matrix, 11

B

Basis, 116, 158
   and coordinates, 139
   and dimension, 125, 159
   and unique representation, 120
   finding basis of a linear space, 161
   of an image, 116, 128, 131
   of a kernel, 127, 131
   of $\mathbb{R}^n$, 132
   standard, 126
Binary digits, 112
Block matrices, 75
   determinant of, 257
   inverse of, 87
   multiplication of, 76
Bounded trajectory, 363

C

Carbon dating, 407
Cardano's formula, 316, 319, 356
Cartesian coordinates, 10, 137
Cauchy–Schwarz inequality, 195, 237
   and angles, 196
   and triangle inequality, 200
Cayley–Hamilton Theorem, 332
Center of mass, 21
Change of basis, 177, 179
Characteristic polynomial, 311
   and algebraic multiplicity, 312

   and its derivative, 318
   of linear differential operator, 426
   of similar matrices, 326
Chebyshev polynomials, 246
Cholesky factorization, 383
Circulant matrix, 356
Classical adjoint, 287
Closed-formula solution
   for discrete dynamical system, 303
   for inverse, 287
   for least-squares approximation, 224
   for linear system, 286
Coefficient matrix
   of a linear system, 11
   of a linear transformation, 41
Column of a matrix, 9
Column space of a matrix, 105
Column vector, 10
Commuting matrices, 73
Complements, 135
Complex eigenvalues, 350
   and determinant, 353
   and rotation–scaling matrices, 352, 359, 416
   and stable equilibrium, 359
   and trace, 353
Complex numbers, 155, 345
   and rotation–scaling, 348
   in polar form, 347
Complex-valued functions, 411
   derivative of, 411
   exponential, 412
Component of a vector, 10
Composite functions, 69
Computational complexity, 89, 96
Concatenation, 323
Conics, 23, 134
Consistent linear system, 25
Continuous least-squares condition, 238
Continuous linear dynamical system, 400
   stability of, 415
   with complex eigenbasis, 416
   with real eigenbasis, 405
Contraction, 56
Coordinates, 139, 158
Coordinate transformation, 159, 167
Coordinate vector, 139, 159
Correlation (coefficient), 198
Cramer's Rule, 286

Cross product
  and polar (QS) factorization, 394
  in $\mathbb{R}^3$, 52, 443, 445
  in $\mathbb{R}^n$, 275
  rules for, 444
Cubic equation (Cardano's formula),
      316, 356
Cubic splines, 19
Cubics, 134
Curve fitting
  with conics, 23, 134
  with cubics, 134

D

Data compression, 393
Data fitting, 225
  multivariate, 228
De Moivre's formula, 348
Determinant, 85, 255
  alternating property of, 251, 265
  and characteristic polynomial,
      310, 311
  and eigenvalues, 315, 353
  and Cramer's rule, 286
  and elementary row operations, 265
  and invertibility, 266
  and Laplace expansion, 268, 270
  and QR factorization, 279
  as area, 85, 278
  as expansion factor, 282
  as volume, 280
  is linear in rows and columns,
      252, 262
  of inverse, 268
  of linear transformation, 272
  of orthogonal matrix, 278
  of permutation matrix, 260
  of product, 267
  of rotation matrix, 278
  of similar matrices, 267
  of 3×3 matrix, 249
  of transpose, 262
  of triangular matrix, 257
  of 2×2 matrix, 85
  patterns and, 255
  Vandermonde, 273
  Weierstrass definition of, 276
Diagonalizable matrices, 333
  and eigenbases, 333
  and powers, 336
  orthogonally, 367, 372
  simultaneously, 342
Diagonal matrix, 9

Diagonal of a matrix, 9
Dilation, 56
Dimension, 125, 159
  and isomorphism, 168
  of image, 129, 165
  of kernel, 129, 165
  of orthogonal complement, 193
Direction field, 401
Discrete linear dynamical system, 303
  and complex eigenvalues, 360, 362
  and stable equilibrium, 358, 359
Distance, 236
Distributive Laws, 75
Domain of a function, 43
Dominant eigenvector, 405
Dot product, 20, 28, 215, 441
  and matrix product, 73, 215
  and product $A\vec{x}$, 29
  rules for, 442
Dynamical system
  (*see* continuous, discrete linear
      dynamical system)

E

Eigenbasis, 322
  and continuous dynamical system, 405
  and diagonalization, 333
  and discrete dynamical system, 303
  and distinct eigenvalues, 324
  and geometric multiplicity, 324
Eigenfunction, 337, 426
Eigenspaces, 320
  and geometric multiplicity, 322
  and principal axes, 381
Eigenvalue(s), 300
  algebraic multiplicity of, 312
  and characteristic polynomial, 311
  and determinant, 315, 353
  and positive (semi)definite
      matrices, 379
  and QR factorization, 316
  and singular values, 387
  and stable equilibrium, 359, 415
  and trace, 315, 353
  complex, 350, 362
  geometric multiplicity of, 322
  of linear transformation, 337
  of orthogonal matrix, 302
  of rotation–scaling matrix, 351
  of similar matrices, 326
  of symmetric matrix, 370
  of triangular matrix, 310
  power method for finding, 355

Eigenvectors, 300
  and linear independence, 323
  dominant, 405
  of symmetric matrix, 369
Elementary matrix, 93
Elementary row operations, 16
  and determinant, 265
  and elementary matrices, 93
Ellipse, 69
  as image of the unit circle, 69, 372, 387
  as level curve of quadratic form, 381
  as trajectory, 362, 416
Equilibrium (state), 358
Equivalence Relation, 146
Error (in least-squares solution), 222
Error-correcting codes, 112
Euler identities, 240
Euler's formula, 413
Euler's theorem, 328
Exotic operations (in a linear space), 172
Expansion factor, 282
Exponential functions, 398
  complex-valued, 412

F

Factorizations,
  Cholesky, 383
  LDL^T, 220, 383
  LDU, 93, 220
  LL^T, 383
  LU, 93
  QR, 207, 208, 279, 316
  QS, 394
  SDS^{-1}, 336
  SDS^T, 367
  $U \Sigma V^T$, 390
Field, 350
Finite dimensional linear space, 162
Flow line, 401
Fourier analysis, 239
Function, 43
Fundamental theorem of algebra, 349
Fundamental theorem of linear
    algebra, 129

G

Gaussian integration, 246
Gauss–Jordan elimination, 17
  and determinant, 265
  and inverse, 82
Geometric multiplicity of an eigenvalue, 322
  and algebraic multiplicity, 326
  and eigenbases, 324

Golden section, 331
Gram–Schmidt process, 206
  and determinant, 278
  and orthogonal diagonalization, 372
  and QR factorization, 207

H

Harmonics, 241
Hilbert space ℓ_2, 201, 235
  and quantum mechanics, 243
Homogeneous linear system, 35
Hyperbola, 381
Hyperplane, 134

I

Identity matrix I_n, 45
Identity transformation, 45
Image of a function, 101
Image of a linear transformation,
    105, 165
  and rank, 129
  is a subspace, 105, 165
  orthogonal complement of, 221
  written as a kernel, 111
Image
  of the unit circle, 69, 372, 387
Imaginary part of a complex number, 346
Implicit function theorem, 276
Inconsistent linear system, 5, 17, 25
  and least-squares, 222
Indefinite matrix, 378
Infinite dimensional linear space, 162
Inner product (space), 234
Input-output analysis, 5, 20, 90, 95, 150
Intermediate value theorem, 68
Intersection of subspaces, 123
  dimension of, 135
Invariant Subspace, 307
Inverse (of a matrix), 80
  and Cramer's rule, 287
  determinant of, 268
  of an orthogonal matrix, 215
  of a product, 83
  of a transpose, 216
  of a 2×2 matrix, 85
  of block matrix, 87
Inversion (in a pattern), 255
Invertible function, 79
Invertible matrix, 80
  and determinant, 266
  and kernel, 109
Isomorphism, 167
  and dimension, 168

K

Kernel, 106, 165
  and invertibility, 109
  and linear independence, 120
  and rank, 129
  dimension of, 129
  is a subspace, 108, 165
Kyle numbers, 131, 320

L

Laplace expansion, 268, 270
LDL^T factorization, 220, 383
LDU factorization, 93, 220
Leading one, 16
Leading variable, 13
Least-squares solutions, 17, 201, 222
  and normal equation, 223
  minimal, 230
Left inverse, 122
Legendre polynomials, 246
Length of a vector, 187, 442
  and orthogonal transformations, 210
Linear combination, 30, 156
  and span, 105
Linear dependence, 116
Linear differential equation, 161, 423
  homogeneous, 423
  order of, 423
  solution of, 425, 433
  solving a first order, 431
  solving a second order, 428, 429
Linear differential operator, 423
  characteristic polynomial of, 426
  eigenfunctions of, 426
  image of, 432
  kernel of, 427
Linear independence
  and dimension, 126
  and kernel, 120
  and relations, 118
  in $\mathbb{R}^n$, 116
  in a linear space, 158
  of eigenvectors, 323
  of orthonormal vectors, 189
Linearity of the determinant, 252, 262
Linear relations, 118
Linear space(s), 154
  basis of, 158
  dimension of, 159
  finding basis of, 161
  finite dimensional, 162
  isomorphic, 167

Linear system
  closed-formula solution for, 286
  consistent, 25
  homogeneous, 35
  inconsistent, 5, 25
  least-squares solutions of, 222
  matrix form of, 32
  minimal solution of, 230
  number of solutions of, 25
  of differential equations: *see* continuous linear
    dynamical system
  unique solution of, 27
  vector form of, 32
  with fewer equations than unknowns, 27
Linear transformation, 41, 44, 48, 165
  image of, 105, 165
  kernel of, 106, 165
  matrix of, 41, 44, 47, 143, 173
Lower triangular matrix, 9
LU factorization, 93
  and principal submatrices, 94

M

Main diagonal of a matrix, 9
Mass-spring system, 422
Matrix, 9
  (for a composite entry such as "zero matrix"
    see zero)
Minimal solution of a linear
    system, 230
Minors of a matrix, 269
Modulus (of a complex number), 346
  of a product, 348
Momentum, 21
Multiplication (of matrices), 72
  and determinant, 267
  column by column, 72
  entry by entry, 73
  is associative, 75
  is noncommutative, 73
  of block matrices, 76

N

Negative feedback loop, 420
Neutral element, 154
Nilpotent matrix, 136, 375
Norm
  of a vector, 187
  in an inner product space, 236
Normal equation, 223
Nullity, 129, 165
Null space, 106

O

Orthogonal complement, 193
    of an image, 221
Orthogonally diagonalizable matrix, 367, 372
Orthogonal matrix, 211, 216
    determinant of, 278
    eigenvalues of, 302
    inverse of, 215
    transpose of, 215
Orthogonal projection, 58, 189, 191,
        225, 237
    and Gram–Schmidt process, 203
    and reflection, 59
    as closest vector, 222
    matrix of, 217, 225
Orthogonal transformation, 210
    and orthonormal bases, 217
    preserves dot product, 218
    preserves right angles, 211
Orthogonal vectors, 187, 236, 443
    and Pythagorean theorem, 194
Orthonormal bases, 189, 192
    and Gram–Schmidt process, 203
    and orthogonal transformations, 212
    and symmetric matrices, 368
Orthonormal vectors, 188
    are linearly independent, 189
Oscillator, 184

P

Parallelepiped, 280
Parallel vectors, 440
Parametrization (of a curve), 102
Partitioned matrices, 76
    (*see* block matrix)
Pattern (in a matrix), 255
    inversion in, 255
    signature of, 255
Permutation matrix, 89
    determinant of, 260
Perpendicular vectors, 187, 236, 443
Phase portrait,
    of continuous system, 400, 419
    of discrete system, 304, 419
    summary, 419
Piecewise continuous function, 239
Pivot, 16
Polar factorization, 394
Polar form (of a complex number), 347
    and powers, 348
    and products, 348
    with Euler's formula, 413

Positive (semi)definite matrix, 378
    and eigenvalues, 379
    and principal submatrices, 379
Positively oriented basis, 290
Power method for finding eigenvalues, 355
Principal axes, 381
Principal submatrices, 94
    and positive definite matrices, 379
Product $A\vec{x}$, 29
    and dot product, 29
    and matrix multiplication, 72
Projection, 56, 67
    (*see also* orthogonal projection)
Pseudo-inverse, 230
Pythagorean theorem, 194, 237

Q

QR factorization, 207
    and Cholesky factorization, 383
    and determinant, 279
    and eigenvalues, 316
    is unique, 219
QS factorization, 394
Quadratic forms, 377
    indefinite, 378
    negative (semi)definite, 378
    positive (semi)definite, 378
    principal axes for, 381
Quaternions, 220, 356, 364
    form a skew field, 356

R

Rank (of a matrix), 26, 129
    and image, 129
    and kernel, 129
    and row space, 136
    and singular values, 389
    of similar matrices, 326
    of transpose, 216
Rank-nullity theorem, 129, 165
Real part of a complex number, 346
Reduced row-echelon form (rref), 16
    and determinant, 265
    and inverse, 80, 82
    and rank, 26
Redundant vector(s), 116, 158
    image and, 116, 131
    kernel and, 131
Reflection, 59
Regular transition matrix, 317, 354
Relations (among vectors), 118
Resonance, 184
Riemann integral, 232, 234

Rotation matrix, 62, 278
Rotation–scaling matrix, 63
   and complex eigenvalues, 351
   and complex numbers, 348
Row of a matrix, 9
Row space, 136
Row vector, 10
Rule of 69, 408

S

Sarrus's rule, 250
Scalar, 12
Scalar multiples of matrices, 28
Scaling, 54
Second derivative test, 383
Secular equation, 308
Separation of variables, 398, 407
Shears, 64
Signature of a pattern, 255
Similar matrices, 145
   and characteristic polynomial, 326
Simultaneously diagonalizable
     matrices, 342
Singular value decomposition
     (SVD), 390
Singular values, 387
   and ellipses, 387
   and rank, 389
Skew field, 356
Skew-symmetric matrices, 214, 382
   determinant of, 274
Smooth function, 157
Space of functions, 155
Span
   in $\mathbb{R}^n$, 105
   in a linear space, 159
Spectral theorem, 368
Spirals, 344, 349, 362, 413, 416, 419
Square matrix, 9
Square-summable sequences, 201, 235
Stable equilibrium, 358
   of continuous system, 415
   of discrete system, 359
Standard basis, 126, 160
State vector, 295
Subspace
   invariant, 307
   of $\mathbb{R}^n$, 113
   of a linear space, 157
Sudoku, 75
Sum of subspaces, 123
   dimension of, 135
Sums of matrices, 28

Symmetric matrices, 214
   are orthogonally diagonalizable, 368
   have real eigenvalues, 370

T

Target space (of a function), 43
Tetrahedron, 92, 150, 289
Theorem of Pythagoras, 194, 237
Trace, 235, 310
   and characteristic polynomial, 311
   and eigenvalues, 315, 353
   and inner product, 235
   of similar matrices, 326
Trajectory, 304, 400
   bounded, 363
Transpose (of a matrix), 214
   and determinant, 262
   and inverse, 216
   and rank, 216
   of an orthogonal matrix, 215
   of a product, 216
Triangle inequality, 200
Triangular matrices, 9
   and determinant, 257
   and eigenvalues, 310
   invertible, 89
Triangulizable matrix, 375
Trigonometric polynomials, 240

U

Unit vector, 187, 443
Upper triangular matrix, 9

V

Vandermonde determinant, 273
Vector(s), 10
   angles between, 196
   addition, 437
   column, 10
   coordinate, 139, 159
   cross product of, 52, 249, 443
   dot product of, 441
   geometric representation of, 439
   length of, 187, 442
   orthogonal, 187, 443
   parallel, 440
   position, 439
   redundant, 116
   row, 10
   rules of vector algebra, 438
   scalar multiplication, 437
   standard representation of, 10, 439
   unit, 187, 443

velocity, 400
  vs. points, 10, 439
  zero, 438
Vector field, 401
Vector form of a linear system, 32
Vector space, 10, 154
  (*see* linear space)
Velocity vector, 400

W

Weierstrass definition of determinant, 276
Wigner semicircle distribution, 246

X

Zero matrix, 9
Zero vector, 438

NAME INDEX

Abel, Niels Henrik, 316
Alcuin of York, 24
al-Kashi, Jamshid, 24
al-Khowarizmi, Mohammed, 1, 18n
al-Ma'mun, Caliph, 1
Archimedes, 53, 331
Argand, Jean Robert, 346n

Banach, Stefan, 163
Berlinski, David, 17
Bézout, Etienne, 272
Binet, Jacques, 272
Burke, Kyle, 131

Cajori, Florian, 272
Cardano, Gerolamo, 316, 319
Cauchy, Augustin-Louis, 195, 255n, 272, 309
Cayley, Arthur, 272, 332
Chebyshev, Pafnuty, 246
Cholesky, André-Louis, 383
Cramer, Gabriel, 135, 272, 286

d'Alembert, Le Rond, 125, 309
Dantzig, Tobias, 345
De Moivre, Abraham, 348
Descartes, René, 118n, 137, 155
De Witt, Johan, 378

Einstein, Albert, 356–357
Euler, Leonhard Paul, 24, 135, 240, 243, 328,
    345, 413

Fermat, Pierre de, 137
Fibonacci (Leonardo of Pisa), 330–331
Fourier, Jean Baptiste Joseph, 239n

Galois, Evariste, 316
Gauss, Carl Friedrich, 17–18, 51, 246, 346n, 350
Gram, Jörgen, 206
Grassmann, Hermann Günther, 125

Hamilton, William Rowan, 220, 332, 356
Hamming, Richard W., 112
Harriot, Thomas, 118n
Heisenberg, Werner, 201, 244
Helmholtz, Hermann von, 309
Hilbert, David, 162, 201, 309

Jacobi, Carl Gustav Jacob, 255n
Jordan, Wilhelm, 17

Kepler, Johannes, 233
Kronecker, Leopold, 255n

Lagrange, Joseph Louis, 309, 364
Lanchester, Frederick William, 409n
Laplace, Pierre-Simon Marquis de, 268n, 272
Legendre, Adrien-Marie, 246
Leibniz, Gottfried Wilhelm von, 272
Leonardo da Vinci, 200
Leontief, Wassily, 5, 150

Mahavira, 24
Mas-Colell, Andreu, 201
Mazur, Barry, 345
Medawar, Peter. B., 64
Mumford, David, 423

Newton, Isaac, 24

Olbers, Wilhelm, 17

Peano, Giuseppe, 154n
Peirce, Benjamin, 413n
Piazzi, Giuseppe, 17
Pythagoras, 194, 243, 272

Riemann, Bernhard, 232, 234, 246
Roosevelt, Franklin D., 90n

Samuelson, Paul E., 364
Sarrus, Pierre Frédéric, 250n
Schläfi, Ludwig, 126
Schmidt, Erhard, 162, 206n
Schrödinger, Erwin, 244
Schwarz, Hermann Amandus, 195
Seki, Kōwa (or Takakazu), 255, 272
Simpson, Thomas, 246
Strang, Gilbert, 393
Strassen, Volker, 96
Sylvester, James Joseph, 9n

Thompson, d'Arcy, 64

Vandermonde, Alexandre-Théophile, 255n, 274

Weierstrass, Karl, 273, 276
Wessel, Caspar, 346n
Weyl, Hermann, 125
Wigner, Eugene Paul, 246
Wilson, Edward O., 285n